# Regional Conference on Acoustics and Vibration (RECAV 2017)



## Enhancing Acoustics and Vibration Research, Education, and Professionalism in Global Community

Journal of Physics: Conference Series
Volume 1075

Bali, Indonesia
27 – 28 November 2017

ISBN: 978-1-5108-7370-4
ISSN: 1742-6588

**Printed from e-media with permission by:**

Curran Associates, Inc.
57 Morehouse Lane
Red Hook, NY 12571

**Some format issues inherent in the e-media version may also appear in this print version.**



Printed by Curran Associates, Inc. (2018)

For permission requests, please contact the Institute of Physics
at the address below.

Institute of Physics
Dirac House, Temple Back
Bristol BS1 6BE UK

Phone: 44 1 17 929 7481
Fax: 44 1 17 920 0979

techtracking@iop.org

**Additional copies of this publication are available from:**

Curran Associates, Inc.
57 Morehouse Lane
Red Hook, NY 12571 USA
Phone: 845-758-0400
Fax: 845-758-2633
Email: curran@proceedings.com
Web: www.proceedings.com

# Table of contents

## Volume 1075

**Regional Conference on Acoustics and Vibration 2017 (RECAV 2017)**

**27–28 November 2017, Bali, Indonesia**

**Accepted papers received: 2 August 2018**
**Published online: 22 October 2018**

### Preface



### Papers

## PREFACE

This book of abstracts is a compilation of the abstracts of the papers contributed to the Regional Conference on Acoustics and Vibration (RECAV) in Denpasar, Bali, Indonesia, being held on November 27-28, 2017. RECAV 2017 is jointly organized by the Association of Acoustics and Vibration Indonesia (AAVI) and Society of Acoustics (Singapore) (SAS) presenting the theme "Enhancing Acoustics and Vibration Research, Education, and Professionalism in Global Community ". It is the intention of this book that participants would be able to see the overview of the papers presented in this conference and to make a plan in attending the sessions with respect to their interest. The range of research in acoustics and vibration and its current developments may also be observed in this book of abstracts which would provide information on various findings that could be of interest to those working in the relevant area. Our great appreciation to all contributions from authors coming from 14 countries for this conference. On behalf of the committee, I would like to thank to all participants and contributors of RECAV 2017 for attending this conference, submitting papers, taking part of the exhibitions, taking part in partnership and sponsorship, and for all kind of supports to make this conference successful. It is our hope that this conference would be enjoyable and provides fruitful interactions between all participants.

Denpasar, November 2017

Benjamin Soenarko
--------------------------------------
Chair of Steering Committee

## EVENT ORGANIZERS

## EVENT PARTNERS

IOP Conf. Series: Journal of Physics: Conf. Series **1075** (2018) 011001 doi:10.1088/1742-6596/1075/1/011001

## REGIONAL CONFERENCE ON ACOUSTIC AND VIBRATION (RECAV) 2017
## SPONSORED BY

## STEERING COMMITTEE

Prof. Dr. Benjamin Soenarko, MSME (Institut Teknologi Bandung)
Ir. Ida Bagus Ardhana Putra, Ph.D (Institut Teknologi Bandung)
Prof. Dr. Ir. Zainal Abidin (Institut Teknologi Bandung)
Ir. Wiratno Argo Asmoro, MSc (Institut Teknologi Sepuluh Nopember)
Dr. Ir. Husein Avicenna Akil, M.Sc. (BPPT)
Dr. Ir. I Gde Komang Merthayasa (Institut Teknologi Bandung)
Drs. Iwan Yahya, M.Si. (Universitas Sebelas Maret)
Dr. Woon Siong Gan (Society of Acoustics Singapore)
Prof. Jeong Guon Ih, (Acoustical Society of Korea)
Ir. Matias Handojo Widhi Budhiantho, M.S.E., Ph.D. (Universitas Kristen Satya Wacana)
Prof. Dr. Ir. Harijono A. Tjokronegoro, DEA. (Institut Teknologi Bandung)
Prof. Ir. Hermawan K. Dipojono MSEE., Ph.D. (Institut Teknologi Bandung)
Dr. Ir. Brian Yulianto (Institut Teknologi Bandung)
Dr. Khoo Boo Cheong (Singapore)
Dr. Venugopalan Pallayil (Singapore)

## ORGANIZING COMMITTEE

**Chairman**
Dr. Eng. Dhany Arifianto (Institut Teknologi Sepuluh Nopember)

**Co-Chairman**
Bagus Tris Atmaja, ST., MT. (Institut Teknologi Sepuluh Nopember)

**Secretary**
Dr. Suyatno (Institut Teknologi Sepuluh Nopember)

**Treasurer**
Dr. Katherine Indriawati (Institut Teknologi Sepuluh Nopember)

IOP Conf. Series: Journal of Physics: Conf. Series **1075** (2018) 011001 doi:10.1088/1742-6596/1075/1/011001

**Event Coordinator**
Dr. rer. nat Aulia Nasution (Institut Teknologi Sepuluh Nopember)
Dr.-Ing. Doty Dewi R. (Institut Teknologi Sepuluh Nopember)

**Technical Program**
Dr. Suyatno (Institut Teknologi Sepuluh Nopember)
Iwan Prasetyo, Ph.D (Institut Teknologi Bandung)
Sentagi Sesotya Utami, Ph.D (Universitas Gadjah Mada)
Ir. R. Sugeng Joko Sarwono, Ph.D (Institut Teknologi Bandung)
Prof. Christina Eviutami Mediastika, Ph.D (Universitas Kristen Petra)
Drs. Iwan Yahya, M.Si. (Universitas Sebelas Maret)
Dr. Eng. Dhany Arifianto (Institut Teknologi Sepuluh Nopember)
Ir. Wiratno Argo Asmoro, MSc (Institut Teknologi Sepuluh Nopember)
Bagus Tris Atmaja, S.T, M.T (Institut Teknologi Sepuluh Nopember)
Prof. Dr. Benjamin Soenarko, MSME (Institut Teknologi Bandung)
Ir. Ida Bagus Ardhana Putra, Ph.D (Institut Teknologi Bandung)
Anugrah Sabdono Sudarsono, Ph.D (Institut Teknologi Bandung)
Dr. Ir. I Gde Komang Merthayasa (Institut Teknologi Bandung)
Prof. Lim Kian Meng, Ph.D (National University of Singapore)
Dr. Woon Siong Gan (Society of Acoustics Singapore)
Prof. H.P. Lee (National University of Singapore)

# Peer review statement

All papers published in this volume of *Journal of Physics: Conference Series* have been peer reviewed through processes administered by the proceedings Editors. Reviews were conducted by expert referees to the professional and scientific standards expected of a proceedings journal published by IOP Publishing.

IOP Conf. Series: Journal of Physics: Conf. Series **1075** (2018) 012001 doi:10.1088/1742-6596/1075/1/012001

# Vibrato correlation to sound quality in violin performance

**T Dimov**[1]

[1] String Dept. Coordinator, Universitas Pelita Harapan, Conservatory of Music, Tangerang, Indonesia

E-mail: tomislav.dimov@uph.edu

**Abstract.** Vibration is the most essential physical, and when applied as *vibrato*, artistic part of the music. When music is performed without *vibrato*, it is depleted to a level of a mere technical sound. The research is explaining why the *vibrato* is important and what is included in the term of artistic *vibrato* in various forms. The analysis is by comparative method of artistically passive and artistically active sound production and the respectful sound spectrum and dynamic properties in attempt to prove that passive sound production is inferior in content of overtones, dynamics and aesthetical qualities. Analyzing only the passive physical and technical properties of given instrumental sound and vibration is not sufficient, but it is necessary to consider the artistic performance and the *vibrato* as a factor and how the artistic *vibrato* reflects on the properties of sound. Samples from live recordings are submitted to a spectral and other analysis in order to establish empirical scientific connection between the physical and the artistic properties of the performance. Significant finding is the direction towards creating software that can verify and approve certain levels of artistic involvement during public performance, or educational practice of music.

## 1. Introduction

Why a skilfully vibrated sound and phrase on the violin captures your attention quickly and affects your mind and heart in a far more exciting way than the straight and not vibrated sound, or tone to be more specific, of a beginner violinist? Being deeply relational, musical instruments (especially string/ bowed instruments) have acoustic properties that are exposed only while there is artistic interaction between them and the performer. How these interactions reflect on the psychological perception and the acoustical properties of the sound while using artistic vibrato is the main object of this research. Violin's instrumental functionality and expression embeds and implies vibration in a variety of forms, functions and meanings that make it one of the best candidates for this analysis. The musical tone is already a vibration by itself with determined: frequency (pitch), power (dynamic), special overtone signature (color) and duration; the four properties in which tone differs from noise. Vibrato is the artistic part of it. When used as technical vibrato, it is generally to connect notes, show metrical significance and bring physical relaxation in playing. When it is used as artistic vibrato, it is part of the dynamic phrasing, culmination, brilliance and emotional expression during the performance. Vibrato brings natural properties to the tone based on the notion that bulk of modulated sounds and tones are produced by living things [1] and the motion of the Earth, while straight, mechanical and long term unmodulated tones and sounds are far less present in nature. We have all experienced the effect when in an airplane after a while we do not percept the sound of the engines although we know it is there, like the noise of the city traffic sound pollution, or noise of the humming air conditioner, fluorescent lightbulb or a fan [2]. Psychologically, the constant unmodulated tone/ sound is perceived as of a mechanical origin and as such is likely to be cancelled by the brain, while the modulated one is perceived as from live origin and is some sort of communication [3]. In that sense, including the communication and recognizing it as part of the quality of the tone through vibrato would be the most significant finding that might change the perception of functionality of the acoustics in general. The artistic vibrato brings aesthetic and expressive properties that are in context of the style and performed musical piece. In the article "The Psychoacoustic secret of vibrato" by Martin Schleske [4] we read: Psychoacoustic experiments have shown that electronically manipulated vibrato that is solely frequency–modulated tends to be perceived as "synthetic" and "vulgar", while pure amplitude modulation of the harmonics (with the frequency modulation artificially re–moved) is perceived as live and natural. It is clear the rejection point of such vibration, but it is not explained the second half of the equation, why the artistic vibration is perceived in a positive way as "live and natural"? The only explanation can be in the inner wiring of the brain and the system of alerting us in perceiving intent of communication.

In the realm of tempo artistic expressiveness is applied through rubato (a tempo–rhythm time sinusoid or tempo–rhythm fluctuation) with short tempo changes, while dynamic is a tool of artistic expression using the physical power of sound. Construction properties of the violin as an instrument, include basics elasticity of the wooden stick of the bow and the bow–hair, which make it to physically vibrate, as well as the sound plates of the violin and the strings that produce the sound, bringing exclusive characteristics to the sound that cannot be repeated except somewhat copied through digital sampling. In pursuit of objectivity, in the research were used six different instruments of a various class of make and sound properties, and only one bow from one maker, thus creating least possible variables in the used hardware that could affect the findings. The violin technique, all of the applied vibrato, phrasing, tempo–rhythm variables and all of the artistic attributes of vibration are in the full discretion of the performer. The analysis was restricted only to the non–vibrato and artistic vibrato sound samples and used was method by comparing artistically passive and artistically active sound production, their respectful sound spectrum and dynamic properties. The hypothesis was that the findings would prove that passive application of sound is far inferior in content of overtones, dynamics and aesthetical qualities. Therefore it was not sufficient to analyse only the passive physical and technical properties of given instrumental sound and vibration, but it was necessary to consider the artistic performance and the artistic vibrato as a factor. Among the other goals was to establish how the applied physic can help the performance process, and how in return the performance process should be part of the applied physics analysis of the sound, thus providing a connection between empirical sound analysis, artistic musical performance and the newly emerging discipline of psycho–acoustics.

## 2. Methods

Half of the samples chosen for analysis included performance with artistic vibrato. Samples from live recordings were presented to a spectral and other analysis in order to establish empirical scientific connection between the physical and the artistic properties of the performance. The recorded samples were obtained in a regular studio using the Voce Vista Video software from the Sygyt Software [5]. For measuring the frequency and the overtones of the violin was used standardized equipment by the ISO: NTi Sound Level Meter XL2 type 1, Earthworks M30 omnidirectional as a measurement microphone, XLR Cable with microphone stand and sound level meter type 1 where the data acquired was using the 1/3 octave frequency band for the Background Noise frequency data. The used microphone are also had flat frequency response and an omnidirectional polar pattern. The distance of the violin with the microphone was 1 meter apart and the tuning of the violins was 440 Hz as referenced in the ISO 16:1975. For production of the sound samples performed by Dr. Tomislav Dimov on the violin, were used the following makes of instruments: Gustav Lutschg (Switzerland), Otakar Spidlen (Czech), Mathias Heinicke (Bohemia), Eugenio Degani (Italy), Svetozar Bogdanovski (Macedonia) and Anonimus (Undetermined). A bow by maker Martin was used in-pair with all of these instruments respectively. As musical samples that would involve artistic vibrato and active grade of engagement during the violinist performance, were used three famous excerpts from the musical literature. For sample 01 were used the first twenty one notes from "Tzigane" Rhapsody for Violin and Orchestra by Maurice Ravel. The properties of this sample are to perform in the lowest register on the violin with all of the pitches on the G string. Sample 02 involved the first seventeen notes from the beginning of the "Devill's Trills" Sonata for Violin and Basso Continuo in g–minor by Giuseppe Tartini. This sample provided the data from the middle range of the violin equally divided between the third and fourth position on the violin and between the A and the D string. Sample 03 consisted of the first nineteen notes of the beginning of the violin solo part of the Concerto for Violin and Orchestra No. 4 in D–Major, K.218 by Wolfgang Amadeus Mozart. For appropriateness of the yielded material for spectral analysis, only the violin solo line was performed without accompaniment. Each of the samples was recorded without any vibrato (non–vibrato) and with artistic vibrato, thus producing a total of 36 samples. From each of the samples were used only notes that yielded clear and energetic vibrato, suitable for analysis. In each of the cases the choice fell to the last note of the sample, namely the twenty first note from the sample 01, the seventeenth note from the sample 02 and nineteenth note from the sample 03. It is noticeably consistent choice which points to probable increasingly artistically charged endings, inspired and intensified after playing the whole sample right before that.

## 3. Result and Discussions

The basic parameters of sound with and without vibrato from six different violins were analysed, their fundamental frequency and the yielded respective overtones. Below are the figures of fundamental frequency and the overtones of each of the violins with vibrato.

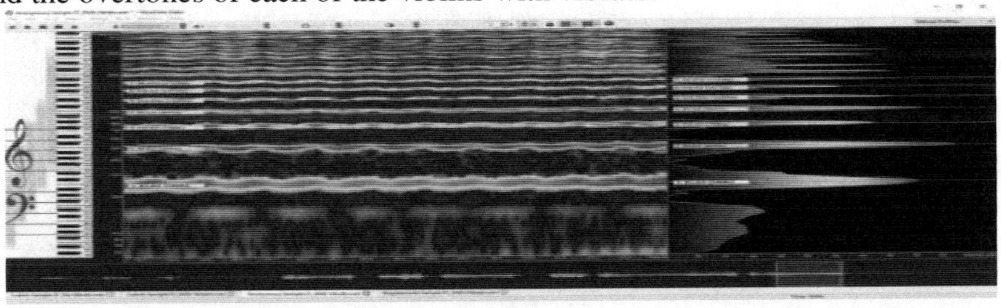

**Figure 1.** In Anonimus sample 01_with *vibrato*.

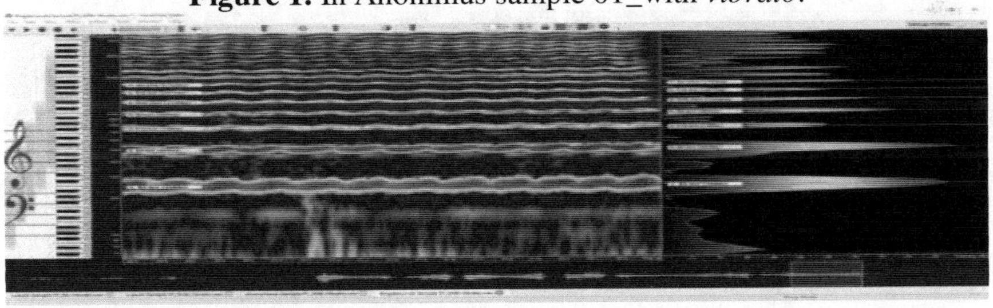

**Figure 2.** Bogdanovski sample 01_with *vibrato*.

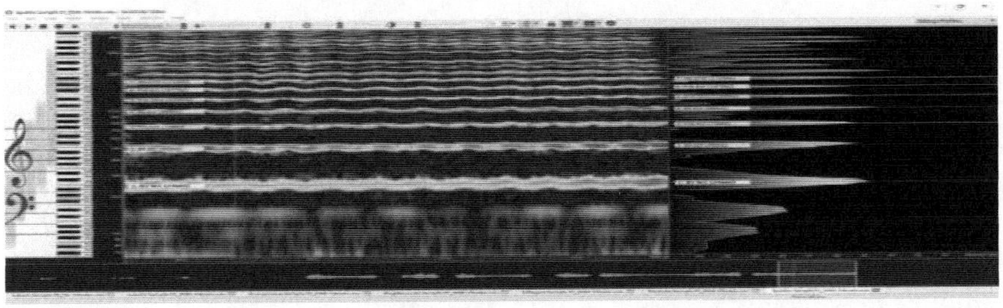

**Figure 3.** Spidlen sample 01_with *vibrato*.

**Figure 4.** Heinicke sample 01_with *vibrato*.

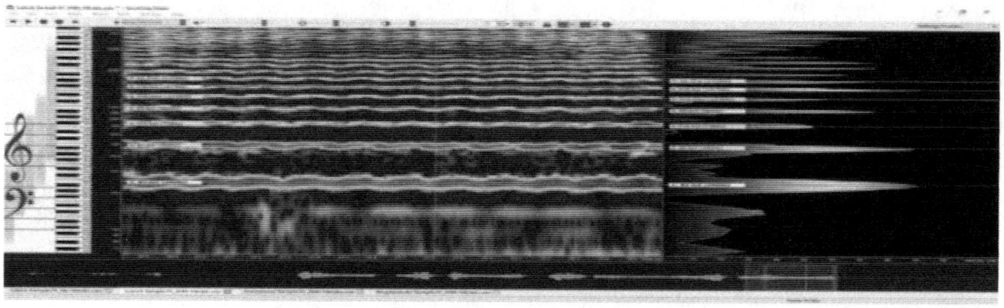

**Figure 5.** Lutschg sample 01_with *vibrato*.

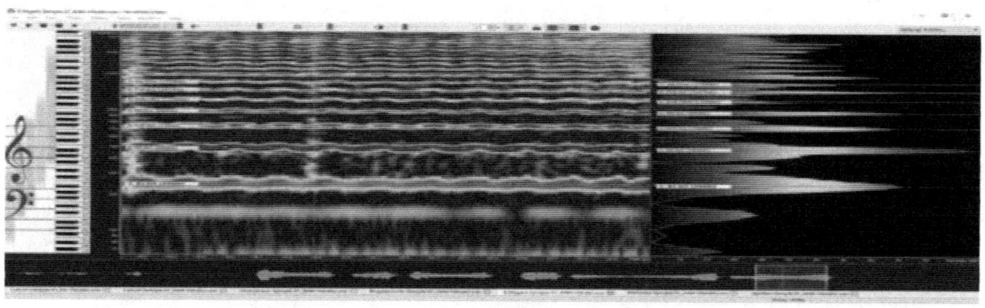

**Figure 6.** Degani sample 01_with *vibrato*.

**Figure 7.** Degani sample 01_ *non–vibrato*.

The notes used as sound samples are the last *vibrato* notes of the each excerpt respectively. From the figures 1 to 6 we can conclude that each of the violins has the exactly same fundamental frequency and the same number of overtones. The sound level of each harmonic has a periodically fluctuating value due to the *vibrato*. The amount of the overtones was the same in the *non–vibrato* samples as seen in figure 7. There is a difference between the *vibrato* samples and the *non–vibrato* samples in sense that *vibrato* samples contain much more artefacts or noise, possibly created by the interference of the fluctuating fundamental and the overtones as seen in figures 6 and 7. This is the area that needs to be researched more in the future.

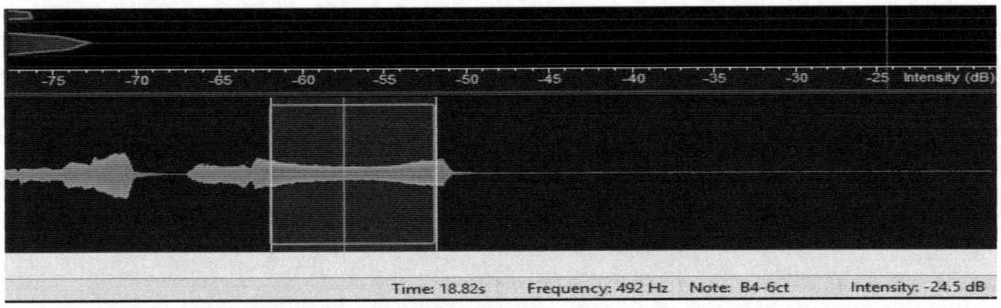


**Figure 8.** Degani sample 01_ *non–vibrato.*

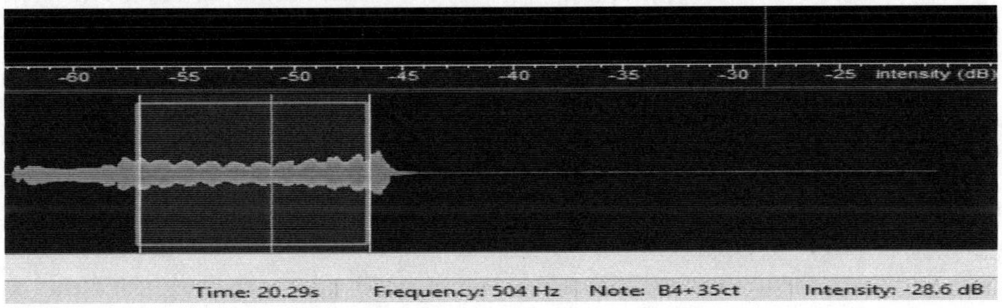


**Figure 9.** Degani sample 01_with *vibrato.*

The biggest loss in decibels was observed between these two samples by Eugenio Degani showed above in figures 8 and 9. The non–*vibrato* sample had intensity of – 24.5 decibels, while the vibrated showed – 28.6 decibels.

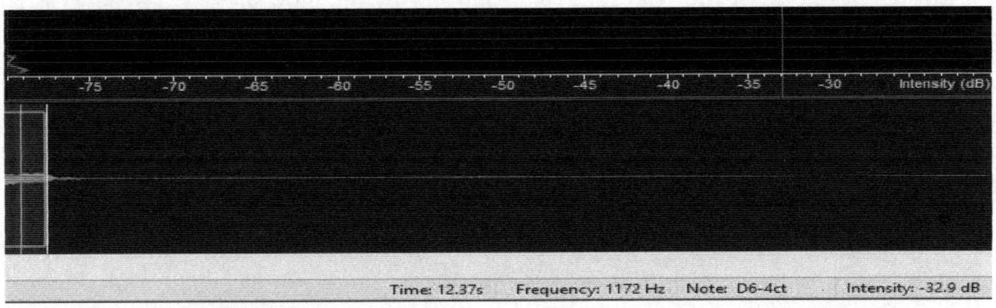


**Figure 10.** Degani sample 03_ *non–vibrato.*

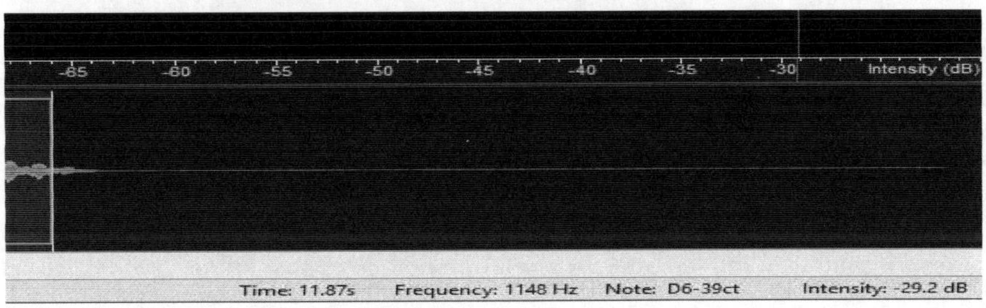


**Figure 11.** Degani sample 03_with *vibrato.*

The biggest gain in intensity was observed between the two samples, also by Eugenio Degani, showed above in figures 10 and 11. The *non–vibrato* sample yielded – 32.9 decibels while the dynamic intensity of the sample with vibrato was at – 29.2 decibels.

Music is a type of communication largely brought by the effects of the artistic *vibrat*o. The results of the research did not yield conclusive empirical results in the amount and extent of overtones. There are results in the field of the dynamic effects of the *vibrato*. The increased power in decibels of the high range samples containing *vibrato* versus the *non–vibrato* samples proved the difference in to the effects of *vibrato* regarding increased brilliance and suggestiveness of the performance. Dynamic increase in the sound projection was noticed in coincidence of the higher tension of the string and the intensifying dynamics of the tone due to the same releasing tension effect while using *vibrato* in high position and in the high range on the violin. The *vibrato* use contributed to less powerful sound during the performance in the lower range of the violin possibly coinciding with the releasing tension effects of the *vibrato* and the lower frequency physical vibration of the violin plates.

## 4. Conclusions

In conclusion of this article it is important to mention that what made this project interesting is the fact that there has been no analogous research of the *vibrato* in correlation to the sound quality. The role of the artefacts in the acoustical completion of the sound in artistic sense is especially interesting and unexplored. This and further exploration of the significance of the *vibrato* should be in the realm of psychoacoustics with analysis of the scanned brain activity using newly developed methods. The findings clearly point in this direction and define the need to further develop psychoacoustics as a whole new discipline within the acoustics. Determination that tone with artistic *vibrato* is more likely to be perceived as communication, is in contribution to a wider perception of the cross disciplinary function of the acoustics in the future. The significance of this research for the music industry and the education system is in setting the ground and the possibility of later creating software that can verify and approve certain levels of artistic involvement during public performances, or educational practices of music. It also adds the artistic dimension in to the acoustic analysis of the sound, in accordance of the most recent scientific trends of interdisciplinary fusion.

## 5. References

[1] Larsen O N and Goller F 1999 Role of syringeal vibrations in bird vocalizations *Biological Sciences* **266** 1429 1609–15
[2] Meyer J 2009 *Acoustics and the Performance of Music* (New York: Springer) ed 5th pp 18–22
[3] Moore B C J 2008 Basic auditory processes involved in the analysis of speech sounds *Biological Sciences* **363** 1493 947–63
[4] Schleske M The psychoacoustic secret of vibrato http://www.schleske.de/en/research/handbook http://www.schleske.de/en/research/handbook-violinacoustics/vibrato-of-the-musician.html
[5] Voce Vista Video Sygyt Software http://www.sygyt.com/en/

# Acoustical characteristics of device system for chasing away birds at Ngurah Rai International Airport Denpasar Bali

**H A Akil[1] and B Basuki[1]**

[1]Research Center For Metrology - Indonesian Institute of Sciences (RCM-LIPI) Komplek PUSPIPTEK, Setu, Tangerang Selatan, Banten.

E-mail: husein.avicenna.akil@lipi.go.id

**Abstract**. Ngurah Rai International Airport Denpasar Bali has two models of the device for chasing away birds these are Mobile Model and Fixed Model. Acoustical characteristics of the device for chasing away birds is determined by two parameters: SPL and Frequency Responses as well as sound propagation in the free field outdoor acoustics, thus to analyze sound properties radiated by the device for chasing away birds must be using a theoretical approach of Inverse Distance Law. Measurement results on Mobile Model that decreasing value of SPL by moving microphone position into the position of double distances from an original distance were somewhat 6 dB, it is nearly agreed with Inverse Distance Law. While measurement results on Fixed Model using the same mechanism of changing the distance the decreasing value of SPL does not comply Inverse Distance Law since the decreasing is about 10 dB. From results of measurement of decreasing SPL, both two models of the device for chasing away birds can be only used effectively to disperse some birds in which their position is very near the speaker that is about 4 - 8 m. Based on a calculation using Inverse Distance Law, the value of SPL at the distance of 1 m from the speaker is below 100 dB. Measurement results on Frequency Spectrum for Mobile Model was dominant at the frequency of 500 Hz to 3150 Hz, while for Fixed Model was dominant at the frequency of 500 Hz to 2500 Hz. At the end, the device for chasing away birds using acoustical method needs a system of sound generator and speaker which can generate the sound wave with very high of sound power level and it has a flat form of frequency responses with a wide range of frequencies. With the high sound power level, the sound will reach very far away this kind of device is called Long Range Acoustic Device (LRAD).

## 1. Introduction

An airport within its facilities must be able to guarantee about the safety of airline especially when the flight is taking off and landing. One of the standard requirements must be provided by the authority of International Airport is a device system for chasing away some birds. This device will be run to dissipate some birds when those birds are roaming in the grass area that is close to the runway aiming to avoid an occurrence of birds strike particularly if a bird gets into the flight engine.

The device system for chasing away bird using acoustical method is an acoustic device consist of the sound generator and speaker, the sound output from the speaker is directed to crowd of birds in order the birds flying away from the airport runway. This kind of acoustic device has been installed in the Ngurah Rai Airport Bali [1] and this paper reports about measurement results of sound characteristic exposed by the device then analysis was made whether the device is quite effective in driving out some birds that are frequently crowded around in a field near the airport runway.

IOP Conf. Series: Journal of Physics: Conf. Series **1075** (2018) 012002 doi:10.1088/1742-6596/1075/1/012002

In the year of 2003, Research Center for Metrology - Indonesian Institute of Sciences, RCM-LIPI (formerly Research Center for Calibration Instrumentation and Metrology) started carrying out a research on applying an acoustical method for driving out the birds in which at that time there were roaming some kinds of birds around Sukarno-Hatta Airport [2]. Since then it was found and anticipated that to drive out the bird effectively it should use sound exposure with a certain frequency, rather than the exposure of alike sound predator, and the frequency value of sound exposure must be arranged to be fitted with the sensitive frequency of the bird hearing [3]. Like a human hearing system in which has a frequency value of hearing sensitivity, this means if someone gets a high-level sound exposure and its sound frequency value is identic with his/her hearing sensitive frequency hence he/she gets a pain inside ears and reacting immediately to close both ears by his/her both hands. This is the human behavior when exposed to noise with the level of a pain threshold, it would also happen to live beings such as birds. When the birds receive a noise exposure with the level of a pain threshold of their hearing system it is expected that they will be flying away in order to avoid a painful noise.

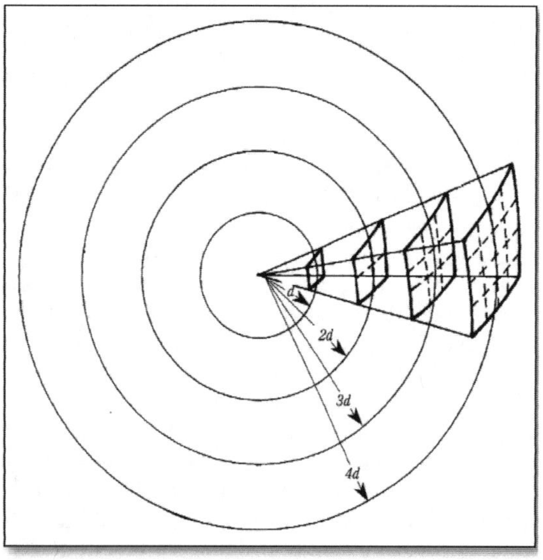


**Figure 1.** In the solid angle shown, the same sound energy is distributed over spherical surfaces of increasing area as $d$ is increased. The intensity of sound is inversely proportional to square of the distance from the point source [4].

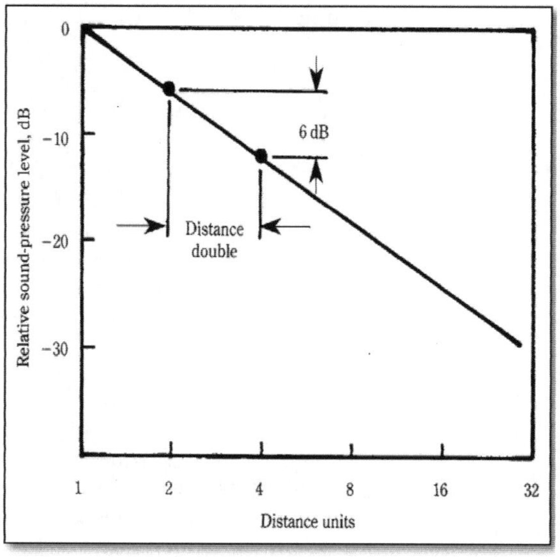


**Figure 2.** The inverse square law of sound intensity becomes the inverse distance law for sound pressure. This means that sound-pressure level is reduced 6 dB for each doubling of the distance [4].

## 2. Theoretical Approach

*2.1 Inverse Square Law and Inverse Distance Law*

A device for chasing away bird using an acoustical method that is attached in many airports is actually a device generating sound energy and radiating it in the free field as the activity of outdoor acoustics. Therefore, the theoretical approach is used in analyzing the affectivity of sound exposure is the theory of sound propagation in an open space free fields. Characteristic of sound propagation in the free field area is illustrated as shown in figure 1. A point sound source radiates sound energy to all direction with a steady sound power, intensity (power per unit area) of sound propagates spherically and uniformly extending with the radius of $d$, $2d$, $3d$, $4d$ and further. The intensity of sound is being decreased with increasing the radius. As the area of sphere surface is $4\pi r^2$, the area of a small segment on the surface of the sphere also varies as the square of the radius. Doubling the distance from $d$ to $2d$ reduces the intensity to 1/4, tripling the distance reduces the intensity to 1/9, and quadrupling the distance reduces the intensity to 1/16 [4]. The intensity of sound is inversely proportional to the square of the distance in a free field following the Inverse Square Law.

The intensity of sound as power per unit area is a difficult parameter to measure. Sound pressure is easily measured. As intensity is proportional to the square of sound pressure, the Inverse Square Law (for intensity) becomes the Inverse Distance Law (for sound pressure). This means sound pressure varies inversely as the first power of the distance. Figure 2 shows the illustration of the Inverse Distance Law for pressure applies in a free field, the sound-pressure level in decibels is plotted against distance and the sound pressure level is decreasing 6 dB per doubling of the distance. This calculation can be shown in equation (1) below, taking example the sound-pressure level *L1* at distance *d1* from a point source is known, the sound-pressure level *L2* at another distance *d2* can be calculated from:

$$L2 = L1 - 20log\frac{d2}{d1}\,\text{dB} \tag{1}$$

Thus, the difference in sound-pressure level between two measuring points that are *d1* and *d2* distance from the source is:

$$L2 - L1 = 20log\frac{d2}{d1}\,\text{dB} \tag{2}$$

*2.2 Frequency Sensitivity of Bird Hearing*

The theoretical approach of acoustics in analyzing the sensitivity of bird hearing system is through the understanding of an acoustical characteristic of a human hearing system since an audible range of bird hearing is almost comparable with the audible range of human hearing system [2], [3], [5]. The most hearing sensitive frequency of birds is a frequency value on the lowest sound level of bird threshold hearing. This means that a sound received by a bird at its hearing sensitive frequency is the most easily heard and it makes a quick reaction from birds although the sound level at its hearing sensitive frequency is categorically very low. On the other side, a sound-pressure level of threshold pain at its hearing sensitive frequency is the lowest value compared with the sound-pressure levels at its other frequencies of threshold pain. When the hearing sensitive frequency of birds is known, it is, therefore, to drive away birds using acoustical technique can be more efficient e.g. to generate sound at a hearing sensitive frequency with the sound power level is somewhat not too high.

A mechanism to get a hearing sensitive frequency of a certain kind of bird so far is to do some measurement and make a recording of sound (whistle) incurred by a certain kind of bird [6], [7]. A certain kind of bird will whistle sound at certain range frequencies and those values of range frequencies are considerably identic with the sound frequencies received by a hearing organ of that certain kind of bird. In other words, an occurred sound frequency spectrum is exactly hearing frequency spectrum of a certain kind bird. From the sound recording, it can be identified a peak of sound spectrum frequency which is later assumed that the peak of the spectrum is an indicator of the highest hearing sensitive of that certain kind of bird. Figure 3 shows an example of the measurement result graphs of the sound whistled by a kind of birds at Lagoon Nusa Dua in the year of 2009. Those birds are similar with the kind of birds that frequently roaming in the area close to the runway at Ngurah Rai Airport. The graphs show that the peak of sound frequency spectrum occurred at low frequencies of 40 Hz and 80 Hz and also occurred at mid frequencies of 1250 Hz and 6300 Hz. Those values of frequencies can be as a reference or assumed that a-hearing-sensitive frequencies of the kind of birds that frequently roaming in the area close to the runway at Ngurah Rai Airport are 40 Hz, 80 Hz, 1250 Hz and 6300 Hz.

The bird audible characteristic is comparable with the human audible characteristic therefore, a bird normal hearing is like a human normal hearing that is human can receive sound normally with SPL under 70 dB. It is, therefore, like a human, a bird will get annoyance when hit by noise exposure of 85 dB [8]. As already mentioned before that if someone gets annoyed by high noise exposure he/she will instantly react avoiding interference, so will do the bird by flying away from a noise.

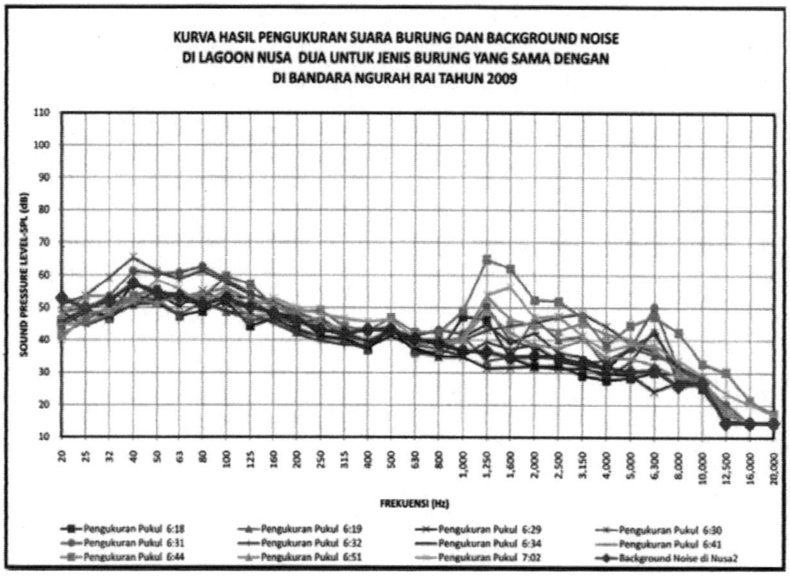


**Figure 3.** A sound frequency spectrum of a certain kind of birds similar with the kind of birds that frequently roaming in the area closed to the runway at Ngurah Rai Airport: the results of measurement taken in 2009 [7].

## 3. Measurement Results and Discussion

Ngurah Rai International Airport has had two models of bird chasing devices that are mobile and fixed models as shown in figure 4. In the mobile model, speakers as a point sound source is installed in the pickup car with the high position of 2 m from the ground, in a fixed-model-speakers installed in a permanent position with the high is only 1 m from the ground and placed in a field area where birds are frequently roaming around this area. As according to a theoretical approach, acoustical characteristic of these devices is determined by two parameters that are the value of SPL and frequency spectrum. Measuring SPL can examine at once the ability of sound source generator and speakers to generate a maximal sound power, and this is the main indicator for sound propagation ability in the free field open space: the higher sound power the further distance of the receiver can receive a sound with significantly high level.

Measurement of SPL at Ngurah Rai Airport was carried out with the shortest distance of receiver (microphone) position from the sound source was 4 m. Then, measurement of decreasing characteristic on SPL by moving receiver position away from a sound source and measurement of the frequency spectrum in order to find a frequency response characteristic of a sound source of each model devices. It was found that a sound exposure of each source from each model device was not a tune sound of single frequency but consisting various frequencies including artificial predator sound i.e. alike hawk voice.

**Figure 4.** Speakers of the device for chasing away birds: mobile model and the fixed model.

**Table 1**. Measurement results of SPL for mobile model and for fixed model

| Microphone position from Speakers (m) | Sound Pressure Level, SPL (dB) | |
|---|---|---|
| | Mobile Model | Fixed Model |
| 4 | 86,6 | 84,0 |
| 32 | 75,7 | 74,8 |
| 64 | 69,6 | 65,0 |
| 128 | 61,7 | 54,0 |

Table 1 shows measurement results of SPL for both two models. For mobile model decreasing SPL further the position of 32 m by moving microphone position at double distances of the last position of microphone that is 32 m to 64 m and then to 128 m decreasing of SPL is approaching the Inverse Distance Law that is somewhat 6dB. While for the fixed model as the same way as did for the mobile model (32 m, 64 m and 128 m) decreasing of SPL value in not meet the Inverse Distance Law that is somewhat 10 dB this is because the placement position of a fixed-model speaker is only 1 m from the ground so that it does not comply a condition of open space free field since the soil and grass on the ground are high sound absorbent.

Measurement results indicated strongly that both two models of the bird chasing devices are much far from minimum requirements needed by a bird chasing device i.e. a sound radiation generated by its device is expected still working effectively to disperse the crowd birds at 100 m position from the device. In order to comply with minimum requirements as well as to meet the Inverse Distance Law, the minimum SPL of the device is ideally 124 dB at the distance of 2 m or about 118 at the distance of 4 m thus it is expected at the distance of 100 m the value of SPL is still higher than 85dB. While for both two models with the condition of an acoustical characteristic as the data shown by the measurement results, the devices will only be effective to disperse the crowd birds when they were very close to the speakers that are 4 - 8 m. Using calculation of the Inverse Distance Law, the value of SPL at the distance of 1 m from speakers for the mobile model and for the fixed model is respectively 98 dB and 96dB.

Measuring a frequency spectrum is also called measuring the frequency characteristic was carried out for both two models. Microphone distances from speakers for each model were 1m for the mobile model and 4 m for the fixed model. The chosen distance is purely based on an environmental condition and the measurement can be carried out conveniently as long as a sound level received by the microphone was far above a background noise level. Figures 5 and 6 are showing the graphs of measurement results of the frequency spectrum for mobile and fixed models. For the mobile model starting from frequency of 500 Hz up to 3150 Hz the form of a spectrum is considerable flat and quite dominant since the sound level average at this frequency range is the highest. While for the fixed model, frequency dominant start from 500 Hz until 2500 Hz. According to measurement results in 2009 (see figure 3) hearing sensitive of the kind of birds that often roaming in the area close to the runway and assuming that the peak of frequency spectrum is a hearing sensitive frequency of the birds so those two models of bird chasing devices would only be effective for sound radiation at the frequency of 1250 Hz.

At last, the above discussion has given a clue about very a simple solution that is to chase away the crowd birds using the acoustical technique it is needed a sound generator and a speaker system that can generate the sound wave with a very high sound power level and a wide flat frequency response. A high-level sound source is automatically able to have far-reaching sound energy. This kind of an acoustic device is usually required for the military activity this means the device must have a specification of a military standard. The device can be used as a non-lethal weapon [9], this kind of sound source is also known as speaker of Long Range Acoustic Device (LRAD) [10] [11]. In many countries such as in the USA, Germany, as well as several other countries in Europe, LARD has been used by the police as equipment. In the event of demonstration, the police use the equipment of LRAD to disperse crowd large group of people rally. An example of a speaker product as equipment of Long Range Acoustic Device (LARD) is shown in figure 7.

| Frequency (Hz) | SPL (dB) |
|---|---|
| 40 | 75.5 |
| 50 | 72.4 |
| 63 | 76.5 |
| 80 | 69.5 |
| 100 | 70.7 |
| 125 | 68.8 |
| 160 | 64.7 |
| 200 | 61.4 |
| 250 | 62.0 |
| 315 | 64.1 |
| 400 | 77.1 |
| 500 | 85.2 |
| 630 | 86.6 |
| 800 | 81.7 |
| 1000 | 81.2 |
| 1250 | 80.6 |
| 1600 | 82.5 |
| 2000 | 81.3 |
| 2500 | 80.9 |
| 3150 | 78.2 |
| 4000 | 70.8 |
| 5000 | 66.6 |
| 6300 | 61.5 |
| 8000 | 57.9 |
| 10000 | 51.5 |

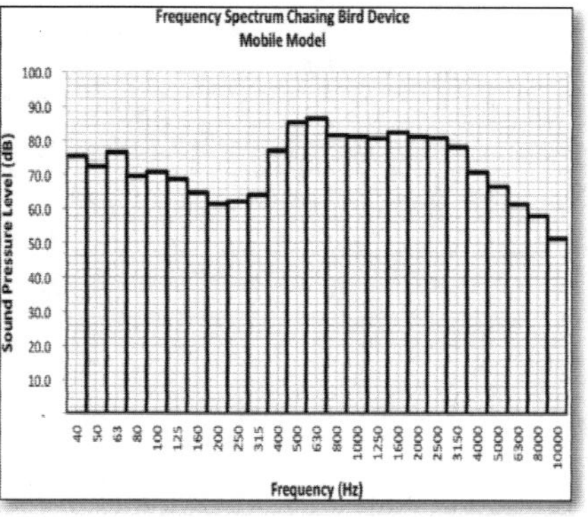


**Figure 5.** The frequency spectrum of a sound source of the mobile model.

| Frequency (Hz) | SPL (dB) |
|---|---|
| 40 | 63.0 |
| 50 | 68.8 |
| 63 | 65.4 |
| 80 | 68.5 |
| 100 | 70.6 |
| 125 | 74.4 |
| 160 | 63.4 |
| 200 | 57.5 |
| 250 | 58.8 |
| 315 | 63.2 |
| 400 | 65.0 |
| 500 | 80.8 |
| 630 | 84.2 |
| 800 | 84.8 |
| 1000 | 92.9 |
| 1250 | 90.9 |
| 1600 | 87.1 |
| 2000 | 79.0 |
| 2500 | 78.1 |
| 3150 | 67.9 |
| 4000 | 57.0 |
| 5000 | 54.8 |
| 6300 | 46.6 |
| 8000 | 42.4 |
| 10000 | 35.5 |

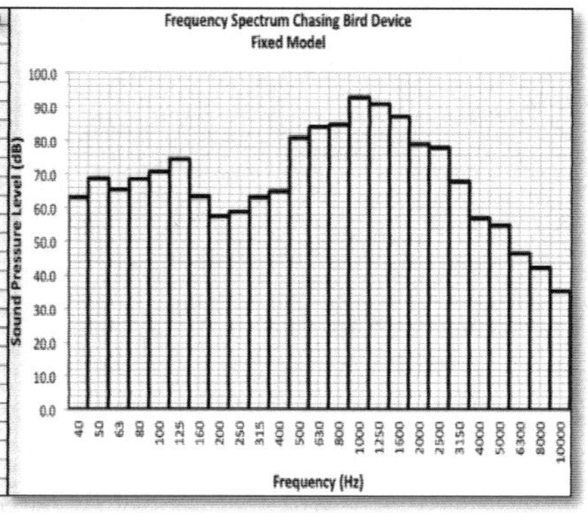


**Figure 6.** The frequency spectrum of a sound source of the fixed model.

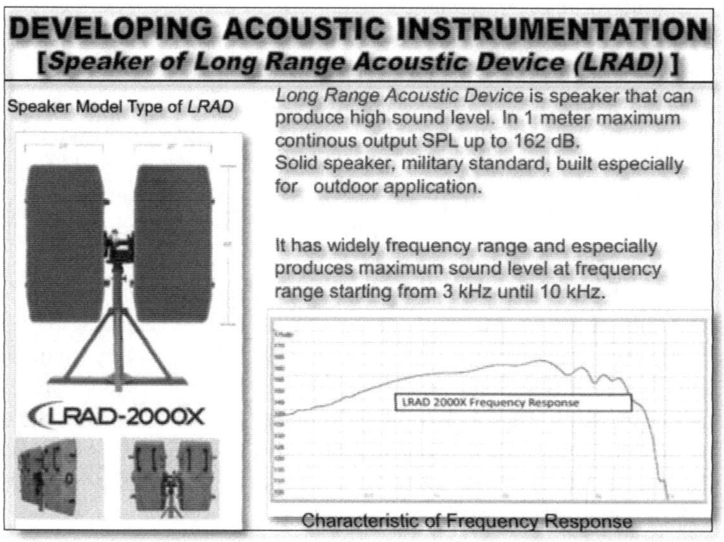


**Figure 7.** Product Example of Speaker: Long Range Acoustic Device (*LRAD*) [11]

## 4. Conclusions

Acoustical characteristics of the bird chasing devices of Ngurah Rai International Airport Denpasar Bali namely mobile model and fixed model both two models have a sound-pressure level (SPL) less than 100 dB at the position of 1 m. With this value, the coverage chasing birds is very short that could only reach birds within the position somewhat 8 m. Dominant frequencies generated by both two models are not quite wide enough, the mobile model starts from 500 Hz up to 3500 Hz and the fixed model is even much narrow that starts from 500 up to 2500 Hz. The development of the bird chasing device using acoustical technique is actually the development of acoustic instrumentation technology i.e. developing sound generator with its speaker arranged to have a capability in generating SPL up to 160 dB in 1 m, much higher of sound power level that means much far away distance covered by sound propagation, so high level sound pressure will still be received by a receiver with very far position from the device.

## 5. References

[1] Pusat Penelitian Metrologi (P2 Metrologi) LIPI, 2016, Laporan Pengukuran Lingkungan di Bandar Udara Internasional Ngurah Rai Bali Tanggal 26 sd 29 Januari 2016: Resume Hasil Pengukuran Birds Strike Control Unit di Bandara Internasional Ngurah Rai Denpasar Bali tanggal 27-28 Januari 2016.

[2] Akil, Husein A., Suwono., Rustandi, Dadang., Sewaka., Basuki, Budhy, 2003, *Metode Akustik Untuk Mengusir Jenis Burung Yang Terdapat di Sekitar Landasan Pacu Bandara Soekarno-Hatta*, Prosiding Seminar Nasional Akustik & Teknik Tata Surya, hal 1 – 10.

[3] Akil, Husein A., Maruf, Ahnan., Suwono., Basuki, Budhy., 2005, *Karakteristik Frekuensi Pendengaran Jenis Burung Yang Berada di Sekitar Landasan Pacu Bandara Juanda Surabaya,* Prosiding Vibrastik, ITS Surabaya.

[4] Everest Alton F., 2001, *Master Handbook of Acou*stics, *Fourth Edition*, McGraw-Hill, page 41-87.

[5] Mayntz, Melissa., update 04/22/17., *Bird Sense – Sight, Hearing, Touch, Taste and Smell, How Birds Use Their 5 Sense,*. The Spruce, https://www.thespruce.com/birds-five-sense-386441.

[6] Akil, Husein. A., 2012, *Identifikasi dan Pengukuran Karakteristik Akustik Jenis Burung Yang Berkeliaran di Kawasan Bandara Polonia/Kuala Namu-Medan dan Sultan Hasanuddin-Makasar*, Instrumentasi Scientific Publication, Volume 36 No. 1, hal. 55-67.

[7] Akil, Husein. A., Basuki, Budhy., 2010, *Analisis Karakteristik Frekuensi Akustik Burung Yang Berkeliaran di Daerah Landasan Pacu Bandara: Soekarno-Hatta, Juanda dan Ngurah Rai*, Instrumentasi Scientific Publication, Volume 34 No. 1, hal. 59-74.

[8] Adams, Melville. S. and McManus, Francis., 1994, *Noise and Noise Law A Practical Approach*, John Wiley & Sons Inc, Colorado Springs, USA., page 39-42.

[9] Altmann J., 2001, *Acoustic Weapon - A Prospective Assessment*, Science & Global Security Vol 9, pp. 165-234.

[10] American Technology Corp., 2005, *Long Range Acoustic Device*, Power Point Presentation, September 2005, San Diego, CA: at http:/www.atcsd.com/

[11] LRAD, *LRAD Product Guide*, https://www.Iradx.com/product_categories/long-range-acoustic-devices/

**Acknowledgements**

The authors would like to express gratitude to the Management Authority, particularly to Safety Management System Team of Ngurah Rai International Airport Denpasar Bali PT Angkasa Pura I for giving us permission to carry out some acoustic measurements. Thanks to the management of Research Center for Metrology (RCM - LIPI) that has given us an opportunity to conduct a research and development on device system against bird strike. Thanks are also delivered to Mr. Akhmad Suwandi and Mr. Rukmana who have made the measurement in the fields.

IOP Conf. Series: Journal of Physics: Conf. Series **1075** (2018) 012003 doi:10.1088/1742-6596/1075/1/012003

# Exploring the biodiversity in Alas Purwo National Park, East Java through soundscape ecology

**G R Setyantho[1], S S Utami[1] and S Hadi[2]**

[1]Department of Nuclear Engineering and Engineering Physics, Universitas Gadjah Mada, Indonesia
[2]Faculty of Biology, Universitas Gadjah Mada, Indonesia

E-mail: gigihrahmandhani@gmail.com

**Abstract**. Biodiversity assessment is becoming one important things to do, to conserve ecosystem. Biodiversity assessment's new method was utilized the nature sounds which is produced by animals in an ecosystem. There are three possibilities sounds sources in an ecosystem, they are: biophonies, geophonies, and anthrophonies, those sound sources will reflect environment acoustics characteristics, called Soundscape. This paper utilized the Soundscape method to predict biodiversity in a forest. Alas Purwo National Park, Banyuwangi, East Java was the location where soundscape data were collected for several days. The soundscape data was processed using R-Statistical Computing Environment software to compute some index values of Bioacoustics Index (BIO), Acoustic Entropy (H), Acoustic Evenness Index (AEI) and Normalized Difference Soundscape Index (NDSI). These indices enabled to characterize the forest in Alas Purwo National Park, East Java, Indonesia for several types of forests selected. The AEI value of 0.69 indicate inequality between frequency bands. The BIO value of 10.79 indicate a forest that is dominant of biotic component. The H value of 0.88 indicate more diverse biotic components. The NDSI value of 0.97 indicate a forest that is dominant with biophonies rather than antrophonies. The results have indicated that some indices (AEI and BIO) could reflect the biodiversity in selected ecosystem, but other indices (H and NDSI) showed anomaly result because of difference landscape between selected ecosystem and ecosystem's weather when data collecting period.

## 1. Introduction

Indonesia has a conservation area that aims to maintain the existence of forest area and its biodiversity. Conservation area is a natural ecosystem to be a shelter for flora and fauna [1]. The more advanced technology, caused human life will be more dynamic. The dynamics of human life covers for residential needs or industrial needs. Frequently, deforestation (forest removal process) become a choice to fulfilled the needs. In 2012-2013, deforestation in Indonesia counted 727,981.2 hectares/year [1]. Of course, with the increasing deforestation rates, the forests in the conservation areas will decrease.

Biodiversity assessment in an ecosystem is a major work for the field of biological conservation. Biodiversity assessment, nowadays, becomes an important work since the widespread deforestation that occurs. This assessment not only assesses the richness of species present in the ecosystem, but also sees changes in species within a given period [2]. Approach methods for biodiversity assessment have been developed before, from Rapid Biodiversity Assessments (RBA) to All-Taxa Biodiversity Inventories (ABTI).

Expeditions undertaken by researchers around the forest to do biodiversity assessment, these methods are inefficient in terms of quality, time and cost [2]. Ordered sounds of animals in a natural ecosystem, are often encountered in a complex natural ecosystem. The sounds can be a source of information to predict the diversity of animal communities (biodiversity) in a forest, if it is processed and analyzed further. Acoustic index in a natural ecosystem is expected to become a new method in biodiversity assessment approach.

The purpose of this paper is to perform soundscape in several types of forest ecosystems in Alas Purwo National Park. The results of the soundscape are processed further so that forest ecosystem's characteristics can be known by using the acoustic index.

Utilization of acoustic data to predict biodiversity in an ecosystem had been done for the last decade. Initially, utilization of acoustic data was performed by Boelman, *et al.* (2007) to estimate the correlation between plant invasion and Aves occurrence in the Hawaiian Rainforest [3]. Then, Sueur, *et al.* (2008) offer a new method for describing biodiversity in Tanzania's Coastal Forest, this method could represent the biodiversity in location and could reveal ecosystem differences based on degrees of vegetation [2]. In contrast to Sueur, Villanueva-Rivera, *et al.* (2011) offers different acoustic data processing based on the frequency domain [4]. The following research, performing acoustic data measurements and data processing as previous studies, in the different location of the case study, namely: Twin Lakes, Michigan [5]; Minas Gerais, Brazil [6]; Queensland, Australia [7][8]. In addition to its relationship to biodiversity, Fuller, *et al.* (2015) examined that utilization of acoustic data could reveal the landscape in the surrounding ecosystem [8].

The index used in soundscape ecology determined biodiversity and landscape conditions of an ecosystem. The Bioacoustics Index (BIO) became the first index introduced by Boelman, *et al.* (2007) [3]. BIO index used, at that time, to measure the spectrum in the Aves' sound frequency range, in the range 2000 Hz to 8000 kHz [3]. In following research, Sueur *et al.* (2008) proposed an index that measure not only on one species of animal, but the entire biotic component of an ecosystem. To measure biodiversity in the ecosystem, Sueur, *et al.* (2008) used index Acoustic Entropy Index (H) and Acoustic Dissimilarity Index (D). Previous research was analyzed acoustic signals in time-domain, then Villanueva-Rivera, *et al.* (2011) analyze the acoustic signal using frequency-domain. This study referred to the hypothesis that each species of biotic component in a natural ecosystem has unique frequencies and unique active times that are different from one another. Villanueva-Rivera, *et al.* (2011) perform calculations to obtain an index to study an ecosystem, example: diversity index (called Acoustic Diversity Index (ADI)), evenness (called Acoustic Evenness Index (AEI)), and dominance over the frequency range. The results showed that the difference of point (location) of data retrieval will result in different index values. This approach method can be used as a material for spatiotemporal comparisons [4].

Pieretti, *et al.* (2011) review the index to measure the songs produced from Aves. This index was called the Acoustic Complexity Index (ACI), this index is claimed to be a new method for observing Aves despite anthrophonies or climate [9]. The previous indexes that proposed by investigators were still investigating the source of sounds derived from animals (biophonies), but Kasten, *et al.* (2012) states that acoustic data could also be a tool for knowing human presence (anthrophonies). Kasten, *et al.* (2012) performed Power Spectral Density (PSD) calculations for biophonies frequency range between 2000 Hz to 8000 Hz and anthrophonies frequency range between 1000 Hz to 2000 Hz. The results of both PSD are then compared and become the index value of The Normalized Difference Soundscape Index (NDSI) [5].

The indices were also examined by Gage, *et al.* (2017) and Fuller, *et al.* (2015). They tried to measure biodiversity in Queensland, Australia [7][8]. Gage, *et al.* (2017) using the indexes (The Normalized Power Spectral Density, ACI, ADI, AEI, BIO, H, and NDSI) for the present soundscape metrics at Samford Ecological Research Facility in Queensland, Australia. The presentation using soundscape metrics will facilitate the correlation, and the results indicate that the number of species (dawn chorus) will correlate with the soundscape power in the range 3000 Hz to 4000 Hz and also correlate with the ADI, AEI, and H indices [7]. The correlation between the index and the number of species still had not fulfilled the main objective of soundscape ecology, which can also describe the landscape of the ecosystem, so Fuller *et al.* (2015) conducted research using the indexes (ACI, BIO, H, ADI, AEI, and NDSI) at 19 different sites. Fuller, *et al.* (2015) correlated between the index results with the landscape at the location of the data. The correlation resulted that the index H, AEI, and NDSI are the best indexes to represent the landscape state of a soundscape, while the BIO and ACI parameters will better represent Aves richness in each forest type [8].

## 2. Methods

This study was conducted using primary data sources. The primary data source was obtained by taking sound recordings in several types of Ecosystem. Sound recordings were process to obtain soundscape ecology index and characterize each ecosystem using those indexes.

### *2.1. Recording Sites*

The location for this research was in Alas Purwo National Park, Banyuwangi, East Java. In this National Park, there are several types of forest ecosystems that exist, namely: Coastal Forest, Mangrove Forest, Lowland Plantation Forest, Bamboo Savana Forest, and Plantation Forest. There are 3 types of forest ecosystem selected for this research, namely: Lowland Forest Ecosystem (tag A), Plantation Forest Ecosystem (tag B), and Coastal Forest Ecosystem (tag C).

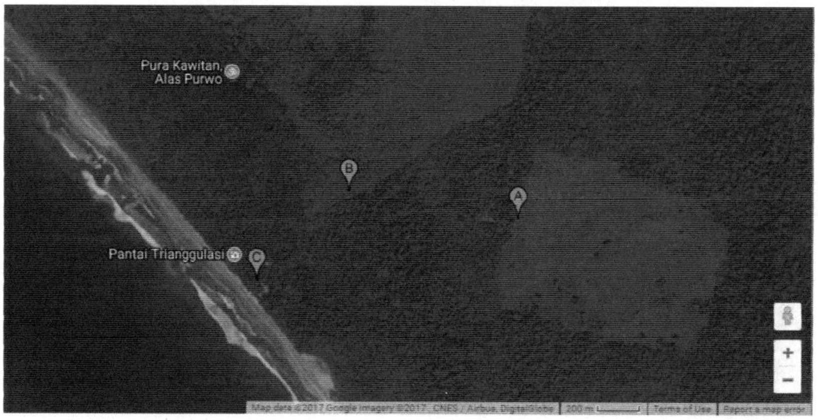


**Figure 1.** Point of recording sites in several types of ecosystem.

### *2.2. Recording Duration and Tools*

Data collecting period was conducted on April 19th, 2017 to April 23rd, 2017. Data collecting would record the environment sounds. For recording, we used mobile phones with external microphone and installed at each data point. The mobile phones had been installed with ARBIMON-Touch applications, this application will allow the device to record for 1-minute for every 10-minute interval. This timing scheme corresponded to the data collecting method used in the previous research.

### *2.3. Recording Duration and Tools*

Data processing use functions in 'seewave' and 'soundecology' that installed in R-Statistical and also Microsoft Excel 2013. R-Statistical allowed to process audio file into the indexes values. To processed such a big data, function multiple_sounds will process all the audio file in a folder at a time. Output from the multiple_sounds function will be a file with .csv (Comma-Separated Values) extension. To visualize the data, Microsoft Excel 2013 processed the .csv file into a graphic for every index (AEI, BIO, H, and NDSI) in each ecosystem.

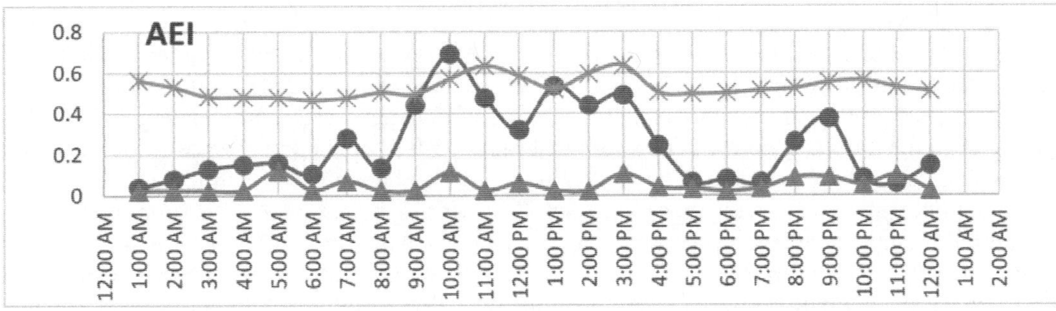


**Figure 2.** AEI in each ecosystem

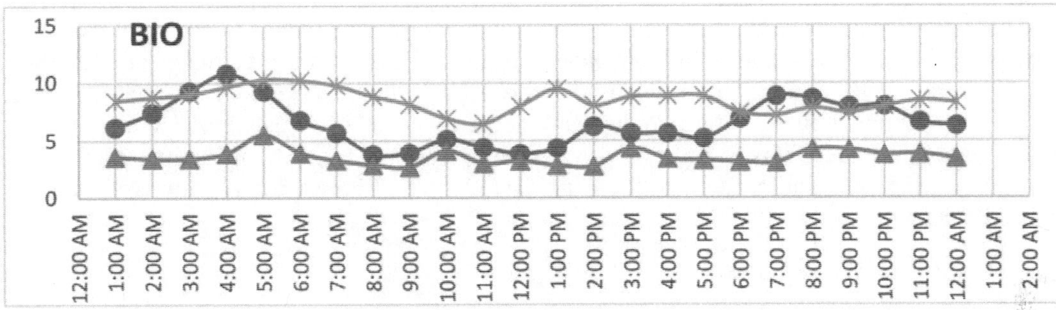


**Figure 3.** BIO in each ecosystem

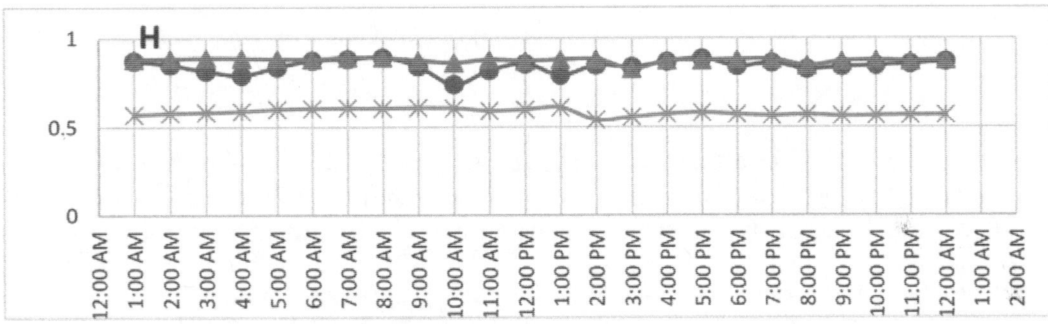


**Figure 4.** H in each ecosystem

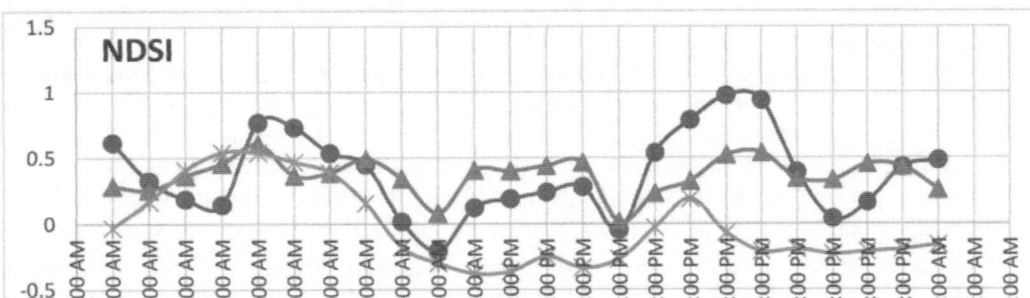


**Figure 5.** Point of recording sites in several types of ecosystem.

Information:

: Index values in plantation forest ecosystem

: Index values in lowland forest ecosystem

: Index values in coastal forest ecosystem

## 3. Result and Discussion

The results showed that the indexes can be used as a tool to predict biodiversity in an ecosystem. The AEI in the Plantation Ecosystem had range of 0.02 to 0.12 indicating the spectrum at each frequency range was equally strong. AEI also indicates that there is a frequency range dominance in Coastal Forest Ecosystem. H in the Lowland Forest Ecosystem had range of 0.74 to 0.89 and tends to fluctuate at any given time, indicating that animals in the ecosystem have a schedule of occurrences. NDSI on the Coastal Forest Ecosystem had a wide range between -0.38 to 0.54 indicating that the sound source varies. BIO monitored the timing of dawn and dusk singing of each ecosystem.

AEI and BIO, can visualize the biodiversity in an ecosystem well. H and NDSI seems to show anomaly data. H, theoretically, will increase proportional with the biological diversity in an ecosystem. H in Plantation Forest Ecosystem was the highest than Lowland Forest Ecosystem and Coastal Forest Ecosystem. Theoretically, biological diversity in Plantation is not bigger than Lowland Forest Ecosystem, so this result seems anomaly. It happens because of Shannon's Index, which the BIO calculation based, have a character proportional to probabilities. Another anomaly data, NDSI result which are zero-values and negative-values. Theoretically, the zero-value of NDSI indicate that biophonies proportional with anthrophonies and negative-value of NDSI indicate that biophonies weaker with anthrophonies. Instead of that theory, zero-values of NDSI happened because of raining when the recording time, and negative-values of NDSI in Coastal Forest Ecosystem happened because of sounds of waves.

## 4. Conclusion

Those indices are interrelated and need to be interpreted simultaneously, to obtain a more comprehensive interpretation result. In addition to biophonies and anthrophonies, the index can also indicate the presence of geophonies. Although the index may describe biodiversity and landscape at the location of the data collection, it is preferable if the description of data collection and voice identification could be done.

## References

[1] Center of Data and Information MLEF, *Statistik Kementerian Lingkungan Hidup dan Kehutanan Tahun 2014*. Ministry of Living Environment and Forestry (MLEF) Republic of Indonesia.

[2] S Pavoine and O Hamerlynck, "Rapid Acoustic Survey for Biodiversity Appraisal," vol. 3, no. 12, 2008.

[3] N T Boelman, G P Asner, P J Hart and R E Martin, "Multi-Trophic Invasion Resistance in Hawaii: Bioacoustics, Field Surveys and Airborne Remote Sensing" *Ecol. Appl.*, vol. 17, no. 8, pp. 2137–2144, 2007.

[4] L J Villanueva-rivera, B C Pijanowski, J Doucette and B Pekin, "A primer of acoustic analysis for landscape ecologists" pp. 1233–1246, 2011.

[5] E P Kasten, S H Gage, J Fox and W Joo, "The remote environmental assessment laboratory's acoustic library: An archive for studying soundscape ecology" *Ecol. Inform.*, vol. 12, pp. 50–67, 2012.

[6] N Pieretti, U Carlo, M Rodrigues and R. J. Young, "Determining Temporal Sampling Schemes for Passive Acoustic Studies in Different Tropical Ecosystems" no. March, 2015.

[7] S H Gage, J Wimmer, T Tarrant and P R Grace, "Acoustic patterns at the Samford Ecological Research Facility in South East Queensland, Australia: The Peri-Urban SuperSite of the Terrestrial Ecosystem Research Network" *Ecol. Inform.*, vol. 38, pp. 62–75, 2017.

[8] S Fuller, A C Axel, D Tucker and S H Gage, "Connecting soundscape to landscape: Which acoustic index best describes landscape configuration?" *Ecol. Indic.*, vol. 58, pp. 207–215, 2015.

[9] N Pieretti, A Farina and D Morri, "A new methodology to infer the singing activity of an avian community: The Acoustic Complexity Index (ACI)" *Ecol. Indic.*, vol. 11, no. 3, pp. 868–873, 2011.

# Current situation of acoustic environment in childcare facilities in Japan: A comprehensive survey in Kumamoto City region

**K Kawai**

Kumamoto University
2-39-1 Kurokami Chuo-ku Kumamoto, 860-8555 Japan

E-mail: kkawai@kumamoto-u.ac.jp

**Abstract**. This is a report on survey results gathered from questionnaires and site visits conducted at licensed child daycare centers in Kumamoto City, a regional core city of approximately 735,000 population, and its adjacent Koshi City (app. 59,000 population). The survey involved 132 centers, and responses were received from 2,075 childcare providers. Results from the site visits indicated that about half of the classes in the centers were acoustically separated by sound insulators such as fixed walls, while the remaining half utilize movable partitions or multiple classes held in a single room. Only 13% of daycare rooms were equipped with sound-absorbing materials. Many childcare providers responsible for children aged three or older, who have more conversation than younger children, reported problems involving throat damage and difficulty conducting conversations. There were indications that using sound-absorbing materials in daycare rooms and separating class spaces by soundproof wall could mitigate these situations.

## 1. Introduction

Child day-care centres, in which children of 0 to 5 years old spend most of their active hours in a day, have had two major problems in terms of the acoustic environment: high sound pressure levels due to children's activity [1], [3] and low speech intelligibility due to the high sound pressure level and excessive reverberation, which could interfere verbal communication of children [4], [5]. It is considered to be possible to alleviate these problems by means of architectural acoustic design. Actually, guidelines with specific values are provided by World Health Organization [6] and Western countries have legislations or standards for the acoustic design of child day-cares [7], [9]. Also, the authors have been conducting on-site experiments by temporarily providing sound-absorbing materials to daycare rooms since 2008 and the results indicated considerably greater noise reduction than that was estimated by the increase of equivalent sound absorption area in the rooms [10].

In Japan, however, there are presently no effective acoustic design standards or guidelines, and there is little research in this area. However, child daycare centers are rapidly proliferating without much consideration given to acoustic environments for children, particularly in designated large cities in response to the growing demand for daycare services as women participate more broadly in society alongside men. In this respect there is an urgent need of acoustic design guidelines for child daycare centers in Japan. Based on this background, the present study undertook a large-scale survey to investigate the current state of acoustic environments in daycare centers.

## 2. Survey overview

This survey comprised inspection of daycare buildings and questionnaires distributed to the daycare administrators and childcare providers. The objective of the inspections was to capture room configurations and partitioning between rooms, as well as any use of sound-absorbing materials. Room configurations in daycare centers are multifarious, in general, and there are many examples of rooms being divided not by fixed walls so as to enable more flexible usage, employing structures that were ineffective in sound insulation such as movable partitions and furniture. Also, reflecting the current lack of design guidelines, it was expected that there was little construction using sound-absorbing materials. Having a grasp on the current state of these factors will provide an important basis for the development of acoustic design guidelines in the future.

The study looked at all of the "licensed nursery schools," the official English expression of nationally licensed daycare centers, in two cities in Kumamoto Prefecture (Kumamoto City, a regional core city with a population of about 735,000, and its adjacent Koshi City, with a population of about 59,000, in 2015.1 currently). In order to survey all of the centers, we obtained the cooperation of nursery school associations in each city and introduced the survey at regular meetings of school directors, distributing complete sets of questionnaires and requesting that they be distributed to and then collected from their childcare providers. We later visited each of the centers and collected the answered questionnaire sheets, and also conducted inspections. This resulted in site inspections of 132 centers out of a total of 140, and more than 2,000 responses collected from childcare providers described in table 1.

Items for site inspections are listed in Table 2. The surveyors needed to inspect each of the sites visually because we expected that personnel on-site might not be knowledgeable about which materials would absorb sound and that there would be various space configurations.

Two questionnaires were created, with one for administrators and one for childcare providers. The questions are listed in Table 3. The providers were asked to answer their subjective evaluation on 3-5 step scales. Administrators were mainly asked about general information about the center.

**Table 1.** Survey overview.

| City | Daycare centers | Surveyed centers | Provider responses |
|---|---|---|---|
| Kumamoto | 122 | 115 | 1809 |
| Koshi | 18 | 17 | 266 |
| Total | 140 | 132 | 2075 |

**Table 2**. Items in site inspection.

| | Inspection items |
|---|---|
| In the center | Room arrangement (type of partition between rooms, number of story), surface materials (sound absorptive or not), class arrangement. |
| Surroundings | Surrounding environment, sound insulation measures. |

**Table 3**. Questions regarding acoustic environment.

| | Question Items |
|---|---|
| Providers | Difficulty of conversing with children, room reverberation, throat damage frequency, noise from other classes, suitability of the room for childcare, disturbance by noise from outside, concern about noise from the childcare center to neighbours. |
| Administrators | Year of building construction and renovations, construction structure, frequency of receiving complaints, numbers of providers and children, class configurations (mixed ages, etc.) Concern about acoustics of the center. |

IOP Conf. Series: Journal of Physics: Conf. Series **1075** (2018) 012004 doi:10.1088/1742-6596/1075/1/012004

## 3. Inspection result

### *3.1. Number of children*

Number of children per center was answered by the administrator of each center described in figure 1. The mode was around 100 children and seven centers had more than 200 children with 320 as the maximum. This result was compared with the official capacity of the centers and it was revealed that 95 % of the centers exceeded the official capacity with half of them exceeding 15-20 %. This is considered to be a reflection of the current demand for child daycare in the cities in Japan.

### *3.2. Types of class environments*

Table 4 summarized the room separation between classes. Based on the result and from a view point of sound insulation, the room separation was categorized into four types as shown in the table. "Common rooms" indicates toilets and closets shared between two rooms located on either side. "Partial openings" are doors or other openings provided to enable access back and forth between two rooms without a hallway being accessed.

The result indicated that 51 % of the classrooms were acoustically separated with fixed walls. Movable partitions, which have poor sound insulating performance in general, were used in 20 % of class borders, and 11 % had partial openings with adjacent rooms that were usually open and did not allow for sound insulation. In 19 % of the cases, single rooms were being used for multiple classes. As a whole, acoustically insulated rooms turned out to be only half.

**Table 4.** Boundaries between classes

| Type | Class Boundaries | Number of class | Ratio (%) | Ratio (%) |
|---|---|---|---|---|
| Separated | Fixed walls | 330 | 41.9 | 51.1 |
| | Shared room (door closed) | 41 | 5.2 | |
| | Partial opening (door closed) | 32 | 4.1 | |
| Movable | Movable partition (closed) | 154 | 19.5 | 19.5 |
| Adjoined | Movable partition (partially open) | 8 | 1.0 | 10.7 |
| | Shared room (door open) | 44 | 5.6 | |
| | Partial opening (door open) | 32 | 4.1 | |
| Same room | Movable partition (open) | 25 | 3.2 | 18.7 |
| | Same room (bookcases, columns, etc. partitioning) | 88 | 11.2 | |
| | Same room (no partitions) | 34 | 4.3 | |
| | Total | 788 | 100 | 100 |

**Table 5.** Numbers of childcare provider responses by age of homeroom children.

| Ages (year) | 0 | 1 | 2 | 3 | 4 | 5 | Mixed ages | Total |
|---|---|---|---|---|---|---|---|---|
| Nos. | 474 | 355 | 374 | 190 | 153 | 150 | 379 | 2075 |

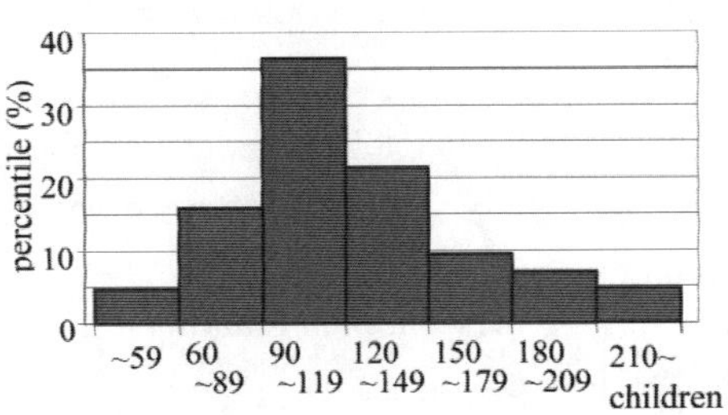


**Figure 1.** Distribution of number of children per center.

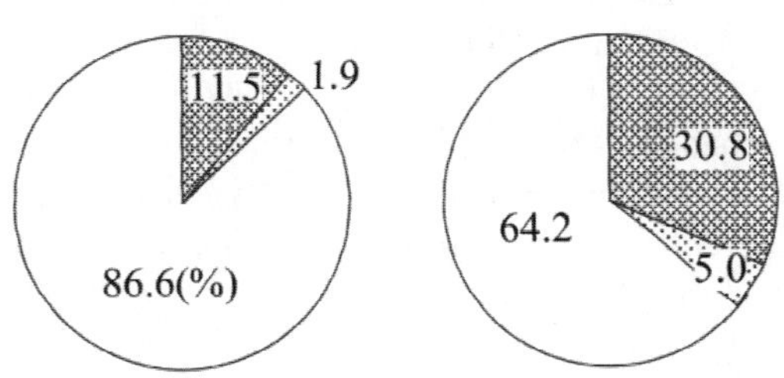


**Figure 2.** Use of sound-absorption in ceilings

IOP Conf. Series: Journal of Physics: Conf. Series **1075** (2018) 012004 doi:10.1088/1742-6596/1075/1/012004

### *3.3. Use of sound-absorbing materials*

Visual inspections of the childcare rooms and play halls in 132 centers were performed to determine whether or not sound-absorbing materials were in use (figure 2). Results indicated that some type of sound-absorbing material was in use in 36 % of the play halls and 13 % of the childcare rooms, and that in most cases, rock wool ceiling board, which is the most commonly used material in Japan because of its fire resistance performance, was used. When we inquired during inspections about the reasons for using acoustical board we were most often told that the architect recommended its use or decided its use without informing the administrator. In many cases, center personnel were not aware that the boards were sound-absorbing.

## 4. Result of questionnaire survey

### *4.1. Totals by ages of children*

The results of responses by childcare providers to the questionnaires are reported here but, due to space limitations, only major results will be presented. There are children from the age of birth up to five years in daycare centers and the activities vary greatly depending on the ages of the children in a class. The results were therefore compiled on the basis of class age. Table 5 shows the number of respondents classified by the age of the children they oversee.

Figure 3 shows the providers evaluations of how easily they could speak with children in the daycare room. The combined ratio for "difficult" and "somewhat difficult" was 4-5% for those taking care of 0 and 1-year-olds, but this number increased to 14-20 % for those minding children aged three or older, who have greater language ability. There was a tendency for an increase in the complaints of throat damages (figure 4), as was the case for ease of communication as the age of the children being supervised increased. In particular, more than 20 % of those caring for five-year-olds answered more than "at least once or twice a week".

Daycare centers receive complaints about noise, and this could be a concern to for childcare providers. The result indicated that such concern increased for whom caring four or five-year-old groups (figure 5). A question in the questionnaire given to administrators inquired about the number of complaints received involving noise disturbances. 92 % of the response was "at most once a year" and the remainder was "a few times a year," with no respondents answering more frequently.

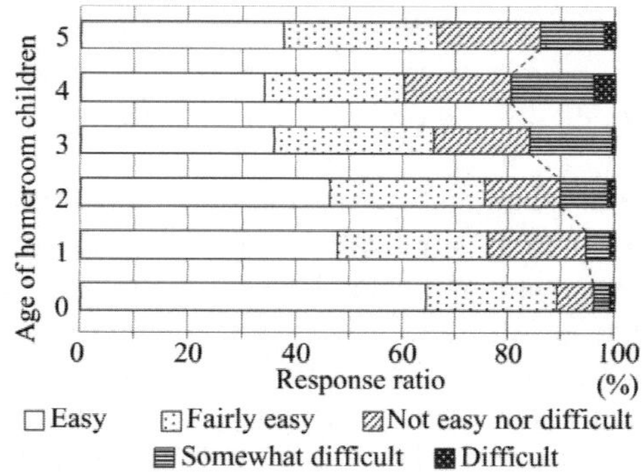


**Figure 3.** Difficulty of conversing in childcare rooms.

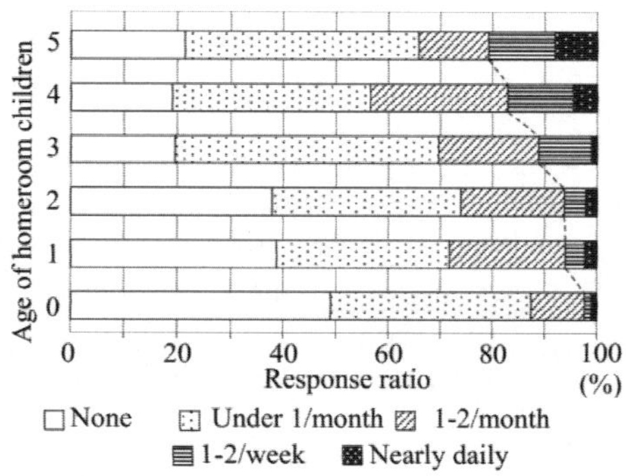


**Figure 4.** Frequency of throat damages

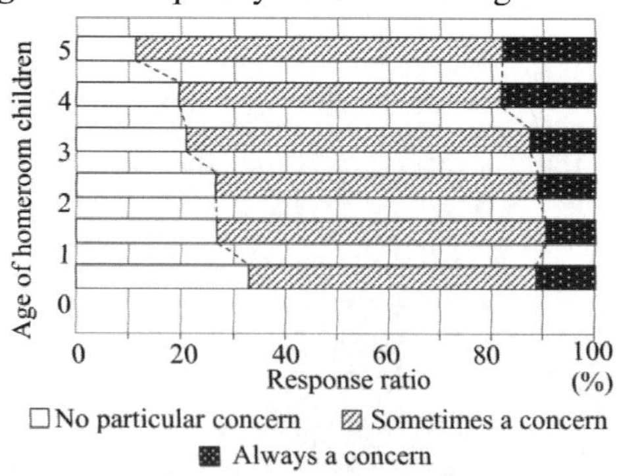


**Figure 5.** Worry about creating noise to neighbours.

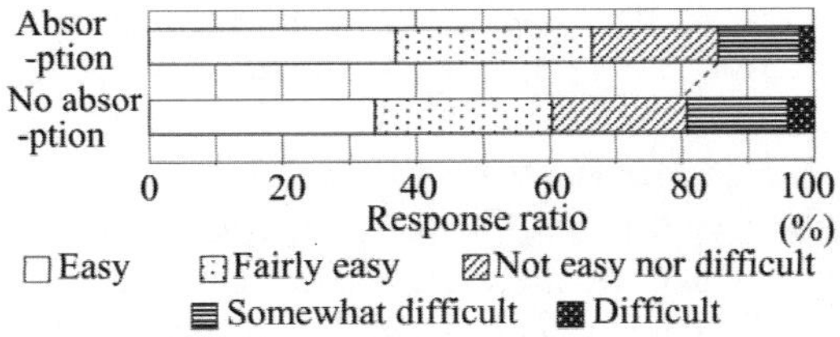


**Figure 6.** Difficulty in conversing in childcare rooms by sound absorption in the room.

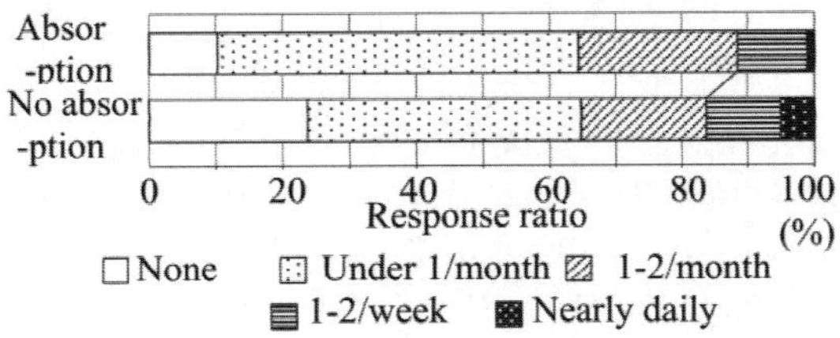


**Figure 7. Frequency of throat damages by sound absorption in the room.**

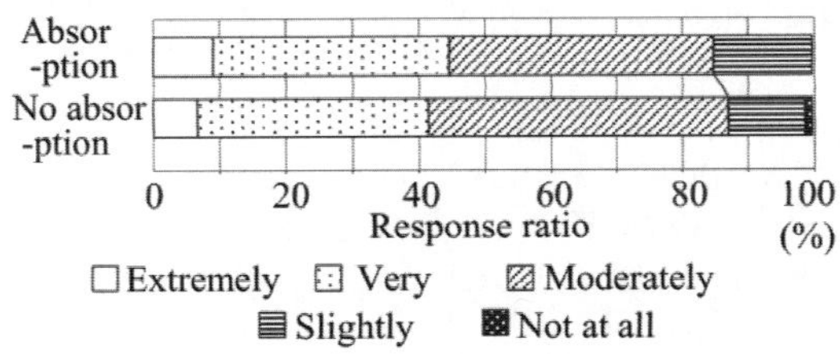


**Figure 8.** Suitability of room for childcare by sound absorption in the room.

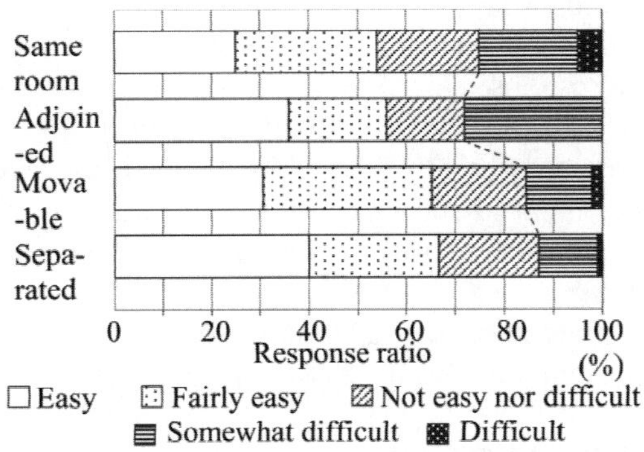


**Figure 9.** Difficulty in conversing in childcare rooms by sound absorption in the room.

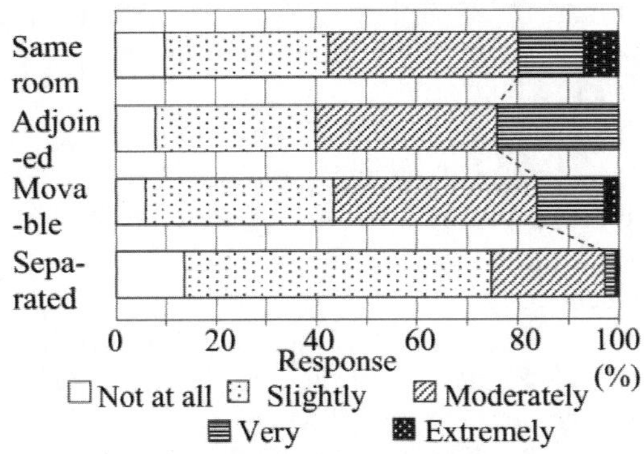


**Figure 10.** Disturbance by noises from other classes by sound absorption in the room.

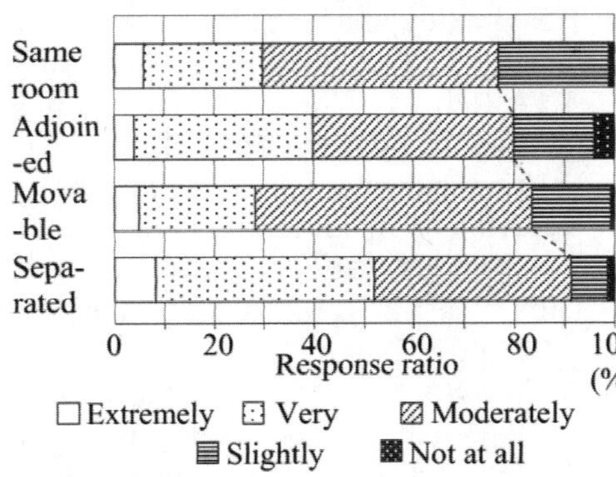


**Figure 11.** Suitability of room for childcare by sound absorption in the room.

### *4.2. Evaluation vs. sound absorption in childcare rooms*

In this survey, whether sound-absorbing material was used and the type of room partition in each childcare room was known by the inspection. Also, the homeroom of each of the childcare provider was known by the questionnaire. Combining these two datasets, it was enabled to divide the providers into groups by the room properties.

Figures 6-8 compiles evaluations of the providers caring three-or more-year-old groups classified according to the use of sound-absorbing materials in their home rooms. In Figures 6 and 7, it is seen that a tendency for a high percentage of childcare providers in rooms lacking sound-absorption to report frequent occurrences of conversational difficulty and throat damages. Since a minority of rooms having sound absorption (the sample with/without sound absorption was around 87 and 485), chi-square tests on the high response groups answering the top 2 categories with/without sound absorption indicated not significantly different. However, the risk ratio of high response without sound absorption was 1.47 and 1.41 for difficulty of conversing and frequent throat damage, respectively, is considerable.

### *4.3. Evaluation vs separating childcare rooms*

As with sound absorption, here the room separation is discussed here and again, totals include only childcare providers caring for children aged three or older (figures 9-11). In the same way as above, high responses are compared by four types of room separation.

The results showed obvious differences in the evaluation. High responses of difficulty of conversing increased when the space of the class was acoustically connected (*same room* or *adjoined*). On the other hand, disturbance by noises from other classes increased even the spaces were separated by movable walls. This indicated that movable walls can reduce speech disturbance but cannot reduce feeling of disturbance between rooms. Possibly reflecting this feeling, the evaluation of not suitable for childcare decreased with the increase of sound insulation (figure 11). These results clearly indicated the advantage of sound insulation.

## 5. Conclusions

As stated above, a survey based on inspections and questionnaires revealed the current state of child daycare centers where little consideration was given to acoustic design regarding space configurations and sound absorption. Combined result of inspection and questionnaire survey also revealed a potential effect of use of sound-absorbing materials in rooms and sound insulated separation between classes.

This study, which is a census of the regional core city region, can be considered to represent the current state of acoustic environments of daycare centers in Japan and to provide foundational material useful in establishing acoustic design guidelines for daycare centers in future.

## References

[1] H. Fujii, Y. Shimura et.al.: Auditory Environment of Kindergarten and Nursery School Part 1-9. Proc. Annual Meeting of Acoustical Society of Japan, 1996-2003 (in Japanese).

[2] L. Grebennikov: Preschool teachers' exposure to classroom noise, Intl. J. Early Years Education 14(1), pp.35-44, 2007.

[3] K.P. Waye et.al: Being in a pre-school sound environment – annoyance and subjective symptoms among personnel and children, Proc. Inter-noise, 2010.

[4] J S Bradley and H Sato, "The intelligibility of speech in elementary school classrooms, J. Acoust. Soc. Am. 123(4), 2078-2086, 2008.

[5] A Astolfi, P Bottalico and G Barbato: Subjective and objective speech intelligibility investigations in primary school classrooms J. Acoust. Soc. Am. 131(1), 247-257, 2012.

[6] WHO: Guidelines for Community Noise, 1999.

[7] Department of Education, U.K.: Building Bulettin 93 acoustic design of schools - performance standards, 2015.

[8] DIN 18041:2016-03, Acoustic quality in rooms - Specifications and instructions for the room acoustic design, 2016.

[9] ANSI/ASA S12.60, American National Standard Acoustical Performance Criteria, Design Requirements, and Guidelines for Schools, 2009 & 2010.

[10] K. Kawai et.al.: Mitigation of noise in nursery classrooms by sound absorption, Part 1-3. Proc. Internoise 2011-2013.

## Acknowledgements

The author appreciates the cooperation of the child daycare centers in Kumamoto and Koshi cities that participated in this survey. This work was supported by JSPS KAKENHI Grant Number 25350947.

IOP Conf. Series: Journal of Physics: Conf. Series **1075** (2018) 012005 doi:10.1088/1742-6596/1075/1/012005

# Analysis of the absolute and the relative transmissibility on the centrifugal pump

**N Falasifah[1], I Taufan[1], D Arifianto[1], and J Susatio[1]**

[1]Departement of Physics Engineering, Institut Teknologi Sepuluh Nopember Surabaya, Keputih, Sukolilo 60111 East Java, Indonesia

E-mail: nihlatulfs@gmail.com

**Abstract**. In this paper, we proposed a new technique on the relative transmissibility which are generally only known as the absolute transmissibility or transmissibility. Relative transmissibility is ratio of the relative deflection amplitude of the isolator to the displacement amplitude imposed at the foundation. Absolute transmissibility is ratio of the vibration amplitude of the equipment to the vibration amplitude of the foundation. Centrifugal pump system which is supported by foundation in the form of A36 steel hollow rectangular has one degree of freedom. This method lead to calculate absolute and relative transmissibility ratio both normal pump and damage pump with variation of spring stiffnes and coefficient damping. Relative transmissibility is significant only in an isolator used to reduce vibration transmitted from a vibrating foundation. It is proven theoretically and experimentally that pump with the A36 steel as the foundation has relative transmissibility less than one, which means that system has a good isolation. The minimum value of spring stiffness which is used of isolation system to have a good isolation is $10^7$ N/m. This research opens opportunities for industry in determining type of material used for machine foundation which is related to the value of spring stiffness.

## 1. Introduction

Mathematical modelling can be applied in a modeling system in the centrifugal pump. It can be used to detect damaged pump as the rotating machine [1]. The damaged pump is caused by inside damaged pump and it because of other vibration around. Pumps which suffered damage to one of its components will result in performance unbalanced and cause vibration anomalies that would negatively affect to other parts. If the source of the vibration around the larger pump will interfere with the performance of the engine at the pump. In almost literature, it is generally only known absolute transmissibility or transmissibility. There is still other type namely the relative transmissibility. The relative transmissibility is the ratio of the relative deflection amplitude of the isolator to the displacement amplitude imposed at the base, while the absolute transmissibility is the ratio of the vibration amplitude of the equipment to the vibration amplitude of the host structure (base).

Both of those transmissibility principles are almost same to find transmitted energy, but the difference is the kind of vibration being compared. The absolute transmissibility is used to determine reduction of transmitted energy conducted by an isolator, whereas the relative transmissibility is used to determine the clearance that transmitted to an isolator. This relative transmissibility would be significant only when an isolator is used to reduce transmitted vibration of a vibrating foundation. The principle of transmissibility can be used to determine the transmitted energy on a structure model of the building with 3 floors as on a thesis titled "Analysis of Dynamic transmissibility as a Feature for Structural

Damage Detection" by Thimoty James Johnson, 2002 [2]. Moreover, the principle of transmissibility also can be used to determine the performance characteristics between linear viscous damping system with pure cubic damping [3].

The principle of vibration transmissibility is used to detect damaged pump in the research which conducted by Anisatul Fauziah [1]. The study states that the transmissibility can be detected using the accelerometer array as evidenced by the change in the diagnosis of damage to the pump that known in the spectrum of the Fast Fourier Transform. From these studies, the principle of transmissibility will be used to analyze the absolute and relative transmissibility in centrifugal pumps and also to determine the approximate value of stiffness (spring stiffness) and the damping coefficient (damping coefficient) of the system.

## 2. Theory

### 2.1. *Centrifugal pump*

Centrifugal pump is a machine that converts fluid kinetic energy into potential energy through a rotating impeller in the casing [4]. Centrifugal pump has an impeller to lift liquid from lower to higher place. Excitation power is given by the pump shaft to rotate the impeller in the liquid, so the liquid in the impeller rotates by impulse blades. The liquid flows from the middle of the impeller to the outside through the channels between the blades because of centrifugal pump arising. It causes the liquid head pressure becomes higher, as well as head speed increases because the liquid is accelerating.

There are several types of damage in pump that can be detected by the level of vibration amplitude at a specific frequency. The types of those damaged are bearing fault, unbalance, and misalignment [5].

### 2.2. *Transmissibility*

Transmissibility is commonly used in the field of vibration isolation. It concerns to minimize or control unexpected transmission and motion from a source to a machine device or other system [6]. There are several types of transmissibility such as force transmissibility, acceleration transmissibility, velocity transmissibility, and displacement transmissibility. Those kind of transmissibility, have the same general equation of transmissibility at one degree of freedom [2]. At displacement transmissibility is generally divided into two types namely transmissibility absolute and relative transmissibility. The absolute transmissibility is the ratio of the vibration amplitude of the equipment to the vibration amplitude of the host structure (base), which should be as small as possible [3][4]. The other characteristic is the relative transmissibility which represents the ratio of the relative deflection amplitude of the isolator to the displacement amplitude imposed at the base [3][4].

The system has a spring stiffness and damping coefficient with Single Degree of Freedom (SDOF). The formula for the absolute and the relative transmissibility ratio for SDOF shown in equation (1) and (2).

$$T_a = \frac{\sqrt{1+4\zeta^2\left(\frac{\omega}{\omega_0}\right)^2}}{\sqrt{\left(1-\left(\frac{\omega}{\omega_0}\right)^2\right)^2+4\zeta^2\left(\frac{\omega}{\omega_0}\right)^2}} \tag{1}$$

$$T_r = \frac{\left(\frac{\omega}{\omega_0}\right)^2}{\sqrt{\left(1-\left(\frac{\omega}{\omega_0}\right)^2\right)^2+4\zeta^2\left(\frac{\omega}{\omega_0}\right)^2}} \tag{2}$$

The value of the damping coefficient, natural frequency, damping ratio, and frequency ratio expressed in equation (3), (4), (5) and (6),

$$c = 2\zeta\sqrt{k.m_{total}} \tag{3}$$

$$\omega_0 = \sqrt{\frac{k}{m_{total}}} \tag{4}$$

$$\zeta = \frac{c}{2\sqrt{k.m_{total}}} \tag{5}$$

$$r = \frac{\omega}{\omega_0} \tag{6}$$

where *Ta* is absolute *transmissibility* ratio (dB), *Tr* is relative transmissibility ratio (dB), *x* is displacement (m), $\omega$ is rotational speed (rad/s), $\omega o$ is natural frequency, *r* is frequency ratio, *c* is damping coefficient (Ns/m), *k* is spring stiffness (N/m), $m_{total}$ is total mass (kg), and $\xi$ is damping ratio.
Isolation is a material used to damper of a vibration. When the value of the frequency ratio is larger than √2 which have absolute transmissibility ratio less than 1, it means that it has a good isolator. When the value of the frequency ratio is less than √2 which have absolute transmissibility ratio more than 1, it means that it has not good isolator. The difference in the value of transmissibility ratio is related to the value of its rotational speed (ω). If the value of rotational speed is small, then it will cause the value of stiffness become smaller so that the absolute transmissibility ratio becomes lower [7].

### 2.3. *Double Clamped Beam*

Beam is a structural component that used as the foundation. Beam is commonly used to support a vertical load, shear stress, and bending moments. There are several types of beam that classified into several categories. One type of category of the beam is doubly clamped beam which is classified based on the type of supporters. Doubly clamped beam is the beam with both ends attached. It is not free to move [8]. The value of spring stiffnes on a doubly clamped beam either in vertical or horizontal axis is shown in the equation (7),

$$k = \frac{2.E.I}{L} \tag{7}$$

where $k$ is spring stiffness (N/m), *E* is modulus of elasticity (Pa), *I* is inertia moment ($m^4$), and *L* is length of beam (m).

### 2.4. *Mass*

Mass is a characteristic of the material that causes inertia. The system that consist of a pump with a beam has the formula as shown in equation (8),

$$m_{total} = m_{pump} + \Omega.m_{beam} \tag{8}$$

where $m_{total}$ is total mass (kg), $m_{pump}$ is mass of each pump (kg), $m_{beam}$ is mass of beam (kg), and Ω is position of centroid.

## 3. Research Methodology

### 3.1. *Taking Data of Mini Plant*

We take the data of centrifugal pump mini plant which type Panasonic GP-129JXK in Vibration and Acoustics Laboratory of Physics Engineering ITS. Those data are masses data from each kind of pump. This centrifugal pump has a value of rotational speed of 3000 rpm (314 rad / s) [9]. Each pump is located on the same foundation that is A36 steel which has a modulus of elasticity E = 200 GPa [10]. This steel is covered by wood. But in this case, wood is ignored. The value of the total mass of the pump foundation is 62.29 kg.

IOP Conf. Series: Journal of Physics: Conf. Series **1075** (2018) 012005 doi:10.1088/1742-6596/1075/1/012005

### *3.2. Plant Modelling.*
Plant modelling of centrifugal pumps are determining the type of pump used in normal pump, pump with 6 grams of damage unbalance, bearing fault, and misalignment of 3 mm. We determine the value of spring stiffness of the beam as the pump foundation that is A36 steel and double clamped beam type using equation (7), the value of the damping coefficient of the pump foundation on the vertical and horizontal axis using equation (3), the total mass of each pump using equation (8), the absolute and relative transmissibility ratio on the vertical and horizontal axis using equation (1) and (2).

**Figure 1.** Beam of pump in outer part [1]

**Figure 2.** Beam of pump in inner part

**Table 1.** Vibration measurement for one pump [1]

| No | Pump Damage | *Ta* (dB) | | |
|---|---|---|---|---|
| | | Vertical | Horizontal | Axial |
| 1 | Unbalance 6 gram.cm | 4.4637 | 6.7306 | 1.6573 |
| | Bearing Fault | 0.9713 | 1.6744 | -2.5261 |
| 2 | Unbalance 27 gram.cm | 0.52213 | -1.7310 | -3.2755 |
| | Misalignment 1 mm | 9.339 | -0.41645 | 0.81151 |
| 3 | Bearing Fault | 3.0104 | -4.9846 | -1.1 |
| | Misalignment 3 mm | 0.63865 | -8.0388 | 2.6465 |

### *3.3. Model variation and validation*
We do the model variations such as the variation of the damping coefficient with constant spring stiffness and the variation of spring stiffness with constant damping coefficient.

Model validation is used to make sure the model representative of the actual plant. Validation of this model used the data measurement that has been done by A Fauziah on research of multi-detection of damaged pump using the accelerometer array [1].

## 4. Data Analysis

### *4.1. Variation of Damping Coefficient with Constan Spring Stiffness*
In the figure 3 and 4, when the value of the damping ratio is less than 1 so that it has the form of a graph which has an amplitude and overshoot as shown in the blue line, red and orange. This system belongs to the type of underdamped. On the other hand, when the value of the damping ratio is greater than one, then the shape of the graph does not have the amplitude and overshoot because this type is the type of overdamped as shown in green and purple lines.

All of the absolute transmissibility ratio is more than one and the relative transmissibility ratio is less than one. The absolute transmissibility ratio value is more than one which means the system does not have good insulation system due to spring stiffness does not function as an insulator because the value of spring stiffness both on the vertical and horizontal axis effect on the value of its natural frequency. The value of its natural frequency is greater than the value of the rotational speed (314 rad/s), so the value of displacement that occurs in the system is transmitted to its mass. In other words, *k* is not functioning as an insulator. But in this case, the relative transmissibility ratio is less than one which means a distance of relative deflection to the isolator quite good. It because of rectangular hollow steel that serves as an insulator. On each pump with the variation of the value of damping coefficient, misalignment 3 mm pump has the greatest values of the absolute and relative transmissibility ratio because misalignment 3 mm pump has the greatest mass pumps among others. The value of this mass is proportional to the transmissibility ratio.

**Table 2.** Total mass in each pump

| Types of pump | Mass (kg) | Total mass (kg) | |
|---|---|---|---|
| | | Vertical axis | Horizontal axis |
| Normal | 5.9 | 18.66945 | 40.7824 |
| *Unbalance* 6 gram | 5.906 | 18.67545 | 40.7884 |
| *Bearing Fault* | 5.9 | 18.66945 | 40.7824 |
| *Misalignment* 3 mm | 6.098 | 18.86745 | 40.9804 |

**Table 3.** The value of $\xi$, *Ta,* and *Tr* misalignment 3 mm pump with variation of damping coefficient in vertical and horizontal axis

| Vertical | | | | Horizontal | | | |
|---|---|---|---|---|---|---|---|
| *c* (Ns/m) | $\Xi$ | *Ta* (dB) | *Tr* (dB) | *c* (Ns/m) | $\xi$ | *Ta* (dB) | *Tr* (dB) |
| $9.57 \times 10^6$ | 11.261 | 1.000176 | 0.000185 | $9.57 \times 10^7$ | 76.408 | 1.000039 | 0.000128 |
| $9.57 \times 10^5$ | 11.261 | 1.000194 | 0.000194 | $9.57 \times 10^6$ | 76.408 | 1.000384 | 0.000403 |
| $9.57 \times 10^4$ | 0.11261 | 1.000194 | 0.000194 | $9.57 \times 10^5$ | 0.76408 | 1.000422 | 0.000422 |
| $9.57 \times 10^3$ | 0.011261 | 1.000194 | 0.000194 | $9.57 \times 10^3$ | 0.076408 | 1.000422 | 0.000422 |
| $9.57 \times 10^2$ | 0.0011261 | 1.000194 | 0.000194 | $9.57 \times 10^4$ | 0.0076408 | 1.000422 | 0.000422 |

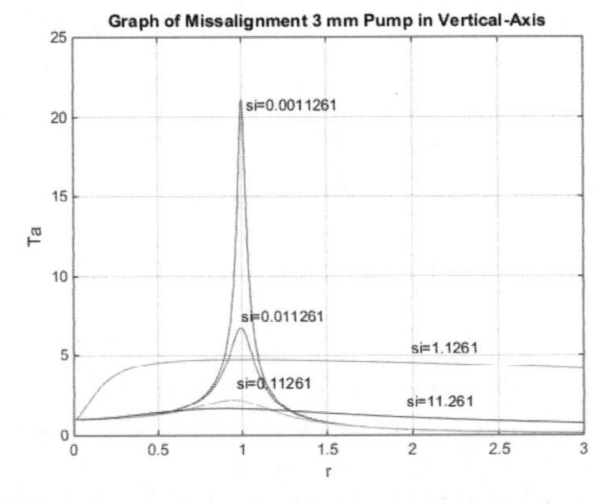


**Figure 3.** Graph of misalignment 3 mm pump with variation damping coefficient in vertical axis

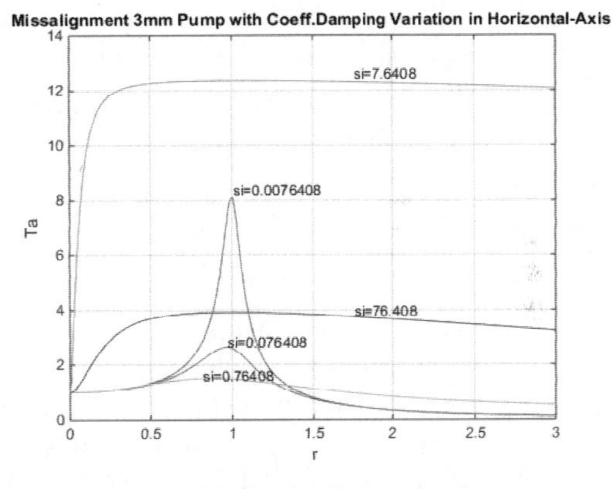


**Figure 4.** Graph of misalignment 3 mm pump with variation damping coefficient in horizontal axis

*4.2. Variation of Spring Stiffness with Constant Damping Coefficient*

On the vertical axis, the value of the damping coefficient is between 1 to $10^9$ Ns/m and the spring stiffness value is less than or equal to $10^5$ N/m to get the value of the frequency ratio $r > \sqrt{2}$. So, the absolute transmissibility ratio value (Ta) is less than 1. When the value of *Ta* is less than 1, it indicates the system has good insulation. In addition, the value of spring stiffness is greater than or equal to $10^6$ N/m to get the value of the frequency ratio $r < \sqrt{2}$ so that the absolute transmissibility ratio value is more than 1. When the value of *Ta* is more than 1, it indicates that insulation used in the system is not good. On the horizontal axis, the value of the damping coefficient is between 1 to $10^9$ Ns/m and the spring stiffness value is less than or equal to $10^6$ N/m to get the value of the frequency ratio r> √2 so the absolute transmissibility ratio value (*Ta*) is less than 1. In addition, the value of spring stiffness is greater than or equal to $10^7$ N/m to get the value of the frequency ratio r <√2 so that the absolute transmissibility ratio value is more than 1. The relative transmissibility in the vertical and horizontal axis will be less than one if it has a minimum value of the spring stiffness of $10^7$ N/m. This means that the deflection distance relative to its isolator is quite good. This is because of A36 steel rectangular hollow that serves as an insulator to reduce the vibrations that transmitted to the foundation. In the spring stiffness variation, the greater the value of its mass, the absolute value and relative transmissibility ratio will be smaller. This occurs because the value of their natural frequencies become smaller and affected on transmissibility ratio.

**Table 4.** The value of $\xi$, *Ta,* and *Tr* misalignment 3 mm pump with variation of spring stiffness in vertical and horizontal axis

| Vertical | | | | Horizontal | | | |
|---|---|---|---|---|---|---|---|
| *k* (N/m) | $\xi$ | *Ta* (dB) | *Tr* (dB) | *k* (N/m) | $\Xi$ | *Ta* (dB) | *Tr* (dB) |
| $10^7$ | 0.00364 | 1.22854 | 0.22854 | $10^8$ | 0.00078 | 1.042106 | 0.042106 |
| $10^6$ | 0.01151 | 1.16224 | 2.16101 | $10^7$ | 0.00247 | 1.677979 | 0.677985 |
| $10^5$ | 0.03640 | 0.05954 | 1.05664 | $10^6$ | 0.00781 | 0.329037 | 1.328822 |
| $10^4$ | 0.11511 | 0.01781 | 1.00526 | $10^5$ | 0.02470 | 0.026598 | 1.025345 |
| $10^3$ | 0.36401 | 0.01689 | 1.00040 | $10^4$ | 0.07811 | 0.008176 | 1.002451 |

## 5. Conclusion

Based on the theoretical and experimental results, this can be concluded several things:

- If the value of the relative transmissibility ratio is less than one, then the required minimum value of spring stiffness is $10^7$ N/m.
- If the absolute value of transmissibility ratio is less than one, the damping coefficient value between 1-$10^9$ Ns/m, then the maximum spring stiffness values is $10^5$ N/m on the vertical axis. On the horizontal axis, the value of the damping coefficient between 10-$10^9$ Ns/m, the maximum spring stiffness values is $10^6$ N/m.

## 6. References

[1] A Fauziah 2014 "Multi-Detection of Damaged Pump Using the Accelerometer Array" Engineering Physics Department, Institut Teknologi Sepuluh Nopember, Surabaya.
[2] T J Johnson 2002 "Analysis of Dynamic Transmissibility as a Feature for structural Damage Detection", Master of Science in Mechanical Engineering, Purdue University.
[3] C M Harris and A G Piersol 2002 Shock and Vibration Handbook, McGraw-Hill, New York, 2012 Rivin E.I: Passive Vibration Isolation, ASME Press, New York.
[4] I Kovacic, Z Milovanovic and M J Brennan 2008 "On the Relative and Absolute Transmissibility of a Vibration Isolation System Subjected to Base Excitation", Journal of Working and Living Environmental Protection vol. 5 no 1 2008 pp. 39 - 48.
[5] T P Agung 2011"Study of Vibration in Maintainability Facility Machine Surabaya Ship", LANTAMAL V.
[6] C M Harris and A G Piersol 2002 "Harris' shock and vibration handbook (5th Edition)" McGraw-Hill.
[7] S S Rao 2004 "Mechanical Vibration 5th Edition" University of Miami: Prentice Hall.
[8] Water Pump Panasonic GP-129JXK, http://www.panasonic.com/id/consumer/home-appliances/water-pumps/non-auto-pump/gp-129jxk.specs.html
[9] MatWeb Material Property Data, "Datasheet ASTM A36 Steel Bar", www.matweb.com.

# The influence of non-audible plural high frequency electrical noise on the playback sound of audio equipment (2nd report)

**N Kimura[1] and T Yoshida[1]**

[1] Department of Electrical Engineering, Tokyo University of science, 6-3-1 Niijuku, Katushika-ku, Tokyo, 125-8585, Japan

E-mail: yoshida@ee.kagu.tus.ac.jp

**Abstract.** Non-audible HF electrical noise is increasing in recent digital audio equipment. Therefore, in our previous study, we measured results showed that inputting non-audible HF electrical noise which has close two frequency component to the analog input of one analog audio amplifier caused intermodulation distortion (IMD) in the audible band of the output of the amplifier. However, the research was insufficient in consideration of sound deterioration difference among some models of audio amplifier. The consideration of dependency on amplitude and frequency of the two HF noise was also insufficient. In this study, in order to investigate the influence of the plural HF noise on the audible band in detail, we measured spectrum of the output of each six models of audio amplifier with inputting two HF sine wave to analog input. The frequency of the HF noise was varied from 100 kHz to 100 MHz, and the amplitude of the HF noise is also set to small amplitude to consider practical condition. From measured results, it was confirmed that immunity against non-audible HF noise is different in each equipment. These results show the importance of consideration of HF characteristic and countermeasure to the HF noise in circuit design of audio equipment.

## 1. Introduction

In recent years, high resolution digital audio equipment with personal computer /network connect ability is increasing. Non-audible high frequency (HF) electrical noise which has various frequency components is increasing more and more due to high performance processor and some radio frequency circuit for wireless connection in these digital audio equipment. There are concerns that these non-audible HF noise may cause bad influence to sound playback. The research on the influence of non-audible HF noise is shortage although development on audio equipment for high-fidelity sound playback is progressing.

Therefore, in our previous study, we investigated the influence of non-audible HF electrical noise on the audio signal of one audio amplifier. The measured results showed that inputting plural non-audible HF electrical noise which has close two frequency component to the analog input of one analog audio amplifier caused intermodulation distortion (IMD) in the audible band of the loudspeaker output of the amplifier [1]. However, in the previous study, we have confirmed this sound degradation phenomenon only by one model of analog audio amplifier. In addition, the amplitude of the HF noise is measured with large amplitude (1 $V_{rms}$) in order to clearly measure the phenomenon.

Therefore, in this study, in order to investigate the influence of the plural HF noise on the audible band in detail, we researched the occurrence and the difference of tendency / characteristics of this sound deterioration phenomenon among audio amplifier models.

In the measurement, non-audible HF sine wave that has two frequency components was generated by mixing the two output signals from the function generator by a high frequency mixer. This HF noise was input to each of three models of audio amplifiers and three models of digital audio amplifiers, and the IMD in the audible band of each loudspeaker output was measured. The frequency of the plural HF noise was varied from 100 kHz to 100 MHz, and the two frequency difference of plural HF noise was varied from 20 Hz to 20 kHz. In addition, we also investigated the amplitude of the plural HF noise set to smaller amplitude (17.78 $mV_{rms}$) to consider practical condition.

## 2. Measurement method of IMD in audible band caused by non-audible plural HF noise

Figure 1 shows the setup for measuring IMD in audible band caused by non-audible plural HF noise. We measure the spectrum of the loudspeaker output of the audio amplifier by the audio analyzer (Audio Precision, APx525) when non-audible HF noise that have two frequency components (plural HF noise) is input to the analog audio input (RCA unbalanced line input).

The plural HF noise is generated by mixing the two output sine waves from the function generator (Tektronix, AFG3252) by the high frequency mixer (R&K, PD010). The frequency of the plural HF noise was selected that the frequency difference between the two sine waves was prime number 2017 Hz, e.g. 100 MHz and 100.002017 MHz, in order to distinguish the frequency component caused by the IMD from buzz noise components. In the experiments shown in section 2.1, 2.2, and 2.3, the amplitude of each HF noise is set to 1 $V_{rms}$ in order to observe the influence remarkably, and that amplitude is set to the range from 0 dBV (1 $V_{rms}$) to -35 dBV (17.78 $mV_{rms}$) to consider practical condition. The gain of the audio amplifier is set to 0 dB at 2017 Hz sine wave.

Actually the function generator outputs somewhat noise also in audible band besides the HF noise. Therefore, in order to observe the targeted noise component caused in the audible band by the plural HF noise, we also measure the noise floor of the measurement apparatus and the targeted audio amplifier when the function generator outputs a DC 0 volt arbitrary waveform signal, i.e. no signal.

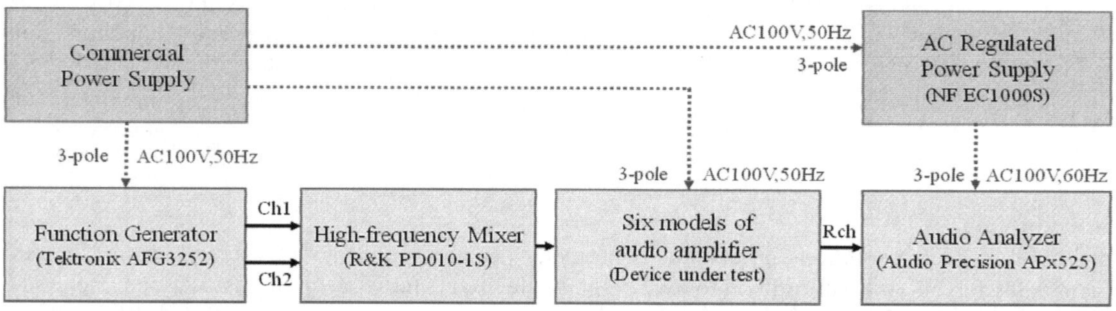


**Figure 1.** Apparatus setup for measuring IMD in audible band on loudspeaker output signal of audio amplifier caused by plural HF noise.

### *2.1. Audio amplifier' model dependency of the IMD in the audible band*

We measure and compare the frequency spectrum in audible band of the loudspeaker output of 3 models of analog audio amplifiers (Model A, B, and C) and 3 models of digital audio amplifiers (Model D, E, and F) in order to consider the audio amplifier' model dependency of the IMD characteristics in audible band caused by non-audible plural HF electrical noise.

### *2.2. Plural HF noise frequency dependency of the IMD in the audible band*

In order to investigate the frequency dependency of the IMD in audible band caused by the plural HF noise which inputs to the audio equipment, we measure spectrum of a loudspeaker output by the audio analyser when the plural HF noise is inputted. In this experiment, frequency of the plural HF noise is varied from 100 kHz + 2017 Hz to 100 MHz + 2017 Hz with 28 logarithmically equal frequency intervals.

*2.3. Plural HF noise frequency difference dependency of the IMD in the audible band*

To investigate the influence of the difference between the frequency of two HF noise component which is inputted to the audio equipment on the IMD, we measure spectrum of a loudspeaker output of the three models of audio amplifiers (Model A, C, F) by the audio analyzer. The frequency difference of the plural HF noise component is varied from 20 Hz to 20 kHz with 200 Hz frequency interval. The base frequency of the plural HF noise is 40 MHz for model A, 600 kHz for model C, and 2 MHz for model F respectively. These frequency was the frequency that the largest IMD are caused in the measurement shown in section 2.2.

*2.4. Plural HF noise amplitude dependency of the IMD in the audible band*

In the measurements in the section 2.1, 2.2 and 2.3, the amplitude of the HF noise was measured with a 0 dBV (1 $V_{rms}$) large amplitude in order to measure the IMD caused by the HF noise clearly. However, since the noise level in the practical condition seems to be considerably low, we measure the IMD caused by the plural HF noise with smaller amplitude. The amplitude of the plural HF noise is varied from 0 dBV (1 $V_{rms}$) to -35 dBV (17.78 $mV_{rms}$) with 5 dBV amplitude intervals.

Measurement below - 30 dBV is performed by inserting 10 dB attenuator (BNC coaxial type, Yuetsu Seiki, band 3 GHz) after the high frequency mixer as shown in fig. 2, since the function generator used in this experiment can output above -25 dBV.

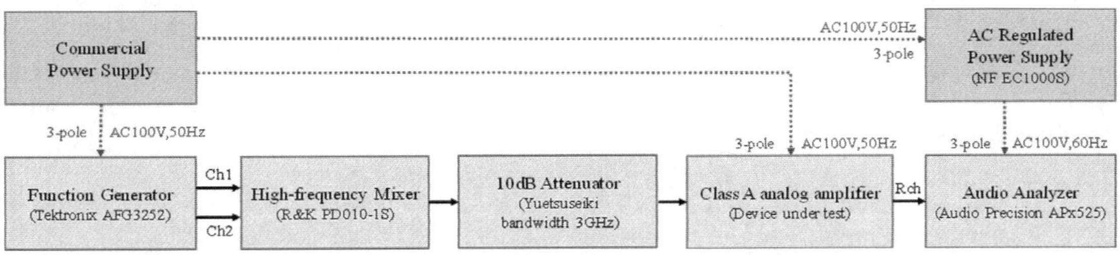


**Figure 2.** Apparatus setup for measuring the IMD caused by plural HF noise these amplitude is below -30dB.

## 3. Experimental Results

*3.1. Audio amplifier' model dependency of the IMD in the audible band*

Figure 3 shows frequency spectrums in audible band of the loudspeaker output of each audio amplifier when the plural HF noise of 100 MHz and 100.002017 MHz was inputted and frequency spectrums of their noise floor. Figure 4 shows the difference spectrum between the spectrum of each audio amplifier when the plural HF noise was inputted and the spectrum of their noise floor.

From figure 3 and 4, it is confirmed that audible band noise (2017 Hz) component is observed in all models of the amplifiers. It is also confirmed that this audible noise (2017 Hz) component is the IMD caused by the inputted plural HF noise, since this frequency 2017Hz is just the difference frequency between 100 MHz and 100.002017 MHz. The magnitude and characteristic of this IMD varies and depends on the model of amplifier greatly. The spectrum from model A has integral order harmonics (2017 Hz, 4034 Hz, 6051 Hz, etc.) of the IMD. In contrast, only fundamental component (2017 Hz) of the IMD was observed from other models.

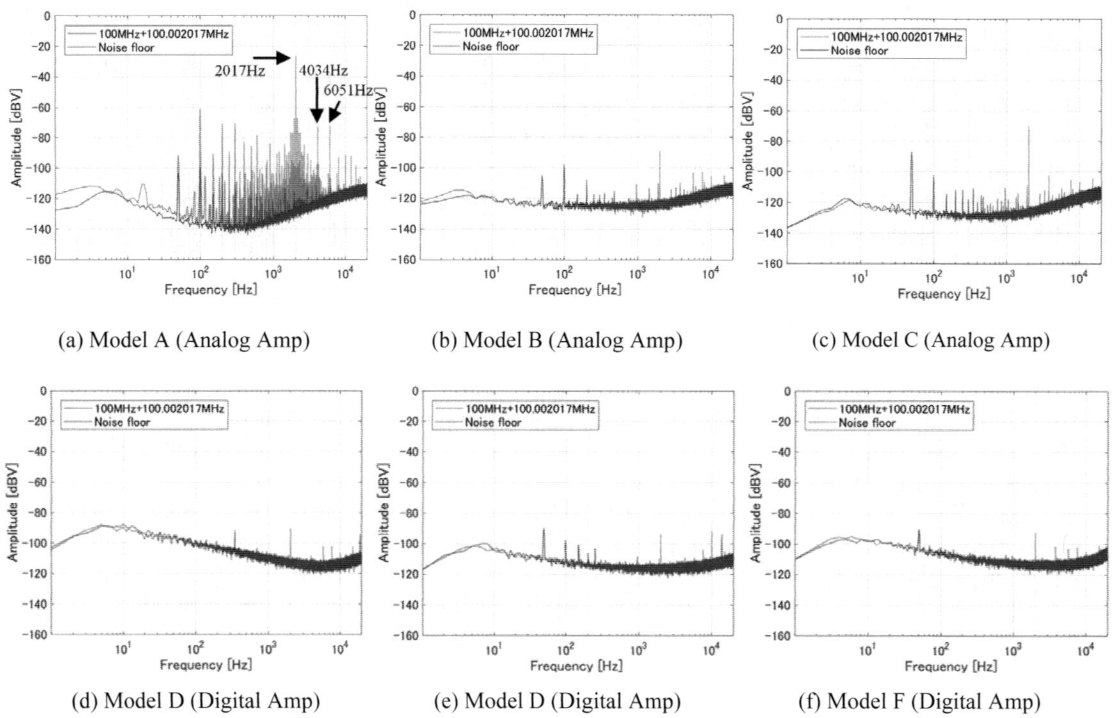


(a) Model A (Analog Amp) (b) Model B (Analog Amp) (c) Model C (Analog Amp)

(d) Model D (Digital Amp) (e) Model D (Digital Amp) (f) Model F (Digital Amp)

**Figure 3.** Frequency spectrums in audible band of the loudspeaker output of each audio amplifier when the plural HF noise of 100 MHz and 100.002017 MHz was inputted and frequency spectrums of their noise floor.

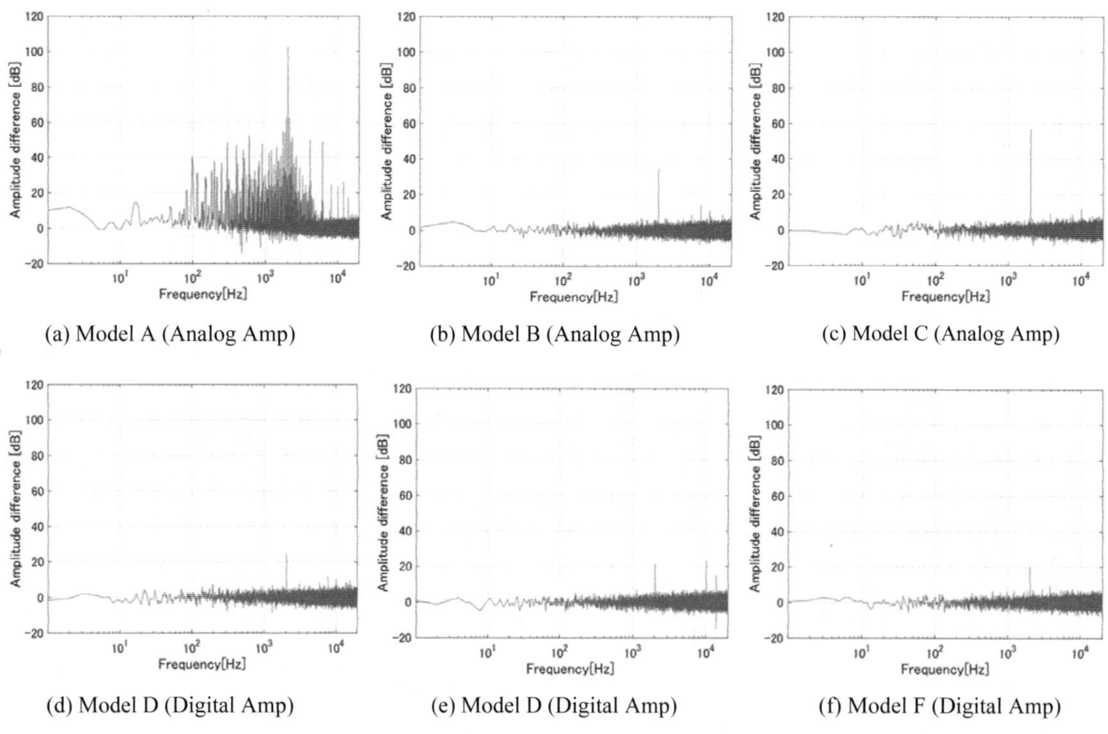


(a) Model A (Analog Amp) (b) Model B (Analog Amp) (c) Model C (Analog Amp)

(d) Model D (Digital Amp) (e) Model D (Digital Amp) (f) Model F (Digital Amp)

**Figure 4.** Difference spectrum between the spectrum of each audio amplifier when the plural HF noise was inputted and the spectrum of their noise floor.

### *3.2. Plural HF noise frequency dependency of the IMD in the audible band*

Figure 5 shows the amplitude of the 2017 Hz component of the IMD at the loudspeaker output of each audio amplifier when the frequency of the plural HF noise is varied from 0.1 MHz to 100 MHz. This amplitude of the IMD is absolute increase of the IMD that is calculated by subtracting amplitude of the noise floor from the IMD caused by the plural HF noise.

From Fig.5, it is found that the amplitude of the IMD depends on the frequency of the plural HF noise and model of the audio amplifier. The IMD maximum amplitude appeared at 40 MHz for model A, 0.1 MHz for model B, 0.3 MHz for model C, 0.1 MHz for model D, and 1 MHz for model E · F, respectively. It is also confirmed that the amplitude of the IMD and their frequency characteristics greatly depend on model of the audio amplifier.

Although in our previous study [1] using only model A amplifier, the IMD in the audible band was getting larger as the frequency of the non-audible HF noise became higher. However, in this experiment, many models (e.g. Model C, E, and F) tended to decrease IMD after their peaks in the range from 1 MHz to 10 MHz. It seems that these differences of characteristics and dependency for each model of the audio amplifier come from design of electrical circuit and structure which affect to characteristics in non-audible HF band.

### *3.3. Plural HF noise frequency dependency of the IMD in the audible band*

Figure 6 shows the amplitude of the IMD at the frequency difference (f2 - f1) of the two sine waves (frequencies f1, f2) of the HF noise which is varied from 20 Hz to 20 kHz. The base frequency of the plural HF noise was 40 MHz for model A, 0.6 MHz for model C, and 2 MHz for model F.

From figure 6, the amplitude of the IMD in the model A and C was almost equal whenever the frequency difference of the plural HF noise was in audible band. In contrast, the amplitude tended to increase slightly as the frequency difference increased in the model F.

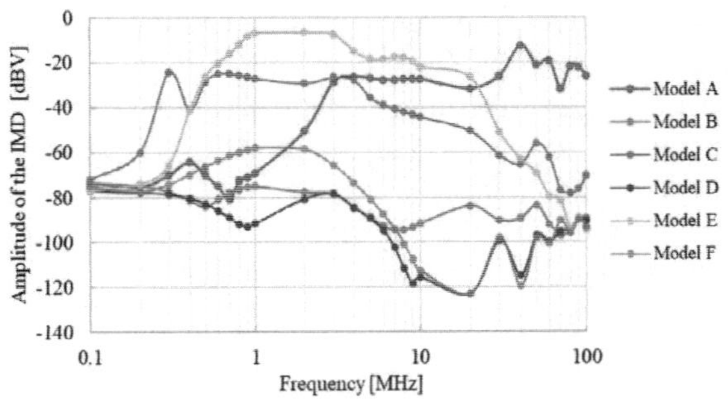


**Figure 5.** Amplitude of the 2017 Hz component of the IMD at the loudspeaker output of each audio amplifier when the frequency of the plural HF noise is varied from 0.1 MHz to 100 MHz.

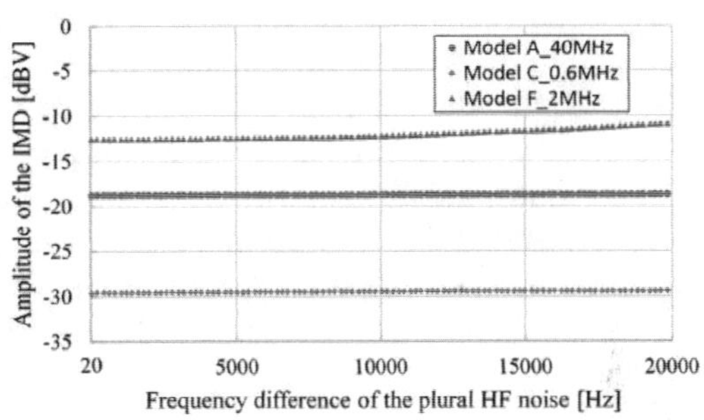


**Figure 6.** Amplitude of the IMD at the frequency difference (f2 - f1) of the two sine waves of the HF noise which is varied from 20 Hz to 20 kHz.

### *3.4. Plural HF noise amplitude dependency of the IMD in the audible band*

Figure 7 shows the amplitude of the 2017 Hz component of the IMD when the amplitude of the plural HF noise is varied from 0 dBV (1 $V_{rms}$) to -35 dBV (17.78 $mV_{rms}$). Figure 8 shows the frequency spectrum of the loudspeaker output of the audio amplifier when the -30dBV plural HF noise was inputted to the model A of the audio amplifier.

From figure 7, it is found that there is a direct relationship between the amplitude of the plural HF noise and the amplitude of the IMD in all three models. The -85 dBV IMD which seems to have some auditory influence is caused in the model A, even if the amplitude of the plural HF noise is small like -30 dBV.

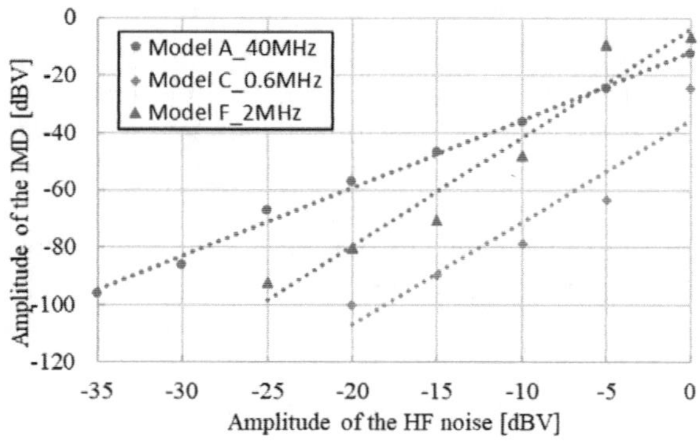


**Figure 7.** Amplitude of the 2017 Hz component of the IMD when the amplitude of the plural HF noise is varied from 0 to -35 dBV.

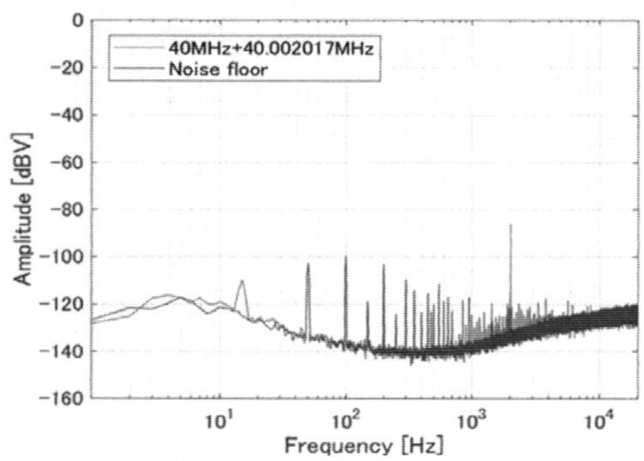


**Figure 8.** Frequency spectrum of the loudspeaker output of the audio amplifier when the -30dBV plural HF noise was inputted to the model A of the audio amplifier.

## 4. Conclusions

In this study, in order to investigate the influence of the non-audible plural HF noise on the audible band in detail, we investigated from the viewpoint of audio amplifier' model dependency of the IMD in the audible band, plural HF noise frequency dependency and frequency difference dependency of the IMD, and plural HF noise amplitude dependency of the IMD.

As a result, we found the followings:

- The amplitude of the IMD caused by the plural HF noise varied depending on the model of the audio amplifier. Their susceptibility is also different in each model.
- The susceptibility to the plural HF noise varied depending on the frequency of the plural HF noise. The frequency characteristics greatly depended on the models.
- The IMD did not almost depend on frequency difference of the plural HF noise.
- Audible levels IMD was caused in some audio amplifiers, even if the amplitude of the plural HF noise was small.

From these results, it was found that non-audible high frequency electrical noise affected the audible band and caused sound quality degradation. These results show the importance of consideration of high frequency characteristic and countermeasure to the high frequency noise in circuit design of audio equipment.

## 5. References

[1] Kobayashi D, Kuwabara T and Yoshida T 2017 The influence of non-audible high frequency noise on the playback sound of audio equipment *Proc. of 2017 IEEE 6th Global Conference on Consumer Electronics (GCCE 2017)* pp 642-644

IOP Conf. Series: Journal of Physics: Conf. Series **1075** (2018) 012007 doi:10.1088/1742-6596/1075/1/012007

# Acoustics perception aspect of Sundanese Celempung's ensemble recording

**J A Simanjuntak[1,2], J Sarwono[1], D Kurniadi[1] and A S Sudarsono[1]**

[1] Engineering Physics, Institut Teknologi Bandung, Indonesia
[2] Conservatory of Music, Universitas Pelita Harapan, Indonesia

E-mail: simanjuntak.jack@gmail.com

**Abstract.** Celempung is a Sundanese traditional instrument from West Java, Indonesia made from bamboo internode. This instrument has three bamboo strings which cut out from the body of celempung itself. Celempung is played by hitting the bamboo strings using bamboo stick and hitting the open section at the end of bamboo by hand. The aim of this study is to determine the important acoustic aspects in celempung ensemble recording technique using semantic categorization. Recordings were made using several stereo recording techniques: XY, ORTF, NOS, Mid-Side (MS), Spaced-pair (AB), Decca Tree, STAAG, and the combination of AB and XY. Nine experts in traditional Sundanese music listened to the recordings using headphone and requested to give the open response regarding the quality of the recordings. Semantic categorization method was implemented to analyze the data using four categories: timbre, spatial, defects, and musical aspects. The study shows that timbre is the most important aspect of recording technique for celempung ensemble. Furthermore, spatial aspects also become the important aspect in evaluating the quality of the recording.

## 1. Introduction

Music recording is a unique and complex process. This process involves the art and science aspects [1]. Music recording is specific because one must embrace two domains that appear contrast, ranging from the sound source of musical instruments, music, and its complexity, the room used its acoustics characteristics as well, and also recording technology. The objective side of music recording includes acoustic aspects of music instruments such as sound dispersion, sound envelope. Then the acoustic aspect parameters of the room used for recording such as reverberation time, ITDG, C80, treble and bass ratio, etc. Also related aspects of microphones such as directivity pattern, frequency response, Stereo Recording Angle (SRA), etc. The art (subjective) side includes music that will be recorded itself from the meaning of the music, the arrangements, the inter-linkage of instruments, and to understand important criteria in a kind of music.

Celempung Sunda, is one of 20 bamboo musical instruments from West Java. In the last eight years, celempung with other Sundanese musical instruments, Karinding, experiencing a revival movement. Various music festivals, performances, music collaborations involving celempung so as to bring up new forms of musical creativities. On the other hand, recorded music has important role for the development of traditional music itself [2]. The recording technique that used affects the sound quality of a music and even opens new possibilities in producing a new sound experience.

In the concept of natural recording [3] one important aspect that determines the quality of the recording is the selection of stereophonic sound recording techniques. There are several stereo recording techniques such as XY, ORTF, NOS, Mid-side (MS), Decca Tree. Each of the technique has different characteristic the implementation must be suitable context for the aim of music recording.

The quality of recorded sound can be evaluated using the categorization from audio attributes. The determination of the categorization is important to determine the important aspects in from sound recording. Moylan conducted a study to develop a system for sound evaluation based on people's perception and determined four attributes aspect that used to evaluate the sound recording: time, pitch, dynamic, sound quality, and spatial properties [4]. Another study also has categorized the sound attributes into four categories: space, defect, timbre, and sound quality [5].

Although the categorization of has been determined, the information about how important those aspect to evaluate the sound recording has not been evaluated. The aim of this study is to determine the important aspect of sound attribute in celempung's music recording based on artist's preference.

## 2. Methodology

### *2.1. Celempung Music Materials and Recording*

There were 24 sample recorded celempungs used for subjective testing. The music sample was a musical composition which was contain of three celempungs which is played live in a performance hall that becomes artist preference.

Recording techniques used in recording live celempung ensemble is a technique XY, ORTF, NOS, Mid-side (MS), Decca Tree, STAAG, and a combination of XY and AB. Recording is performed using a 48 kHz and 24 bit sampling rate. The variation of aspects used in these techniques include angular variants between microphones, microphone height, and distance of the microphone to the sound source.

Two other techniques used are techniques that are a combination of pure stereo recording techniques. STAAG is a recording technique that uses two opposite ORTF [6] . One pair captures the sound from the sound source while the other one is facing away from the sound source to capture the ambience of the room. One more technique used is a combination of XY and AB with variable level changes on AB technique 0, -5, and -10 dB at mixing.

Similarly done on the MS recording, which the microphone side level setting was 0 dB, -10 dB.

### *2.2. Test Setup and Subjects*

Tests performed using AKG HD280 headphones Pro to remove the effect of the room from the experiment. Music was played using laptop via audio interface Steinberg UR22.

Results of the recording will be tested to the respondents. They were asked to make a numerical judgments of a rating scale from 24 samples of recordings with a sample duration of 90 seconds. The rating scale used is the 5-scale rating. The smallest numbers 1 to 5 successively represents bad, poor, fair, good, and excellent.

Subjective tests were conducted on nine respondents, divided into groups of experts and non-experts. The expert group consisted of six respondents, while the non-expert group consisted of three respondents. The group of experts has qualified experience in the field of Sundanese musical art more than 25 years while the non-expert group has less than 10 years' experience.

### *2.3. Interview*

After an assessment of 24 samples of celempung recording given by the respondent, then conducted the interview. The purpose of this interview is to obtain the subjective criteria of both groups in providing an assessment of the quality of sounds of celempung sounds.

Interviews were conducted on 3 respondents from expert groups and 3 respondents from non-expert groups. Interviews are conducted in a semi-structured manner, by asking what the criteria for assessment of a sample record are and what the reason for giving an excellent or bad assessment of a sample of celempung recordings.

The interview was recorded using a handy recorder for subsequent transcription of the recording. Interview results were analyzed using Word Cloud technique. World cloud is a visual representation of the text data used to describe keywords. Visible visuals are presented by color and font size to describe a certain level of text data. Voyant was performed to analyze the text.

Word Cloud results will be filtered and grouped according to the four categories of space, defects, timbre, and quality (Table 1). This method is excellent useful in studying and classification of artists' attributes regarding sound quality of celempung's recording.

**Table 1.** Categorization and sound quality attributes [5]

| Category | Meaning | Attribute |
|---|---|---|
| Defects | Relating to distortion, hiss, hum, etc. | Noise<br>Distortion<br>Background noise<br>Hum<br>Hiss<br>Disruption |
| Space | Relating to the three-dimensional nature of the sound sources and environments | Depth<br>Reverberation<br>Width<br>Distance<br>Localization<br>Spatial distributions<br>Spatialization<br>Envelopment<br>Immersion |
| Timbre | Relating to the tone color | Brightness<br>Tone color<br>Clarity<br>Hardness<br>Equalization<br>Richness |
| Quality/ miscellaneous | Relating to the remaining properties | Homogeneity<br>Stability<br>Sharpness<br>Realism<br>Fidelity<br>Dynamics |

## 3. Result and Discussion

The most frequent words that used during the interview by the subjects, assumed as the most important words that define the aspect the quality of recording celempung. The result of word cloud analysis of expert respondents is shown in Figure 1. The most frequent word used by Subject A is “gema” which means reverberation. The other word frequently used by Subject A are: “gemanya” (the reverberation), “ting” (the sound of highest note of celempung), “jenger” (boxy vocals sound), ”halimpu” (soft, balance in low and high frequency), “hempeung” (unbright), “nyaring” (loud) ,”jauh” (far) ,“intensitas” (intensity), “sop”(soft), “cempreng”, “tung”, “teungkeur” (dull), “warna” (color), “soft”, “saweur” (harsh metal sound), and “murni” (pure). Subject B uses the word “warna” (color), “jelas” (clear) and "kemurnian" (purity) as the most frequent word. Subject C use the word “bersih” (clean) and “timbre” as the most frequent word.

From Subject A, it was found that some words such as "gema" (reverberation) and "jauh" (far) are the word representing the category of space. The subject explained that those two words must be considered because it will decrease the recording quality of the celempung. This result shows that the most important aspect in the evaluation of celempung's music recording is related with the aspect of timbre.

Another interesting finding that comes from the Subject A was he uses the terminology (*heumpeung, halimpu, jenger, teungkeur, saweur, cempreng)* that commonly used to explain the timbre characteristic of vocal quality in Sundanese Karawitan such as *Cianjuran* and *Kiliningan.*

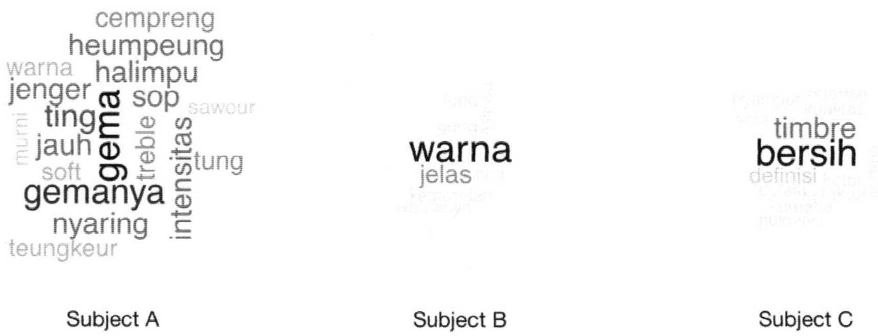


**Figure 1.** Word Cloud Group of Expert

The result of the expert subjects is also consistent with the non-expert subjects. Figure 2 shows the Word Cloud of the non-expert subjects. Subject D use the word "gema" (reverberation),"detail", "jelas" (clear) ,"asli"(original), "jauh"(far), dan "mirip" (similar) mostly in his interview. Subject E mostly use the word "bright" and "bek" (refering to low frequency impression) in his interview and Subject F mostly use the word "jelas" (clear) and "keseimbangan" (balance).

Further analysis is conducted to determine the definition of good recording. The use of the word "kemurnian" (purity) and "kebersihan" (cleaness) by Subject B and Subject C indicates the preference of the sound recording to be as similar as the actual sound without any coloration. The word "bright" and "jelas" (clear) that used by Subject E and F respectively, also indicates the clarity of the instrument in the recording. The clarity aspect also consistent with the terminology "gema" (reverberation) used by Subject A and D. They do not like the reverberation in the recording because the reverberation makes the instrument sound unclear and different than the actual instrument.

Almost all subjects describe the criteria for evaluating the recording of celempung's ensamble related to the tone color, brightness, and clarity. According to the Table 2.1, these attributes are included in the aspect of timbre. Even if there is an explaination of reverb and space, both of these are mentioned because both affect the timbre of the recording.

**Figure 2.** Word Cloud Group of Non-Expert

## 4. Conclusions

This is the first study to identify the subjective criteria that used by artists to determine sound quality of Sundanese celempung's recording. This study has examined the factors which are thought to contribute to determine sound quality aspect of celempung sunda. The result of this study indicates that timbre is most important aspect in celempung's music recording. Spatial aspects also become the important aspect in evaluating the quality of the recording, although the importance seems to be not as high as the timbre aspect. The result is consistent between the expert and non-expert subjects.

Further analysis shows that the respondents prefer the recordings that sound similarly to the actual instrument. Although the spatial aspects emerge in the description of the sound quality, the effect seems to be related with the timbre of the instruments. The respondents do not prefer the recoding in reverberant space because the reverberation affects the clarity and coloration of the instrument.

## 5. References

[1] Rumsey F and McCormick T 2014 *Sound and Recording* (Hoboken: Focal Press)

[2] Suryadi S 2014 The recording industry and 'regional' culture in Indonesia: *the case of Minangkabau* (Netherlands: Leiden University)

[3] Romblom D, Guastavino C, and King R 2012 A comparison of recording, rendering, and reproduction techniques for multichannel spatial audio *Audio Engineering Society Convention 133*

[4] Moylan M 2007 Understanding and crafting the mix: *the art of recording*, ed 2 (Amsterdam, Boston: Elsevier/ Focal Press)

[5] Le Bagousse S, Paquier M and Colomes C 2014 Categorization of Sound Attributes for Audio Quality Assessment—A Lexical Study *J. Audio Eng. Soc.* vol. 62 no. 11 pp. 736–747

[6] Tagg J 2012 A Microphone Technique for Improved Stereo Image, Spatial Realism, and Mixing Flexibility: *STAGG (Stereo Technique for Augmented Ambience Gradient) Audio Engineering Society Convention 133*

**Acknowledgments**

Thank you to Ivan Stefanus, Kevin Leonardo, Bapak Iyon Supiyono, Deara, Arul, and Ari (Institut Seni Budaya Indonesia, Bandung).

IOP Conf. Series: Journal of Physics: Conf. Series **1075** (2018) 012008 doi:10.1088/1742-6596/1075/1/012008

# Performance of underwater source separation using narrowband and broadband signal at mini underwater test tank with varied in salinity and temperature

**B H Purwojatmiko[1], E Cahyaningtyas[1], D Arifianto[1] and W A Asmoro [1]**

[1]Department of Engineering Physics, Institut Teknologi Sepuluh Nopember, Surabaya, Indonesia

E-mail: h.purwojatmiko@gmail.com

**Abstract**. SONAR is one of underwater communication by using acoustic wave which having smaller attenuation value in underwater than others wave. However, SONAR received mixing information from the source caused by noise from ship, offshore activity and seismic noise. Therefore, in this paper we propose a method using Independent Component Analysis (ICA) to separate the mixing signal to get the desired signal. Sound source separation is performed at mini underwater test tank in Vibration and Acoustics Laboratory using the same number of source and sensor configuration. The variables are varied based on salinity, temperature and sensor displacement changes. The separation process was carried out using TDICA, FDICA, and MSICA methods. From the mixed sound recording results obtained the best separation results using FDICA and MSICA methods for each variable change. The average MSE score was 0.0890 and 0.0899 for the FDICA and MSICA methods, while for the TDICA method the average score of MSE 0.1692 was obtained. Along with the increase in distance and changes in propagation medium, the quality of the separation is decreased.

## 1. Introduction

SONAR is one of the application of underwater communication using acoustic wave which having smaller attenuation value in underwater than others wave. However, SONAR received mixing information from the source caused by noise from ship, offshore activity and seismic noise. Therefore, in this paper we propose a technique to separate and distinguish between the information signal and the unwanted signals received by the receiver. BSS (Blind Source Separation) is a technique to model the selection of signal from the mixed signal where the initial condition is unknown. In the previous research has been conducted the application of sound separation using ICA (Independent Component Analysis) to simulate the separation of underwater acoustics signals varied in several source and algorithm. The result shows the cross-correlation value in 0.98 which shows that the mixed signal can be separated [8]. Also, the separation of mixed signal has been conducted in semi-anechoic underwater test tank by using TDICA (Time Domain Independent Component Analysis) method. The simulation shows the better result with MSE value about $3.8\text{x}10^{-8}$, while in the separation of mixed signal in underwater test tank shows lower result in MSE that is 0.01 below the simulation result [9].

Hence, in this paper we will discuss the method to separate unknown mixing signals by using narrowband and broadband signal as a stimulus in the mixing signal with varied of temperature changes

IOP Conf. Series: Journal of Physics: Conf. Series **1075** (2018) 012008 doi:10.1088/1742-6596/1075/1/012008

and salinity. The experiment will be conducted in semi-anechoic mini underwater test tank in Vibrastic Laboratory, Department of Engineering Physics, ITS, Indonesia.

This paper contains of four sections, the first section will describe the background and purpose of the research. Section 2 explains the method used in this research, section 3 describes the results of separation signal, and ends with the conclusion and the future work.

## 2. Method

### *2.1. Test Signals*

In this experiment, according to ICA estimation, the test signal should be independent and non-Gaussian. The Gaussian distribution allowed only on one of the test signal. The independency of the mixed test signal can be known using covariance value of two signal. While kurtosis is used to examine non-Gaussianity of the test signal. Table 1. showed sourced sound mixture with multi input and multi output. Narrowband and broadband signal was used as source signal with five variations of mixed signal. Signal tone signal with frequency 500 Hz and 800 Hz are categorized as narrowband signal. Signal of ship noise, sonar ping and broadband (pure-tone with frequency 500 Hz, 700 Hz, 900 Hz, 1100 Hz and 1300 Hz which generated simultaneously) are categorized in broadband signal.

**Table 1.** Sound Source Multi Input Multi Output

| Signal 1 | Signal 2 | |
|---|---|---|
| Pure-tone 500 Hz | Pure-tone 800Hz | Narrowband |
| Pure-tone 800Hz | Ship Noise | |
| Ship Noise | Sonar Pinging | Broadband |
| Broadband | Ship Noise | Broadband |
| Sonar Pinging | Pure-tone 800Hz | |

### *2.2. Experimental Set-up*

The experiment conduct at test tank dimension is 2 x 1 x 1 in meter from tempered glass and covered inside by corrugated sponge to reduce the echo and possible reflection from signal that we generated. In this experiment, we are using hydrophone array as sensor. Hydrophone array configuration will achieve spatial selection, strengthen source propagation from a particular direction and weaken the source propagation from the other direction. The position of hydrophone array on mini underwater test tank should be considered to avoid spatial aliasing. Distance of each hydrophone is determined by equation (1), $d < \frac{\lambda_{\min}}{2}$ $\boldsymbol{d}$ is the distance of array component, while $\boldsymbol{\lambda_{min}}$ is the minimum wavelength from the maximum signal frequency. Then, the sensor's distance should be smaller than the wavelength from upper and lower frequency limits.

This experiment used 500 Hz as the lower frequency and 1300 Hz for upper frequency. The allowed distance of the sensor according to wave propagation in the mini test tank should be smaller than 45 cm. Therefore, we used three variations in each hydrophone distance that is 10cm, 20cm, 30cm. The sampling frequency depends on Nyquist Theorem to prevent aliasing.

The mixed sound source recording configuration was using two speakers and two hydrophones set as figure 1. There used three distance variation between speaker and hydrophone in 80 cm, 100 cm and 150 cm. Referring to Indonesian shallow sea water characteristic, salinity level and temperature difference is considered in this experiment. We used 3.5% salinity level and with temperature of 9°C as same as Indonesian shallow sea water characteristic. In this experiment, we also performed variation in temperature difference and salinity level to know the changing characteristic of the method used. The temperature difference variation was using 9°C, 13°C, 16°C, 19°C, 22°C, 15°C, and 28°C and 3.1%, 3.2%, 3.3%, 3.4%, and 3.5% was used for the salinity level. Duration of the data is 4 second with sampling frequency 16.000 Hz.

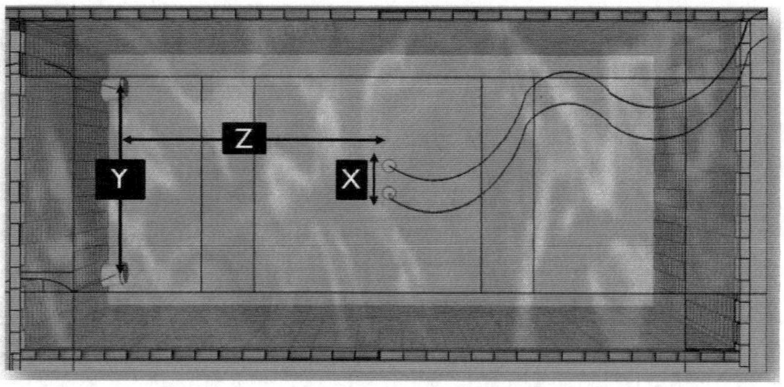


**Figure 1.** underwater test tank top view and hydrophone configuration. X is the distance between the sensor array. Y is the two speakers distance. Z is distance between speaker and sensor

### *2.3. Independent Component Analysis (ICA)*

Data processing for separate the mixed signal was done by using ICA method. Before the ICA estimation phase, there are pre-processing stage which consist of centering stage and whitening stage. The algorithm model for sound separation mixture as $X(t) = AS(t)$ where $X(t) = (x_1(t),...,x_j(t))$ is observation signal, $A$ is environment mixing matrix and $S(t) = (s_1(t),...,s_N(t))$ is source. Then for gain $S(t)$ the model become, $Y(t) = WX(t)$ where $Y(t)$ represent source expectation $(Y(t) \approx S(t))$, $W$ is environment de-mixing matrix $(W \approx A^{-1})$ and $X(t)$ is observation signal.

Time-Domain ICA (TDICA). Separation of mixed sound using TDICA is done by directly give the mixed input signal which recorded in the anechoic test tank. The input and the reconstructed signal will remain in the time domain (t), illustration of TDICA method can be seen in figure 2. TDICA method will minimize the nonnegative cost-function which will be minimum if cross-correlation value is zero [7]. The estimated of separated signal can be obtained by multiplying the observation signal with the $w(z)$ filter.

Frequency-Domain ICA (FDICA). The separation using FDICA method is done by transforming the input mixed signal from time domain into frequency domain as illustrated in figure 3. The transformation was using Fast Fourier Transform (FFT). Once obtained the reconstructed signal of the separation result, then returned into time domain using Inverse Fast Fourier Transform (IFFT). Separation of signal with FDICA can simplify the sum of convolution into instantaneous sums, it makes the separation process will easier to achieve convergence during the iteration.

Multistage ICA (MSICA). The MSICA method for separate the signal is complete by combining processes from FDICA and TDICA. The mixed signal used as input will be processed first by FDICA, after that the FDICA separation result will be input for TDICA. Then the output of separated signal from TDICA will be the estimated signal of MSICA.

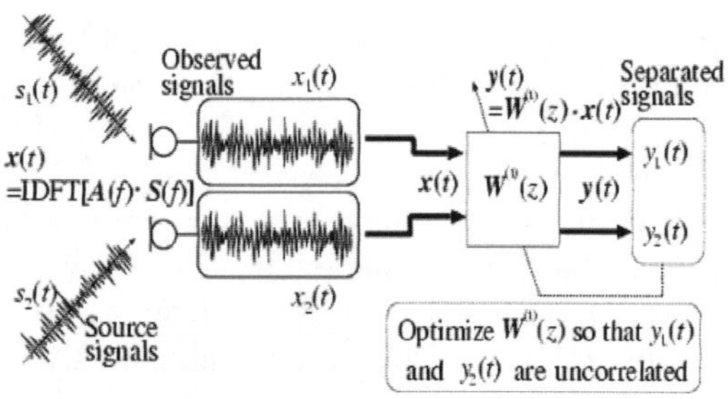


**Figure 2.** Block diagram of TDICA method.
(Source: Saruwatari et al., 2001)

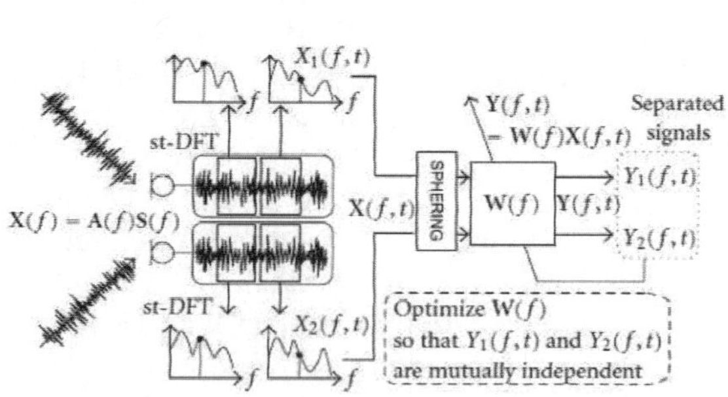


**Figure 3.** Block diagram of FDICA method.
(Source: Saruwatari et al., 2001)

In order to evaluate the estimated of separated signal, here used mean square error (MSE) and signal to interference ratio (SIR). The smallest MSE value means the estimated signal is more similar with the original signal. While the greater SIR value gives better estimated signal quality.

## 3. Result and Discussion

### *3.1. Results of Data Processing*
The requirement of ICA method for estimated signal is independent and non-Gaussian test signal. The independence of test signal performed by finding its covariance value, the closer covariance value of two test signal to zero means that the signal has strong independence characteristic. The independence test result shown that almost all test signal meet the independence assumption. The highest covariance value is found between signal sonar ping and 800 Hz tone while the smallest value is in signal ship noise and broadband signal. Other conditions besides the signal must be independent also not included in the Gaussian distribution. It performed by the value of kurtosis and negentropy of each test signal. Signals that have a kurtosis value equal to three and a negentropy value close to zero will not be used. The negentropy calculation result showed that the signal fulfill the non-gaussian criteria. The test signal should meet both ICA criteria to make an optimal signal separation.

### *3.2. Mixed Sound Separation*
Pre-processing stage consist of centering and whitening. The centering process was given to input signal $x$. Then, the determination of whitening matrix was obtained by give the value of eigen value and eigen vector on covariance matrix. The separation of mixed sound in this study is grouped by method, distance variation and propagation medium change. The evaluation of the separation performance can be seen from the average MSE value of the overall signal estimation. MSE is one way of knowing the quality of an estimation signal compared to the original signal. Figures 4 and 5 show the mean comparison of MSE based on salinity medium and temperature change with variation in the separation method, the distance between the sensor and the distance between the sensor and the source. In terms of separation method, separation performance using TDICA method is worse than FDICA and MSICA.

In figure 4 and figure 5, the MSE value tend to be higher along with the increase of distance between sensors and the distance between sensors and source that is equal to 0.199. The best performance for TDICA, FDICA and MSICA method achieved in the distance between sensor of 10 cm and with distance between the sensor and source in 80 cm for both on salinity of 3.1% and in temperature of 9°C. The further distance between the sensor and the distance between sensor and source will give bigger MSE value. The MSE value for both FDICA and MSICA method only have slightly difference, but in TDICA method give big difference as the distance between the sensor and source increases. This is because of the result of TDICA affected by the delay time for about 0.0014 second which obtained from the data retrieval.

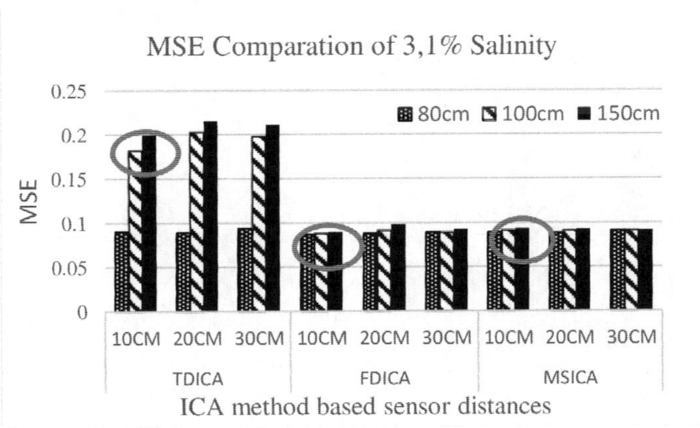


**Figure 4.** MSE comparison of TDICA, FDICA and MSICA with 3.1% salinity and variation in sensor distances

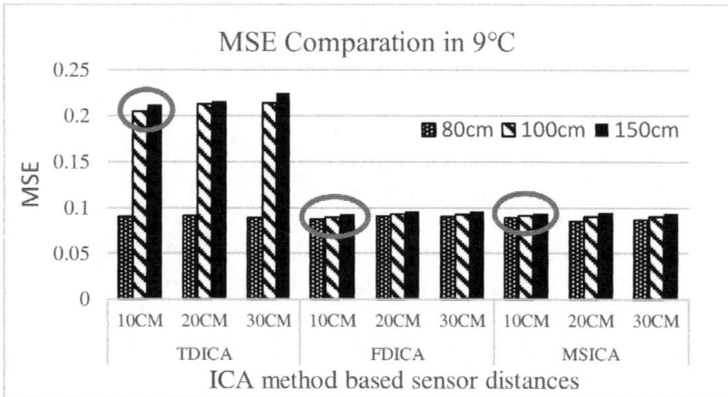


**Figure 5.** MSE comparison of TDICA, FDICA and MSICA with 3.1% salinity and variation in sensor distances

The separation of mixed signal, sonar ping signal and ship noise signal, have the best MSE value among the other mixed signals. Figure 6 show the MSE value comparison of the mixed signal with variation in distance between sensor and source an also with variation in salinity level of 3.1% to 3.5%. While in figure 7 can be seen the MSE value comparison of the mixed signal for sensor and source difference and variation in temperature change. Both using TDICA method. It can be seen that the performance of separation signal getting worse as the salinity level and temperature difference increase, shown by the MSE value getting bigger. This caused by the medium density which affecting the velocity of the sound, so there is a delay time for sound from source to sensor. The increase of salinity level will cause the sound propagation velocity decreasing. The same result also obtained in every temperature change, from the lowest temperature to near room temperature, which give increasing MSE value. With the increase of the mean value of MSE for TDICA method on each change of propagation medium and the difference of distance between sensor and source indicates that the separation quality decreases for each variable change. The quality decrease also occurs almost every mixed signal used as well as other separation methods.

The separation of mixed sounds in the anechoic test tank uses several test signals that are mixed in different propagation mediums. Salinity and temperature changes are the control variables in this study where salinity and temperature changes are conditioned near the true state of the ocean. In the case of this mixed sound separation using three separation methods that are TDICA, FDICA and MSICA. In some test signals used there are signals that cannot be completely separated, but some mixing components can be removed. Of the three methods of separation of quality obtained from each distance either between sensors and sensors to the source has a quality difference. The TDICA method has a decreasing quality when there is a change in distance and variables as shown in Figure. 5 and 6.

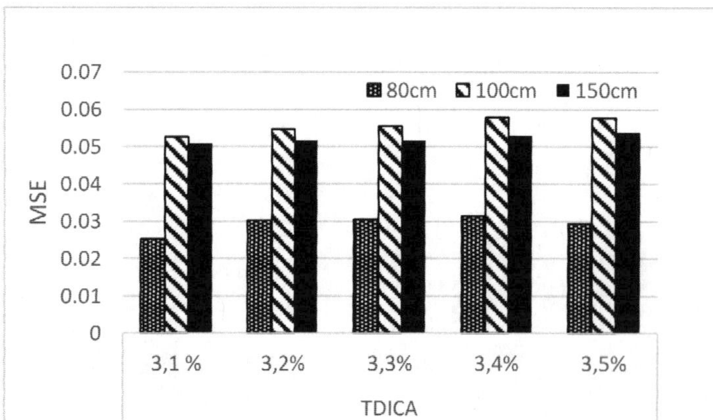


**Figure 6.** MSE Comparison of mixed signal using TDICA method with variation in salinity level

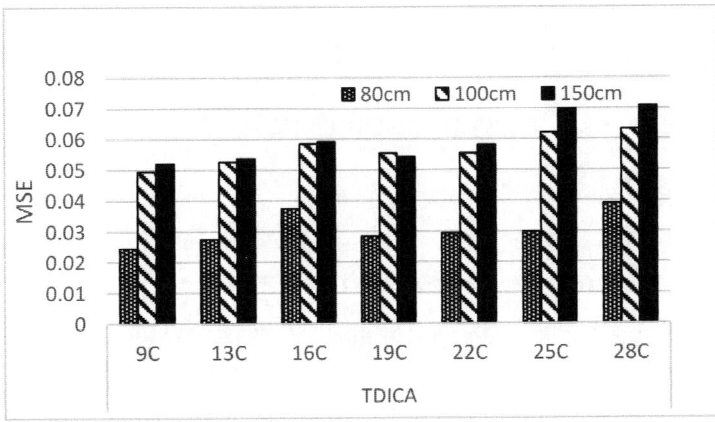


**Figure 7.** MSE Comparison of mixed signal using TDICA method with variation in temperature difference

## 4. Conclusion

In this paper, we performed separation of mixed signal with variation in propagation medium and position of the sensor in anechoic underwater test tank. Among the methods used in separation, the FDICA method is superior to other methods. The average value of MSE FDICA was 0.0890, while for TDICA and MSICA methods were 0.1692 and 0.0899, respectively. Changes in some variables such as distance and medium propagation can affect the performance of mixed separations. The further distance between the sensor will decrease the separation result. And also, the higher salinity level and temperature difference give separation result decrease.

For the future work will be conducted the experiment with variation in the angle of the sensor and source. Then used other algorithms to provide approximate direction of the signal arrival and increase signal level which reduced due to noise and reverberation.

## References

[1] Urick R J 1983 Principle of Underwater Sound 3rd Edition (McGraw-Hill).

[2] Naik G R and Kumar D K 2011 An Overview of Independent Component Analysis and its Application. Informatica p 35

[3] Cochard N and Lacoume J L 2000 Underwater Acoustic Noise Measurement in Test Tank, IEEE Journal of Oceanic Engineering, vol.25, no.-1,October 2000

[4] Pedersen M S and Larsen J A survey of Convolutive Blind Source Separation Methods, Springer Handbook on Speech Processing and Speech Communication.

[5] Yang Y and Zhang Di 2011 Noise Source Separation based on the Blind Source Separation, Control and Decision Conference (CCDC) 2011 Chinese

[6] Makino S and Sawada H 2005 Blind Source Separation of Convolutive Mixtures of Speech in Frequency Domain, IEICE Trans Fundamentals, vol.e88-A, No. 7 July 2005.

[7] Nishikawa T, Saruwatari H and Shikano K April 2003 Blind Source Separation of Acoustic Signal Based on Multistage ICA Combining Frequency Domain ICA and Time Domain ICA. IEICE Trans Fundamentals, Vol.E86-A, No.4

[8] N P Yuwono, D Arifianto, E Widjiati and Wirawan. "Underwater sound propagation characteristics at mini underwater test tank with varied salinity and temperature." *2014 6th International Conference on Information Technology and Electrical Engineering (ICITEE)* (2014): 1-5.

[9] Hyvarinen A, Karhunen J dan Oja E 2001 "Independent Component Analysis: Basic Independent Component Analysis", John Willey & Sons, New York.

[10] Saruwatari H, Kawamura T and Shikano K 2001 "Blind Source Separation for Speech based on Fast-Convergence algorithm wih ICA and Beamforming." *7th European Conference on Speech Communication and Technology*. pp. 846-858

IOP Conf. Series: Journal of Physics: Conf. Series **1075** (2018) 012009 doi:10.1088/1742-6596/1075/1/012009

# Experimental investigation of fabric-based micro perforated panel absorber

**G Desendra[1], M N Hermanto[1], I Prasetiyo[1] and D R Adhika[1]**

[1]Acoustic Laboratory, Engineering Physics, Institut Teknologi Bandung, Ganesa 10, Bandung 40132, Indonesia

Email: gradi.desendra@gmail.com

**Abstract**. The most popular sound absorbers are made of porous material. Although porous material provides decent sound absorption, their fibers may harm health and environment. Thus, micro perforated panels (MPP) are considered an alternative option to deal with the health and environment issues. However, manufacturing minute holes of the MPP in Indonesia is not easy. As an alternative to MPP, this research uses microstructure of available materials on the market. The inherent microstructure potentials of fabric as MPP material will be investigated. Two kinds of woven fabric, namely Alabama and Lisburn, also one kind of denim fabric are employed. Absorption coefficient of the fabrics is measured using impedance tube with various air cavities. Measurement results show that all of the three fabrics are potential to be used as sound absorbers, as the absorption coefficient of the fabrics are greater than 0.5 at various frequency bands. For further investigation, the geometrical properties of the fabrics are measured using Scanning Electron Microscope. From the measured microstructures, woven fabrics are predicted to be more feasible for further development as the microstructures are more predictable compared to the denim fabric. For woven fabric themselves, Alabama shows greater performance than Lisburn in terms of absorption coefficient. It is predicted that the fibrous holes for Alabama made the resistive part become more apparent rather than Lisburn that has clearer holes, thus increasing its potential as sound absorber.

## 1. Introduction

Porous absorber has been widely used for a long time, but nowadays the use of Micro Perforated Panel (MPP) absorber is being considered as an alternative because MPP absorber poses no environmental and health issues like most porous absorbers do [1]. The basic construction of MPP absorber consists of a thin flat panel that has sub-millimeter holes backed by a rigid wall with an air space between them, and none of these components produce fibrous waste that harms health and environment. Based from this fact, MPP absorber is considered to be a basis for the next generations of sound absorbing constructions [2], and many researches are being conducted regarding MPP absorbers technology. For instance, MPP absorbers have been used when a robust sound absorber was needed for severe environments such as on airplane [2] and for low frequency sound absorption in room acoustics such as glazing [9-10].

For some developing countries, mass production of these micro perforated panels with minute holes may be very costly to produce because of insufficient manufacturing technologies. As an attempt to eliminate the usage of high-end technologies, fabric - which is considered to be analogous to MPP structure - will be investigated. Hypothetically, the yarns serve as the panel and the gaps between the yarns serve as the minute holes. In order to prove this prediction, the sound absorption coefficients of several fabrics would be investigated using impedance tube with transfer matrix method. The measurement results determine whether the predetermined fabrics proved feasible to be used as low-cost micro perforated panel absorber.

In this study, results show that the three tested fabrics have the potential to be used as a resonator-based sound absorber. Despite their potential, it is still not clear yet whether the absorption phenomenon is mainly contributed by inertial effect of the fabric vibrations or by the viscous-thermal effect of the fabric holes backed by air cavity. In order to provide more insight to the mechanism, morphology investigation using scanning electron microscope (SEM) is conducted. The morphology structure contributes in explaining the difference of sound absorption coefficient amplitude and the presumptions of the fabric utilization as MPP absorber. There are some constraints for a panel to be considered MPP such as hole diameter should be less than 1mm (d<1mm), panel thickness should be less than 1mm (t<1mm), corresponding panel should be rigid and backed with air cavity, and the perforation ratio should be less than 19.6%. The selected fabrics are presumed to fulfil the prerequisites in this study, and validated later on throughout the experiment. Further parametric survey for mathematical modeling of these fabrics may reinforce the presumption of fabric as MPP absorber, but is yet to be done.

## 2. Methods and Materials

### *2.1. Fabric Materials*

Ideally, the MPP absorber proposed by Maa assumed being rigid. Therefore, the strength of the knitting structure is strongly considered as the fabrics are stretched to fulfill the rigid MPP assumption. Other than fulfilling the assumption, the rigidity of the fabric is mandatory in the impedance tube mounting system. There are three types of fabric based on the integration structure of the yarns to determine the feasibility of the fabric utilization as MPP.

The first type of fabric is non-woven textile, presented in Figure 1(a). The figure shows that non-woven fabric has no clear structure in its build, as the composition may consist of recycled materials and other yarns that are tangled one to each other randomly. This structure results in bulkier build and seemingly harder to do parametric survey, and it is not preferably investigated as an alternative to MPP absorber.

Knitted fabric is the second type, and the morphological structure represented in Figure 1(b). It is visually clearer than non-woven fabric in terms of parameter determination from the recurring perforated morphology structure. This fact implies knitted fabric as a promising alternative as the fabric-based MPP absorber, but knitted fabrics consists of only one yarn stitched in such manner or integration from many coils of yarn, leaving the structure very stretchy.

The last type is woven fabric represented in Figure 1(c). The hole-yarn pattern resembles knitted fabric, but differs in the structure's strength. The figure shows that yarns are adjacent to other yarns, thus increasing the static friction between the yarns. Therefore, the main focus on this study is related to woven fabric as the potential MPP substitute.

**Figure 1.** Type of Fabrics (a) Non-Woven[3] (b) Knitted[4] (c) Woven[5]

In a previous experiment conducted by Nico [6] a kind of denim fabric was investigated and modeled using membrane absorber approach, but the correlation between the measured absorption coefficient and model-predicted absorption coefficient show a significant gap. The membrane model approach suspected to be the reason, and this study will provide another basis of MPP absorber investigation by scanning the surface of the denim fabric.

### *2.2. Experimental Setup*

The absorption coefficient of fabric based MPP was measured using impedance tube as per ISO 10534–2 [7]. The schematic diagram of impedance tube could be seen in Figure 2 in this system. First, the values of relative humidity and environment temperature were measured and involved in the measurement to add corrective values of the results. Also, two microphones were calibrated using 114dB white noise to ensure the accuracy of the measurement. The next step, each respective MPP was placed inside impedance tube with 15mm air cavity. White noise was generated as the input through speaker inside the impedance tube. This noise propagated through 3 cm and 10 cm diameter tubes using two calibrated microphones. These microphones obtained sound pressure level data in the impedance tube in order to get the normal incidence absorption coefficient of the MPP using built -in analyzer. From this dimension of tubes, the normal incidence absorption coefficient between 64 to 6300 Hz thus could be obtained. Measurements of the absorption coefficient were conducted three times each tube. The duration of each measurement lasted 120 seconds to ensure the steady state response of the system (60 seconds each configuration of microphone). Lastly, the absorption coefficients of three measurements were averaged, resulting in one set of measurement. The experiment was conducted on two other different samples of the same fabric, as the data presented on this study consists of three sets of measurements for each kind of fabric. The presented data is assumed to be accepted statistically as the variance of fabric properties could vary very randomly.

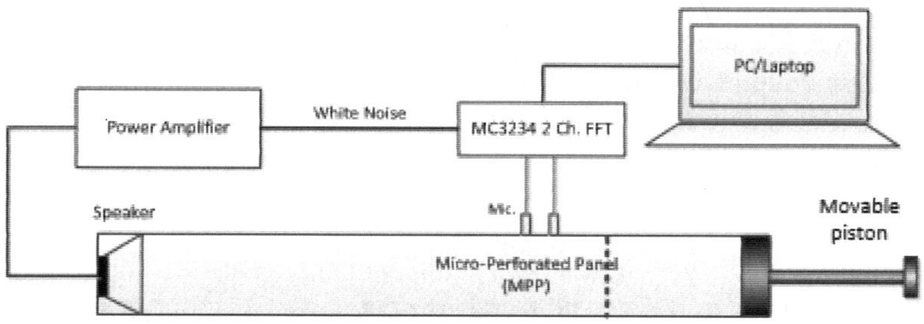


**Figure 2.** Schematic Diagram of Measurement Using Impedance Tube

To further explain the absorption coefficient difference and classifying the fabric, measurements using electron microscope were conducted with 35 times magnification. This high-end microscope provided an accurate and reliable result, especially if parametrical surveys in terms of fabric's perforation diameter and yarn thickness were to be done. In this study, the presented images are random samplings from the predetermined fabrics (see Figure 4) . The presented figures are assumed to be able represent the whole fabrics' morphology structure in this investigation phase.

## 3. Measurement Results and Discussion

There were 3 different woven fabrics measured using impedance tube, namely Alabama, Lisburn, and Denim. The results showed all 3 woven fabrics showed a different absorption coefficient, which could be seen in Figure 3.

Figure 3 shows the absorption coefficient of each fabric at 64 to 6300Hz. The absorption coefficient fluctuations at low frequencies are likely to be caused by the vibrational modes of the mounted fabric in the impedance tube. But, the environmental moisture and temperature has little impact on the result as each measurement showed similar output values.

The difference in the absorption coefficient can be explained by its surface morphology that has been observed using SEM. The Alabama fabric has a small hole between its yarn and amorphous yarn pattern that can be seen in Figure 4(a). These small holes and amorphous yarn pattern are predicted to be the reason why Alabama fabric performed better (with absorption coefficient close to 1) compared to the Lisburn and Denim fabric.

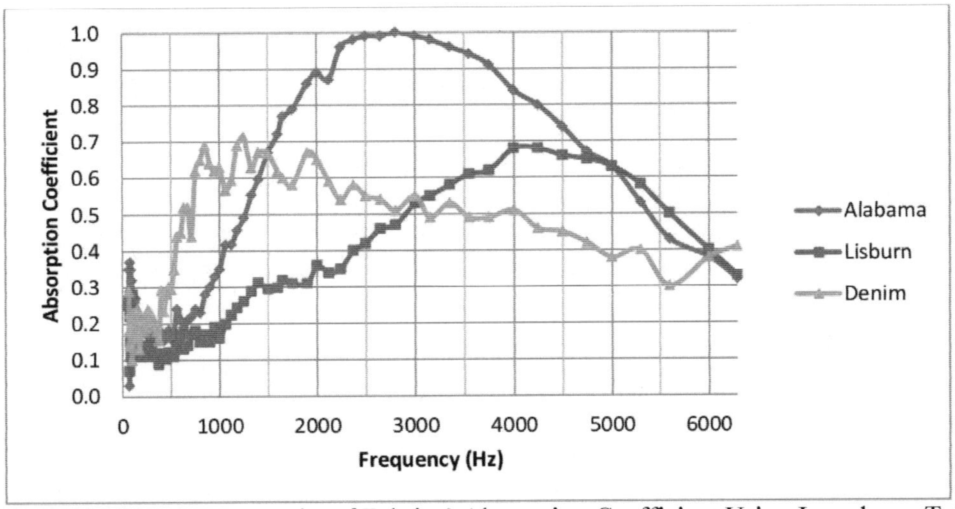


**Figure 3.** Measurement Results of Fabrics' Absorption Coefficient Using Impedance Tube

For Lisburn, its hole can be seen clearly and the yarn pattern is well-shaped. Because of its bigger hole, the coefficient absorption is lower than Alabama (about 0.7) which can be seen in Figure 4(b). As for Denim, its surface morphology is different from Alabama and Lisburn. In Figure 4(c), the yarn has oblique pattern and its hole cannot be seen clearly. The coefficient absorption of Denim is analogous with Lisburn but has a different frequency resonance. The frequency resonance of each fabric is affected by its stiffness. Therefore, all 3 fabrics have a different absorption coefficient on a different frequency resonance using the same air cavity in the investigation.

(a) (b) (c)

**Figure 4.** Surface Morphology of Fabrics Observed Using SEM (a) Alabama (b) Lisburn (c) Denim

Error analysis investigation is not explained in this study as the main aim for this study is to investigate the potential of the three fabrics as MPP-like absorber. It would be crucial to include error analysis in the model validation phase, as the measurement result could differ from the model output that could be used, namely Maa model [2] or Johnson-Allard model [8]. The input parameters for MPP-like structure models such as diameter of the holes, distances of the holes, and the panel's thickness may vary broadly as the mounting process and random sampling of the fabric can result in different surface geometry thus resulting in different absorption performance.

## 4. Conclusion

In conclusion, based on ISO 10534-2, the conducted measurement result showed that the 3 fabrics provided a good result of absorption coefficient. The 3 fabrics absorption coefficient amplitudes showed values larger than 0.5 using 15mm cavity thus could be potentially implemented as sound absorbers. Despite being so, the surface morphology observation using SEM had unseen holes and oblique patterns on Denim fabric, therefore only Alabama and Lisburn fabric were considered to be further investigated as MPP absorbers. The three sets of measurements were assumed to provide sufficient insight regarding the potential of these fabrics, but for further analysis such as absorption modelling, more complicated analysis on the data variance and error should be elaborated.

Therefore, further studies could focus on parametrical survey of fabrics (i.e. perforation diameter and distance between perforations) and model validation on these MPP-like structured fabrics using Maa Model, Johnson-Allard Model or other MPP models.

**References**

[1] U. Ackermann, H. V. Fuchs, and N. Rambausek, 'Sound absorbers of a novel membrane construction', *Appl. Acoust.*, vol. 25, no. 3, pp. 197–215, Jan. 1988.
[2] D.-Y. Maa, 'Potential of microperforated panel absorber', *J. Acoust. Soc. Am.*, vol. 104, no. 5, pp. 2861–2866, Oct. 1998.
[3] F. Suvari, Y. Ulcay, B. Maze, and B. Pourdeyhimi, 'Acoustical absorptive properties of spunbonded nonwovens made from islands-in-the-sea bicomponent filaments', *J. Text. Inst.*, 2013.
[4] 'Garment Tear-Down', *Teresa Flour Lamb*, 01-Feb-2016. [Online]. Available: http://104.131.25.1/garment-tear-down/. [Accessed: 11-Sep-2017].
[5] 'Scanning electron microscopy, Liverpool museums'. [Online]. Available: http://www.liverpoolmuseums.org.uk/conservation/departments/science/scanning-electron-microscopy.aspx. [Accessed: 11-Sep-2017].
[6] A. Nico, 'PENELITIAN KARAKTERISTIK KOEFISIEN ABSORBSI SUARA DARI PANEL KAIN', Institut Teknologi Bandung, 2007.
[7] 'ISO 10534-2:1998 - Acoustics -- Determination of sound absorption coefficient and impedance in impedance tubes -- Part 2: Transfer-function method'. [Online]. Available: https://www.iso.org/standard/22851.html. [Accessed: 11-Jun-2017].
[8] N. Atalla, F. Sgard, 'Modeling of perforated plates and screens using rigid frame porous models', *Journal of Sound and Vibration* 303 pp. 195–208, Mar. 2007.
[9] J. F. Allard, N. Atalla, 'Propagation of Sound in Porous Media : Modelling Sound Absorbing Materials', *John Wiley & Sons,Ltd.*, 2009.
[10] T. J. Cox, P. D'Antonio, 'Acoustic Absorbers and Diffusers : Theory, Design, and Application', *Taylor & Francis*, 2009.

**Acknoledgement**

The Authors would like to thank Engineering Physics Research Group and Laboratory for providing the funding through P3MI scheme

IOP Conf. Series: Journal of Physics: Conf. Series **1075** (2018) 012010 doi:10.1088/1742-6596/1075/1/012010

# Drill string nonlinear vibrations: experimental studies and finite-element analysis

**M Kanzari[1], I M Shahin[1, 2], M Y Alqaradawi[1] and B Balachandran[3]**

[1]Department of Mechanical and Industrial Engineering, College of Engineering, Qatar University, Doha 2713, Qatar

[2]Department of Mechanical Engineering, Shoubra Faculty of Engineering, Benha University, Egypt

[3]Department of Mechanical Engineering, University of Maryland, College Park, Maryland, USA

E-mail: mkanzari@qu.edu.qa

**Abstract**. In the present work, nonlinear vibrations of drill strings are studied. A laboratory scale arrangement consisting of a flexible rod with rotor-stator system is used to study the system dynamics. A finite-element model has been developed to study the system dynamics. Qualitative changes in the system dynamics are examined with respect to variation in the rotation speed, with a focus on whirling and stick-slip motions. The finite -element model results are compared with experimental results. It is believed that the developed finite-element model can be important for the design analysis of drill strings as well as understanding the system dynamics.

## 1. Introduction

Drill strings are assemblies of slender structured pipe segments, which are used especially in drilling operations to transfer power from the rotating motor, at the surface, to the drill bit at the bottom. Due to interactions with the accumulation of cuttings in a wellbore as well as the wellbore, a drill string can experience several problems during operations [1]. One of them is nonlinear vibrations, which can affect the different system components and lead to drilling operation failures. In general, capturing nonlinear effects is not an easy task, and in the current context, this makes the prediction of drill-string dynamics difficult.

Over the last decade, challenging research studies involving experimental and numerical methods have been undertaken to explore drilling operations and understand associated stick-slip whirl interactions. For instance, several finite -element-models (FEM) have been developed and used to predict system response. Klocke *et al.* [2] predicted the cutting forces by using a three-dimensional thermomechanical coupled FE method for carbide gun drills. Su *et al.*[3] used commercial FE analysis software to carry out finite volume simulations of titanium alloy drilling for different feed rates and drilling speeds. Fu *et al.* [4] modelled the transient response of string performances during gas drilling. Gardner and Dornfeld [5] showed that it is possible to capture needed drilling parameters on a FEM based software called DEFORM. Experimental validation of FEM studies has been discussed by Kheireddine *et al.* [6]. Overall, from a review of previous studies, the authors have formed the opinion that FEM based modelling efforts to explain the phenomenon of stick-slip motions and whirling involving torsional and lateral motions have not received much attention. Furthermore, experimental efforts in conjunction with FEM studies have been rather limited. This gap is addressed in the present work.

In the present study, an experimentally validated three-dimensional FEM model for a rotor enclosed within a stator and subjected to dry friction is presented. Predictions of lateral vibrations are made and compared to experimental observations. Different operating condition are examined.

## 2. System Description

One of the main components of the bottom hole part of a drilling system is the drill string that is comprised of three parts, namely, drill pipes, bottom hole assembly, and drill bit, as presented Figure.1.

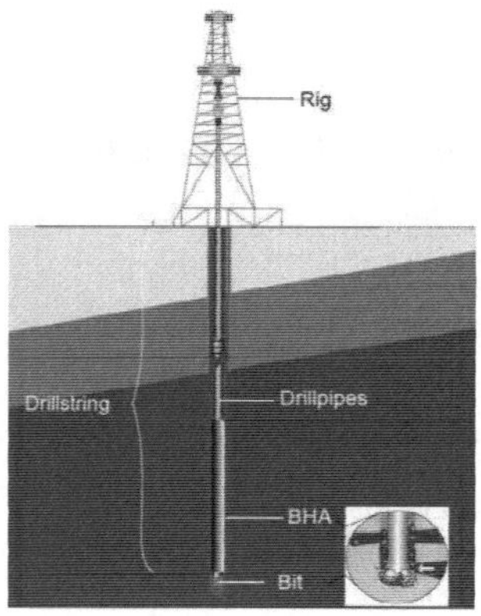


**Figure 1.** Schematic of vertical drilling system.

The drill string can be said to be the most critical part of a drilling system since it is used to transmit torque and drilling mud to the bit, exert weight on bit, and control trajectory of the bit deep below the Earth's surface. The drill bit is used to cut through rock and other formations to create holes. The top drive is principally composed of rotary table, which is used to provide the drilling torque to the drill string and through it to the drill bit. The drill string system is surrounded by soil and rock. Structurally, a drill string can be considered as a long, flexible and rotating core embedded into an outer shell or simplistically as a rotor enclosed within a stator subjected to dry friction. This structure is the subject of the author' investigation, which is presented next.

## 3. Finite Element Analysis and Experimental Validation

### *3.1 Experimental design and arrangement*

During operations, a drill string may undergo forward whirling without any contact with the surroundings, forward whirling with contact and stick-slip motions, or backward whirling with contact. For furthering the understanding of drill string dynamics and exploring schemes for controlling them, a 24:1 laboratory scale version of a drill string section has been constructed. As presented Figure. 2 and 3, a rotor-stator system is considered: The drill string is represented by long aluminium rod with an unbalanced aluminium disc at the bottom.

The system is driven at the top by a servomotor with controlled rotation speed. The borehole is simulated by an outer shell. The experimental system parameter values and operating conditions are shown in Table 1.

IOP Conf. Series: Journal of Physics: Conf. Series **1075** (2018) 012010 doi:10.1088/1742-6596/1075/1/012010

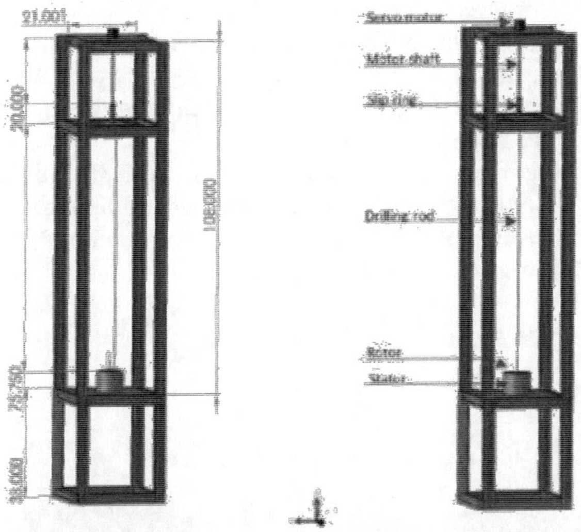


**Figure 2.** Three-dimensional design of experimental apparatus.

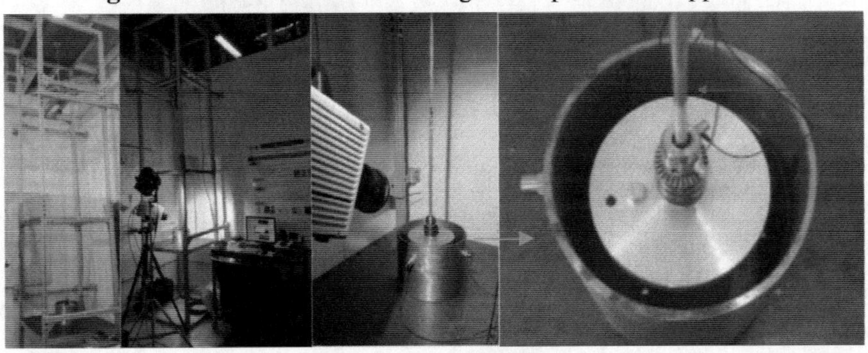

**Figure 3.** Experimental realization of rotary drilling system at Qatar University.

**Table 1.** Parameter values for experimental studies.

| System Variables & Parameters | Notation | Value | Units |
|---|---|---|---|
| Rotor diameter | $D_r$ | 0.15 | m |
| Stator diameter | $D_s$ | 0.18 | m |
| Drill string length | L | 1.92 | m |
| Mass of rotor | $m_r$ | 1.1 | Kg |
| Unbalanced mass on rotor | $m_m$ | 0.06 | Kg |
| Initial rotor position | $\rho_0$ | $10^{-3}$ | m |
| Rotation speed | $\Omega$ | 60-120 | RPM |

Strain gauges were used at different spatial locations along the drill string and accelerometers were fixed to the rotor to measure accelerations in the lateral direction. Experimental data of rotor lateral vibrations and torsional strain were recorded for different rotor drive speeds. The studied motions included whirling. By integration of the obtained acceleration time histories, velocity of the rotor is obtained.

*3.2 Finite element analysis: Result and comparisons*

Rotor motions were examined for driving speeds in the range of 60 to 200 RPM. A finite-element model of the drill string was first constructed by using LS DYNA software. Adaptive explicit meshing is performed. The assembly is composed of four FEM parts, including the servomotor, the drill string, the rotor and the outer shell, as presented in Figure 4.

The drill string, rotor, and stator are described with 2.5, 4.6, and 6.321 millions of nodes, respectively. The resulting model is used to study the non-linear response for different rotation speeds. The nonlinear response of the structure has been analyzed with finite element method. Qualitative comparisons are made between experimental results and FEM based results.

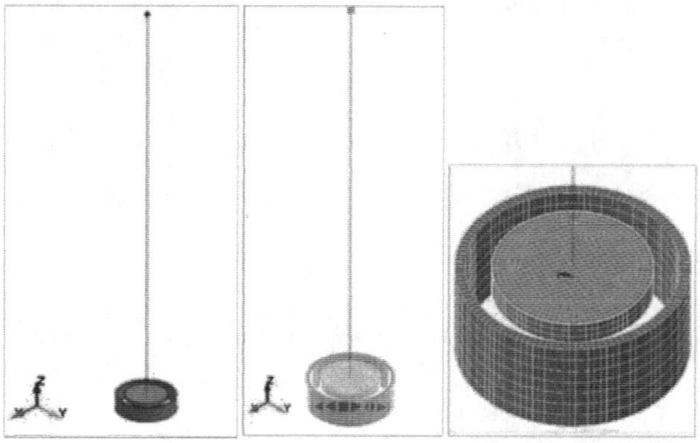

**Figure 4.** FEM representation of system components.

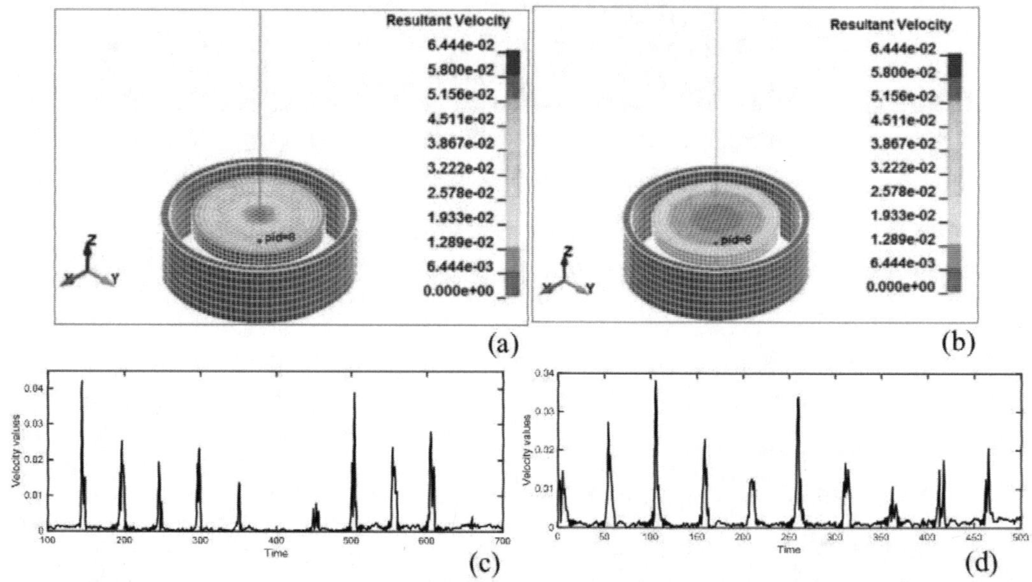


**Figure 5.** Rotor response velocity variation with drive speed. (a) FE analysis at Ω=60 RPM. (b) FE analysis at Ω=100 RPM. (c) Experimentally observed rotor velocity response at Ω=60 RPM. (d) Experimentally observed rotor velocity response at Ω=100 RPM.

At low driving speeds, the simulation results suggest that the lateral vibrations have irregular characteristics. Rolling and bouncing phases of motion dominate, leading to the irregular characteristics observed. When the vibrations have regular characteristics, the rotor motion is found to be stable and the drill-string motions are closer to the radial center of the system. As the driving speed is increased, the lateral vibrations are found to be excited further and sliding motion phases are also observed, as found in previous work[7]. The FEM based simulation results are found in to be good qualitative agreement with the expermental results. When the driving veloity is enhanced, these vibrations are converted to regular lateral oscillations over time. Rotor is more stable and drillstring motions are closer to the radial centre. For both cases, the FE analysis results compared well with the experimental studies (see Figure 6). For high driving speed, lateral vibrations are excited, sliding motion with rotor-stator contact results on dry friction permanent whirl[7]. Torsinal and lateral motions are strongly coupled and backward whirl also occurs for certain rotation speeds. When only the first torsion mode of vibration is excited, the strain measurements are directly promotinal to the torsion displacement[8]. Strain measurements made on the drill string close to the rotor-stator system are presented in Figure 7.

At low driving speeds, the drill-string motion is dominated by torsion motion, with associated impacts of the rotor with the stator, as shown in Figure 8a. For intermediate rotation speed values, the drill-string lateral vibrations are excited, which result in bending of the system.. For high driving speeds, lateral and torsional motions are coupled and the rotor undergoes whirling motions, as seen in Figure 7b and confirmed by the EM based results of Figure 8d.

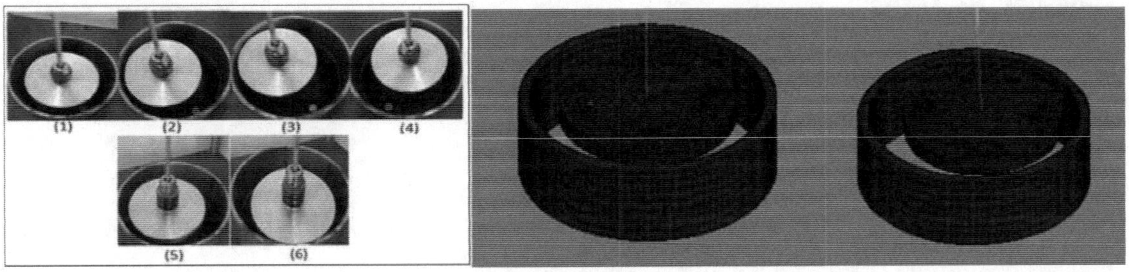


(a) (b)

**Figure 6.** Rotor motions illustrating sticking and sliding phase at Ω=200 RPM. (a) Experimental observation. (b) FEM results.

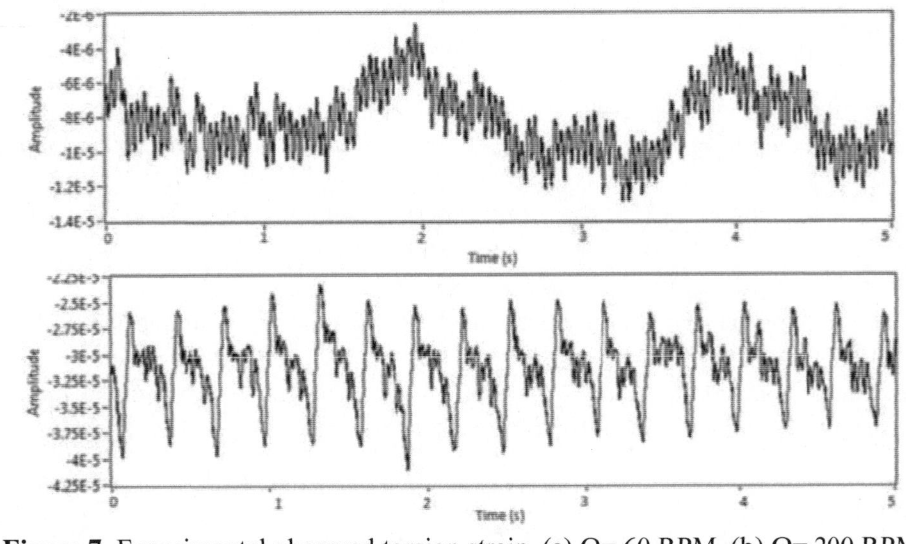


**Figure 7.** Experimental observed torsion strain. (a) Ω= 60 RPM. (b) Ω= 200 RPM.

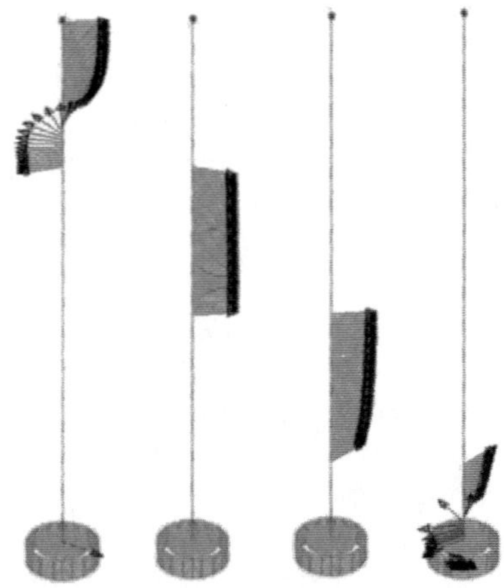

**Figure 8.** System lateral displacement. (a) Ω= 60 RPM, (b) Ω= 10 0 RPM, (c) Ω= 140 RPM, a nd (d) Ω=200 RPM.

## 4. Concluding Remaks

Dynamic of a section of a drill- string within a wellbore has been studied as a rotor -stator system through experiments and finite-element model-based simulations. The experimental results and simulation results are in good qualitative agreement. The results suggest that intermediate driving speeds may be preferable comparable to low or high driving speeds.

## References

[1] Spanos, P. D., Chevallier, A. M., Politis, N. P., and Payne, M . L., 2003. Oil and gas well drilling: a vibrations perspecti e, *The Shock Vibr. Digest* **35** 85-103.

[2] Klocke, F., Abouridouane, M., Gerschwiler, K., and Lung, D., 2011. 3D modelling and simulation of gun drilling, *Adv. Mater. Res*, **223** 12-19.

[3] Su, Y., Chen, D. D., and Gong, L., 2015. 3D Finite Element Analysis of Drilling of Ti-6 Al-4V Alloy, *Int. Conf. on Co mput. Inf. Syst. Ind. Appl. (Thailand),* 07-911.

[4] Jianhong, F., Kexiong, S., Zhi, Z., Dezhi, Z., Fei, L., Xin, X. and Xin, Z., 2012. Stress Analysis on Drilling String Vibration in Gas Drilling, *Int. Conf. Future Enrgy, Env. Mater, Enrgy Procedia* **16,** 1264-1268.

[5] Gardner, J. and Dornfeld, D., 2006. Finite Element Modeling of Drilling Using DEFORM, *Consortium on Deburring and Edge Finishing*, Laboratory for Manufacturing and Sustainability, UC Berkeley.

[6] Kheireddine, A. H., Ammouria, A. H., Lu, T., Jawahir, I. S., and Hamade, R. F., 2013. An FEM Analysis with Experimental Validation to Study the Hardness of In-Process Cryogenically Cooled Drilled Holes in Mg AZ31b, *Procedia CIRP* **8**, 588 -593.

[7] Huang, D. G., 2007. Characteristics of torsional vibrations of a shaft with unbalance. *J Sound Vib.* **308**, 692–698.

[8] Vlajic, N., Liao, C. M., Karki, H., and Balachandran, B., 2014. Stick-slip motions of a rotor-stator system, *ASME J. Vib. Acoust* **136**, 021005-1-021005-8.

## Acknowledgement

This work was made possible by NPRP grant # [7-083-2-041] from the Qatar National Research Fund (a member of Qatar Foundation). The statements made here in are solely the responsibility of the authors.

# Vibration Transmissibility Measurement on Multi Pump Damage Detection Using Accelerometer Array

**Anisatul Fauziyah, Try Rahadi Sulistomo, Mifta Nur Farid, Dhany Arifianto**

Dept. of Engineering Physics, Faculty of Industrial Technology, Institut Teknologi Sepuluh Nopember, Kampus ITS Sukolilo, Surabaya 60111, Indonesia

E-mail: nisafauziyah91@gmail.com

**Abstract.** Pump damage detection methods (based on vibration) are normally used on a pump located on a single foundation. However, when more than one pump located on a single foundation, the vibration transmissibility at one pump with other pump vibrations will affect each other. In this case, we can not use pump damage detection methods normally. Hence, this research used accelerometer array to know the effect of vibration transmissibility to the result of pump damage detection method. Using Fast Fourier Transform (FFT) analysis we known that the effect of vibration transmissibility on the diagnosis of detection of pump damage is destructive superposition occurs in the configuration of misalignment & unbalance pump (-7.7296 dB) and misalignment & bearing fault pump (-8.037 dB). While the constructive superposition occurs in the configuration of unbalance & bearing fault pump (3.6719 dB). The conclusion is transmissibility ratio from accelerometer array measurement could be used to determine the superposition of vibration between two different pump which mounted on same supporting base.

## 1. Introduction

In Industrial process, Installed machine have integrated work mechanism. Hence, once the machine fault occurred, whole industrial process may be impacted. To maintain the machines in good condition, good maintenance strategies need to be implemented. Predictive maintenance is one of the maintenance strategy that often be implemented in industry, especially for rotating machine. In this maintenance strategy, it is compulsory to monitor machine condition overtime, thus it is often be called condition-based maintenance[1].

In rotating machine, condition of the machine is determined through the vibration analysis. Thus, by having preliminary condition analysis, machine could maintain in proper schedule and treated correctly[1]. However, when there is more than one machine being operated and mounted in same base, which in this study we call it multi-pump, the vibration of one machine will transmitted to the other, thus the collected vibration data of one machine could be superposition of the other machine. This may result in wrong interpretation of machine condition, and continue to wrong corrective treatment. This paper shows the evaluation of the effect of vibration transmissibility of two machines that mounted in same base.

## 2. Vibration Analysis

Vibration analysis is one of the methods for implementing predictive maintenance. This method has been widely applied in industrial field. Vibration of machine represented by its acceleration, althoughit may also being state in term of velocity or displacement. Since vibration is oscillating motion, there are three main features that being utilized in the analysis. Those are amplitude, frequency, and phase. Amplitude of vibration indicated the level of machine damage. Increasing in vibration amplitude indicated increasing of machine severity. Amplitude of vibration stated in root mean square value of the recorded vibration, as shown in equation [?].

$$f_{rms} = \sqrt{\frac{1}{T}\int_0^T (f(t))^2\, dt} \tag{1}$$

where $f(t)$ is time domain signal, $T$ is length of data.

Frequency of vibration often being corresponded to the failure mode of rotating machine. Each of the failure modes have different frequency characteristic. Frequency of vibration can be acquired by transforming the time domain signal to frequency domain signal using Fourier Transform.

$$F(j\omega) = \int_{-\infty}^{\infty} f(t)e^{-j\omega t} dt \tag{2}$$

where $\omega = 2\pi f$. In this paper FFT algorithm implemented to acquired frequency domain signal.

## 3. Failure Mode of Rotating Machine

Rotating machine assembled from different kind of parts. Wrong installation and defect of rotating machine part may affected the vibration frequency and amplitude of rotating machine. In this paper 3 kinds of failure mode will be analyzed, which is unbalance, misalignment, bearing fault, and mechanical looseness.

(1) Unbalance
Unbalance is the condition where the centre of gravity is not aligned with the axis of rotation[2]. Unbalance condition can be recognize when peak of frequency exist in $1 \times RPM$ of radial vibration (horizontal and vertical direction)[3].

(2) Misalignment
Misalignment is the condition when rotating shaft is not aligned. It may be result from bended shaft or improper coupling. Peak frequency of this failure exist on 1 to $3 \times RPM$[3]. This failure can be detected by shifted phase of the vibration.

(3) Bearing Fault
This failure mode detected by peak in high frequency (may exist until $10 \times RPM$)[3]. This failure mode occurred due to defect or wear of bearing. Additional measurement position in variation of depth need to be add in order to have understanding major source of scattering.

## 4. Transmissibility

Rotating machine vibration could propagate through the supported base. If there are two or more than one machine operating, this propagated vibration may added to the other machine. Hence the measured vibration of that machine is the result of superposition between the machine own vibration and other machine vibration which propagate through the base[4]. The ratio of machine vibration force which transmitted to the supported base to its actual force is called transmissibility factor ($T$),

$$T = 10\log_{10}\left(\frac{S}{B}\right) \tag{3}$$

where $S$ is the total summation of machine vibration and transmitted vibration and $B$ is machine vibration when operating alone[5].

IOP Conf. Series: Journal of Physics: Conf. Series **1075** (2018) 012011 doi:10.1088/1742-6596/1075/1/012011

## 5. Experiment Setup

Two pumps with different failure mode will be mounted in same supporting base with the distance of 15.5 cm as shown in Figure 1. Pumps operated in 3000 RPM (50Hz). Three accelerometer were placed in three different axis (vertical, horizontal and axial) to measure vibration of the pump. The combination of two pumps with different failure mode can be seen in Table 1.

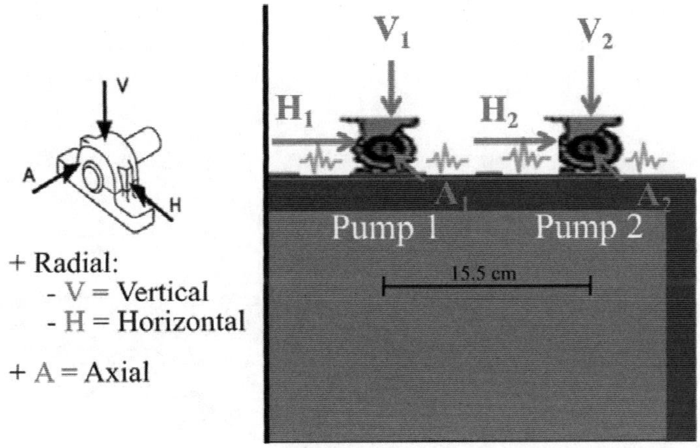


**Figure 1.** Configuration of sensors

**Table 1.** Variation of two different pumps

| No | Pump 1 | Pump 2 |
|---|---|---|
| 1 | Unbalance 18g, 1.5 cm | Misalignment 1 mm |
| 2 | Unbalance 18g, 1.5 cm | Misalignment 2 mm |
| 3 | Unbalance 18g, 1.5 cm | Bearing fault |
| 4 | Bearing fault | Misalignment 1 mm |
| 5 | Bearing fault | Misalignment 2 mm |

## 6. Result and Discussion

The results of transmissibility factor was shown in Table 2. The negative sign indicated that there was destructive superposition, since acquired signal pump vibration less powerful than the vibration where the pump operation solely. The positive sign indicated opposite event, thus superposition was constructive.

In constructive superposition case, recorded signal behave like a pump with 2 failure mode. Figure 2a shown the recorded vibration signal of bearing fault pump, which operated along with the unbalance pump. Transmissibility of this configuration is 3.6 dB (configuration 3, pump 2 vertical axis), thus it is belong to constructive superposition. The amplitude of first harmonic frequency increased when compared to the baseline signal. This event may lead to wrong analysis of machine condition failure mode, the pump which only have bearing fault also contain vibration of nearby pump which have unbalance.

In destructive superposition case, the wave cancel out each other, thus made the recorded signal have lower power of vibration. Figure 2b shows the spectrum of pump 2 (configuration 2, horizontal axis). The transmissibility value of this configuration is -7.729 dB. The amplitude of

**Table 2.** Transmissibility factor of two different pumps

| No. | Transmissibility (dB) | | | | | |
|---|---|---|---|---|---|---|
| | Pump 1 | | | Pump 2 | | |
| | $V_1$ | $H_1$ | $A_1$ | $V_2$ | $H_2$ | $A_2$ |
| 1 | -0.354 | -0.558 | -0.731 | 10.378 | -2.798 | 1.705 |
| 2 | 1.244 | -0.548 | -1.180 | 1.764 | -7.729 | 2.841 |
| 3 | -0.036 | -0.908 | -1.249 | 3.672 | 0.385 | 4.265 |
| 4 | 0.006 | 0.068 | 5.456 | 5.025 | -2.347 | -2.879 |
| 5 | 0.541 | 0.501 | 5.720 | -0.129 | -8.037 | 4.489 |

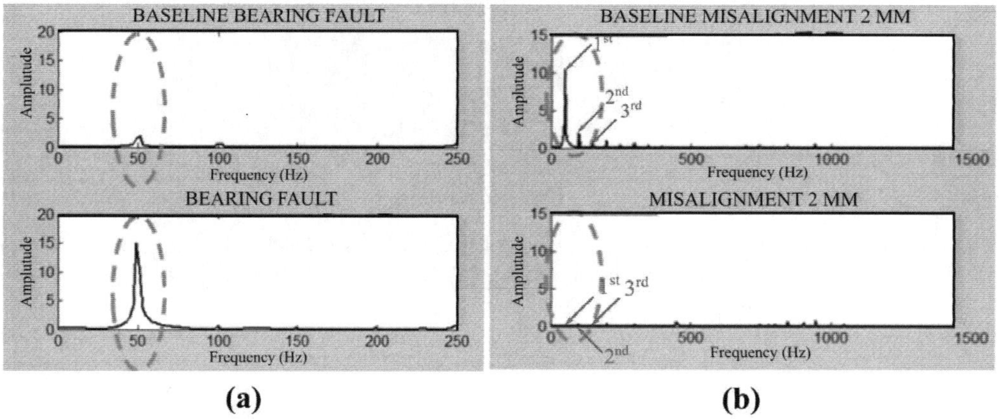


**Figure 2.** Vibration spectrum of (a) bearing fault pump and unbalance, (b) misalignment 2mm pump and bearing fault

1st, 2nd and 3rd harmonic frequency were decrease. This may lead to wrong analysis of machine failure severity.

## 7. Conclusion

(1) Transmissibility ratio from accelerometer array measurement could be used to determine the superposition of vibration between two different pump which mounted on same supporting base.

(2) Destructive superposition occurs in the configuration of misalignment & unbalance pump (-7.7296 dB) and misalignment & bearing fault pump (-8.037 dB). While the constructive superposition occurs in the configuration of unbalance & bearing fault pump (3.6719 dB).

## References

[1] ISO 13373 - Condition monitoring and diagnostics of machines - Vibration condition monitoring.
[2] Girdhar, Pares.2004. Practical machinery vibration and analysis & Predictive Maintenance. Oxford: Newnes Inc.
[3] Tri P., Agung, 2011. Pembelajaran Vibrasi Bengkel Mesin Fasilitas Pemeliharaan Kapal Surabaya, LANTAMAL V.
[4] Saktyo,Haris. 1997. Simulasi Superposisi Getaran dengan menggunakan Metode FFT. ITS
[5] Amir Hossein, 2013 Aplication of Transmissibility Measure in Estimation of Modal Parameters for a Structure. Blekinge Institute of Technology. Sweden

IOP Conf. Series: Journal of Physics: Conf. Series **1075** (2018) 012012 doi:10.1088/1742-6596/1075/1/012012

# Audio Signal Deterioration Caused by Propagation Noise between Audio Equipment (2nd report)

**Takahiro Fujino and Takahiro Yoshida**

Department of Electrical Engineering, Tokyo University of Science,
6-3-1 Niijuku, Katsushika-ku, Tokyo 125-8585, JAPAN

E-mail: yoshida@ee.kagu.tus.ac.jp

**Abstract**. Sound quality of music playback may be deteriorated due to various complex factors such as power supply state of the audio equipment. Therefore, in our previous study, we partly confirmed the mechanism that the power supply states affected alternative-current (AC)-potential difference and propagation noise between the equipment, which influenced audio-signal intermodulation distortion (IMD) on one of personal-computer (PC)-audio system. However, the research was insufficient in the number of measured audio equipment and measuring setup. In this study, we increased the model of digital-to-analog converter (DAC) and audio amplifier used for measurement and improved the measurement method to reduce the influence of measuring equipment. We measured ground AC potential, propagation noise, and the analog audio signal output of the amplifier. Their measurements were done in changing combination of audio equipment and each power supply states to analyze the relationship among these three domains. As a result, it is found that each domain depended on the model, combination, and power supply states of each audio equipment. There is a strong positive correlation between the ground AC potential and the propagation noise. There is also a positive correlation between the propagation noise and the audio-signal IMD in some combination of audio equipment.

## 1. Introduction

In recent years, personal-computer (PC)-audio systems that use the PC and digital-to-converters (DAC) have become popular, and digital-audio technologies are now in progress to improve high-quality-sound playback. However, in an actual music playback environment using audio equipment, sound quality of music playback may be deteriorated due to various complex and interrelated factors, such as the state and polarity of each alternative current (AC) power supply in each equipment [1],[2], the operating state of the PC [3], and so on. In these previous study [1]~[3], focusing on the sound quality deterioration due to jitter, and the mechanism of generation has not been elucidated. Even today, it is necessary to tune audio equipment and systems subjectively based on the knowledge and experience of the acoustic technicians in order to play with best performance of the audio equipment.

We considered one reason of remaining such troublesome situation is that the mechanisms leading to the degradation of sound quality have yet to be elucidated. In addition, another reason is that acoustic-measurement methods and objective evaluation criteria to describe sound-quality deterioration have not been specified. To address these situation, we considered one factor of sound-quality deterioration mechanism as AC potential differences between audio equipment caused by the power supply in each piece of equipment, and the noise loop due to indirect coupling and the resulting propagation of electromagnetic noise between parts of the equipment. These propagation noise can modify the analog audio output signal and thereby perturb the resulting subjective auditory sensation [4].

Therefore, in our previous study, we measured the ground AC potential at the USB ground of the PC using a digital multimeter, the propagation noise current at the USB ground line between the PC and the DAC using an oscilloscope with a differential probe, and the analog audio signal at the loudspeaker output of the amplifier using an audio analyzer, as a function of the type and polarity of each power supply on each equipment on one of PC-audio system [5]. This audio system for the measurement consists of PC, universal-serial-bus (USB) connected DAC, and audio amplifier for the purpose of verifying the focused the sound-quality deterioration mechanism. The result of the previous study showed that the amplitude and shape of the propagation-noise waveforms on the USB ground between the PC and the DAC differed and depended on PC-power supply state, i.e., battery powered or AC power supply (AC adapter) powered. In particular, the 50 Hz noise component (hum noise) from the AC power supply was remarkable. Furthermore, power supply state affected the 50 Hz component and its harmonics distortion in the audio signal output of one amplifier. In addition, power supply state also affected the IMD components around the fundamental wave of the playback single sine wave. There was a positive correlation between the ground AC potential, the propagation noise, and the audio-signal IMD. However, the previous research was insufficient in the number of measured audio equipment and measurement setup, because the targeted audio equipment used in the previous measurement was only one set. In addition, there was a problem that the results might be affected by a model of measurement equipment.

Therefore, in this study, we measured by changing combination of audio equipment consisted of 3 models of DAC and 6 models of amplifier and each power supply states and we analyzed among those 3 domains in order to elucidate that sound deterioration mechanism more detail. In this study, the measurement was done by improved measuring method to reduce the influence of measuring equipment.

## 2. Measurement setup and conditions

Figure 1 shows the setup for measuring the ground AC potential, the propagation noise, and the audio signal. This system can accommodate different types and polarity of power supplies for each device. In our previous study, one model of DAC and digital amplifier was used. In this study, we measured them using 18 combinations of audio equipment consisted of 3 models of DAC (model A, B, and C) and 6 models of amplifier (3 models of digital amplifier (model a, b, and c) and 3 model of analog amplifier (model d, e, and f)).

The AC potential of the equipment is the potential difference between the ground of the USB output of the PC (Apple, MacBook air) for playback and the laboratory ground. This was measured by a digital multimeter (Tektronix, DMM4040) in AC voltage mode with PC for playback, DAC, and amplifier are all connected and powered on. The current related to the propagation noise was measured using a Tektronix oscilloscope (DPO7254, bandwidth 2.5 GHz) equipped with a differential probe (Tektronix,

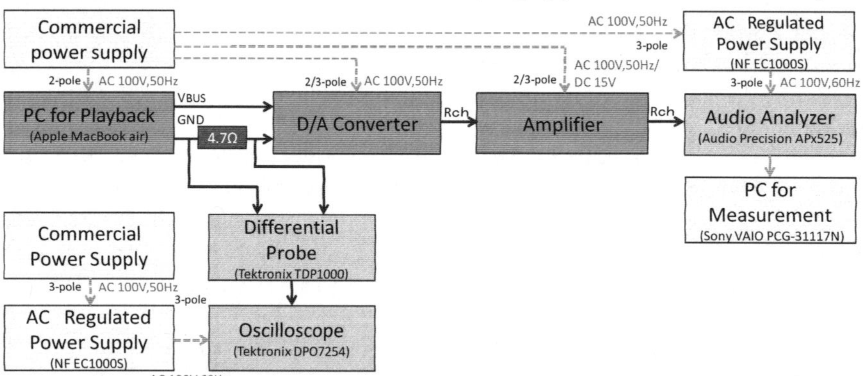


**Figure 1.** Measuring setup for measurement of the ground AC potential, the propagation noise, and the analog audio output signal.

IOP Conf. Series: Journal of Physics: Conf. Series **1075** (2018) 012012 doi:10.1088/1742-6596/1075/1/012012

TDP1000, bandwidth 1 GHz) by a resistor (4.7 Ω) inserted into the ground line of the USB connecting the PC and the DAC. Its resistance value was set at balanced value considered of sensitivity and insertion influence. The oscilloscope operated at 1 MHz sampling frequency, 1 MSample record length, and 1 second measurement duration.

These measurements of the ground AC potential and the propagation noise were measured without connecting an audio analyzer in order to avoid influences by measuring instrument on measured results. The reason is that an audio analyzer makes another indirect connection such as ground loop and causes some influences on measured results because an audio analyzer is active device that has a power supply. In contrast, connecting loudspeakers to the output of the amplifier does not make influences on measured results because loudspeakers are passive device that have not a power supply. These influences were experimentally confirmed in our experiment.

The analog audio signal was measured at the loudspeaker output of the amplifier using an audio analyzer (Audio Precision, APx525) and battery powered notebook-PC (Sony, VAIO type-Z) only for control the audio analyzer. This measurement applied a fast Fourier transform (FFT) to analyze the audio spectrum. This was measured 5 times for each measurement condition, and the average value was calculated. A 10.011 kHz sine wave (WAV file format, 44.1 kHz sampling frequency, 24-bit quantization) was used as a test signal for the measurement. The amplitude of playback sine wave is set to 1.0 $V_{rms}$ by setting the gain of each amplifier. We used music-player-software PlayPcmWin in WASAPI push mode on Microsoft Windows 8.1 to play this WAV file. The specifications of the FFT analysis of the audio spectrum was 48 kHz sampling frequency, 192 kSample record length, and 4 seconds measurement duration. The measurements were done after each device warmed up for over two hours.

Table 1 lists the combinations of power-supply states that can affect the AC potential of the equipment and the noise loop. There are 3 kinds of power supply states of the PC for playback and 2 power supply state of the DAC and amplifier, i.e., there are total 12 combinations of power supply states. This is mainly focused on the influence due to noise and noise-loop change derived from the power supply. The output of the DAC and the amplifier were connected only the right channel of stereo 2 channels.

Japanese commercial AC 100 V power and audio-equipment-power source is normally 2 poles, i.e., Line and Neutral. We defined the original polarity of the power supply as "Polarity A" and the opposite polarity, i.e., the line and neutral were reversed, as "Polarity B". The line and the neutral were reversed also on an audio equipment which had 3 poles AC inlet. The commercial AC-power-line frequency in eastern Japan was 50 Hz. The audio analyzer was driven by a regulated AC power supply at 60 Hz in order to distinguish whether the frequency components of the measured spectrum of the audio signal was caused by the interaction with audio analyzer or only by the interaction within the audio equipment.

**Table 1.** Experimental condition about the combination of the power state of each equipment.

| Equipment | PC | DAC | Amplifier |
|---|---|---|---|
| Power supply state | · Internal battery<br>· AC Adapter (Polarity A)<br>· AC Adapter (Polarity B) | · AC Adapter (Polarity A)<br>· AC Adapter (Polarity B) | · AC Adapter (Polarity A)<br>· AC Adapter (Polarity B) |

## 3. Measurement results and consideration

### *3.1.* Audio equipment's model dependency of the relationship between the AC potential and the propagation noise

Figure 2 shows the relationship between the ground AC potential and the amplitude of the propagation noise amplitude on the USB ground. The 216 measurement results for all combinations of the 12 power supply states and the 18 combinations of equipment are plotted in this figure. An approximate straight line of the 216 measurement results is also plotted in this figure. Plotted colour means the combination of the equipment. The amplitude of the propagation noise is averaged peak-to-peak amplitude of the 100-point-moving-averaged waveform in order to reduce the random noise.

From figure 2, it is confirmed that the ground AC potential and the amplitude of the propagation noise varies greatly with the combinations of audio equipment and their power supply states. The relationship is a direct proportion and strong positive correlation represented by single approximate straight line.

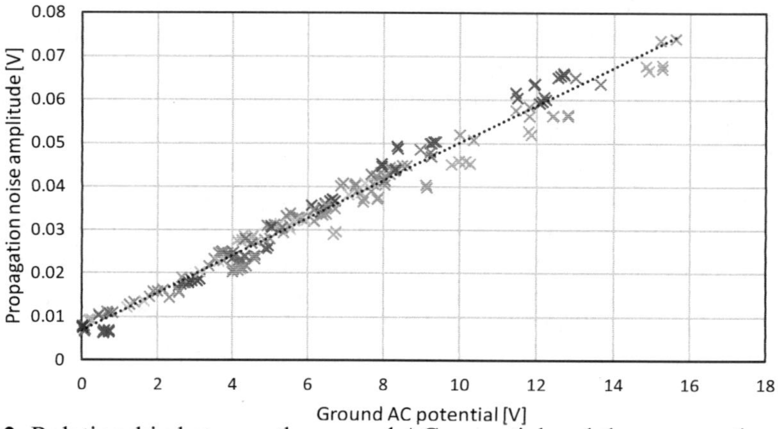


**Figure 2.** Relationship between the ground AC potential and the propagation noise amplitude on the USB ground.

Figure 3 shows comparison of the propagation noise waveforms through the USB ground line in different power condition of the PC. The black line in this figure shows maximum case of the propagation noise waveform, and the gray line shows minimum case. From figure 3, it is found that 50 Hz noise components from AC power differs mainly by the power supply states.

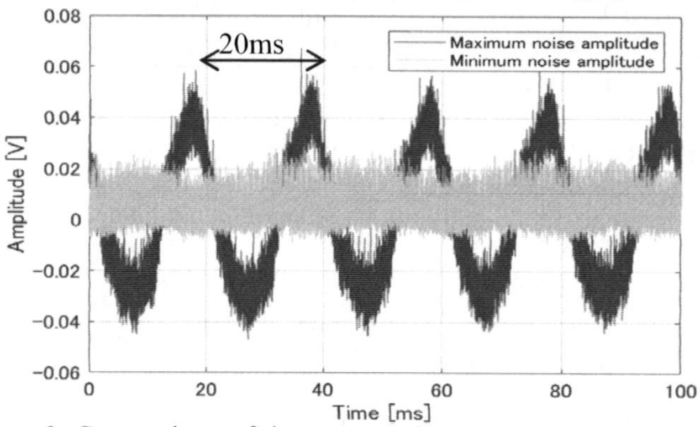


**Figure 3.** Comparison of the propagation noise waveforms through the USB ground line in different power condition of the PC.

*3.2. Audio equipment's dependency of the relationship between the propagation noise and the audio signal output*

Figure 4 shows the audio equipment dependency of the relationship between the propagation noise amplitude and the total value of the IMD of the audio signal output. The 216 measurement results for all combinations of the 12 power supply states and the 18 combinations of equipment are plotted in this figure. An approximate straight line of each combination of the equipment (18 combinations) is also shown in this figure. Plotted colour means the combination of the equipment. The total IMD that seems to affect the auditory sensation is calculated as follows,

$$\mathrm{IMD}_{Total}\ [\mathrm{dBV}] = 20\log_{10}\sum_{n=1}^{5} A(f_1 \pm f_2) \tag{1}$$

where A(f) is the IMD amplitude at frequency f. In this measurement, $f_1$ = 10.011 kHz, $f_2$ = 50 Hz, and the IMD amplitude sums from 1st order to 5th order.

From figure 4, it is found that the audible IMD of the audio signal output at near frequency of the playback sound depend on the combinations of the equipment. The influences of the combination of the power supply state that has correlation with the propagation noise amplitude on the total IMD are different among the combinations of the equipment. The 5 of all 18 combinations of equipment had a positive correlation between the propagation noise amplitude and the total IMD. The other 13 combinations of equipment did not have the correlation, i.e., these equipments were not influenced by the combination, the power supply state, and propagation noise. In addition, there was no combinations of equipment with negative correlation. From this results, it seems that it is desirable to set the power supply states to less propagation noise and lower ground AC potential of audio equipment.

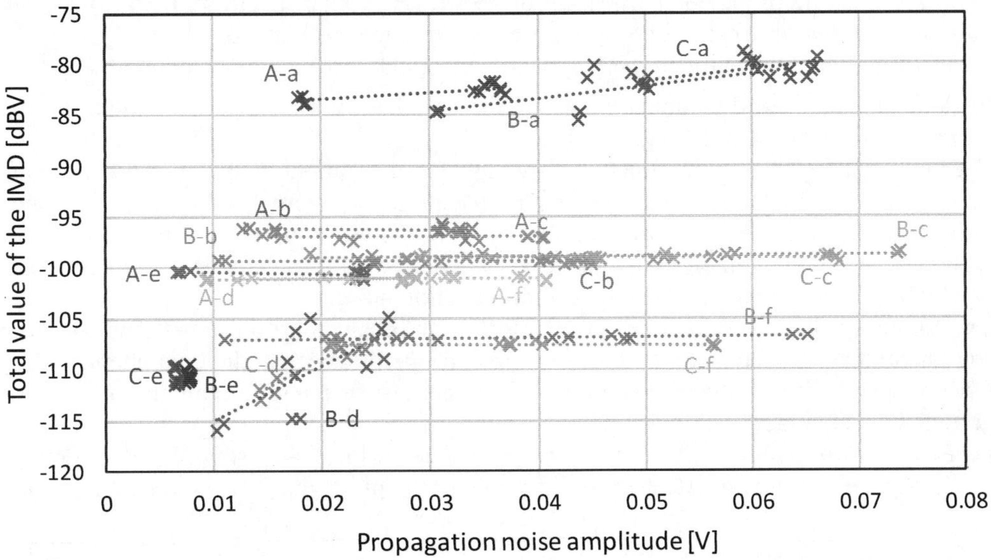


**Figure 4.** Relationship between the propagation noise and the total value of the IMD of the audio signal output. (Example: 'A-b' means DAC model A, Amplifier model b.)

Figure 5 shows an example of the IMD part of FFT spectrum of the loudspeaker output of an amplifier at two power supply states in the same combination of the equipment. This figure is magnified view around the playback 10.011 kHz sine wave component. Magnified view if the IMD part of FFT spectrum of the loudspeaker output of an amplifier around the playback 10.011 kHz sine wave.

From figure 5, it is found that the occurrence of the audible IMD between the playback signal and hum noise varies greatly with dependence on the power supply states.

## 4. Conclusion

In this study, in order to elucidate the audio-signal deterioration mechanism, we did additional research by increasing the model of DAC and amplifier used for measurement, and improved the measurement method.

From experimental results, we found and confirmed followings;

1. The ground AC potential and the propagation noise depend on the combination of equipment and the power supply states. The audio signal IMD partly depend on the combinations of equipment and the power supply state.
2. In particular, there is a strong positive correlation between the ground AC potential and the propagation noise in all conditions of the combinations of equipment and the power supply states.

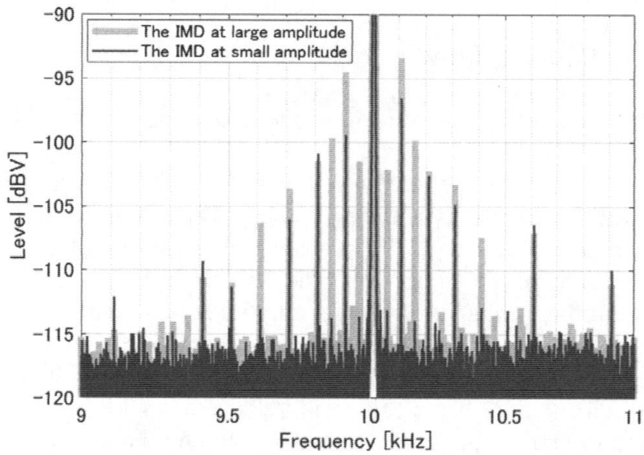


**Figure 5.** An example of the comparison of the IMD part of FFT spectrum of the loudspeaker output of an amplifier at two power supply states on the same combination of the equipment. (Magnified view around the playback 10.011 kHz sine wave component.)

3. In some equipment and combination of equipment, there is a positive correlation between the propagation noise and audio-signal IMD.
4. The proposed sound-deterioration mechanism in our previous study that the power supply states affects the AC potential difference and the propagation noise between the equipment, and which influences the audio-signal IMD, occurs in some audio equipment and their combination. However, there are some equipment and some combinations that is not influences by the combination, the power supply state, and propagation noise.
5. The sound-quality deterioration mechanism with dependent on the combination of the equipment may be one factor that causes compatibility between audio equipment.
6. It seems that it is desirable to set the power supply states to less propagation noise and lower ground AC potential of audio equipment.

In future work, we will identify the conditions under which this sound-quality deterioration mechanism is applied, and investigate another influence of the propagation noise on audio signal.

## References

[1] Mitui M, Ishikawa T, Dang J, and Miyahara M 2008 Investigation of deterioration of sound quality concerned with deep *KANSEI* in digital audio system –Structure and inprovement of deteriolation of sound quality caused by jitter- *Proc. of journal of Japan Society of Kansei Engineering (JSKE2008)* Vol 7 No 4 pp 759-764

[2] Fuyuki S, Kobayashi Y, Ishikawa T, and Miyahara M 1998 The deteriolation of high order sensations in digital audio signal causeb by jitter –A sharp loss of "Atmospheric" and (deepness)- *Proc. of the Institute of Electronics, Information and Communication Engineers (IEICE1998) technical report* EA97-104 pp 9-16 (In Japanese)

[3] Kaneko K and Mitui M 2012 A study for improvement of the sound quality in PC andio system –Relation between the hardware surrounding of PC and jitter of a digital signal- *Proc. of the Institute of Image Information and Television Engineers (ITE2012) technical report* Vol 36 No 54 ME2012-142 (Dec. 2012) (In Japanese)

[4] Tagaya S and Yoshida T 2015 Effect of AC coupling of ground connection between audio equipment on propagation noise suppression and Inter-Modulation Distortion noise *Proc. of the Institute of Electronics, Information and Communication Engineers (IEICE2015) technical report* Vol 115 No 377 pp 19-23 (In Japanese)

[5] Fujino T and Yoshida T 2017 A consideration of mechanism of audio signal deterioration caused by propagation noise between audio equipment Proc. of 2017 Asia-Pacific International Symposium on Electromagnetic Compatibility (APEMC2017) pp 155-157 (In Japanese)

IOP Conf. Series: Journal of Physics: Conf. Series **1075** (2018) 012013 doi:10.1088/1742-6596/1075/1/012013

# Underwater Source Separation Using Multi-Stage Independent Component Analysis in Semi-Anechoic Water Tank

**Ridhwan Juniarga Pribadi, Mifta Nur Farid, Wiratno Argo Asmoro, Wirawan, Endang Widjiati, Dhany Arifianto**

Dept. of Engineering Physics, Faculty of Industrial Technology, Institut Teknologi Sepuluh Nopember, Kampus ITS Sukolilo, Surabaya 60111, Indonesia

E-mail: `ridhwan.juniarga@gmail.com`

**Abstract.** The experiment of source separation in sea water environment often faced difficulties especially for controlling certain variables. One of the upcoming solution to simplify sea water environment was to conduct the experiment in the anechoic water tank. To evaluate the effects of the experiment performed in the water tank, the comparison of source separation using linear mixture and in-tank mixture will be analyzed. The source separation was performed in three different methods, Time Domain Independent Component Analysis (TDICA), Frequency Domain Independent Component Analysis (FDICA), and Multi Stage Independent Component Analysis (MSICA). The experiment result showing that compared to the linear mixture source, the performance of source separation degrade significantly while performed on the water tank, indicated by the MSE value which increased from $3^{1}0^{-8}$ to $6^{1}0^{-2}$. This result showing that the source mixture that occurred in water tank, much more complex compared to the linear mixture.

## 1. Introduction

Acoustic wave played important role in underwater communication.In terms of propagation and power attenuation, it performed better than electromagnetic wave in medium of water. However, the application of acoustic signals in ocean encountered many problems, such as limitation of bandwidth, huge noise environment, complexity of the medium itself, and the impact to surrounding marine biological life, thus required further processing in order to be used properly. Considering those level of complexity, a technique that is able to simplify the process of propagation modelling, as well as the reconstruction of the mixed sound was required.

References [1] proposed one method of solving this problem by constructing a marine miniature in an anechoic water tank. By using this kind of tank as test-bed the fluctuation of important variables such as temperature, and salinity can be controlled and multipath channel can be eliminated.

By having proper experiment place, the modelling of propagation will be simplified, then Independent Component Analysis (ICA) will take place of sources separation. ICA is one of the method which estimate sources signal without any prior information about sources itself. One of the common method of ICA is Fast ICA, which in this paper will be called Time-Domain ICA (TDICA). The process of source mixing are formulated as common wave superposition (linear mixture). The drawback of this method came when facing signal which mixed in reverberant condition (convolutive mixture). In those condition, the propagation of source signal also impacted by the room impulse response, thus the assumption of independence and number of sources are less then sensors will collapse. Another method of ICA is to perform separation under frequency domain, which called Frequency Domain ICA (FDICA). This method will allow convolutive mixture to be simplify as linear mixture through Fourier Transform. However this methods also have drawback since the assumption of signal independence will collapse in each sub band. To overcome this problem, [2] proposed to combine both of this methods in sequence, which called Multistage ICA (MSICA). In this method, signal estimation will be done in two stages, first in FDICA then followed by TDICA.

This paper will evaluated the quality of sources estimation using three of mentioned methods. The experiment conducted in water tank which belong to Vibration and Acoustic Laboratory, Institut Teknologi Sepuluh Nopember. The Performances of estimated signal will be evaluated in Mean Square Error (MSE) value by comparation to baseline signal. The MSE value of each method will be evaluated as function of distance of source and sensors.

## 2. Determination of Test Signals

There are two condition that have to be met by source signals. First, they have to be independent each other. It means that the first signal doesnt contain any information of another signal. Second, the statistical distribution of the two sources signal are not gaussian distribution [3].

Using two preliminary requirement above, the signals below (Table 1) will be treated as the source signals of the experiment.

**Table 1.** Source signal

| No. | Signal | Note |
|---|---|---|
| 1. | Puretone 500 Hz | Narrowband |
| 2. | Puretone 600 Hz | Narrowband |
| 3. | Puretone 700 Hz | Narrowband |
| 4. | Puretone 800 Hz | Narrowband |
| 5. | Puretone 900 Hz | Narrowband |
| 6. | Puretone 1 kHz | Narrowband |
| 7. | Ship noise | Wideband stationary |
| 8. | Sonar pinging | Wideband Non-stationary |

Each of the signal were recorded in 8000 Hz sampling frequency, 16 bit per sample, and have 4 seconds duration.

## 3. Experiment Setup

The Experiment was conducting in water tank which have dimension of 2 m length, 1 m width, and 1 m height. The Interior wall of tank was covered with absorber, to decrease the amount of reflection. The tank was filled with salt water which have salinity of 3.5% and temperature of $25^{o}$C. The Velocity of sound in this condition is 1468.37 m/s.

The configuration of sources and sensor can be seen in Figure 1. There are three variations of the distance between the hydrophone-speakers are 85 cm, 1 m, and 1.5 m. The number of sensors is two, so does the sources. The distance of each sensor is 15 cm, while the distance of each speaker is 55 cm.

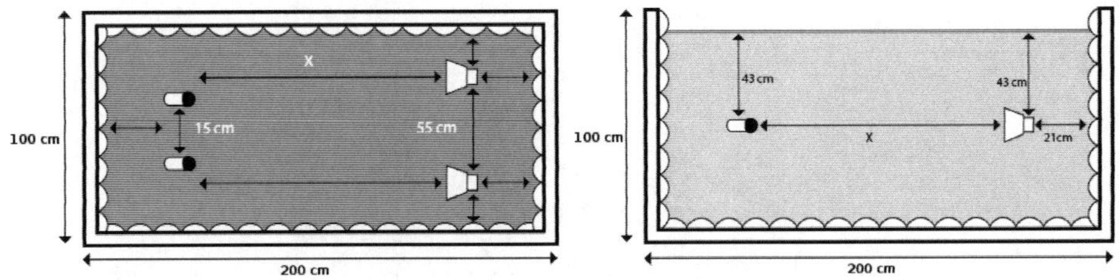


**Figure 1.** Experiment Setup. Side view (right) and top view (left)

## 4. Time Domain Independent Component Analysis (TDICA)

In the case of linear mixture, mixed signal is just sum or superposition of each source followed by some gain. This assumption can be formulated as $X(t) = A \cdot S(t)$, where $X(t)$ is vector matrix of mixed signal, $S(t)$ is vector matrix of source signal, and $A$ is mixing matrix of signal. Using above equation, then the source can be estimated by finding the inverse of $A$.

Since ICA using iteration process to estimate inverse of A, which quite time consuming if it is applied in real time world, there exist some preprocessing method in order to shorten the iteration process. First step is centering the signal. This step intended to make zero-mean signal observation using equation, $X_c(t) = X(t) - E(X)$, where $X_c(t)$ is centered vector signal, and $E(X)$ is the signal expectation.

After signal have been centered, it will be get to whitening process. This step intend to make recorded signal uncorrelated each other by using Eigenvalue Decomposition, $V = \frac{E}{\sqrt{D}}E^T$. Where $V$ is whitening matrix, $E$ is orthogonal matrix of eigenvector of $E(XX^T)$, and $D$ is diagonal matrix of its eigenvalues. After $V$ acquired, whitened signal can be obtained by using equation, $X_w(t) = VX_c(t)$

After signal being preprocessed, the estimation of source signal will be accomplished using ICA algorithm. In this experiment, algorithm of [4] was applied to estimate source signal. The principle of this algorithm was to minimize the entropy of estimated signal, since the low entropy is one of the feature of independence signal, and high entropy is the characteristic of gaussian random variable.

$$L = \sum_{t=1}^{T}\sum_{i=1}^{n} log f_i \left(w_i^T x(t)\right) \; + \; T log|detW| \tag{1}$$

Demixing matrix is optimized using this epoch, $W^+ = W + \mu\left[I + f(X)X^T\right]W$, where $g(y)$ is non linear function, in this paper $Tanh(y)$ was used. Then the estimation signal acquired by using equation, $Y(t) = WX_w(t)$.

After estmation source $Y(t)$ acquired, it will be compared to the baseline signal to evaluate the quality of source separation using equation (2).

$$MSE = \frac{1}{M}\sum_{m-1}^{M}\left(s[m] - y[m]\right)^2 \tag{2}$$

Figure 2 shows the score of estimated signal using TDICA. Lower value of MSE indicated better separation performance.

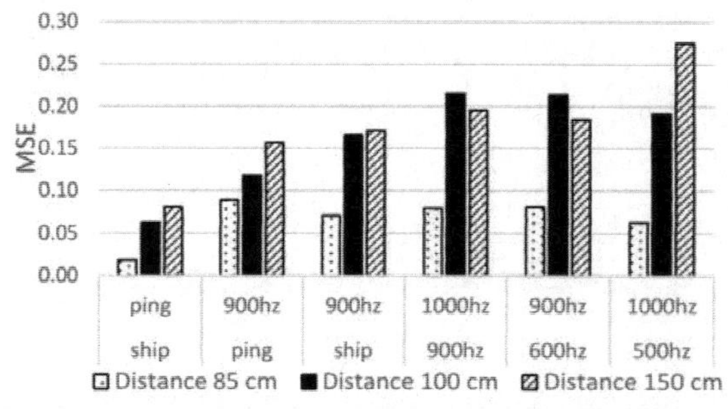


**Figure 2.** Result of mixed signal separation using TDICA indicated by its MSE.

## 5. Frequency Domain Independent Component Analysis (FDICA)

If the mixing process is not just simple superposition of each sources, in other word the signal recording performed in reverberant condition, convolutive mixture model have to be taken. This model accommodate room impulse response as the Finite Impulse Response (FIR) filter of mixed signal.

It is shown in equation, $X(t) = \sum_{n=0}^{N-1} A^{(t)}(n)S(t-n)$, where $N$ is the length of FIR filter. If the length of reverberation is 0.5 s and our recording frequency sampling is 8000 Hz, the length of FIR is 4000 taps, thus the number of demixing matrix W(n) that need to be estimated is just same in number. In order to simplify the calculation process, [5] proposed to tansforming this equation in frequency domain, $X(f) = A(f)S(f)$.

In this paper, this is achieved by performing ICA after signal being transformed in frequency domain using Fast Fourier Transform (FFT) algorithm. After demixing matrix estimated so does the signal estimation in frequency domain, it will be transformed back to time domain using Inverse Fast Fourier Transform (IFFT), then the quality of estimation signal will be acquired by comparing to baseline signal. The MSE value of each FDICA estimation signal are shown in Table 3.

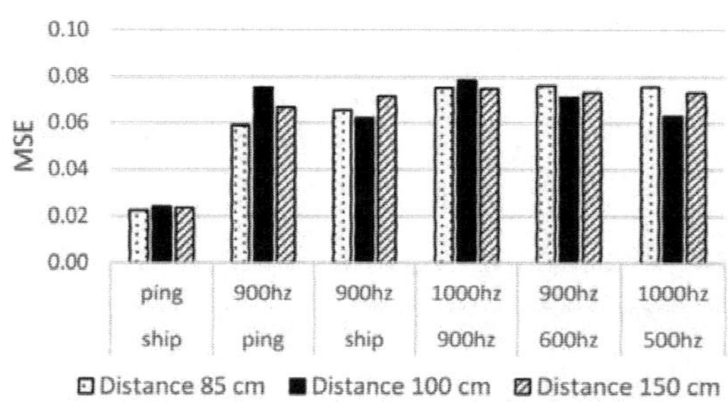


**Figure 3.** Result of mixed signal separation using FDICA indicated by its MSE.

## 6. Multistage Independent Component Analysis (MSICA)

Although FDICA can resolve the complexity of convolutive mixture, it also have some disadvantage. When the signal being transformed into frequency domain, it lost the independence assumption (since the independence assumption is statistical feature that exist in time domain), thus the separation performance will be saturated before reaching sufficient performance.

In order to resolve this problem, [2] proposed new algorithm to combine FDICA and TDICA which called MSICA. In this method, FDICA and TDICA are performed sequentially. FDICA was performed in the first stage, then estimated signal of FDICA will be feed as the input of second stage TDICA. The MSE value of each MSICA estimation signal are shown in Table 4.

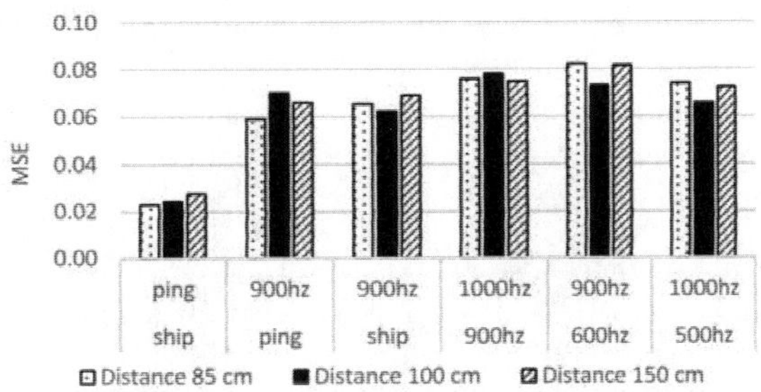


**Figure 4.** Result of mixed signal separation using MSICA indicated by its MSE.

## 7. Discussion

Comparison of the performance of each method can be seen in Figure 5. Total MSE of of each signal estimation is averaged to obtain the MSE of each method.

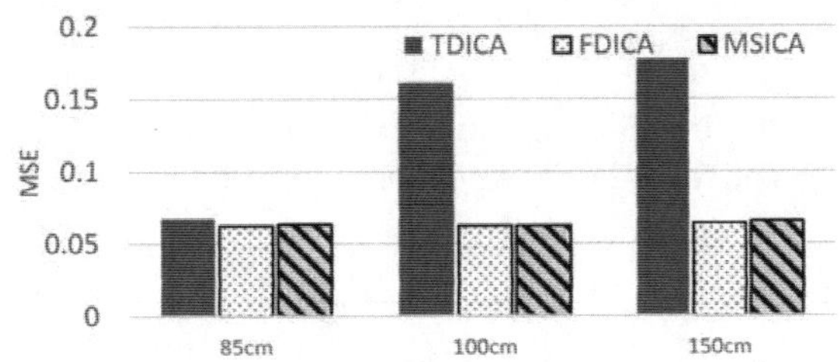


**Figure 5.** Comparison of the average value of the MSE Each separation method. MSICA and FDICA has identical score. In contrast, the performance of the method TDICA decreases as the increasing of distance between sources and sensors

In the distance of 85cm, TDICA, FDICA, and MSICA have similar performances, but eventually as the distance increase, performance of TDICA was decrease. FDICA and MSICA gave stable result of MSE value. There are two explanation of this event. First, the performance of source separation decrease as the intensity of signal get attenuated by propagation, in other words the SNR of signal play big role. Second, as the distance increase, the ratio of indirect sound to direct sound got bigger due to the scattering. Since the FDICA and MSICA shown stable performances, as shown in Figure 5, the first assumption got weak, and second assumption is more dependable.

The reason of this hypothesis was the changing of signal distribution in the addition of distance as shown in Figure 6. It is show that the distribution of both signal tends to be Gaussian for the increasing of distance. According to the central limit theorem, the distribution of independent component summation, tends to be Gaussian distribution, hence the assumption of instantaneous mixture collapsed.

In underwater communication the scattering occurred when the sound wave faces different temperature and salinity due to the increasing water pressure by depth of water. Since the tank had homogen salinity and temperature, the scattering suspected came from the surface of the water. The surface of the water act as a boundary between two different medium, which is water and air. This two different medium has different characteristic of sound propagation. This made the sound wave that came to the surface will be scattered. Another suspect of scattering is due to reflection of the wall. Since the wall have been covered by the absorber, the indirect sound wave that came from this event wont be bigger than from the surface.

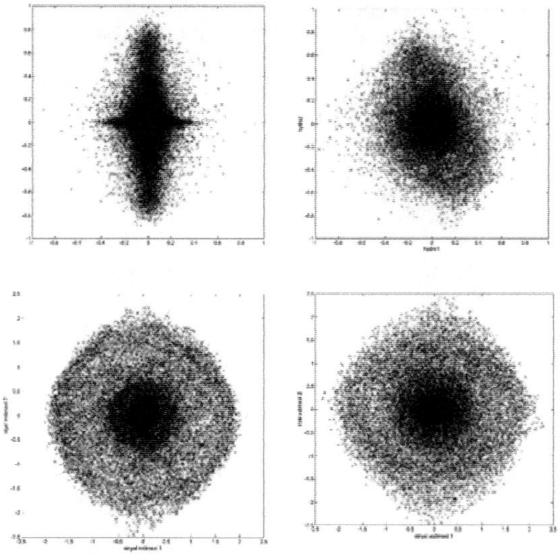

**Figure 6.** Distribution of Testing Signal Ship dan Ping; baseline (top-left); distance 85 cm (top-right); distance 100 cm (bottom-left); distance 150 cm(bottom-right).

## 8. Conclusion

(1) Pre-filtering is required in data acquistision stage to eliminate the effect of harmonic distortion generated by the speaker.

(2) The water tank impulse response can be accounted while modelling the convolutive mixture, in order to get better simmulation.

(3) Additional measurement position in variation of depth need to be add in order to have understanding major source of scattering.

## References

[1] Cochard, N., Lacoume, J. L. 2000 Underwater Acoustic Noise Measurement in Test Tank.IEEE Journal of Oceanic Engineering, 25(1).

[2] Nishikawa, T., Saruwatari, H., dan Shikano, K. 2003. "Blind Source Separation of Acoustic Signals Based on Multistage ICA Combining Frequency-Domain ICA and Time-Domain ICA". IEICE Trans. Fundamentals, E86-A (4)

[3] Hyvarinen, A., Karhunen, J., dan Oja, E. 2001, "Independent Component Analysis: Basic Independent Component Analysis".NewYork: John Willey & Sons.

[4] Bell, Sejnowski, 1995. An Information-Maximization Approach to Blind Separation and Blind Deconvolution, Neural computation, vol. 7, no. 6, pp. 11291159.

[5] Kawamoto, M., Matsuoka, K., dan Ohnishi, N. 1998. "A method of blind separation for convolved non-stationary signals". Elsevier Neurocomputing 22 :157-171

# High-speed Vision System for Dynamic Structural Distributed Displacement Analysis

**Zulhaj Aliansyah[1], Mingjun Jiang[1], Takeshi Takaki[1], and Idaku Ishii[1]**
Robotics Laboratory, Graduate School of Engineering, Hiroshima University
E-mail: `aliansya@robotics.hiroshima-u.ac.jp`

**Abstract.** Structural health monitoring (SHM) has become prominent topic in the field of civil engineering in detecting structural damage, estimating remaining service life of structure, optimizing decision making process on maintenance, and aiding performance-based structure design methodology. Dynamics measurement is one of important techniques to diagnose deterioration of structures by inspecting their dynamics properties, mainly derived from displacement time-series from multiple measurement points. Various conventional distributed displacement measurement techniques of civil structures currently in place have been limited due to various constraints: inadequate accuracy, overly-localized measurement point, measurement drift, robustness issue on field usage, structural obstruction, and reliability for long-term installation. Optical technique utilizing LED markers has been proposed as an alternative for dynamic distributed displacement measurement, providing visual cue and and easily-identifiable marker position. The technique incorporates high-speed camera system and telephoto lens to capture both lateral and vertical displacement of bridge members, LED markers to cope with the insufficient incident light over long distance and small camera aperture, and computer vision system to calculate multiple markers' displacement simultaneously. Further analysis was conducted to evaluate structure natural vibration frequencies using FFT (Fast Fourier Transporm) peak-picking method.

## 1. Optical displacement measurement

Structures are subject to repetitive and temporary external forces from wind, current, quakes, and load in addition to its inherent weight. These forces may cause either small or large displacement of structure's members. In the effort in detecting structural damage, estimate remaining service life of the structure, scheduling maintenance, and improving design methodology, structural health monitoring is widely employed on civil structures [1]. Dynamics measurement is one of technique to diagnose deterioration of structures, mainly derived from distributed displacement measurement. Furthermore, evaluation on natural vibration frequencies of the structure may tell vibration characteristics.

Various distributed displacement measurement techniques currently in place have been limited due to several constraints: inadequate accuracy, precision, and both spatial and temporal resolution, measurement drift, ease of practical setup, and reliability for long-term installation. Vision measurement provides alternative for contact-less distributed displacement measurement over some distance. The increasing availability of high-performance vision-based systems provides the potential for developing adaptable/reconfigurable monitoring systems of complex deformation field measured at selectable locations and orientations [2].

IOP Conf. Series: Journal of Physics: Conf. Series **1075** (2018) 012014 doi:10.1088/1742-6596/1075/1/012014

**Figure 1.** Truss structure with LEDs attached on a single lower chord of the truss.

Several implemented vision-based systems for displacement measurement were still limited. Target-less vision-based displacement sensor system was proposed by extracting and tracking feature points in video stream [3]. Multiple displacement point approach can be achieved using image key-points [4], but these methods largely depend on low noise image and crisp edges. Given that only feature point was extracted from a small area and edge detection is susceptible to focal blur, the system isn't suitable for distributed measurement points with different frontal distances, long distance measurement, and low ambient lighting condition. Optical motion tracking system using surveillance camera was proposed to measure induced motion in the event of an earthquake [5], but the system is limited due to low frame rate, and low pixel resolution of the surveillance camera, and in the event of an actual earthquake, inherent camera self-vibration resulting in no steady measurement reference point.

A more complex system incorporating two cameras equipped with telephoto lens and non-coplanar control points markers with Affine camera modelling has been proposed to measure tri-axis displacement and rotation of fixed monitoring zone of a structure [6]. Considering the extended period of operation, the geometry of components may change due to temperature effects and camera frame synchronization had yet been taken into account.

Optical system with telephoto lens, focal length extender, and active lighting device close to the camera may develop noticeable inaccuracy due to camera head vibration and heat wave from the lighting device, thus operating time of the lighting device needs to be limited [7]. This renders such system only suitable for structures with routine periodical loading condition such as railway bridge.

The contribution of this work is to capture lateral and vertical displacements of high-contrast markers installed on a structure with different distances and estimating natural vibration frequencies of the structure.

## 2. Truss vibration experiment with arbitrary forcing action

### *2.1. Experiment structure*

Bridge model used for the experiment is Warren truss structure with L-beam aluminum (25 mm flanges, 1.2 mm thickness) for end posts, top and bottom chords, and flat aluminum strips (15 mm width, 2 mm thickness) for the struts and diagonal members. The bridge model incorporates no deck and bracing. The span, height, and width of the truss are 4 m, 21 cm, and 33 cm respectively. LED markers are 5 mm with white color, installed with 25 cm distance between one another. Figure 1 shows 16 LED makers installed on a single side of lower chord of the truss. Excitation force was applied to the center of the span on both bottom chords with arbitrary direction.

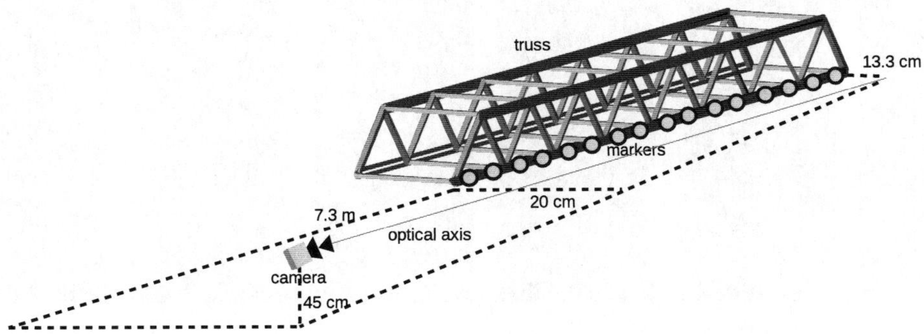


**Figure 2.** Concept overview of vision-based distributed displacement measurement experiment.

### 2.2. Vision system settings

Focal length, aperture, and focus distance need to be adjusted proportionally to cover the whole span of the truss within the depth of field of the camera. The equation to the depth of field is given as: $D_p = D_f - \frac{\epsilon N D_f^2}{f^2 + \epsilon N D_f}$, $D_d = D_f + \frac{\epsilon N D_f^2}{f^2 - \epsilon N D_f}$ (1)

where $D_p$, $D_d$, and $D_f$ are depth of field near distance, far distance, and focus distance respectively. $N$ is aperture number, $f$ is focal length, and $\epsilon$ is blur diameter. Optical settings were 9.3 m focus distance with F/32 aperture at 410 mm focal length which resulted in depth of field of 4.178 m with 22 px acceptable blur diameter, covering the whole span of the truss.

Camera head was set to capture vibration of the truss at 180 fps with 10 seconds duration. Summary of vision system specification is shown in Table 1. Figure 2 shows general overview of the experiment setup.

**Table 1.** Vision system specification.

| Camera | | Lens | | Computer | |
|---|---|---|---|---|---|
| Ximea MQ042MG-C | | SIGMA APO HSM 300-800 | | Dell Latitude 3540 | |
| Resolution | 2048x1024 px | Focal length | 410 mm | Memory | 8 GB |
| Sensor | 5.5 $\mu$m pitch | Focus distance | 9.3 m | OS | Ubuntu 14.04.1 |
| Exposure, frame rate | 5 ms, 180 fps | Aperture | F/32 | CV library | OpenCV 2.4.8 |

### 2.3. Scaling factor

Scaling/calibration factor gives estimate between vision-system resolution in pixels and actual physical dimension. For setups with non-perpendicular lens optical axis between camera and markers, scaling factors for lateral and vertical direction need to be estimated separately [8]. Calibration was done taking into account sensor size and distance between camera and marker.

To eliminate occlusion between markers, the camera system was placed slightly away from the truss axis both in lateral and vertical directions, resulting in a slanted angle between markers and the optical axis of the camera, thus scaling factor will be compensated following the equation: $SF_v = \frac{11.27 D_m}{f\cos(\pi - atan(D_v/D_m))}$, $SF_l = \frac{11.27 D_m}{f\cos(\pi - atan(D_l/D_m))}$ (2)

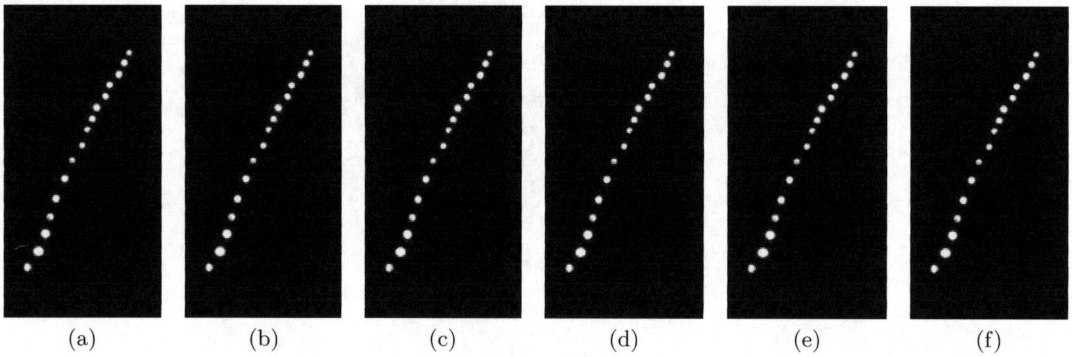

**Figure 3.** Captured frames of sixteen markers showing minute displacements between $t$=3 s (a) to $t$= 5.5 s (f) with 0.5 s interval between each other.

where $D_f$, $D_v$, $D_l$, and $D_m$ are frontal distance, vertical distance, lateral distance, and marker distance (Phytagorean resultant of $D_f$, $D_v$, and $D_l$). $SF_v$ and $SF_l$ are scaling factor in vertical and lateral direction respectively.

Camera head was put 45 cm higher than the chord on which all sixteen LED markers were installed with slanted lateral angle between camera optical axis and bridge axis, resulting in lateral distance of 20 cm between Marker1 and optical axis and 13.3 cm between Marker16 and optical axis. These frontal, lateral, and vertical distances between camera axis and LED markers were taken into account to calculate the appropriate scaling factor of each individual marker.

## 3. Captured marker movement and extracted displacement

Pixel-wise movement of each marker was translated to actual displacement using its respective scaling factors after subtracted from its steady-state value. Each marker's centroid was extracted using OpenCV's `cvblob()` blob detection function then multiplied to its respective scaling factor to obtain distributed displacements of the structure after firstly substracted from its steady state value.

Figure 3 shows six frames capturing minute pixelwise movement of markers from $t$=3 s to $t$= 5.5 s with 0.5 s interval. Figure 4 shows the truss' chord displacement from the same time span. Figure 5 shows Marker9 displacement, positioned at the center of the truss' span, over the course of 10 s capture duration and highlight between $t$=3 s to t=$t$=4 s.

## 4. Vibration natural frequency analysis

FFT analysis was conducted to evaluate natural frequencies of the structure vibration. Peak-picking method was applied to frequency response of Marker9 vibration to extract up to fifth mode of vibration. Only 1024 samples ($t$=2.5 s to $t$=8.183 s) was selected in order to avoid adding initial transient response to the natural frequency analysis. Figure 6 shows the structure's frequency response both lateral and vertical directions. Table 2 summaries the natural vibration frequencies in both lateral and vertical directions.

Unique natural vibration frequency was found on 8.6133 Hz for lateral direction and 19.6875 Hz for vertical direction with the rest of the natural vibration frequencies are most likely due to material characteristics. The higher unique natural vibration frequency on vertical direction compared to one of lateral direction shows that the truss is more rigid vertically.

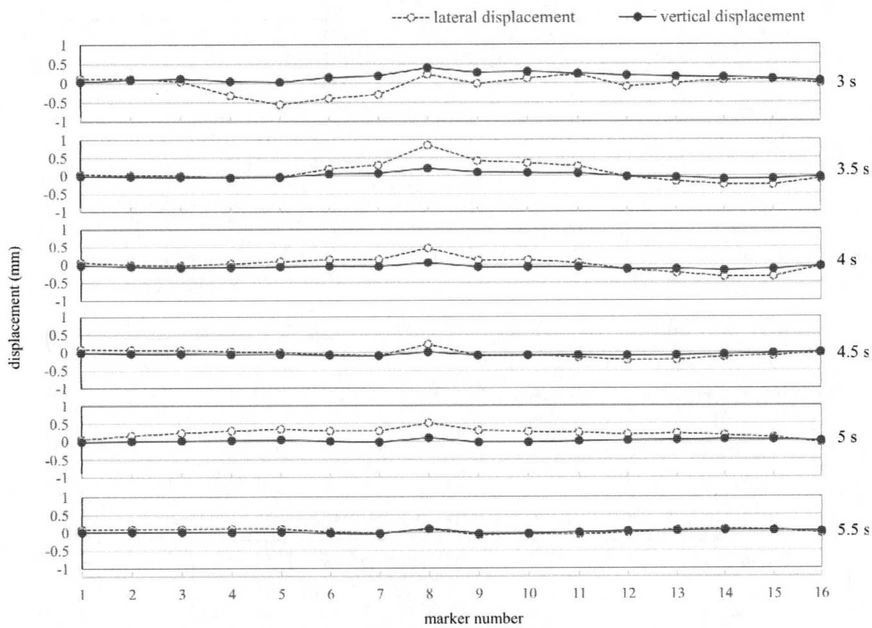


**Figure 4.** Chord displacement between $t$=3 s to $t$= 5.5 s.

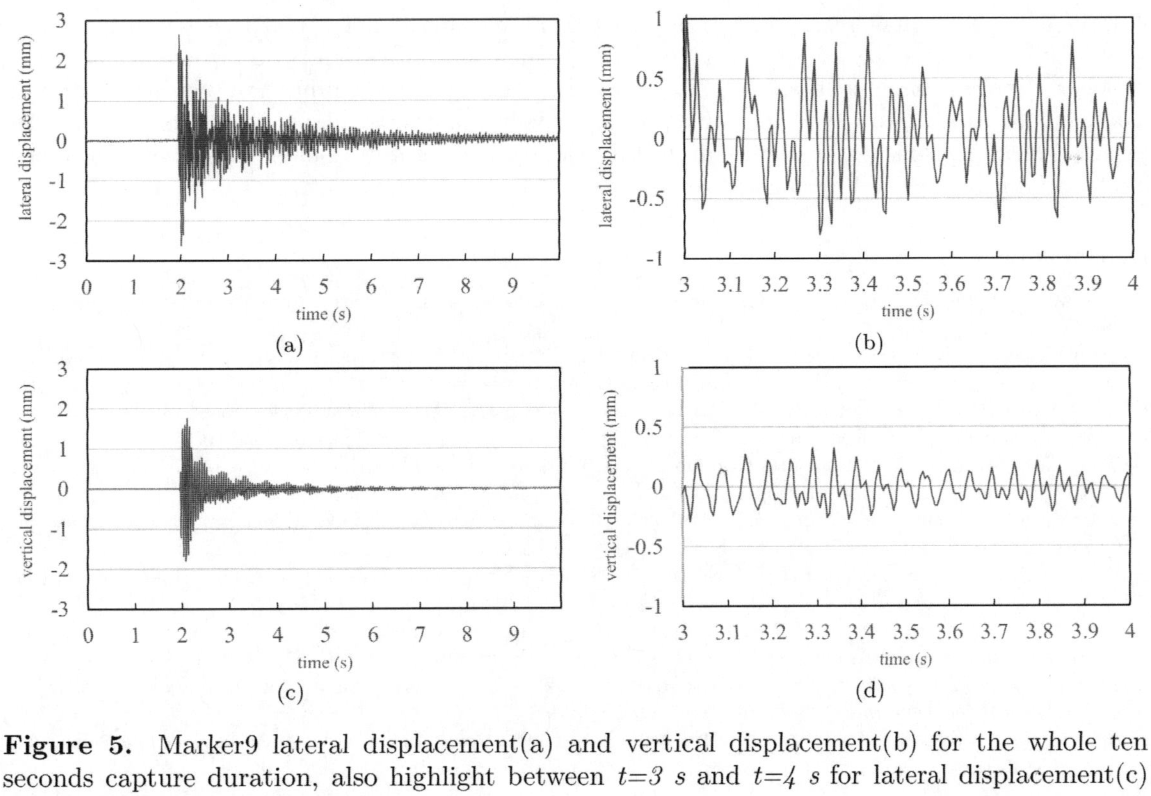


**Figure 5.** Marker9 lateral displacement(a) and vertical displacement(b) for the whole ten seconds capture duration, also highlight between $t=3$ $s$ and $t=4$ $s$ for lateral displacement(c) and vertical displacement(d).

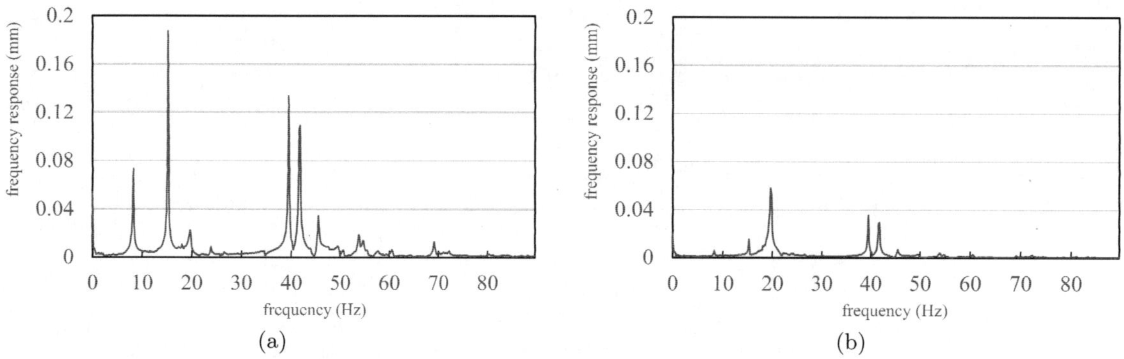


**Figure 6.** Normalized frequency response of the truss in lateral direction (a) and vertical direction (b).

**Table 2.** Natural vibration frequencies of the truss.

| Direction | Natural frequencies (Hz) | | | | |
|---|---|---|---|---|---|
| | Mode 1 | Mode 2 | Mode 3 | Mode 4 | Mode 5 |
| Lateral | 8.6133 | 15.2930 | 39.5508 | 41.8359 | 45.5273 |
| Vertical | 15.2930 | 19.6875 | 39.5508 | 41.8359 | 45.5273 |

## 5. Conclusion

High-speed (180 fps) vision system, concurrently measuring the lateral and vertical displacement of multiple markers, has been proposed for structural dynamics analysis of truss structure. FFT peak-picking analysis was conducted to evaluate struture's natural vibration frequencies.

Several improvements could be made to the experiment: calibration against external displacement sensors to verify measured displacement accuracy and precision, longer capture duration reaching steady state again after excitation to find for any residual plastic displacement and to obtain higher FFT resolution.

## 6. References

[1] Liu, M, Frangopol DM, Kim S 2009 Bridge system performance assessment from structural health monitoring *J. Struct. Eng.* **135** 733-42

[2] Wahbeh AM, Caffrey JP, Masri SF 2013 A vision-based approach for the direct measurement of displacements in vibrating systems *Smart Mater. and Struct.* **12** 785-94

[3] Insub C, JunHee K, Donghyun K 2016 A target-less vision-based displacement sensor based on image convex hull optimization for measuring the dynamic response of building structures *Sensors* **16** 2085

[4] Tung K and Catbas FN Catbas 2017 Completely contactless structural health monitoring of real-life structures using cameras and computer vision *Struct. Control Health Monit.* **24** e1852

[5] Katsuhisa K, Miyamoto Y, Akihiro K, Makoto O 2005 Monitoring of earthquake induced motions and damage with optical motion tracking *Smart Mater. Struct.* **14** 32-38

[6] Santos CA, Costa CO, Batista J 2016 A vision-based system for measuring the displacements of large structures *Mech. Sys. and Signal Process.* **72-73** 678-94

[7] Ribeiro D, Calada R, Ferreira J, Martins T 2014 Non-contact measurement of the dynamic displacement of railway bridges using an advanced video-based system *Engineering Structures* **75** 164-80

[8] Feng D and Feng MQ 2017 Experimental validation of cost-effective vision-based structural health monitoring *Mechanical Systems and Signal Processing* **88** 199-211

[9] Salawu S and Williams C 1995 Review of full-scale dynamic testing of bridge structures *Engineering Structures* **17** 113-21

# Objective and subjective acoustics measurement of audience seating areas in a medium size auditorium

**L F Sari[1], S S Utami[1] and J Sarwono[2]**

[1] Department of Nuclear Engineering and Engineering Physics, Universitas Gadjah Mada, Yogyakarta, Jl. Grafika No. 2, Yogyakarta, INA
[2] Department of Engineering Physics, Institut Teknologi Bandung, INA

E-mail: lisa.fita.s@mail.ugm.ac.id

**Abstract**. Several consideration of the acoustical quality of a multi-purpose hall should be taken into account and therefore, the audiences must be fully aware of the acoustical consequences in deciding their seating positions. In this paper, objective measurement and subjective assessment in the medium capacity, Driyarkara multi-purpose hall, has been investigated. Room impulse responses (RIR) were taken for several monaural parameter distribution including G, EDT, and C80 at 16 positions referring to ISO 3382-1:2009. The signals were convolved with anechoic sound of musical instruments to reproduce acoustic stimuli for subjects to listen. Stimuli data is reduced to 8 measurement' positions, which indicated the noticeable difference. These were generated to reproduce binaural sounds for further subjective assessment. The distribution of binaural parameters were observed in 250 Hz, 1 kHz, and 4 kHz to obtain JND (Just Noticeable Difference). The result reveal that this hall contains 3 different acoustic zones with the range of acoustics parameters of G: -8.91 dB – 1.07 dB, EDT: 1.12 s – 1.52 s, C50: -0.11 dB – 5.05 dB, C80: -3.70 dB – 2.03 dB, D50: 31.00 – 60.71, Ts: 62.75 ms – 106.50 ms, IACC: 0.43 s – 0.82 s, Treble Ratio: 0.67 – 0.96 and Bass Ratio: 0.81 – 2.80. Furthermore, this result correlates with the JND value for G: 2.26 dB, Ts: 17.5 ms, Treble ratio: 0.2, Bass Ratio: 0.03 and D50: 16.67.

## 1. Introduction

Auditorium have many functions, some of the most familiar functions are for meeting room and for having performances. Auditorium as a performing space should have a capacity of volume per seat minimum 6.2 $m^3$ and maximum 10.8 $m^3$ [1]. Music performance space or concert hall emphasizes on distribution of sound in the entire hall. Commonly closed performance space produce low background noise. Suyatno et al [2] studied the acoustic parameters for gamelan performances in Pendopo Mangkunegaran Surakarta. The result of background noise shows in the main hall, pendopo terrace and outside area pendopo are 54 dB, 62 dB and 71 dB, respectively. The pavilion that has a roof with pyramid shape, greatly affects the sound distribution inside the main hall. The largest reverberation time inside the pendopo at the center of the main hall is 2.2 s. This is influenced by the distribution of reflected sounds from the geometry of the roof and its material.

The same phenomenon occurs in an atrium that has a large capacity with solid material used for the roof. These architectural elements created acoustic discomfort. The results of computer simulations show that micro-perforated panels can improve the noise level and reverberation time [3] where the reverberation time decreased around 1 to 2 s. A research done by Marc Aretz and Raf Orlowski[4] reported that changes in the area of absorption not only change the value of the reverberation time but

IOP Conf. Series: Journal of Physics: Conf. Series **1075** (2018) 012015 doi:10.1088/1742-6596/1075/1/012015

also the value of G. Concert hall will produce a good sound distribution when reverberation time is between 1.6 s to 2 s [5]. Reverberation time also affects the conductor and the orchestra player in their ability to hear the clarity of the tone of each phrase from musical composition [5].

Measurement technics and just noticeable difference (JND) value is provided in ISO 3382-1:2009 [6]. A study reported a JND value for reverberation time that is larger than the value in ISO 3382-1, which is 24.5 with 6.9% in standard deviation [7]. Distribution of acoustic parameters can be presented with area mapping using data at points that are suitable enough to see the differences. Taeko Akama et al [8] suggested that it is not about the larger amount of measurement points that will give one the mapping discrepancies but by enough measurement points. This is to insure a best visualization in presenting the acoustics parameters distribution. To get enough measurement points, Taeko Akama also suggested to do some measurements with different amount of points. Other study reported by Mike Barron [9], found that the value of G (strength) is not constant. As the distance between sound source and receiver increases, the G value will decrease. As the distance gets farther away from the sound source, the G value will decrease more.

To evaluate the quality of a performance space, subjective evaluation is also needed besides of acoustics parameter measurements Acoustic stimuli helps respondents to identify their favorite seats rather than using visual stimuli only [10]. The combination of these two stimuli shall produce a more suitable assessment. Angelo Farina reported the index for subjective evaluation to assess concert hall and opera houses [11]. The indices are obtained by the regression value more than 0.3 in every experiment of the comparison between objective parameters and subjective index. Case studied in this research is the Driyarkara Auditorium in Yogyakarta, Indonesia. Driyarkara Auditorium was designed specifically for speech and music purposes without electrical sound system. With capacity more than 1000 audiences, this auditorium is required to fulfill the acoustic needs for the entire audience seats.

## 2. Research Methodology

Auditorium Driyarkara, Yogyakarta as a multipurpose-hall is functionalized as a graduation ceremony building and also rent as a concert hall. It is a medium size auditorium with a capacity of 974 seats at the first floor and 126 seats at the second floor. Design of the wall is covered by diffusor panels and absorber material. Structure of the wall is arranged with a zig-zag pattern. In front of the stage, a huge diffuser panel is installed to scatter the sound. This condition affects the audience seating area by providing sound reflections that are coming from many surfaces.

Equipment of this research is balloon with 30 cm diameters as the sound source, omnidirectional microphone BSWA MPA416 as the receiver, soundcard, laptop and headphone. Preliminary research was done to obtain positions of measurement that represent the acoustics profile of the auditorium. Objective measurement was done by recording at 16 seats (A1, A8, B15, B22, G1, G9 G18, M26, S1, S8, S18, and S26; see Figure 1 and 2). The recorded Room impulse response (RIR) was then convolved with a dry sound from guitar instrument following the steps of auralization. The auralization signals are then used as the acoustic stimuli for the subjective evaluation. In the subjective evaluation, a comparison of 2 sounds at three directions (vertical, horizontal and diagonal) was performed. Every comparison consist of three questions; (1) Which sound is louder?; (2) Which sound is clearer?; (3) Which sound is more reverb?. The answer options are (a) sound 1; (B) sound 2; (C) No differences perceived. The result of the objective measurement is the percentage difference of three parameters (G, RT, and C80). The result of the subjective evaluation is compared with the parameters obtained from the objective measurements for positions with significant differences.

**Figure 1.** Top View Audience Seat

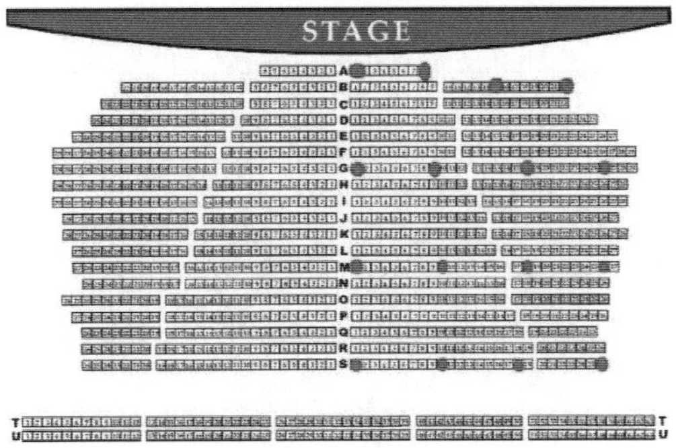


**Figure 2.** Seat Plan

### *2.1. Objective Measurement*
Binaural recording was done at 8 seats (C5, C16, H7, H18, M8, M22, R9, and R22) based on the previous preliminary result. Audacity was used as the software for recording the RIR with each 7 seconds. The height of the source (balloon) and microphone receiver are 1.5 m and 1.2 m respectively. Microphone mounted on a person's head was used to obtain a binaural recording. Data of the recording file was save in .wav format. Before generating the acoustics parameters, the recording file had to be trimmed to delete the early portions of the in order to minimalize error in the energy decay curve. In addition, "trimming" was done to reduce the discrepancies between the reverberation time values obtained from measurement using dodecahedron loudspeaker and balloon as the sound sources. Acoustic parameters analyzed are the G (strength), EDT (Early Decay Time), C80, D50, C50, Ts, IACC, Treble Ratio and Bass Ratio. Room Impulse Response (RIR) in 8 eats convolved with dry sound from guitar instrument by GratisVolver from CATTacoustics. The result of convolution become acoustics stimuli on subjective evaluation.

### *2.2. Subjective Evaluation*
Subjective evaluation was done using questionnaire with semantic scale rating from -3 to +3 with zero as neutral value. Respondents were listening to 8 acoustics stimuli through headphones and were giving the assessment based on the subjective parameters, which represents acoustic parameters in the objective measurement. Subjective parameters are pleasant-unpleasant, round-sharp, soft-hard, diffuse-localizable, detached-enveloping, dry-reverberant, treble booster-treble reduced, bass boosted-bass reduced, and quiet-loud. Respondents were selected based on several criteria which are students that have taken acoustics courses, or musicians or those that have experienced attending an indoor live concert.

## 3. Results and discussion
Result of the objective measurement is shown in Table 1. Many positions are having values there are under the recommended JND in ISO 3382-1:2009. The result should be validated with the respondent's perception in order to observe their listening quality of the sound happening in the auditorium.

### *3.1. Subjective Evaluation*
Subjective evaluation was done with 81 people (17 until 27 years old), contain of 50 males and 31 females also 32 musicians and 48 non-musicians. Respondents should closed at least one of the requirements; 1. Have been attending acoustics class in university, 2. Musicians, or 3. Have been watching indoor live concert. Result of the subjective evaluation is shown in Figure 3, with the axes are (1) pleasant-unpleasant, (2) round-sharp, (3) soft-hard, (4) diffuse-localizable, (5) detached-enveloping, (6) dry-reverberant, (7) treble booster-treble reduced, (8) bass boosted-bass reduced, and (9) quiet-loud. The result of rating scale shows that respondents have a tendency to give the assessment at -2 till +2. It is shown that we can actually reduce the rating scale to 5 scale only from -2 to +2.

**Table 1.** Objective measurement result

| | Frek (Hz) | C5 | C16 | H7 | H18 | M8 | M22 | R9 | R22 |
|---|---|---|---|---|---|---|---|---|---|
| G (dB) | 250 | -3.17 | -0.45 | -5.45 | 0.73 | -5.11 | 1.13 | -5.29 | -7.82 |
| | 1000 | -5.39 | -5.69 | -6.23 | -4.25 | -5.80 | -3.76 | -6.20 | -8.61 |
| | 4000 | 1.83 | 0.66 | 4.07 | 6.53 | 2.37 | 3.48 | 2.15 | -2.07 |
| EDT (s) | 250 | 1.81 | 1.32 | 1.89 | 0.88 | 1.81 | 1.16 | 1.73 | 1.77 |
| | 1000 | 1.50 | 1.30 | 1.33 | 1.36 | 1.35 | 1.35 | 1.40 | 1.31 |
| | 4000 | 1.00 | 1.17 | 0.99 | 1.08 | 1.14 | 1.10 | 0.83 | 1.11 |
| C80 | 250 | -4.40 | 2.76 | -4.24 | 3.89 | -2.67 | 5.33 | -3.51 | -1.40 |
| | 1000 | 2.06 | 2.12 | 3.57 | 1.75 | 4.63 | -0.49 | 2.57 | 4.54 |
| | 4000 | 5.10 | 5.75 | 8.20 | 5.70 | 5.77 | 3.13 | 6.76 | 5.89 |
| Ts | 250 | 158.00 | 91.00 | 165.33 | 85.67 | 153.33 | 75.67 | 152.00 | 136.33 |
| | 1000 | 78.67 | 91.00 | 72.00 | 88.67 | 61.33 | 105.00 | 92.67 | 68.33 |
| | 4000 | 57.67 | 48.67 | 40.67 | 45.00 | 47.33 | 69.33 | 57.67 | 55.67 |
| D50 | 250 | 17.69 | 45.59 | 18.27 | 37.57 | 15.19 | 65.67 | 15.32 | 27.50 |
| | 1000 | 53.23 | 46.54 | 52.43 | 49.79 | 56.42 | 34.63 | 39.94 | 59.47 |
| | 4000 | 61.13 | 71.10 | 64.32 | 69.77 | 63.80 | 52.26 | 65.69 | 66.16 |
| IACC | 250 | 0.95 | 0.85 | 0.92 | 0.86 | 0.95 | 0.91 | 0.94 | 0.94 |
| | 1000 | 0.84 | 0.42 | 0.74 | 0.55 | 0.69 | 0.39 | 0.42 | 0.40 |
| | 4000 | 0.56 | 0.59 | 0.53 | 0.49 | 0.69 | 0.29 | 0.50 | 0.41 |
| TR | - | 0.74 | 0.96 | 0.84 | 0.86 | 0.91 | 0.82 | 0.67 | 0.88 |
| BR | - | 1.03 | 1.19 | 1.53 | 0.85 | 1.36 | 0.81 | 1.16 | 1.39 |

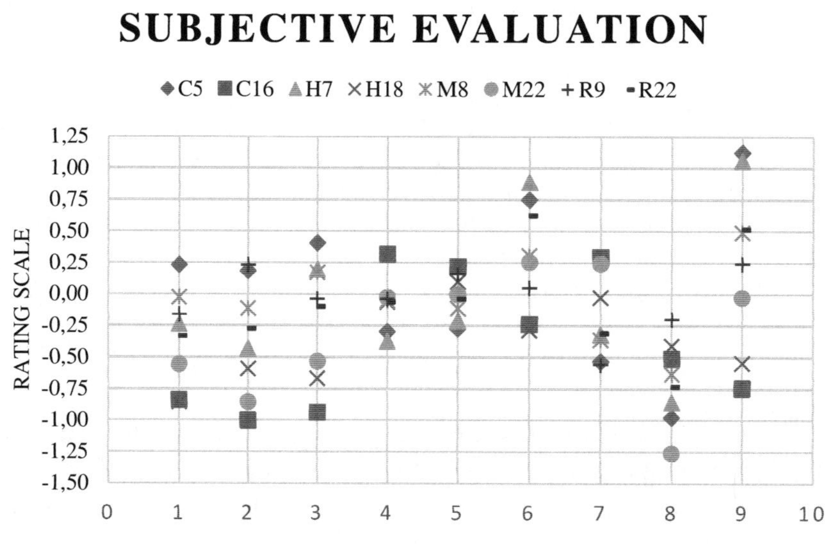


**Figure 3.** Subjective assessment result.

### *3.2. Just Noticeable Difference*

Objective measurement result in Table 1 converts to subjective scale and compares with the subjective evaluation. This comparison obtained the percentage value of respondents who gave the same perception with the objective measurement rating. The Yellow color represent the maximum percentage of the rating from respondent in subjective evaluation. The blue color is the result of converting the objective measurement result. Meanwhile, the green color shows the same result between objective measurement and subjective evaluation. From the 9 comparison, we can obtained the value of JND for G, D50, Ts, Bass Ratio and Treble Ratio as 2.26 dB, 16.68 %, 17.5 ms, 0.03 and 0.2 respectively.

**Table 2.** Comparison objective vs subjective

| Points/ Scale | -3 | -2 | -1 | 0 | 1 | 2 | 3 |
|---|---|---|---|---|---|---|---|
| H7 | 4% | 26% | 17% | 12% | 26% | 14% | 1% |
| C5 | 0% | 12% | 17% | 23% | 31% | 14% | 2% |
| R22 | 1% | 21% | 27% | 23% | 15% | 12% | 0% |
| M22 | 7% | 27% | 22% | 16% | 15% | 9% | 4% |
| H18 | 7% | 30% | 27% | 17% | 14% | 5% | 0% |
| M8 | 1% | 14% | 27% | 16% | 28% | 14% | 0% |
| C16 | 16% | 33% | 14% | 10% | 14% | 10% | 4% |
| R9 | 5% | 15% | 27% | 14% | 25% | 14% | 1% |

### *3.3. Acoustic Zoning*

Regression analysis was used to obtain parameters which has the highest correlation. The distribution of the parameters is to predict the location of the most convenient seat, acoustically speaking. The regression analysis shows that G, EDT, Ts, C80, D50, C50, IACC, Treble Ratio, Bass Ratio and IACC are 0.42, 0.32, 0.24, 0.06, 0.08, 0.26, 0.23, 0.06 and 0.18 respectively. From this result only G and EDT which have regression value above 0.30. The distribution of G and EDT at low (250 Hz), mid (1000 Hz), and high (4000 Hz) were analyzed and obtained the acoustic zoning (see Figure 4). The result indicated that middle seats have the highest listening quality. Hypothetically, these middle seats are receiving the most sound reflections from all surfaces that enable to support the listening quality. Meanwhile, the fair listening quality are seats at the front rows. Seats that are far away from stage have the poor listening quality.

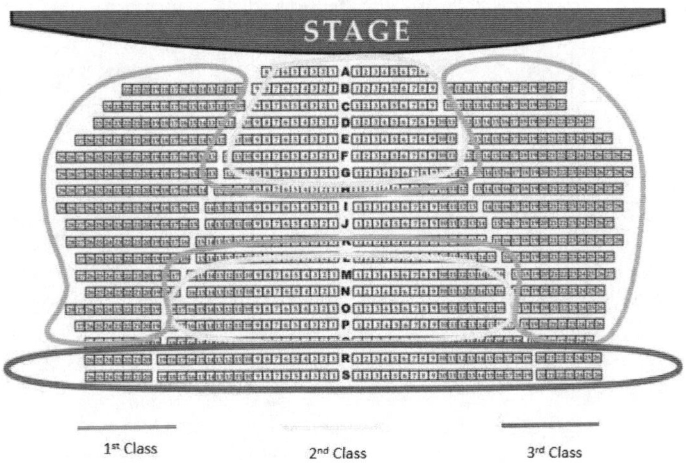


**Figure 4.** Acoustics zoning.

## 4. Conclusion

Using the objective measurement, we obtained range of values of acoustics parameter of G, EDT, C80, C50, D50, Ts, IACC, Treble Ratio, and Bass Ratio are (-8.91 to 1.07 dB), (1.12 to 1.52 s), (-0.11 to 5.05 dB), (-3.70 to 2.30 dB), (31.00 to 60.71), (62.75 to 106.50 ms), (0.42 to 0.82 s), (0.67 to 0.96), and (0.81 to 2.80). Comparison between the results from objective measurement and subjective evaluation give JND values for G, D50, Ts, Bass Ratio and Treble Ratio of 2.26 dB, 16.68 %, 17.5 ms, 0.03 and 0.2

respectively. In addition, based on the regression result of G (0.42) and EDT (0.32), we obtained three acoustics 3 zones (see Figure 4) of different listening quality where middle seats are the best, front seats are the intermediate class and the worst listening quality are seats that are further away from the stage.

## 5. References

[1] Ingegneria D, Ferrara U and Saragat V 2003 On the Effect of Domed Ceiling in Worship Spaces: *A Scale Model Study of a Mosque* vol 10 no 2 pp 117–133

[2] Tjokronegoro H A, Merthayasa I G N and Supanggah R 2015 Acoustic parameter for Javanese Gamelan Performance in Pendopo Mangkunegaran Surakarta *Procedia - Soc. Behav. Sci.* vol 184 pp 322–327

[3] Urbán D, Zrneková J, Za P, Maywald C and Rychtáriková M 2016 Acoustic comfort in atria covered by novel structural skins vol 155 pp 361–368

[4] Aretz M and Orlowski R 2209 Sound strength and reverberation time in small concert halls *Appl. Acoust.* vol 70 no 8 pp 1099–1110

[5] Beranek L L 2010 Concert halls and opera houses: music, acoustics, and architecture (New York: Springer)

[6] ISO 3382-1 2009 Acoustics - Measurement of Room Acoustic parameters (International Standard)

[7] Blevins M G, Buck A T, Peng Z and Wang L M 2013 Quantifying the just noticeable difference of reverberation time with band-limited noise centered around 1000 Hz using a transformed up-down adaptive method

[8] Akama T, Suzuki H and Omoto A 2010 Distribution of selected monaural acoustical parameters in concert halls *Appl. Acoust.* vol 71 no 6 pp 564–577

[9] Barron M 2012 Taking account of loudness constancy for the loudness criterion for concert halls *Appl. Acoust.* vol 73 no 11 pp 1185–1189

[10] Kawase S 2013 Factors influencing audience seat selection in a concert hall: *A comparison between music majors and nonmusic majors J. Environ. Psychol.* vol 36 pp 305–315

[11] Farina A 2001 Acoustic quality of theatres: *correlations between experimental measures and subjective evaluations Appl. Acoust.* vol 62 no 8 pp 889–916

# Application of Acoustic Doppler Current Profiler (ADCP) to Observe Diel Vertical Migration of Zooplankton

**Angga Dwinovantyo[1], Henry M. Manik[2]*, Tri Prartono[2], Susilohadi[3]**

[1]Graduate School of Marine Technology, PMDSU Batch II, Bogor Agricultural University, Bogor 16680, Indonesia

[2]Department of Marine Science and Technology, Faculty of Fisheries and Marine Sciences, Bogor Agricultural University, Bogor 16680, Indonesia

[3]Marine Geological Institute, Ministry of Energy and Mineral Resources Republic of Indonesia, Bandung 40174, Indonesia

Corresponding Author: henrymanik@ipb.ac.id

**Abstract**. Study of zooplankton behavior was observed because of their importance as a source energy for small fish and other organisms in the food chain. The aim of this study was to investigate diel vertical migration (DVM) of zooplankton using high frequency acoustic scattering techniques from an acoustic Doppler current profiler (ADCP) instrument. The DVM was observed by utilizing mean volume backscattering strength (MVBS) from the recorded ADCP's echo intensities. MVBS data were obtained using a bottom-moored, upward looking 750 kHz ADCP at 15 m depth in Lembeh Strait, North Sulawesi, Indonesia. In this work, we simultaneously presented ADCP quantifications, zooplankton net samples, and oceanographic data (tidal and water current). The movement of zooplankton was shallower during the night and deeper during the day with changes of MVBS and ranged between -65 and -80 dB. From laboratory analysis, it was found that crustacean had the highest abundance and biomass, dominated at different time during the observation. However, further research should be performed with specific species using measured sound speed and density of zooplankton and laboratory analysis for grazing angle measurement should be conducted in order to improve accurate results of specific zooplankton.

## 1. Introduction

In the field of marine technology, hydroacoustic is one of the advanced technology that can detect and quantify various underwater objects, for example on detection of presence and abundance of zooplankton in the seawaters. The use of acoustic technology for zooplankton research is important because zooplankton is a link between primary producers with large and small carnivores that can affect the complexity of food chains in aquatic ecosystems. Previous studies that utilize hydroacoustic technology for the detection of zooplankton include monitoring the abundance of zooplankton in aquatic area [1], vertical migration of zooplankton (diel vertical migration) and its variation [2], zooplankton detection and its relation to the sound scattering layer [3]. Furthermore, acoustics can be used as a combination of models [4], as well as acoustic signal processing to predict the presence of zooplankton with a specific target strength (TS) [5, 6].

One approach has been the use of acoustic Doppler current profiler (ADCP) instrument for zooplankton detection. In ADCP, the response of scattering objects to different sound signals produces different results in terms of the average scattering volume strength (SV). The difference in mean volume backscattering strength may be used as a reference to differentiate zooplankton from other scatterers [7]. Estimation of zooplankton abundance can be done by processing backscattering values from acoustic devices, and mathematical inversion techniques was used to convert the recording results of acoustic data as counts unit to the mean volume backscattering strength (MVBS) in dB units.

The backscattering strength values in dB units can be used as a conversion value into the size and abundance distributions of biological sound scatterers [8].

The main problem in zooplankton analysis in seawaters currently is need water sampling. The water samples should be taken at different depths and at any given time to obtain the distribution of zooplankton in the water column. Studying the movement and behavior of zooplankton, requires continuous MVBS data, so the method by which this water sampling needs to be evaluated. The acoustic approach can be a solution because it can produce continuous data on a time scale and wide space. The aim of this research was conducted to test the validation and ability of acoustic method can be used to detect quantitatively the existence of zooplankton and test the precision of zooplankton measurement result obtained from ADCP to result obtained from laboratory analysis. In this paper, the diel vertical migration of zooplankton with special emphasis in its response to change in daylight and midnight time was observed using ADCP based on the mean volume backscattering strength.

## 2. Method

### *2.1. Research Location*

The collection of water samples and ADCP data were conducted in Lembeh Strait, North Sulawesi, Indonesia (1°28'2.8800" N - 125°14'6.3240" E) in April 2016. Data processing and analysis were done at Laboratory of Ocean Acoustic Data Computation and Sonar System, Department of Marine Science and Technology, Faculty of Fisheries and Marine Sciences, Bogor Agricultural University, Indonesia in December 2016 – July 2017.

### *2.2. Equipment and Preparation Setup*

ADCP SonTek Argonaut- XR with 750 kHz frequency was deployed in Lembeh Strait at 15 m of depth with 0.5 m resolution, 20 minutes recording interval data, and total 30 bins. Tides data were recorded using Tidemaster Tide Gauge. For seawater sampling, plankton net was used with mesh size 50 μm and was taken from 5 different time and depth. The water samples were preserved using Lugol's solution.

### *2.3. Mean Volume Backscattering Strength (MVBS) Computation*

Acoustic data used as an echo intensity unit data extracted from ADCP, corrected against transmission losses to eliminate the effect of distance and sound absorption [9]. Acoustic waves in the water will propagate with an attenuation and geometrical spreading in the water column. Sound attenuation due to scatterers like zooplankton, were consist of scattering suspended particles, viscous and thermal absorption. The acoustic response in the detection of zooplankton relative with their physical characteristics, including shape, size, type, and body response that reflects acoustic signals [10]. The corrected echo intensity data were then analyzed and compared with the results using the model. A ping from ADCP has several stages in transmission, propagation, backscattering, returning echo, and reception. The transmission losses in propagation process can be expressed as follows

$$2TL = 20 \log_{10} R + 2 \alpha R \quad (1)$$

where $2TL$ is transmission losses in two-way travel time (dB), $\alpha$ is absorption coefficient (dB km$^{-1}$), $R$ is the distance of the ADCP transducer to the measured layer (m).

The algorithm validation is important to estimate the volume backscattering value in the water column for later Diel Vertical Migration (DVM) pattern of the zooplankton. The volume backscattering strength can be simplified as follows [11]:

$$SV = 10\log [N\sigma / 4\pi V] = 10\log(N\sigma_{bs} / V) \quad (2)$$

where $Sv$ is scattering volume (dB), $\sigma$ is the average value of backscattering cross section on the number of $N$ scatter on sample volume $V$, and $V$ can be calculated as:

$$V = (c\tau/2)r^2\psi \quad (3)$$

Based on Eq. 2, $c$ is the speed of sound in the medium (m $s^{-1}$), $\tau$ is the pulse duration (ms), and $\psi$ is beam angle $\psi = \int b^2 d\Omega$.

The value of $b^2$ is the result of the beam pattern, which is typically denoted by a unit of relative intensity. The backscattering strength from scatterers in each depth (bin) depends on the size and density of zooplankton. These factors were correlated with the target strength, which obtained by [12]:

$$TS = 10\ log[\sigma / 4\pi r_0^2] = 10\ log(\sigma_{bs} / r_0^2) \quad (4)$$

Using the following expression for $TS$, the $\sigma$ defined as backscattering cross section ($m^{-2}$) and ro is reference distance (m). The second form of the equation, $\sigma_{bs}$ represents the differential of the backscattering cross section, which is different from $\sigma$ derived from $4\pi$ factor. An acoustic signal that affects multiple targets in a volume area (density), the size of the scattering strength characteristics is considered the total scattering volume. Using the following formula in Eq. 2 and 4, and assuming the numerical density of zooplankton is proportional to the individual of zooplankton, then it can be written as [13] $SV = 10 \log \rho + TS$, where $\rho$ is density of zooplankton (ind $m^{-3}$). The data processing is the conversion of data in the form of intensity echo from ADCP to scattering volume. In laboratory measurements, the zooplankton abundance analysis data were compared with acoustic measurements. All of the experiment setup is shown in Figure 1.

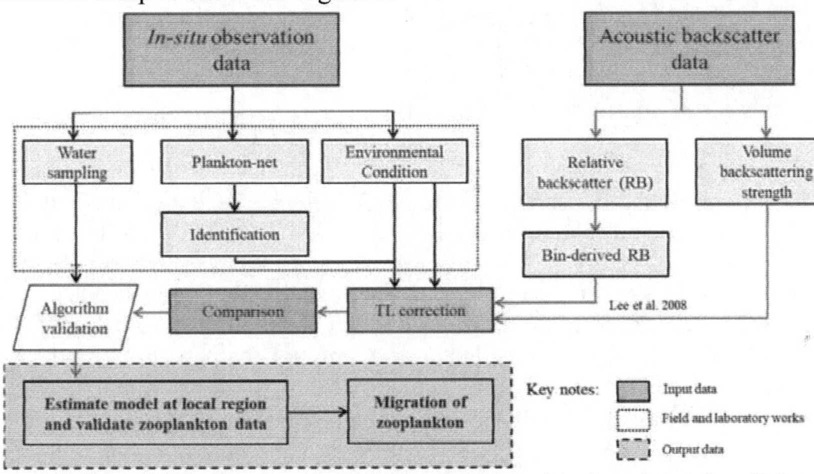


**Figure 1.** Research flowchart of data acquisition and processing for measuring diel vertical migration of zooplankton

*2.4. Laboratory-based Zooplankton Analysis*

Zooplankton sampling was done at several points of the station using plankton-net with stratified random sampling method at 2 m vertical depth and three sampling times. The abundance of zooplankton was calculated using Sedgwick -Rafter Counting Method with sweep technique as an individual unit of $m^{-3}$. Abundance value used the modification equation of APHA method. Diversity of species was observed using the Shannon-Wiener diversity index, the generalization of the type using the Evenness levelness index formula [14].

**3. Results and Discussion**

Field observation in the area of ADCP SonTek is at a depth of 18 meters. The observation area has strong currents, as it is located in a narrow gap between Sulawesi Island and Lembeh Island. The absorption coefficient ($\alpha$) found was 0.1676 dB $m^{-1}$ and the average sound speed from ADCP was 1519.23 m $s^{-1}$. Conversion results show the time series of scattering volume in the observed area.

Scattering volume measured from ADCP was a parameter often used in underwater acoustic to measure and detected a medium to scatter acoustic waves direction back towards the source, producing consequent echo [15]. The highest scattering volume value was in the range of -57 dB and the lowest ranged in -89 dB (Figure 2). Based on the time series data, the highest scattering volume value was at a depth of 12 meters to 16 meters. In the columns and near the surface, the scattering volume value was relatively lower when compared with the near-bottom area which ranges from -58 dB to -75 dB. An important information based on the echogram is vertical distribution and migration.

Based on the echogram, movement of zooplankton were up and down in the water column over a daily cycle at 12-16 meters. Based on the echo tracing analysis on the echogram, at night, zooplankton move to upper layer with speed 0.127 m $h^{-1}$. It was thought that zooplankton migration was triggered by the relative changes of temperature, because zooplankton have low swimming capability [16]. But there was no time series data of temperature in the stratified layer of depth.

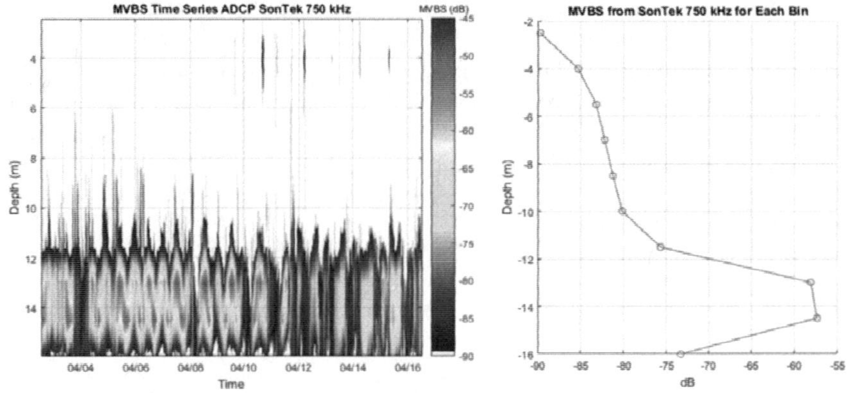


(a) (b)

**Figure 2.** Time series data of scattering volume (SV) from ADCP measurement. It was shown that the zooplankton were located and moved at 12 – 15 m of depth

Based on laboratory analysis and net sampling, it was revealed that crustacean zooplankton is the cause of the strong sound scattering in the water column at the near bottom area. Diel vertical migration of zooplankton is commonly considered an adaptation with warm temperature at day and cold temperature at night. Beside of that, the availability of light from the sun plays an important role of the phenomenon, especially for crustacean migration [17].

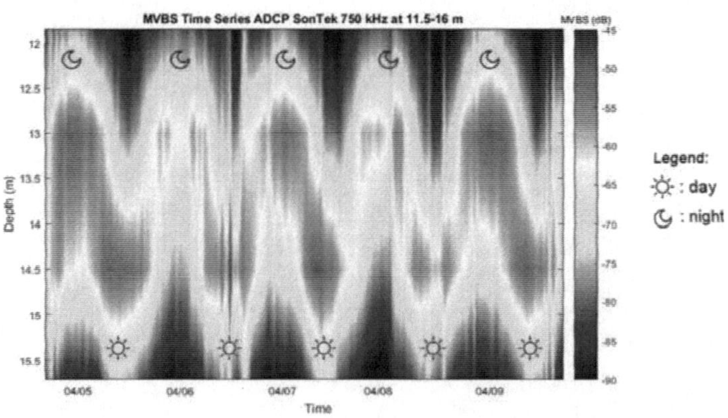


**Figure 3.** Diel vertical migration of zooplankton at the water column from an upward looking ADCP in daily periods. During the night, zooplankton was moved to higher depth compared to sunlight

IOP Conf. Series: Journal of Physics: Conf. Series **1075** (2018) 012016 doi:10.1088/1742-6596/1075/1/012016

Measurements of zooplankton with ADCP was reasonably well applied to aquatic conditions with relatively similar in-situ concentration ranges against the estimated data from ADCP. The value of the echo intensity is proportional to the zooplankton abundance of the scattering material in the water column.

Most study demonstrate that specific species of zooplankton was found in the ocean (for example Antartic Krill; *Euphasia superba*) [6], but from our research it was found that the heterospecies consists of protozoa, crustaceae, and polychaeta (Table 1), not only single species. The laboratory observation was from five days sampling in the deployed ADCP area. This unresolved issue is need to be verified through specific target strength (TS) measurement at the laboratory for the further research, especially measuring TS for each zooplankton found (Table 1) for determining zooplankton abundance.

**Table 1.** Abundance of zooplankton at each depth at 13.5 - 15 m based on laboratory analysis for 5 day observation

| Organism | Time | | | | |
|---|---|---|---|---|---|
| | 05/04/16 | 06/04/16 | 07/04/16 | 08/04/16 | 09/04/16 |
| **PROTOZOA** | | | | | |
| *Amphorella* sp. | 101 | 0 | 0 | 201 | 101 |
| *Codonelopsis* sp. | 0 | 101 | 101 | 201 | 201 |
| *Globoratalia* sp. | 0 | 0 | 201 | 0 | 0 |
| *Eutintinnus* sp. | 0 | 0 | 0 | 101 | 0 |
| *Tintinnopsis* sp. | 0 | 0 | 0 | 0 | 301 |
| **CRUSTACEAE** | | | | | |
| Nauplius (stadia) | 702 | 201 | 0 | 1,103 | 1,304 |
| *Paracalanus* sp. | 0 | 0 | 201 | 101 | 101 |
| *Oithona* sp. | 0 | 0 | 101 | 101 | 0 |
| *Corycaeus* sp. | 0 | 0 | 301 | 0 | 0 |
| **Polychaeta larvae** | 0 | 0 | 0 | 0 | 101 |
| The number of Taxa | 2 | 2 | 5 | 6 | 6 |
| Abundance (ind/m$^3$) | 803 | 302 | 905 | 1,808 | 2,109 |
| Diversity Index | 0.38 | 0.64 | 1.52 | 1.27 | 1.24 |
| Uniformity Index | 0.55 | 0.92 | 0.95 | 0.71 | 0.69 |
| Dominance Index | 0.78 | 0.55 | 0.23 | 0.41 | 0.42 |

Plankton calculation using Ln
Method: Two-replication census

## 4. Conclusion

This study is very useful for monitoring zooplankton especially it's abundance and behavior (diel vertical migration; DVM) in the ocean. For zooplankton monitoring, ADCP can provide continuous backscattering signal data with long time observation. Some of the obstacles in research were yet to be able to distinguish the size of zooplankton from acoustic signal, the type of material reflector near the seabed that reflects the highest acoustic signal, and determined species of zooplankton based on backscattered signal. Further research needs to be continued to distinguishing between acoustic signal from zooplankton and suspended sediments and measuring target strength of each zooplankton found.

## Competing Interests

The authors declare that there are no competing interests.

## References


[1] Iida K, Mukai T, Hwang D 1996 Relationship between acoustic backscattering strength and density of zooplankton in the sound-scattering layer *ICES Journal of Marine Science* **53** 507-512Cisewski B, Strass

[2] V H, Rhein M, Kragefsky S 2010 Seasonal variation of diel vertical migration of zooplankton from ADCP backscatter time series data in the Lazaref Sea, Antartica *Deep-Sea Research I* **57** 78-94

[3] Lee K, Mukai T, Lee DJ, Iida K 2014 Classification of sound-scattering layers using swimming speed estimated by acoustic Doppler current profiler *Fisheries Science* **80** 1-11

[4] Jech J M, et al. 2015 Comparisons among ten models of acoustic backscattering used in aquatic ecosystem research *The Journal of Acoustic Society America* **138** 3742 – 3764

[5] Smith J N, Ressler P H, Warren J D 2013 A distorted wave Born approximation target strength model for Bering Sea euphausiids *ICES Journal of Marine Science* **70** 204–214

[6] Manik H M 2015 Acoustic observation of zooplankton using high frequency sonar *Indonesian Journal of Marine Sciences* **20** 61-72

[7] Sawada K, Mukai T, Fukuda Y, Matsuura T 2016 Comparison of zooplankton density estimated by acoustic inversion method and net sampling *The Journal of the Acoustical Society of America* **140** 3243-3243

[8] Harris R, et al. 2000 ICES Zooplankton Methodology Manual. London (UK): Academic Pr. 669 pp

[9] Dwinovantyo A, et al. 2017 Estimation of suspended sediment concentration from Acoustic Doppler Current Profiler (ADCP) instrument: A case study of Lembeh Strait, North Sulawesi *IOP Conference Series: Earth Environment Science* **54** 1-8

[10] Benoit-Bird K J 2009 The effects of scattering-layer composition, animal size, and numerical density on the frequency response of volume backscatter *ICES Journal of Marine Science* **66** 582–593.

[11] Maclennan D N, Simmonds E J 2005 Fisheries Acoustic. Blackwell Science, Chapman and Hall, Oxford, London, England.

[12] Chu D, Wiebe P H, Copley N 2000 Inference of material properties of zooplankton from acoustic and resistivity measurement *ICES Journal of Marine Science* **57** 1128-1142

[13] Hannachi M S, Abdallah L B, Marrakchi O 2004 Acoustic identification of small pelagic fish species: target strength analysis and school descriptor classification. *Med Sud Med Technical Documents* **5** 1-10

[14] Rao K R 2017 Zooplankton diversity and seasonal variations in Thandava reservoir, Visakhapatnam, India *International Journal of Fisheries and Aquatic Studies* **5** 90-97

[15] van Haren H, Compton T J 2013 Diel vertical migration in deep sea plankton is finely tuned to latitudinal and seasonal day length. *PLoS ONE* **8** 1-8

[16] Armengol X, et al. 2012 Pseudo-diel vertical migration in zooplankton: a whole-lake 15N tracer experiment *Journal of Plankton Research* **34** 976-986

[17] Stéphane M, et al. 2001 Diel vertical and horizontal distribution of crustacean zooplankton and young of the year fish in a sub-alpine lake: an approach based on high frequency sampling *Journal of Plankton Research* **23** 1041–1060,


## Acknowledgments

The author would like to thank the Directorate General of Higher Education (DIKTI) Indonesia, for financially supporting this research through PMDSU Research Grant Scheme Year II (No. 136/SP2H/LT/DRPM/IV/2017). This work is also supported by Marine Geological Institute, Ministry of Energy and Mineral Resources, Bandung, Indonesia.

# Measurement and Visualization System of Eigenmode Shapes Considering Orthogonal Modes on Circular Membrane

**Eri Zempo[1], Naoto Wakatsuki[1,2*], Koichi Mizutani[1,2], and Yuka Maeda[1,2]**

[1] Graduate School of Systems and Information Engineering, University of Tsukuba, 1-1-1 Tennodai, Tsukuba, Ibaraki, 305-8573 Japan
[2] Faculty of Engineering, Information and Systems, University of Tsukuba, 1-1-1 Tennodai, Tsukuba, Ibaraki, 305-8573 Japan

E-mail: wakatuki@iit.tsukuba.ac.jp

**Abstract.** This work proposes a visualization system of eigenmode shapes of a tapped circular membrane of membranophones. Under tapped membrane vibration, it is not enough to divide modes in frequency domain, because of the existence of eignmodes whose frequencies are close to each other including orthogonal modes. Therefore, we structured this visualization system of eigenmode shapes, which could divide the orthogonal modes appearing at slightly different frequencies. This system firstly measures the sound pressures emitted from the vibration on tapped circular membrane of membranophone with a circular microphone array placed in proximity to the membrane. Secondly, it extracts the circumferential shape of each eigenfrequency. In this system, dominant frequencies of estimated power spectrum density obtained by Yule-Walker method are considered as approximate eigenfrequences of this vibration. Multiple eigenmodes whose azimuthal mode order is different, are divided by Fourier series expansion of the circumferential shapes of each eigenfrequency. In addition, orthogonal modes, which have the same mode order and slightly different frequencies are divided by considering the orthogonality of mode shapes, sine and cosine functions. Lastly, the system estimates the eigenmode shape of the whole membrane using the Bessel function of the first kind correspond to each mode order as radial shape, and shows them to users. It is expected that user could recognize the unbalance of tension distribution intuitively using this system. It is also expected the system to be applied to the tuning assistive system of the tuning of circular membranophone.

## 1. Introduction

Membranophones are musical instruments, which emit sounds by the vibration of membranes. Most of them have circular membranes, called "head", whose tension distribution has an effect on the sound of the instruments, as the eigenmode frequencies and these compositions in the vibration are changed. "Tuning" is the process to adjust of the tension distribution. Players follow this process mainly before their performances.

Schematic views of snare drum are shown in Fig. 1 as an example of membranophones. The edges of heads are caught between a hoop and a cylindrical drum shell. Heads are more stretched over the edge of the shell, when you tighten the "tuning rods", which are kinds of screws, and

IOP Conf. Series: Journal of Physics: Conf. Series **1075** (2018) 012017 doi:10.1088/1742-6596/1075/1/012017

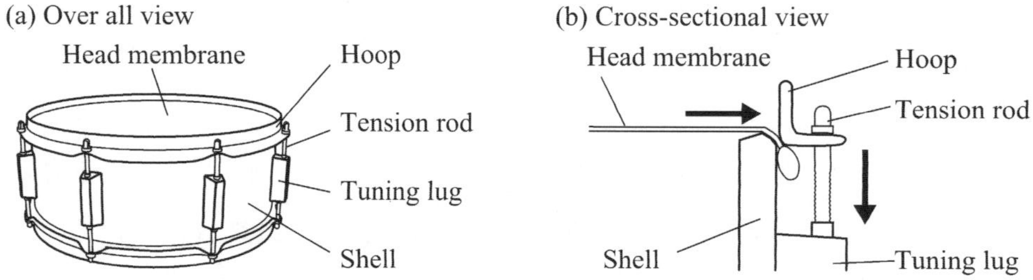


**Figure 1.** Schematic views of snare drum: Tightening tension rods make the head membrane stretched.

the hoop is pulled toward "tuning lug". A membrane, usually has from six to ten tuning rods. The tension distribution of the head is determined by controlling the tensions of the rods in the tuning process. In the common practical method, the player adjusts each rod little by little, and guess the tension distribution of the head by hearing the pitches of the sounds emitted by tapping several points[1]. However, it is not easy for beginners to master these processes as there are many difficult points in them.

On the other hand, it is known that non-uniform tension makes the eigenmode shapes non-symmetric[2, 3]. This means visualizing the mode shapes might help the players to guess the tension distribution. To assist tuning or training that processes, we proposed the system, which immidiately visualizes the eigenmode shape of the vibration on tapped circular membranes of membranophones[4]. In the system, the membrane vibration is measured by means of sound pressures acquired by a circular microphone array located adjacent to the membrane, and the peaks of the estimated power spectral density (PSD) of the sound are regarded as the eigenfrequencies. Each component of eigenfrequency is extracted by synchronous detection, and shown in a few tenth seconds to users.

However, it is also known that orthogonal modes on the circular membrane under non-uniform tension may cause beat in the sound; such states are likely to occur in the tuning process[2, 3]. Such a beat was observed as is in the system mentioned above, however, multiple modes including orthogonal modes should be separated for further assistance of tuning. Thus, in this paper, we applied the Fourier analysis not only to the temporal waveform, but also to the spatial distribution in azimuthal coordinate, with the aim to extract the eigenmodes considering adjacent orthogonal modes, and show their mode shapes.

## 2. Method of measurement and visualization eigenmode shapes

### *2.1. Ideal membrane theory*

It is well known that the harmonic vibration $z(t)$ on ideal circular membrane under uniform tension is given by:

$$z(r, \phi, t) = R(r)\,\Phi(\phi)e^{j\omega t} \,, \tag{1}$$

where,

$$\Phi(\phi) = Ae^{\pm jm\phi} \,, \tag{2}$$
$$R(r) = J_m(k_{mn}r) \,, \tag{3}$$

for polar coordinates $(r, \phi)$, where $J_m$ are Bessel functions of first kind, of order $m$, and $m$ are the mode order of the radial coordinate for the vibration[5]. $k_{mn}$ are the solution for the equation $J_m(k_{mn}r) = 0$, correlated, from lowest to highest, to azimuthal mode order $n$ for the vibration.

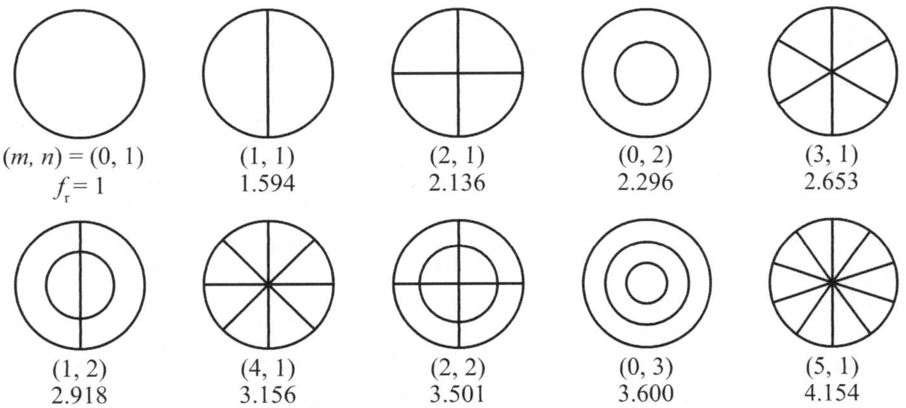


**Figure 2.** Mode shape and frequency-rate to $(0, 1)$ fundamental mode on ideal membrane under uniform tension

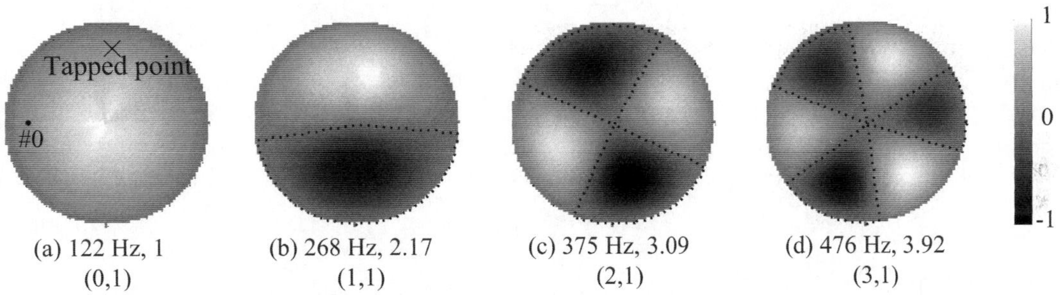


**Figure 3.** An example of the estimated eigenmode shapes: to calculate the estimated eigenmode shape for each frequency, $f_{\hat{d}}$, the measurement values, $\varphi_{\hat{d}}(i)$, and the Bessel functions which have appropriate orders, are used as azimuthal and radical shapes, respectively.

Figure 2 shows the first ten eigenmode shapes on the ideal membranes. Diameters and concentric lines in the figure represent nodal lines for each mode. $(m, n)$ indicates mode order, $f_{\mathrm{r}}$ is the frequency relative to the $(0, 1)$ mode. The vibration on circular membrane is the superposition of these eigenmodes. Every $(m, n)$ mode except for $m = 0$, consists of a pair of orthogonal modes. These have the same eigenfrequency on the ideal membrane under uniform tension, which is called "degenerated", while the frequencies of these orthogonal modes are split into different frequencies and they cause beat sounds under non-uniform tension. In our system, eigenmode shapes are shown by using and reflecting these knowledges.

### *2.2. Measurement of vibration and extraction of eigenmodes*

Let $z_i(l)$ as the offset-eliminated sound pressures of the vibration of the head, acquired by a circular microphone array located adjacent to the membrane, where $i$ and $l$ indicate the channel index of microphones and the sample index, respectively. The dominant frequencies of the estimated PSD, obtained by Yule-Walker method, are regarded as the approximate eigenfrequencies, $f_{\hat{d}}$, and let them from lower to higher, as $f_1, f_2, \cdots$. Figure 3 shows a typical eigenmode shapes estimated by the process written above.

Next, the azimuthal eigenmode shapes on each eigenfrequencies, $f_{\hat{d}}$, are extracted. These shapes are calculated by the internal product of sine function, whose frequency is $f_{\hat{d}}$, and the temporal waveform of each channel as,

$$\varphi_{\hat{d}}(i) = \sum_{l} z_i(l) e^{j2\pi f_{\hat{d}} l} \; . \tag{4}$$

The real part of the left member in eq. 4, $\mathrm{Re}[\varphi_{\hat{\mathrm{d}}}(\mathrm{i})]$, denotes the estimated azimuthal eigenmode shapes.

In addition, multiple modes including orthogonal modes would be observed by Fourier analysis to azimuthal coordinate. This method is based on that the measurements are periodic to azimuthal coordinate, as the microphone array located in a circle, and especially that the azimuthal eigenmode shape is represented as sine function[6].

By applying Fourier series expansion to the estimated azimuthal eigenmode shapes extracted in the former process, $\varphi_{\hat{d}}(i)$, multiple modes, which have different azimuthal mode order $m$, can be separated as,

$$\tilde{\varphi}_{\hat{d}m} = \sum_{i} \varphi_{\hat{d}}(i) e^{j2\pi m i} \; . \tag{5}$$

If multiple of $\tilde{\varphi}_{\hat{d}m}$ which have enough value for each $\hat{d}$ exist, it is thought as the multiple eigenmodes exsit around at the frequency $f_{\hat{d}}$, and these modes might be observed in $\varphi_{\hat{d}}(i)$ extracted above.

If the orthogonal modes have slightly different frequencies to each other, two eigenfrequencies, $f_{\hat{d};\mathrm{a}}$ and $f_{\hat{d};\mathrm{b}}$, can be found as,

$$f_{\hat{d};\mathrm{a,b}} = \arg_{f_{\hat{d}}} \max \left[ \sum_{i} z_i(l) e^{j2\pi m (i + \frac{f_{\mathrm{si}}}{360^\circ} \theta_0)} \right] , \tag{6}$$

where, $f_{\mathrm{si}}$ and $\theta_0$ denote the number of the microphones and the initial angle of the sine function, respectively. Through these processes, the pair of orthogonal mode can be separated by discrete Fourier transform even if their eigenfrequency is very close to each other. This is because their mode shape is orthogonal, that is, the inner product of their mode shape is equal to zero.

## 3. Experimental data on 13 inch Tom-Tom

### *3.1. Settings on the experiment*

We measured the vibration of a head on 13 inch Tom-Tom as an example of membranophones. The diameter of this Tom-Tom is 318 mm, and the depth is 279 mm. A head is stretched over

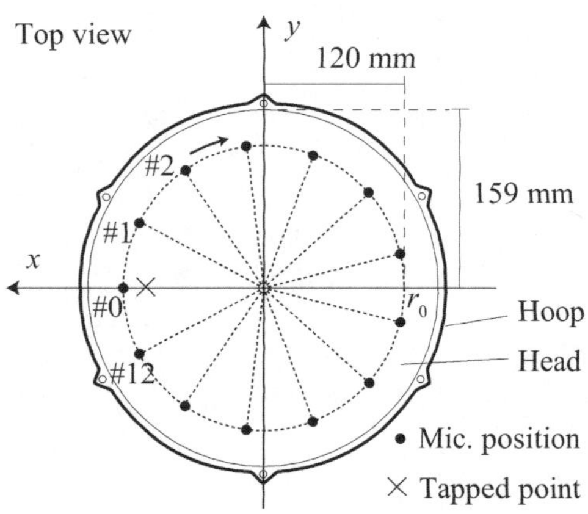


**Figure 4.** Measurement Points: Microphones are circularly arranged at regular intervals.

IOP Conf. Series: Journal of Physics: Conf. Series **1075** (2018) 012017 doi:10.1088/1742-6596/1075/1/012017

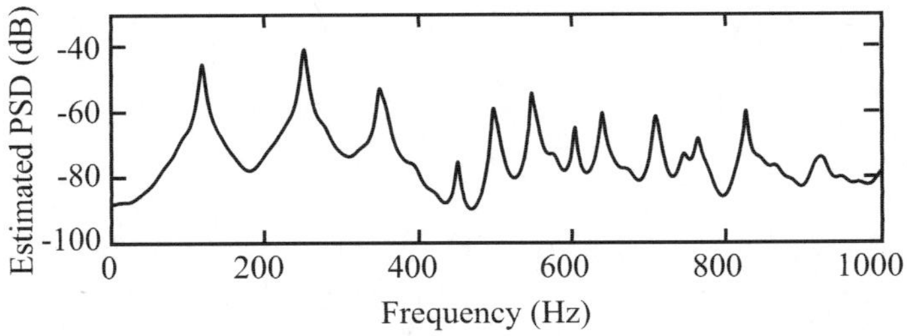


**Figure 5.** Estimated PSD by Yule-Walker method, on #0.

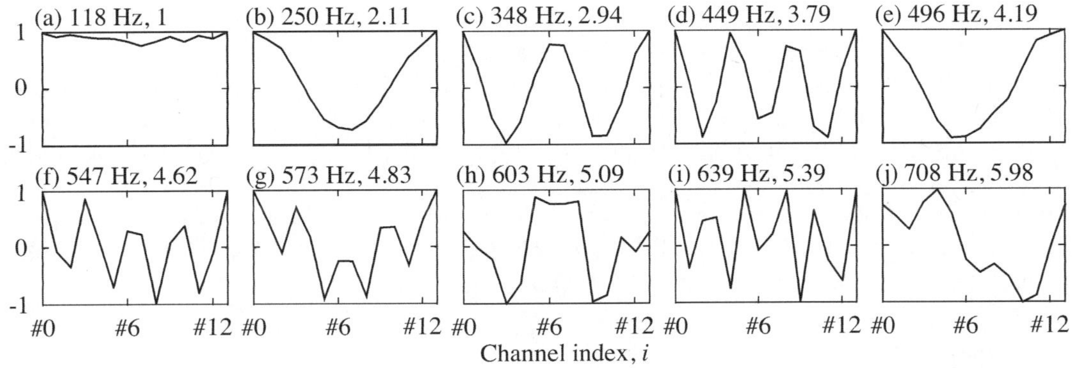


**Figure 6.** The estimated azimuthal eigenmode shapes extracted by Fourier analysis to the temporal waveform

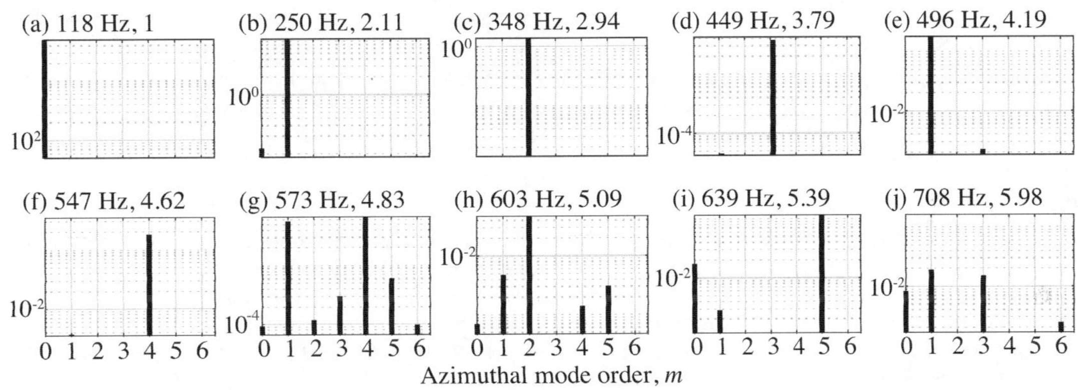


**Figure 7.** Components of each mode order $m$, for each frequencies extracted by Fourier analysis to the azimuthal

on only one side, to observe the vibration of a single membrane. The microphone $\#0 - 12$ are arranged circularly, inside of the shell, and located adjacent to the membrane with the sampling angular frequency, $f_{\text{si}} = 13$, which is equal to the number of the microphones. These microphone position is shown in Fig. 4. Sound pressures by each microphones are measured in 5 sec with the sampling frequency $f_{\text{sl}} = 30$ kHz.

### *3.2. Results of the experiment*

Figure 5 shows the estimated PSD obtained by Yule-Walker method. The estimated dominant frequency, $f_{\{1,2,\cdots,10\}} = \{118, 250, 348, 449, 496, 547, 573, 603, 639, 708\}(\text{Hz})$ were extracted as $f_{\text{d}}$ from the lowest.

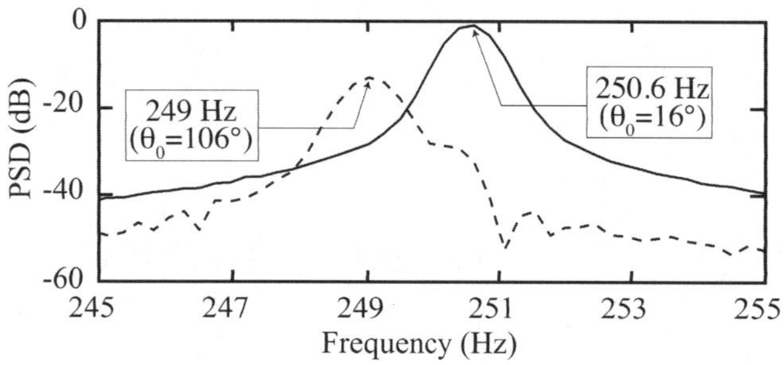


**Figure 8.** Observation of the orthogonal modes of $(1,1)$: the peaks of PSD appear at slightly different frequencies.

The azimuthal mode shapes correspond to each peak in Fig. 5, are shown in Fig. 6. Some figures such as (b), (c), (e), have curves like sine functions, while other such as (g),(i), draw intricate curves, relatively. In addition, Fig. 7 shows the modes such as (g), (i) have multiple modes at close frequencies.

Figure 8 shows the extraction orthogonal modes on (1,1). (1,1) mode on this vibration has two orthogonal modes; one has a nodal line on $\phi = \theta_0 = 16°$ with eigenfrequency 250.6 Hz, the other has a nodal line on $\phi = \theta_0 = 106°$ with eigenfrequency 249 Hz.

## 4. Conclusion

In this work, with the aim to extract and visualise eigenmode shapes of the vibration on tapped circular membranes, we proposed the system to extract the multiple modes including orthogonal mode appearing in close frequencies. by applying the Fourier analysis not only to the temporal waveform, but also to the spatial distribution in azimuthal coordinate. Besides, we measured the vibration of the membrane on 13 inch Tom-Tom, and showed the multiple modes could be extracted. The system for visualizing eigenmode shapes utilizing this method is expected to help users to tune the tension of the membrane of membranophones, and also to learn the tuning process.

## References

[1] R. Toulson, C. C. Crigny, P. Robinson and P. Richardson (2009). The perception and importance of drum tuning in live performance and music production. Journal on the Art of Record Production **4** 1.
[2] R. Worland (2008). Drum tuning: an experimental analysis of membrane modes under non-uniform tension. Proceedings of Meeting on Acoustics **5**.
[3] R. Worland (2010). Nomal modes of a musical drumhead under non-uniform tension. The Journal of the Acoustical Society of America **127** (1): 525-533.
[4] E. Zempo, N. Wakatsuki, K. Mizutani, and Y. Maeda (2017). Measurement and Visualization System of Vibration Mode Shape on Tapped Membrane. Proceedings of the Technical Committee on Musical Acoustics of the Acoustical Society of Japan, **36** (3): 1-6 (in Japanese).
[5] N. Fletcher and T Rossing (1998). "The Physics of Musical Instruments", Springer .
[6] T. Yudasaka, N. Wakatsuki, and K. Mizutani (2014). Acoustic model of cymbal considering geometrical nonlinearity. The Journal of the Acoustical Society of Japan, **70** (7): 362-370 (in Japanese).

# Acoustical measurement and questionnaire survey on the comfortableness of conversation in tavern

**K Harada[1] and K Kawai[1]**

[1]Graduate School of Science and Technology Kumamoto University

E-mail: kazunori.harada.108@gmail.com

**Abstract.** In the spaces such as taverns and cafes, there are more than one group in a space, and each group is conversing within the group. These spaces are defined as "multi-group conversation spaces (MGCS)". The purpose of this study is to find the optimal acoustic environment in taverns as an example of MGCS regarding with the comfortableness of conversation. For this purpose, acoustic measurement and questionnaire survey were conducted in two types of taverns, semi-separated room type and mono-space type. In the acoustic measurement, unoccupied room impulse responses and sound pressure levels in the operating time were measured. In the questionnaire survey, questions of subjective evaluations such as ease of conversation and importance of speech privacy were asked to guests. As a result, the noise level of mono space type in the busy time was 76-77 dB and higher than that of semi-separated room type (69-73 dB), and this difference seemed to be caused by the higher level of background music in the mono space type. The result of questionnaire survey showed that guests of semi-separated room type answered higher importance of privacy in tavern than those in the mono space type. It is also suggested that importance of privacy might affect client's preference of the type of tavern.

## 1. Introduction

In the spaces such as taverns and cafes, there are more than one group in a space, and each group is conversing within the group. In our previous study, these spaces were defined as "multi-group conversation spaces (MGCS)". Optimum acoustics environments and acoustics design has been less discussed for MGCS than spaces that require speech privacy such as hospitals or pharmacies. It is assumed that there are many factors to be evaluated for acoustics in MGCS, such as: comfortableness within groups, privacy between groups, liveliness, calmness, and easiness of ordering foods or drinks. It should be necessary for the acoustic design of MGCS to understand the relationship between these factors and acoustic characteristics.

Thus, in this study, an acoustic measurement and a questionnaire survey were conducted in two different types of tavern to investigate actual acoustic situation of tavern as a basic step.

## 2. Survey

### *2.1. Surveyed tavern*

Two taverns were selected for the survey. One was semi-separated room type (T1) where tables were set up like separated rooms by partial walls of about 1.6m height. All the tables of T1 were regarded as acoustically connected. Another was mono-space type (T2) without partition. Plans of the taverns are shown in Fig. 1. The ceiling height of both taverns was about 2.6 m. Sound absorbing material were not installed to the surfaces of both of the taverns. Only a carpet put in a part of a passage floor and two or three cushions placed at every table in T1 were regarded as sound absorptive. We divided T1 into 11 sections and T2 into 7 sections for the survey. Sections 1 and 10 in T1 can be connected by removing the panel between them and "10c" denotes the connected condition.

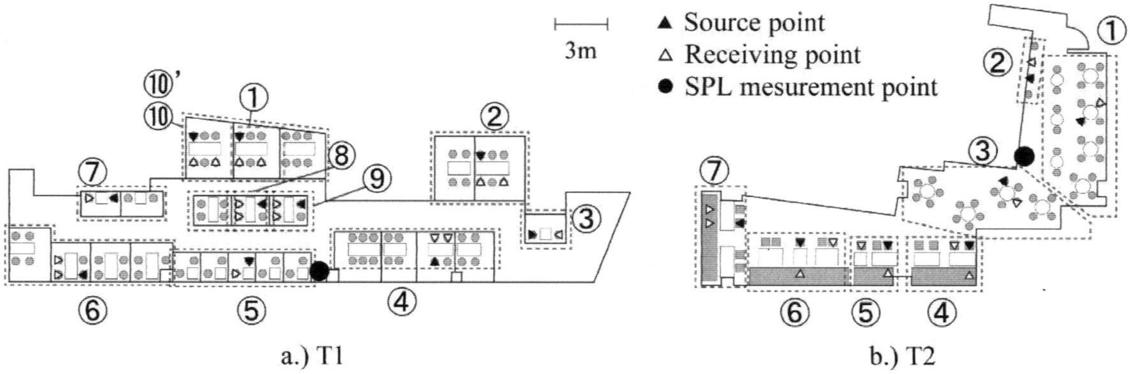


**Figure 1.** Plans of surveyed taverns and measurement point.

**Table 1.** Subject attribute.

| Tavern | Gender | Spacing 20-24 | 25-34 | 35-44 | 45-54 | 55- | Total |
|---|---|---|---|---|---|---|---|
| T1 | Male | 7 | 16 | 10 | 2 | 7 | 42 |
| | Female | 14 | 12 | 7 | 5 | 7 | 45 |
| T2 | Male | 12 | 15 | 6 | 11 | 3 | 47 |
| | Female | 8 | 11 | 19 | 5 | 1 | 35 |
| | blank | | | | | 1 | 1 |

*2.2. Reverberation time*

In each of the sections, one seat was set as a source point and another one or two seats were set as receiving points assuming a conversation partner, and impulse responses were measured using time-stretched pulse.

*2.3. Sound pressure level*

A-weighted sound pressure level was measured at a point near the center of each of the taverns for 3 hours from just after the opening of the tavern without guests until busy period. At the same time number of guests was counted separately for two areas of near tables (1, 4, 5, 8, 9, 10 of T1, and 1, 2, 3 of T2) and far tables (others), which were divided by about 8 m from the measurement point.

*2.4. Questionnaire survey*

A questionnaire survey was conducted for the guests. The questionnaire sheet consisted of five questions asking impressions at the table, such as: importance of privacy regarding others' eye and conversation voices, evaluation and preference of liveliness or calmness, easiness of conversation, kinds of distracting sounds, and two face items of age and gender. The questionnaire were handed to the guests when they made an order at the first time and collected at the first serving because it would be as difficult to request answering after drinking alcohol. At the same time, occupied tables around the respondents were recorded by the researcher to grasp how many conversing groups existed at the time of the answer. The survey was conducted on five normal days and busy two days (weekend or holiday) for both of the taverns. In total 87 (male 42, female 45) in T1, and 83 (male 47, female 35, blank 1) in T2 (Table 1) answers were collected.

## 3. Results

### *3.1. Reverberation time*

Reverberation times were calculated from the measured impulse responses. The average T30 (500-2000 Hz) in each section were 0.2 - 0.4 s in T1 and 0.4 - 0.5 s in T2 (Fig. 2). These values seemed to be short considering the volumes of each of the spaces. The reason might be that they were an elongated shape with many tables and chairs and, furthermore, T1 was separated into small sections.

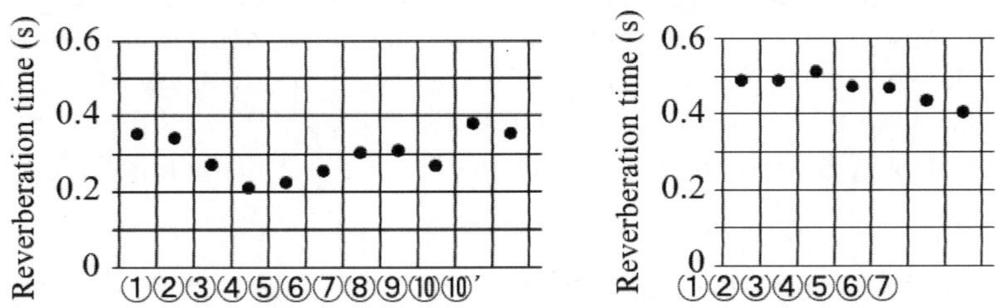


a.) T1 b.) T2

**Figure 2.** Reverberation time of each section.

### *3.2. Sound pressure level*

**Table2.** Percentile sound pressure level of T1

| Index | Time zone | | | | | |
|---|---|---|---|---|---|---|
| | 17:30 -18:00 | 18:00 -18:30 | 18:30 -19:00 | 19:00 -19:30 | 19:30 -20:00 | 20:00 -20:30 |
| $L_{95}$ | 43.7 | 45.0 | 46.2 | 47.8 | 53.2 | 55.1 |
| $L_{90}$ | 44.7 | 46.2 | 47.3 | 48.9 | 54.3 | 56.1 |
| $L_{50}$ | 50.0 | 50.4 | 50.7 | 52.2 | 58.6 | 60.3 |
| $L_{10}$ | 54.1 | 54.2 | 54.5 | 56.9 | 65.5 | 70.1 |
| $L_5$ | 55.2 | 55.3 | 55.5 | 58.7 | 69.0 | 72.7 |
| $L_{Aeq}$ | 51.1 | 52.0 | 52.2 | 54.9 | 63.5 | 66.3 |
| Number of guests | | | | | | |
| All | 0 | 4 | 6 | 14 | 40 | 40 |
| Far | 0 | 4 | 6 | 6 | 6 | 6 |
| Near | 0 | 0 | 0 | 8 | 34 | 34 |

**Table3.** Percentile sound pressure level of T2

| Index | Time zone | | | | | |
|---|---|---|---|---|---|---|
| | 18:00 -18:30 | 18:30 -19:00 | 19:00 -19:30 | 19:30 -20:00 | 20:00 -20:30 | 20:30 -21:00 |
| $L_{95}$ | 55.7 | 54.4 | 56.8 | 58.5 | 62.9 | 64.5 |
| $L_{90}$ | 57.5 | 56.5 | 58.3 | 60.2 | 64.1 | 65.5 |
| $L_{50}$ | 61.7 | 61.3 | 63.6 | 64.3 | 68.4 | 69.9 |
| $L_{10}$ | 66.6 | 64.9 | 67.4 | 68.6 | 73.7 | 75.3 |
| $L_5$ | 69.2 | 66.5 | 68.5 | 70.2 | 75.8 | 76.9 |
| $L_{Aeq}$ | 64.1 | 63.2 | 64.8 | 66.0 | 71.0 | 72.2 |
| Number of guests | | | | | | |
| All | 0 | 0 | 6 | 15 | 16 | 19 |
| Far | 0 | 0 | 4 | 12 | 10 | 12 |
| Near | 0 | 0 | 2 | 3 | 6 | 7 |

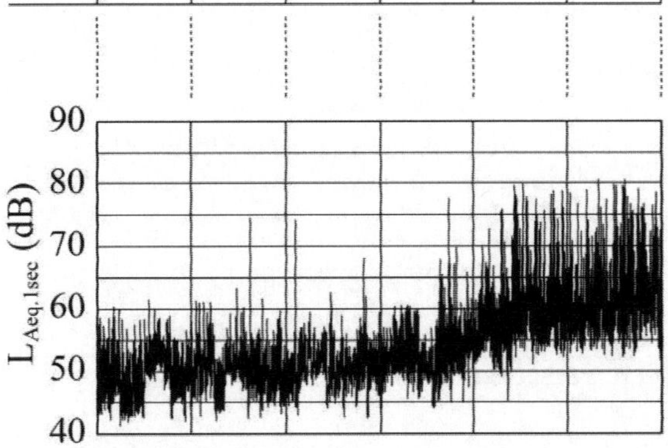


**Figure 3.** Sound pressure level of T1

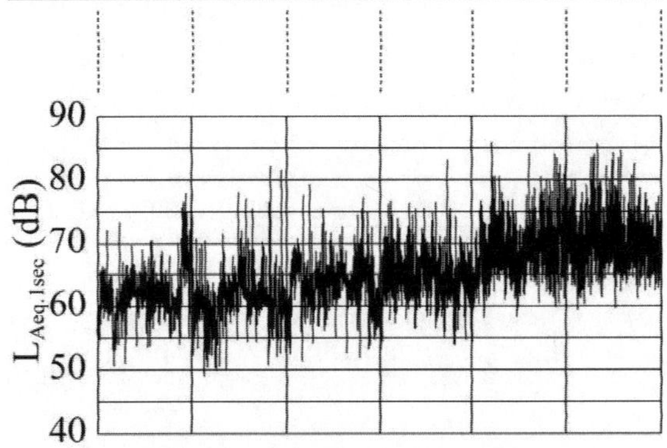


**Figure 4.** Sound pressure level of T2

Table 2 and Table 3 show the percentile sound pressure levels (Lx), equivalent sound pressure levels (LAeq), and numbers of guests in the tavern by 30 minutes, and Figs. 3 and 4 show the time series of 1 second LAeq. The major sound sources were mostly conversation and background music (BGM).

The SPLs of the period without guest could be regarded as the background noise level. The value of each tavern was 44 - 55 dB in T1 (17:30-18:00), and 54 - 69 dB in T2 (18:00-19:00) in 90 % range of time. This difference seems to come from a higher volume of BGM in T2 than that of T1.

There were rather few guests in the *near area* in the first 120 minutes in both of the taverns and the LAeq didn't change much and after that the LAeq raised by about 5 dB with the increase of guest in the near area. This result implied that the speech in *far area* didn't affect the noise level at the measurement point probably because the voice level was lower than that of BGM. This also implied that speech of several meter distant from a table couldn't be heard at the table. It is assumed that this increase of SPL was caused by conversation as the researcher observed that the volume of BGM was not raised with the increase of the guest. Moreover, it was suggested that the conversation voice level in T2 was louder than that in T1 because the LAeq in this busy period in T2 was 4-6 dB louder than that of T1 and the number of guest sat on the near chairs in T2 was 5 times less than those in T1.

### *3.3. Questionnaire survey*

Respondents answered "very important" and "rather important" were about a half in T1 and about 23% in T2 (Fig. 5) to the question of importance of privacy. This result suggested that a preference of a guest might reflect his/ her choice of a tavern. As for the evaluation of liveliness/ calmness of the tavern, the difference of evaluation between the two types clearly appears in the result, and more than half of the respondents in T1 evaluated "calm" (Fig. 6). The easiness of conversation was analysed by the number of other occupied tables at the answering time (Fig. 7). In the result T1 received better evaluation than T2 as a whole and the evaluation shifted to toward worse evaluation under the condition of three or more occupied tables. The question about distracting sounds were multiple choice questions from four kinds of sounds shown in Fig. 8. As a result, major answers were "surrounding conversation (45 %)" and nothing (41 %)" in T1 and "nothing (70 %)" and "BGM (20 %)" in T2. The result in T1 seemed to be a reflection of the preference of the respondents who put importance on the privacy. The answer to BGM in T2 was higher than that in T1 (13 %) and this is in accordance with the actual volume of BGM. Also, the high ratio "nothing" in T2 suggested a possibility that the BGM masked other sounds.

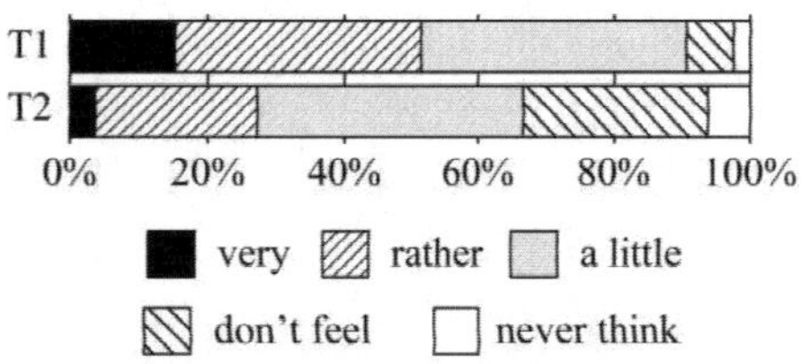


**Figure 5.** Importance of privacy in tavern

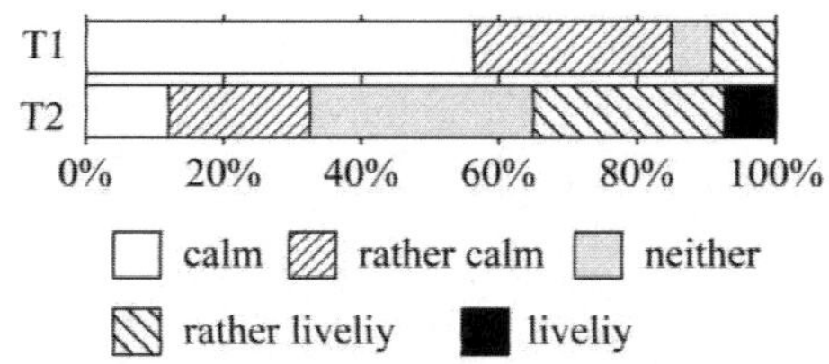


**Figure 6.** Evaluation of liveliness or calmness

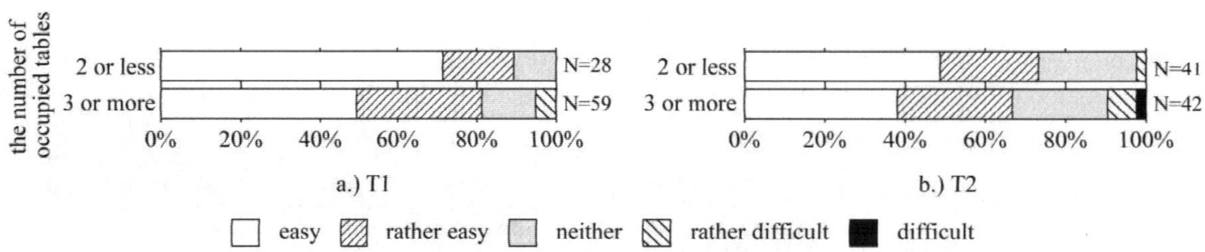


**Figure 7.** easiness of conversation

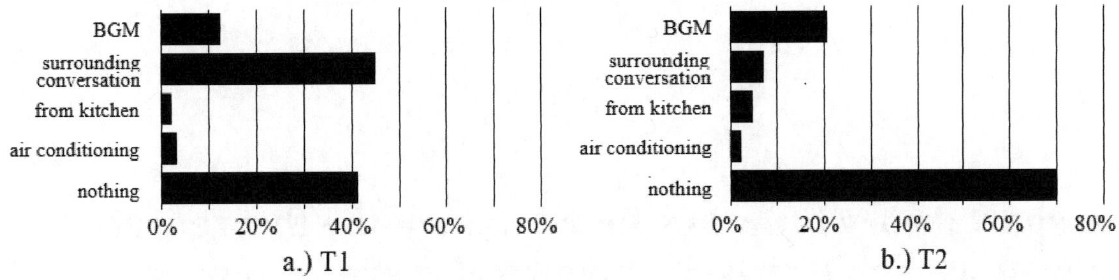


**Figure 8.** Kinds of distracting sound.

## 4. Conclusions

In this study, an acoustic measurements and questionnaire survey was conducted for two different types of taverns to clarify the actual condition of the acoustic environment in taverns. Major results are that the conversation level of the mono-space tavern was higher than the semi-separated room tavern probably because of the volume of the BGM and possibly affected by the longer room reverberation, and that the evaluation of importance of privacy may influence the guests' choice of the type of tavern.

In further studies, more investigation for actual conditions of tavern is necessary and, additionally, experimental studies on easiness of conversation such as balancing speech intelligibility between the table inside and between tables would be needed.

## 5. References

[1] Hiraguri Y, Ueda M, Kawai K and Yano 2014 T. Fundamental investigation of easy to talk at multigroup communication space *ICBEN 2014*

### Acknowledgments

We thank the taverns for their helpful cooperation in this research.

# Transport theory is the key foundation of the theoretical design of metamaterials-metamaterial is artificial phase transition

**W S Gan**[1]

[1] Acoustical Technologise Singapore Pte Ltd, 5 Derbyshire Road, #04-05, Singapore 309461,Singapore.

E-mail: wsgan5@gmail.com

**Abstract.** This is a revisit of my PhD thesis work is which I coined the term transport theory. In 1966. This is the first introduction of transport theory into condensed matter physics. Transport theory is today the key foundation of the theoretical design of materials. Since metamaterial is a new material, naturally it will fall under this category of work. Recently I discovered that metamaterial is in fact, artificial phase transition. Permeability, permittivity, effective bulk modulus, effective mass density are transport properties. The transport property also shows a singularity behavior at the critical point of phase transition in metamaterial same as that of the singularity behavior of transport property during condensation at phase transition. The successful fabrication of quantum metamaterial producing artificial superconductivity and metamaterial interpretation of high temperature superconductivity further confirms that metamaterial is artificial phase transition. Being artificial phase transition, one is able to control the degree of phase transition by tuning the resonance frequency of the transport properties.

## 1. PhD work revisited --introduction of transport theory into condensed matter physics

In 1966 W. S. Gan coined the term transport theory during his PhD work in the physics department of Imperial College London. His PhD thesis (1969, Imperial College London) Transport Theory in Magnetoacoustics [1] is the first to introduce transport theory into condensed matter physics. Since then transport theory has undergone tremendous development in linear transport theory, electronic transport theory, nonlinear transport theory, quantum transport theory etc. Today transport theory is the key foundation of the theoretical materials design [2]. It is the most important theory in condensed matter physics. The status of transport theory in condensed matter physics is equivalent to that of Yang Mills Theory [3] in particle physics. Also in 1967 Philip Warren Anderson and Volke Heine coined the term condensed matter physics when they changed the name of the solid state theory group to condensed matter physics group at Cavendish Lab, Cambridge. , to combine liquid state physics with solid state physics and to reflect the important role of phase transition. Today condensed matter physics group has the largest membership in the American Physical Society. Thus his PhD thesis also played a role in the founding of the field of condensed matter physics.

IOP Conf. Series: Journal of Physics: Conf. Series **1075** (2018) 012019 doi:10.1088/1742-6596/1075/1/012019

## 2. Discovery of metamaterial as phase transition –metamaterial is a branch of condensed matter physics

In this paper, my PhD thesis will be revisited. It will be illustrated that metamaterial in fact is a phase transition and metamaterial is a branch of condensed matter physics. The double negativity of permeability and permeability of electromagnetic metamaterial and double negativity of the effective bulk modulus and the effective mass density of the acoustic metamaterial is phase transition from the positive phase material to the negative phase material. Permeability, permittivity, bulk modulus and mass density are transport properties. Based on the backbone of the transport properties, one is able to discover new forms of metamaterials beyond the electromagnetic metamaterial and the acoustic metamaterial. Metamaterial is a branch of condensed matter physics Metamaterial is a transport phenomenon. Hence transport theory which describes transport properties and transport phenomena is the key foundation of the theoretical design of metamaterials. Metamaterial is a new material and transport theory is the key foundation of the theoretical design of new materials. Besides this, one does not have to use analogy to extend electromagnetic metamaterial to acoustic metamaterial. One can base on transport properties, the root of the problem.

## 3. Singularity behavior of transport properties at the critical point of phase transition

The description of metamaterial as a phase transition can be further confirmed by the singularity behavior of the transport properties of permeability [4], permittivity [5] and effective bulk modulus [6] during the critical point of phase transition, during the resonance frequency. (Please refer to the plots in Figs. 1, 2, 3, 4) There is a common behavior of the transport properties: permeability, permittivity, effective bulk modulus at the critical point of phase transition that at this critical point of resonance frequency they will have a sudden increase in value followed by a sudden drop to a huge negative value and then followed by a gradual rise in value in the negative region. The metamaterial is used in the phase transition of high temperature superconductivity further confirmed that metamaterial is a phase transition.

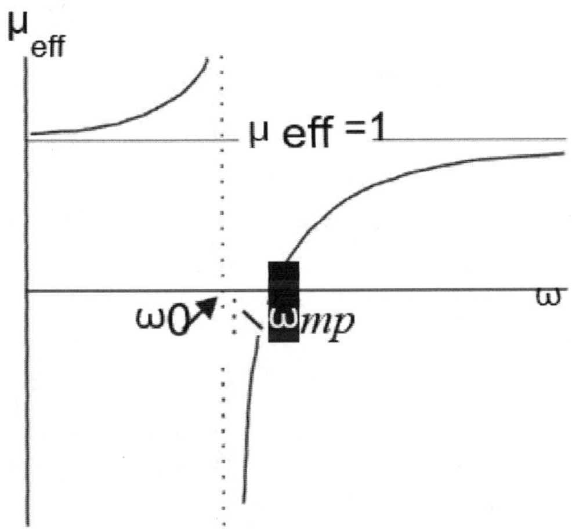


**Figure 1.** Singularity behavior of permeability in magnetism at the critical point of phase transition [4].

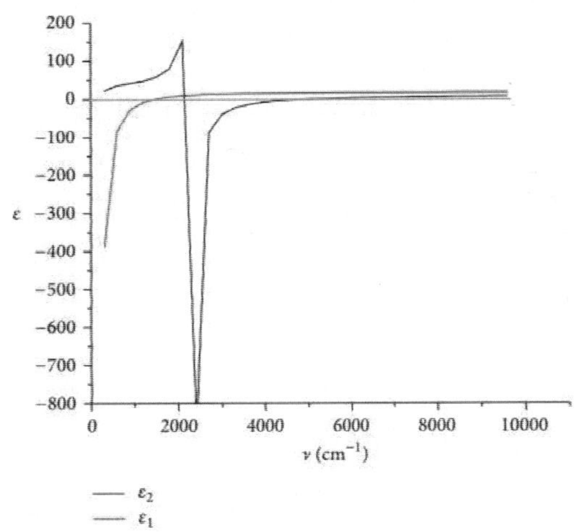


**Figure 2.** Singularity behavior of permittivity in high temperature superconductivity at the critical point of phase transition [5].

IOP Conf. Series: Journal of Physics: Conf. Series **1075** (2018) 012019 doi:10.1088/1742-6596/1075/1/012019

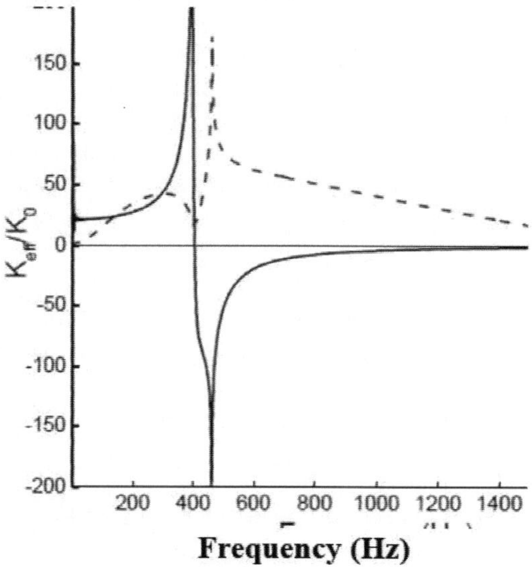


**Figure 3.** Singularity behavior of the bulk modulus at the critical point of phase transition or resonance frequency [6].

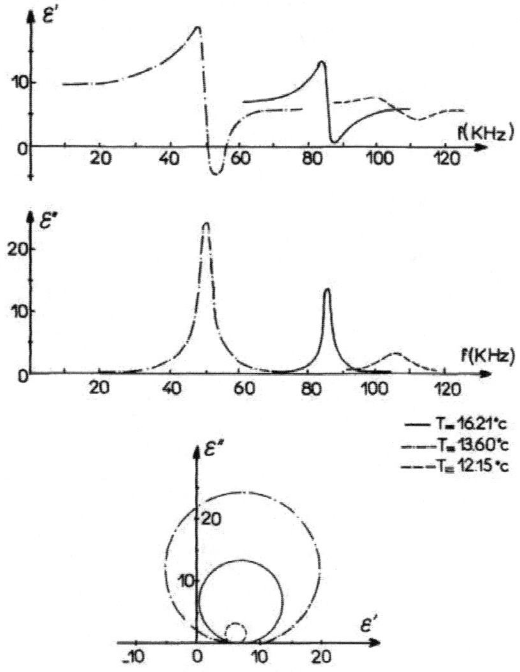


**Figure 4.** Typical piezoelectric resonance curves and corresponding Cole-Cole diagram for a low frequency acoustic resonance mode observed in the sample ∝ above and below the transition temperature = 13.43$^0$C [7].

## 4. Discovery of other forms of metamaterials beyond electromagnetic and acoustic metamaterials

In this paper, based on transport properties, two new forms of metamaterials are introduced. One is the artificial piezoelectricity and another is artificial ferromagnetism based on the transport properties of dielectric constant and dipole moment respectively. The dielectric constant of the piezoelectricity is found to have the singularity mentioned in the previous paragraph [7]. An immediate fabrication of this form of metamaterial is to use the split ring resonator (SRR) as the unit cell of the piezoelectric metamaterial. The material of the unit cell will be made of piezoelectric material. This allows one to manipulate and control the piezoelectricity. Piezoelectricity provides the physical basis for almost all practical applications of acoustic fields. This is because they provide an effective means for electrically generating and detecting acoustic vibrations.

I am in the process of investigating the singularity behavior of the dipole moment, the transport property of ferromagnetism versus frequencies during phase transition, the critical point or the resonance frequency.

## 5. Discovery of other forms of metamaterials beyond electromagnetic and acoustic metamaterials

The double negativity of electromagnetic metamaterial and acoustic metamaterial and the singularity of the dielectric response function of high temperature superconductivity are first examples of the use of transport properties to discover new forms of metamaterials.

### *5.1. Artificial elasticity*

The singularity behavior of the effective bulk modulus at the resonance frequency or critical point of phase transition can be exploited to fabricate acoustic metamaterial [6]. Here there is a sudden increase of the effective bulk modulus to a very high value at the resonance frequency followed by a sudden drop to very high negative value and a gradual increase in value in the negative region. This singularity behavior can be exploited to fabricate acoustic metamaterial using the geometric structure of split ring resonator (SRR) with the unit cell made of elastic material. This enables the control and manipulation of the elasticity of the material. This is artificial **elasticity.**

### *5.2. Artificial magnetism*

Pendry et al [4]'s paper on magnetism from conductors and enhanced nonlinear phenomena is an example of artificial magnetism. Here the permeability which is a transport property shows a singularity behaviour at the critical point of phase transition with a sudden rise at the resonance frequency to positive infinite value followed by a sudden drop to negative infinite value and then a gradual rise in the negative region. The negative values of permeability is used to fabricate the negative electromagnetic metamaterial using the SRR geometric structure as the unit cell of the metamaterial and the material of the unit cell is made of magnetic material. This enables the achievement of an electromagnetic metamaterial experimentally and the manipulation and control of magnetism. This is artificial magnetism.

### *5.3. Artificial high temperature superconductivity*

Smolyaninov et al [5] proposed that high temperature superconductivity can be achieved using metamaterial. This is because dielectric response function governing electron-electron interaction can be used to increase the critical temperature of high temperature superconductivity according to Kirzhnits et al [8]. It is found that the dielectric response function which is a transport property has singularity behavior (hyperbolic shape) at the resonance frequency or the critical point of phase transition. The dielectric response function versus frequencies plot shows a sudden increase to high positive value and then a sudden drop to large negative value followed by a gradual increase in value in the negative region. This behavior can be exploited in the fabrication of high temperature superconductor by using a geometric structure of SRR for the unit cell made of high temperature superconductor. This enables one to manipulate and control high temperature superconductivity. Hence the name high temperature superconductivity metamaterial.

### *5.4. Artificial piezoelectricity*

Piezoelectricity is an important phenomenon in acoustics. It provides the physical basis for almost all practical applications of acoustic fields. This is because they provide an effective means for electrically generating and detecting acoustic vibrations. It is found by Legrand et al [7] that the permittivity also has a singularity behavior at the resonance frequency or point of phase transition with a sudden increase to a very high value followed by a sudden drop to negative value then a gradual increase in value in the negative region. This singularity behavior can be used to fabricate artificial piezoelectricity by using a SRR as geometric structure of the unit cell made of piezoelectric material. This will enable the manipulation and control piezoelectricity. This is artificial piezoelectricity.

### *5.5. Artificial ferromagnetism*

In ferromagnetism, according to Joseph Mayer [9], there is some sort of mathematical singularity at the condensation point. The transport property of ferromagnetism is magnetization or dipole moment. It is of interest to investigate the singularity behavior of the dipole moment at the critical point of phase transition or condensation point. There will be a temperature dependence of the dipole moment besides the frequencies dependence. One will have to plot the dipole moment versus the frequencies at a series of temperatures. There will be no singularity behavior of the dipole moment versus frequencies plot at other temperatures. However, at the critical temperature of second order phase transition, there will be a hyperbolic shape behavior of the dipole moment versus dipole moment versus frequencies plot at other temperatures. However, at the critical temperature of second order phase transition, there will be a hyperbolic shape behavior of the dipole moment versus frequencies plot similar to that of the cases of the artificial elasticity, artificial magnetism, and artificial piezoelectricity.

## 6. Metamaterial as phase transition as breakthrough to a new world of artificial materials

Double negativity electromagnetic metamaterial and double negativity acoustic metamaterial are manifestations of artificial magnetism and artificial elasticity. This will be a starting point for the exploration of various new artificial materials based on metamaterial and phase transition. Immediate examples are artificial piezoelectricity and artificial ferromagnetism. This will open to a new world of new materials based on other transport properties besides permeability, permittivity, effective bulk modulus and effective mass density.

## 7. Phase transition approach to metamaterial

Phase transition approach to metamaterial will enable a deeper understanding of metamaterial. The advantage of using metamaterial in phase transition is to enable to adjust the degree of phase transition from full phase transition to partial phase transition by tuning the resonance frequencies of the transport property as phase transition arises from the singularity behavior of the transport property at the critical point.

## 8. Conclusions

Transport theory has come a long way in tremendous development in condensed matter physics since the term was coined by me in 1966. It has now become the key foundation of theoretical materials design. Hence it has become the most important theory in condensed matter physics. Since metamaterial is a new material, it will naturally fall under the category of design using the transport theory.

## 9. References

[1] Gan W S 1969 Transport Theory in Magnetoacoustics: *PhD thesis* (Imperial College London: Google Book)
[2] Shenoy V B 2006 Transport Theory, *Lecture notes* (SERC School on Condensed Matter Physics)
[3] Yang C N and Mills R 1954 Conservation of isotopic spin and isotopic gauge invariance, *Phys.* Rev 96 (1) pp 191-195
[4] Pendry J B, Holden A J, Robbins, D J and Stewart W J 1999 Magnetism from conductors & enhanced nonlinear phenomena *IEEE Transaction on Microwave Theory and Techniques* vol 47 issue 11 pp 2075-2084
[5] Smolyaninov and Smolyaninova V N 2014 Is there a metamaterial route to high temperature superconductivity *arXiv.org,arxiv.1311.3277*

[6] Sharma B, and Sun C T Acoustic metamaterial with negative modulus and double negative structure *arxiv.org/pdf/1501.02833.5*
[7] Legrand J F 1982 Ferroelastic and ferroelectric phase transition in a molecular crystal: tanance 2 Phenomenological model and piezoelectric resonance study of the soft acoustic mode, *J. Physique* **43** pp 1099-1116
[8] Kirzhnits D A, Maksimov E G and Khomskii D I 1973 The dependence of superconductivity in terms of dielectric response function *J. Low. Temp. Phys.* **10** 79
[9] Mayer J E and Mayer M G 1940 Statistical Mechanics (Wiley)

IOP Conf. Series: Journal of Physics: Conf. Series **1075** (2018) 012020 doi:10.1088/1742-6596/1075/1/012020

# Acoustic emission technique as an adaptive biomarker in integrity analysis of knee joint

**T I Khan[1], M Kusumoto[1], Y Nakamura[1], S Ide[2] and T Yoshimura[3]**

[1] Department of Advanced Technology Fusion, Saga University, Saga 840-8502, Japan.
[2] Department of Orthopaedic Surgery, Saga University, Saga 849-8501, Japan.
[3] Daishin co., Ltd, Higashioka-machi, Saga 840-2223, Japan.

E-mail: khan@me.saga-u.ac.jp

**Abstract.** Acoustic emission (AE) technique has been applied as an adaptive biomarker for evaluating the disorder of knee joint. Integrity analysis of knee joint involves a detail study of several anatomical parts of knee joint like bones, cartilage, tendons etc. Any damage of these anatomical parts causes several knee diseases like osteoarthritis (OA). The incidence of knee OA, a widely manifested knee disease, particularly in aging society, increases due to some damages in the cartilage of knee joint. The major concern of this disease (OA) is its incurability at its matured stage. However, early detection for adopting appropriate measures can reduce the risk of this disease [1], [2]. The present investigation focuses on the dynamical behavioral characterization of knee joint for its integrity analysis with acoustic emission (AE) characterizing parametric features. AE signals have been collected from different positions of tibia, patella, femur etc. by AE sensors with adaptive frequency bands for getting sufficient information about the condition of cartilage in knee joint. Data has been collected from the participants with different age groups without any knee problems as well as participants with knee diseases. All data have been clarified effectively to identify the proposed technique as an adaptive biomarker for monitoring knee condition in early stage of OA effectively.

## 1. Introduction

The knee joint is anatomically made up of three bones and various ligaments. Femur (the thigh bone), tibia (the shin bone), and patella (the kneecap) are the major components of this joint [1]. Knee joint provides necessary supports to the skeleton for allowing them to be flexible in movements. The motion control as well as the protection of the knee joint is done by several muscles and ligaments [2]. These ligaments of the knee ensure that the body-weight must be transmitted through the knee axis for minimizing the amount of wear and tear on the cartilage inside the knee [3].

Bones in knee joint are receiving cushiony supports by cartilage, synovial membrane and fluid inside the joint [4]. Moreover, muscles and ligaments provide the appropriate forces and strength to this joint for its stable movements. However, due to getting ages, the quality of bones including the cushiony items are degenerated. Furthermore, the quality as well as the quantity of muscle and synovial tissues are also degraded due to aging and thus, the balance movements of knee joint are decreased. Therefore, the surface roughness of the articular cartilage is increased and the risk of happening the osteoarthritis (OA) is also increased [5], [6].

The dynamic analysis of knee joint consists of the integrity analysis of knee's anatomical parts for several movements of sitting and standing. Major objectives of this study are to develop a simple bio-marker for clarifying the initiation of knee disorder responsible for OA due to aging, and also to diagnosis the matured OA in knee joint. The disease of OA increases with age due to some damages in cartilage of knee joint [7], [8].

Potential methods like X-ray, magnetic resonance imaging (MRI) etc. are presently using for clinical diagnosis of knee diseases. However, externally inserted high energy to the body for diagnosis of these technique and their static sensitivity with high cost make them unpleasant to the patient. In the contrary, proposed acoustic emission (AE) technique is considered safe and user-friendly diagnosis of knee joint with low cost in dynamic analysis mode.

In the present research, integrity analysis of knee joint is planned to be conducted by applying AE technique to young aged, middle aged and old aged volunteers with no knee pain or knee problems (self-declared) as well as knee patients in dynamic movements (sit-stand-sit) of their knees. In the next section, experimental methodology is discussed, after which results and discussion and finally conclusions and references are presented.

## 2. Experimental Methodology

According to the objectives of the present research, four AE sensors are attached to the knee joint. For avoiding noise from muscle, tendon etc. sensors are placed to nearest to the knee bones as two to femur and tow to tibia according to the experimental observations [9]. The positioning of sensors is mentioned in Figure 1(a) by numbering as (1) ~ (4) to a knee model (3B Scientific GmbH, Germany). AE sensors are attached to the mentioned positions of knee joint with high elastic medical tape for avoiding undesirable noise during sit-stand-sit movements. Furthermore, coupling gel is used between the surface of the sensor and the contact place of the knee for keeping continuous contact of the sensing surface to the anatomical site of the knee. For getting an angular position of the knee during the movement, electronic goniometer is attached to the knee with double-stick tape. In one set of movement (sit-stand-sit) the goniometer angle is recorded as 180 degrees. Ten sets of movement are considered as 1 cycle, while 1 minute is taken as intermittent rest time. For AE signals 1MHz and for goniometer 100 Hz are used as sampling frequencies.

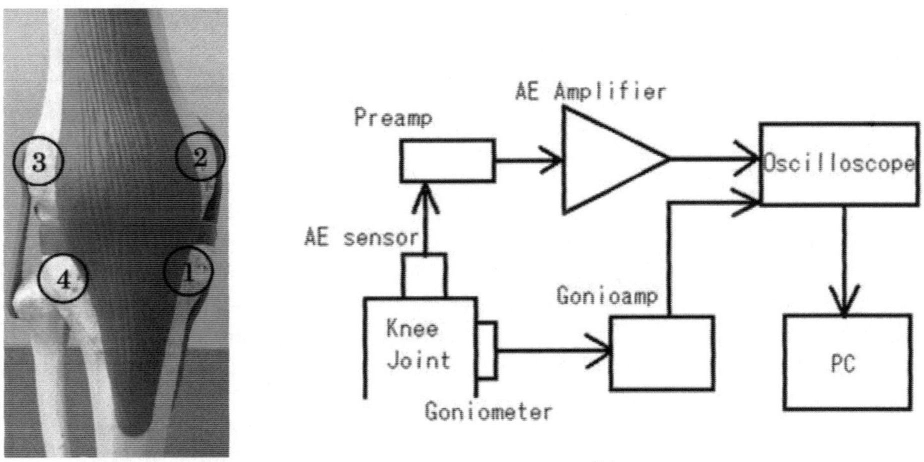


(a) (b)

**Figure 1.** Experimental apparatus of AE integrity analysis of knee joint. (a) AE sensor positions, (b) schematics of AE signal acquisition set up.

Acquisition of AE signals from knee joint has been designed according to the schematic views as shown in Figure 1 (b). An array (2x2) of AE sensors (R6□, Physical Acoustics Corporation) with an operating frequency range of 35 to 100 kHz and resonant frequency of 55 is used in the experiment. The sensors are connected to a digital oscilloscope through pre- and main amplifiers and finally the data has been transferred to the personal computer (PC). For the acquisition of angular movements, a two channel goniometer (SG150, Biometrics Ltd.) is also connected to the oscilloscope as shown in the figure and the data has been recorded into the PC similarly.

## 3. Experimental Results and Discussion

Experimental results for the integrity analysis of knee joint have been focused in two major divisions. One division consists of different aged participants without any self-realizing knee problems. Another division has been consisted with knee patients who have been taking knee surgery for their knee damage with osteoarthritis or other similar knee disease. Again, participants without any knee problem have been grouped into three based on their ages. For example, young participants with below the age of 30 years old, are defined as 20 yrs. group and is coded as A. Similarly, middle aged people with near 40 years old or above, however, below 60 years old are grouped into 40 yrs. group and is coded as B and the participants with 60 years or above are grouped into 60 yrs. and is coded as C. All of these volunteer participants are self-declared participants with no knee problems. Major objectives of these experiments are to identify the bio-marker for the identification of knee problem initiation stage. However, the objectives of experiments with group D participants is to diagnosis of knee problems with damage intensity and position as well. Detail specifications of these participants are summarized in Table 1.

**Table 1.** Participants Specification and Characterization.

| Participants | Code name | Spacing |
|---|---|---|
| 20 yrs. Group | A | 10 |
| 40 yrs. Group | B | 14 |
| 60 yrs. Group | C | 14 |
| Patients of knee problem | D | 11 |

Experimental results are evaluated based on the damage intensity and the intensity has been identified with maximum amplitude of AE hits, based on percent of occurrence of AE hits, and the concentration angle of AE hits in a complete cycle of knee extension-flexion dynamic motion generated by one sit-stand-sit movement. Detail explanation of these biometric parameters are given below.

### *3.1. Maximum amplitude distribution as bio-marker of knee evaluation*

As mentioned in Table 1, 49 volunteer participants joint the present experiment, out of which 10 people for A, 14 people as each group of B and C, and 11 patients as D. The participants of A to C are self-declared healthy people and never went to the hospital for their knee pain. However, the participants of D group are those who possess knee disease of OA. They took knee operations following day of joining to our experiments. As the intensity of AE signals depends on the intensity of transient elastic wave generated from sudden redistribution of stresses, in the present experiment, AE signals are generated similarly due to the frictional effects of cartilage during sit-stand-sit movements of the people. Therefore, surface roughness is a major criterion in increasing the signal amplitudes or increasing signal intensity. Thus, in the present experiment, participant who has damage to knee cartilage (OA patients), generates increased friction force and thus, shows the significantly increased amplitudes in AE hits during sit-stand-sit dynamic motion. Similarly, participants of C group, also exhibit increasing signal amplitude compared with 20 yrs. or 40 yrs. groups of participants, however, it is much less than D group. Similar significance is also found in group B, although, the increasing level of amplitude is very small (around 2 dB). Thus, it can be concluded that the surface condition of knee cartilage has been started degrading in group B, when the condition of young people is considered healthy knees. Of course, a lot of observations regarding the declaration of threshold values (for considering young people as standard healthy knee) are needed to be investigated with many participants. The total results of the maximum amplitude distribution for A, B, C, D participants are shown in Figure 2. Average values of all participants in each group are represented in the graph. Increasing of signal amplitudes with increasing the participants ages are indicated by arrow indication as well.

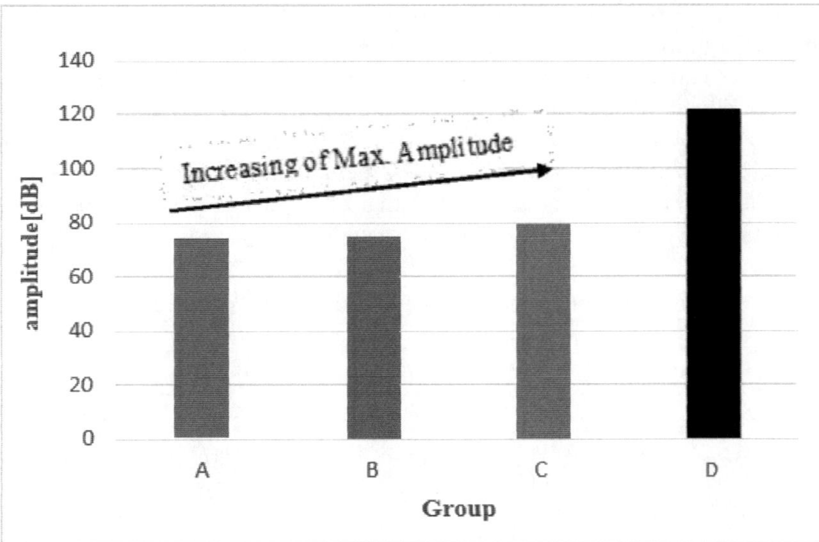


**Figure 2.** Maximum amplitude distribution of different aged participants and OA patients.

*3.2. Maximum amplitude distribution as bio-marker of knee evaluation*

The percent occurrence of AE hits is defined as the percent of the ratio of total required sit-stand-sit movements for a fixed number of AE hits in one cycle. As explained in above that the rate of transient AE hit generation increases due to surface condition of knee cartilage decreases. Experimental results show that increasing of aging increasing the percent occurrence of AE hits and it increases more for OA patients. Thus, according to the observation, percent occurrence of AE hit works as an acceptable bio-marker for knee integrity analysis.

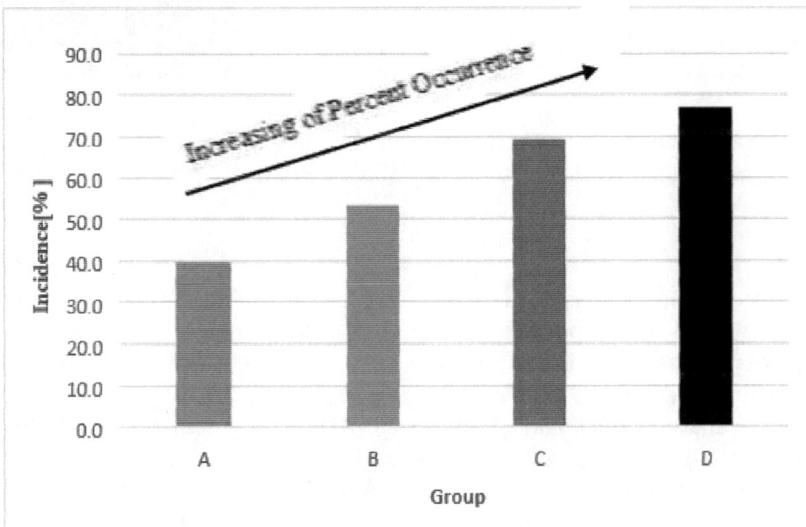


**Figure 3.** Percent occurrence distribution of AE hits for different aged participants and OA patients.

*3.3. Concentrated distribution of AE hits as bio-marker of damage position in knee joint*

The position of cartilage damage of knee joint can be identified by analyzing its angular position as well. Therefore, in the present research angular position of knee movements has been clarified by goniometer. MRI technique has been applied and examined the damage position of OA participants (patients). It is found that AE hits are also concentrated to the position of knee joint. It is found in Table 2 and in Figure 4 that AE hit data are concentrated than other non-damage places. In Table 2, it is found that 50% AE hits concentrated to 0 to 30 degrees and 90 to 60 degrees of knee surface. MRI data also shows the damage location at 60 to 90 degrees (at 0 to 30 degrees, the damage location is obscured due to standing image). Based on this observation, it is also found that concentrated distribution of AE hits can be considered as a bio-marker in knee integrity analysis as well.

**Table 2.** Percent concentration of AE hits in knee angles.

| Angle range ( ° ) | Occurrence ( % ) |
|---|---|
| 90 - 60 | 50.0 |
| 60 - 30 | 0.0 |
| 30 - 0 | 50.0 |

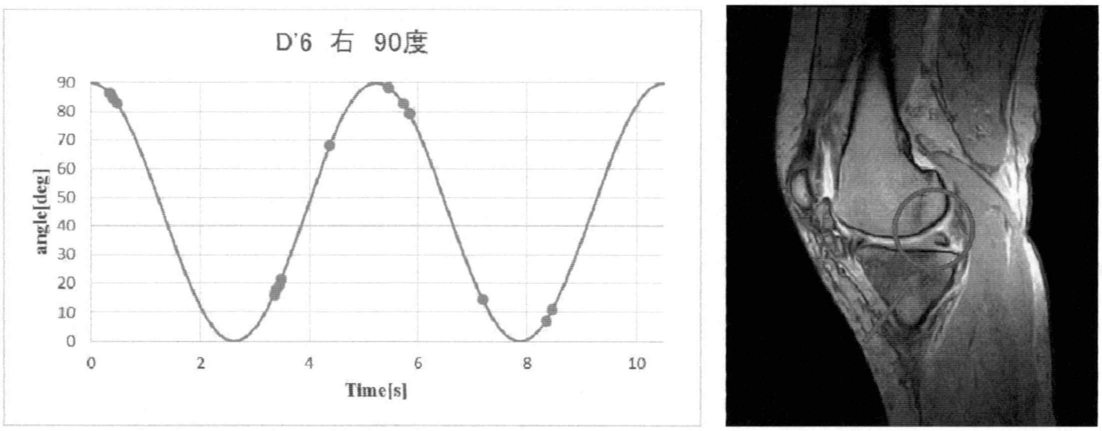


(a) (b)

**Figure 4.** Concentrated distribution of AE hits and the knee integrity evaluation by MRI imaging. (a) Indicates the concentrated AE hits distribution showing damage positions, (b) MRI imaging of an OA knee.

## 4. Conclusions

The integrity of knee joint has been identified by AE technique. Three bio-markers for knee osteoarthritis (OA) diagnosis based on AE technique have been explained with experimental validations. According to the objectives, two major findings have been clarified by the proposed system. Degeneration of knee cartilage by aging have been successfully identified by the proposed AE technique. Furthermore, integrity analysis of damaged knee by OA is also discussed satisfactorily. Maximum amplitude distribution of AE hits, percent occurrence of AE hits and concentrated distribution of AE hits are presented as three successful bio-markers in the AE diagnosis system of knee problems including OA. Damage places of knee joint are also identified by clarifying AE hits with respect to the angular position evaluation by goniometer. MRI examinations are also employed successfully in the present research for the validation of the proposed technique of integrity analysis of knee problems by AE technique.

## 5. References

[1] Robert F, Engebretsen A H, Thuan V, Steinar J, Fred A W and Lars E 2007 *J. Bone Joint Surg Am*, **89**, 2000
[2] Shrivastava S and Prakash R 2009 *J. Biomedical Science and Engineering* ***2*** 144
[3] Ian S 2004 *J Sports Med* ***38*** 526
[4] Hurley M V 1999 *Rheum Dis Clin North Am* **25** 283
[5] Eckhoff D G, Bach J M Spitzer V M et al. 2005 *J. Bone Joint Surg Am* **87**(2) 71
[6] Archibeck M J White R E Jr 2002 *J. Bone Joint Surg Am.* ***84*** 1719
[7] Thienpont E, Schwab P E and Fennema P 2014 *Bone Joint J.* ***96***-B 1052
[8] Guccione A A, Felson D T, Anderson J J 1994 *Am. J. Public Health* **84** 351
[9] Khan T I, Harino Y 2016 *J. Multimodal User Interfaces* (springer) **10** 319

## Acknowledgments

This research is supported by the Grants-in-Aid for Scientific Research (KAKENHI 17K06266) from the MEXT of the Government of Japan. Authors are grateful to all volunteer participants and staffs to the research.

IOP Conf. Series: Journal of Physics: Conf. Series **1075** (2018) 012021 doi:10.1088/1742-6596/1075/1/012021

# Acoustic signal of Javanese 56 cm brass kempul

**M H W Budhiantho[1] and D Salim[2]**

[1] Department of Computer System Engineering, Matana University, Gading Serpong, Tangerang, Indonesia

E-mail: matias.matana@gmail.com

**Abstract.** The 56 cm Brass Kempul Gong is regarded to produce better musical sound compared to its sibling one that made of iron. This study aims to investigate the spectral and temporal properties of that Gong. Acoustic measurements were conducted and sounds analyzed using the PC program, ARTA. Both frequency and time domain signals were explored and compared to a similar kempul made of iron. The fundamental frequency, which decays much more slowly than the other harmonics, starts with a lower, increasing frequency. The wave-like sound of the Gong is due to signal behavior that resembles the primary and secondary mistuned beats of its second and third harmonic frequencies.

## 1. Introduction

Kempul is the second largest Javanese gong which hang in a crossbar supported by two side posts as shown in Figure 1. A kempul may be crafted from iron as well as brass. Traditionally a good gamelan (the metal section) was made of brass. However because of budget concern the iron made gamelan became popular as well. [1], [2] It is believed that brass kempul has musical sound than the iron sibling. There have been lacks of studies in the past that attempt to describe the spectrum of gongs both theoretically and experimentally to better comprehend such analyses. Previously we have investigated an iron made kempul gong [3-6].

This paper considers the spectral and temporal properties of brass kempul to examine the spectral harmonics and partials of theses gongs, and how they are put together to produce the continuous beats.

Generally a gong has about a dozen prominent exponentially decaying partials, with some component frequency ratios that closely correspond to harmonics and other that are inharmonic. [7-10] Many of the partials have a slow amplitude and frequency modulation of a few Hertz, and a faster modulation around 20 Hz resulting from multiple components falling into the same harmonic bin. The studies in [12-13] indicated that gongs show a pronounced nonlinear behavior. The pitch of the larger gong glides downward as much as three semitones after striking, whereas that of the smaller gong glides upward by about two semitones. Rossing et al. [14] investigated some vibrational modes of the larger gongs. Some of the modes were confined pretty much to the flat inner portion, some to sloping shoulders, and some involve considerable motion in both parts. When the gong was hit near the center, the central modes clearly dominate the sound. When the gong was hit lightly on the shoulder, the lowest mode is heard. It was also found that the Gong Kempul possesses a degree of harmonicity, making it sounds with targeted pitch.

**Figure 1.** The gong kempul.

## 2. Experimental Investigation

### *2.1. Measurement Setup*

A controlled gong striker was utilized to exert a controlled impact force upon the kempul's boss, as seen in Fig. 2. A measurement ECM 8000 condenser microphone acquired the acoustic signal by near field measurement from behind the boss, externally powered by a Samson phantom power. A sound card then interfaced and digitized this signal in order that computers are able to recognize. We recorded and analyzed the 48 kHz sampled acoustic signal using ARTA PC Software.

**Figure 2.** Gong Striker.

### *2.2. Brass kempul partials*

From the brass kempul smoothed spectrum we found with strong consistency from 7 trials that there were at least 9 partial in a span from the fundamental to the third harmonic (91.59 Hz to 274.84 Hz) as shown in Figure 3 and Table 1, compared to the only 6 partials of the iron kempul measured previously [3]. The harmonic partials interestingly be equally spaced with three inharmonic partials in between. Regularity of the harmonic partials leads to the clearer pitched sound of the brass kempul. This may explain why brass kempul has better musical sound compare to its iron sibling.

IOP Conf. Series: Journal of Physics: Conf. Series **1075** (2018) 012021 doi:10.1088/1742-6596/1075/1/012021

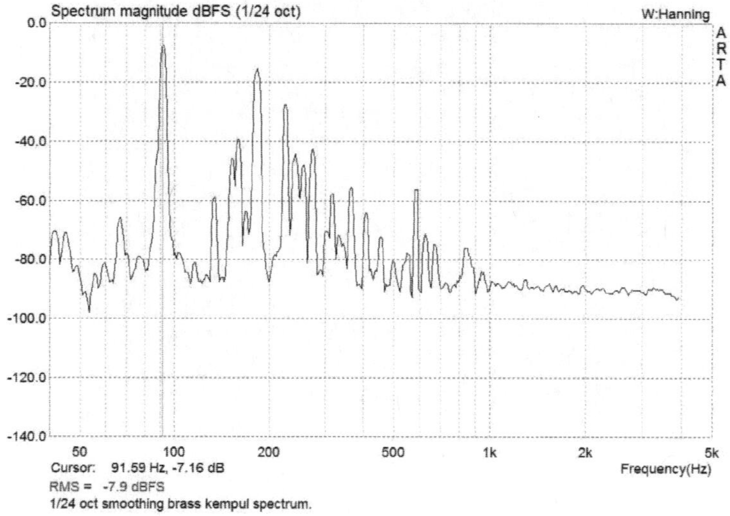


**Figure 3.** 1/24 oct smoothed brass kempul spectrum.

**Table 1.** A slightly more complex table with a narrow caption.

| Partials (Hz) | Harmonics (Hz) | Freq. ratio to fundamental | Primary beats (Hz) | Mistuned harmony beats (Hz) |
|---|---|---|---|---|
| 91.59 | Fundamental | | | |
| 132.92 | | 1.45 | 41.33 | |
| 151.44 | | 1.65 | 18.52 | |
| 158.65 | | 1.73 | 7.21 | |
| 182.44 | 2nd | 1.99 | 23.79 | 0.74 |
| 224.97 | | 2.46 | 42.53 | |
| 241.24 | | 2.63 | 16.27 | |
| 256.29 | | 2.80 | 15.05 | |
| 274.84 | 3rd | 3.00 | 18.55 | -0.07 |

### *2.3. Mistuned Harmonics*

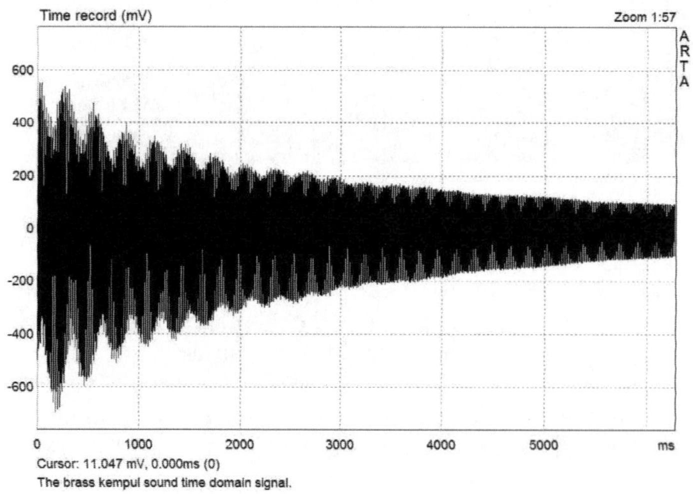


**Figure 4.** The brass kempul sound time domain signal.

IOP Conf. Series: Journal of Physics: Conf. Series **1075** (2018) 012021 doi:10.1088/1742-6596/1075/1/012021

The slowly decaying sound of the kempul that can be seen in Figure 4 opens up the possibility of those partials to interfere, resulting in various beats. The harmonic and or inharmonic partials that occur in fairly close frequency beat together and form the roaring sound which is often associated with Bima's laughter (Bima is one of the Pandawa brothers in the story of Mahabarata).

There are two kinds of beats, i.e. primary and second–order or secondary beats. The primary beats occur between closely existing partials in the spectrum, while the secondary beats occur between a tone and its mistuned harmonics [7].

When there are two discrete tones with frequency $f_2$ higher than $f_1$ and $n$ is an integer number, then those two tones will produce a beat frequency of $f = f_2 - nf_1$. For $n = 1$, we obtain primary beat, for $n>1$ we obtain second-order or mistuned harmonic beats. Since, if $f_2 = n f_1$ we get exact harmonically related tones.

Primary beat:

$$\text{delta} f = f_2 - f_1 \quad (1)$$

Second-order mistuned $n$ harmonic beats:

$$\text{delta} f = f_2 - nf1 \quad (2)$$

A beat is perceptible as a fluctuating sound if its frequency is less than 10 Hz [8]. And, in the case of this brass kempul, we found one primary beat of 7.21 Hz and two mistuned 2nd harmonic at 0.74 Hz and 3rd harmonic at – 0.07 Hz.

We investigate further into the kempul wave form at the beginning, mid and tail section of the signal. We found that the signal repetition period drifted from 87.91 Hz to longer lasting 89.89 Hz that span approximately 1.98 Hz.

Looking at the energy decay curve as shown in Figure 5 we can see that energy fluctuation was prominent in the 250 Hz octave band with periodic peak interval of 598.5 ms or, 1.6 Hz in frequency. This one octave band filter contained the 2nd and 3rd harmonics of the kempul signal.

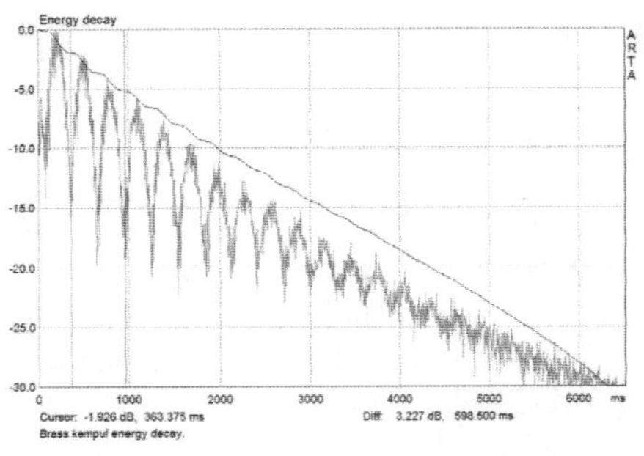


**Figure 5.** Brass kempul energy decay.

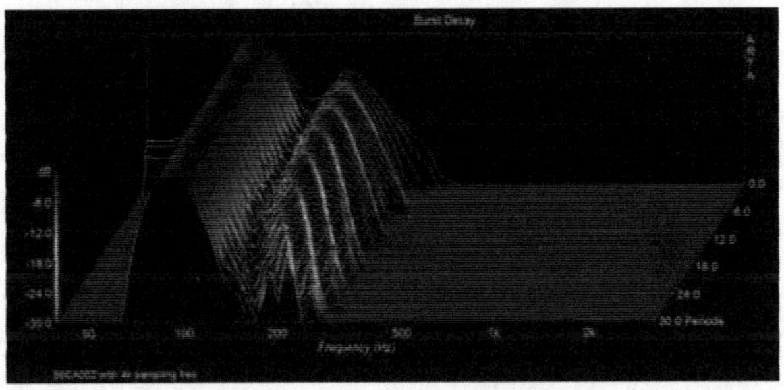

**Figure 6.** Brass kempul burst decay.

Figure 6 shows the burst decay graph to investigate the most prominent partials along a certain time intervals at the opening and tail of the sound, respectively. A complex Morlet wavelet analytic signal is used in convolution with kempul time domain *wavelet scalogram*, represents the envelope of the shaped burst response decay. A complex Morlet wavelet analytic signal is defined as:

$$w(t) = e^{\frac{-t^2}{\tau^2}}(\cos\omega_0 t + j\ \sin\omega_0 t) \quad (3)$$

The Fourier transform of the Morlet wavelet is equal to:

$$W(\omega) = e^{-(\omega-\omega_0)^2\tau^2/4} \quad (4)$$

It can be seen that in Figure 6, the fundamental frequency at 91.59 Hz decays much more slowly than the other harmonics and increases in dominance as the other partials weaken in decaying tone. We can also see that signal energy 2$^{nd}$ and 3$^{rd}$ fluctuate and decreasing with time.

## 3. Comparison to iron kempul

A short impulse can be represented as the instantaneous and gentle strike on the boss of the kempul. Thus, an impulse-generated response was recorded to analyze in the time domain. Figure 4 shows the impulse response of the kempul within about 8000 ms in which the sound gradually vanishes. The time plot shows fluctuation indicating at least seven prominent peak on the signal (Javanese: ombak pitu).

With regard to the behaviors of the iron kempul that we studied previously [3-6], The brass kempul acoustic sound as shown in Figure 4 clearly has longer sustain sound compares to its iron sibling as depicted here in Figure 7 for a direct comparison.

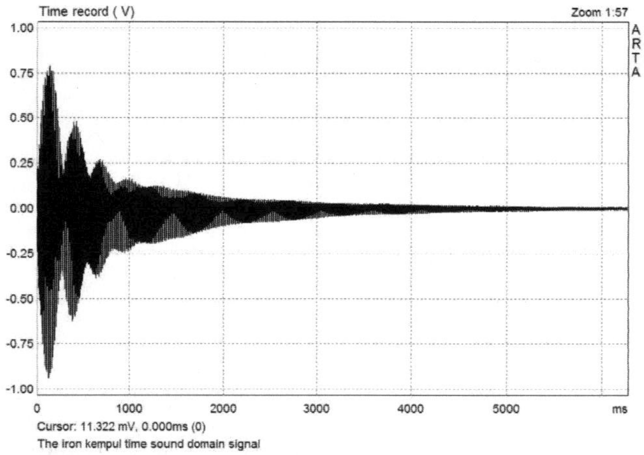


**Figure 7.** The iron kempul sound time domain signal [3].

While wave sound of the brass kempul has approximately 7 waves compare to approximately 3 wave only produced by the iron one.

Harmonic partials of brass kempul were more regularly spaced between inharmonic partials, this produced better pitched sound compares to the iron kempul that we studied previously.

The burst decay graph of brass kempul in figure 6, shows that its energy fluctuation was occur at combined 2$^{nd}$ and 3$^{rd}$ mistuned harmonics that finally sifted at the 3$^{rd}$ harmonic. The iron kempul burst decay was related mainly to the 2$^{nd}$ mistuned beat, as can be seen from Figure 7 below.

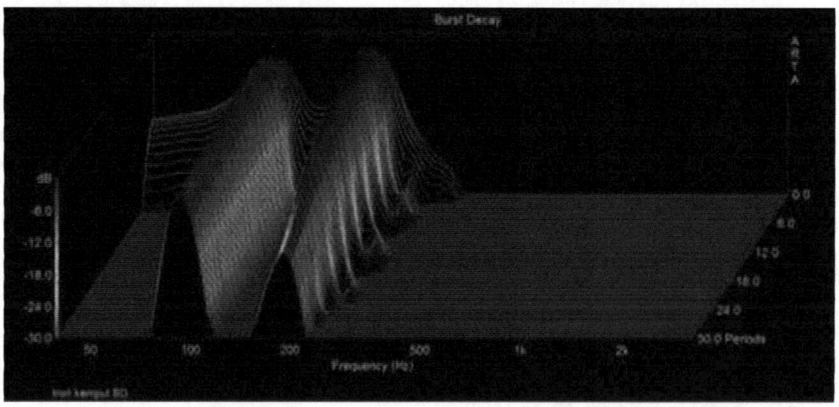


**Figure 8.** Iron kempul burst decay [3].

In the effort of tracing the origin of the wave sound this work found that the mistuned 2nd harmonic frequency that was 0.74 Hz does not agree with the fluctuation frequency in the energy decay that was around 1.6 Hz.

It may be that the kempul itself can be considered as non-linear system. A nonlinear system may produce higher harmonic of its input excitation. If we considered 0.74 Hz as the excitation input then 1.6 Hz is in the vicinity of its 2nd harmonic. And, that human tactile beats perception becomes more sensitive as the vibration [9] frequency of stimulation increases.

Perception of beats by human hearing and brain are another area that need to be study since. Hearing is a sensation and just like every other sensation it is an esoteric process. One is not able to decouple and independently vary any of the variables of interest (however those might be defined) and even if one could, the principle of superposition, in general, does not apply [10-11].

## 4. Conclusions

The 56 cm brass kempul has a nature of harmonicity with fundamental frequency at 91.59 Hz, while the second, third, mistuned harmonics are, respectively, at 182.44 Hz, 274.84, with three partials lie between each harmonics. The primary and mistuned harmonics beats together resulted in a kempul *ombak pitu* (7 waves) sound. This experimental investigation shows that brass kempul produces longer sustaining and cleaner pitched therefore has partially support some believes that the brass kempul produces better sound to it inexpensive iron sibling. However it has not able to showed conclusively the origin of the wave sound of the kempul.

## 5. References

[1] Sorrell N 1990 A Guide to The Gamelan, *1st ed* (Portland: Amadeus Press)
[2] Capwell C 2004 The Music of Indonesia *Excursions In World Music* 4 ed. (Bohlman P V Upper Saddle River: Pearson Prentice Hall) pp 141-170
[3] Budhiantho M H W and Dewantoro G 2013 The Spectral and Temporal Description of Javanese Gong Kempul *International Conference on Information Technology and Electrical Engineering* (Yogyakarta) pp 300-304
[4] Budhiantho M H W and Dewantoro G 2014 Javanese Gong Wave Signals *Proceedings of Meetings on Acoustics* (San Francisco) pp 1-8
[5] Budhiantho M H W and Dewantoro G 2014 Homomorphic Filtering for Extracting Javanese Gong Wave Signals *International Conference on Telecommunication System, Services, and Application* (Bali)
[6] Dewantoro G and Budhiantho M H W 2015 Javanese Gong Acoustics and Its Modeling using Finite Element Method *TELKOMNIKA Indonesian Journal of Electrical Engineering* Vol 14 No 1 pp 147-153
[7] Rossing T D 1982 The Science of Sound (Addison-Wesley) pp 43-44
[8] Truax B 1999 Handbook for Acoustic Ecology, *2nd ed.* (Cambridge Street Publishing)
[9] Kyung L and Kwon 2012 Effect of frequency difference on sensitivity of beats perception *Exp Brain Res* **216** pp 11– 19
[10] https://en.wikibooks.org/wiki/Engineering_Acoustics/The_Hu man_Ear_and_Sound_Perception
[11] Herrera L and Velasco 2015 Neural Networks for Beat Perception in Musical Rhythm *HYPOTHESIS AND THEORY*

**Acknowledgments**
This research was partly supported by Matana University, 2017.

# Performance evaluation of a back-to-back speaker system using a low-cost single board computer as a digital signal processor

**Irwansyah[1], Raphaël Masson[2], Naoyuki Kuse[1], Tsuyoshi Usagawa[1]**

[1]Graduate School of Science and Technology, Computer Science and Electrical Engineering, Kumamoto University, 2-39-1 Kurokami, Chuo-ku, Kumamoto 860-8555, Japan

[2]École Nationale Supérieure d'Ingénieurs de Caen & Centre de Recherche Caen, France 14000

E-mail: irwan@hicc.cs.kumamoto-u.ac.jp, raphael.masson@ecole.ensicaen.fr, naoyuki@hicc.cs.kumamoto-u.ac.jp, tuie@cs.kumamoto-u.ac.jp

**Abstract.** A back-to-back speaker system with a user-designed filter results in a low-frequency cardioid-like radiation pattern where a back-facing loudspeaker is used to suppress the radiation in the rear direction. This paper presents an application of a low-cost single board computer as a digital signal processor on a back-to-back speaker system to obtain a cardioid-like radiation pattern in the low-frequency range. The system consists of two loudspeakers and a US$7 Orange Pi Zero PC. The loudspeakers are closely spaced and mounted in a back-to-back arrangement. A cancellation filter, which is identified in advance in an anechoic chamber using the filtered-x least mean square (FxLMS) algorithm, is used in the Orange Pi Zero PC. A buffering technique is used to make the PC act as a real-time-like signal processing unit. The performance evaluation was conducted under two conditions which were offline and online filtering. In offline filtering, sounds with and without cancellation were pre-prepared. On the other hand, in online filtering, the sound was simultaneously filtered and played. This paper also indicates conditions, for which the system exhibits great performance.

## 1. Introduction

Designs of back-to-back speaker systems that produce sound in the front direction while keeping the energy of the sound radiated towards rear directions low have been studied intensively because of their useful applications [1, 2, 3]. When designing alarm systems with multiple loudspeakers to be installed in hallways, it is recommended to use loudspeakers with a directional characteristic [4]. Directional loudspeakers would benefit people in emergency situations. This enables them to follow a particular direction to a safe exit determined by the louder sound. Moreover, concerning sound intelligibility, a directional loudspeaker is more preferable [5].

One of the methods for designing a directional speaker system is based on a combination of two loudspeakers, a microphone, and a signal processing unit [2, 3]. Simply mounting two closely spaced loudspeakers back-to-back can result in a low-frequency cardioid-like radiation pattern. The pattern is obtained by utilizing the back-facing loudspeaker, with an appropriate filter, to produce an anti-sound signal. This signal is used to suppress the radiation from the front loudspeaker at the location of the microphone in the rear direction. By observing the output of the loudspeakers at the microphone, the cancellation filter can be identified as a result of the filtered-x least mean square (FxLMS) algorithm.

**Figure 1.** Orange Pi Zero PC as an audio signal processing unit

This paper presents an example of the application of a low-cost single board computer as a digital signal processor (DSP) in the back-to-back speaker system. A cheap single board computer such as Raspberry Pi was introduced in 2012, and nowadays there are so many single board computers that run Linux and are quite powerful and cheap. Since it is equipped with a Wi-Fi module that means it can communicate and remotely control two ordinary loudspeakers to achieve an intelligent speaker system with a directional characteristic. Moreover, if the speaker system is part of an "Internet of things" (IoT) network, using a customizable and programmable small computer would be of increasing convenience and flexibility when transmitting a sound to only selected loudspeakers in the network.

## 2. Speaker System Configuration

### *2.1. Signal Processing Unit and Software*

The signal processing unit to which we interfaced loudspeakers is an Orange Pi Zero PC [6] as shown in Fig. 1. This is a small computer with the dimensions of 48 mm x 45 mm and weight of only 26 g. It can run the latest Ubuntu Xenial 16.04 operating system (OS) stored on a MicroSD card. To operate, the computer is supplied with only a 5VDC/1A source, and thus it does not consume a lot of power. The Orange Pi Zero PC was chosen because it is popular, flexible, and inexpensive (costs US$7–10) computer based on ARM Cortex A7 Quad-core 1.2 GHz with 256/512 MB DDR3. The GPU of the PC is a Mali-400MP2 with 600 MHz clock speed. Equipped with Wi-Fi module and Ethernet port which allows easy access to the PC through SSH remote login. The Python programming language which has a wide variety of audio signal processing libraries is also installed by default.

### *2.2. Active Control Technique*

Active control technique is a method for reducing unwanted sound by estimating a cancellation filter, which is utilized by a secondary sound source to produce an anti-sound signal. The technique works well to suppress sounds at low frequency (<1000 Hz) [7]. Figure 2(a) shows the schematic diagram of the filtered-x least mean square (FxLMS) algorithm used to obtain a low-frequency cardioid-like radiation pattern. The pattern is achieved by utilizing the back-facing loudspeaker to produce the anti-sound signal, *anti* $d(n)$, as a result of the cancellation filter $W(z)$. Before identifying the filter $W(z)$, the transfer functions $P(z)$ and $H(z)$ from the front and back loudspeakers to the cancellation point are first estimated by presenting a time stretched pulse signal (TSP) over the loudspeakers [8]. The cancellation filter $W(z)$ is then calculated based on the schematic diagram of the FxLMS algorithm where the filter $C(z)$ is assumed to be the estimation of the $H(z)$. Once the filter $W(z)$ is obtained, it is used by the Orange Pi Zero PC to filter the input signal $x(n)$ that goes to the back-facing loudspeaker as shown in Fig. 2(b).

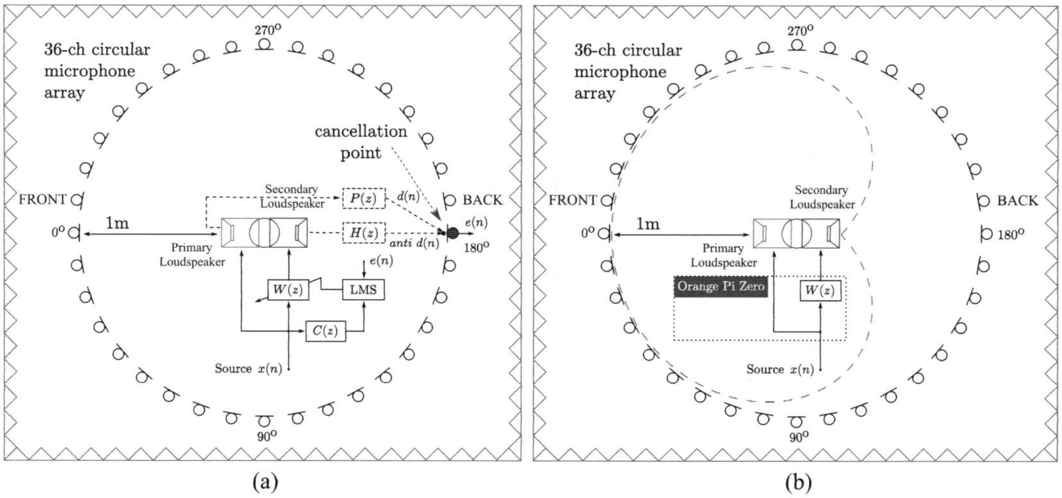


**Figure 2.** Schematic diagram of (a) the FxLMS algorithm and (b) online filtering using an Orange Pi Zero PC as a digital signal processor.

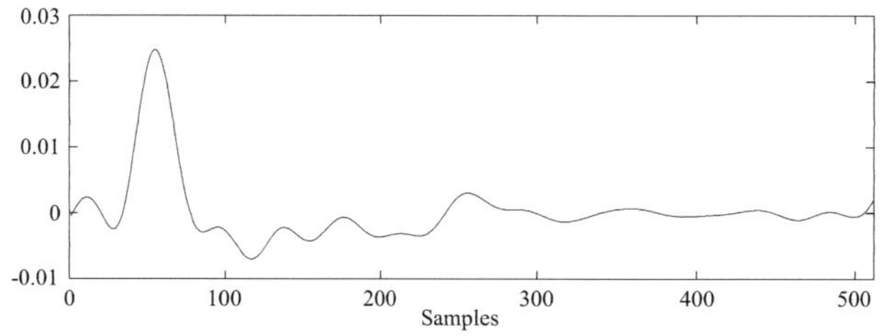


**Figure 3.** Estimated coefficients of the cancellation filter $W(z)$

### 2.3. Online filtering and Measurement Setup

A buffering technique is used to make the Orange Pi Zero PC act as a real-time-like signal processing unit. As it is well known in audio processing, a sufficient buffer size is an important parameter for playing an audio signal. If a buffer size is too small, crackles or audio dropouts will happen. On the other hand, if a larger buffer size is used, it will increase audio latency. Moreover, a larger filter tap length also results in increasing the latency. In this paper, after preliminary experiments with the PC, we chose a buffer size of 256 samples and a filter tap length of 512 taps. A sampling frequency of 44100 Hz was used. The main Python libraries used for audio signal processing are "pyalsaaudio" [9] and "scipy" [10]. In addition, the loudspeaker used in the measurement was TASCAM-VL-S3 with a size of 0.11m(W) x 0.17m(H) x 0.138m(D). Polar radiation patterns of the back-to-back speaker system were measured using a 36-ch circular microphone array proposed in [11] as shown in Fig. 2(b). The microphone used was Audio Technica AT899.

## 3. Results and Discussion

In order to obtain a cardioid-like radiation pattern, the cancellation filter $W(z)$ used by the back-facing loudspeaker to produce an anti-sound signal in the rear direction should be estimated in advance. Figure 3 shows coefficients of the cancellation filter $W(z)$ estimated in the anechoic chamber based on the schematic diagram of the FxLMS algorithm shown in Fig. 2(a). The coefficients were used for implementation on the Orange Pi Zero PC to evaluate the performance of the PC as a digital signal processor (DSP).

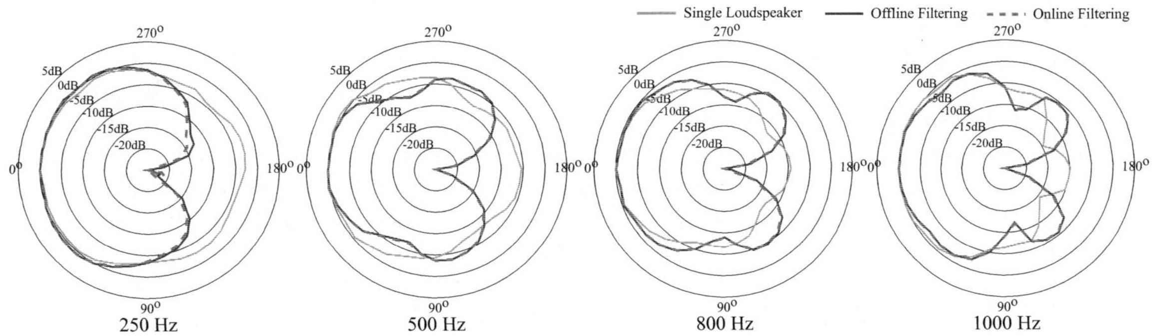


**Figure 4.** Polar radiation patterns

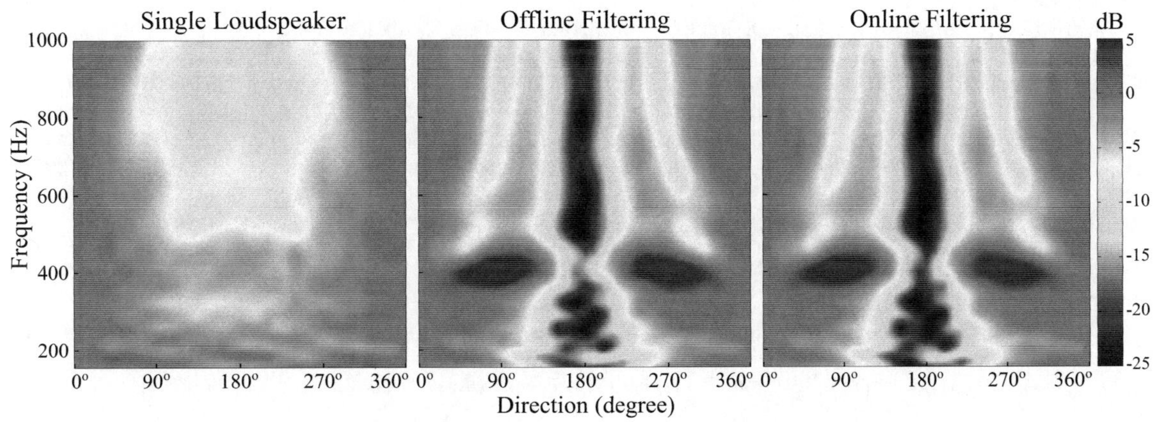


**Figure 5.** Radiation patterns as a function of direction and frequency

For the performance evaluation, the experiment was conducted under two conditions namely, offline and online filtering. In offline filtering, sounds with and without cancellation were pre-prepared. On the other hand, in online filtering, the sound was simultaneously filtered and played by the PC. Figure 4 shows polar radiation patterns in a low-frequency range for a single loudspeaker, offline and online filtering. As shown in the comparison between the green line and either blue or red line in the figure, it can be seen that cardioid-like radiation patterns could be obtained using the cancellation filter $W(z)$ for either offline or online filtering. In addition, the continuous filtering using a buffering technique, which actually created a delay for the online condition, has less effect on the low-frequency cardioid-like radiation patterns. As a result, very similar radiation patterns were obtained for both offline and online conditions.

Figure 5 shows the overall performance of polar radiation patterns as a function of direction and frequency. From the figure, it can be seen clearly that, in the direction of 180°, a significant attenuation of around 20 dB was achieved over a frequency range of 150 to 1000 Hz for both offline and online filtering. The front-to-back ratios over a frequency range of 150 to 1000 Hz were $22 \pm 3$ dB and $23 \pm 3$ dB for offline and online conditions, respectively. Generally, it can be said that a low-cost single board computer acted as a DSP in the back-to-back speaker system works well to achieve a cardioid-like radiation pattern in a low-frequency range.

## 4. Conclusions

This paper presented the application of a low-cost single board computer (SBC) as a digital signal processor (DSP) on a back-to-back speaker system to obtain a cardioid-like radiation pattern in the low-frequency range. The DSP to which the back-to-back speaker was interfaced is a US$7 Orange Pi Zero PC. The coefficient filter which results in the cardioid-like radiation pattern was estimated in advance and later used in the PC. Polar radiation pattern measurement was conducted under two conditions which were offline and online filtering. Experimental results showed that, for a frequency range of 150 Hz to 1 kHz, the front-to-back ratios were 22 $\pm$ 3 dB and 23 $\pm$ 3 dB for offline and online filtering, respectively. The cardioid-like radiation pattern for both conditions was also confirmed. In general, it can be said that a low-cost single board computer can act as a DSP in a back-to-back speaker system and works well to achieve a cardioid-like radiation pattern in the low-frequency range.

## References

[1] M. M. Boone and O. Ouweltjes., "Design of a loudspeaker system with a low-frequency cardiodlike radiation pattern," J. Audio. Eng. Soc., vol. 45, issue 9, pp. 702-707, 1997.
[2] I. Kakuhari, H. Hashimoto, and K. Terai, "Development of a loudspeaker system with a unidirectional radiation pattern in a speech frequency range," J. Acoust. Soc. Jpn., vol. 21, issue 6, pp. 369-373, 2000.
[3] V. C. Poekoel, K. Hira, Y. Chisaki, and T. Usagawa, "Unidirectional sound signage for speech frequency range using multiple-loudspeaker reproduction system," Open J. Acoust., vol. 3, issue 4, pp. 120-126, 2013.
[4] L. D. Tice, "Directional speaker system," US Patent No. 8,457,324., 2013.
[5] K. D. Jacob, "Subjective and predictive measures of speech intelligibility–The role of loudspeaker directivity," J. Audio Eng. Soc., vol 32, issue 12, pp. 950-955, 1985.
[6] http://www.orangepi.org/orangepizero/ (accessed on June 10, 2017)
[7] W. F. Druyvesteyn, and J. Garas, "Personal sound," J. Audio Eng. Soc.,45(9): 685-701, 1997.
[8] Y. Suzuki, F. Asano, H. Y. Kim, and T. Sone, "An optimum computergenerated pulse signal suitable for the measurement of very long impulse responses", J. Acoust. Soc. Am, vol. 97, pp. 1119-1123, 1995.
[9] https://pypi.python.org/pypi/pyalsaaudio (accessed on September 14, 2017)
[10] https://pypi.python.org/pypi/scipy (accessed on September 14, 2017)
[11] K. Hira, Y. Chisaki, and T. Usagawa, "Development of a directivity characteristics measurement system using a circular microphone array," Proc ISITIA, 2014.

IOP Conf. Series: Journal of Physics: Conf. Series **1075** (2018) 012023 doi:10.1088/1742-6596/1075/1/012023

# Flanking floor impact sound insulation in cross laminated timber model building for experiment

**A Hiramitsu[1], T Otsuru[2], R Tomiku[2] and K Harada[3]**

[1] Equipment Standards Division, Building Department, National Institute of Land Infrastructure Management, 1 Tachihara, Tsukuba-City, Ibaraki, 305-0802, Japan
[2] Department of Architecture and Mechatronics, Faculty of Engineering, Oita University, 700 Dannoharu, Oita, 870-1192, Japan
[3] Wood Structure Prom Inc., 5F, Adoresu Bldg., 2-2-19 Akasaka, Minato-ku, Tokyo, 107-0052, Japan Type the author addresses here

E-mail: hiramitsu-a92ta@mlit.go.jp

**Abstract.** The Act for Promotion of Use of Wood in Public Buildings (Law No. 36 of 2010 of Japan) was enforced in 2010. The promotion of the use of wood can contribute to the prevention of global warming etc. A public building in a low layer is supposed to attempt making to timber construction by this act. Moreover it is expected that the building such as apartment houses comes to be made from timber construction. For effective use of the Japanese wood, CLT (Cross Laminated Timber), which had been used in Europe, was standardized by JAS (Japanese Agricultural Standard) as a building material in 2014. Among the objections and the troubles of buildings in Japan, the sound insulation performance is one of the most serious issues, especially floor impact sound insulation. We, however, have little knowledge about sound insulation of CLT building. Therefore we built a 3-story CLT model building for experiment and investigated the floor impact sound insulation and the flanking floor impact sound insulation. This paper presents the results of floor impact sound insulation and the surfaces vibration in a sound-receiving room. The results showed the influences of damping material, ceiling and floating on the floor impact sound insulation, and the floating floor showed possible improvements in the heavy-weight and the light-weight floor impact sound insulation of CLT building. The vibration velocity level of the ceiling was the largest. However, it is necessary to decrease the vibration velocity level of the sound-receiving room's wall for improvement of the heavy-weight floor impact sound insulation by using damping material.

## 1. Introduction

*The Act for Promotion of Use of Wood in Public Buildings* (Law No. 36 of 2010 of Japan) was enforced in 2010. The purpose of this act is proper maintenance of the forest and an improvement of the rate of self-sufficiency of wood. The promotion of the use of wood can contribute to the prevention of global warming etc. A public building in a low layer is supposed to attempt making to timber construction. Moreover it is expected that the building such as apartment houses comes to be made from timber construction. For effective use of the Japanese wood, CLT (Cross laminated Timber), which had been used in Europe, was standardized by JAS (Japanese Agricultural Standard) as a building material in 2014. Among the objections and the troubles of buildings in apartment houses of Japan, the sound insulation performance is one of the most serious issues, especially floor impact sound insulation. A timber building usually has inferior performance of floor impact sound insulation than a concrete construction. Zeitler et al. investigated the impact sound insulation of cross laminated timber floors [1]. Author reported the effect of floating floor (dry double system floor) on the floor impact sound insulation in timber building [2]. We, however, have little knowledge about sound insulation of cross laminated timber building. Therefore we built a 3-story cross laminated timber model building for experiment and investigated the floor impact sound insulation and the flanking floor impact sound insulation. This paper presents the results of floor impact sound insulation and the surface vibration in a sound-receiving room.

### *1.1. Building frame*

Figure 1 shows the cross-sectional view of the cross laminated timber model building. One floor of this building measures 3.64 x 3.18 m (10.4 $m^2$ in a floor area) and the entire building is three-storied. H steels were set on the concrete earth floor and 1st floor was finished the floating floor. The 2nd and 3rd floor were 5-ply 150 mm thickness CLT panel. All walls were 3-ply 90 mm thickness CLT panel. As for the ceiling height, the 2nd floor was 2.3 m, the 1st floor was 2.7 m. The floor is jointed the wall by bolts and angles, it was structure to be proof against earthquakes.

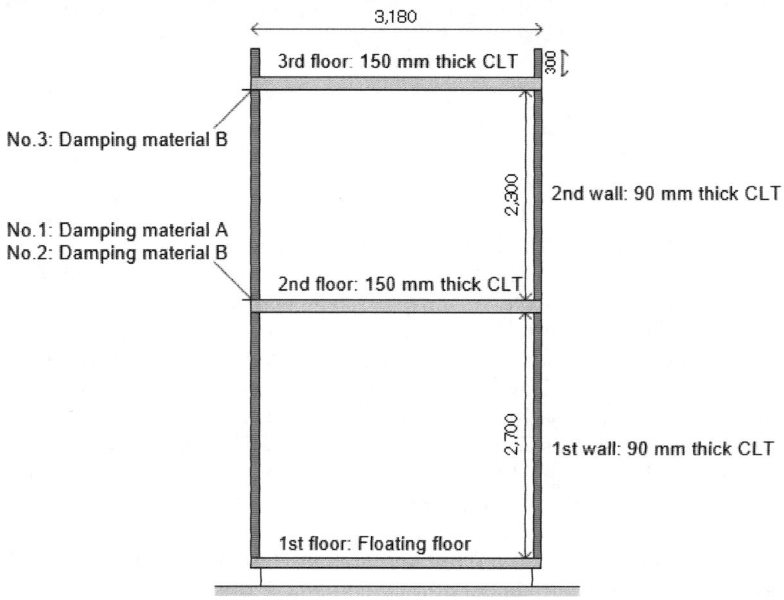


**Figure 1.** Cross-sectional view of the cross laminated timber model building.

### *1.2. Specifications of floor section*

In order to investigate the influences of damping material, ceiling and floating floor on floor impact sound insulation for cross laminated timber building, the floor section specifications of the specimen were determined. Table 1 shows the outline of section specifications of 11 specimens. Specimen No. 1 as reference specimen was only building frame without damping material, ceiling and floor topping.
Specimens No. 1, 2 and 3 were No. 0 with damping material. The damping material is polyurethane elastomer foam of 12.5 mm thickness. As for each Young's modulus of static load, damping material A is approximately 8.0 N/mm2, damping B is approximately 1.8 N/mm2. Specimen No. 1 was No. 0 with damping material A between 2nd wall and 2nd floor, Specimen No. 2 was No. 0 with damping material B between 2nd wall and 2nd floor, and Specimen No. 3 was No. 0 with damping material B between 3rd floor and 2nd wall.

Specimens No. 4 was No. 0 with suspended ceiling. The celling was constructed in 9.5 mm thickness gypsum board with substrate light gauge steel. Specimens No. 5, 6 and 7 were No. 0 with floating floor, and specimens No. 8, 9 and 10 were No. 4 with floating floor. Figure 2 shows the cross-sectional view of floating floors. The floating floor is designed by supporting legs with rubber vibration isolators supported the particle boards etc., and contained air layers. The topping of floating floors was changed with and without 8 mm thick asphalt sheet.

**Table 1.** Outline of floor section specifications of specimen.

| Damping material | Ceiling | Floor topping |
|---|---|---|
| None | None | None |
| Damping material A (between 2nd floor wall and 2nd floor) | None | None |
| Damping material B (between 2nd floor wall and 2nd floor) | None | None |
| Damping material B (between 3rd floor and 2nd floor wall) | None | None |
| None | Gypsum board, t=9.5 mm | None |
| None | None | Floating floor A |
| None | None | Floating floor B |
| None | None | Floating floor C |
| None | Gypsum board, t=9.5 mm | Floating floor A |
| None | Gypsum board, t=9.5 mm | Floating floor B |
| None | Gypsum board, t=9.5 mm | Floating floor C |

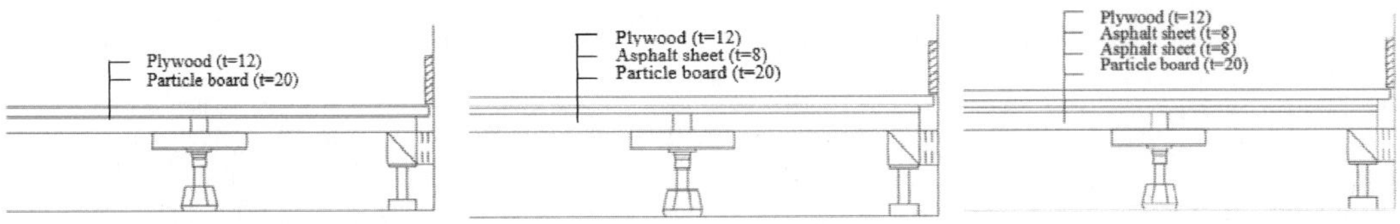


**Figure 2.** Cross-sectional view of floating floors. (Left: Floating floor A, Center: Floating floor B, Right: Floating floor C)

## 2. Measurements

The floor impact sound insulations were measured in conformity with JIS A 1418-2 [3] for the heavy-weight floor impact sound insulation and JIS A 1418-1 [4] for the light-weight floor impact sound insulation. The impact sources for measurement of the heavy-weight floor impact sound were a car-tire source (a Bang machine) and a rubber ball source, and of the light-weight floor impact sound was a tapping machine. The impact positions were five positions, and the sound-receiving positions were five positions. The floor impact sound insulations were evaluated with JIS A 1419-2 [5].

The heavy-weight floor impact sound insulation would be closely related to the indoor surface vibration in the sound-receiving room. Therefore the vibration velocity levels were measured. The vibration velocity levels (reference value 0 dB = 5 x 10-8 m/s) of the ceiling, four walls (north-side, south-side, east-side and west-side) and floors of the sound-receiving room were measured in response to the rubber ball source impacts on the floor of the sound source room. The impact position was centre position which was similar to those used in the measurement of floor impact sound. For each surface, the maximum vibration velocity energy were measured in every third octave band frequency.

## 3. Results and Discussions

### *3.1. Floor impact sound insulation*

Figure 3 shows the measurement results of the floor impact sound insulation; the 3rd floor was the sound-source room and the 2nd floor was the sound-receiving room. As for specimen No. 0, the heavy-weight floor impact sound insulation using the car-tire source was Lr-90, the heavy-weight floor impact sound insulation using the rubber ball source was $L_r$-80 and the light-weight floor impact sound insulation was $L_r$-105. These floor impact sound insulations were exceedingly low performance for actual buildings. Specimen No. 10 yielded high performance, showing $L_r$-75 in the heavy-weight floor impact sound insulation using car-tire source, $L_r$-65 in the heavy-weight floor impact sound insulation using rubber ball and $L_r$-70 in the light-weight floor impact sound insulation.

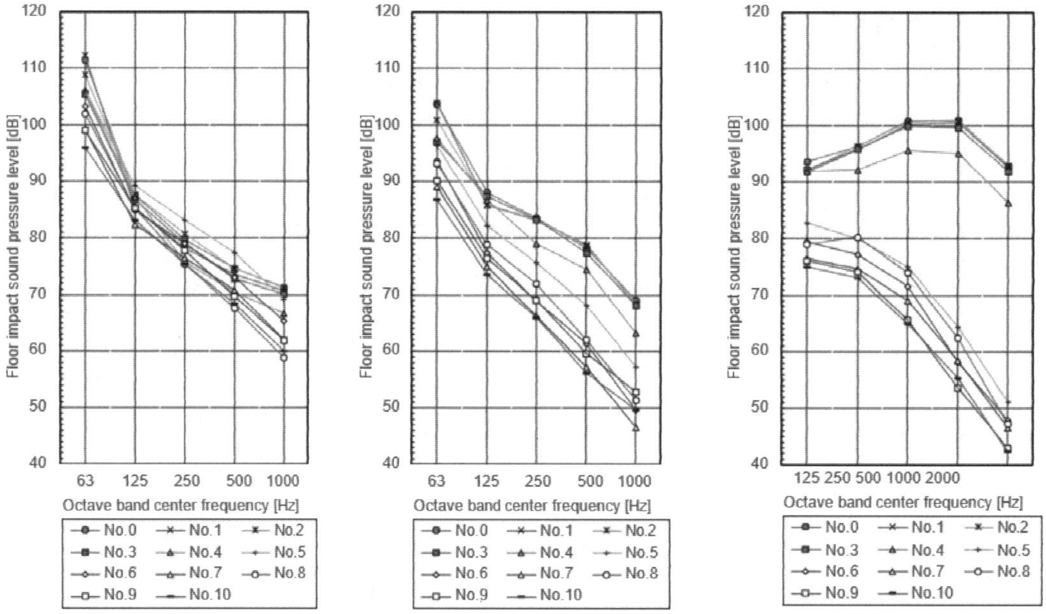


**Figure 3.** Measurement results of the floor impact sound insulation of the cross laminated timber model building. (Sound-source room: 3rd floor, Sound-receiving room: 2nd floor) (Left: Car-tire source, Center: Rubber ball source, Right: Tapping machine).

### *3.2. Effects of damping material, ceiling and floating floor*

To confirm the effects of the damping material, ceiling and floating floor on the floor impact sound insulation, the floor impact sound pressure level differences from specimen No. 0 were calculated. Figure 4 shows the improvements of floor impact sound pressure levels relative to that of specimen No. 0. The positive values indicate that the room has a higher floor impact sound insulation than specimen No. 0.

There were few level differences in the 63 Hz octave band between car-tire source and rubber ball source. In contact, the level differences of the rubber ball were higher than of the car-tire source in the 125 Hz octave band or higher frequency.

There are few changes of the level difference by the damping material (specimen No. 1, 2 and 3) in the 125 Hz octave band or high frequency regardless of the floor impact source. The floor impact sound insulations were increased in the 63 Hz octave band by damping material. In addition, an effect of damping material was the highest in specimen No. 3. The differences were 6.0 dB in the case of the car-tire source, and 6.6 dB in the case of the rubber ball source.

The value of the effect of the ceiling (specimen No. 4) in all octave bands were approximately 5 dB regardless of the floor impact source. From the above results, the ceiling has the effect on heavy-weight and light-weight floor impact sound insulation of cross laminated timber building.

The effect of the floating floor (specimen No. 5, 6, 7, 8, 9 and 10) on the heavy-weight and the light- weight floor impact sound insulation. Furthermore the floor impact sound insulation was high by mass on the topping of the floating floor. It was a performance gain of approximately 3 dB a layer in the asphalt sheet in the floor impact sound insulation.

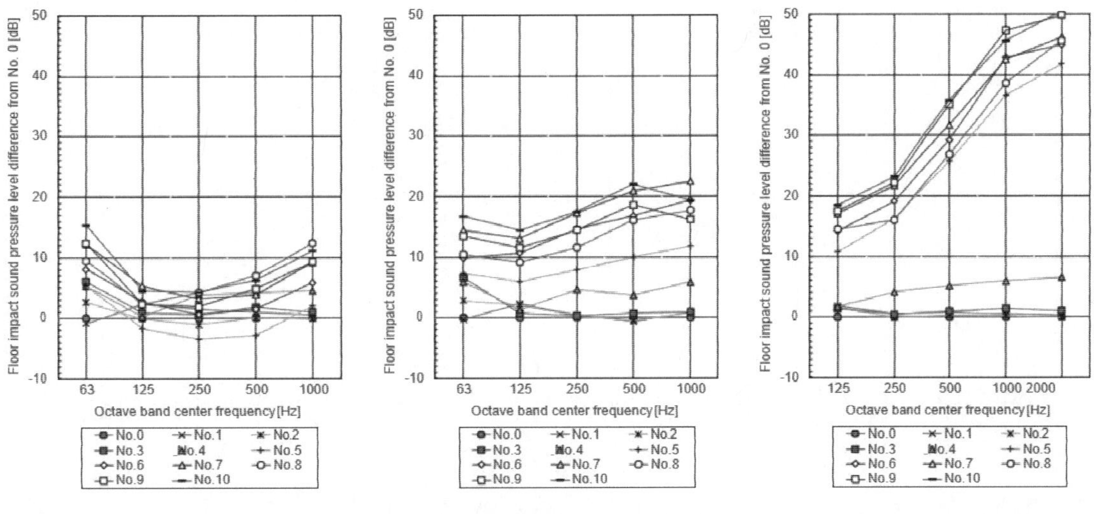


**Figure 4.** Improvements of floor impact sound pressure levels from specimen No. 0. (Left: Car-tire source, Center: Rubber ball source, Right: Tapping machine).

*3.3. Flanking floor impact sound insulation*

Figure 5 and 6 show the measurement results of vibration velocity levels when rubber ball source impact on room's center position. The vibration velocity level measurement results of specimens No. 0, 1, 2 and 3 in Figure 5 shows that the vibration velocity level of the ceiling surface hardly changes even when a damping material was inserted. On the other hand, the wall surface vibration velocity level of specimen No. 3 is small near the 63 Hz octave band. This indicates that the damping material between the floor and the floor reduces vibration propagation to the wall and reduces the radiated sound from the wall surface (the flanking floor impact sound). The vibration velocity level measurement results of the specimen's No. 0, 8 and 10 in Figure 6 show that the vibration level is reduced in the frequency range lower than the 63 Hz octave band and higher than the 80 Hz octave band by the construction of the floating floor. Furthermore, although the effect of reducing the radiated sound from the ceiling surface is small, the radiated sound from the wall surface is greatly reduced by inserting the asphalt sheet on the floating floor, and the effect of the asphalt sheet on the upper surface material is appearing.

Based on the above results, the floor impact sound insulation performance of cross laminated timber building is often not able to ignore the radiated sound (the flanking floor impact sound) from the wall surface.

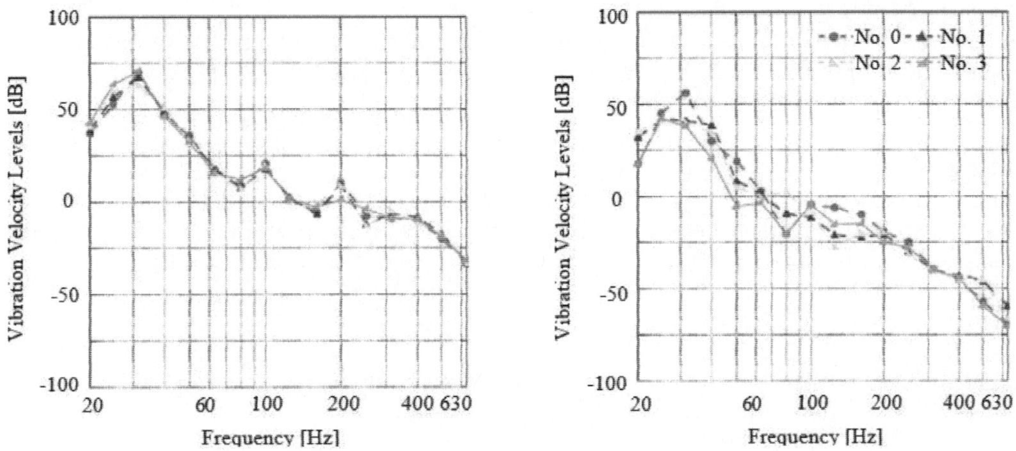


**Figure 5.** Measurement results of vibration velocity levels 1. (Left: Ceiling, Right: West wall).

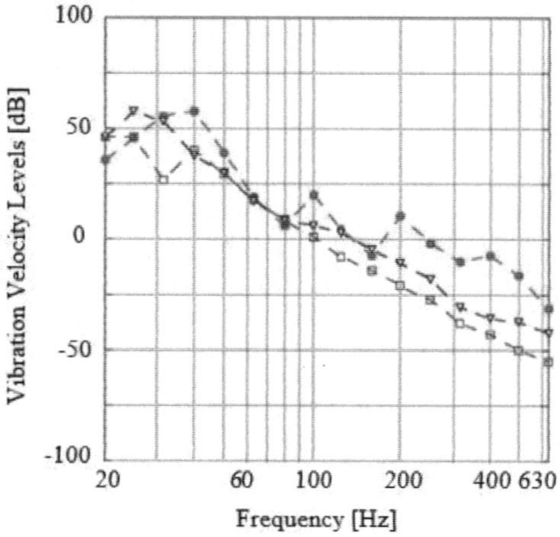


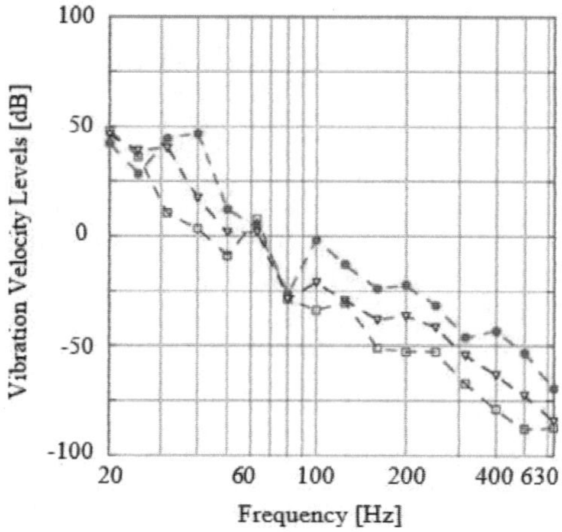


**Figure 6.** Measurement results of vibration velocity levels 2. (Left: Ceiling, Right: West wall).

## 4. Conclusions

This paper presents the results of floor impact sound insulation and the vibration of surfaces of sound-receiving room of the 3-story cross laminated timber building. The results of the floor impact sound insulation suggest the influences of the damping material, ceiling and floating on the floor impact sound insulation of cross laminated timber building. We conclude that the floating floor showed possible improvements in the heavy-weight and the light-weight floor impact sound insulation of cross laminated timber building. In addition, we showed the possibility of reducing the radiated sound from the wall surface (the flanking floor impact sound) by using damping material.

## 5. References

[1] Zeitler B, Schoenwald S and Sabourin I 2014 Direct impact sound insulation of cross laminate timber floors with and without topping *Proc INTER-NOISE 14*

[2] Hiramitsu A 2013 Floor impact sound insulation of wooden three-story school building for full-scale fire experiment *Proc INTER-NOISE 13*

[3] Acoustics - Measurement of floor impact sound insulation of buildings – Part 2: Method using standard heavy impact sources. (2000). Japanese Industrial Standard JIS A 1418-2: 2000, Japanese Industrial Standards Committee.

[4] Acoustics - Measurement of floor impact sound insulation of buildings – Part 1: Method using standard light impact source. (2000). Japanese Industrial Standard JIS A 1418-1: 2000, Japanese Industrial Standards Committee.

[5] Acoustics - Rating of sound insulation in buildings and of building elements – Part 2: Floor impact sound insulation. (2000). Japanese Industrial Standard JIS A 1419-2: 2000, Japanese Industrial Standards Committee.

## Acknowledgments

The authors would like to thank to Mr. Yuta Katsuyama (former master's course student of Oita Univ.) for their continuous contribution to this research. This research was conducted as a project of CLT Building Promotion Council of Kochi. A part of this research was supported by JSPS KAKENHI Grant Number 17K06690.

# Vibro-acoustic characterization of Angklung bamboo tube

**T Aditanoyo**[1]

[1]Department of Engineering Physics, Institut Teknologi Bandung, Indonesia

Email: teguh.aditanoyo@gmail.com

**Abstract.** Angklung is one of the most popular Indonesian traditional bamboo musical instruments. In this paper, bamboo tubes were harvested from West Java, which were sorted and crafted by an angklung-maker maestro. This paper is focused on finding the correlation of those bamboo tubes both in acoustic characteristic and mechanical vibration. There are three data that are measured simultaneously: force from the exciting impact hammer, normal acceleration from accelerometer, and sound pressure from a microphone. Accordingly, bamboo tubes' mechanical admittance and acoustic sound pressure level was calculated. In terms of frequency response, it was found that the admittance and SPL of each tube are well correlated. All of the tubes have inharmonic frequency overtones with the first overtone 2.17 times higher than its fundamental frequency.

## 1. Introduction

Bamboo is a kind of plants that easy to find in Indonesia. It has been used in various ways including as material for traditional musical instruments. When it comes to traditional musical instrument and bamboo, Indonesians will surely mention the idiophone of angklung as this instrument is very popular. It is usually made of two bamboo tubes, which one is tuned an octave higher than the other one, bound in a bamboo frame as depicted at Figure 1. Each of the tubes has unique shape as they seem like a tube with a tongue at its opening. People play angklung mostly in a group and each member rattles their notes in turn.

**Figure 1.** Illustration of an angklung.

The basic acoustic and vibration properties of bamboo has been generally described and compared to other wood [1], [2]. Researchers also have determined some acoustical characteristics of angklung. From empirical experiment, Zainal [3] determined angklung's fundamental frequency from its length

IOP Conf. Series: Journal of Physics: Conf. Series **1075** (2018) 012024 doi:10.1088/1742-6596/1075/1/012024

and diameter of the air resonator with an overtone at 2.45 times of the fundamental frequency. Sudarsono and Merthayasa [4] have made remarkable findings about angklung as they made complete analysis including spectral, temporal, and spatial characteristics of angklung. In terms of overtones, their conclusion does not agree with Zainal's as they stated 1.44 $_0$, 2 $_0$, 3.47 $_0$, 6.31 $_0$ and 7 $_0$ with $_0$ for the fundamental frequency. Despite of different overtones pattern, both researches agree that angklung has inharmonic overtones. Inharmonic is the opposite term of harmonic, which overtones are integer's multiplication of the fundamental frequency.

This paper explains the vibration feature of angklung bamboo tube and its relation to the acoutiscal radiation in frequency domain. It is expected that this paper gives appropriate guidance for angklung maker in the future.

## 2. Methods

There were 10 angklung bamboo tubes tuned at note A4 440 Hz and 8 tubes tuned at note A5 880 Hz, an octave higher. All of these tubes were made of temen hitam bamboo (*Gigantochloa Atroviolacea Widjaja*), which is known as the best material for making angklung [5], from Sukabumi and Majalengka, West Java, Indonesia. Handiman Diratmasasmita, who is acknowledged as angklung maker maestro, carefully crafted all the tubes so that these could be a benchmark as favorable angklung tubes.

Each tube was hung by a rope through its hole at the tongue to imitate free-free boundary condition. The excitation was done by striking the lower edge of tube with an impact hammer. The tip of the hammer was hardened plastic to provide broadband frequency impulse. An accelerometer was glued at the other side and measured the vibration response of the tube normal to the tube surface. The measured both force and acceleration are collected by an oscilloscope with sampling frequency of 10,000 Hz. Simultaneously, a flat-response omnidirectional microphone with a distance of 1 m from the tube in the same direction of the accelerometer recorded the radiated sound pressure. The audio frequency sampling was 44,100 Hz. The experiment set, including the impact point and the position of the accelerometer, is illustrated at Figure 2. The measurement repeated 5 times for each tube. All of the measurements have been done in an anechoic chamber to avoid any acoustical distortion due to room surface reflection.

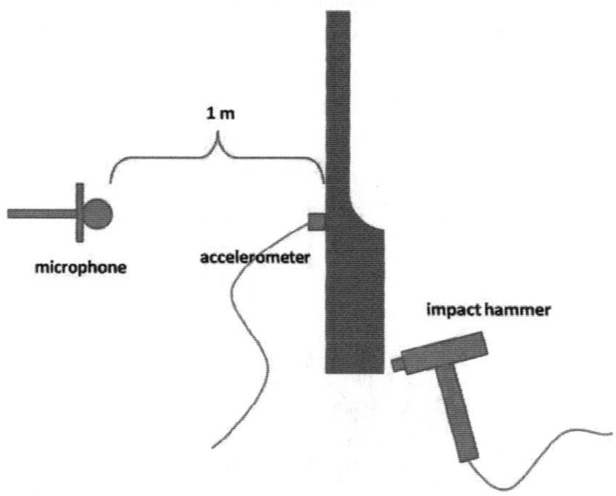


**Figure 2.** Physical experiment set illustration.

## 3. Results and Discussion

All of the discussions in this subtopic are in scope of spectral parameter. Thus, all of the measured data had to be transformed into frequency domain by DFT with FFT algorithm. The acceleration data were integrated into velocity and divided by the force data so that yielded mechanical admittance. The pressure data were just converted into SPL. Then, 5 data of each tube were averaged. The FFT size for both admittance and SPL was $2^{14}$ so that the frequency sensitivity is 0.6 Hz for admittance and 2.9 Hz for SPL.

Some of all the results from tubes tuned at 880 Hz can be found at Figure 3 as well as tubes tuned at 440 Hz are at Figure 4. All of the remaining results are tabulated at Table 1 for tubes tuned at 880 Hz and Table 2 for tubes tuned at 440 Hz. First to take note is admittance peak located near 4000 Hz in each graphs is occurred due to the unknown filtering effect of the impact hammer electrical instrument, and nothing to do with the tube vibration characteristic. This frequency dip of the impact hammer that took place in each measurement as can be seen in Figure 5.

**Table 1.** Peak frequency of angklung bamboo tube tuned at 880 Hz.

| Tube Label | Frequency (Hz)/ SPL Peak (dB) | | | | | | | | Frequency (Hz)/ Admittance Peak (m/N.s) | |
|---|---|---|---|---|---|---|---|---|---|---|
| | $f_0$ | | $f_1$ | | $f_2$ | | $f_3$ | | $f_0$ | |
| 1a | 881.8 | 81.17 | 1916 | 57.36 | 2194 | 49.59 | - | - | 888.3 | 0.6426 |
| 2a | 881.8 | 70.34 | 1881 | 47.75 | 2180 | 38.16 | - | - | 877.3 | 0.5230 |
| 3a | 878.9 | 75.38 | 1843 | 44.88 | 2121 | 51.27 | 2581 | 41.80 | 890.0 | 1.100 |
| 4a | 878.9 | 74.36 | 1986 | 50.75 | - | - | - | - | 865.7 | 0.1885 |
| 5a | 878.9 | 66.05 | - | - | - | - | - | - | 884.7 | 2.532 |
| 6a | 884.8 | 63.78 | - | - | 2051 | 45.36 | - | - | 890.8 | 0.6540 |
| 7a | 878.9 | 71.45 | 1904 | 42.50 | 2136 | 52.80 | - | | 889.5 | 1.017 |
| 8a | 878.9 | 71.45 | - | - | 2083 | 42.27 | 2525 | 38.49 | 884.7 | 0.847 |

**Table 2.** Peak frequency of angklung bamboo tube tuned at 440 Hz.

| Tube Label | Frequency (Hz)/ SPL Peak (dB) | | | | | | | | | |
|---|---|---|---|---|---|---|---|---|---|---|
| | $f_0$ | | $f_1$ | | $f_2$ | | $f_3$ | | $f_4$ | |
| 1b | 439.5 | 78.23 | 919.9 | 48.64 | 1148 | 58.25 | 1269 | 52.26 | 1781 | 42.94 |
| 2b | 439.5 | 77.10 | 940.4 | 42.40 | 1128 | 49.34 | 1277 | 43.33 | 1781 | 37.06 |
| 3b | 436.5 | 69.79 | - | - | 1163 | 53.85 | 1254 | 49.50 | 1875 | 36.97 |
| 4b | 439.5 | 74.68 | - | - | 1189 | 55.71 | - | - | - | - |
| 5b | 439.5 | 76.38 | 966.8 | 36.01 | 1125 | 47.83 | - | - | 1781 | 36.10 |
| 6b | 439.5 | 73.16 | 960.9 | 39.66 | 1116 | 50.00 | 1271 | 50.11 | 1849 | 49.55 |
| 7b | 442.4 | 76.01 | - | - | - | - | 1233 | 63.27 | 1682 | 43.81 |
| 8b | 442.4 | 78.34 | - | - | - | - | 1213 | 57.17 | 1857 | 45.00 |
| 9b | 445.3 | 78.83 | 1014 | 43.08 | - | - | 1216 | 56.15 | 1875 | 42.54 |
| 10b | 439.5 | 77.92 | - | - | 1157 | 60.63 | 1266 | 54.85 | 1720 | 43.73 |

| Tube Label | Frequency (Hz)/ Admittance Peak (m/N.s) | | | | | | | |
|---|---|---|---|---|---|---|---|---|
| | $f_0$ | | $f_1$ | | $f_2$ | | $f_3$ | |
| 1b | 440.8 | 0.7239 | - | - | 1132 | 0.0557 | - | - |
| 2b | 440.2 | 0.6023 | 933.6 | 0.0237 | 1118 | 0.0735 | - | - |
| 3b | 432.0 | 1.179 | - | - | - | - | - | - |
| 4b | 443.9 | 0.8449 | - | - | 1185 | 0.5568 | - | - |
| 5b | 448.1 | 1.500 | - | - | 1120 | 0.2590 | - | - |
| 6b | 438.4 | 1.274 | - | - | - | - | - | - |
| 7b | 444.5 | 0.8989 | - | - | - | - | 1245 | 0.2115 |
| 8b | 440.2 | 0.6696 | - | - | - | - | 1208 | 0.1477 |
| 9b | 440.2 | 0.4553 | - | - | - | - | 1225 | 0.0203 |
| 10b | 437.1 | 0.9339 | - | - | 1146 | 0.2347 | - | - |

All of the fundamentals frequencies of each tube are laid in the frequency where they were tuned for, both around the 440 Hz and the 880 Hz. In addition, all of the fundamental frequencies have the highest both SPL and admittance compared to the overtones. It indicates strong correlation between the vibration response and the radiated sound pressure in terms of frequency and magnitude.

In contrast of being able to spot on easily, the overtones were defined subjectively by the peak's figure of merit. For example, the SPL peak around 900 Hz in tube 2b can be clearly seen so that we mark it as an overtone (see Figure 4); while it is hard to define the SPL peak throughout frequency area in tube 7b. The admittance graphs are a bit noisy so an overtone is marked by a small hump with less noise in its vicinity; not merely marked by a single frequency spike. SPL peaks and admittance peaks in overtones, specifically at 440 Hz tubes, also marked at similar frequency as it support the strong correlation between vibration and sound pressure.

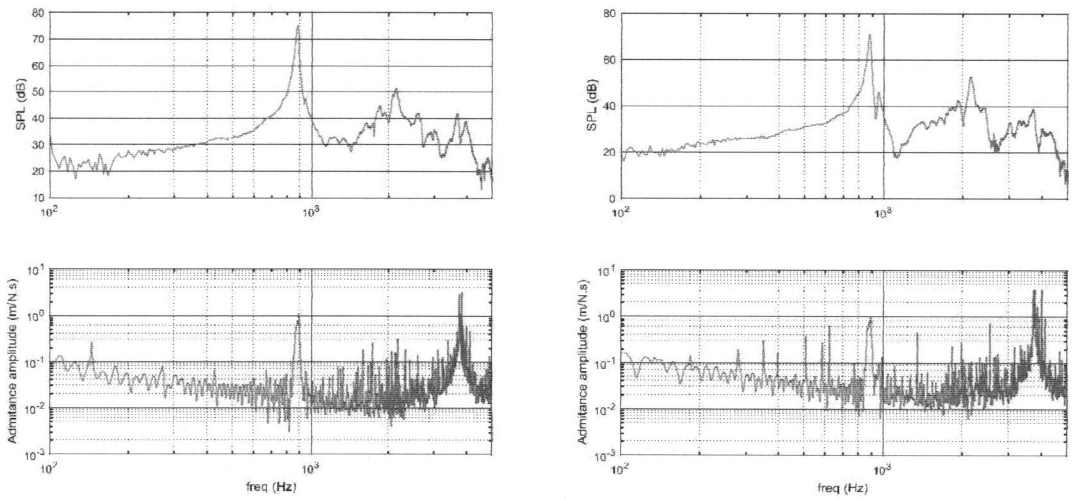


(a) (b)

**Figure 3.** SPL and admittance of angklung bamboo tube tuned at 880 Hz labelled as (a) 3a and (b) 7a.

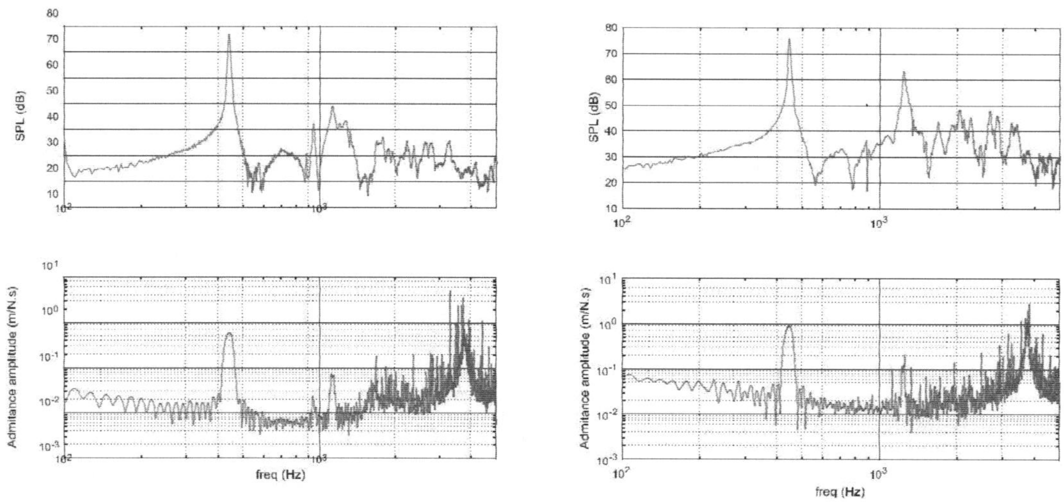


(a) (b)

**Figure 4.** SPL and admittance of angklung bamboo tube tuned at 440 Hz labelled as (a) 2b and (b) 7b.

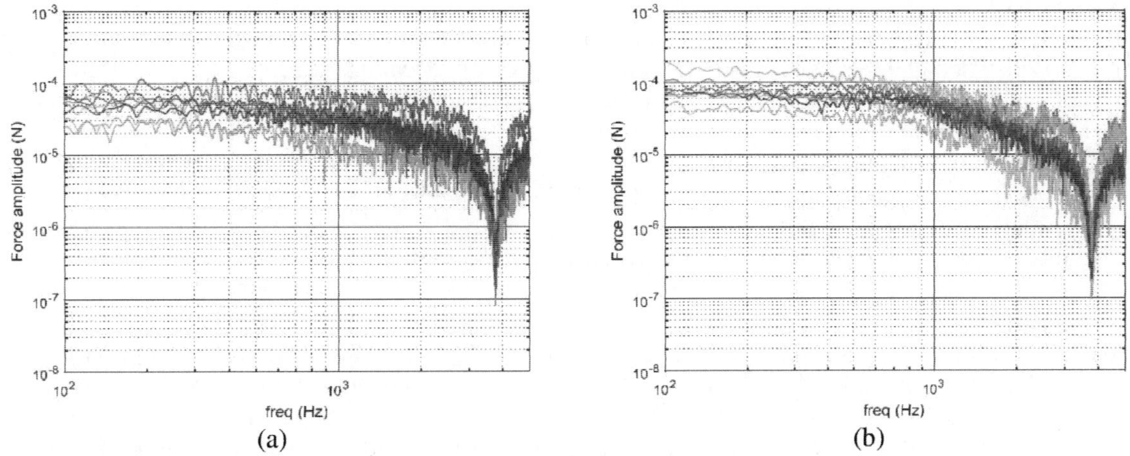


**Figure 5.** Force magnitude frequency spectrum from the measurement of tubes tuned at (a) 880 Hz and (b) 440 Hz.

For the 880 Hz tuned tubes (see Table 1), generally there are three overtones in the SPL results while no overtones can observed in the admittance results. Theoretically, admittance shall be obtained even if the mechanical system was excited with small force or vibrating in the slightest displacement. In reality, small force or acceleration signal could have the same magnitude with the noise thus some of admittance peaks cannot be distinguished. Moreover, the higher the overtones' frequency, the weaker the magnitude would be, hence only the admittance peaks of the fundamental frequency of each tube are observable. From the SPL results, the first and second overtones are quite high in the value, up to 50 dB. In the other hand, the third overtone is barely occurred with only two tubes define this overtone. Taking account these three overtones, the overtones frequency pattern is around $f_0$; $2.17f_0$; $2.42f_0$; $2.90f_0$.

Overtones for the 440 Hz tune tubes (see Table 2) are more observable with four overtones for the SPL results and three overtones for the admittance results. It is interesting that the first overtones at around 960 Hz are barely occurred and have smaller magnitude, both SPL and admittance, compared to the second and third overtones. The second and third overtones are located respectively at around 1140 Hz and 1240 Hz. The admittance peaks of each tube at the fourth overtones are no longer recognized from the graphs while the SPL peaks are still perceivable at around 1800 Hz. The overtones frequency pattern for this 440 Hz tubes is around $f_0$; $2.17f_0$; $2.60f_0$; $2.83f_0$; $4.09f_0$.

It is interesting that the results of bamboos with two different fundamental frequencies share the same multiplication number 2.17 $_0$ for the first overtone. The correlation may indicate the criteria of good angklung tube. In the other hand, the second overtones and above it have different number. Another thing to be concern about is overtones pattern found in this experiment doesn't agree with the findings of Zainal [3] nor Sudarsono and Merthayasa [4].

## 4. Conclusion

In angklung bamboo tube, vibration and radiated sound have strong correlation. Both peaks from SPL and admittance are located at the similar frequency and have the same magnitude profile, with the highest at the fundamental frequency. The first inharmonic overtone of angklung tube is found at 2.17 f0, with f0 as the fundamental frequency. The second overtones and above are spot on different multiplication factor while still maintaining the inharmonic characteristic.

More angklung bamboo tube samples are needed so that the measured characteristic can represent angklung in general. The explanation of why tubes with same fundamentals frequency can have varying inharmonic patterns will be discussed in the future paper. The factors that affect overtones shift will be described as it will be very useful for making precise timbre of angklung bamboo tube.

## 5. References

[1] Wegst U G 2008 Bamboo and wood in musical instruments *Annu. Rev. Mater. Res.* vol 38
[2] Yoshikawa S and Walthan C 2014 Woods for wooden musical instruments
[3] Zainal M R, Samad S A, Hussain A and Azhari C H 2009 Pitch and timbre determination of the angklung *Am. J. Appl. Sci.* vol 6 no 1 pp 24
[4] Sudarsono A S and Merthayasa I G N 2013 Acoustic analysis from pentatonic Angklung *Proceedings of Meetings on Acoustics ICA2013* vol 19 pp 035078
[5] Widjaja E A 1980 Angklung and other West Javanese bamboo musical instruments *Bamboo research in Asia: proceedings of a workshop held in Singapore*

## Acknowledgments

The author sincerely gives thanks to Mr. Joko Sarwono and Mr. Eko Mursito Budi who supervised this angklung bamboo project; Mr. Handiman Diratmasasmita who has craved all of the bamboo tubes; Megarini Hersaputri, Ivan Stefanus, Hizkia Natanael, and Kevin Leonardo who helped a lot during the data measurement. Without those kind-hearted people, this paper would have not even existed.

# Effect of Movement on Positioning Accuracy in a Transponder-based Acoustical Positioning

**Hirokazu Iwaya[1], Satoki Ogiso[2], Koichi Mizutani*[1,3], Tadashi Ebihara[1,3], and Naoto Wakatsuki[1,3]**

[1]Graduate School of Systems and Information Engineering, University of Tsukuba, 1-1-1 Tennodai, Tsukuba, Ibaraki, 305-8573 Japan

[2]Ph. D. Program in Empowerment Informatics, School of Integrative and Global Majors, University of Tsukuba, 1-1-1 Tennodai, Tsukuba, Ibaraki, 305-8573 Japan

[3]Faculty of Engineering, Information and Systems, University of Tsukuba, 1-1-1 Tennodai, Tsukuba, Ibaraki, 305-8573 Japan

E-mail: mizutani@iit.tsukuba.ac.jp

**Abstract.** A transponder-based positioning method using audible sounds for a large-scale area positioning has been proposed by the authors. The proposed method has two characteristics; it employs audible sounds instead of ultrasounds to measure long distance by using audible sounds, and it achieves positioning without clock synchronization among devices which is required additional equipment such as radio wave. However, positioning of the moving object (terminal) is still challenging because the proposed method requires longer measurement time than other methods since it measures the round-trip time of flight (TOF) among the terminal and the anchors (transponders). In this paper, we evaluated the effect of the terminal movement on positioning in simulations and experiments. From simulations and experimental results, it was found that the effect of Doppler and the terminal movement cause random error and offset in TOF/distance, respectively. Consequently, it was found that our proposed method is sensitive to the terminal movement.

## 1. Introduction

Indoor positioning is a technique to localize people/objects and it is utilized for various location services such as people navigation and robot automation. Generally, indoor positioning is achieved by using acoustical waves [1–6], laser light [7], radio waves [8], infrared light [9], geomagnetic field [10], and inertial measurement units [10]. Among them, acoustical positioning is cost effective, and it is used to achieve for small-scale area localization such as a corridor/room. These techniques commonly use ultrasounds higher than 20 kHz, and achieve accurate positioning by measuring time of flight (TOF) of ultrasound. However, ultrasound attenuates rapidly in air, and it is not suitable for large-scale localization. Hence, it is difficult to achieve positioning in large-scale area by ultrasound means. In addition, existing positioning methods commonly require clock synchronization among devices in the measurement of TOF, resulting in complicated system.

To achieve simple and cost effective large-scale localization, we have proposed a transponder-based positioning method using audible sounds lower than 10 kHz [11]. The proposed method is achieved as follows [Fig. 1(a)];

(i) A terminal transmits audible sounds (request signals) to transponders of known position.
(ii) The transponders detect the request signal from received sound. After a constant delay time, the transponders transmit response signal of audible sounds (response signal) to the terminal shortly.

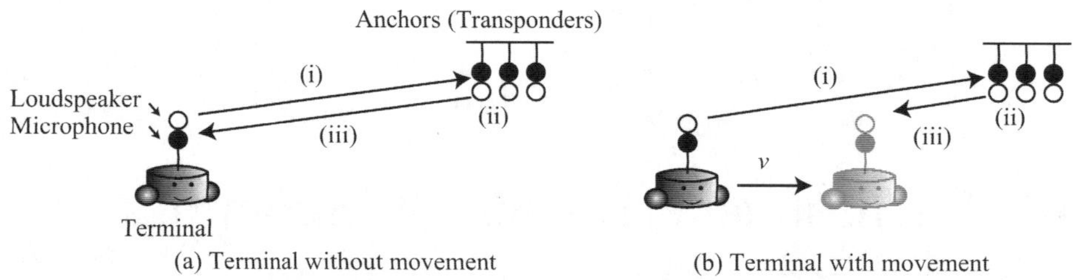


**Figure 1.** Transponder-based acoustical positioning.

(iii) The terminal receives the response signals and measures round-trip time of flights (TOFs) among the terminal and the transponders.

The proposed method has two characteristics; it employs audible sounds instead of ultrasounds to measure long distance, and it achieves positioning without clock synchronization among devices which is required additional equipment such as radio wave. We tested this method in experiments, and found that the proposed method can achieve accurate positioning (error: within 0.1 m).

However, positioning of a moving terminal is still challenging because the proposed method requires longer measurement time than other methods since it measures round-trip TOF among the terminal and the transponders [Fig. 1(b)]. Moreover, the terminal movement causes an effect of Doppler that may affect the measurement of TOF. Hence, the effect of the movement on positioning accuracy in the transponder-based positioning should be clarified.

In this paper, we evaluate the effect of the terminal movement on positioning in simulations and experiments. This paper consists of four sections. In section 2, we overview transponder-based acoustical positioning and analyze effects of the terminal movement on positioning. In section 3, we perform simulations and experiments. In section 4, we conclude our works.

## 2. Transponder-based positioning and effects of terminal movement

### *2.1. A overview of transponder-based positioning*

The proposed method measures round-trip TOFs among the terminal and transponders. A schematic view of the measurement procedure of TOF is shown in Fig. 2 [11]. Firstly, the terminal transmits a request signal to the transponder #$i$ at a specific time $t_{\mathrm{T}i}$. The request and response signals are modulated by maximum-length sequence (M-sequence) signals that have good cross-correlation property [6]. The unique code of M-sequence is assigned each transponder. Secondly, the transponder receives the request signal and calculates a cross-correlation function between the received request signal and the assigned request signal. When the cross-correlation function exceeds a threshold, the transponder detects the request signal and immediately transmits the response signal to the terminal. Note that a delay time $w$ (that is the transponder timing interval of recording the request signal and the response signal) is kept constant. Finally, the terminal receives the response signal and calculates the cross-correlation function between the received waveform and the assigned response signal. Then, TOF between the terminal and the transponder #$i$ ($t_i$) is calculated by peak detection of the cross-correlation function. If the terminal received the response signal at $t_{\mathrm{R}i}$, $t_i$ is determined by the relationship among $t_{\mathrm{T}i}$, $t_{\mathrm{R}i}$, and $w$,

$$t_i = \frac{t_{\mathrm{R}i} - t_{\mathrm{T}i} - w}{2}. \tag{1}$$

From the above method, the terminal can measure the round-trip TOFs. After the measurement of multiple TOFs, the terminal positioning is performed by using TOFs and the positions of the transponders. In following, we assume a case of three transponders existing at $\boldsymbol{x}_i = (x_i, y_i, z_i)^{\mathrm{T}}, (i = 1, 2, 3)$. In this case, the relationship of known information [position of transponder #$i$, ($\boldsymbol{x}_i$), and the

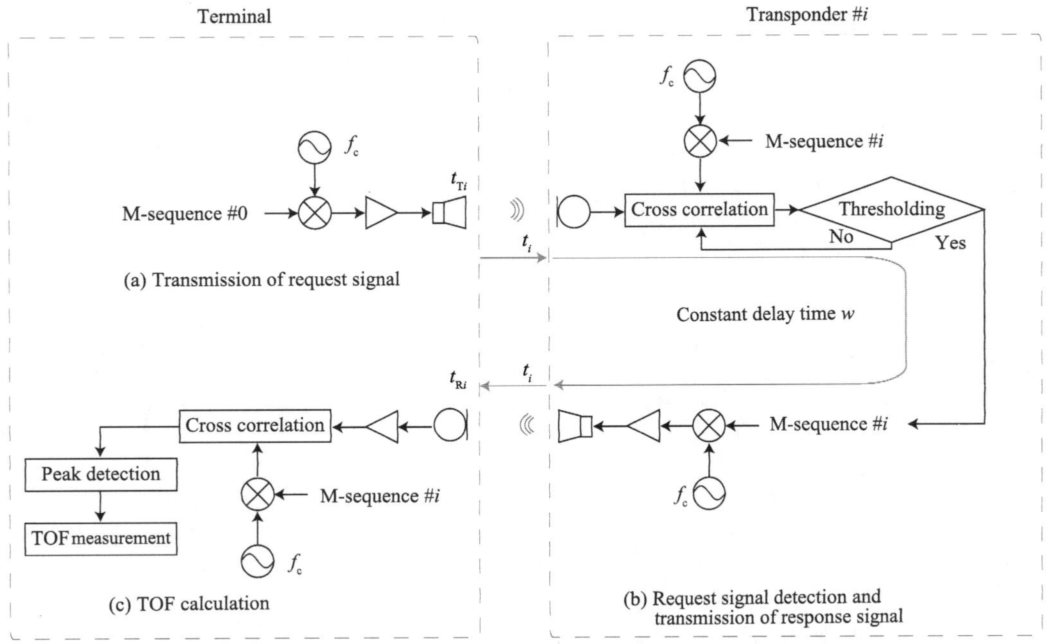


**Figure 2.** Schematic view of measurement procedure of TOF [11].

distance between the terminal and transponder #i, ($\rho_i$)] and unknown information [position of terminal, $\boldsymbol{x} = (x, y, z)^{\mathrm{T}}$ ] is expressed as

$$\begin{aligned} \rho_i &= ct_i, \\ \rho_i^2 &= (x_i - x)^2 + (y_i - y)^2 + (z_i - z)^2, (i = 1, 2, 3). \end{aligned} \tag{2}$$

Because Eq. (2) is a nonlinear equation, the terminal solves Eq. (2) using iterative methods such as the Newton-Raphson method and obtains $\boldsymbol{x} = (x, y, z)^{\mathrm{T}}$, as well as existing positioning schemes.

### *2.2. Effects of terminal movement*

The transponder-based positioning method assumes that the positioning is performed in a static/quasi-static environment. However, if we consider positioning in a dynamic environment, the following two problems may become an issue. The first problem is the increase of random error in the measurement of TOF. In detail, when the terminal transmits the request signal or receives the response signals, the effect of Doppler is caused in both signals by the terminal movement. Because the effect of Doppler stretches the request and the response signals in time, it brings deterioration on the cross-correlation characteristic. In other words, the peak-to-average-power-ratio (PAPR) of the cross-correlation function becomes small as the effect of Doppler increases. This means that the measurement of TOF becomes sensitive to the noise, so that the random error increases. Secondly, the terminal movement may cause offset in the measurement of round-trip TOF because the position of the terminal changes during the measurement of TOF. The TOF obtained from this measurement may be different from both of the transmission of the request time and the reception of the response time.

## 3. Simulations and Experiments

In this section, we tested the effects of the terminal movement on positioning accuracy in the transponder-based acoustical positioning by simulations and experiments. In detail, we firstly evaluated the effect of

IOP Conf. Series: Journal of Physics: Conf. Series **1075** (2018) 012025 doi:10.1088/1742-6596/1075/1/012025

**Table 1.** Parameters used in simulations and experiments.

| | |
|---|---|
| Sampling frequency | 50 kHz |
| Carrier frequency | 5 kHz |
| M - sequence order | $7^{th}$ |
| Chip time | 0.2 ms |
| Moving velocity | 0 to 1.2 m/s |

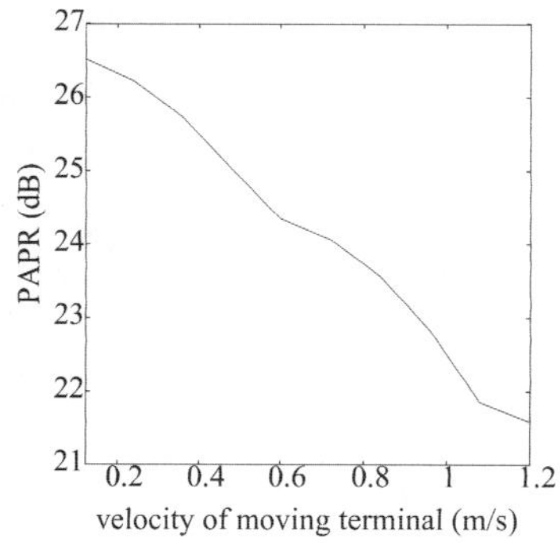


**Figure 3.** Simulations of velocity of moving terminal versus PAPR.

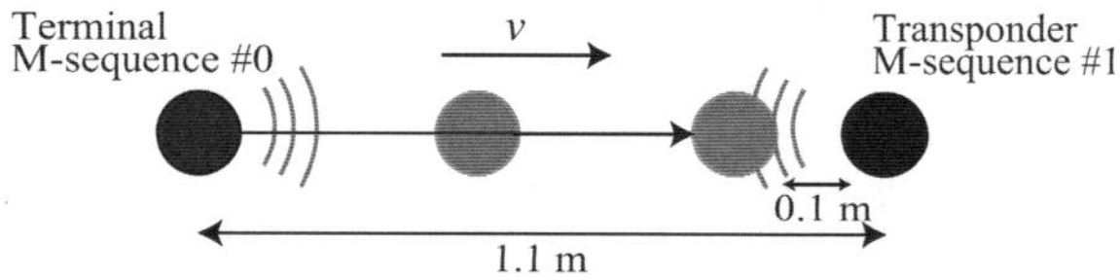


**Figure 4.** Schematic view of experiment set up.

Doppler on the measurement of TOF in simulations. Moreover, we also evaluated the effect of the terminal movement on the measurement of TOF in experiments.

### *3.1. Simulations*

Table1 shows parameters used in simulations. We generated a request signal by modulating carrier frequency of 5 kHz by M-sequence. We calculated the received response signal by resampling the transmitted request signal using the velocity information of the terminal. Then, we calculated a cross-correlation function between the original and resampled signals. The PAPR of the cross-correlation function becomes

$$PAPR = \frac{\max(f(\tau)^2)}{\mathrm{mean}(f(\tau)^2)}, \tag{3}$$

where $f(\tau)$ is the cross-correlation function and $\tau$ is the time lag between signals. Figure 3 shows the simulation results. In this figure, the horizontal axis shows the velocity of the terminal and the vertical axis shows PAPR. As shown in Fig. 3, as the velocity of the terminal increases, the cross-correlation function peak decreases. Hence, it is expected that the random error of TOF will increase as the velocity of the terminal increases.

### *3.2. Experiments*

In the simulations, it was expected that the random error of TOF increases as the effect of Doppler increases. In the experiments, we also evaluate the effect of Doppler on the measurement of TOF.

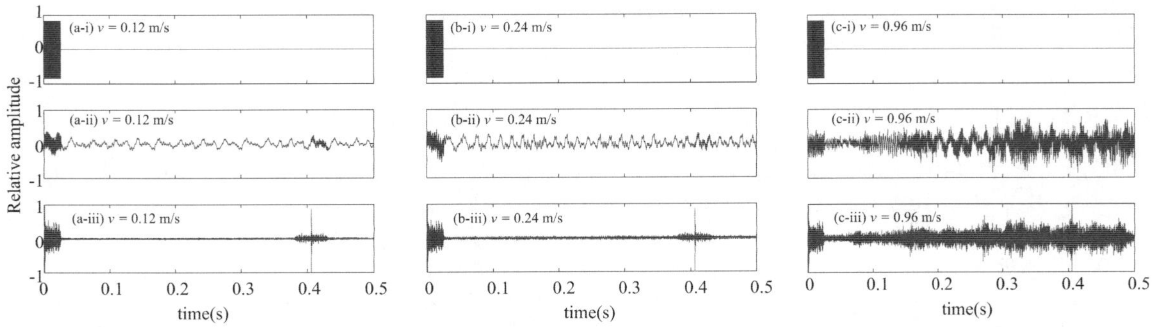


**Figure 5.** Examples of signal processing; (i) transmitted response signal, (ii) received response signal, (iii) cross-correlation between (ii) and (i), (a) $v$ = 0.12 m/s, (b) $v$ = 0.24 m/s, and (c) $v$ = 0.96 m/s.

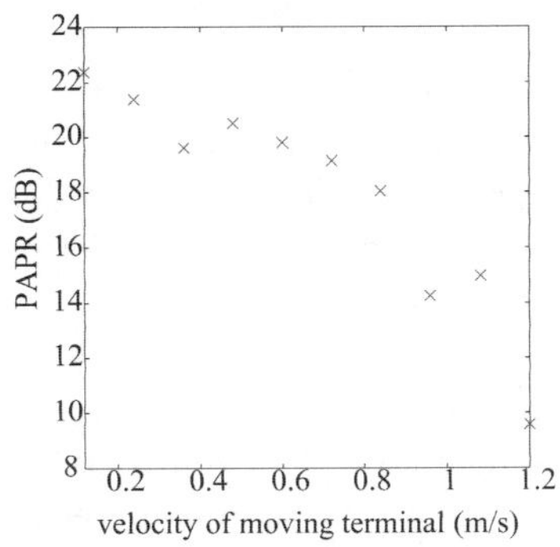


**Figure 6.** Relationships between velocity of moving terminal and PAPR.

Moreover, we evaluate the effect of the terminal movement on the measurement offset of TOF.

Figure 4 shows the experimental set up in an anechoic chamber of 3 m (W) × 3 m (D) × 3 m (H) size. Loudspeakers (Fostex, P650K), omnidirectional microphones (DB Products Limited, C9767), A-D/D-A converters (National Instruments, USB-6221 and 6259), and personal computers (PCs) were used in both a terminal and a transponder. The terminal and the transponder were set on a linear actuator (Misumi, RSB2, 1.2 m). Transmitting signals and receiving signals were processed using a measurement software (National Instruments, LabVIEW) on PC. Acoustical parameters such as TOF and terminal position were calculated by a software (MathWorks, MATLAB) on PC. We used parameters shown in Table 1 as well as simulations. Using above parameters, ranging experiments were performed. The terminal approaches the transponder at various velocities $v$ (0.12 to 1.20 m/s, every 0.12 m/s) and transmits/receives signal 10 times from the fixed transponder.

Figure 5 shows examples of signal processing result. As shown in Figs. 5 (a-iii), 5(b-iii), and 5(c-iii), the cross-correlation functions between the transmitted and received response signals have a peak. However, as the velocity of the moving terminal increases, the PAPR tends to decrease as the velocity increases.

Figure 6 shows the relationships between the velocity of the moving terminal and the PAPR of the cross-correlation function obtained from the experiments. As shown in this figure, the velocity of the terminal increases, the PAPR also decreases as well as the simulation results. Hence, it is expected that the random error of TOF increases as the velocity of the moving terminal increases.

The relationships between the measured distance and the true distance is shown in Fig. 7. The horizontal axis shows the true distance between the terminal and the transponder (when the terminal transmits the request signal). The vertical axis shows the measured distance obtained from the

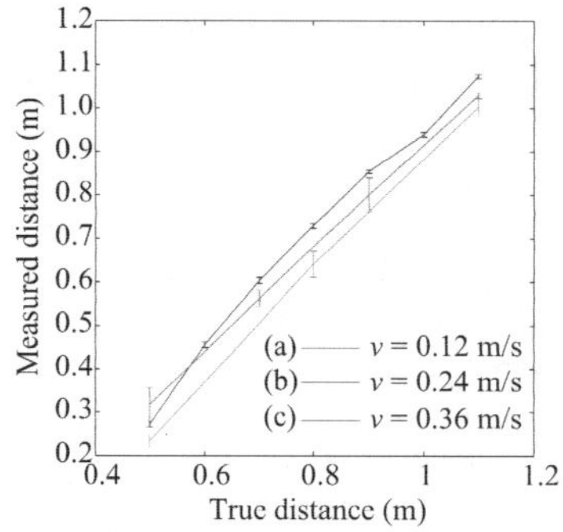


**Figure 7.** Relationships between measured distance and true distance in various velocity of terminal movement; (a) $v$ = 0.12 m/s, (b) $v$ = 0.24 m/s, (c) $v$ = 0.36 m/s.

measurement of TOF. Focusing on the random error, as the velocity of the moving terminal increases, the variance also increases. Focusing on the measurement offset of TOF, the velocity of the moving terminal increases, the offset value tends to large. From the above experiment results, we found that the both affect on the terminal movement.

## 4. Conclusion

To achieve simple and cost effective positioning in large-scale area, we proposed a transponder-based positioning. We evaluated the effects of the terminal movement on positioning in the simulations and the experiments. From the simulations, we found that the measurement error of TOF increases as the effect of Doppler increases. From the experiments, we observed same tendency as the simulations. Focusing on the random error of TOF, as the velocity of the moving terminal increases, the variance also increases. Focusing on the measurement offset of TOF, the velocity of the moving terminal increases, the offset value tends to large. The obtained results suggest that the proposed method can achieve positioning accurately when the velocity of the moving terminal is low. On the other hand, it was also found that as the velocity of the moving terminal becomes higher, additional signal processing that cancels the effect of the Doppler and offset of TOF may be required. Positioning experiments using multiple transponders in the large-scale indoor environment is one of our future work.

## References

[1] Widodo S, Shiigi T, Hayashi N, Kikuchi H, Yanagida K, Nakatsuchi Y, Ogawa Y and Kondo N 2013 *Robotics* **2** 36–53
[2] Priyantha N B, Miu A K, Balakrishnan H and Teller S 2001 *Proceedings of the 7th annual international conference on Mobile computing and networking* (ACM) pp 1–14
[3] Ogiso S, Kawagishi T, Mizutani K, Wakatsuki N and Zempo K 2015 *ROBOMECH Journal* **2** 12
[4] Hashizume H, Kaneko A, Sugano Y, Yatani K and Sugimoto M 2005 *Proc. TENCON 2005-2005 IEEE Region 10 Conference* (IEEE) pp 1–6
[5] Mandal A, Lopes C V, Givargis T, Haghighat A, Jurdak R and Baldi P 2005 *Proc. Second IEEE Consumer Communications and Networking Conference, 2005. CCNC. 2005* (IEEE) pp 348–353
[6] Hirata S, Kurosawa M K and Katagiri T 2008 *IEICE Transactions on Fundamentals of Electronics, Communications and Computer Sciences* **91** 1031–1037
[7] Quintana B, Prieto S A, Adán A and Bosché F 2016 *Indoor Positioning and Indoor Navigation (IPIN), 2016 International Conference on* (IEEE) pp 1–8
[8] Zhang W, Hua X, Yu K, Qiu W and Zhang S 2016 *Indoor Positioning and Indoor Navigation (IPIN), 2016 International Conference on* (IEEE) pp 1–5
[9] Sakai N, Zempo K, Mizutani K and Wakatsuki N 2016 *Indoor Positioning and Indoor Navigation (IPIN), 2016 International Conference on* (IEEE) pp 1–4
[10] Hostettler R and Särkkä S 2016 *Indoor Positioning and Indoor Navigation (IPIN), 2016 International Conference on* (IEEE) pp 1–8
[11] Iwaya H, Mizutani K, Ebihara T and Wakatsuki N 2017 *Japanese Journal of Applied Physics* **56** 07JC07

# Experimental acoustic modal analysis of an automotive cabin: challenges and solutions

**G. Accardo [1], P. Chiariotti [2], B. Cornelis [1], M. El-Kafafy [3,4], B. Peeters [1], K. Janssens [1], M. Martarelli [5]**

[1] Siemens Industry Software NV, Interleuvenlaan 68, B-3001 Leuven, Belgium
[2] Università Politecnica delle Marche, Dipartimento di Ingegneria Industriale e Scienze Matematiche, Via Brecce Bianche 12, 60131 Ancona, Italy
[3] Vrije Universiteit Brussel, Acoustic and Vibration Research Group, Pleinlaan 2B, B-1050 Brussels, Belgium
[4] Helwan University, Cairo, Egypt
[5] Università degli Studi eCampus, Via Isimbardi 10, 22060 Novedrate (CO), Italy

E-mail: giampiero.accardo@siemens.com

**Abstract.** In this paper, a full Acoustic Modal Analysis (AMA) procedure to improve the CAE predictions of the car interior noise level is proposed. Some of the challenges that can be experienced during such an analysis are described and new solutions to face them are proposed. Particular AMA challenges range from the arrangement of the experimental setup to the post-processing analysis. Since a large number of microphones are needed, a smart localization procedure, which automatically determines the microphone three dimensional (3-D) positions and dramatically reduces the setup time, is presented herein. Furthermore, the need for a large number of sound sources spread across the cavity to assure a homogeneous sound field makes modal parameter estimation a nontrivial task. Traditional modal parameter estimators have indeed proven not to be effective in cases where many input excitation locations have to be used. Hence, a more suitable estimator, the Maximum Likelihood Modal Model-based (ML-MM) method, will be employed for such an analysis.

## 1. Introduction

Nowadays, the automotive industry is asked to fulfil ever more demanding requirements for noise reduction and passenger comfort. Design engineers are asked to face the big challenge of reducing in-vehicle noise and improving passengers acoustic experience by keeping the intervention costs to a minimum. It is clear that the adoption of a numerical method to perform vibration and acoustic analyses is attracting increasing attention because of its merits in saving costs and time. Nevertheless, it is obvious that the effectiveness of this approach greatly depends on the accuracy of the predictions made using such models. So, in order to guarantee reliable simulations of the interior sound field of a vehicle cabin, experimental Acoustic Modal Analysis (AMA) can be considered a must-be-performed step, since it allows for validation and updating of these numerical models, and improvement of the overall modelling know-how.

In this paper, a full procedure to perform an experimental acoustic modal survey of an automotive cabin is proposed, with the goal of providing useful and practical guidelines. The particular challenges of such an analysis, which range from the test preparation to the post-processing analysis, will be shown, and different solutions will be presented.

IOP Conf. Series: Journal of Physics: Conf. Series **1075** (2018) 012026 doi:10.1088/1742-6596/1075/1/012026

## 2. Formulation of the acoustic problem

Consider a cavity of volume $\mathcal{V}$ enclosed by a surface $\Omega = \partial\mathcal{V}$ and excited by a point monopole of volume velocity per unit volume $q$, located at the arbitrary point $\mathsf{r}_0$. If the surface $\Omega$ is acoustically rigid, the governing equation of the system is [3]:

$$\nabla^2 p(\mathsf{r}, t) - \frac{1}{c^2}\frac{\partial^2 p}{\partial t^2}(\mathsf{r}, t) = -\rho\frac{\partial q}{\partial t}\delta(\mathsf{r} - \mathsf{r}_0), \quad (1)$$

where $p$ is the acoustic pressure, which is a function of space $\mathsf{r}$ and time $t$, $\nabla^2$ is the Laplace operator, $c$ is the speed of sound and $\rho$ the density of the medium, and the source function is represented mathematically by a delta function.
Assuming now that a number of point monopoles of known volume velocity per unit volume are placed in the cavity and the sound pressure across the volume is sampled at an appropriate number of points, it can be shown that the continuous wave equation (1) can then be substituted by its discrete equivalent:

$$\mathsf{M}_F\ddot{\mathsf{p}} + \mathsf{C}_F\dot{\mathsf{p}} + \mathsf{K}_F\mathsf{p} = -\rho\dot{\mathsf{q}}, \quad (2)$$

where, to preserve the analogy with a structural finite element model, the matrix $\mathsf{M}_F$ is called the acoustic mass matrix; the matrix $\mathsf{C}_F$ is the acoustic damping matrix; the matrix $\mathsf{K}_F$ is called the acoustic stiffness matrix. In view of this equivalence between acoustics and structural dynamics, it can be concluded that the classical modal parameter estimation approach can be followed also in the acoustic modal analysis case.

## 3. Test preparation, test model creation and setup

In this section, the measurement setup and equipment needed for typical AMA tests will be described. Details on the sound sources and their position, on the number of sensors, their spatial distribution, and their mounting inside the cavity will be given.

### *3.1. Sound sources*

Calibrated volume velocity sources are necessary to measure acoustic FRFs that are required in AMA tests. The sound sources have to be omnidirectional and have a negligible size in order not to influence the field, especially in the higher frequency range. So the need exists for a dedicated source that is compact, omnidirectional and capable of generating high noise levels. The LMS Qsources Low-Frequency Monopole Sound Source (Q-MED) can fulfil such requirements (Fig. 1). It is a unique monopole sound source that has been developed to acquire acoustic and vibro-acoustic FRFs accurately without disturbing the acoustic behaviour of the passenger compartment.

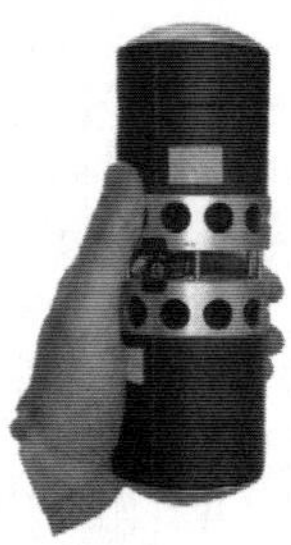

Figure 1: LMS Qsources low-frequency monopole sound source

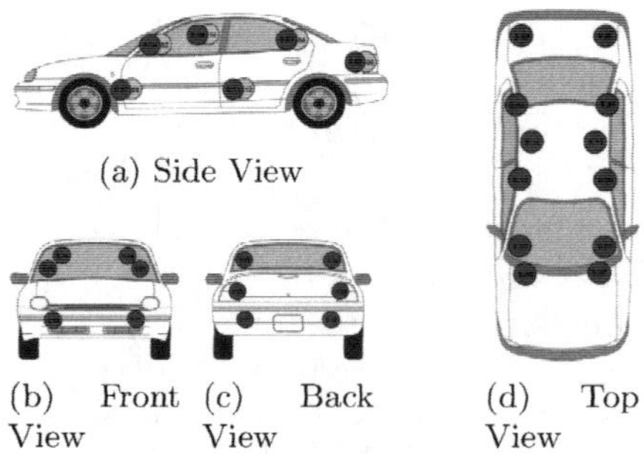


Figure 2: Reference distribution

As shown in [1, 4, 5], an appropriate source distribution over the entire cabin is required to properly estimate the acoustic modes. With reference to the analysis reported in [1], where up to 12 volume velocity sources were set in geometrically symmetric locations, close to the edges, corners and at the maximum amplitude locations to avoid nodal lines and excite close to pressure maxima on the boundaries, a typical distribution of sound sources is displayed in Fig. 2. Too few sources do not allow for correct identification of the mode shapes as exciter-location-dependent mode shape distortions are clearly visible [1]. For this reason, it is highly advisable to use a rather large number of sources and source locations.

### *3.2. Sensor Placement*

In order to guarantee a good description of the mode shapes and a sufficient number of degrees of freedom for the updating, a sensor layout as the one depicted in Fig. 3 is usually employed.

Figure 3: Wireframe model

Sensors need to be uniformly spread across the whole cabin, even in extreme positions, such as in foot regions, between the windshield and the dashboard, and in the hat shelf region. With reference to the test case reported in [1], more than 500 measurement points are required. Determining the exact location of so many sensors by hand would be extremely cumbersome. Hence, automatic methods would be essential for drastically reducing the setup time and localizing microphones in a smarter way. For this purpose, a fast, accurate and cost-effective procedure has been developed and validated [6–8].

#### *3.2.1. Smart localization procedure*

The microphone localization procedure employed is based on multilateration: the distances between at least four sources (anchors), whose coordinates are known or estimated *a priori*, and a microphone (target) are utilized to determine the unknown position of the microphone in three-dimensional (3-D) space. However, due to the complex structure and obstructions typical of a car cabin (e.g., seats, dashboard, etc.), the direct path may be obstructed, a so-called Non-Line-Of-Sight (NLOS) condition. As a consequence, acoustic range estimates based on the time-of-arrival (TOA) may have an erroneous positive bias, i.e., the signal arrives at a microphone through reflections instead of through the direct (shortest) path.

The proposed method copes with the problem of NLOS through an identification and discard (IAD) algorithm: the erroneous NLOS measurements are detected and pruned, so that the microphones are localized using the LOS distances only, hence yielding more accurate 3-D localization results. For further details, the reader can refer to [6].

For multilateration, the sound source should ideally be a monopole. It means that, within the frequency range of interest, the characteristic length of the source must be smaller than the minimum wavelength, and the source itself must be always omnidirectional. For this purpose, a new, dedicated and compact LMS Qsources volume source has been utilized (Fig. 4). These LMS Qsources volume velocity sources are designed to be used in a high frequency band (1 to 20 kHz), which is an important bandwidth to ensure accurate TOA measurements [9].

Figure 4: New compact LMS Qsources volume velocity source

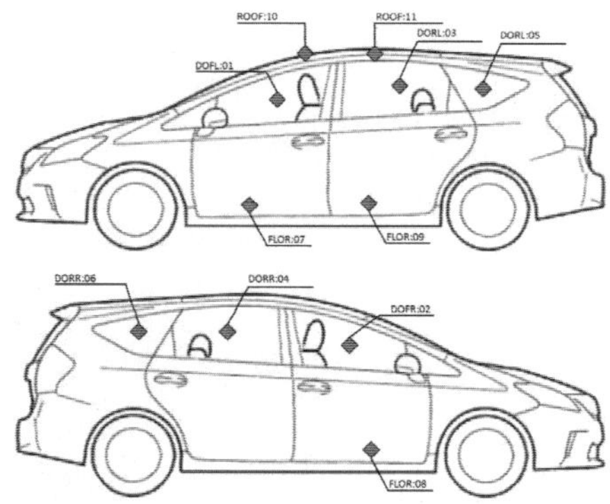


Figure 5: Source Distribution

The anchors should be placed in strategic positions so as to localize the largest number of microphones. With reference to the experimental case reported in [8], where microphones were placed in critical positions, an example of anchors configuration is illustrated[1] in Fig. 5. During the measurements, the averaged temperature of the environment must be recorded in order to calibrate the speed of sound, whereby the temperature is assumed to be constant throughout the cabin.
The linear frequency modulated (chirp) signal have been used for this application.
In order to have a qualitative idea of the effectiveness of the method, the coordinates from the CAD model are assumed as a reference. The discrepancies between the (inaccurate) CAD positions and the acoustically estimated coordinates do not allow for an absolute localization error quantification, but they are sufficient for demonstrating the effectiveness of the approach in a complex scenario, such as a car cabin.
In Fig. 6, a comparison is made between a localization algorithm were the pruning is not applied (i.e. NLOS distances remain present), versus the considered localization algorithm where erroneous distances are identified and discarded[2]. As visible in Fig. 6, the application of the NLOS IAD algorithm is not only effective, but also essential for a correct localization of all microphones.

## 4. Modal Parameter Identification

It has been observed in [1, 4, 5, 10] that it is quite challenging for classical modal parameter estimation methods to curve-fit an FRF matrix with so many columns (12 references, as reported in § 3.1); typically, not all references are well fitted for a particular sensor location. Therefore, there is a need for a new solver capable to overcome such a difficulty. The Maximum Likelihood Modal Model-based (ML-MM) modal parameter estimator [2] has been proven to be more suitable for such a kind of data [1, 10]. A brief description of the estimator is reported in the follow.

[1] The model of the car is for illustrative purposes only.
[2] The results have been obtained by using all the 11 sources, imposing a threshold $\zeta = 2$ cm, and using as initial guess the centre of gravity of the anchors.

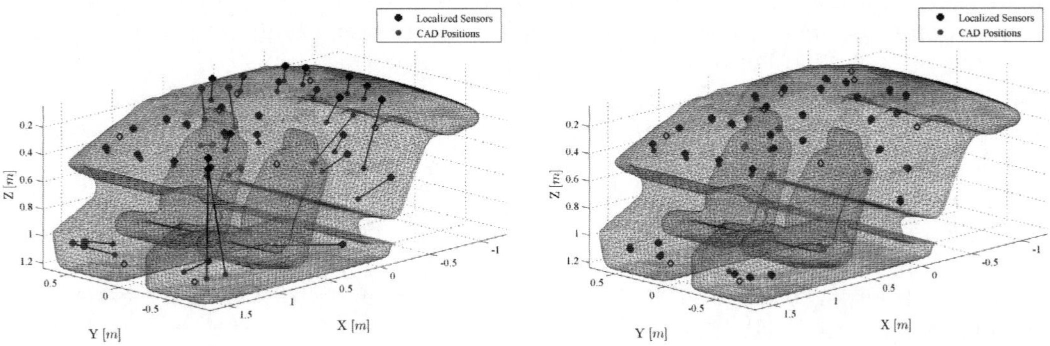


(a) (b)

Figure 6: Localized microphones without (a) and with (b) the application of the NLOS IAD algorithm

*4.1. Maximum Likelihood Modal Model-based method*

The so-called ML-MM method is a multiple-input, multiple output (MIMO) frequency-domain estimator providing global estimates of the modal model parameters. Since it is an iterative method based on solving a non-linear optimization problem, initial values for the modal model parameters (i.e., poles, participation factors, mode shapes, lower and upper residuals) are needed to start the optimization process. In the first step, the Polymax method [?] is applied to the FRFs to obtain the initial estimates for the poles and the participation factors of the physical modes within the analysis band. Then, initial values for the mode shapes and the lower and upper residuals are estimated in a complementary step using the so-called Least-Squares Frequency Domain (LSFD) estimator [11]. In the next step, once the initial values for the entire modal model parameters are obtained, the ML-MM solver starts minimizing the error between the modal model equation and the measured data in a maximum-likelihood sense. Assuming the different measured FRFs to be uncorrelated, the ML-MM cost function to be minimized can be formulated as:

$$K_{\text{ML-MM}}(\theta) = \sum_{o=1}^{N_o}\sum_{i=1}^{N_i}\sum_{k=1}^{N_f} \frac{\left|H_{oi}(\omega_k) - \hat{H}_{oi}(\theta,\omega_k)\right|^2}{\sigma^2_{H_{oi}}(\omega_k)}, \tag{3}$$

where $N_0$ is the number of outputs, $N_i$ the number of inputs, $N_f$ the number of frequency lines, $\omega_k = 2\pi f_k$ the circular frequency at frequency $f_k$ [Hz], $H_{oi}(\omega_k) \in \mathbb{C}$ is the measured FRF, $\hat{H}_{oi}(\theta,\omega_k) \in \mathbb{C}$ the modelled FRF, and $\sigma^2_{H_{oi}}(\omega_k) = \text{var}[H_{oi}(\omega_k)] \in \mathbb{R}$.

Assuming volume acceleration FRFs, $\hat{\mathsf{H}}(\theta,\omega_k) \in \mathbb{C}^{N_o \times N_i}$ can be represented using the modal model formulation [11]:

$$\hat{\mathsf{H}}(\theta,\omega_k) = \sum_{r=1}^{N_m}\left(\frac{\phi_r \mathsf{l}_r}{j\omega_k - \lambda_r} + \frac{\phi_r^* \mathsf{l}_r^*}{j\omega_k - \lambda_r^*}\right) + \frac{\text{LR}}{(j\omega_k)^2} + \text{UR}, \tag{4}$$

where $N_m$ is the number of the identified modes, $\phi_r \in \mathbb{C}^{N_o \times 1}$ is the $r$-th mode shape, $\lambda_r$ is the $r$-th pole, $(\bullet)^*$ stands for the complex conjugate of a complex number, $\mathsf{l}_r \in \mathbb{C}^{1 \times N_i}$ is the $r$-th participation factors vector, $\text{LR} \in \mathbb{R}^{N_o \times N_i}$ and $\text{UR} \in \mathbb{R}^{N_o \times N_i}$ are the lower and upper residual terms used to compensate for the out-of-band modes, and $\theta$ is the parameters vector (i.e., $\theta = \{\phi_r, \mathsf{l}_r, \lambda_r, \text{LR}, \text{UR}\}$). The maximum likelihood estimates of $\theta$ are obtained by using a Gauss-Newton optimization. Furthermore, to ensure convergence, the Gauss-Newton

IOP Conf. Series: Journal of Physics: Conf. Series **1075** (2018) 012026 doi:10.1088/1742-6596/1075/1/012026

optimization is implemented together with the Levenberg-Marquardt approach, which forces the cost function to decrease. More details about the ML-MM method are presented in [2].

### *4.2. Considerations and results*

The ML-MM method has been proven to outperform more classical modal parameter estimators with such a kind of data.
With reference to the test case in [1], the initial values generated by applying the Polymax method to the full $526 \times 12$ FRF matrix were improved by applying the ML-MM method. The analysis was stopped after 20 iterations. Nine pure acoustic modes are well identified in the frequency range from 0-200 Hz (Fig. 8). The initial mean fitting error between measured FRFs and Polymax synthesized FRFs was around 9%. The mean fitting error after applying the ML-MM method reduced to only 2%. This improved overall curve fit is illustrated using two typical elements from the full FRF matrix in Fig. 7.

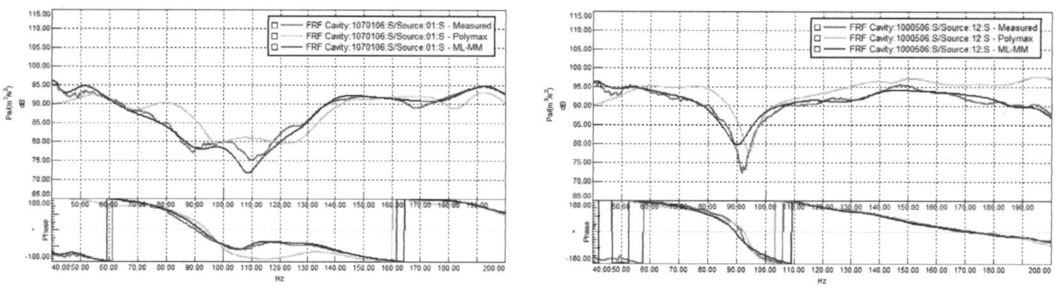


Figure 7: Improved FRF curve-fitting quality shown for two typical FRFs; - Measured (red), Polymax synthesis (green), ML-MM synthesis (blue)

## 5. Conclusions

A full Acoustic Modal Analysis (AMA) procedure to improve the CAE prediction of the the car interior noise level has been proposed and successively validated. The challenges of typical AMA tests are described in details. The huge amount of sensors required to have a good description of the acoustic cavity, and the large number of sound sources to properly excite the cavity of a car cabin make such tests extremely time-consuming and demanding. In order to have a good description of the dynamic behaviour of the system, many sensors are indeed placed inside the cabin. In such a complex environment, determining the microphone positions by hand is not only tedious, but also inaccurate and cumbersome. Furthermore, where many input excitation locations have to be used, it has been observed that traditional modal parameter estimators have proven not to be effective.
In order to face such typical challenges, different solutions are proposed. Firstly, a smart approach capable to automatically localize microphones in such a complex scenario is presented. The method is based on acoustic distance measurements between a microphone and (at least 4) sources. With the introduction of novel algorithms coping with reflections and non-line-of-sight issues, the localization procedure has been proven to be effective, providing reliable results and drastically reducing the measurement set-up time. Secondly, modal parameters are estimated in the frequency range between 40 and 200 Hz, by applying a new modal parameter estimation method, the so-called ML-MM estimator. Nine acoustically dominant modes are identified. Although Polymax still yields good modal parameter estimates, ML-MM provides superior FRF synthesis results and, hence, more reliable values.

IOP Conf. Series: Journal of Physics: Conf. Series **1075** (2018) 012026 doi:10.1088/1742-6596/1075/1/012026

| # | Numerical Modes | # | Experimental Modes | Mode Shapes (Frequency, Damping) |
|---|---|---|---|---|
| 2 |  | 2 |  | I Longitudinal (51.06 Hz, 13.82%) |
| 3 |  | 3 |  | I Longitudinal & Rigid-Body Trunk (81.44 Hz, 14.94%) |
| 4 |  | 4 |  | I Lateral (97.24 Hz, 10.65%) |
| 5 | 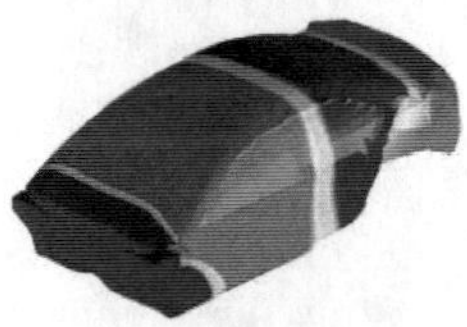 | 5 |  | II Longitudinal & Rigid-Body Trunk (137.79 Hz, 7.25%) |
| 7 |  | 6 |  | I Vertical (148.66 Hz, 13.70%) |
| 6 |  | 7 |  | I Longitudinal & I Lateral (149.34 Hz, 6.57%) |
| 8 |  | 8 |  | I Longitudinal & I Lateral & I Lateral Trunk (150.74 Hz, 11.79%) |
| 9 |  | 9 | 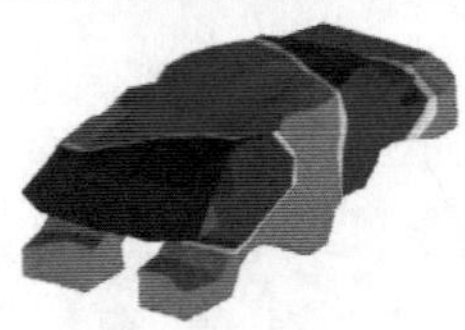 | III Longitudinal (195.13 Hz, 6.05%) |

Figure 8: Identified mode shapes using 12 references (rigid-body mode not shown)

**Acknowledgements**

This work was supported by the Marie Curie ITN project ENHANCED (joined Experimental and Numerical methods for HumAN CEntered interior noise Design). The project has received funding from the European Union Seventh Framework Programme under grant agreement No. 606800. The whole consortium is gratefully acknowledged.

## References

[1] G. Accardo, M. El-kafafy, B. Peeters, F. Bianciardi, D. Brandolisio, K. Janssens, M. Martarelli, *Experimental Acoustic Modal Analysis of an Automotive Cabin*, Sound & Vibration, May (2015), pp. 10-18.

[2] M. El-Kafafy, T. De Troyer, P. Guillaume, *Fast maximum-likelihood identification of modal parameters with uncertainty intervals: A modal model formulation with enhanced residual term*, Mechanical Systems and Signal Processing, Vol. 48, No. 12, Elsevier (2014), pp. 49-66.

[3] F. Fahy, P. Gardonio, *Sound and Structural Vibration. Radiation, Transmission and Response*, Elsevier Ltd., 2nd edition (2007).

[4] T. Yoshimura, M. Saito, S. Maruyama, S. Iba, *Modal analysis of automotive cabin by multiple acoustic excitation*, in *Proceedings of the 25th International Conference on Noise and Vibration Engineering (ISMA2012)*, Leuven, Belgium, September (2012).

[5] B. Peeters, M. El-kafafy, G. Accardo, T. Knechten, K. Janssens, J. Lau, L. Gielen, *Automotive cabin characterization by acoustic modal analysis*, in *Proceedings of the 2014 JSAE Annual Congress (Spring)*, Pacifico Yokohama, Japan, May (2014).

[6] G. Accardo, B. Cornelis, K. Janssens, B. Peeters, *Automatic microphone localization inside a car cabin for experimental acoustic modal analysis*, in *Proceedings of the 22nd International Congress on Sound and Vibration (ICSV 22)*, Florence, Italy, July (2015).

[7] G. Accardo, B. Cornelis, P. Chiariotti, K. Janssens, B. Peeters, M. Martarelli, *A non-line-of-sight identification algorithm in automatic microphone localization for experimental acoustic modal analysis*, in *Proceedings of the 44th International Congress and Exposition on Noise Control Engineering (Inter-Noise 2015)*, San Francisco, California, USA, August (2015).

[8] G. Accardo, B. Cornelis, K. Janssens, B. Peeters, *Smart localization of microphones inside an automotive cabin*, in *Proceedings of the 2016 JSAE Annual Congress (Spring)*, Pacifico Yokohama, Japan, May (2016).

[9] K. Lauterbach, A. Ehrenfried, L. Koop, S. Loose, *Procedure for the Accurate Phase Calibration of a Microphone Array*, in *Proceedings of the 15th AIAA/CEAS Aeroacoustics Conference (30th AIAA Aeroacoustics Conference)*, Miami, Florida, USA, May (2009).

[10] M. El-kafafy, G. Accardo, B. Peeters, K. Janssens, T. De Troyer, P. Guillaume, *A Fast Maximum Likelihood-Based Estimation of a Modal Model*, in *Proceedings of the 33rd International Modal Analysis Conference (IMAC XXXIII)*, Orlando, Florida, USA, February (2015).

[11] W. Heylen, S. Lammens, P. Sas, *Modal Analysis Theory and Testing*, Katholieke Universiteit Leuven, Departement Werktuigkunde, Leuven (2012).

IOP Conf. Series: Journal of Physics: Conf. Series **1075** (2018) 012027 doi:10.1088/1742-6596/1075/1/012027

# Dynamical mechanical characterization of a nanostructured vibration damping layer

**JP Arenas[1], JL Castaño[1] and ML Auad[2]**

[1] Institute of Acoustics, University Austral of Chile, PO Box 567, Edif. 6000 Campus Miraflores, Valdivia, Chile
[2] Polymer and Fiber Engineering Department, College of Engineering, Auburn University, 103 Textile Building, Auburn, AL, USA

E-mail: jparenas@uach.cl

**Abstract.** A vibration damping layer can dissipate relatively large amounts of energy. A known effect of this dissipation of energy is the reduction of the response of a structure to sound and vibration excitation and the reduction of the transmission of sound through structures at high frequencies. Added damping also increases the impedances of structures at their resonances and thus may improve the effectiveness of vibration isolation at these resonances. It has been observed that the molecular behavior of new nanomaterial composites may have significant influence on their shock and vibration insulation properties. Dynamical mechanical characterization of these nanostructured materials is important for determining the potential application as a viscoelastic vibration damping layer. In this work, inclusion of several nanostructures to thermostable polyurethane has been made and their main elastic properties have been measured using current standard methods. It is observed that specific amounts of nanostructures can improve the impact damping capabilities of the material, with the advantage of reduced thickness and weight as compared to traditional materials.

## 1. Introduction

One of the recent experimental and research objectives is to improve the thermal, electrical and mechanical performance of a polymer by adding nanostructures. They include reinforcement nanotubes, nanoparticles (NP) and nanofibers that, depending on their characteristics, improve either the physical or chemical polymer properties [1,2]. The advances in nanotechnology have provided several new materials with diverse mechanical properties which can be used in both acoustic and vibration applications. In particular, the molecular behavior of these new nanocomposites may have a significant influence on their impact and shock absorption [3].

This work aims to evaluate, through vibration and mechanical testing, nanostructured materials with different types of nanoclusters at different mass quantities. The main objective is to identify nanostructured materials for their potential use as viscoelastic layers to reduce impact noise in floating slabs.

## 2. Nanostructured damping layers

To synthesize the nanostructured samples, four types of clay were used: cloisite 15A, cloisite 30B, hectorite, and laponite. Initially, the clays were dried for 8 hours in an oven to 110°C. A polymer matrix thermostable polyurethane (PU 455-203 # 1) was used and proportions of 0.5%, 1%, 2%, 5%

and 10% in mass for each clay where chosen to synthesize the films. Subsequently, the polymer in pellet form was mixed with tetrahydrofuran (THF) and N, N'-dimethylformamide (DMF), with mass ratio of 20% for the polymer, THF to 40% m/m and DMF to 40% m/m. The dried clay was then added to the mixture and samples were shaken during 30 min in an intensive mixer. Subsequently, the samples were mixed again during a period of 12 hours in a magnetic stirrer.

After the mixing process, the stirring process was repeated for 30 min in the intensive mixer. Then, the samples were agitated by applying sound energy during 30 min in a sonifier Branson with 10% intermittent amplitude as well as pulsed (30 s on, 30 s off). Finally, the samples were subjected to 30 min of stirring in the intensive mixer and, again, in the sonifier during 30 min. Samples were poured into teflon molds and cured in an oven at 45° C for 12 h. Additionally, single wall (SWCNT) and multiwall (MWCNT) carbon nanotubes samples with concentrations of 0.5% and 21% m/m were made for comparison purposes. The process of mixing and drying was the same for the case of samples with NP and no additional chemical modifications were made.

## 3. Experimental results

In order to evaluate the mechanical properties of the nanostructured materials, a dynamic mechanical analysis (DMA) was used. The working principle of DMA is to apply a controlled sinusoidal deformation or strain to a sample of knows geometry [4]. Thus, to study the mechanical properties through the DMA, it is possible to separate the main mechanical properties of a material into two parts: the elastic modulus corresponding to the elastic response under an applied strain, and the loss modulus, which indicates the damped response to the applied strain. Finally, the loss factor shows the exchange rate between the loss modulus and elastic modulus [5]. In addition, the procedure defined in ISO 9052-1 [6] was used to evaluate the dynamic stiffness of the material samples. It is known that the noise reduction properties of a viscoelastic layer are related to its dynamic stiffness [7].

### *3.1. Mechanical test using DMA*

After evaporating THF and DMF solvents from the samples in the oven, both density and thickness measurements of each nanostructured film were performed. Subsequently, tests were carried out using a commercial dynamic mechanical analyzer (DMA). The samples were cut with dimensions of $7 \pm 2$ cm long, $0.5 \pm 0.1$ cm wide with a thickness of $0.35 \pm 0.05$ mm. Furthermore, to have more reliable values, five samples of each nanostructured polymer were measured for each experiment.

To evaluate the Young's modulus E of the samples, mechanical traction on the DMA test was performed for each nanostructured film under a constant extension rate of 10 mm/min at room temperature. In addition, all samples were brought to the point of breakage. After obtaining their responses against static tensile strain, the Young's modulus was extracted from the linear elastic section of each sample.

To evaluate the dynamic mechanical properties of polymers (dynamic modulus and loss factor), the experimental conditions were a constant room temperature and a frequency sweep signal from 0.1 Hz to 90 Hz.

Table 1 shows the increase of the Young's modulus of each nanostructured sample in relation to the pure polymer matrix thermostable polyurethane ($E = 1.542 \pm 0.183$ MPa). We notice that E increases with an increase in the amount of NP in all samples (except for cloisite 30B which remains constant from amounts larger than 2% m/m). The increase in Young's modulus is caused by the exfoliation that the material experiences and the bonds generated between the organic and inorganic nanoclay. Adding NP to the polymer matrix causes the high-E of NP to modify the low-E of the polymer matrix, which results in an increase of the Young's modulus of the nanostructured material. Thus, as reported previously, the greater the amount of NP in the PU, the higher the E in the nanostructured film [8]. However, at higher values of E, the fragility of the material increases (which increases with the slope of the elastic region of the static tensile curves), which causes breakage more easily [4].

Figure 1 shows the values of loss factor as function of frequency for all samples of NP and CNT with 0.5% m/m concentration. It is noted the apparent increase in the loss factor for SWCNT and MWCNT samples. This is due to the mobility of the CNTs and their Van der Waals bonds with the polymer chains [9]. All loss factor curves showed a slight drop in the slope due to the viscoelastic behavior of materials generated by an increase in the elastic behavior [10]. When the frequency is increased, the vibration of the viscoelastic material is inhibited by the polymeric chains, resulting in a stiffer material.

**Table 1.** Increase in percentage of the Young's modulus of each nanostructured sample in relation to the pure polymer matrix thermostable polyurethane (E = 1.542 ± 0.183 MPa).

| Material's concentration (% in mass) | Cloisite 15A (%) | Cloisite 30B (%) | Laponite (%) | Hectorite (%) |
|---|---|---|---|---|
| 0.5 | 110 | 47.4 | 86.7 | 101.6 |
| 1.0 | 295 | 66.3 | 96.9 | 164.7 |
| 2.0 | 340.9 | 100.6 | 135.2 | 164.5 |
| 5.0 | 984.1 | 83.4 | 227 | 657.5 |
| 10.0 | 4646.5 | 92 | 2495.1 | 1820.1 |

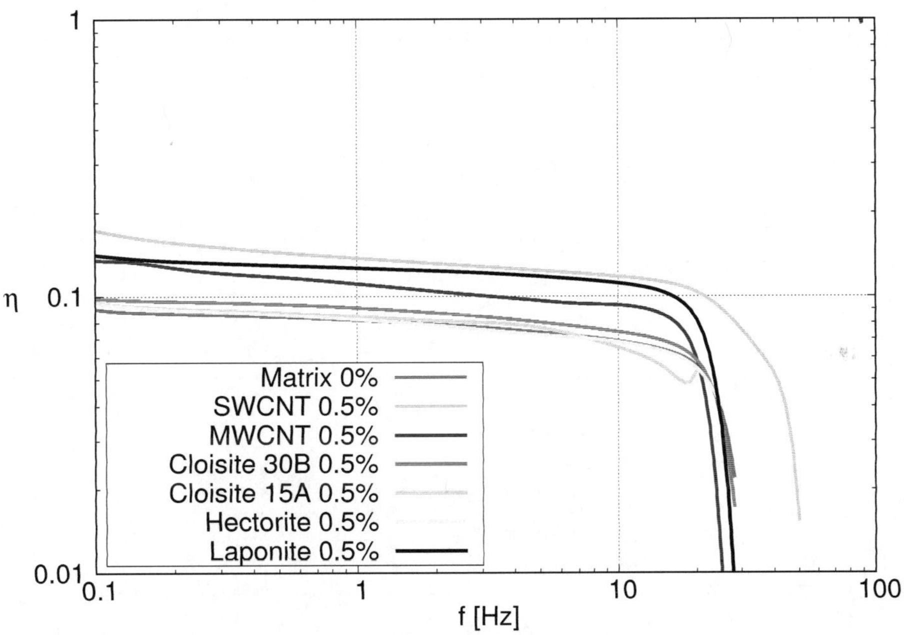


**Figure 1.** Loss factor η versus frequency from 0.1 Hz to 90 Hz under constant temperature of nanostructured polymers at 0.5% concentration of added nanostructures.

### *3.2. Dynamic stiffness test*

Based on the standard ISO 9052-1 [6], which specifies the process for determining the dynamic stiffness, 8 ± 0.5 kg-steel plate of dimensions 200 ± 3 mm × 200 ± 3 mm was used. The samples were placed and centered on two heavy marble plates 400 mm x 400 mm which avoided vibration interference measurement between the nanostructured films and direct surface. Data acquisition and frequency responses were performed by a FFT real-time signal analyzer connected to a laptop via a LAN port. The sensors used were an accelerometer and an impact modal hammer (with a steel, rubber and hard plastic end). Handling the data acquisition process and data analysis was performed by specialized software. Results obtained for single layer of these nanostructured materials have been published before by the authors [11].

Figure 2 shows the results of the dynamic stiffness using nanostructured materials made up of 10 layers, resulting in approximately 3.5 mm thickness for each sample. We observe that the stiffness changes with an increase in the concentration of NP present in the samples. In addition, the behavior of the dynamic stiffness of materials at 0.5% m/m concentration is like the polymer matrix. However, increases of NP beyond 1% m/m for all samples show a dissimilar behavior between the different nanostructures.

For samples with cloisite 15A, 30B and hectorite with NP ≥ 2% m/m the dynamic stiffness presents a slight fall as the amount of NP is increased. Laponite has a remarkable fall in the dynamic stiffness value below the corresponding value of pure polyurethane. In laponite films, there is a clear decrease in dynamic stiffness for 5% and 10%. These films demonstrate the best stress forces distribution between the laponite and the polymer matrix. It is presumed that the laponite morphology can intervene in the increase of energy dissipation because this nanoclay has the highest loss factors.

Finally, by comparing the values of the results obtained from nanostructured materials, it is noted in Fig. 2 that the dynamic stiffness values are in some cases comparable to those of common sound insulation materials such as glass and mineral wool with thickness of 13 mm and 33 mm respectively, but requiring much smaller thicknesses.

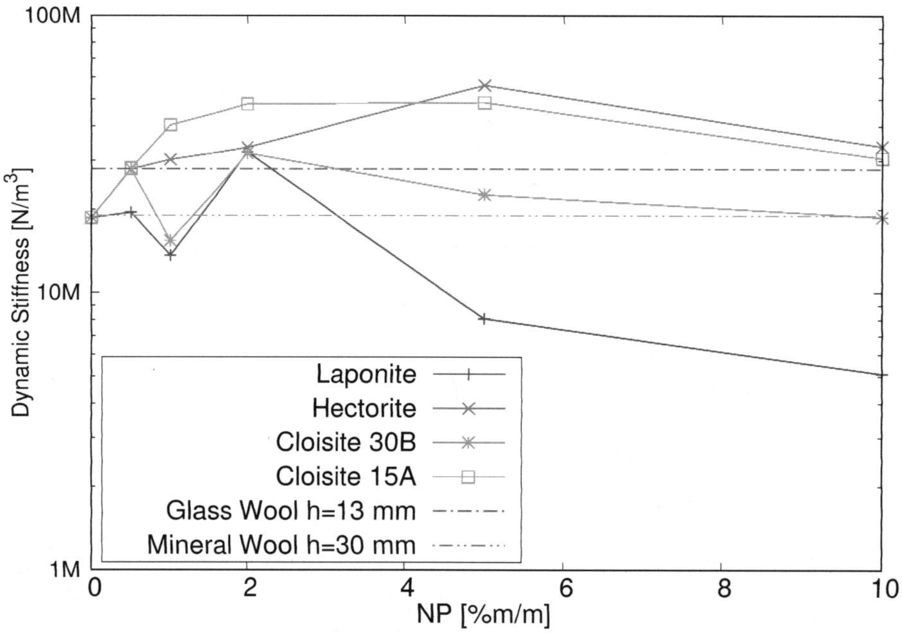


**Figure 2.** Test results of dynamic stiffness for nanostructured samples composed of 10 layers. The lines show the value of dynamic stiffness for glass wool with thickness 13 mm and mineral wool with 30 mm.

By linking the high loss factor values with lower values of E (implying a low dynamic stiffness) will allow one to obtain a material with good impact sound reduction properties. In all materials, the amount of NP is proportional to E and the loss factor. Therefore, if there is a change in the amount of NP in the polymer matrix this will change both its loss factor $\eta$ and their respective dynamic stiffness. It has been shown [12,13] that the index of impact sound reduction $\Delta L_W$, in dB, for a floating slab of mass per unit area $m$ (kg/m$^2$) which is mounted on a resilient layer of dynamic stiffness $s$ (MN/m$^3$), can be estimated by the equation

$$\Delta L_W = 18 + 15\log\frac{m}{s}. \quad (1)$$

Table 2 reports the results of the index of impact sound reduction of a 6 cm thick homogenous concrete (144 kg/m$^2$) floating slab with 10 elastic nanostructured layers. The layers are made up of the nanostructured polymer matrix with added laponite nanoclay for 5% m/m and 10% m/m concentrations. The results are compared with those obtained using typical glass and mineral wool [14].

For the particular case of laponite, dynamic stiffness is less than that of the polymer matrix for samples with NP $\geq$ 5%, which shows that although there is a high E, the loss factor indicates that nanoparticles are involved in the dissipation of vibration energy more efficiently. This is observed with the reduction in dynamic stiffness and, consequently, in increasing $\Delta L_W$. We notice that in the case of laponite with 5% and 10% m/m concentration, the index of impact sound reduction of this 3.5 mm thick layer is 5.6 dB and 8.5 dB greater, respectively, than that of a 30 mm thick mineral wool [14].

**Table 2.** Properties and acoustic performance of nanostructured samples at two different concentrations and two common materials used as viscoelastic layers in a 6 cm thick homogeneous concrete floating slab (144 kg/m$^2$).

| Material | Density (kg/m$^3$) | Thickness (mm) | Dynamic stiffness (MN/m$^3$) | $\Delta L_W$ (dB) |
|---|---|---|---|---|
| Laponite (5% m/m) | 780 | 3.5 | 8 | 36.8 |
| Laponite (10% m/m) | 784 | 3.5 | 5.1 | 39.7 |
| Cloisite 30B (5% m/m) | 1433 | 3.5 | 22.6 | 30.0 |
| Cloisite 30B (10% m/m) | 871 | 3.5 | 18.8 | 31.3 |
| Glass wool | 36 | 13 | 28 | 28.7 |
| Mineral wool | 140 | 30 | 19 | 31.2 |

## 4. Conclusions

In this study, the effects of nanostructures cloisite 15A, cloisite 30B, hectorite, and laponite on a matrix of thermostable polyurethane with various concentrations of nanostructures were determined. Based on the experimental results, it can be concluded from the mechanical characterization that the loss factor and Young's modulus increase with the amount of nanoparticles added. Furthermore, it was found that there is a strong dependence on the type of nanoparticle added to the polymer matrix.

Material characterization by ISO 9052-1 pointed out that materials may be used as components in damping applications when percentages of NP are higher than 5% m/m. The lowest dynamic stiffness found was for laponite with 10% m/m concentration, with a reduction in stiffness of 72% when compared to the polymer matrix without nanostructures.

The index of impact sound reduction of a floating slab with a viscoelastic layer was also determined from the dynamic stiffness values, and the index of impact noise reduction for thin films of 3.5 mm thickness was obtained. Experimental results obtained for Laponite and cloisite 30B at higher proportions than 5% m/m have indicated that these materials can be used as viscoelastic layers for construction of floating slabs. These results suggest that the use of nanostructured materials may be an alternative to the use of common commercial materials and, in some cases, enhance the impact noise isolation with lower thicknesses.

Further work is planned using SEM to study other parameters of the nanostructured materials, such as their morphology and NP distribution, in order to better explain the physics behind the results presented here.

**Acknowledgments**
This work is supported by CONICYT-FONDECYT No 1171110.

IOP Conf. Series: Journal of Physics: Conf. Series **1075** (2018) 012027 doi:10.1088/1742-6596/1075/1/012027

## References

[1] Karger-Kocsis J and Stoyko F 2009 *Nano-and Micro-Mechanics of Polymer Blends and Composites* (Cincinnati: Hanser Publications)

[2] Sinha Ray S and Okamoto M 2003 Polymer/layered silicate nanocomposites: A review from preparation to processing *Progr. Poly. Sc.* **28** 1539–1641.

[3] Corsaro RD and Sperling LH 1990 *Sound and Vibration Damping with Polymers* ed LH Sperling and RD Corsaro (American Chemical Society) pp 5–22

[4] de Saja-Saenz JA 2002 *Introduction to the Physics of Materials* (in Spanish, Valladolid: Universidad de Valladolid)

[5] Crocker M J 2007 *Handbook of Noise and Vibration Control* (Hoboken, NJ: John Wiley & Sons).

[6] ISO 9052-1 1989 *Acoustics - Determination of Dynamic Stiffness - Part 1: Materials used Under Floating Floors in Dwellings* (Geneve: International Standadization Organization)

[7] Maderuelo-Sanz R, Martín-Castizo M and Vílchez-Gómez R 2011 The performance of resilient layers made from recycled rubber fluff for impact noise reduction *Appl. Acoust.* **72** 823–828

[8] Richardson TB 2009 *Nanoreinforced Shape Memory Polyurethane* PhD Thesis (Auburn, AL: Auburn University)

[9] Kireitseu M, Hui D and Tomlinson G 2008 Advanced shock-resistant and vibration damping of nanoparticle-reinforced composite material *Composites Part B: Engn.* **39** 128–138.

[10] Menard KP 1999 *Dynamic Mechanical Analysis: a Practical Introduction* (New York: CRC Press).

[11] Castaño JL, Arenas JP and ML Auad 2016 Study on nanostructured thermostable polyurethane for passive damping applications *Proc. 23$^{rd}$ Int. Cong. on Sound and Vibration* (Athens: IIAV)

[12] Brosio E 1980 Footstep noise protection: floor performance and evaluation problems (in Italian) *Rivista Italiana di Acustica* **IV** 1

[13] Asdrubali F and D'Alessandro F 2011 Impact sound insulation and viscoelastic properties of resilient materials made from recycled tyre granules *Int. J. Acoust. Vib.* **16** 119–125

[14] Hopkins C 2007 *Sound Insulation* (Amsterdam: Butterworth-Heinemann)

# Analysis for active isolation of the equipment on flexible beam structure

**H E Oh[1], J M Ku[1], D H Lee[1], C S Hong[2,*] and W B Jeong[1]**

[1]School of Mechanical Engineering, Pusan National University, Busan, South Korea
[2]School of Mechanical Engineering, Ulsan College, Ulsan, South Korea

E-mail : cshong@uc.ac.kr

**Abstract**. This paper dealt with the active isolation of vibration. Many home appliance and vehicles in operation generate the vibration for several reasons. Due to the vibration propagated to the structure, it has a negative influence of the structure such as breakage and decreased reliability. As basic study on vibration isolation, the analytical model is as follows. The analytical model consists of the equipment as source of vibration and the flexible beam to be controlled. The equipment is assumed by rigid body and excited by a concentrated force as disturbance. And the equipment and the beam are connected to spring located at both ends the equipment. The actuators are located slightly away from spring. To minimize the vibration propagated to the beam structure through the equipment, the controller was designed by using the multi-channel Wiener filter after calculating the transfer path through the theoretical formula. The disturbance was given as harmonic components. The propagated vibration to the beam caused by the disturbance was controlled using a designed controller. Finally, the numerical results for the displacement obtained at each point of the beam structure with and without control were carried out and compared.

## 1. Introduction

When many home appliance and vehicles operate, the vibration occurs. Here, the vibration propagates through the structure, which has a negative influence on life of the structure [1]. Therefore, it is necessary to study how to isolate the propagated vibration. To isolate the propagation vibration, there is a method of the passive damping and a method of redesigning the vibration transmission path of the structure [2]. However, the method of the passive damping is only a control of high frequency band, and the redesign of the structure is costly. Thus, active vibration control (AVC) technique is used as a solution to these problems [3].

In this study, the control was implemented on a model in which the equipment with disturbance connected with spring on the flexible beam supporting both ends. The object to be controlled is a flexible beam. The controller was designed using a Wiener filter for isolate the vibration transmitted from the equipment. Using the designed filter, the displacement with and without control were carried out and compared.

## 2. System modelling

### *2.1. Analytical model*

Figure 1 shows the analytical model. This model consists of the equipment as source of vibration and the flexible beam as receiver to be controlled. The equipment is assumed by rigid body and excited a concentrated force as disturbance. The equipment and flexible beam is connected to spring located at both ends of the equipment.

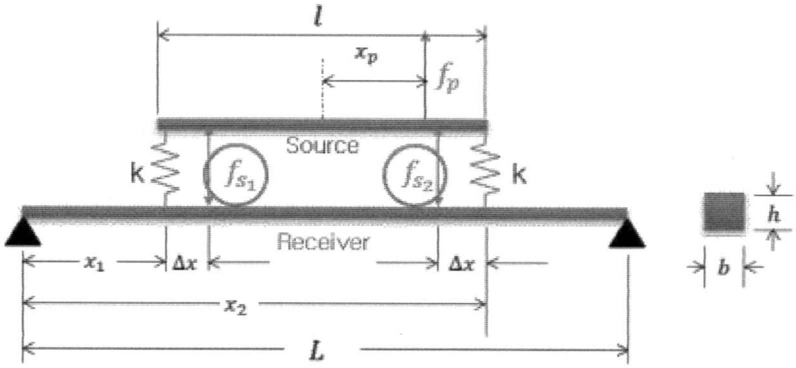


**Figure 1.** Analytical model

Here, $l$ and $L$ are the length of receiver and source, respectively. And $k$ is the spring constant, $x_1$, $x_2$ are the location of spring on receiver, $f_{s_1}$, $f_{s_2}$ are the controlled force.

### *2.2. Governing equation*

The governing equation of analytical model can be obtained through the free body diagram. Figure 2 shows the free body diagram of analytical model.

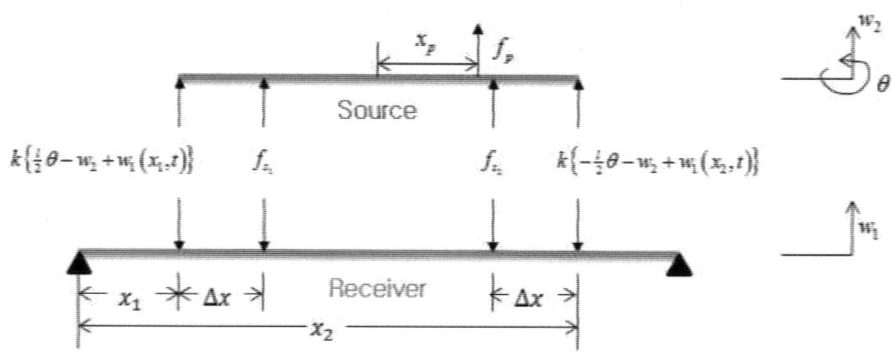


**Figure 2.** Free body diagram of analytical model

Receiver was assumed the Euler-Bernoulli beam. The equation of motion of transverse vibration of a uniform cross section beam under concentrated force is as follows.

$$EI\frac{\partial^4 w_1(x,t)}{\partial x^4}+\rho A\frac{\partial^2 w_1(x,t)}{\partial t^2}=f(x,t) \quad (1)$$

Here, $E$ is the elastic modulus, $I$ is the moment of inertia of beam section, $\rho$ is the density, $A$ is cross section of beam and $f$ is the external force per unit length.

In this case, the displacement of beam is expressed by the sum of the eigenmodes up to N-th order, as follows.

$$w_1(x,t)=\sum_{n=1}^{N} d_n(\omega)\phi_n(x)e^{j\omega t} \tag{2}$$

Here, $d_n(\omega)$ is the n-th order modal displacement by frequency and $\phi_n(x)$ is the n-th order eigenmode.

Using Fourier transform, Equation (2) in frequency domain can be expressed.

$$w_1(x,\omega)=\sum_{n=1}^{N} d_n(\omega)\phi_n(x) \tag{3}$$

In equation (3), if beam length is $L$ and the boundary condition of beam is simply supported, normalized eigenmodes can be expressed.

$$\phi_n(x)=\sqrt{2}\sin\left(\frac{n\pi x}{L}\right) \tag{4}$$

The force on the receiver is assumed to be applied to the $x_1$ and $x_2$ concentrated loads. And it consists of force transmitted through the source and controlled force $f_{s_1}$, $f_{s_2}$. Therefore, the external force can be expressed.

$$f(x,t)=\left[k\left\{w_2-\frac{l}{2}\theta-w_1(x_1,t)\right\}-f_{s_1}\right]\delta(x-x_1)+\left[k\left\{w_2+\frac{l}{2}\theta-w_1(x_2,t)\right\}-f_{s_1}\right]\delta(x-x_2) \tag{5}$$

Here, $w_1$ is the transverse degree of freedom of receiver, $w_2$ and $\theta$ are the transverse and rotational degree of freedom, respectively.

Substituting equation (3) and equation (5) into equation (1), it can be expressed as follows.

$$\begin{aligned}\omega_m^2 a_m-\omega^2 a_m &= -\frac{k}{\rho A}\left\{\sum_{n=1}^{N} d_n(\omega)\phi_1(x_1)\right\}\phi_m(x_1)-\frac{k}{\rho A}\left\{\sum_{n=1}^{N} d_n(\omega)\phi_1(x_2)\right\}\phi_m(x_2)\\ &+\frac{k}{\rho A}\{\phi_m(x_1)+\phi_m(x_2)\}W_2-\frac{k}{\rho A}\{\phi_m(x_1)-\phi_m(x_2)\}\Theta\\ &-\frac{1}{\rho A}F_{s_1}\phi_m(x_1)-\frac{1}{\rho A}F_{s_2}\phi_m(x_2)\end{aligned} \tag{6}$$

Here, $m$ is the arbitrary order left by the mode orthogonality, $W_2$, $\Theta$, $F_{s_1}$ and $F_{s_2}$ are variables obtained by Fourier transforming $w_2$, $\theta$, $f_{s_1}$ and $f_{s_2}$.

The equation of motion of source can be expressed using Figure 2.

$$m\ddot{w}_2+2kw_2-k\{w_1(x_1,t)+w_1(x_2,t)\}=f_p+f_{s_1}+f_{s_2} \tag{7}$$

$$\frac{ml^2}{12}\ddot{\theta}+\frac{kl^2}{2}\theta+\frac{kl}{2}\{w_1(x_1,t)-w_1(x_2,t)\}=f_p x_p-f_{s_1}\frac{l}{2}+f_{s_2}\frac{l}{2} \tag{8}$$

Here, equation (7) and equation (8) are transverse and rotational equation of motion, respectively.

Substituting equation (3) into equation (7) and equation (8), it can be expressed as follows.

$$\left(-\omega^2 m+2k\right)W_2-k\left\{\sum_{n=1}^{N} d_n(\omega)\phi_1(x_1)+\sum_{n=1}^{N} d_n(\omega)\phi_1(x_2)\right\}=F_p+F_{s_1}+F_{s_2} \tag{9}$$

$$\left(-\omega^2\frac{ml^2}{12}+\frac{kl^2}{2}\right)\Theta+\frac{kl}{2}\left\{\sum_{n=1}^{N}d_n(\omega)\phi_1(x_1)-\sum_{n=1}^{N}d_n(\omega)\phi_1(x_2)\right\}=F_px_p-F_{s_1}\frac{l}{2}+F_{s_2}\frac{l}{2} \quad (10)$$

## 3. Controller design

The controller was designed using the Wiener filter. The Wiener filter one of ways of obtaining the coefficient of fir filter, it calculates the coefficients to minimizes the mean square error between the estimated random process and the desired process.

In this study, the control was performed using SIMO (Single-input multiple-output) Wiener filter. A schematic diagram is shown in figure 3.

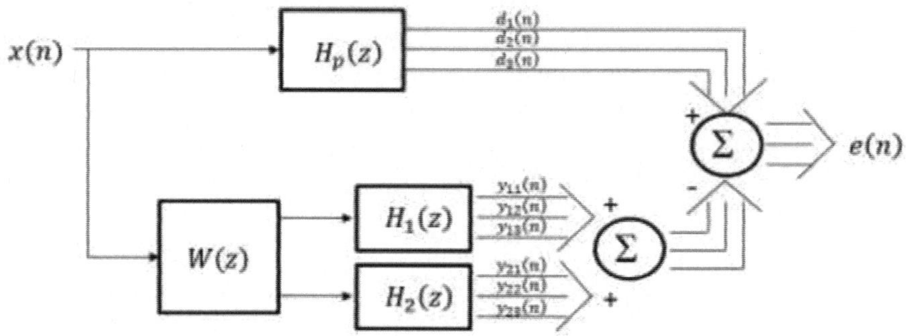


**Figure 3.** Schematic diagram using SIMO Wiener filter

Here, $x$ is the input force, $W$ is the controller (filter coefficient), $H$ is the transfer function between force and displacement, $d$ is the displacement before control, $y$ is the displacement for control and $e$ is the displacement after control.

The filter coefficient $W$ can be obtained by minimizing the mean square of the $e$.

$$\begin{aligned}e^2(n)=&\left[d_1(n)-\{y_{11}(n)+y_{21}(n)+\cdots+y_{M1}(n)\}\right]^2+\\&\cdots+\left[d_L(n)-\{y_{1L}(n)+y_{2L}(n)+\cdots+y_{ML}(n)\}\right]^2\end{aligned} \quad (12)$$

where,

$$y_{ij}=\bar{x}_{ij}^T W_i=\bar{x}^T H_i$$

Here, $M$ is the number of controlling force, $L$ is the number of transducer to measure displacement. To minimize the $e$, the equation (14) was used.

$$\frac{\partial e^2(n)}{\partial w_1}=\frac{\partial e^2(n)}{\partial w_2}=\cdots=\frac{\partial e^2(n)}{\partial w_M}=0 \quad (13)$$

Thus, filter coefficient $W$ was calculated by the following equation (14).

$$\begin{bmatrix}w_1\\w_2\\\vdots\\w_M\end{bmatrix}=\begin{bmatrix}\sum_{i=1}^{L}T_{1_i1_i} & \sum_{i=1}^{L}T_{1i2i} & \cdots & \sum_{i=1}^{L}T_{1_iM_i}\\ \sum_{i=1}^{L}T_{1_i2_i} & \sum_{i=1}^{L}T_{2_i2_i} & & \sum_{i=1}^{L}T_{2_iM_i}\\ \vdots & & \ddots & \\ \sum_{i=1}^{L}T_{1iMi} & \sum_{i=1}^{L}T_{2_iM_i} & & \sum_{i=1}^{L}T_{M_iM_i}\end{bmatrix}^{-1}\begin{bmatrix}\sum_{i=1}^{L}R_{p_ix_{1i}}\\ \sum_{i=1}^{L}R_{p_ix_{2i}}\\ \vdots\\ \sum_{i=1}^{L}R_{p_ix_{Mi}}\end{bmatrix} \quad (14)$$

IOP Conf. Series: Journal of Physics: Conf. Series **1075** (2018) 012029 doi:10.1088/1742-6596/1075/1/012029

where,

$$T_{a_i b_i} = \begin{bmatrix} R_{x_{ai}x_{bi}}(0) & R_{x_{ai}x_{bi}}(1) & \cdots & R_{x_{ai}x_{bi}}(J-1) \\ R_{x_{ai}x_{bi}}(1) & R_{x_{ai}x_{bi}}(0) & & R_{x_{ai}x_{bi}}(J-2) \\ \vdots & & \ddots & \\ R_{x_{ai}x_{bi}}(J-1) & R_{x_{ai}x_{bi}}(J-2) & \cdots & R_{x_{ai}x_{bi}}(0) \end{bmatrix}$$

Here, J is the filter length, T is the Toeplitz matrix, R is the cross correlation,

## 4. Analysis

### 4.1. Analysis condition

The specification and properties of analytical model are as follows.

**Table 1.** Mechanical properties and geometric data of source and receiver

| Parameter | Value | Unit |
|---|---|---|
| Length | 1.0 | m |
| Width | 8.6 | mm |
| Thickness | 8.6 | mm |
| Density | 7580 | Kg/m3 |
| Elastic modulus | 2.07e11 | Pa |

The length of equipment was assumed to be $0.4\,L$. And the input force was assumed to be white noise at $0.2\,l$ on source. And location of spring on receiver $x_{s_1}$, $x_{s_2}$ were assumed to be $0.1\,L$, $0.7\,L$, respectively. The time interval for analysis was 0.1 seconds, and the record length was 60 seconds. And Wiener filter length is 2000.

### 4.2. Results

Figure 4 shows the displacement before and after at $0.3\,L$ on receiver. When the control is performed, it can be confirmed that the displacement is reduced compared the displacement before control. And Figure 5 shows the displacement after Fourier transform of result of Figure 4.It can be confirmed that the largest control effect occurs in the frequency band where the displacement is the largest.

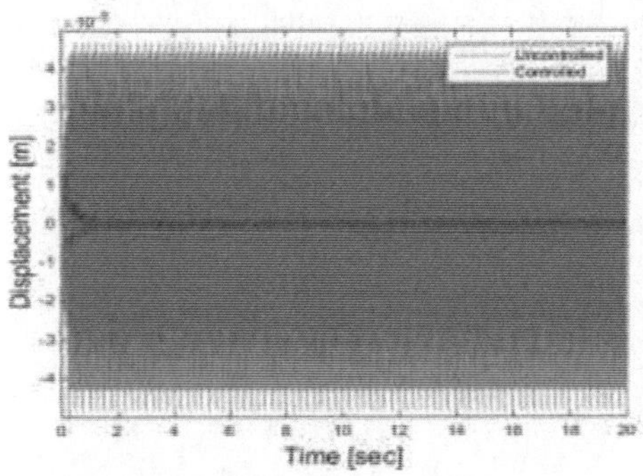


**Figure 4.** Displacement with/without control (time domain)

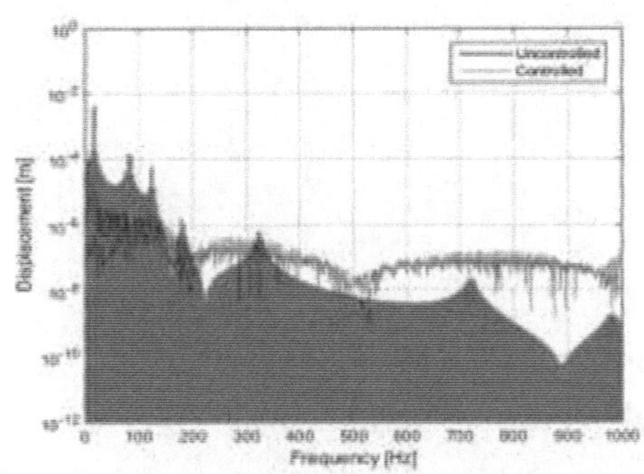


**Figure 5.** Displacement with/without control (frequency domain)

## 5. Conclusion

The control analysis was carried out to isolate the vibration transmitted the equipment connected with flexible beam supported both ends. To design the controller, the SIMO Wiener filter was used.

As a result after analysis, it was confirmed that the displacement was reduced and the largest control effect is occurred in the frequency band where the displacement is the largest. Later, it will test on actual model by using the analytical results.

## References

[1] Minson J S and Powell C A 1985 Review of recent research on interior noise of propeller aircraft. *J. of Aircraft* **22** pp 931-949

[2] Shin C J, Hong C S and Jeong W B 2008 Active Vibration Control of Cantilever Beams Using PZT Actuators. *Transactions of the KSNVE* **18** pp 1293-1300.

[3] Moheimani S R, Halim D and Fleming A J 2003 Spatial control of vibration: *theory and experiments* Vol. 10 (World scientific)

# Design of half-car active suspension system for passenger riding comfort

**Y Susatio[1], L Oktaviana[1], N K Rizki[1], E Listijorini[2], T R Biyanto[3]**

[1]Vibration and Acoustic Laboratory, Departement of Engineering Physics, Institut Teknologi Sepuluh Nopember, Surabaya, Indonesia
[2]Department of Mechanical Engineering, Universitas Sultan Ageng Tirtayasa
[3]Instrumentation and Control Laboratory, Departement of Engineering Physics, Institut Teknologi Sepuluh Nopember, Surabaya, Indonesia

E-mail: trb@ep.its.ac.id and trbiyanto@gmail.com

**Abstract**. In this paper, the active suspension system is designed to reduce vibration on the passenger seat caused by change in road-surface shapes or disturbance. PID control is utilized in the design of this active suspension system, and Direct Synthesis method is proposed to tune PID parameters. Direct Synthesis method optimize PID tuning parameters to achieve desired output response in different road surface. In this study, three interference signals are used as disturbance from surface road i.e step, impulse, and sinus. The aim of the suspension control design is to obtain the control parameters that conform to ISO 2631. The simulation results show that value of $\tau_c$ 0.0002 for step disturbance provide $a_{RMS}$ 0.086 m/s$^2$ and damping ratio 0.21. For impulse input the value of $\tau_c$ 0.002 provide $a_{RMS}$ 0.112 m/s$^2$ and damping ratio of 0.22. For sinus input value of $\tau_c$ 0.00005 provide $a_{RMS}$ 0.09 m/s$^2$.

## 1. Introduction

Nowadays, riding comfort becomes the main focus in the automotive industries. The passengers feel the vibration effects due to interaction of the vehicle and road surfaces. Many researchers said that vibration can degrade human fitness due to discomfort and fatigue [1]. There are several standards about whole body vibration that aim for restricting the allowable magnitude of vibration on human body such as ISO 2631 [2].

The vehicle suspension system is responsible for ride comfort and road holding as the suspension carries the vehicle body and transmit all forces between the body and the road. A classical car suspension or passive suspension system consists of a spring (coil spring, air spring or leaf spring) and a damping element. The spring and damping coefficients are chosen according to comfort, road holding, and handling specifications. However, conventional suspensions can achieve a trade-off between ride comfort and road holding since their spring and damping coefficients cannot be adaptively tuned according to driving efforts and road conditions. They can achieve good ride comfort and road holding only under the designed conditions. To avoid the tradeoff, intelligent suspension systems have been investigated since the 1980s with the development of microprocessor, sensor, and actuator technologies; their spring and damping coefficients can be controlled. The active suspension usually requires a substantial amount of external energy to generate the required control forces. As

IOP Conf. Series: Journal of Physics: Conf. Series **1075** (2018) 012030 doi:10.1088/1742-6596/1075/1/012030

intelligent suspension systems, active suspensions belong to the controlled suspension system, which consists of sensors, controllers, plants and actuators. Therefor the active suspension can improve riding comfort against vibration due to road surfaces change [3].

Recent years, a lot of studies about control method used in active suspension have been engaged. PID controller still widely used because of its simple structure and robust performance in linear model system. The application of PID on active suspension has been developed by many researchers and the result shows the response of active suspension of sprung mass decrease overshoot peak as compare with its passive suspension [4] [5]. However, it is necessary to build controller which will bear with any road changes. In this research Direct Synthesis method is used to optimize PID tuning parameters to achieve desired output response in different road surfaces. The presented controller is tested with three road surfaces i.e step, impulse and sinus. The robustness of control system will performed within simulation in MATLAB SIMULINK.

## 2. Mathematical Modelling

The 5DOF active suspension shown in Figure 1. The suspension, tire, passenger seat are modeled by linear springs in parallel with dampers. The $Y_s$ expresses seat displacement while Y expresses sprung mass vertical displacement (bounce) at the center of gravity and Θ is the pitch displacement of sprung mass. Front and rare tire displacement are denoted by $Y_1$ and $Y_2$. The device U set at the front suspension as the actuator which can be controlled. The Z1 and Z2 are the disturbance which hit the vehicle at front and rare tire.

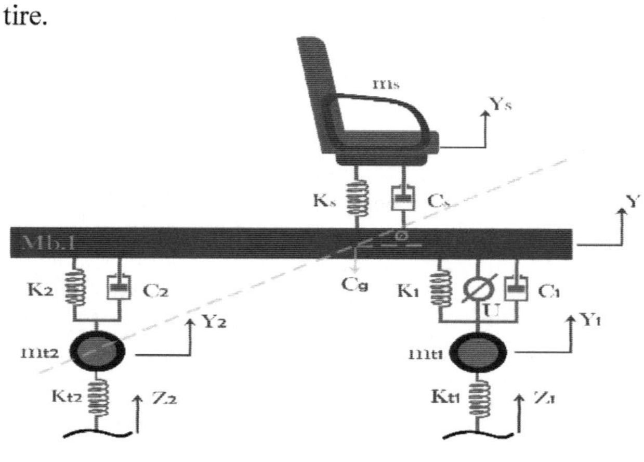


**Figure 1.** Half-car active suspension model

The equation of motion of half-car active suspension system is being carried out by using Newton second law of motion and written bellow as

$$m_s\ddot{y}_s + c_s(\dot{y}_s + a\dot{\theta} - \dot{y}) + k_s(y_s + a\theta - y) = 0 \quad (1)$$

$$m_b\ddot{y} + (c_1 + c_2 + c_s)\dot{y} - k_2y_2 - k_sy_s + (c_1l1 - c_2l2 - c_sa)\dot{\theta} - c_1\dot{y}_1 - c_2\dot{y}_2 + (k_1 + k_2 + k_s)y + (k_1l1 - k_2l2 - k_sa)\theta - c_s\dot{y} - k_1y_1 = U \quad (2)$$

$$I\ddot{\theta} + (c_1l1 - c_2l2 + c_sa)\dot{y} - k_say_s - c_sa\dot{y} + (c_2l2^2 + c_1l1^2 + c_sa^2)\dot{\theta} + c_1l1\dot{y}_1 + c_2l2\dot{y}_2 + k_2l2y_2 + (k_2l2 - k_1l1 + k_sa)y + k_1l1y_1 + (k_2l2^2 + k_1l1^2 + k_sa^2)\theta = Ul1 \quad (3)$$

$$m_{t2}\ddot{y}_2 - k_2y + k_2l2\theta - c_2\dot{y} + c_2l2\dot{\theta} + c_2\dot{y}_2 + (k_2 + k_{t2})y_2 = k_{t2}z_2 \quad (4)$$

$$m_{t1}\ddot{y}_1 + c_1\dot{y} - k_1x_b - k_{sf}l2\theta + (k_{sf} + k_{tf})y - c_1l1\dot{\theta} + c_1\dot{y}_1 = k_{t1}z_1 + U \quad (5)$$

In order to optimize the simulation and to analyze the half-car suspension behavior, the parameters are input as follow:

**Table 1.** Suspension parameter of half-car

| **Description** | **Value** |
|---|---|
| Driver Seat Mass (ms) | 75 Kg |
| Sprung Mass (mb) | 505.1 Kg |
| Front unsprung mass (mt1) | 28.58 Kg |
| Rare unsprung mass (mt2) | 54.3 Kg |
| Moment of inertia of the vehicle (I) | 651 Kg.m |
| Rare body length from the CG (l2) | 1.468 m |
| Front body length from the CG (l1) | 1.098 m |
| Driver seat length from the CG (a) | 0.7 m |
| Spring stiffness of front and rare tire (kt1, kt2) | 155900 N/m |
| Spring stiffness of front and rare unsprung mass (k1, k2) | 15000 N/m |
| Damping coefficient of front nad rare unsprung mass (c1, c2) | 1828 N.s/m |
| Spring stiffness of driver seat (ks) | 15000 N/m |
| Damping coefficient of driver seat (cs) | 150 N.s/m |

*2.1. Riding Comfort Standard Condition*

ISO 2631 has been recognized as minimum standard for whole body vibration on passenger as described below:

**Table 2.** ISO 2631 standard for riding comfort

| **RMS (Root Mean Square) Acceleration** | **Ride comfort Criteria** |
|---|---|
| $a<0.315$ m/s$^2$ | Comfortable |
| $0.315<a<0.63$ m/s$^2$ | A little Uncomfortable |
| $0.5<a<1$m/s$^2$ | Fairy Uncomfortable |
| $0.8<a<1.6$m/s$^2$ | Uncomfortable |
| $1.25<a<2.5$m/s$^2$ | Very Uncomfortable |
| $a>2.5$m/s$^2$ | Extremely Uncomfortable |

Another riding comfort standard comes from Optimum G, an England automotive consultant, claims that passenger vehicles generally use a damping ratio ($\xi$) of approximately 0.2- 0.25 for maximizing riding comfort [6]. Furthermore, the controlled responses should be conformed to those riding standards.

## 3. PID Tuned by Direct Synthesis

The first step to tune PID parameters using direct synthesis method is recognizing open loop behaviour of the system. Previous study has successfully arranged the equation to tune Kp, Ki, and Kd on the second order system shown in the equation (6)-(8) [7]:

$$kc = (\tau_1 + \tau_2)/k_p\tau_c \tag{6}$$

$$\tau_i = (\tau_1 + \tau_2) \tag{7}$$

$$\tau_d = \tau_1\tau_2/(\tau_1 + \tau_2) \tag{8}$$

$\tau_c$ is the parameter in direct synthesis method that can be changed to reach the best response of the system. When the disturbance shape change, system may be unsteady and unrobust. Therefor in this present attempt $\tau_c$ parameter used to build an adaptive control system to any different road surfaces. Table 3, table 4, and table 5 show PID gain parameters for step input, impulse input and sinus input with several varieties of $\tau_c$. The range of $\tau_c$ would be different for each input shapes since this value depend on the settling time and gain of each system.

**Table 3**. Gain parameters for step input

| Gain | $\tau_c$ | | | | | |
|---|---|---|---|---|---|---|
| parameter | 1 | 0.1 | 0.01 | 0.001 | 1E-04 | 2E-04 |
| Kp | 9 | 93 | 927 | 9268 | 92680 | 46340 |
| Ki | 3 | 34 | 336 | 3358 | 33580 | 16790 |
| Kd | 6 | 64 | 639 | 6395 | 63949 | 31974 |

**Table 4.** Gain parameters for impulse input

| Gain | $\tau_c$ | | | | |
|---|---|---|---|---|---|
| parameters | 1 | 0.1 | 0.01 | 0.001 | 0.002 |
| Kp | 39.5 | 395.3 | 3952.6 | 39526.4 | 19763.2 |
| Ki | 9.1 | 91.1 | 910.7 | 9107.5 | 4553.7 |
| Kd | 42.9 | 428.9 | 4288.6 | 42886.2 | 21443.1 |

**Table 5.** Gain parameters for sinus input

| Gain | $\tau_c$ | | |
|---|---|---|---|
| parameters | 0.001 | 0.0001 | 0.00005 |
| Kc | 10174.42 | 101744.2 | 203488.4 |
| Ki | 4069.767 | 40697.67 | 81395.35 |
| Kd | 6359.012 | 63590.12 | 127180.2 |

## 4. MATLAB simulation and result analysis

The MATLAB SIMULINK simulation is prepared for the passenger seat displacement and acceleration responses to analyse the achievement of riding comfort. The goals of this study will show the comparison between active suspension system response and its passive suspension for each road shapes.

### *4.1. Step road profile*

The first road profile is step input with 0.08 m high. Figure 2 shows passenger seat displacement response while figure 3 shows passenger seat acceleration response.

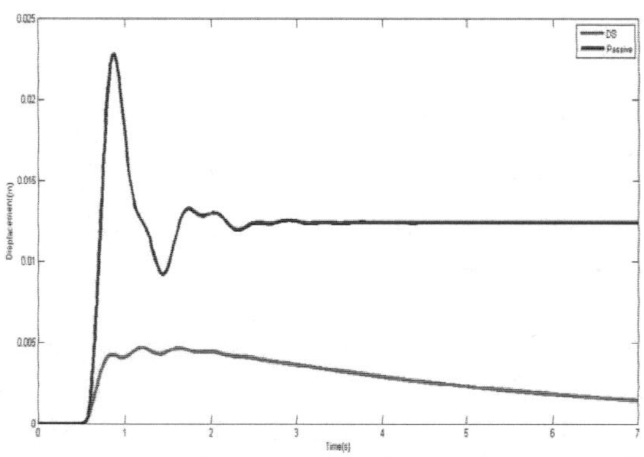

**Figure 2.** Seat displacement response due to step input with $\tau_c$ value 0.0002

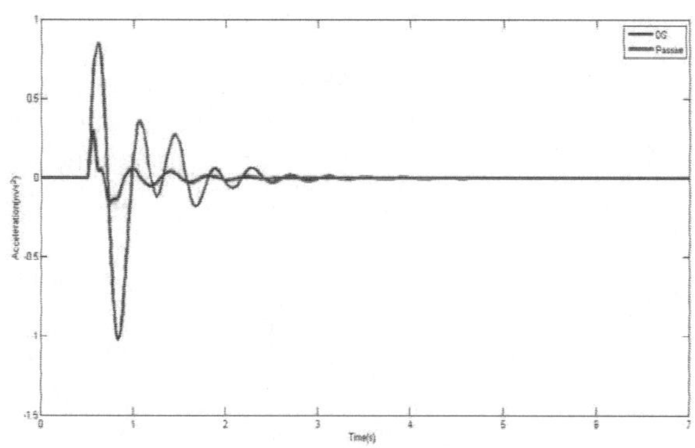

**Figure 3.** Seat acceleration response due to step input with $\tau_c$ value 0.0002

Figures show the maximum overshoot of the respon from 0.0227 m of passive to 0.0021 m of active, and lessens the $a_{RMS}$ from 0.39 m/s$^2$ to 0.08 m/s$^2$ and gives the controlled response with 0.21 damping ratio. So that the active response meets all qualifications of riding comfort standard.

### *4.2. Impulse road profile*

The second road profile formed as impulse with 0.08 m high and hit by the car for 0.06 seconds while it is traveling at 20 km/hour. Figure 4 shows passenger seat displacement response and figure 5 shows acceleration response.

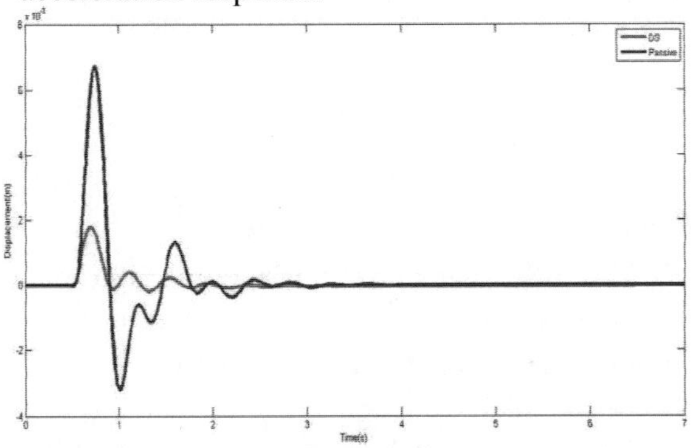


**Figure 4.** Seat displacement response due to impulse input with $\tau_c$ value 0.002

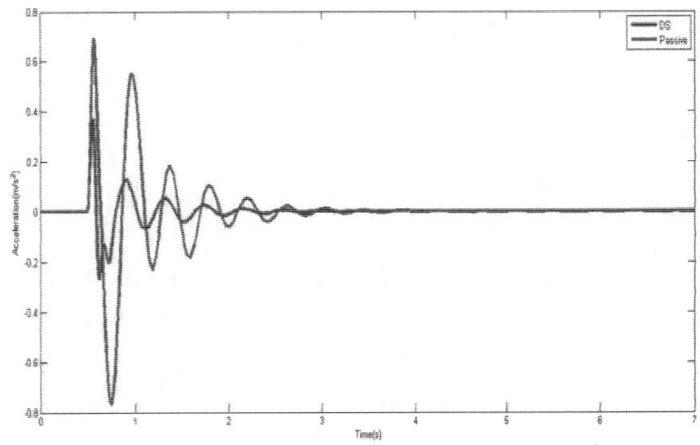


**Figure 5.** Seat acceleration response due to step impulse with $\tau_c$ value 0.002

Figures observe that the result of active suspension system degrade maximum overshoot from 0.087 m of passive to 0.0012 m of active. Active suspension system with direct synthesis obtain $a_{RMS}$ 0.112 m/s$^2$ and damping ratio ($\xi$) 0.22, these two results absolutely meet riding comfort standard.

### *4.3 Sinus road profile*

The third road profile formed as sinus wave. It assumed that the road is approximated as sinusoidal in cross section with amplitude of 0.035 m and the wavelength ($\lambda$) of 5 m. The car is traveling at 20 km/hour.

The period and cycle frequency for the harmonic input is as follow

$$Z1 = 0.035 \sin 17.27t \tag{9}$$

Figure 6 shows passenger seat displacement response and figure 7 shows passenger seat acceleration response.

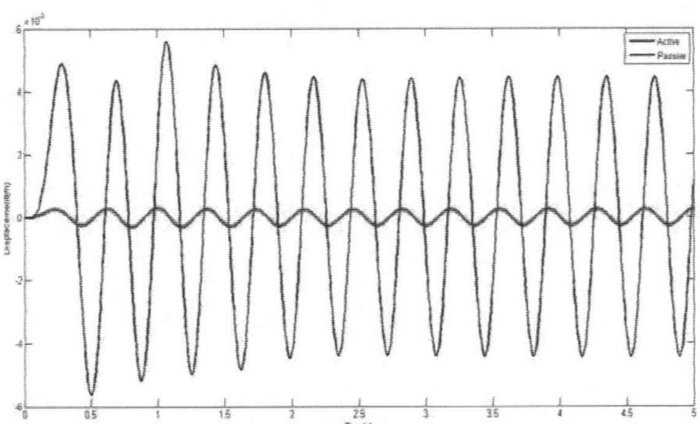


**Figure 6**. Seat displacement response due to sinus input with $\tau_c$ value 0.00005

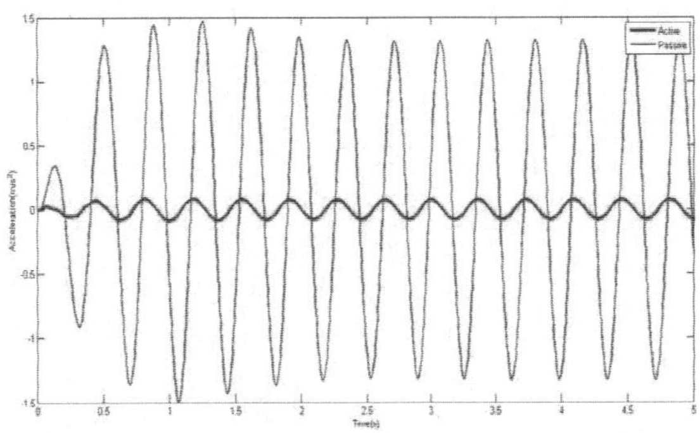


**Figure 7.** Seat acceleration response due to sinus input with $\tau_c$ value 0.00005

Figures observe that active suspension gives the better result. With maximum overshoot depression from 0.010 m of passive to 0.003 to active. And reach $a_{RMS}$ 0.099 m/s$^2$. This result successfully conform to ISO 2631 standard.

## 5. Conclusion

This research presents that direct synthesis is the appropriate method for active suspension system. $a_{RMS}$ and damping ration obtained by active suspension system can adapted to three different road profiles and fulfill the requirements of passenger car riding comfort standard. It can be surely denoted that direct synthesis is the robust control strategy for automobile model

## References

[1] Gagorowski A 2010 Simulation Study on Stiffness of Suspension Seat in The Aspect of the Vibration Assesment Affecting a Vehicle Driver *Logistics and Transport* vol **11** no 2 pp 55-62,

[2] ISO 2631 1997 Mechanical Vibration and Shock Evaluation of Human Body Exposure to Whole Body Vibration: *General Requirement* (International Organization of Standarization)

[3] Honghai L, Huijun G and Ping L 2013 Handbook of Vehicle Suspension Control System, (London: The Institution of Engineering and Technology) pp 1

[4] Al-Mutar W H and Abdalla T Y 2015 Quarter Car Active Suspension System Control Using PID Controller tuned by PSO *Iraq J. Electrical and Electronic Engineering* vol **11** no 2 pp 151-154

[5] Al-amir H S and Al-zahra A T 2014 Design a Robust PID Controller of an Active Suspension System *International Journal of Mechanical and Mechatronics Engineering* vol **14** no 2 pp 122-127

[6] Brisson S and Giaraffa M Tech and Tip Spring & Damper Ep Four *Optimum G, Denvor.co*

[7] Emad A 2015 Chapter 4. PID Control Design *Advanced Topic in Process Control* (Riyadh, Saudi Arabia, King Saud University) pp 1-27

IOP Conf. Series: Journal of Physics: Conf. Series **1075** (2018) 012031 doi:10.1088/1742-6596/1075/1/012031

# Sound quality degradation of audio equipment by hum noise induced by non-audible high frequency electrical noise

**D Kobayashi[1] and T Yoshida[1]**

[1] Department of Electrical Engineering, Tokyo University of Science, 6-3-1 Niijuku, Katsushika-ku, Tokyo, 125-8585, Japan

E-mail: yoshida@ee.kagu.tus.ac.jp

**Abstract.** In recent audio equipment, induction and inflow of non-audible high frequency (HF) electrical noise to audio equipment is increasing more and more due to implementation of personal computer / network connectivity and high-resolution capability in digital audio equipment. In this study, in order to investigate another causes of influence of non-audible HF electric noise on playback sound, we measured a loudspeaker output of an audio amplifier by the audio analyser with inputting single HF sine wave from a function generator to analog input of the audio amplifier. From measured results, it was found that the power of harmonics of 60 Hz (hum noise) on the loudspeaker output of some audio amplifiers increased when single HF sine wave was inputted. Tendency of the increasing of the hum noise component depended on the frequency of the inputted sine wave and model of audio amplifier. These results show the importance of consideration of HF characteristic and countermeasure to the HF noise also in circuit design and tuning of audio equipment.

## 1. Introduction

In recent years, the progress of research and development on audio equipment leads to high-fidelity sound playback. However, equipment design and tuning are still done subjectively based on the knowledge and experience of the acoustic technicians even today. We think the one cause of this situation is that the playback-sound-quality-deterioration mechanism due to complex and interrelated additional factors, e.g. the operating state of the PC, the state and polarity of each AC power supply [1] in the equipment, cable materials, etc., has not been elucidated. To address this situation, we focused on non-audible high frequency (HF) electrical noise because the HF noise that audio equipment is concerned is increasing more and more due to implementation of personal computer / network connectivity and high resolution capability in recent digital audio equipment.

In our previous study, we confirmed that the HF electrical noise that had "plural" frequency components cause their intermodulation distortion (IMD) in audible frequency band [2]. However, the factors and modes of the sound quality degradation caused by non-audible HF noise were not considered. In this study, in order to investigate sound quality degradation caused by non-audible "single" HF electric noise on playback sound, we measured the loudspeaker output of 5 models of audio amplifier using the audio analyzer when inputting single HF sine wave to analog audio input of the audio amplifier in the frequency range from 1 MHz to 240 MHz.

IOP Conf. Series: Journal of Physics: Conf. Series **1075** (2018) 012031 doi:10.1088/1742-6596/1075/1/012031

## 2. Measurement method of sound quality degradation caused by non-audible single HF noise

*2.1 Distortion characteristics of audio output signal caused by non-audible single HF noise*
Figure 1 shows apparatus setup for measuring sound quality degradation in audible band of loudspeaker output signal of audio amplifier cause by single HF noise. We measure the frequency spectrum of the loudspeaker output of one model of the audio amplifier (Class A analog amplifier) with the audio analyzer (Audio Precision, APx525) when single non-audible HF sine wave is inputted from a function generator (Tektronix, AFG3252) to the right channel of an analog audio input (LINE IN) of the audio amplifier. The specifications of the FFT analysis is 48 kHz sampling frequency, 1 second measurement duration, and 3 measured spectrum averaging. The amplitude of the HF sine wave input to the audio amplifier is 1 $V_{rms}$ and the frequency is 60 MHz. The gain of the audio amplifier is set to 0 dB at the 1 kHz sine wave. The power source of the function generator is a commercial AC power supply (100V, 50 Hz). The power source of the audio amplifier is a regulated AC power supply (NF EC1000S, 100 V, 50 Hz, or 60 Hz) connected with 3 poles cable, i.e. Line, Neutral, and Grounding. To prevent noise propagation to audio amplifier from measurement equipment, power source of the audio analyzer is DC / AC inverter (100 V, 50 Hz AC output) and battery (DC 48 V).

In order to confirm the increase of hum noise is caused not in measurement equipment, e.g. the function generator and the audio analyzer but in the audio amplifier under test, we measure the frequency spectrum of the function generator output. In addition, we also measure the noise floor of measuring system and the audio amplifier by outputting 0 V DC signal (equivalent for silent signal) from the function generator.

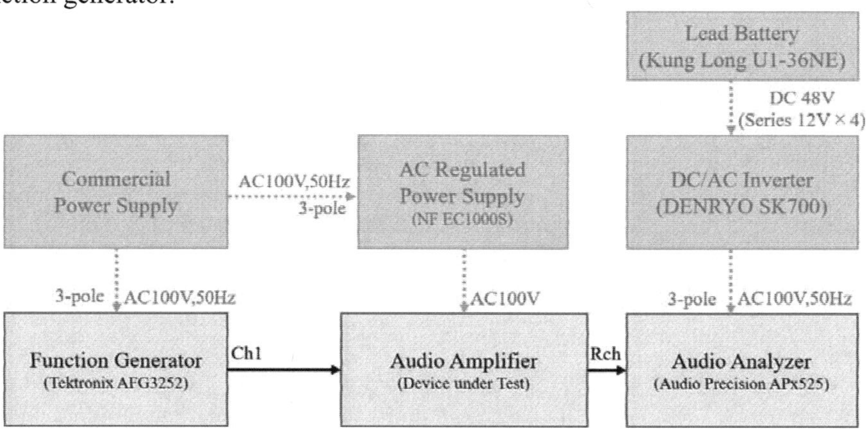


**Figure 1.** Apparatus set up for measuring sound quality degradation in audible band of loudspeaker output signal of audio amplifier caused by single HF noise.

*2.2 Frequency and audio amplifier models dependency of the increase of hum noise*
In order to research the frequency dependency and audio amplifier models dependency of the increase of hum noise in audible band caused by the non-audible single HF noise, we measure and compare the frequency spectrum of loudspeaker output of the 5 models of audio amplifiers when the frequency of the HF noise is changed 0.5 MHz each in the range from 1 MHz to 240 MHz. The 5 models of audio amplifiers consist of 3 models of analog amplifiers (model A, C, and D) and 2 models of digital amplifiers (model B and E). The amplifier models A and D has three poles, and the other amplifier, i.e. model B, C, and E, and has two poles. The audio amplifier with 3 poles inlet is connected to the regulated AC power supply with 3 poles power cable. Only the audio amplifier is driven by a regulated AC power supply at 60 Hz in order to distinguish the hum noise source. To evaluate the amount of increase hum noise quantitatively, calculate the sum of the harmonics of the hum noise up to the 5th order (300 Hz).

## 3. Measurement result of sound quality degradation caused by non-audible single HF noise

*3.1 Distortion characteristics of audio output signal caused by non-audible single HF noise*

Figure 2 shows the frequency spectrum in audible band of the loudspeaker output of the audio amplifier (model A) when the HF sine wave (1 $V_{rms}$, 60 MHz) is inputted to the analog audio input of the audio amplifier, and the noise floor of measured system and the audio amplifier.

From figure 2, it is confirmed that some frequency components such as harmonics of the 60 Hz hum noise increase remarkably in audible band, even though inputted noise is non-audible high frequency. The amplitude of the 60 Hz component increased 47 dBV and that of the 120 Hz increased 72 dBV.

Figure 3 shows the increase of the hum noise when the 60 MHz single HF noise was inputted. We measured and compared for the three cases that the audio amplifier was AC 100 V 50 Hz powered and AC 100 V 60 Hz powered, and the output of the function generator was directly measured with the audio analyzer not via the audio amplifier. The increase of the hum noise is difference between the sum of amplitude of the hum noise harmonics up to $5^{th}$ order and the amplitude of the noise floor (without the HF noise inputted) at the frequency of the hum noise harmonics.

From figure 3, it is confirmed that the 50 Hz component and its harmonics are increased when the audio amplifier is AC 50 Hz powered, and the 60 Hz component and its harmonics are increased when the audio amplifier is AC 60 Hz powered. In the case that not via the audio amplifier, the hum noise was not increased even though the single HF noise was outputted from the function generator. It is found that these hum noise of the loudspeaker output caused by the single HF noise had relation with internal power supply of the audio amplifier.

From these results, it is found that the non-audible HF noise inputted to audio amplifier increases, hum noise, and degrades sound quality of audio equipment.

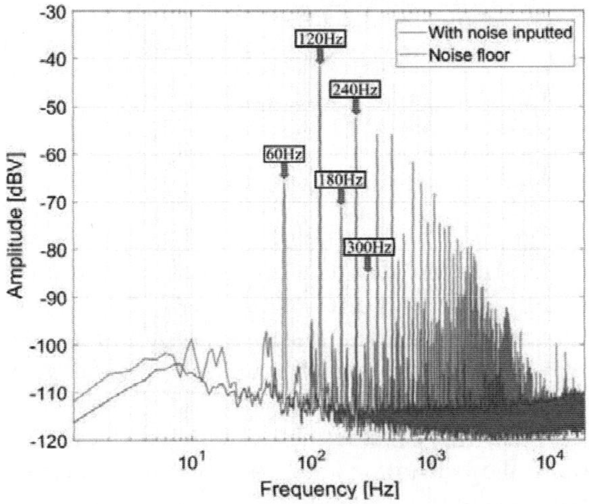


**Figure 2.** Frequency spectrum in audible band of the loudspeaker output of the audio amplifier when the 60 MHz single HF noise.

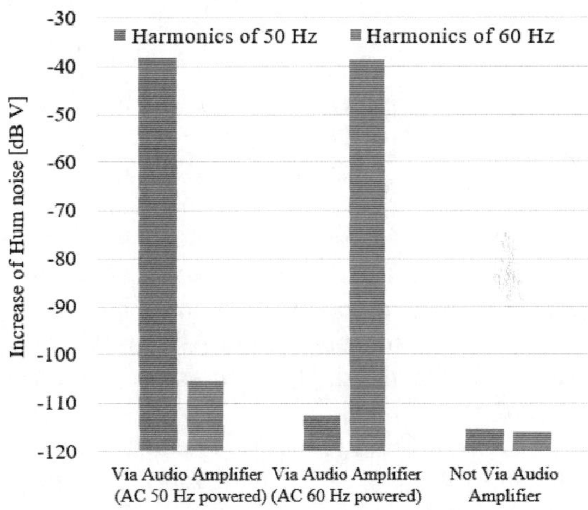


**Figure 3.** Hum noise increase when single HF noise of 60 MHz was inputting, audio amplifier is powered by AC 50 Hz, powered by AC 60Hz, or not via the audio amplifier.

*3.2 Frequency and audio amplifier models dependency of the increase of hum noise*

Figure 4 shows the increase of the hum noise of the 5 models of the amplifiers, caused by the single HF noise in the range of 1 MHz to 240 MHz. The audio amplifier was AC 60Hz powered in this measurement.

From figure 4, it is found that the susceptibility to the single HF noise to the single HF noise depends on the models of audio amplifier and the frequency of the HF noise. With the frequency range and audio amplifier models in this measurement, it is confirmed that the increase of the hum noise is caused in 2 models out of 5 models of the amplifier. The maximum increase of the harmonics of 60 Hz was 27.4 $mV_{rms}$ (-31.3 dBV) in the case that 64.0 MHz single HF noise was inputted to the model an audio amplifier. The harmonics of 60 Hz was also increased 7.89 $mV_{rms}$ (-42.1 dBV) in the case that 15.5 MHz single HF noise was inputted to the model B audio amplifier. However, the hum noise didn't increase in the rest 3 models in these frequency range of the single HF noise. These susceptibility did not depend on the amplification circuit architecture, i.e. analog amplifier or digital amplifier.

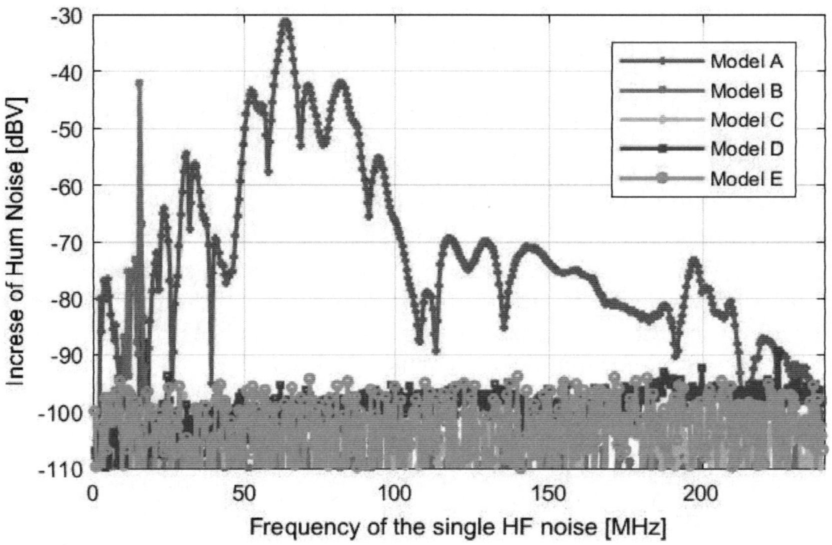


**Figure 4.** Increase of the hum noise of the 5 models of the amplifier caused by the single HF.

## 4. Conclusion

In this study, we investigated influence of non-audible "single" HF electric noise on playback sound of 5 models of audio amplifier in the frequency range from 1 MHz to 240 MHz. From measured results, we confirmed that non-audible single HF noise inputted to audio amplifier increased the hum noise related to the frequency of the AC power supply in the audio amplifier. It was also found that the susceptibility to the single HF noise depended on the models of audio amplifier and the frequency of the HF noise.

The results mean that circuit design and countermeasures with considering characteristics even in higher frequency over audible band is necessary to avoid the bad influence in audio equipment and connected audio equipment, particularly in modern electro-acoustic equipment with many HF noise source.Elucidation of the cause of the sound quality degradation mechanism which is found in this study is our future work.

## References

[1] Fujino T and Yoshida T 2017 A consideration of mechanism of audio signal deterioration caused by propagation noise between audio equipment *Proc. of 2017 Asia-Pacific International Symposium on Electromagnetic Compatibility (APEMC2017)* pp 155-157

[2] Kobayashi D, Kuwabara T and Yoshida T 2017 The influence of non-audible high frequency noise on the playback sound of audio equipment *Proc. of 2017 IEEE 6th Global Conference on Consumer Electronics (GCCE 2017)*

IOP Conf. Series: Journal of Physics: Conf. Series **1075** (2018) 012032 doi:10.1088/1742-6596/1075/1/012032

# Fractal diffusers as noise barrier top-edge devices

**Prachee Priyadarshinee[1], Kian Meng Lim[2] and Heow Pueh Lee[3]**

Department of Mechanical Engineering, National University of Singapore

E-mail: a0135595@u.nus.edu[1], mpelimkm@nus.edu.sg[2], mpeleehp@nus.edu.sg[3]

**Abstract.** With every passing decade, the issue of noise pollution has worsened in urban areas and the usefulness of noise barriers have increased. To improve the effectiveness of noise barriers in limited frequency bandwidth, a few researchers have reported that quadratic residue diffusers as noise barrier top-profiles have shown promising noise attenuation, albeit in their narrow frequency bandwidth. It has been reported for room acoustics that the frequency bandwidth of quadratic residue diffusers may significantly be extended by a nested or fractal design. Therefore, the aim of this work is to investigate the potential of fractal acoustic diffusers for noise barrier application. Upon conducting a numerical analysis of 1-D fractal diffuser as noise barrier top-profile using finite element method, it is observed that fractal diffusers outperformed T-barriers in both the shadow zone and the transition zone. Hence, fractal acoustic diffusers are promising candidates for our application as these have good diffusion properties in the integrated frequency bandwidth. If properly designed, the fractal diffuser top-edge device can be used to enhance noise attenuation over extended frequency bandwidth.

## 1. Introduction

The effect of top-edge of noise barriers in diffracted sound propagation into the shadow zone is known from past studies. Therefore, for the past several decades, extensive works have been published about the various designs of top-profiles in passive noise barriers. Shroeder diffusers are one such top-profile that have been reported by Monazzam and Lam[1], and Fard et. *al* [2] [3] to show improved attenuation over designs like T, Y, arrow, etc.

Schroeder diffusers use number-sequences to form surfaces with series of reflecting wells with the same width and different depths, separated by thin fins. The well depths are based on the residues of modulo arithmetic, which equate to fractional wavelength and wavelets of different phases. In 1979, Schroeder postulated that theoretically, the quadratic residue diffuser attempts to distribute the incident waves by scattering them into many wavelets. The scattered wavelets have definite phase relationships among them due the fact that each wavelet involved in scattering process experiences a different path length when it is reflected from the wells. The path length difference introduces phase differences which cause a wide angular range of diffusion [4] [5] [6]. The acoustic pressure at any receiver location is given by the interference between all the reflected waves generated with different phases, diffracted and scattered sound. However, useful diffusion can be achieved in a narrow frequency bandwidth, which is limited at high frequencies by the well width and at low frequencies by the maximum depth. This limitation led to construction of nested or fractal diffusers, wherein the surface of the parent diffuser is integrated with progressively scaled-down diffusers. The surface consisted of nested self-similar diffusers, each generation of diffuser was designed for a specific frequency bandwidth [7]. Konnert et. *al* [8], Cox et. *al* [7], Landerman et. *al*, Lock et. *al* [9] have reported that good diffusion can be achieved in the integrated bandwidth of the diffusers in a closed room. As fractal diffusers have been proven to improve the sound quality in room acoustics over the integrated bandwidth, an investigation has been conducted in this paper to study and report the performance of fractal diffusers as top-edge device of noise barriers.

### *1.1. Design parameters of diffuser*

In this work, the variation of the depth of the wells is only in one direction, the resultant diffusers being called 1D Schroeder diffusers. The diffusers studied here are designed based on quadratic residue sequence and are hence called quadratic residue diffusers(QRD), as shown in Figure 1(a).

The well depths for a QRD are the non-negative residues or remainders of $n^2$ modulo N. Here, N is a prime number and n is the well position, i.e., 0,1,2,3,...,N-1. Specification of bandwidth is limited by the fact that if the wavelength is too large in comparison with well dimensions, the the diffuser surface may act like a plain surface. Also, if the frequency is too high, achieving good diffusion may be difficult due to complex modes coming into play. Bandwidth is defined as the difference between design frequency $f_0$ and maximum frequency $f_{max}$, where maximum frequency $f_{max}$ is limited by the well width such that

$$f_{max} = \frac{c}{2w} \tag{1}$$

The depth sequence $d_n$ is therefore given by

$$d_n = \frac{s_n \lambda_0}{2N} \tag{2}$$

where $s_n = n^2$ mod N. $\lambda_0$ corresponds to design frequency $f_0$ and is given by Equation 3.

$$\lambda_0 = \frac{c}{f_0} \tag{3}$$

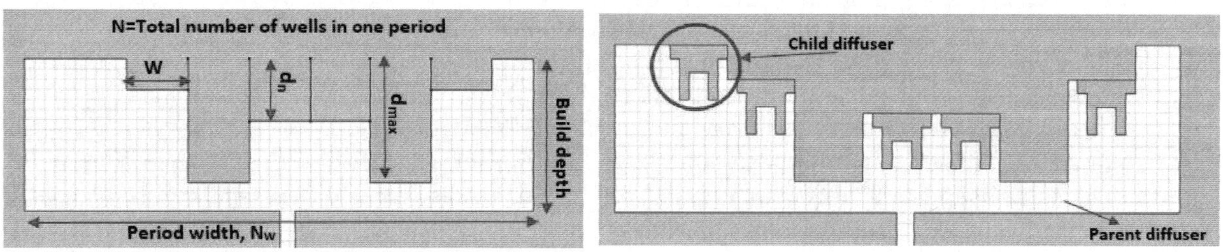


(a) (b)

Figure 1: (a) Design variables of a quadratic residue diffuser and (b) illustration of parent and child diffusers.

## 2. Methodology

The acoustics solver of the commercial software Comsol Multiphysics 5.3, which uses finite element method to solve the Helmholtz equation, was used to obtain results for this numerical investigation. The problem has been solved in the two-dimensional frequency domain solver, and the exterior domain is modelled as air.

A schematic of the finite element model is shown in figure 2(a), the boundary conditions imposed at all exterior boundaries of the domain are the radiation boundary conditions, to ensure that the scattered waves are outgoing waves and satisfy Sommerfeld's radiation condition. The surfaces of all barriers is assumed to be sound reflective, therefore, sound hard boundary

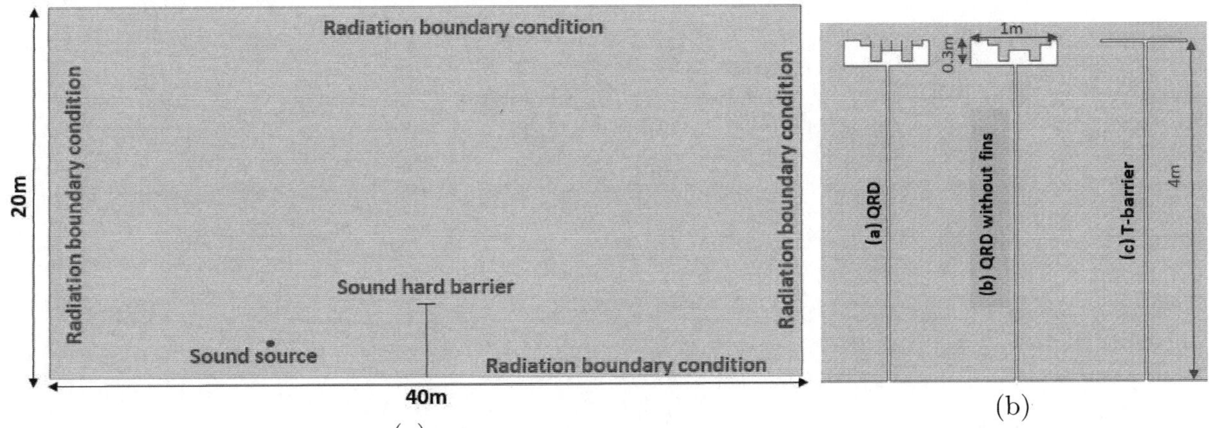


Figure 2: (a) A schematic of finite element model, and (b) Illustration of some designs of top-edge devices.

conditions are imposed in the finite element model. The monopole source in 2D represents a line source in 3D. In this study, this acoustic source is 5m away from the barrier. As shown in Figure 2(b), the overall height of the barriers is 4m, thickness of the stem is 0.03m and the width-span of the top profile is 1m for all barrier configurations. The maximum depth of the QRD is 0.3m.

Figure 1(b) depicts a nested or fractal diffuser, which consists of a parent diffuser and a child diffuser. The parent diffuser, designed for lower frequencies is denoted as LFD( acronym for low frequency diffuser) in this paper. Similarly, as the child diffuser is designed for higher frequencies it is referred to as HFD(high frequency diffuser). For the purpose of comparison between child diffusers, the parent diffusers are the same in every model. The parent-diffuser common for all fractal designs analyzed in this paper has the number of wells N as 7, design frequency $f_0$ of 400Hz, width of each well w as 12cm, and maximum depth $d_{max}$ as 0.25m.

Keeping in mind the undulations in spatial pressure response, the values represent an average over three receivers spaced 0.1m apart. Sound field was measured at receiver locations behind the barrier, in the geometrical shadow zone and near the shadow boundary, as can be seen in Figure 3. The geometrical shadow boundary is an imaginary boundary obtained by drawing a line connecting the source and the top of the barrier, shielding the receivers from direct incident waves. In this study, the shadow boundary calculated by geometry is at $123^0$.

## 3. Results and discussions

The objective of this study is to investigate whether the sound measured is lower when a noise barrier with fractal diffuser as top-profile is in place than when shielded by a T-barrier. To do so, measurements are conducted in the shadow zone and near the shadow boundary. A schematic of receivers is represented in the Figure 3, the lines $60^0$ and $90^0$ being in the shadow zone whereas the line $120^0$ being near the shadow boundary.

The results are reported in terms of additional attenuation achieved against a reference T-barrier.

$$Additional\ attenuation = SPL_{T-\ barrier} - SPL_{model-\ barrier}$$

where SPL stands for sound pressure level in dB. Therefore, positive additional attenuation

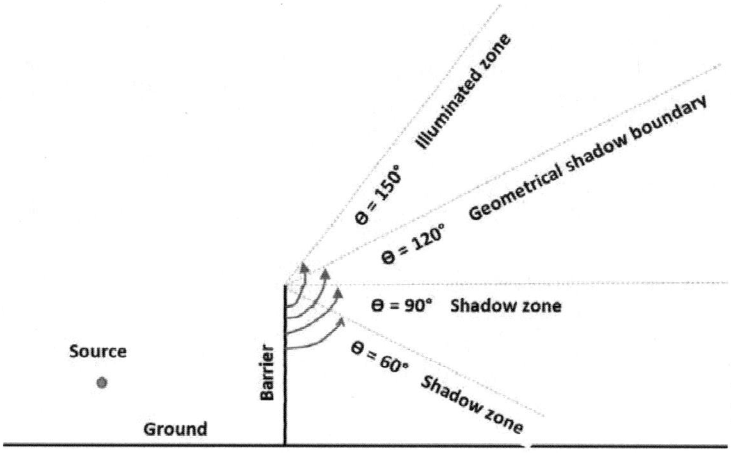


Figure 3: Schematic of receiver positions for sound measurement.

implies that our test cases are performing better than T-barrier.

Keeping the max depth and total width fixed, several designs of nested diffusers featuring a variation of number sequence, N and design frequency, $f_0$ were studied in this exercise. The selected results shown here are compare the designs LFD, $model_A$ and $model_B$. LFD denotes QRD without fins, whereas $model_A$ and $model_B$ are nested diffusers , as shown in Figure 2(b). The child diffuser in both $model_A$ and $model_B$ have N7 sequence but design frequency of 1000 Hz and 800 Hz, respectively. The design parameters of LFD can be found in section 2.

The lowest design frequency, of the parent diffusers, is 400Hz and hence the results are presented from 400Hz onwards.

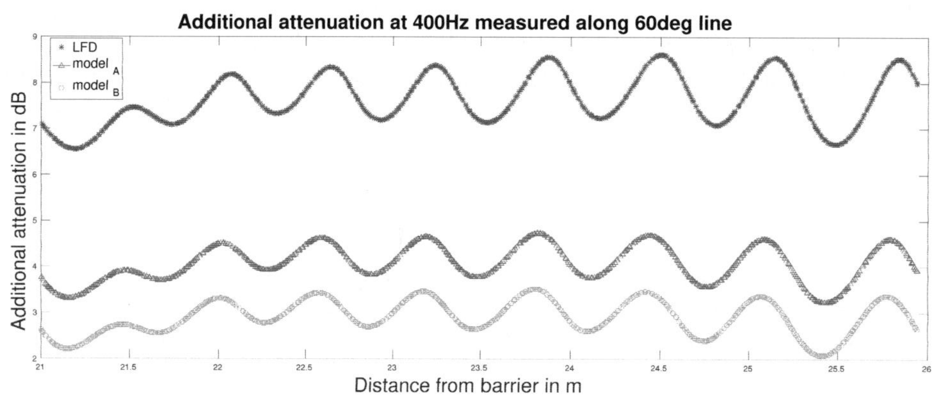


Figure 4: Additional attenuation along the $60^0$ line at 400Hz

As seen in Figure 4, the blue line assigned to model LFD has higher attenuation than the other two designs. Although the results are presented only along the $60^0$ line, this characteristic was observed along $60^0$, $90^0$ and $120^0$ lines, i.e., in the shadow zone as well as near the shadow boundary. This behaviour could be attributed to the difference in the diffusion pattern of a parent diffuser with and without the inclusion of child diffusers, which have a diffraction pattern of their own. However, as it can be seen in the figures below, this behaviour is observed only at the design frequency of 400Hz of LFD.

As seen in the Figures 5(a) and 5(b), the red and yellow lines denoting $model_A$ and $model_B$ have an average of 6dB more attenuation than the blue line denoting the barrier LFD. These

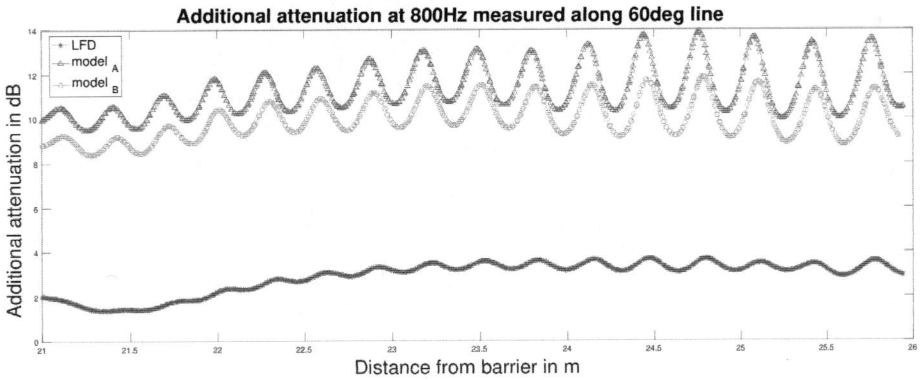


(a)

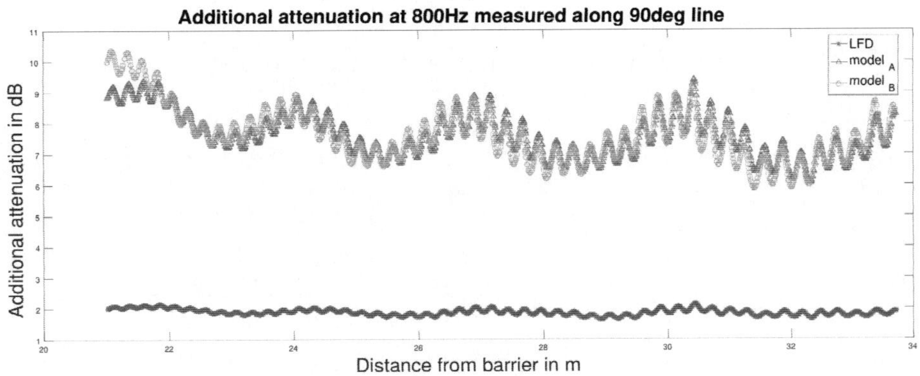


(b)

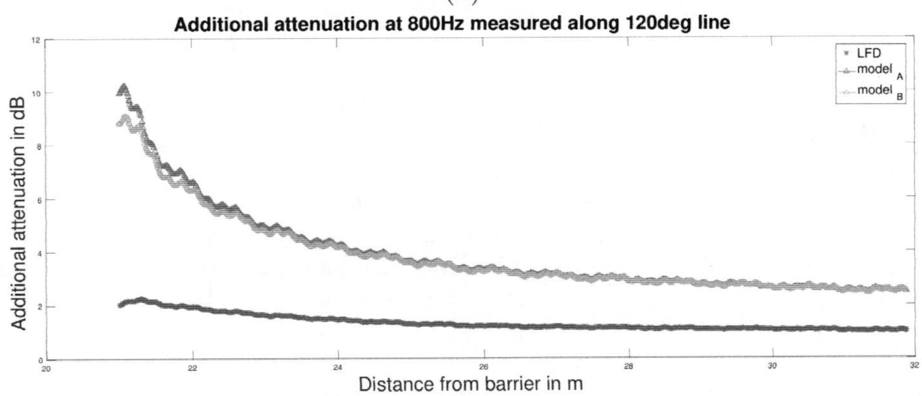


(c)

Figure 5: Additional attenuation at 800Hz along (a) $60^0$, (b) $90^0$, and $120^0$ lines.

figures support the claim in this sample frequency size that at multiples of design frequency of parent and child diffusers, i.e., at 800, 1000, 1200, 1600 and 2000Hz, $model_A$ and $model_B$ outperformed the LFD model. Therefore, at these frequencies, the response of the barriers having fractal diffusers as top-profiles is similar to their response in Figure 5.

For further proof, this response has been clearly captured in Figure 6. In the Figure 6, instead of attenuation, the sound pressure levels are directly plotted 5.5m behind the barrier at a height of 4m. This receiver position is a representative of other positions tested and in the shadow zone

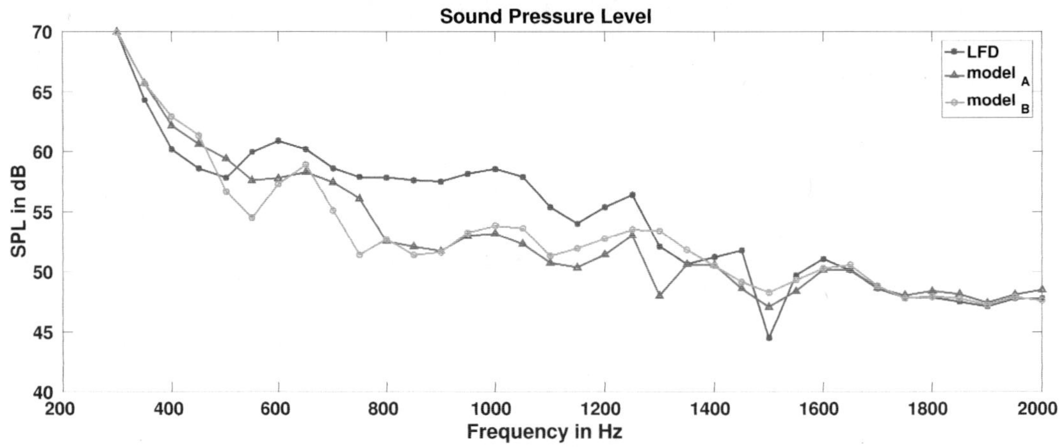


Figure 6: Sound pressure level with respect to frequency.

and the has similar response. Thus, higher SPL implies that the barrier is performing worse than others and vice versa. The SPL of LFD, $model_A$ and $model_B$ with respect to frequency has been plotted at a resolution of 50Hz. Closer to the design frequency of 400Hz, the LFD barrier has lower SPL than the other two models. However, it can be seen that between 550Hz and 1650Hz, with the exception of 1500Hz, the barriers with fractal diffusers as their top-profiles have lower SPL.From 1700Hz onwards, the difference is negligible and all three seem to perform as well as each other.

However, there is an uncertainty of numerical measurement of $\pm$1dB and hence any similar gain in attenuation may not be considered significant.

## 4. Conclusions

Nested diffusers offered promising response and improved the performance of barriers over simple quadratic residue diffusers at the tested frequencies. T-barrier has the lowest attenuation in the shadow zone and near the geometric shadow boundary when compared with the quadratic residue diffuser and the fractal diffusers at all tested frequencies. The difference is particularly evident at the design frequencies of diffusers and integer multiples of design frequencies. The diffusive noise barrier caps have an average additional attenuation of 6dB over the T-barrier at these frequencies.

## References

[1] Monazzam M 2006 *Journal of Environmental Health Science & Engineering* **3** 229–238
[2] Naderzadeh M, Monazzam M R, Nassiri P and Fard S M B 2011 *Applied Acoustics* **72** 393–398
[3] Fard S M, Kessissoglou N, Samuels S and Burgess M 2013 *Proceedings of Meetings on Acoustics* vol 19 (Acoustical Society of America) p 040081
[4] D'Antonio P and Konnert J H 1984 *Journal of the Audio Engineering Society* **32** 228–238
[5] Schroeder M R 1979 *The Journal of the Acoustical Society of America* **65** 958–963
[6] Schroeder M R 1980 *Physics Today* **33** 24–30
[7] Cox T J and D'antonio P 2009 *Acoustic absorbers and diffusers: theory, design and application* (Crc Press)
[8] D'Antonio P and Konnert J H 1984 *Journal of the Audio Engineering Society* **32** 228–238
[9] Lock A and Holloway D 2015 *Acoustics 2015 Hunter Valley* pp 1–10

# The relationship between sound source and urban soundscape

**A S Sudarsono, N P A Nitidara and J Sarwono**

Department of Engineering Physics, Institut Teknologi Bandung

Email: anugrah@tf.itb.ac.id, amd.nitidara@students.itb.ac.id, jsarwono@tf.itb.ac.id

**Abstract**. Noise analysis of urban area is usually conducted according to noise level measurement. The noise level measurement concept only considers how loud the overall acoustic environment but not how people perceived each sound source. The soundscape concept has become a new alternative to analyse acoustic environment. This concept tries to understand the perception of the acoustic environment from the interaction between the sound sources and people who use the space. This study tries to understand the relationship between sound sources in an urban area with people's perception related with three feeling: Relaxation, Dynamic, and Communication/Communality. An experiment was conducted using 26 simulated urban acoustic environments. The stimuli were composed using six sound sources: The sound of hubbub, urban traffic, construction noise, music, event sounds, and water fountain. The analysis was conducted using forward logistic regression. The perception of Relaxation is affected by the sound of music, urban traffic, and construction noise. The perception of Dynamic is affected by the sound of hubbub, urban traffic, and event sounds. The perception of Communication is affected by the sound of hubbub and water fountain. This result indicates that the perception of acoustic environment are related to specific sound source inside the environment.

## 1. Introduction

Nowadays, noise measurement methods are still widely used to rate the sound in urban areas. Noise measurement methods only rate sounds according to noise level, without consideration of discrete sound objects [1]. The soundscape method attempts to understand the perception of a soundscape from the interaction of the sound objects inside it, hoping to create a better auditory environment by using positive sound objects and working more with people's perceptions rather than simply the overall sound level measurement [2]. Using the noise measurement approach, as long as the sound level of a soundscape exceeds a certain level, it will be considered an unpleasant soundscape. With the soundscape approach, however, the focus is on how people who use the space feel about the soundscape and their interaction with sound objects in the soundscape. Researchers have therefore begun to investigate the factors affecting the perception of soundscapes, and one of the factors being investigated is the interaction between sound objects and the rating of the soundscape [1].

A soundscape environment is formed by the sound objects inside it. Sound object in here is taken as a source of sound in a soundscape including behaviours (movement and position in space), distribution in time (event sound or background sound), sound level, and interactions with the environment. Some studies have tried to find the relationship between these sound objects and the perception of the soundscape as a whole [1,3–5]. These studies have grouped sound objects according to how they are perceived, and not how they affect the overall perception of the soundscape. In addition, the previous studies have not clearly defined which specific sound objects significantly affect the perception of a soundscape, nor how specific sound objects affect the rating of soundscape compared to other sound objects.

This study tries to understand the relationship between sound objects in a soundscape to the perception of soundscape representing with three perceptions: Relaxation, Dynamic, and Communication/Communality.

## 2. Method

The experiment was conducted using simulated soundscape composed using six sound sources: the sound of hubbub, music, water fountain, traffic noise, construction noise, and the event sound source. The description of the sound sources is shown in table 1. Variation of sound level are also given. The length of each stimulus was one minutes. The stimulus of the experiment is shown in table 2. The experiment was conducted in a listening room and reproduced using two-dimensional reproduction system which can reproduce immersive sound including the position and movement of sound source in space.

Sound objects in a soundscape are considered to affect the ratings of the total soundscape; for example, natural sound is believed to increase the perception of pleasantness [3,4], whereas human sound is considered to increase perceptions of eventfulness. In this study, the natural sound is represented by the sound of water fountain, and human sound is represented using the sound of hubbub.

Mechanical or artificial sounds correspond with unpleasant ratings of a soundscape [1]. The sound of mechanical sound is represented by the sound of construction noise and traffic noise. The sound of music was also selected because there are evidence showing the relationship between music and relaxation [6,7].

Twenty-one participants (15 male, 6 female) joined the experiment voluntarily. The entire participants were undergraduate and master students (22-27 years old, M = 24.6, SD = 2.38) of Engineering Physics Department, ITB with the background of engineering and acoustic.

The experiment was conducted individually. Each of participants gave the rating of each stimulus (which was played in random order) according to three perceptions: Relaxation, Dynamic, and Communication. The perceptions were selected according to the study of soundscape dimension in urban area [8,9]. The questionnaire used 11 points Likert scale and made using Microsoft Excel as shown in Figure 1.

## 3. Results and discussion

Analysis of the data is conducted using two approaches: Stepwise Linear Regression and Stepwise Logistic Regression. The logistic regression is calculated using the output of one/zero so the data need to be transform first before this regression is implemented. The Linear Regression is evaluated from the value of Pearson's Correlation Coefficient (R) from the models, shown in Table 3, while the Logistic Regression is evaluated from the value of percentage of correct prediction of the model, shown in Table 4.

The linear model gives Pearson's coefficient correlation (R) between 0.475-0.635. The highest correlation is shown by the linear model for Communication indicating a good prediction of the model. The logistic model can predict the data between 63.9%-81.4%. The highest model is shown by the model for the perception of Communication. This result is consistent with the linear model and the model also indicates a good prediction.

Further analysis is conducted from the linear and logistic models as shown in Table 5 and 6. The linear and logistic model shows similar characteristic. The models show that there are correlation between the sound source and the perception of the acoustic environment.

The perception of relaxation is related to the sound of music, traffic noise, and construction noise. The negative value of Beta (B) represents the relation to the perception of comfortable while the positive value Beta (B) represent the relation to the perception of uncomfortable. The sound of music related to the perception of comfortable and the sound of traffic noise and construction related to the perception of uncomfortable.

The perception of dynamic is related to the sound of hubbub, event sound, and traffic noise. The entire significant sound sources make the soundscape perceived as dynamic soundscape (indicated by the negative value of B).

The perception of communication is related to the sound of hubbub and water fountain. The sound of hubbub makes the soundscape to be perceived as communal (indicated by the negative value of B) while the sound of water fountain is related to the perception of private (indicated by the negative value of B).

**Table 1.** Sound source description.

| No | Sound Source | Description |
|---|---|---|
| 1 | hubbub | the sound of people talking in the background |
| 2 | music | the sound of people playing relaxing music with string instrument |
| 3 | water fountain | the sound of water fountain in the background |
| 4 | traffic noise | the sound of traffic noise in the background |
| 5 | construction noise | the sound of construction noise in the background |
| 6 | event sound | the combination of the sound which occurs once and in the short period. It consist of: the sound of footstep, the sound of trolley bag, bike being pushed, and bird chirping |

**Table 2.** Stimulus used in the experiment[a]

| No | Sound Source 1 | Sound Source 2 | Sound Source 3 |
|---|---|---|---|
| 1 | People 0 | Music 0 | |
| 2 | People 0 | Fountain 0 | |
| 3 | People 0 | Music -10 | |
| 4 | People 0 | Fountain -10 | |
| 5 | People -10 | Music 0 | |
| 6 | People -10 | Fountain 0 | |
| 7 | People 0 | Music 0 | Event |
| 8 | People 0 | Fountain 0 | Event |
| 9 | People 0 | Music -10 | Event |
| 10 | People 0 | Fountain -10 | Event |
| 11 | People -10 | Music 0 | Event |
| 12 | People -10 | Fountain 0 | Event |
| 13 | People 0 | | Event |
| 14 | Music 0 | | Event |
| 15 | Fountain 0 | | Event |
| 16 | Traffic 0 | Fountain 0 | |
| 17 | Traffic 0 | Music 0 | |
| 18 | Traffic 0 | People 0 | |
| 19 | Traffic 0 | | |
| 20 | Construction 0 | Fountain 0 | |
| 21 | Construction 0 | Music 0 | |
| 22 | Construction 0 | People 0 | |
| 23 | Construction 0 | | |
| 24 | Fountain 0 | | |
| 25 | Music 0 | | |
| 26 | People 0 | | |

[a]0 indicates that the sound was reproduced at the actual sound level and -10 indicates that the sound was reproduced 10 dB below the actual level

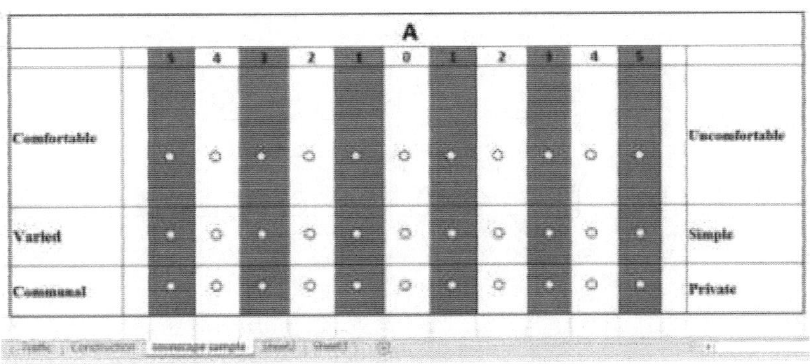


**Figure 1.** Questionnaire used in the experiment.

**Table 3.** Correlation coefficient of linear model

| Perception | R |
|---|---|
| Relaxation | .475 |
| Dynamic | .337 |
| Communication | .635 |

**Table 4.** Percentage correct of logistic model

| Perception | Percentage Correct |
|---|---|
| Relaxation | 68.3% |
| Dynamic | 63.9% |
| Communication | 81.4% |

**Table 5.** Soundscape linear model according to the perception of relaxation, dynamic, and communication

| **Relaxation** | B | Std. Error | Standardized Beta | Sig. |
|---|---|---|---|---|
| Constant | .020 | .144 | | .891 |
| Music | -.022 | .003 | -.295 | .000 |
| Traffic | .030 | .004 | .285 | .000 |
| Construction | .022 | .004 | .218 | .000 |

| **Dynamic** | B | Std. Error | Standardized Beta | Sig. |
|---|---|---|---|---|
| Constant | .365 | .184 | | .048 |
| Traffic | -.034 | .004 | -.354 | .000 |
| Event | -.726 | .212 | -.148 | .001 |
| Hubbub | -.010 | .003 | -.143 | .001 |

| **Communication** | B | Std. Error | Standardized Beta | Sig. |
|---|---|---|---|---|
| Constant | .772 | .176 | | .000 |
| Hubbub | -.053 | .003 | -.608 | .000 |
| Fountain | .014 | .003 | .151 | .000 |

**Table 6.** Soundscape logistic model according to the perception of relaxation, dynamic, and communication

| **Relaxation** | B | Sig. | Exp(B) |
|---|---|---|---|
| Music | -.018 | .000 | .983 |
| Traffic | .025 | .000 | 1.025 |
| Construction | .018 | .000 | 1.018 |
| Constant | -.115 | .380 | .891 |

| **Dynamic** | B | Sig. | Exp(B) |
|---|---|---|---|
| Hubbub | -.006 | .035 | .994 |
| Event | -.417 | .037 | .659 |
| Traffic | -.024 | .000 | .976 |
| Constant | .042 | .808 | 1.042 |

| **Communication** | B | Sig. | Exp(B) |
|---|---|---|---|
| Hubbub | -.040 | .000 | .961 |
| Fountain | .011 | .002 | 1.011 |
| Constant | .163 | .329 | 1.177 |

## 4. Conclusion

This study has confirmed that there is a relationship between the perception of acoustic environment (according to the perception of relaxation, dynamic, and communication) and the specific sound source. The perception models have been developed using linear regression (Pearson's correlation between 0.337-0.635) and logistic regression (Percentage Correct between 63.9%-81.4%). The models from both analysis shows consistent result. The perception of relaxation is affected by the sound the sound of music, traffic noise, and construction noise. The perception of dynamic is affected by the sound of hubbub, event sound, and traffic noise. The perception of communication is affected by the sound of hubbub and water fountain.

## References

[1] Hall D A, Irwin A, Edmonson-Jones M, Phillips S and Poxon J E W 2013 An exploratory evaluation of perceptual, psychoacoustic and acoustical properties of urban soundscape *Appl. Acous.* 74 248-54

[2] Davies, Adams, Bruce, Cain, Carly, Carlyle, Cusack, Hume, Jennings and Plack 2007 The positive soundscape project *19th International Congress on Acoustic* (Madrid)

[3] Axelsson Ö, Nilsson M E and Berglund B 2010 A principal component model of soundscape perception. *J. Acoust. Soc. Am.* 128 2836-46

[4] Brambilla G, Gallo V, Asdrubali F and D'Alessandro F 2013 The perceived quality of soundscape in three urban parks in Rome. *J. Acoust. Soc. Am.* 134 832-9

[5] Dubois D, Guastavino C and Raimbault M 2006 A cognitive approach to urban soundscape: using verbal data to access everyday life auditory categories. *Acta Acust. United with Acust.* 92 865-74

[6] Stratton V N and Zalanowski A H 1984 The relationship between music, degree of liking, and self-reported relaxation. *J. Music Ther.* 21 184-92

[7] David W B and Thaut M H 1989 The influence of preferred relaxing music on measures of state anxiety, relaxation, and physiological responses 1. *J. Music Ther.* 26 168-87

[8] Kang and Zhang 2002 Semantic differential analysis on the soundscape of urban public spaces. *J. Acoust. Soc. Am.* 112 2435

[9] Sudarsono A S, Lam Y W and Davies W J 2016 The effect of sound level perception of reproduced soundscape. *Appl. Acoust.* 110 53-60

## Acknowledgment

The author would like to thank Asahi Glass Foundation for their funding of this research.

# An Indonesian speech synthesis system considering low-power event before vowel-starting-syllable

**I Setiawan and T Hirai**

Arcadia, Inc.

Minoh Sunplaza 1, 7F
6-3-1 Minoh, Minoh-city, Osaka 562-0001, Japan

Email: ivan@arcadia.co.jp

**Abstract**. In our Indonesian concatenative speech synthesis system, there existed a linking issue between vowel-starting-syllable with its preceding phoneme. Unlike other languages, in Indonesian the power of the speech signal decreases (sometime it is a pause) in this kind of boundary. Phonemes before and after the boundary are uttered separately. For example, the underlined phonemes before and after the boundary in "buah apel (apple fruit)" are not to be uttered continuously, but should be separated. This is not the case in English, where the underlined phonemes in "an apple" are linked. We did not treat this kind of low-power event ("lpow") explicitly, such that the lpow generated indirectly from the syllable boundary information, is sometime too short, resulting in the above linking issue. In this paper, we propose to explicitly treat the lpow. The lpow is treated similarly with phoneme during the model training, so that it is appropriately generated during the synthesis. We confirmed that the synthesized speech is more natural by the introduction of lpow.

## 1. Introduction

Arcadia, Inc. provides services to deliver disaster prevention/reduction messages through e-mail, FAX, telephone, and public announcement speakers. Thus, speech synthesis system with a high broadcast-quality is required.

An Indonesian language (Indonesian: "Bahasa Indonesia") speech synthesis system has been proposed by Sakti et al [1]. Since this system is a statistical parametric synthesizer that synthesizes speech from feature vector, quality improvement by the adoption of concatenative synthesis method [2, 3], was required.

We have already developed an Indonesian concatenative speech synthesis in [4]. By utilizing a pronunciation dictionary and pronunciation rules, this speech synthesizer can differentiate between the pronunciation of [@] and [e] (X-SAMPA [5]) which are represented with identical "e" character in the text. The system in [4] considered only the phoneme substitution cost and concatenative distortion, but did not include F0 feature when computing target cost during unit selection. This resulted in a rising end -of- sentence intonation in some synthesized speech of declarative sentences (so it sounds like interrogative sentences) which is caused by units with relatively high F0 was selected.

An HMM-based model training method to obtain a model that generates feature vector target based on linguistic information such as phoneme environment (environment where the phoneme appears, such as adjacent phonemes), number of phonemes in phrases, etc., is already released and available for public use [6]. We incorporated this method to provide a model with a correct F0 target, so that units with appropriate F0 are selected [7]. Rising intonation was solved, however, linking of vowel-starting-syllable/word with the preceding syllable/word during pronunciation occurred.

This paper presents our solution to the linking issue. In section 2, we briefly explain about the concatenative speech synthesis. Then, the main theme in this paper, i.e. the linking issue and its solution by introducing low-power event, are described in section 3. Section 4 describes our Indonesian speech corpus and some narration styles that we found while constructing the corpus. Some experiments to check the effect of the low-power event and narration style classification are shown in section 5, and we provide our summary in section 6.

## 2. Indonesian concatenative speech synthesis system

This section briefly describes the Indonesian concatenative speech synthesis that we have reported in [4] and [7].

As in many systems, there are 2 parts in the system, that is, the analysis part and the synthesis part. The analysis part, or more widely known as speech corpus building, is done only once during the system construction. On the other hand, the synthesis part is called each time a text needs to be synthesized as a speech waveform.

In the corpus building, speech with transcription (text) is analyzed. Transcription texts are converted to phoneme sequences, then a phoneme acoustic model trained using speech and the phoneme sequences, is generated. A forced alignment technique was used to obtain the start/end point of each phoneme in the speech. When the analysis finished, we obtain the corpus, the acoustic model, the F0 control model, and the duration model. The corpus stores phoneme environment, start/end point, and acoustic features (F0, power, cepstrum) of each phoneme. The model will be used to estimate acoustic features from the phoneme environment.

To synthesize text to a speech, first the input text is converted to phoneme sequence. Then, the acoustic features (target features, F0 etc.) are generated using the trained model. Let $s$ be sequence of unit, then an appropriate unit sequence $s^*$ which has features similar to the target, i.e., low target cost $C_t(s)$, and shows the least concatenative distortion, i.e., low concatenation cost $C_c(s)$, was selected from the database and was concatenated into a waveform. Written formally:

$$s^* = argmin_s\{w_t C_t(s) + w_c C_c(s)\} \quad (1)$$

where $w_t$ and $w_c$ are weights that control the contribution of the target cost and the concatenation cost to the overall cost, respectively.

In the system in [4], the target cost only considered the phoneme environment disagreement cost (phoneme substitution cost) . In the system in [7], the target cost also considered the F0 target cost. Both implementations have their own issues, but in comparison with the statistical parametric synthesis which gives muffled sound, the concatenative synthesis produces a crisp and clearer sound, satisfying the broadcast quality requirement for a public announcement system.

## 3. Low-power event

When saying Indonesian word or syllable starting with vowel, it is natural to put a brief pause or decrease the speech signal power, before the vowel. For example, phrases like "buah apel (apple fruit)" and "bawah atas (bottom, top)" is better to utter as [buah ap@l] and [bawah atas], instead of [bua hap@l] and [bawa hatas] (a space inside the phoneme sequence, showed inside square brackets "[]", implies a brief pause or decreasing of the speech signal power). The latter case is what we refer as the linking issue, where the vowel (the first phoneme, the vowel [a] of the second word) is linked to the final phoneme of the preceding word (the final phoneme [h] of the first word).

The linking issue occurs not only before vowel-starting-word, but also before vowel-starting-syllable, especially when uttering abbreviations. For example, "SNI (Standar Nasional Indonesia, Indonesian National Standard)" and "PMI (Palang Merah Indonesia, Indonesian Red Cross)" which should utter as [es en i] and [pe em i], might be incorrectly synthesized as [e se ni] and [pe emi] due to the linking issue.

The linking issue may or may not occur in our previous system described in [7], depends on the existence of the phrases in the corpus. If it is part of the training data, then this issue does not happen in our previous system. If it is not, then the linking will happen because the brief stop or the decrease in the speech signal power before vowel -starting-word/syllable was not explicitly treated. In this paper, we propose to explicitly treat this event by introducing a low- power event "lpow" to indicate a brief pause or a decrease in the speech signal power. Beware that lpow is not a pause event like a comma or a period, neither it is a phoneme. Since the lpow event is phonetically meaningful and can be considered during the model training, an lpow event is expected to be generated before the vowel-starting-syllable when using the trained model.

Incorporating the lpow event, "bawah atas", "SNI", and "PMI" will have phoneme sequences of [bawah lpow atas], [es lpow en lpow i], and [pe lpow em lpow i], respectively.

## 4. Indonesian speech corpus

Indonesian speech data were recorded in 44.1 kHz sampling and 16-bit integer wav-format, from a female native Indonesian speaker who teaches Indonesian language and has been living in Japan over 20 years. Phrases included proper nouns and text samples from Indonesian language learning books, newspaper articles, and TV scripts, which covered almost all possible Indonesian syllables. There were 4,490 phrases lasting a total of 5.1 hours. Pauses were notated in the transcription by listening to the recorded speech.

During corpus checking, we noticed that in our corpus there are 2 types of end-of-sentence's intonation for question (interrogative sentence). These 2 types of intonation were due to the narration styles, i.e., prose (read speech) and dialog (conversational speech). The prose style question ends with a rising intonation as shown in figure 1 (the text is "Siapa anda? (Who are you?)"), while the dialog style question ends with a long and flat intonation as shown in figure 2 (the text is "Wanita yang rambutnya pendek itu siapa? (Who is that short-haired woman?)"). Phoneme boundaries in both figures 1 and 2 are obtained by forced alignment.

Since rising and flat intonations will result in a different F0 feature, we manually classify and assigned different type of question mark for each style in the corpus; and we expect to obtain a better F0 model.

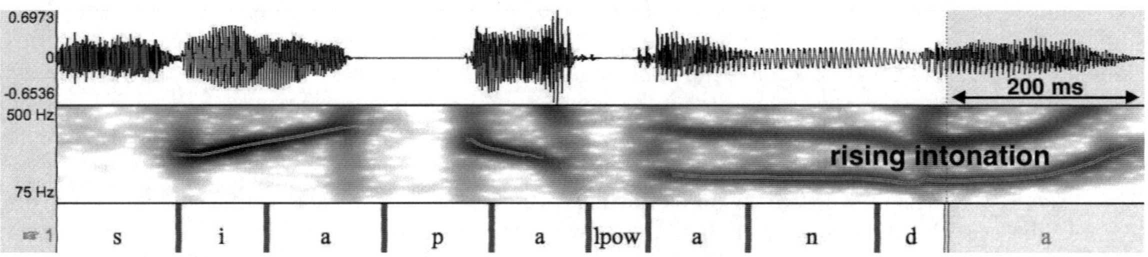


**Figure 1.** Prose style question "Siapa anda?" in corpus. Top to bottom: waveform, F0 pattern on narrow band spectrum, and phonemes with boundaries. Rising intonation in the shaded part.

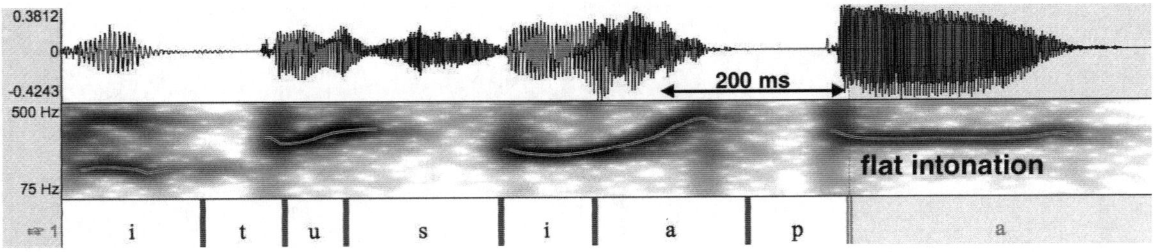


**Figure 2.** Dialog style question "Wanita yang rambutnya pendek itu siapa?" in corpus. Top to bottom: waveform ("itu siapa?" part), F0 pattern on narrow band spectrum, and phonemes with boundaries. Flat intonation in the shaded part.

## 5. Experiments and results

Two examples where low-power event improves the naturalness of the synthesized speech are shown below. Then, we also mention the effect of classifying type of question sentence in the corpus.

### *5.1. Effect of low-power event "lpow" in the case where the lpow exists before a WORD starting with vowel*

The input text is "bawah atas" (phoneme sequence: [bawah atas]). A vowel-starting-word "atas (top)" is following the first word "bawah (bottom)". Figure 3 and 4 shows the full waveform of the synthesized "bawah atas" before and after the introduction of lpow event, respectively.

In figure 3, it is clear that the first phoneme [a] of the succeeding word is linked to the preceding word's final phoneme [h]. The speech was sound as [bawahatas]. Actually, the phoneme [h] and [a] were both selected from the continuing units [ha] in the word "kejahatan (crime)" in the corpus, where [h] and [a] are both in the same syllable.

In figure 4, the lpow correctly separates the [bawah] and [atas] such that the speech was clearly sound as [bawah atas]. The lpow here resembles a brief pause.

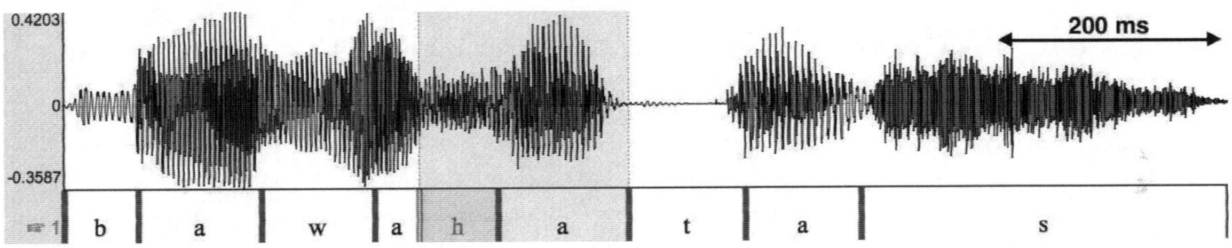


**Figure 3.** Full waveform of the synthesized "bawah atas" before the introduction of lpow. Top: waveform, bottom: phonemes with boundaries. Linking of [ha] in the shaded part.

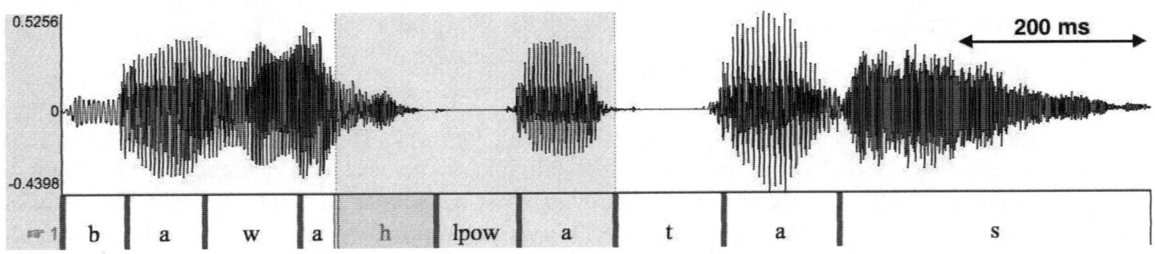


**Figure 4.** Full waveform of the synthesized "bawah atas" after the introduction of lpow. Top: waveform, bottom: phonemes with boundaries. An appropriate lpow is selected, thus no linking of [h] and [a] in the shaded part.

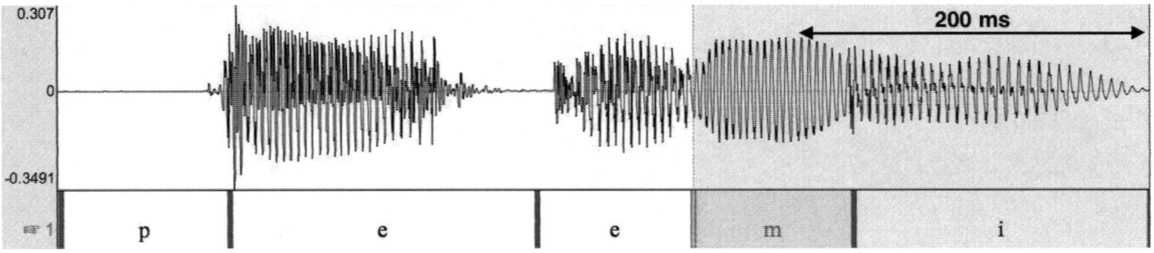


**Figure 5.** Waveform of the synthesized "PMI" before the introduction of lpow. Top: waveform, bottom: phonemes with boundaries. Linking of [mi] in the shaded part.

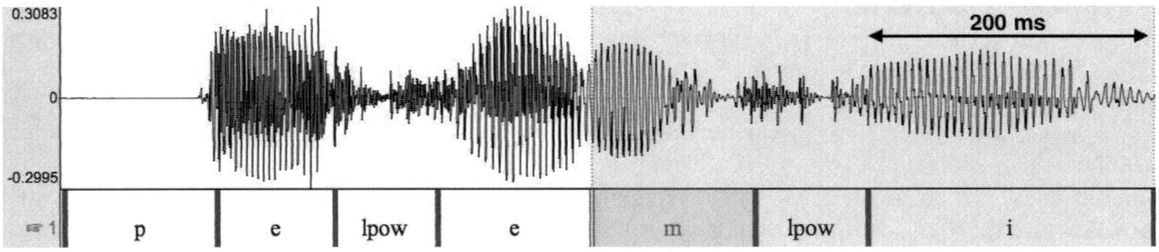


**Figure 6.** Waveform of the synthesized "PMI" after the introduction of lpow. Top: waveform, bottom: phonemes with boundaries. An appropriate lpow is selected, thus no linking of [m] and [i] in the shaded part.

*5.2. Effect of low-power event "lpow" in the case where the lpow exists before a SYLLABLE starting with vowel*

The input text is "Palang Merah Indonesia disingkat PMI. (Indonesian Red Cross is abbreviated as PMI.)", and we pay attention to the vowel- starting-syllable [i] in the abbreviation "PMI [pe em i]". Figure 5 and 6 shows the waveform of the synthesized "PMI" before and after the introduction of lpow event, respectively.

In figure 5, the [e] of [pe] and its succeeding [e] of [em] are separated because there is such a unit in the corpus just by coincidence. The [i], however, is linked to the preceding [m], such that the speech was sound as [pe emi]. Similar to the case in the previous example, the phoneme [m] and [i] were both selected from the continuing units [mi] in the word "kami (us)" in the corpus, where [m] and [i] are both in the same syllable.

In figure 6, lpow correctly separates the [em] and the [i] such that the speech was clearly sound as [em i], instead of [emi]. The lpow here is a low-power speech signal, not a brief pause.

*5.3. Effect of classifying types of question in the corpus*

In section 4, we have described the classification of 2 types of question sentence. However, we have found no good evidence whether this classification really improves the naturalness of the question speech synthesis. This is probably due to the small number of the question sentence in the corpus, which are only 121 (prose style questions: 71, dialog style questions: 51; One phrase includes two question sentences with different type of question.) out of 4,490 phrases. The trained model related to question might be unstable.

## 6. Summary

We have introduced the low-power event before vowel-starting-word/syllable and showed through examples that its effect is as expected. We also differentiate type of question sentence based on its end-of-sentence intonation, however no improvement was observed which maybe due to the small number of question sentences in the corpus.

## References

[1] Sakti S, Maia R, Sakai S, Shimizu T and Nakamura S 2008 Development of HMM-based Indonesian Speech Synthesis *Oriental COCOSDA*

[2] Kawai H, Toda T, Ni J, Tsuzaki M and Tokuda K 2004 XIMERA: A New TTS from ATR Based on Corpus-Based Technologies *ISCA 5th Speech Synthesis Workshop*

[3] Gonzalvo X, Tazari S, Chan C, Becker M, Gutkin A and Silen H 2016 Recent Advances in Google Real-time HMM-driven Unit Selection Synthesizer *INTERSPEECH*

[4] Hirai T and Setiawan I 2016 An Indonesian Concatenative Speech Synthesis System *5th Joint Meeting of ASA and Acoustical Society of Japan (ASJ)*

[5] https://ja.wikipedia.org/wiki/X-SAMPA

[6] http://hts.sp.nitech.ac.jp/

[7] Hirai T and Setiawan I 2017 An Indonesian Concatenative Speech Synthesis System (in Japanese) *ASJ Autumn Meeting*

## Acknowledgment

The authors wish to thank all the members of Arcadia, Inc. for all their supports during this work.

# A method for selecting a proper modulation technique for the parametric acoustic array

**F Farias and W Abdulla**

Department of Electrical and Computer Engineering, University of Auckland, New Zealand

E-mail: ffar475@aucklanduni.ac.nz

**Abstract.** The implementation of the parametric acoustic array has invariably being based on Berktay's far-field solution, despite its limitations. In addition to the problems due to the approximations of the model, recent studies suggest that the frequency response of the transducers add substantial errors in the demodulation process, which are not considered by Berktay's equation. This paper introduces a method to overcome the frequency response problem of the ultrasonic transducers, in order to reduce the harmonic distortion produced by the parametric acoustic array, by using a suitable modulation technique. The frequency response of a parametric loudspeaker has been acoustically measured. The characteristics of this response have been applied as a filter to the signal to be demodulated, using Berktay's far-field solution. A modulation technique has been proposed, taking into consideration the frequency response of the parametric loudspeaker. The results show an improvement in the estimation of the Total Harmonic Distortion (THD) when the parametric loudspeaker frequency response is considered in the modelling. Moreover, the experimentally measured Total Harmonic Distortion of the proposed modulation technique is consistent with the predictions of the model.

## 1. Introduction

Westervelt described the theory of the parametric acoustic array in 1963 [1]. Additionally, Berktay derived a quasi-linear approach in 1965 [2], which introduced a far-field solution describing a self-demodulation process in the propagation field.

From early experimentations of the parametric acoustic array in liquids until the latest development of parametric loudspeakers [3, 4], Berktay's solution has remained to be the main model in the implementation of the parametric acoustic array. However, minor amendments have been done to this model [5], despite the limitations of the equation describing the parametric acoustic array. In addition to inaccuracy due to the approximations in the origin of the model, the bandwidth limitations of the transducers affect the demodulation process, by increasing the harmonic distortion [6, 7].

This paper introduces a method to consider the frequency response of the ultrasonic transducer to obtain a suitable modulation technique, in order to lower the harmonic distortion. The rest of the paper is structured as follows: Section 2 describes Westervelt and Berktay's equations, Section 3 describes the method proposed and the experimental setup, Section 4 shows the results, and Section 5 contains the conclusions and future work.

## 2. Westervelt equation

The parametric acoustic array was first described by Westervelt as an approximation of the full second order wave equation, for the cases where local nonlinear effects are insignificant compared to cumulative nonlinear effects:

$$\Box^2 p + \frac{\delta}{c_0^4}\frac{\partial^3 p}{\partial t^3} = -\frac{\beta}{\rho_0 c_0^4}\frac{\partial^2 p^2}{\partial t^2} \tag{1}$$

Where $p$ is pressure, $\rho_0$ is the mass density, $\Box^2 = \nabla^2 - \frac{1}{c_0^2}\frac{\partial^2}{\partial t^2}$ is the d'Alambertian operator, $\beta = 1 + \frac{B}{2A}$ is the coefficient of nonlinearity, and $\delta$ is the diffusivity of sound. A quasi-linear solution has been considered to solve Westervelt equation:

$$p = p_1 + p_2 \tag{2}$$

This solution includes a linear part $p_1$ and a nonlinear correction $p_2$.

### 2.1. Berktay's far-field solution

In 1965, a quasi-linear solution was implemented by Berktay [2], in the form of equation 2. Berktay described the parametric acoustic array as a self-demodulation process in the propagation medium. If the source is considered as a modulated signal from a circular piston with radius $a$, the source condition is [8]:

$$p(r, 0, t) = p_0 f(t) H(a - r) \tag{3}$$

Where $p_0$ is the characteristic source pressure, $H$ is the unit step function and $f(t)$ is the time-varying signal, defined by:

$$f(t) = E(t) sin[\omega_0 t + \phi(t)] \tag{4}$$

Where $E(t)$ is the envelope and $\phi(t)$ is the phase, both are slowly varying functions compared to $sin(\omega_0 t)$.

Westervelt's analysis was based on strong absorption criterion [9], so nonlinear effects are limited to the near-field. To abide with this assumption, the following condition should be considered:

$$\alpha_0 z_0 \geq 1 \tag{5}$$

Where $\alpha_0$ is the absorption coefficient at frequency $\omega_0$, and $z_0 = \frac{1}{2}k_0 a^2$ is the Rayleigh distance at this frequency, which establishes the limit between near-field and far-field.

For solving Westervelt equation, the quasi-linear approach of equation 2 is considered, so the following form of $p_1$ is proposed, which includes exponential attenuation [8]:

$$p_1(r, z, t) \approx p_0 e^{\alpha(t) z} E(t) sin[\omega_0 t + \phi(t)] H(a - r) \tag{6}$$

In a fluid, $\alpha(t) = [\Omega(t)/\omega_0]^2 \alpha_0$.

After solving Westervelt equation, and considering strong absorption, the nonlinear part of equation 2 is obtained:

$$p_2(0, z, t) = \frac{\beta p_0^2 a^2}{16 \rho_0 c_0^4 z}\frac{\partial^2}{\partial t^2} E^2(t) \tag{7}$$

Which is Berktay's far-field solution [2]. This equation is valid in the far-field, as several assumptions were made in this regard. The audible sound wave is proportional to the second derivative of the square of the envelope.

The nonlinear interaction is expected to be limited by the near-field, although it can extend over the far-field in some cases. The propagation of the primary ultrasonic waves is limited by the absorption of the medium.

## 3. Method proposed

Since Berktay's far-field solution establishes the propagated wave demodulates in air, the implementation of the parametric acoustic array considers a modulation stage. This stage is based on the form given by equation 4, which corresponds to amplitude modulation.

Westervelt and Berktay's analyses are based on assumptions and criteria which are not always met. One example is the case of the strong absorption criterion given by equation 5, so the far-field solution becomes inaccurate. Moreover, the characteristics of the parametric loudspeaker are not considered in the theoretical model, so the bandwidth limitations of the transducers have not been taken into consideration for the selection of the proper modulation technique. The bandwidth effect has already been described in the demodulation process [6, 7].

Different studies have proposed several modulation techniques [6, 10], all of them based on Berktay's solution, but the frequency response of the transducers has not been considered. Since these techniques are based on the generation of additional frequencies to the primary waves, the frequency limitations of the transducers have a deep impact in the modulation stage.

To overcome the inaccuracies due to the physical characteristics of the ultrasonic transducers, a new method for selecting a proper modulation technique is proposed. It consists in measuring the frequency response of the parametric array loudspeaker before selecting a suitable modulation technique. Additionally, the second-order nonlinear generation is calculated using Berktay's far-field solution. Finally, a filter describing the frequency response of the parametric loudspeaker is applied to the signal to be demodulated, in order to include the frequency limitations of the transducer.

### *3.1. Implementation*

The parametric loudspeaker used for this experiment was a NICERA AS195. This is an array of 195 ultrasonic transducers of 10 mm of diameter, with a resonance frequency of 40 kHz, as expressed by the manufacturer. The frequency response of the array of transducers was acoustically measured in an isolated silent room, at 1.5 meters from the source, using a Brüel & Kjær 2670 microphone. A sweep from 20 kHz to 70 kHz with a total duration of 20 seconds was used for obtaining the frequency response. The result of this test is shown in figure 1.

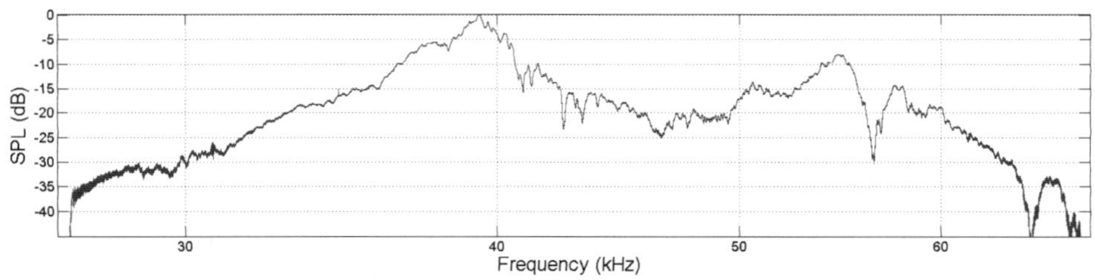


**Figure 1.** Frequency Response of NICERA AS195 array, measured at 1.5 m.

The resonance frequency of the array of transducers is 39440 Hz, as observed from figure 1. The cut-off frequencies are 38440 and 40060 Hz, as a -3 dB attenuation is observed at those points. Subsequently, a band-pass digital filter is designed in Matlab, based on a Van Dyke's model describing ultrasonic transducers [11].

Secondly, seven different modulation techniques were implemented using Matlab. Five of these techniques have been previously presented [6], which are: Quadrature Amplitude Modulation (QAM), Double Sideband Amplitude Modulation (DSBAM), Square-root Amplitude Modulation (SRAM) and Modified Amplitude Modulation (MAM2 and MAM3). Additionally, two more modulation techniques have been introduced, in order to compare the

IOP Conf. Series: Journal of Physics: Conf. Series **1075** (2018) 012035 doi:10.1088/1742-6596/1075/1/012035

effects of considering the frequency response in the process of developing a specific technique. These techniques are based on a QAM scheme, which also serves to produce MAM techniques [6]. The scheme used is shown in figure 2:

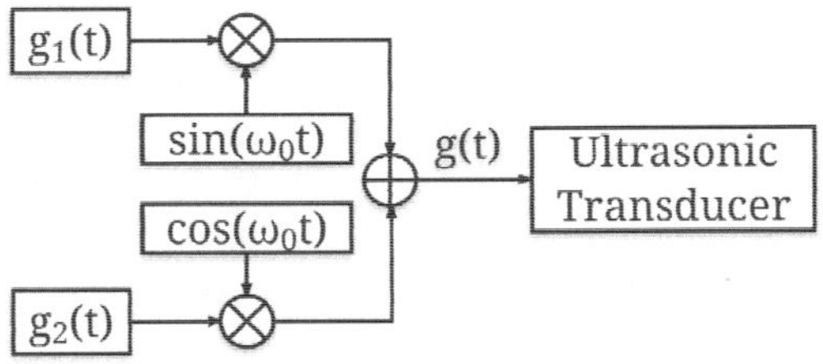


**Figure 2.** Quadrature Amplitude Modulation Scheme.

In all cases, the first signal is $g_1(t) = 1 + mg(t)$ and the second signal $g_2(t)$ is specific to each technique. For the development of MAM from QAM, $g_2(t) = 1 - m^2g^2(t)$ was used, as it would eliminate the harmonics generated by $g_1(t)$, according to Berktay's solution. The final forms of MAM2 an MAM3 are obtained using a Taylor series approximation [6]. Our proposal for the second signal has considered the addition of distortion in a smaller amount, in order to eliminate the distortion produced by $g_1(t)$, but considering the frequency response of the transducer.

For this purpose, $g_2(t) = 1 - \frac{m^2g^2(t)}{2}$ was selected as the second envelope for the first proposal, and $g_2(t) = 1 - \frac{m^3g^3(t)}{3}$ for the second one. The first proposal adds up to the fourth harmonic but in a small amount, and the second proposal adds up to the third harmonic, but in a higher level, compared to the previous technique. Table 1 summarizes the envelopes of all used modulation techniques.

**Table 1.** Envelopes of the modulation techniques used.

| Modulation Technique | Envelope |
|---|---|
| QAM | $\sqrt{(1+mg(t))^2 + 1 - m^2g^2(t)}$ |
| DSBAM | $1 + mg(t)$ |
| SRAM | $\sqrt{1+mg(t)}$ |
| MAM2 | $\sqrt{2}\sqrt{1 + mg(t) + \frac{1}{16}m^6g^6(t) + \frac{1}{128}m^8g^8(t)}$ |
| MAM3 | $\sqrt{2}\sqrt{1 + mg(t) + \frac{5}{128}m^8g^8(t) + \frac{1}{128}m^{10}g^{10}(t) + \frac{1}{512}m^{12}g^{12}(t)}$ |
| Proposal 1 | $\sqrt{2 + 2mg(t) + \frac{1}{4}m^4g^4(t)}$ |
| Proposal 2 | $\sqrt{2 + 2mg(t) + m^2g^2(t) - \frac{2}{3}m^3g^3(t) + \frac{1}{9}m^6g^6(t)}$ |

If a sinusoidal input is considered, the second derivative of the squared envelope of the first modulation technique proposed is:

$$\frac{\partial^2 E_1(t)}{\partial t^2} = -2msin(t) + \frac{1}{2}m^4cos(2t) - \frac{1}{2}m^4cos(4t) \quad (8)$$

Which shows the generation of the second and fourth harmonic. The harmonic generation of the second proposal is given by:

$$\frac{\partial^2 E_2(t)}{\partial t^2} \approx \frac{m^3+4m}{2} sin(t) + 2m^2 cos(2t) + \frac{3}{2} m^3 cos(3t) \tag{9}$$

An approximation was necessary in this case, since higher harmonics were present but with a very low level. Subsequently, all seven modulation techniques were tested using the same setup as for the frequency response measurement. In this case the acoustic measurement was done at 3.6 meters from the source, in order to obtain data comparable to Berktay's far-field solution results. Only one point was measured, due to limitations of the room.

Finally, total harmonic distortion (THD) was calculated for the data measured, along with the Berktay's far-field solution predicted by equation 7, for each modulation technique. Additionally, the filter describing the frequency response of the transducers array was applied to all seven modulation techniques, and their Berktay's far-field solution was computed. The THD was calculated according to equation 10:

$$THD = \sqrt{\frac{P_2^2 + P_3^2 + P_4^2 + \cdots + P_{n-1}^2 + P_n^2}{P_1^2 + P_2^2 + P_3^2 + P_4^2 + \cdots + P_{n-1}^2 + P_n^2}} \times 100 \tag{10}$$

Where $P_1$ is the fundamental frequency and $P_2$ to $P_n$ the harmonics.

## 4. Results

The Total Harmonic Distortion calculated for the data measured and for Berktay's solution computed with and without the filter is presented in table 2, for all modulation techniques used.

**Table 2.** Total Harmonic Distortion of computed and measured data (%).

| Modulation Technique | Berktay (No filter) | Berktay (Filter) | Measurement |
|---|---|---|---|
| QAM | 0.44 | 17.53 | 27.78 |
| DSBAM | 44.59 | 33.31 | 43.48 |
| SRAM | 0.74 | 9.38 | 8.29 |
| MAM2 | 1.05 | 9.44 | 6.23 |
| MAM3 | 0.75 | 9.39 | 6.21 |
| Proposal 1 | 4.43 | 17.89 | 13.83 |
| Proposal 2 | 12.58 | 16.44 | 18.10 |

In most cases the harmonic distortion predicted by Berktay's solution is very low compared to the measurements, except for DSBAM. When the filter is applied, higher THD is obtained, as more harmonics appear due to the frequency response of the transducers. In general, a better approximation to harmonics generation is found when a filter is applied to the modulated signal.

Only DSBAM shows a more accurate Berktay's far-field solution without the filter applied. This is because the second harmonic of this modulation technique is very high and it is almost the sole contribution towards Total Harmonic Distortion. The filter applied attenuates more of the second harmonic than anything else, so the results for DSBAM become less accurate. A similar case occurs with QAM, where the second harmonic is also very high, but not as important as for DSBAM, and also the third harmonic contributes towards THD calculation.

Finally, the proposed techniques have a close THD value, 13.83% for the first one, and 18.10% for the second one, according to the measurements. Although Berktay's far-field solution initially shows a bigger difference between the two, 4.43% for the first one, and 12.58% for the second one, the frequency response affects especially to the first proposal. In all experiments, the parametric loudspeaker is capable of producing up to the third harmonic of the fundamental demodulated wave (1 kHz), and the first proposed technique includes up to the fourth harmonic. The inability to produce the expected higher harmonics causes the generation of lower harmonics, which ultimately increases the THD. Although proposal one seems to be much less distorted than proposal two, the difference is much closer at the end.

## 5. Conclusions and future work

A new method for selecting an appropriate modulation technique has been proposed. This method increases the accuracy of Berktay's solution, by including the characteristics of the transducers used to implement the parametric acoustic array. With a more accurate far-field solution it is possible to select an appropriate modulation technique for the specific transducer. In six of the seven cases studied, the harmonic generation predicted was more accurate when the filter was applied, with an increase of up to 17 % of THD, in the case of QAM. Despite the good results, there are some challenges to improve accuracy. The generation of the second harmonic in many cases was weak when the method was applied, so it needs to be improved. Additionally, more measurement points should be considered in future work, so the propagation effect is taken into consideration. The use of a different model, like the Khokhlov-Zabolotskaya-Kuznetsov equation would be suitable, as it includes absorption, diffraction and nonlinearity effects of sound beams. The proposed method could be used as a tool to describe THD and anticipate the undesired harmonic generation, so other mitigation actions can be taken.

## Acknowledgments

This work is funded by the National Commission for Scientific and Technological Research of Chile (CONICYT), Becas Chile program.

## References

[1] Westervelt P J 1963 Parametric Acoustic Array *J. Acoust. Soc. Am* **35** 535-7
[2] Berktay H O 1965 Possible exploitation of non-linear acoustics in underwater transmitting application *J Sound Vib* **2** 435-61
[3] Bennett M B and Blackstock D T 1975 Parametric array in air *J. Acoust. Soc. Am* **57** 562-8
[4] Gan W S, Yang J and Kamakura T 2012 A review of parametric acoustic array in air *Appl. Acoust.* **73** 1211-9
[5] Merklinger H M 1973 Fundamental frequency component of a finite amplitude plane wave *J. Acoust. Soc. Am* **54** 1760-1
[6] Tan E L, Ji P and Gan W S 2010 On preprocessing techniques for bandlimited parametric loudspeakers *Appl. Acoust.* **71** 486-492
[7] Shi C and Kajikawa Y 2015 Ultrasound-to-ultrasound volterra filter identification of the parametric array loudspeaker *Proc. 20th Int. Conf. Digital Sig. Process.* pp. 1-4
[8] Averkiou M A, Lee Y, and Hamilton M F 1993 Self-demodulation of amplitude and frequency-modulated pulses in a thermoviscous fluid *J. Acoust. Soc. Am* **94** 2876-83
[9] Farias F and Abdulla W 2015 On rayleigh distance and absorption length of parametric loudspeakers *Proc. 2015 APSIPA Annual Summit and Conf.)* pp. 1262-5
[10] Liew Y H 2002 *Signal processing techniques for sound reproduction in parametric arrays* (Nanyang Technological University, School of Electrical and Electronic Engineering)
[11] Jordan T and Ounaies Z 2001 Piezoelectric ceramics characterization *NASA Langley Research Center*

# Measurement of bottom backscattering strength using single-beam echosounder

**S Solikin[1], H M Manik[2*], S Pujiyati[2] and S Susilohadi[3]**

[1] Graduate School of Marine Technology, PMDSU Program Batch II, Jl. Agatis Kampus IPB Dramaga, Bogor Agricultural University, Bogor 16680, Indonesia
[2] Department of Marine Science and Technology, Faculty of Fisheries and Marine Science, Jl. Agatis Kampus IPB Dramaga, Bogor Agricultural University, Bogor 16680, Indonesia
[3] Centre for Research and Development of Marine Geology, Ministry of Energy and Mineral Resources, Jl. Dr. Djunjunan No. 236 Pasteur, Bandung, Indonesia

*henrymanik@ipb.ac.id

**Abstract**. A quantified single beam echosounder was applied to detect and quantify the bottom echoes. Bottom backscattering strength ($S_B$) is necessary to be quantified for accurate seabed mapping and classification. The $S_B$ can be measured by Lambert's Law that enabled to interpret the bottom echo and to measure the $S_B$ values. The research was conducted in Tidung Island, Jakarta which had several seabed types, such as coral reef, seagrass, sand, silt, clay, mud, and mixed sediment. Acoustic data was acquired using SIMRAD EK15 with the frequency of 200 kHz and sediment grain size was analyzed using Wentworth grain size classification. The results show that the backscattering of each seabed type is different. Sediment with greater mean diameter had greater $S_B$ value, and seabed with rougher surface had greater $S_B$ value as well.

## 1. Introduction

Nowadays, accurate marine seafloor mapping is needed due to increased marine activity, such as dredging, oil and gas exploration, and geological research and marine morphology. The conventional method of mapping the seafloor is to take sediment samples directly, but this method costs a great deal and a very long time. The new method developed is using sound waves (acoustics) that can cover a wider area in the not too distant future [1]. Several studies have succeeded in developing algorithms to classify and map the seafloor using single beam, split beam, multibeam, and side scan sonar [2-6].

Underwater acoustics are highly dependent on frequency (frequency dependence) ranging from 10 Hz to 1 MHz. The smaller the frequency used, the acoustic waves will penetrate deeper into the bottom layer of the water, and vice versa. This is caused by high frequency that will be more quickly absorbed by the sea water medium [7].

The seafloor has the characteristic to reflect and scatter the sound waves. The acoustic method, which in principle is using sound waves, can be used to obtain information on the basic types of waters. The sea in Tidung Island is classified as shallow water (average 30 m) with very diverse ecosystems, including coral reef and seagrass ecosystems. The seabed types in the sea of the Tidung Islands are very diverse, ranging from the reefs, gravel, sand, silt, and clay [8].

This research used Lambert's law to model the backscattering strength from the sample sediments in Tidung Island. There is some support that backscattering strength follows a Lambert's law approximation, especially for low grazing angles [9-10]. The model from the Lambert's law will be compared with other theories, such as Peturbation Theory and Kirchoff Approximation. Furthermore, reflection coefficient is also modeled to relate with the physical parameters of the sediment (mean grain size).

## 2. Methodology

The research was conducted in Tidung Island and Bintan Island. The data collected consist of the acoustic data and sediment samples as ground truth data. The acoustic data was acquired using SIMRAD EK15 single-beam echosounder with the frequency of 200 kHz, while the sediment samples was collected using sediment grab. Figure 1 shows the location of the data acquisition.

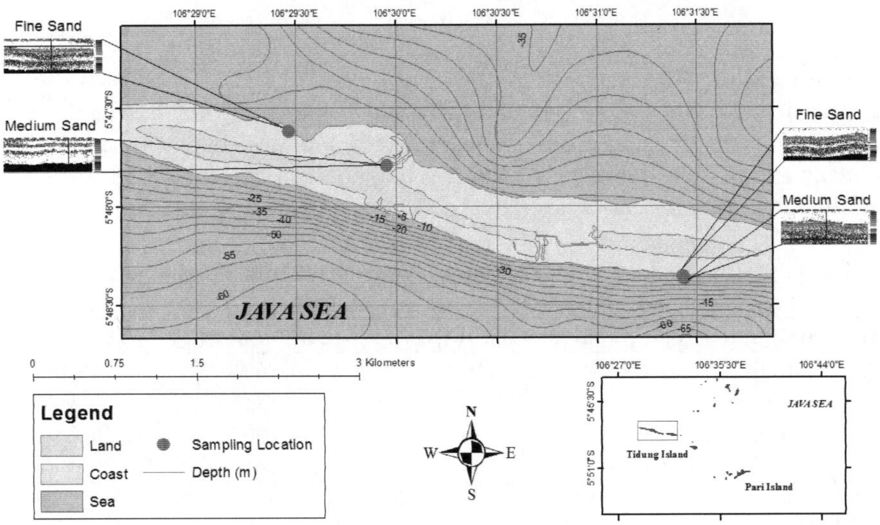


Fig. 1 The location of the data acquisition in Tidung Island. Sediment samples was collected in 4 stations symbolized with red circle. The sediment samples is classified into two classes, medium sand and fine sand

The grain size is a sediment descriptor. It is usually measured in logarithmic unit and denoted $M_z$. [6]

$$M_z = -\log_2 \frac{d}{d_0} = -3.32 \log_{10} \frac{d}{d_0} \qquad (1)$$

Where $d$ is the mean grain size (or "diameter"), $d_0$ is the reference length (1 mm), and the units of $M_z$ are denoted φ. The sediment names are taken from the Wentworth's classification schemes that divided the sediment into 5 primary classes; boulders, pebbles, sand, silt, and clay.

The acoustic data was acquired along the island to see the distribution of the SS. The bottom scattering of the seabed surface is computed by Lambert's law [11]. The backscattering strength of the bottom varies with the grazing angle, sound frequency, and the bottom type induced roughness. The Lambert's law relationship between the backscattering strength and the grazing angle fits to many experimental data satisfactorily accurate for angles below 60°. Consequently, the backscattering strength can be described by Lambert's law and an empirically specified scattering coefficient, e.g

$$S_B = K(f, bt) + 10\ \log(\sin^2(\theta)) \qquad 0 \leq \theta \leq 60°, \qquad (2)$$

Where $S_B$ denotes bottom backscattering strength, $K(f, bt)$ denotes the scattering coefficient depending on the frequency of sound $f$ and bottom type $bt$, and $\theta$ denotes the grazing angle. The bottom type parameter is defined as follows

| | |
|---|---|
| $bt = 1$ | mud |
| $bt = 1$ | sand |
| $bt = 1$ | gravel |
| $bt = 1$ | rock |

In principle $bt$ can be any real number satisfying

$$1 \leq \theta \leq 4$$

## 3. Results and Discussion

The sediment samples are analysed to get their physical parameters (mean grain size). From the 4 samples, there are two samples categorized as medium sand and other two samples categorized as fine sand with the mean grain size is between 138.96-159.10 µm. The physical parameters and the acoustic characteristic of the four samples can be seen on Table 1.

Table 1. The physical (mean grain size) and acoustical parameters (backscattering strength and reflection coefficient value) of the four sediment samples

| Station | Sediment Type | Mean Grain (µm) | Mean Grain (φ) | $S_B$ (dB) | R |
|---|---|---|---|---|---|
| 1 | Medium sand | 159.10 | 2.65 | -21.89 | 0.2162 |
| 2 | Fine sand | 138.96 | 2.85 | -25.52 | 0.1966 |
| 3 | Fine sand | 154.60 | 2.69 | -25.76 | 0.2119 |
| 4 | Medium sand | 142.21 | 2.81 | -20.63 | 0.1998 |

Table 1 showed that the medium sand has greater backscattering strength value than the fine sand, following with the reflection coefficient value as well. $S_B$ denotes the bottom backscattering strength and $R$ denotes the reflection coefficient value. The mean grain size gives contribution to the backscattering value and the reflection coefficient, but not in all case. It depends on other physical parameters of the sediment, such as density, porosity, etc [12]

There is a correlation between the physical parameters and the acoustical parameters of the sediment. The physical parameters will affect the acoustical parameters. Least square polynomial is used to relate these parameters [13]. Figure 2 shows the relation between the mean grain size and the reflection coefficient value of the four samples of sediment.

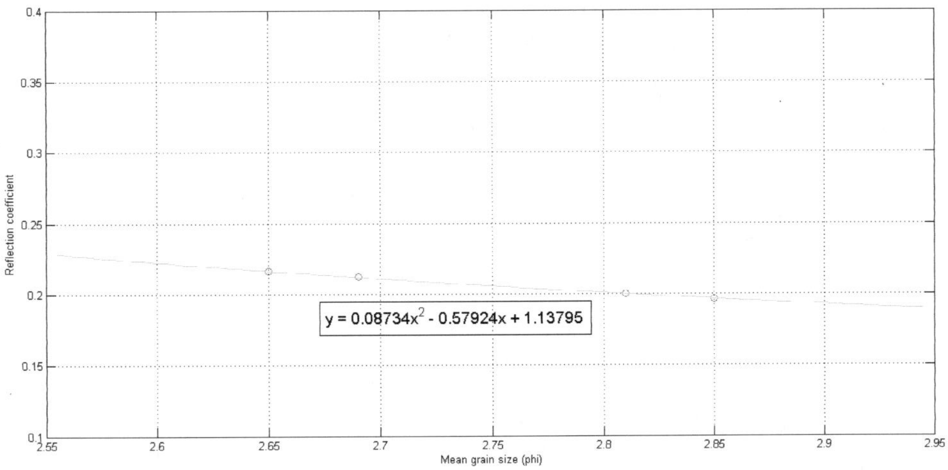


Fig. 2 The relation between mean grain size (μm) and the reflection coefficient. By least square polynomials, the relation is $0.008734\ x^2 - 0.57924\ x + 1.13795$. The regression equation fits the curve well

The least square polynomials is implemented in this case to see the non-linear relation between the mean grain size and the reflection coefficient value. The regression equation got is $0.008734\ x^2 - 0.57924\ x + 1.13795$. It has the quadratic regression that shows the relation is a non-linear relation.

The echogram of the 4 stations can be seen in Figure 3. The white column indicates the water column and the red one below the water column is the bottom of the ocean.

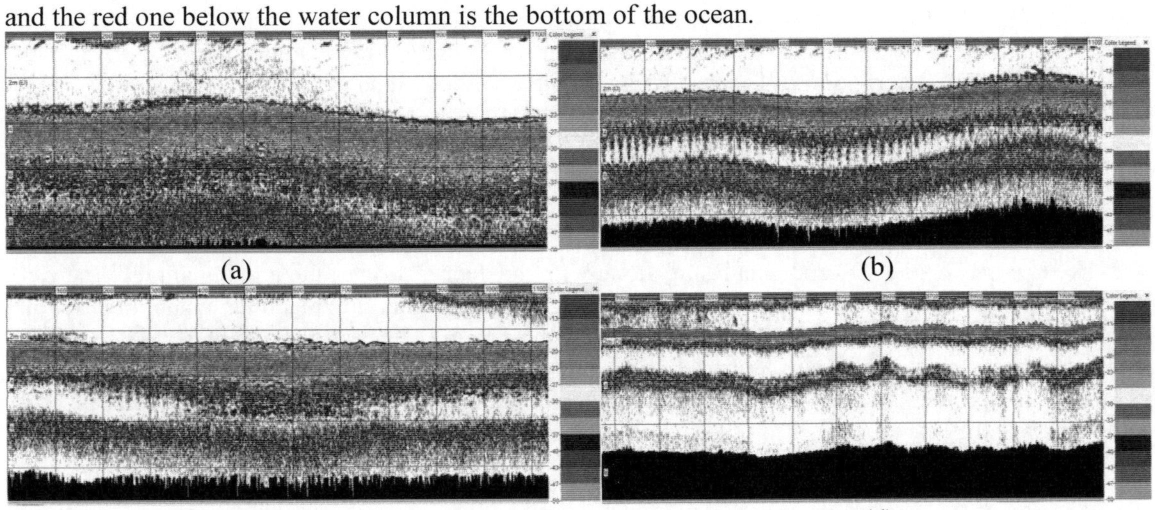


Fig. 3 The echogram of the four sample sediments. Fig 3a is the station 1, 3b is the station 2, 3c is the station 3, and 3d is the station 4. The color bars beside the echogram show the backscattering strength value of the bottom.

Figure 3 showed that the bottom backscattering strength value from the 4 stations is not much different. The backscattering strength is around -25 – (-20) dB and it shows similarities between the 4 stations. The similarities are caused by the sediment's mean grain size. Different type of sediment will cause the differences of the sediment's grain size, such as gravel will have greater grain size than sand, and sand will have greater grain size than silt and clay. The further research is to find the differences between the gravel, sand, silt, and clay.

The backscattering strength value for ocean sediment in Tidung Island is modeled using Lambert's law and compared with other theory, Peturbation Theory (PT) and Kirchoff Approximation (KA). Figure 4 showed the bottom backscattering strength model.

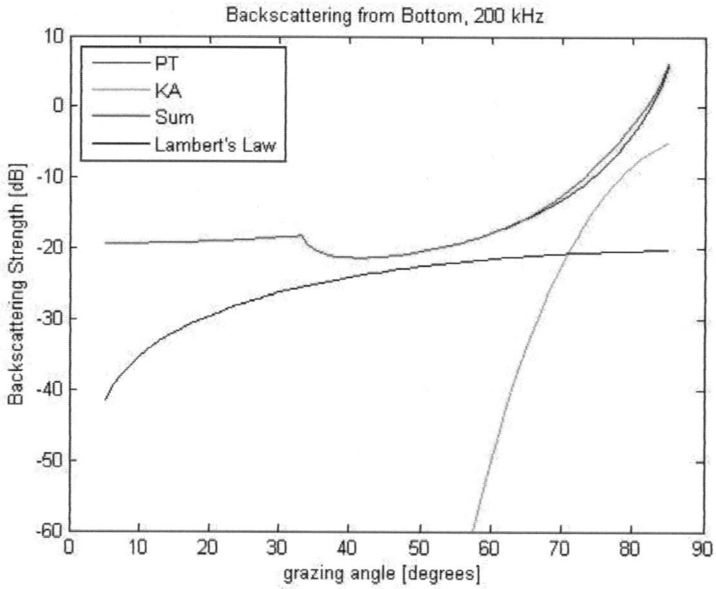


Fig. 4 Bottom backscattering strength model using Lambert's law model (black line) and compared with other models, Peturbation Theory (PT) (blue line) and Kirchoff Approximation (KA) (green line) with the frequency of 200 kHz single-beam echosounder

Figure 4 showed the bottom backscattering strength value against the grazing angle. Lambert's law model showed a stable pattern from grazing angle 0° to 90°. The highest value for backscatter is on nadir line (grazing angle 90°) and decreasing along with the degradation of the grazing angle. Whereas, other theories showed a fluctuating pattern and only effective in higher grazing angle value. In Kirchoff Approximation the backscattering strength value only exist in grazing angle 60° to 90° and decreasing drastically. So, it shows that the Lambert's law is more effective than other theories, but only in this research. This argumentation might not be implemented in other researches depends on the characteristic of the ocean sediment.

## 4. Conclusion

The sediment samples in Tidung Island is classified into two classes, medium sand and fine sand with the backscattering strength value is ranged between -25.76 – (-20.63) dB and the reflection coefficient is ranged between 0.1966 – 0.2162. The backscattering strength model used in this research, Lambert's law, is more effective to use than other theories, Peturbation Theory and Kirchoff Approximation. Lambert's law can be used in lower grazing angle and more stable than other two theories.

**Acknowledments**

The authors would like to thank the Directorate General of Higher Education (DIKTI) Indonesia for financially supporting this research through PMDSU Research Grant Scheme Year II.

**References**

[1] Manik HM. 2011. Underwater acoustic detection and signal processing near the seabed. In Nikolai Kolev, editor. Sonar Systems; ISBN: 978-953-307-345-3, InTech.

[2] van Walree PA, Ainslie MA, Simons DG. 2006. Mean grain size mapping with single-beam echo sounders. J. Acoust. Soc. Am. 120(5), 2555-2566.

[3] Mamede R, Rodrigues AM, Freitas R, Quintino V. 2015. Single beam acoustic variablitiy associated with seabed habitats. Journal of Sea Research, 100: 152-159.

[4] Cutter GR and Demer DA. 2013. Seabed classification using surface backscattering strength versus acoustic frequency and incidence angle measured with vertical, split-beam echosounders. ICES Journal of Marine Science, 71(4): 882-894.

[5] Zhi H, Siwabessy J, Nichol S, Anderson T, Brooke B. 2013. Predictive mapping of seabed cover types using angular response curves of multibeam backscatter data; testing different feature analysis approaches. Continental Shelf Research, 61-62: 12-22.

[6] APL-UW. 1995. High-Frequency Ocean Environmental Acoustic Models Handbook. University of Washington, Seattle.

[7] Penrose JD, Siwabessy PJW, Gavrilov A, Parnum I, Hamilton LJ, Bickers A, Brooke B, Ryan DA, Kennedv P. 2005. "Acoustic Techniques for Seabed Classification." Report prepared for the CRC for Coastal Zone Estuary and Waterway Management.

[8] Pujiyati S, Hartati S, dan Priyono, W. 2010. Efek Butiran, Kekasaran, dan Kekerasan Dasar Perairan terhadap Nilai Hambur Balik Hasil Deteksi Hidroakustik. E-Jurnal Ilmu dan Teknologi Kelautan Tropis. 2(1):59-67.

[9] Lamarche G, Lurton X, Verdier AL, Augustin JM. 2011. Quantitative characterization of seafloor substrate and bedforms using advanced processing of multibeam backscatter-application to Cook Strait. 31. Continental Shelf Research, New Zealand, pp. S93-S109.

[10] Yu J, Henrys SA, Brown C, Marsh I, Duffy G. 2015. A combined boundary integral and Lambert's Law method for modelling multibeam backscatter data from the seafloor. Continental Shelf Research, 103: 60-69.

[11] Brekhovskikh LM and Lysanov YP. 2003. Fundamentals of Ocean Acoustics, 3rd ed.

[12] Manik HM. 2012. Seabed identification and characterization using sonar. Advance in Acoustics and Vibration, vol. 2012, Article ID 532458, 5 pages.

[13] Matthews JH and Fink KD. 1999. Numerical Methods using MATLAB. Prentice Hall, Upper Saddle River, NJ.

# Methods to reduce noise level due to environmental changes: A case study in Universitas Gadjah Mada

**S S Utami[1], D D Avoressi[2], R J Yanti[1], K Zakiya[1] and H Sutanta[3]**

[1] Department of Nuclear Engineering and Engineering Physics, Universitas Gadjah Mada, Yogyakarta, Indonesia
[2] Department of Architectural and Planning, Universitas Gadjah Mada, Yogyakarta, Indonesia
[3] Department of Geodetic Engineering, Universitas Gadjah Mada, Yogyakarta, Indonesia

sentagi@ugm.ac.id

**Abstract**. A sound level mapping in Universitas Gadjah Mada (UGM) was done in 2015. At the end of 2016, there were changes in UGM's infrastructure and operational management at several areas. This research aims to find the effect of environmental changes toward sound level through sound level mapping. Sound level was recorded by H6 Zoom Recorder at three allocated times, in the morning (6 - 8 AM), at noon (11 AM – 1 PM), and in the afternoon (4 – 6 PM). The sound level mapping in 2017 has shown a decrease of sound level within the range of 2 dBA to 12 dBA. This decrease is indicated in the noise map by the dominance of blue color in the color gradation. The conclusion refers to examples of significant infrastructure and management changes that can reduce sound level such as the usage of ticket or identity card and arrangement of parking area and public space.

**Keywords: sound level mapping, sound level, infrastructure, management**

## 1. Introduction

Based on the theory of city growth, Spiro Kostof (1991) proposed that the analogy of element of city shaper is human body organ, where free space refers to lung, city center as the heart, and streets as the arterial tissue in the body. The high population growth leads to the change of city shaper elements, as what is currently happening in Yogyakarta. The changes of city shaper element can be observed through both visual and auditory perception. The changes can be visualized directly from the building shape and streets transformation. The changes of element can also be observed as audio-related changes of vehicle's noise.

Sound level mapping can be used to indicate the effect of changes of city shaper element towards the auditory perception. Sound level mapping is a map with color gradation based on the noise level distribution of an area. The same sound levels are described in contour lines and the overall sound level contour map will describe the areas of different sound levels [1]. Studies about sound level mapping have been developed worldwide. Sukru Dursun et al. (2006) conducted a sound level mapping in Konya, Turkey [2], Kang-Ting Tsai et al. (2009) created sound level mapping in Taiwan [3], while Tara P McAlexander et al. (2015) made sound level mapping in New York City [4]. The sound level maps are also have been built in other areas such as in China [5], Korea [6], Brazil [7], [8], and Hong Kong [9]. Our previous sound level mapping in Universitas Gadjah Mada (UGM), conducted in 2015, is an example of similar study in Indonesia. Sound level mapping in UGM was created to find the distribution of sound level from traffic noise in a university area [10].

At the end of 2016 until early 2017, street infrastructure and operational management has changed in several areas of UGM which caused changes in the number and type of vehicles in this area as compared to the previous year. A different sound level and sound level mapping is expected to happen. The paper provides a description of such changes and further analysis of its impact to the sound level distribution in the campus area. The paper also provides several ideas that can reduce the sound level due to traffic.

## 2. Research Methodology

### *2.1. Research Location*

The environmental changes and sound level distribution from the condition in 2015 to 2017 of three different areas in UGM were observed. The first location was the area of UGM Mosque and its surrounding. In the past two years this area has gone through significant changes, mostly changes of the environmental elements. Therefore, the traffic pattern also changed. Data were collected at 17 measurement points. The second location was the surrounding of an outdoor food park that has undergone some management changes with 14 measurement points. Seventeen measurement points were taken at the third location, which is the area around the university hospital, RSUP Dr. Sardjito.

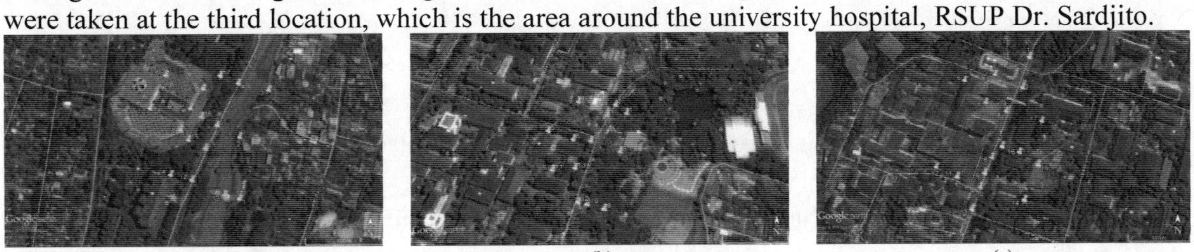

(a) First Location: Area of UGM Mosque
(b) Second Location: Area of Food park
(c) Third Location: Area near the Hospital

**Figure 1.** Measurement points.

### *2.2. Interpretation of Sound Level in Sound Level Map*

The acoustic data recorded by the measurement tool, H6 Zoom Recorder positioned at a height of 1.5 m above the ground referring to a human ear level [2], [3], [11], were processed to obtain sound level in the unit of dB(A). Sound levels for each point are plotted as contour lines using Surfer 11 software, creating a sound level map. This sound level map display color gradation, which describes sound level distribution of the area.

## 3. Result and Discussion

### *3.1. Changes of Infrastructure and Management*

There are three case-studies discussed in this paper based on the selected locations within UGM that have gone through infrastructure and management changes. The first discussion is the area of UGM Mosque. The main changes of this area from the condition in 2015 to the current condition in 2017 are in the street elements. In 2015, there existed a street divider in front of the UGM Mosque, which created a narrow street and therefore, only limited vehicle could pass this area. Other elements were vegetation on the side of the street positioned close to the building's gate or the mosque main entrance gate. In 2017, the vegetation was removed and their plantation area was replaced by a side-walk for pedestrians and bicyclists. The street divider was also removed, which made the street seems slightly wider even though this is not true.

IOP Conf. Series: Journal of Physics: Conf. Series **1075** (2018) 012037 doi:10.1088/1742-6596/1075/1/012037

The second case studied applied managerial changes, which took place in the surrounding of the food park area. The managerial changes included utilization of ticket to enter the university property. This entrance system had not been applied in 2015. This entrance system involves the use of identity card as a pass clearance for faculty members, staff, or students as. If identity card is not available, ticket would be used as the pass clearance.

The area around the RSUP Dr. Sardjito hospital is the last location selected as the case study. The different environment condition between 2015 and 2017 in this location involves physical and managerial changes. In 2015, the walkway and street side in front of the hospital is used as a street market for street sellers and as parking area for motorbikes. It is considered a crowded street especially from 6 AM to 6 PM. A market was built to relocate the street sellers and the side walk is now officially used for pedestrians. A street divider is also employed to avoid sudden turn and prevent the use of the street side as parking area.

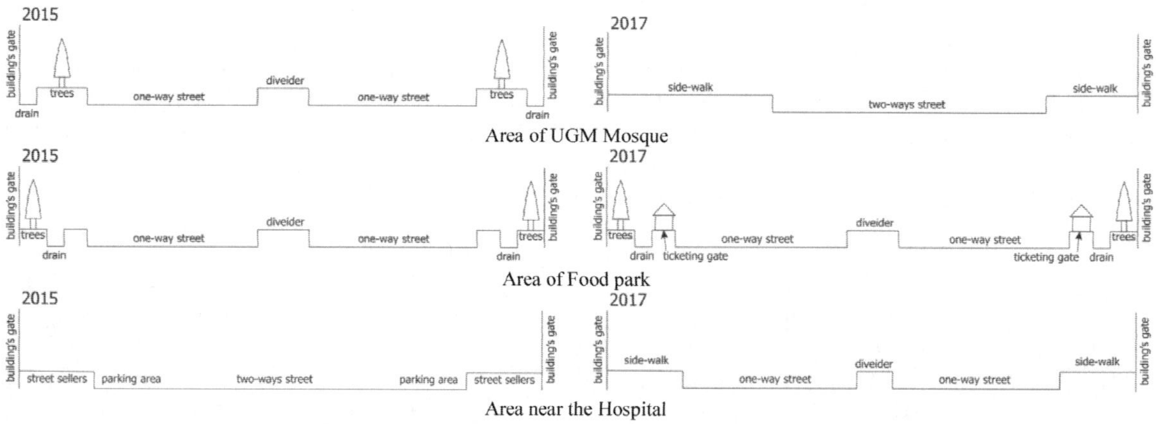


**Figure 2.** Street infrastructure and management changes.

*3.2. Comparison of Sound Level Mapping in 2015 and 2017*

A recent sound level mapping created in 2017 is used as a comparison to the sound level mapping created in 2015. The previous research in 2015 has shown a sound level map that classified to three zones. However in this paper we only discuss on sound level map not sound zoning. Further analysis on this comparison of the sound level distributions enables to see the changes that already occurred in the past two years due to physical and managerial changes.

*3.2.1. Physical Changes.* The physical changes of the street elements lead to the change of sound level distribution in area of UGM Mosque. The detail of sound level distribution differences between 2015 and 2017 for this location is shown in figure 3.

There is a dominance of bluish color in the sound level mapping of 2017, where blue indicates a lower sound level within the range of around 45 – 55 dBA for dark blue, 56 – 59 dBA for blue, and 60 – 64 dBA for light blue. Meanwhile, a wider area of yellowish to orange color is shown in the 2015 sound level mapping, indicating a larger area of a higher sound level of around 65 – 75 dBA. For all the three measurement times, in the morning, noon, and afternoon, nearly all of the sound level value of each measurement point decreased by 2-10 dB from 2015 to 2017.

The changes of the street infrastructure also caused the type of vehicle passing through this street to differ from those passing in the previous year. In 2015, buses were seen passing this street, while in 2017 there is nearly no bus passing through this street. In addition, because of the divider, the streets become narrower and for some way it led to less number of vehicles passing by.

*3.2.2. Managerial Changes.* The significant sound level mapping changes between 2015 and 2017 is due to management changes is in the area where pass clearance is applied by employing ticket or identity card. Description of the changes is provided in figure 4.

Figure 4 shows that the sound levels in the food park area decreased when compared to the 2015 measurement, especially in the morning. The majority of the sound level mapping contains bluish color, indicating a lower sound level in the morning. Even the range of the lowest sound level in the morning in 2015 and 2017 reached 12 dB. Similar condition is occurring at noon where the range of the lowest and highest sound level in 2015 and 2017 reached 5 dB and 7 dB, respectively. However, in the afternoon, the lowest sound level in 2015 is higher than the sound level in 2017, although the sound level mapping is still dominated by blue color.

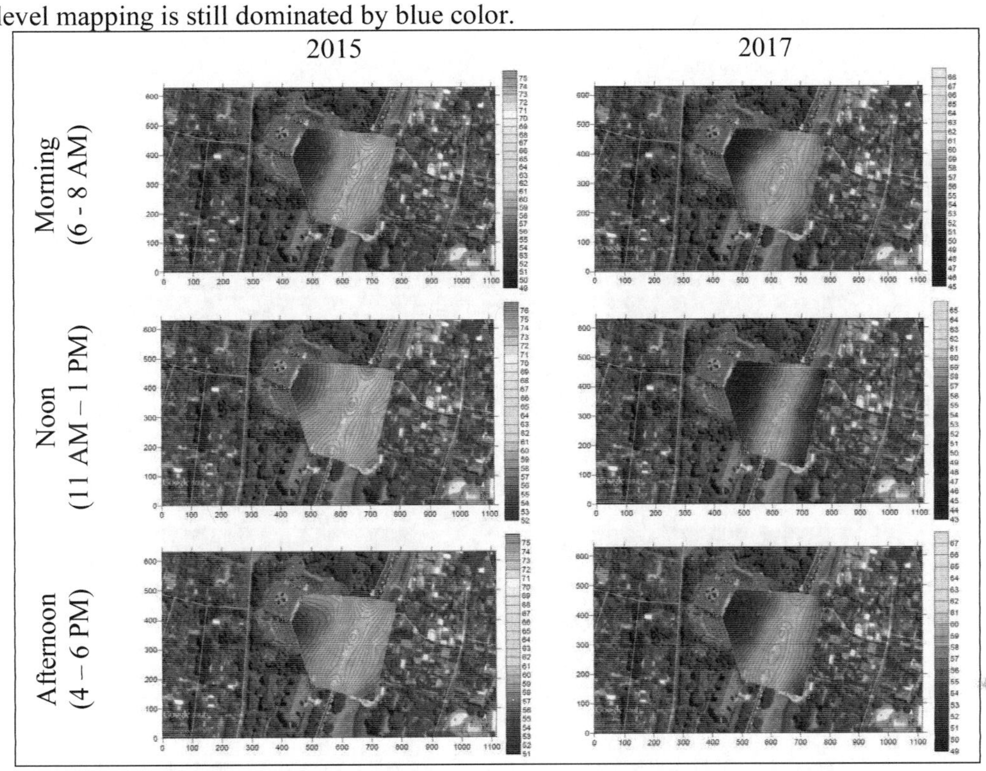


**Figure 3.** Sound level mapping in mosque area due to street infrastructure changes.

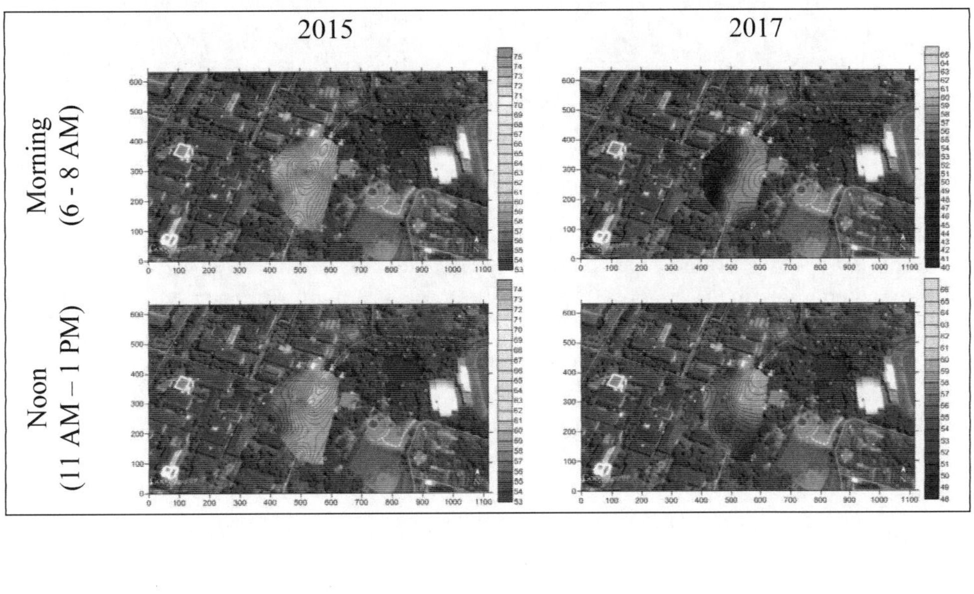

IOP Conf. Series: Journal of Physics: Conf. Series **1075** (2018) 012037 doi:10.1088/1742-6596/1075/1/012037

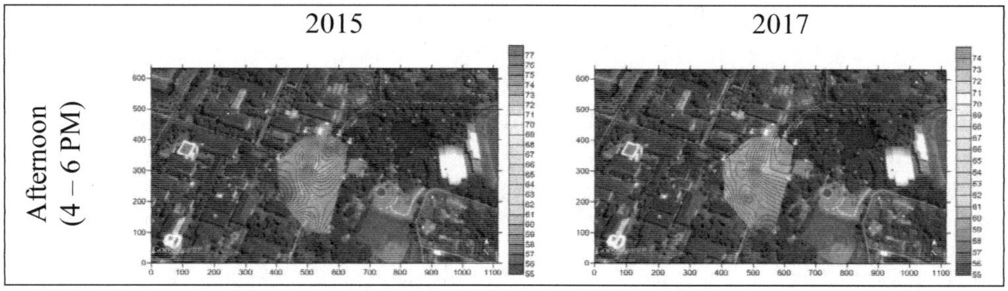


**Figure 4.** Sound level mapping in food park area due to managerial changes.

.The operational management changes made it hard for mass transportation to enter the college area so there have been no bus passing through this street.

*3.2.3. Physical and Managerial Changes.* The area where a combination of physical and managerial changes was applied is in the area near the hospital. The street infrastructure changes and relocation of street sellers and parking area led to the change in noise level from 2015 to 2017 (see figure 5).

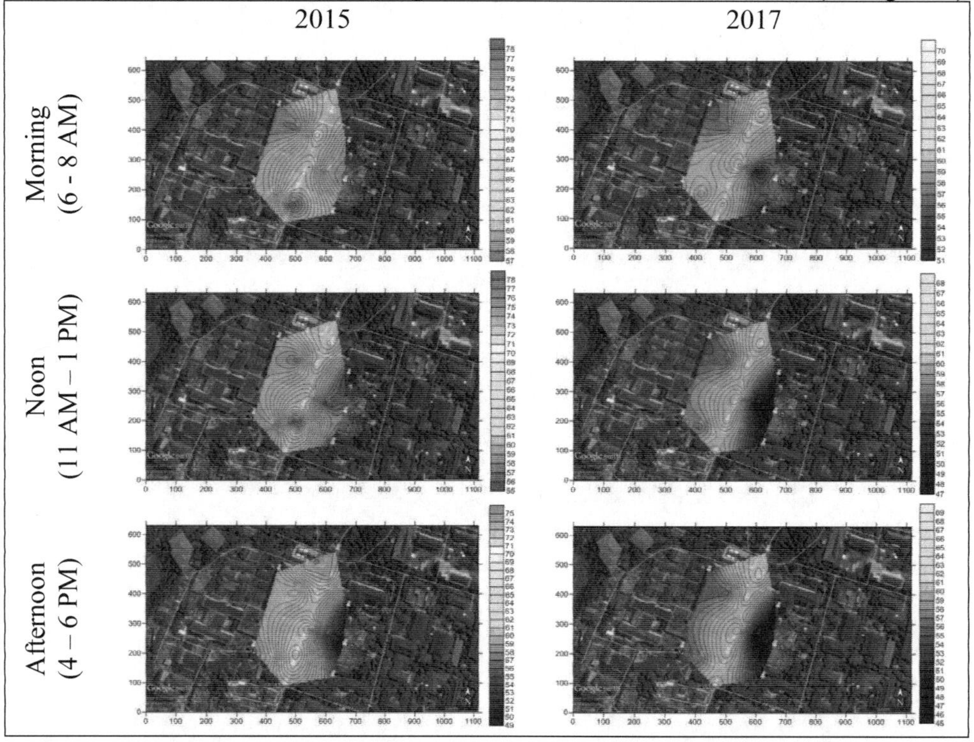


**Figure 5.** Sound level mapping in hospital area due to infrastructure and managerial changes.

Figure 5 shows that the sound level mapping in 2017 significantly changed with the domination of blue color indicating a lower sound level distribution in most of this area. The range of the lowest to highest sound level changes in 2015 and 2017 for three measurement times varied from 3 dB to 10 dB.

The street infrastructure change did not affect the mass transportation route in this area. The relocation of street sellers and parking area is a policy that created significant changes since the street is not as crowded as it was in 2015. Visually the street also looks neater and people walking through the street especially heading to the hospital are more comfortable to use the walkway.

Based on three of case studies, the result of sound level mapping in 2017 has shown a decrease of sound level within the range of 2 dB(A) to 12 dB(A). This decrease is indicated in the sound level map by the dominance of blue color in the color gradation.

## 4. Conclusion

Physical and managerial changes applied on an area will affect the sound level distribution of the area and can be identified through sound level mapping. Sound level mapping were conducted in 2015 and 2017 which shows that the sound level decreased in 2017 within the range of 2 dB to 12dB.

From the calculation, the most impact of sound level decrease is resulted from the use of pass clearance to enter the university by ticketing or identity card. This can be claimed as the most effective strategy to reduce noise level in an area since public transportation will have only limited access to enter this area. The transmission loss value due to the combination of physical and managerial changes of the surrounding area of the hospital near university has resulted in a decrease in the sound level. However, public transportation, particularly bus, is still passing through the narrower street given the street divider and therefore it remains still a street with a high density of vehicle.

## References

[1] NoiseMap, "What is Noise Mapping?," 2008. [Online]. Available: http://www.noisemap.ltd.uk/home/what is noise mapping.html. [Accessed: 15-Mar-2017].

[2] Ş. Dursun, C. Ö. İ. R, H. Karabörk, and S. Koçak, "Noise pollution and map of Konya city in Turkey #," *J. Int. Environ. Appl. Sci.*, vol. 1, pp. 63–72, 2006.

[3] K. T. Tsai, M. Der Lin, and Y. H. Chen, "Noise mapping in urban environments: A Taiwan study," *Appl. Acoust.*, vol. 70, no. 7, pp. 964–972, 2009.

[4] T. P. McAlexander, R. R. M. Gershon, and R. L. Neitzel, "Street-level noise in an urban setting: assessment and contribution to personal exposure.," *Environ. Health*, vol. 14, p. 18, 2015.

[5] B. Wang and J. Kang, "Effects of urban morphology on the traffic noise distribution through noise mapping : A comparative study between UK and China," *Appl. Acoust.*, vol. 72, no. 8, pp. 556–568, 2011.

[6] J. Lee, J. Gu, H. Park, H. Yun, S. Kim, W. Lee, J. Han, and J. Cha, "Estimation of Populations Exposed to Road Traffic Noise in Districts of Seoul Metropolitan Area of Korea," no. 14, pp. 2729–2740, 2014.

[7] F. Augusto and D. N. Castro, "Noise Mapping of Densely Populated Neighborhoods - Example of Copacabana," pp. 1–13, 2007.

[8] P. Henrique, T. Zannin, D. Queiroz, and D. S. Ana, "Noise mapping at different stages of a freeway redevelopment project – A case study in Brazil," *Appl. Acoust.*, vol. 72, no. 8, pp. 479–486, 2011.

[9] C. Law, C. Lee, A. S. Lui, M. K. Yeung, and K. Lam, "Advancement of three-dimensional noise mapping in Hong Kong," *Appl. Acoust.*, vol. 72, no. 8, pp. 534–543, 2011.

[10] S. S. Utami, D. D. Avoressi, K. Zakiya, and H. Sutanta, "Sound Level Mapping Using Geographic Information System (GIS) To Optimize a Green Campus Environment Quality," *ARPN J. Eng. Appl. Sci.*, vol. 11, no. 6, pp. 4058–4064, 2016.

[11] M. A. Zytoon, "Opportunities for Environmental Noise Mapping in Saudi Arabia : A Case of Traffic Noise Annoyance in an Urban Area in Jeddah City," 2016.

IOP Conf. Series: Journal of Physics: Conf. Series **1075** (2018) 012038 doi:10.1088/1742-6596/1075/1/012038

# Prediction and simulation of internal train noise resulted by different speed and air conditioning unit

**B.T. Atmaja, M.F. Puabdillah, M.N. Farid, W.A. Asmoro**

Department of Engineering Physics, Faculty of Industrial Technology, Institut Teknologi Sepuluh Nopember, Surabaya 60111, Indonesia

E-mail: bagus@ep.its.ac.id

**Abstract.** Railway noise is a problem not only for the train passenger inside but also from public and environment outside the train. The noise of railway is contribution from the noise of wheel, track and the friction of both. Other components such as electrical instruments, mechanical equipment and structure construction add the noise level of the train. This noise can be predicted as the function of the train velocity. By knowing, the noise level at certain speed, the noise at another speed can estimated accurately. In this research we obtained an error value of estimated noise level about 2.6 dB and 0.375 dB in the center of train car.

## 1. Introduction

Train noise is one of the oldest problem occurred acoustics noise. The environment impact of railway noise enforces the need to build such noise controls like tunnel and noise barrier. Not only for environment, but also passenger inside the train cabin are impacted by train noise. So, the railway noise can be divided into two: external/environmental noise and internal/interior noise. The noise inside the train can be resulted from many sources: rolling noise (friction of wheel and rail with its sleeper) and train's equipment noise (traction motor, inverter, compressor and air conditioning). If the train moves at speed above 320 km/h, the aerodynamic noise also contributes to total railway noise.

This paper evaluated internal train noise, especially when speed is changed, and internal train noise caused by air conditioning unit. For the last mentioned noise, the shape of air conditioning diffuser and the different between experiment and simulation are evaluated. The first diffuser uses circle shape ($KA_1$) and the second diffuser uses square shape ($KA_2$). These two different shape of diffuser give two similar sound distribution but different air flow.

The train models used for this research are KRL-KFW and Soekarno-Hatta airport train. The result of this research is expected can be used to reduce the noise of future train by suppressing the dominant noise sources and designing better noise absorber materials and enclosure.

## 2. Internal railbound noise as function of speed

Railway noise can be predicted by its speed. It means, if we knew emitted noise at certain speed, noise level at another speed can be predicted accurately. The emitted rolling noise from a train usually is taken proportionally to the logarithm of speed. The following equation is used for this research [3].

$$L_p = L_{p0} + N \log_{10} \frac{V}{V_0} \tag{1}$$

The equation above is is usually used for rolling noise prediction, in this case, we used it for internal train noise prediction. $L_{p0}$ is the sound level at reference speed $V_0$. $N$ is the value of speed exponent, it is determined from linear regression, generally from 20 to 35, we use $N = 20$ for this research. The lower $N$ value is used for lower train speed, while higher $N$ value is for higher speed ( $>$ 300 km/h) [4].

For this case, the internal noise measurement was conducted on train speed at 37 km/h. The obtained noise level is 69 dB. For the those data, we would like to predict internal train noise at 70 km/h as we also have the real experiment data for that speed. By using $N = 20$, we obtain the following noise prediction,

$$\begin{aligned} L_p &= L_{po} + N.\log\left(\frac{V}{V_{p0}}\right) \\ &= 69 + 20\log\left(\frac{70}{37}\right) \\ &= 74.5 \ \ dBA \end{aligned}$$

For the real measurement, we got noise level 74.125 dB. So the comparison between prediction and experiment can b e summarized on the table 1.

**Table 1.** Comparison of noise prediction and measurement inside railbound vehicle (study case: KRL KFW)

| Prediction (dB) | Measurement (dB) | Error (dB) |
|---|---|---|
| 74.5 | 74.125 | 0.375 |

## 3. Train air conditioning noise with different diffuser shape

Air conditioning unit is one of noise sources inside a railbound vehicle. The noise simulation on air conditioning unit is desired to get noise level resulted by AC's components, i.e compressor, centrifugal fan, axial fan, and streamline in ducting. The measurement to verify the simulation was conducted beneath AC, the cabin room where passenger stays inside.

An air conditioner is heat an exchanger system, a modification of the cooling system which can be classified into two purposes: cooling system to refresh the air or (1) a cooling system for process [2]. Noise resulted from AC comes from its components. The biggest contribution of AC noise comes from compressors an exhaust fan. When in operation, the fan should be operated on 70 % to 80 % of maximum capacity. A small fan operated at higher capacity could result an additional noise of 5 dB [2]. The noise from air flow then transmitted to cabin passenger through exhaust path, ducting and diffuser. Noise from structure borne dominantly appears on lower frequency while air flow noise appears on higher frequency [3].

The two shapes of AC diffuser was evaluated: circle and square. The train AC prototype used for this research was AC-4001 model that have been developed by national company. The first

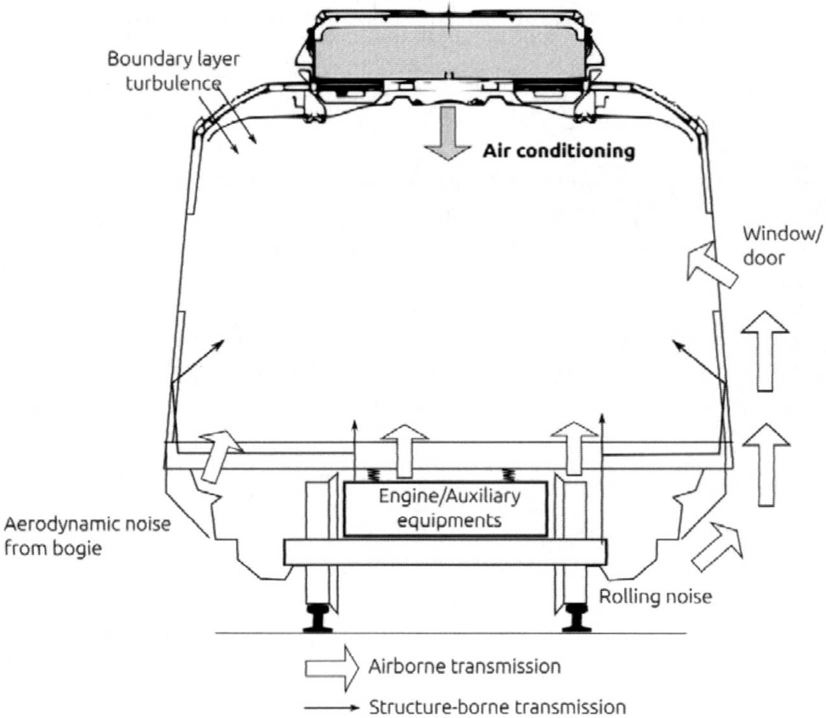


**Figure 1.** Noise sources inside the train, adapted from [3]

circle shape diffuser is the existing model while the square one is another shape for comparison that previously considered by the company. The circle shape diffuser AC is coded as $KA_1$ and the square shape diffuser AC is $KA_2$.

### *3.1. Methods*

The following specification was made during simulation using computational fluid dynamics. The ACI-4001 consist of two symetrical AC system with one centrifugal fan output.

(i) Output debit (each) = 90 m 3 /min
(ii) Return air debit by AC = 64 m 3 /min
(iii) Cooling capacity = 40000 kcal/h
(iv) Evaporator coil temperature output = 27 o C
(v) Humidity = 65 %

### *3.2. Simulation steps*

To simulate AC noise inside railbound vehicle, a 3D drawing of train and AC unit have been built according to its geometry. Then, the following steps have been conducted.

(i) Pre-processing
In this step, the train cabin was modeled in real dimension along with its parameters.
(ii) Meshing
This step follows the previous step i.e. the modeled train cabin was meshed (divided) into smaller elements. The mesh geometry is 20 $mm$ with cutcell method.

(iii) Solver
This step is including determination of model scale, iteration time, and viscous models used to obtain speed and pressure of air from AC that can be used for noise level calculation. The simulation parameter for the solver is as follows.
(a) Operating condition,
1. $P = 291\ Pa$
2. $g = 9.8\ m/s^2$
3. $\rho = 1.2257\ kg/m^3$
(b) Boundary conditon,
1. Velocity Inlet; $V_{ref} = 12.764\ m/s$
2. Pressure Outlet; Pressure gauge = 0 Pa

(iv) Post-processing
This step is result of CFD simulation that displayed air speed contour with the streamline.

### *3.3. Simulation Result*

From the steps in method and provided data in AC specification, the simulation resulted some findings between diffusre $KA_1$ and diffuser $KA_2$ i.e comparison of speed, comparison of pressure and noise level and comparison of streamline distribution. We will discuss those variables one by one.

The speed comparison between $KA_1$ and $KA_2$ can be seen on table 2. For the each diffuser model, $KA_1$ and $KA_2$, the three viscous models give the similar result. The average result for each model shows the the second diffuser model, $KA_2$, with square shape output has slightly faster speed. However the speed different is very small that hardly can be perceived by the passenger.

**Table 2.** Air speed on 1.2 m high above the train floor

| Diffuser model | Statistics | Air speed (m/s) K-Epsilon Standard | K-Epsilon Realizable | K-Omega SST |
|---|---|---|---|---|
| $KA_1$ | Max | 3.99 | 3.88 | 4.267 |
| | Min | 0.00706 | 0.00792 | 0.004069 |
| | Average | 0.61 | 0.659 | 0.6701 |
| $KA_2$ | Max | 4.31 | 3.825 | 4.32 |
| | Min | 0.00464 | 0.021 | 0.042 |
| | Average | 0.658 | 0.7049 | 0.741 |

The next investigation for the different AC diffuser shape is its pressure that can be converted to noise level. Again, the simulation shows the similar result between $KA_1$ and $KA_2$. Table 3 show the different pressure level resulted by diffuser $KA_1$ and $KA_2$.

The pressure resulted from two different shape of diffuser can be converted to the resulted noise level. Table 4 shows the comparison between both. The measurement position is located based on ISO standard [5].

**Table 3.** Pressure comparison on 1.2 m high above the train floor

| Diffuser model | Statistics | Pressure (Pa) K-Epsilon Standard | K-Epsilon Realizable | K-Omega SST |
|---|---|---|---|---|
| $KA_1$ | Max | 431.624 | 426.755 | 427.372 |
| | Min | 424.544 | 419.177 | 420.331 |
| | Average | 425.95 | 420.753 | 421.83 |
| $KA_2$ | Max | 432.064 | 427.52 | 426.569 |
| | Min | 425.021 | 420.691 | 419.298 |
| | Average | 422.162 | 422.162 | 420.541 |

**Table 4.** Comparison of Noise level

| Position | $KA_1$ | $KA_2$ |
|---|---|---|
| ducting | 70 dB | 70 dB |
| 1.2 m | 8 dB | 6.15 dB |
| 1.6 m | 14.5 dB | 16 dB |

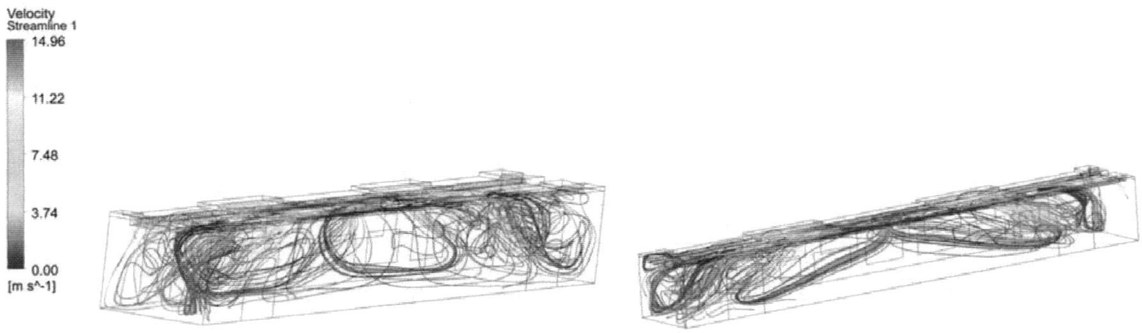


**Figure 2.** Comparison of streamline between $KA_1$ (left) and $KA_2$ (right)

Similar to the result of the pressure and the air speed of different diffuser shape model, the noise level is also similar for both diffuser model. It can be concluded in this stage that either model of diffuser can be applied to AC unit of the train. However, the last evaluation about streamline shows the different between both.

The result of simulation for streamline of the air flow from AC unit can be shown in figure 3.3. The numerical values of streamline shows the different each other. In figure 3.3 left side, it is clearly shown that CFD simulation resulted spread evenly of the air flow streamline by using first model diffuser, the circle shape. The output air from ducting through diffuser is distributed into train cabin. The right side shows distribution of streamline resulted by using second diffuser, the square one. The right side shows less broadly distribution especially in the center of cabin. For the $KA_1$ diffuser model, average speed for the streamline is 3.25 m/s and for the $KA_2$ is 4.43 m/s. This result shows the superiority of circle diffuser model over square model.

**Table 5.** Simulation Vs experimental result at the center of cabin

| CFD simulation (dB) | Measurement (dB) | Error (dB) |
|---|---|---|
| 19.3 | 21.9 | 2.6 |
| 22.6 | 23.1 | 0.5 |

### *3.4. Experimental result*
Four points at the center of train cabin was simulated to get the noise value of each point that varied from 19.3 dB to 22.6 dB. By measuring the real noise value when AC is on and other components is off and then logarithmically subtracted with the ambient noise, the noise from AC at real situation can be measured. Table 5 shows the different between simulation and experimental result.

## 4. Conclusion
It can be concluded based on the study of train noise analysis both for different speed and different AC unit diffuser as follows:

- Train noise inside the cabin can be predicted by using rolling noise prediction equation. By using $N = 20$ we obtain small error for noise prediction at 70 km/h from known data at 37 km/h.
- The circle shape diffuser for AC unit has more benefits compared to the square unit. The streamline of circle shape diffuser spreads more evenly compared to square shape diffuser with similar air speed, pressure and noise level.

## 5. Acknowledgment
Authors would like to thank to Institut Teknologi Sepuluh Nopember for funding of this research with the grant No. 810/PKS/ITS/2017 and also to PT. INKA (Industri Kereta Api) Madiun as research collaborator who provide data and prototype of AC Unit along with train cabin for this research.

## References
[1] AHSRAE 1999 *Application Handbook, Ch. 46: Sound and Vibration Control.*
[2] GÃijngÃűr, F.E., 2003 *Computer aided noise prediction in heating, ventilating and air conditioning systems* (Doctoral dissertation, METU)
[3] Thompson D 2008 *Railway noise and vibration: mechanisms, modelling and means of control* (Amsterdam: Elsevier) p 13
[4] Xiaoan G 2006 Railway environmental noise control in China Journal of Sound and Vibration 293 1078âĂŞ1085 (Journal of Sound and Vibration 293) (Amsterdam: Elsevier) p 1078—1085
[5] Orczyk, M et. al. 2016 Assessment of noise inside a tram during a ride and at a standstill ICSV23 (Athens)

IOP Conf. Series: Journal of Physics: Conf. Series **1075** (2018) 012039 doi:10.1088/1742-6596/1075/1/012039

# Management at corporate level of noise exposure assessment and control for Petrobras refining

**P PISCHEDDA[1], K CORMIER[2] and I MOHAMED[3]**

[1] Asia Managing Director, ACOEM Group
[2] Metravib Business Unit Managing Director, ACOEM Group
[3] Support Manager, ACOEM Group

E-mail: patrice.pischedda@acoemgroup.com

**Abstract.** This article presents the work carried out as part of a project aiming to regulate the noise exposure of workers of PETROBRAS refining units. This is a corporate project carried out over a three years' period (2012 to 2015) by the Supply business unit of the Brazilian company. The uniqueness of this project is due to its scale. Indeed, this encompassed twelve refineries located throughout Brazil. The corporate approach to deal with noise exposure risk leads to an overall data consolidation allowed by a standardized methodology. The methodology used is split into four phases. First, a long but necessary planning stage, then an exhaustive assessment of the current noise situation of production areas was carried out during the first twelve months of the project. This stage allowed the involvement of several important tools used to identify and rank noise exposure issues. This lead to noise maps of every industrial area and an acoustic database of equipment exceeding regulatory limits (sound power level and directivity). Third, every industrial area was modelled into noise prediction software compliant with ISO 9613-2 standard. Finally, acoustic damping solutions were designed and proportioned by size of equipment. This mixed approach of measurement and numerical simulation mobilized a dedicated ACOEM project team comprising of eight acoustic experts. Principal achievement of this project, concluded in 2015, is a clear vision of the noise control priorities allowed by standardized methodology on all refineries in a short time. This achievement leads to a coherent set of controls for the future investments to be done to mitigate noise exposure risks. In addition to highlighting the benefits of integrated Health Safety and Environment (HSE) management, this work provides an understanding on the use of virtual computing doses from propagating acoustic model.

## 1. Introduction

The concern with workers exposed to high noise levels is an issue of constant importance. Such exposure can lead to unhealthy work environment and consequently be a potential cause of sleep and cardiovascular problems [1].

Based on this, an analysis and evaluation of the acoustical environment of 12 operational refineries and 4 refineries in the establishment phase were carried out. The size of this project is unique in a global context – analysis and simulation of environmental noise of 265 industrial zones in a total investigated area of 1.900.000 m$^2$. For each zone, a noise map was generated, helping the identification of noise sources of nearly 3500 equipment/installations. The mobilization of skilled technicians, detailed determination of required measuring equipment, meticulous logistic planning and

IOP Conf. Series: Journal of Physics: Conf. Series **1075** (2018) 012039 doi:10.1088/1742-6596/1075/1/012039

the quality in the analysis of collected data and acoustic simulations are amongst the factors leading to a successful project.

## 2. Development

The project was divided into 4 phases in accordance with its unique requirements. An initial planning phase where required resources and equipment were defined, a field campaign where data was collected, a diagnostic phase where acoustic computational simulation took place and a final stage including a conceptual and a detailed noise mitigation plan.

### *2.1. Resources and equipment*

The first challenge was defining the required resources for the realization of the proposed scope within 36 months – the demanding logistics for field works which took more than 250 days and mobilization of a single measuring team of a total distance greater than 15000 km. The project planning considered all logistics, technical and operational requirements. The field campaign survey team was comprised of 6 technicians equipped with 6 sound level meters and concluded their measurements in 12 months.

### *2.2. Field campaign*

Field campaign had two main objectives: mapping of sound pressure levels of each zone and determination of sound power levels of sources where sound pressure levels were measured above 80 dB(A).

*2.2.1. Mapping of noise levels.* Noise maps were calculated, depicting the iso-phonic contours calculated from sound pressure levels measured at 1.5 m above ground in a 5 x 5 m squared grid. Exception to the proposed grid occurred in accordance to the spatial noise gradient, a courser or finer grid was adopted per the local needs and the access to measurement areas. The adopted typical grid resolution was defined also with basis on the dimensions and distances of equipment in the refining plants, allowing source identifications on the final calculated noise maps.

*2.2.2. Characterization of noise sources.* The noise sources were identified with the noise maps of each zone. The ISO 3746 was employed in determination of sound power of sources. The standard defines the conditions and methods for calculating the sound power from sound pressure levels. Figure 1 depicts the typical noise measurement arrangements for calculating the sound power levels of the sources.

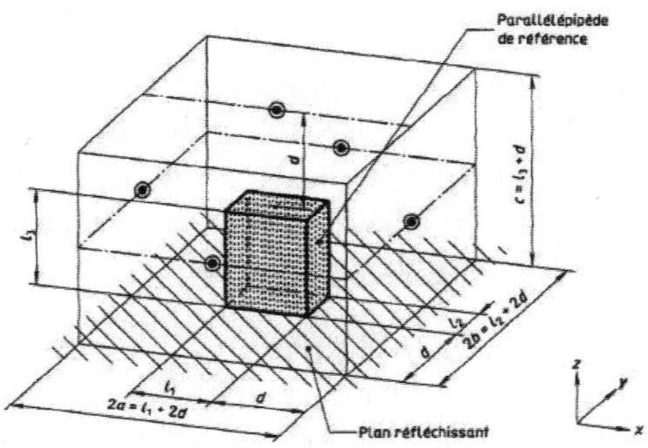


(a)

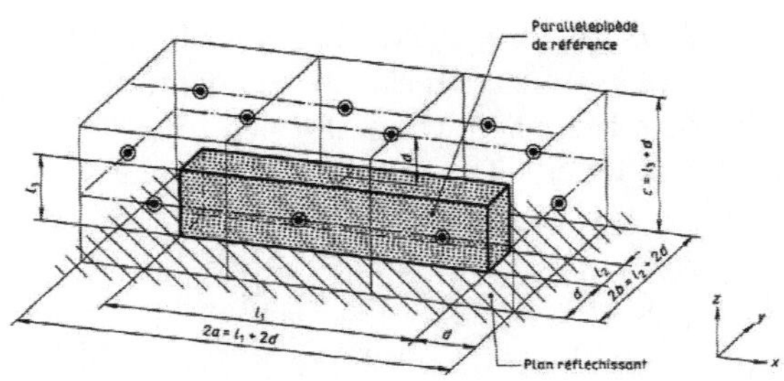


(b)

**Figure 1.** Measurement points for determining the sound power levels of (a) small and (b) large machines [2].

Adapting of the employed methodology was necessary due to operational constraints – in some areas the elevated background noise could not be isolated from the machine of interest. In these cases, the measurements were conducted in smaller grids, which means the measurement locations were closer to the source, nevertheless without reaching the source near-field [3]. Consequently, an overestimation of sound power levels of up to 3 dB(A) can incur, in this case the noise assessment plan assumes a conservative criterion for noise control. Additionally, the computational noise models were adjusted and recalibrated with the sound pressure level measurements to minimize the risk of overestimation of noise levels.

### *2.3. Computational simulation and conceptual mitigation plan*

Computational models in the CadnaA software were designed using the sound power of equipment determined in the simulation of the refinery sites to provide an initial mitigation plan. CadnaA's prediction model is based on the ray-tracing method to calculate attenuation in the propagation of acoustical waves. In case of multiple source models, such as this study, the contribution of individual sources is added up assuming uncorrelated sources. The calculation protocol was in accordance to ISO 9613-2, including the parameters for insertion loss of obstacles, air and ground absorption and source sound directivity [4,5].

The sound directivity of noise sources was adjusted in the computational model taking into consideration the data collected in the field and the calculated noise map. The mitigation plan takes into consideration, that the sound pressure levels at 1 m of equipment should be limited to 90 dB(A) in accordance with the customers' requirements (Petrobras).

Typical noise mitigation for equipment which did not fulfill noise requirements was proposed taking into consideration the type of equipment (pumps, motors, compressors etc.) and the required noise attenuation. Undesired noise sources related to maintenance issues (such as steam leakages in the pipelines) were not considered in the simulation. This is provided that the corrective maintenance was included in the mitigation plan.

The validation of necessary noise attenuation was assessed in CadnaA using and iterative process. A global attenuation was introduced at each source' façade and noise maps were recalculated, allowing the assessment of the impact of each mitigation measure and its combined results in the acoustic field.

### *2.4. Detailed mitigation plan*

In this stage, a basic plan of noise attenuation measures was presented, indicating the spectral attenuation of employed materials. An analysis and selection of potential suppliers capable of providing the indicated noise solutions was also part of the scope.

Based on the response of consulted potential suppliers, a final cost estimative was calculated and a revalidation of noise maps were conducted considering the guaranteed attenuation informed by potential suppliers.

## 3. Results

As of now, the concluded phases include the field campaign, characterization of noise sources, computational simulation and conceptual mitigation plan.

The field campaign was performed from October 2012 to November 2013, on an average pace of 1 refinery per month. The project management was successful in allowing all activities to be concluded within the expected time, causing no delays to the customer and respecting all health and safety requirements imposed by the activities.

From the collected grid measurements in the field, the noise maps were calculated with the aid of CadnaA noise mapping software using ISO9613-2 standard for the 265 operational units investigated. The noise maps provided crucial information in the identification of relevant noise sources in each unit. The figure 2 below illustrates a noise map calculated in this phase.

In figure 2, it is possible to identify some of the more critical noise sources in the unit, helping the selection of noise sources to be further characterized. The selected sources were investigated by determining its equivalent noise level, $L_{eq}$ at a specific distance in the boundary of the source, as described in 3.1.2.

In the initial mitigation plan phase, the results obtained with the noise map were compared with the computational model of the noise sources of the plant. In most cases the resulting analysis indicated a direct coherence between the maps, nevertheless in some cases regions depicting sound pressure levels above 90 dB(A) were only present in the maps calculated from the simulated model which can be attributed to different operation condition of machines along the measuring day, in other words, the sound sources may not be perfectly stationary.

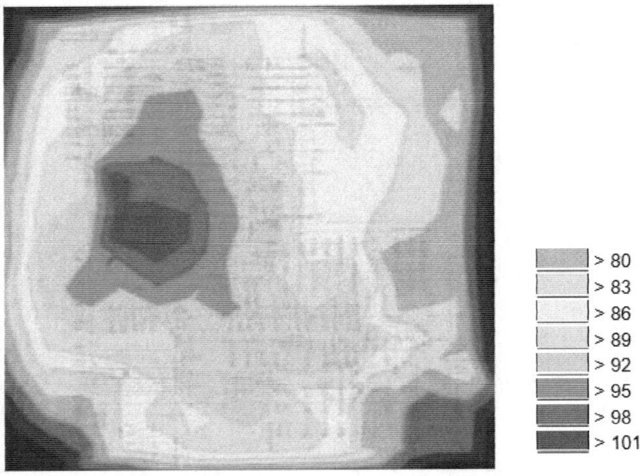


**Figure 2.** Noise map of operational unit zone [6].

In this phase of the study, the contribution of undesired sources (pipeline leakages) were considered for adjusting the simulated model with respect to the measured map. Later on, their contribution was ignored in the simulated model.

The figure 3 below shows the noise map obtained from the computational model of the same zone as figure 2.

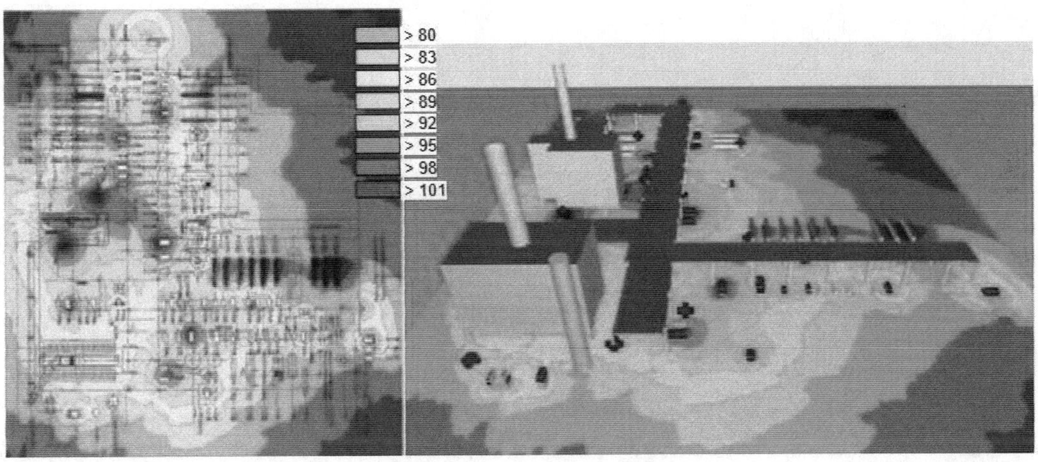


**Figure 3.** Computational noise simulation, (a) 2D Representation (b) 3D Representation [6].

In the noise simulation model, all noise sources were calibrated with respect to measured values in the refinery, including the directivity of the sources. The sound propagation was computed via CadnaA, in accordance with ISO 9613-2, following the equation 1:

$$L_{fT} = L_w + D_c - A \quad (1)$$

Where: $L_{fT}$ is the sound pressure level at receiver; $L_w$ is the sound power level of the source in octave bands; $D_c$ is the directivity correction index and A is the attenuation from source to receptor in octave bands.

IOP Conf. Series: Journal of Physics: Conf. Series **1075** (2018) 012039 doi:10.1088/1742-6596/1075/1/012039

For noise sources exceeding the defined noise limit, the mitigation measures were based on the type of equipment:

**Table 1.** Standard noise control measures for several types of equipment [7].

| Mitigation measure | Equipment |
|---|---|
| Noise enclosure | Electric motor |
| Low noise fan | Electric motor |
| Lagging | Pipeline and valves |
| Acoustic blanket | Air compressors |
| Noise enclosure | Air compressors |
| Silencer | Exhausting |
| Acoustic barrier | Furnace Oven |

The presented solutions were adapted per the required attenuation levels and the equipment's constraints regarding installation of mitigation measures. In some due to the lack of space or the proximity of other equipment, the mitigation measures could not be implemented. An initial cost estimate could be determined based on the sources dimensions and CAD of the entire plant.

The use of noise control measure in the computational model of an oven is illustrated below. The oven's wall was extended around the burners, acting as a noise barrier. The cost of this noise control measure was estimated based on the cost of wall per square meter.

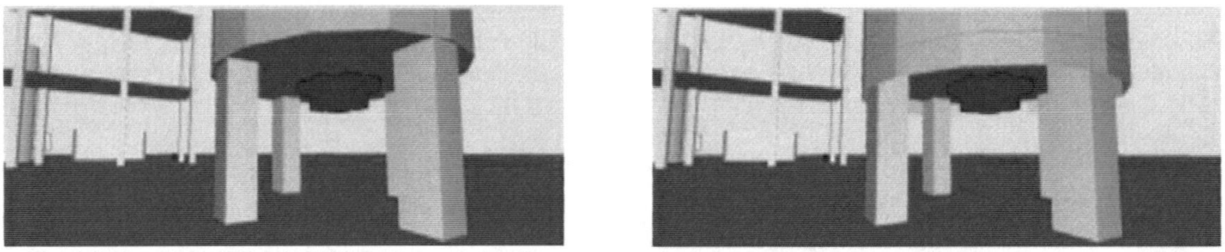

(a) (b)

**Figure 4.** The figures present the bases of the oven with burners depicted in blue. In (a) original installation and extended wall acting as a barrier for noise from [6].

To verify the initial mitigation plan, a noise map is calculated considering the applicable noise control measures. The overall result of the combine mitigation plan can then be appreciated.

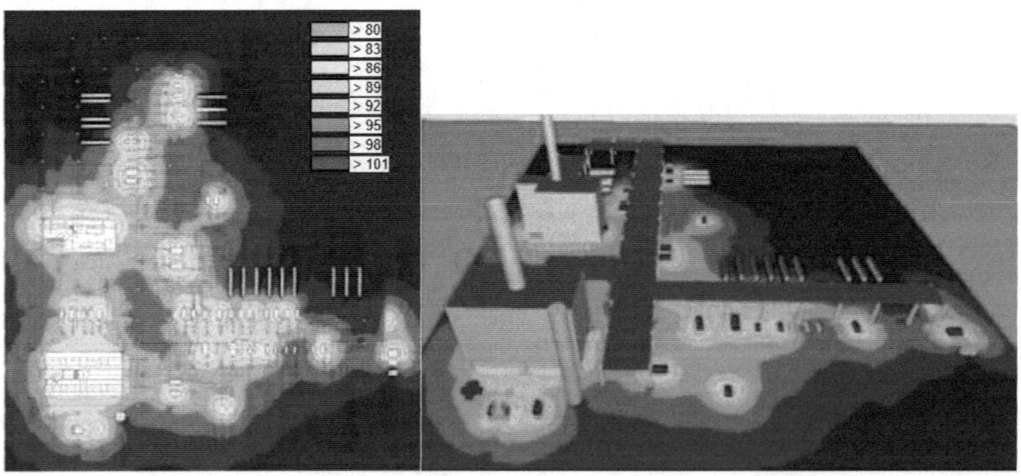


**Figure 5.** Map of noise simulation including the mitigation measures [6].

The methodology described above for mapping, modelling and mitigating the environmental and noise exposures defines a structured management framework to be applied to all 265 zones. Based on a noise contribution ranking and evaluation of the effectiveness of the noise control of individual sources, the mitigation plan could be determined and focused in the areas with higher circulation of workers.

## 4. Conclusions

The stages of sound pressure measurements, noise mapping, characterization of noise sources and the noise modelling were concluded within schedule, including the initial mitigation plan.

The software CadnaA was proven to be a decisive tool in the project, allowing in depth investigations and conclusive results. It has allowed the calculation of noise maps from sound pressure and the construction of a computational noise model representing all the sources of interest. The results from the computational model could be compared to the noise maps and fine-tuned in terms of source level and directivity. In possession of a representative noise model of the refinery, the contribution of individual sources, ranking and assessment of mitigation effectiveness could be performed – proving valuable information in defining a complete mitigation plan to all zones.

The goal of the mitigation plan was to reduce the risk of noise exposure of workers and to reduce the environmental impact of assessed areas.

The developed framework for managing environmental and noise exposure was proven to be well adapted to the requirements and the characteristics of this project despite their bring no precedents – Activities were concluded within schedule and provided conclusive results for defining a noise mitigation plan. Such frameworks can be extended to other large industrial plants were noise control is often a concern either in legal or occupation safety terms.

## 5. References

[1] CHALMERS UNIVERSITY OF TECHNOLOGY 2012 Division of Applied Acoustics (Disponível) *em:<http://www.ta.chalmers.se/education.php?page=mst_role.>Acessado em 26/05/2014*

[2] INTERNATIONAL ORGANIZATION FOR STANDARDIZATION **ISO 3746**: Determination of source power levels and sound energy levels of noise source using sound pressure- Survey method using on enveloping measurement surface over a reflecting plane (Genebra: ISO, 1996)

[3] BIES D A and HANSEN C H 1996 Engineering noise control and theory and practice (E & F N Spon, Londres)

[4] DATAKUSTIK GMBH 2009 CadnaA Reference Manual (Greinfeberg)

[5] INTERNATIONAL ORGANIZATION FOR STANDARDIZATION **ISO 9613-2**:Acoustics: Attenuation of sound during propagation outdoors- Part 2: General Method of Calculation (Genebra: ISO, 1996)

[6] ACOEM 2013 Acervo de relatórios técnicos (São Paulo)

[7] CROCKER M J 2007 Handbook of noise and vibration control (John Wiler & Sons, Nova Jersey)

# Acoustics of noise, vibration, and harshness

**W S Gan**[1]

[1]Acoustical Technologies Singapore Pte Ltd,33 Oxford Road, #04-03, Kentish Court, Singapore 218816, Singapore.

E-mail: wsgan5@gmail.com

**Abstract.** In this paper, we will go to a more fundamental level of treating acoustics. We introduced the gauge invariance into the acoustic wave equation of motion in 2007. This is an important milestone in the history of gauge invariance. The earliest gauge invariance is shown by Maxwell's electromagnetic wave equations. The root of acoustics originated from fluid mechanics. We show that the usual form of acoustic wave equation for fluids is in fact the linearized form of the Navier Stokes equations. Hence there is in fact, no need to use acoustic analogy to derive the formula for the jet noise which was done by James Lighthill. In this paper we also treat finite amplitude sound wave and finite amplitude vibration using the deductive approach. This is a generalization of the linear theory to the nonlinear theory using relativistic and curvilinear spacetime treatment. This produces the linear case as a special case of the nonlinear case. The usual method is the reverse which starts from the linear theory and extend to the nonlinear theory. This will miss out some useful information. The treatment of vibration will need to incorporate the theory of elasticity. This paper is also the first introduction of singularities into vibration. Singularities is mathematically more sophisticated than resonance.

## 1. Introduction

We extend the gauge invariance property of the Maxwell's equation for electromagnetic theory to the acoustic field equations. This is an important milestone in the history of gauge invariance. The earliest gauge invariance is used in Maxwell's equations of electromagnetic wave equations. Gauge invariance which includes symmetries is a basic property of field theory which covers strong nuclear forces, electromagnetic force, and gravitational force. In extending gauge invariance approach to acoustic fields, it will be a more sophisticated approach than the vector theory of acoustic fields and will include also the characteristics of the vector theory. We will also show that vector theory is a subset of gauge theory. We address the symmetry properties of the acoustic field equations, and interpretation of the inhomogeneous wave equation in terms of gauge invariance. Gauge invariance has long been applied to electromagnetic wave theory. Due to the similarities between electromagnetic waves and acoustic waves, we feel that the interpretation of acoustic fields in terms of gauge invariance will provide more understanding of the acoustic fields and throw lights on new potential applications. This paper is the first application of gauge invariance to acoustic fields.

## 2. Gauge invariance formulation of the inhomogeneous acoustic wave equation

We will start with the derivation of the acoustic Lorentz gauge condition. First, we show that for electromagnetic waves, the Lorentz gauge condition gives rise to the equation of continuity. In electromagnetic theory, the Lorentz gauge condition is given as

$$\nabla.\vec{A} + \frac{1}{c^2}\frac{\partial\phi}{\partial t} = 0 \tag{1}$$

where A = vector potential and φ = scalar potential.

The potential theory is used here. By substituting in the retarded potential for A and φ, and with simplification, one can obtain:

$$\overrightarrow{\nabla.\mathrm{j}} + \frac{\partial\rho}{\partial t} = 0 \tag{2}$$

which is the equation of continuity.

On the other hand, the acoustic equation of continuity is:

$$\nabla(\rho V) + \frac{\partial\rho}{\partial t} = 0 \tag{3}$$

with just replacing j by ρV which is correct with charge density being analogous to the mass density.

So, one can use equation (1) as the acoustic Lorentz gauge condition where c = sound velocity. From the electromagnetic theory, the electric field $\vec{E}$ and the magnetic field $\vec{B}$ are expressed in terms of potential functions as:

$$\vec{E} = -\nabla\phi - \frac{\partial\vec{A}}{\partial t} \tag{4}$$

and

$$\vec{B} = \nabla \times \vec{A} \tag{5}$$

These potentials are invariant under gauge transformations. Here one will use the analogy of $\vec{E}$ as $\vec{T}$ the acoustic stress field.

The usual acoustic wave equations are expressed in terms of the potentials A and ϕ. This gives only partial aspects of the acoustic fields $\vec{V}$ and $\vec{T}$.

From equation (4), one can write

$$\vec{T} = -\nabla\phi - \frac{\partial\vec{A}}{\partial t} \tag{6}$$

From Maxwell's equation

$$\nabla.\vec{E} = 4\pi\rho \tag{7}$$

For the acoustic field

$$\nabla.\vec{T} = 4\pi\rho \tag{8}$$

Substituting equation (6) into equation (8), one has

$$\nabla.\left(-\nabla\phi - \frac{\partial\vec{A}}{\partial t}\right) = 4\pi\rho \tag{9}$$

But Lorentz gauge condition, equation (1),

Thus,

$$-\nabla^2\phi + \frac{1}{c^2}\frac{\partial^2}{\partial t^2}\phi = 4\pi\rho \tag{10}$$

which is the inhomogeneous acoustic wave equation in ϕ and ρ = field density.

One can also express the velocity field in terms of the scalar potential and vector potential as

$$\vec{V} = \nabla\phi + \nabla \times \vec{A} \tag{11}$$

The Christoffel equation for an isotropic medium with plane wave solutions and harmonic time variation can be written as

IOP Conf. Series: Journal of Physics: Conf. Series **1075** (2018) 012040 doi:10.1088/1742-6596/1075/1/012040

$$c_{11}\nabla(\nabla.\vec{V}) - c_{44}\,\nabla \times \nabla \times \vec{V} = \rho\frac{\partial^2}{\partial t^2}\vec{V} \tag{12}$$

Substitution of equation (11) into equation (12) gives

$$\nabla(c_{11}\nabla^2\phi - \rho\frac{\partial^2}{\partial t^2}\phi) - \nabla \times (c_{44}\nabla \times \nabla \times \vec{A} + \rho\frac{\partial^2}{\partial t^2}\vec{A}) = 0 \tag{13}$$

For the second term, the quantity in brackets is set equal to the gradient of an arbitrary function f. The vector potential $\vec{A}$ can thus be taken as a solution to the vector wave equation

$$\nabla^2\vec{A} - \frac{1}{V_s^2}\frac{\partial^2}{\partial t^2}\vec{A} = 0 \tag{14}$$

The first term in equation (13) is made zero by simply by requiring that the scalar potential $\phi$ satisfy the scalar equation

$$\nabla^2\phi - \frac{1}{V_l^2}\frac{\partial^2}{\partial t^2}\phi = 0 \tag{15}$$

So, one has obtained homogeneous sound wave equation in A and $\phi$.

This shows that the acoustic fields $\vec{T}$ and $\vec{V}$ can be expressed in terms of the solutions of the homogeneous acoustic wsave equations A and $\phi$. A and $\phi$ in turn are only partial aspects of $\vec{T}$ and $\vec{V}$.

The Lorentz gauge condition has the advantage of introducing complete symmetry between the scalar and vector potentials, i.e., it makes both potentials satisfy the same wave equation as. that obeyed by the fields. Equations (14) and (15) are a symmetrical set of equations.

By using the analogy of momentum density $\vec{P}$ as equivalent to $\vec{B}$, one has

$$\vec{T} = -\nabla\phi - \frac{\partial\vec{A}}{\partial t} \tag{16}$$

and

$$\vec{P} = \nabla \times \vec{A} \tag{17}$$

Equations (16) and (17) are unchanged by transformations of the type

$$A' = A - \nabla\phi \tag{18}$$

$$\phi' = \phi + \frac{\partial\phi}{\partial t} \tag{19}$$

These transformations are usually known as gauge transformations, and a physical law that is invariant under such a transformation is said to be gauge invariant.

Gauge invariance or gauge field theory is a more sophisticated treatment of acoustic fields than vector theory. It gives a more accurate solution for the velocity field and the stress field than those obtained from the inhomogeneous acoustic wave equations. It also gives the difference between the strain field and the velocity field which previously all treat them as wave functions. It provides better understanding of the stress field and the velocity field.

## 3. The root of acoustics is fluid mechanics-there is no need to use acoustic analogy

Acoustics for fluids originated from fluid mechanics. The acoustic wave equation for fluids is in fact the linearized form of the Navier Stokes equation. Hence there is no need to use the acoustic analogy to use the Navier Stokes equation to derive the formula for the jet noise as what James Lighthill [1] did.

The Navier Stokes equations consist of two equations:

1) The continuity equation. This equation is derived through the hypothesis of conservation of mass. In its typical form, it can be written down as

$$\frac{D\rho}{Dt} + \rho\nabla.\mathrm{u} = 0 \tag{20}$$

where $\frac{D}{Dt} = \frac{\partial}{\partial t} + u_j \frac{\partial}{\partial x^j}$ known as the material derivative, and ρ the density of the fluid.

2) Momentum equation. The fluid must also conserve momentum. We ensure this by requiring that

$$\frac{D}{Dt}u_j = \frac{DT_{ij}}{Dx_j} - \rho g_j \tag{21}$$

with

$$T_{ij} = -\mathrm{P}\,\delta_{ij} + 2\mu e_{ij} + \lambda e_{mm}\delta_{ij} \tag{22}$$

where u is the fluid velocity, $T_{ij}$ is the stress tensor, P is the pressure, $g_j$ is the constant gravitational acceleration, $\delta_{ij}$ is the unity matrix, $e_{mm}$ is the strain tensor, with μ and λ are fluid dependent scalars.

If u is incompressible, (∇. u = 0) these equations reduce to the Navier-Stokes Equations. On the other hand, usually the acoustic wave equation is derived as follows [2] as what is done by B. A. Auld [2]:

These equations are used. The first is the equation of motion:

$$\nabla.\vec{T} = \rho\frac{\partial}{\partial t}\mathrm{u} - \mathrm{F} \tag{23}$$

where F = body force and the second equation are the strain-displacement relation:

$$\mathrm{S} = \nabla_s \mathrm{L} \tag{24}$$

where S = strain, and L = displacement.

One will realize that equation (21) is equivalent to equation (23) and equation (22) and equivalent to equation (24). The acoustic wave equation of motion is obtained by the combination of equation (23) and (24). This shows that the Navier Stokes equation and the acoustic wave equation of motion are equivalent. And they have the same origin.

## 4. Nonlinear acoustics on the curvilinear spacetime- a deductive approach

Curvilinear spacetime is a coordinates system that satisfy the general covariance requirements of all physics equations. All physics equations must remain the same form on whatever type of coordinate system being used. The curvilinear coordinate system was used by Albert Einstein in his general theory of relativity [3]. He introduced the curvilinear coordinates to analyse the general theory of relativity because besides fulfilling the general covariance requirement, it also describes accurately the curvilinear path of the gravitational force. Curvilinear spacetime is useful for explaining intense sound fields due to their curvilinear paths and hence will be suitable for the treatment of nonlinear acoustics. This is a more accurate approach and appropriate for the design of extremely sensitive equipment.

The use of curvilinear spacetime will enable a deductive approach to a nonlinear acoustics problem. The usual treatment of nonlinear acoustics is the proceeding from the to the general; from the special, linear case to the more general, non-linear case. This will tend to miss out some useful information's. But it has become clear in recent years that in many cases it is advantageous to reverse this procedure. That is, to proceed from the general to the particular. Take an example, to regard the case of a material with linear constitutive relations as a special case of a more general class of materials.

## 5. Treatment of vibration and introduction of singularities into vibration problems

To understand vibration, one needs to incorporate in the theory of elasticity. For example, for the treatments of vibration in structures, in shells, in plates. This is because one has to deal with the stress-strain relationships.

Resonance is a very common phenomenon in vibration problems and the usual treatment on resonance is to deal with the solution of the free vibration and forced vibration partial differential equations. Singularities is not a shortcoming of theories but is evident as the pattern of nature. Singularities play an important role in several branches of physics such as the Big Bang and the origin of the universe, singularities in second order phase transition, singularities in the Yang Mills theory of Standard Model of particle physics and so on.

Mathematically, singularities are more sophisticated than resonance. Usually it is only mentioned that the amplitude reaches a peak during resonance but not mentioned that in fact the amplitude tends to move towards infinity and causes instability during resonance. The figure below shows the amplitude of vibration tends towards infinity at the resonance frequency of a driven damped simple harmonic oscillator. Here the term disastrous resonance is used instead of the term singularities.

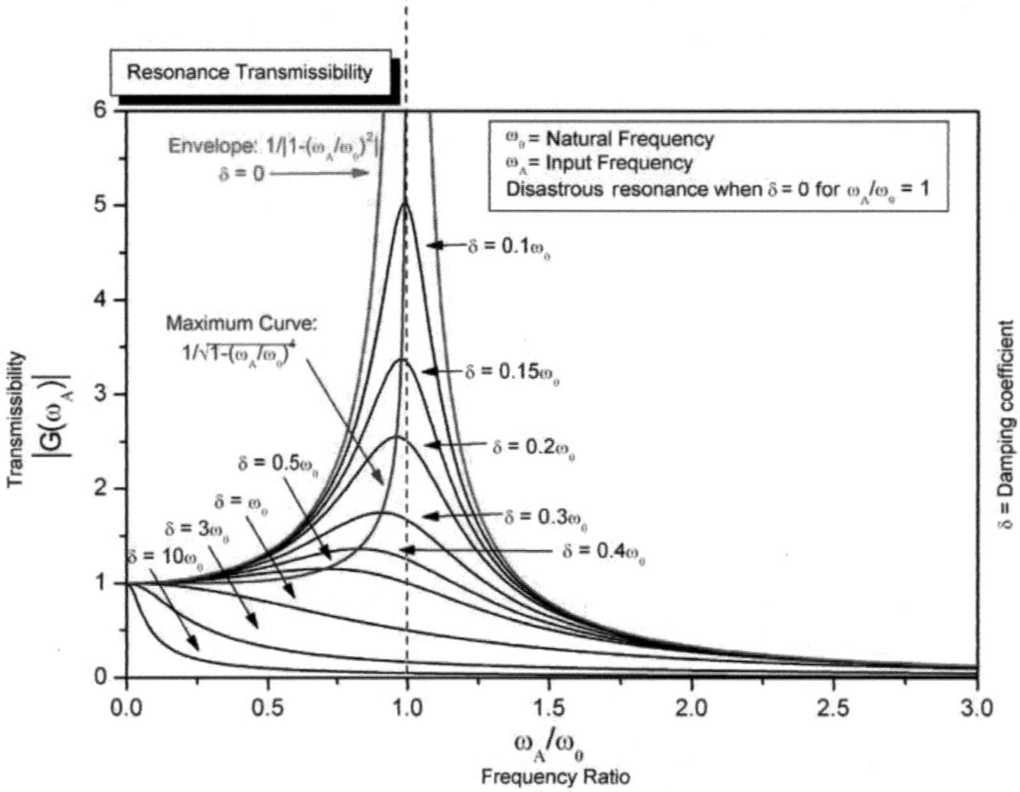


**Figure 1.** The amplitude of vibration tends towards infinity at the resonance frequency of a driven damped simple harmonic oscillator.

## 6. References


[1] Lighthill, M J 1952 On Sound Generated Aerodynamically I General Theory *Proc. R. Soc.* (London) A 211 pp 564-587

[2] Auld, B A 1990 Acoustic Fields and Waves in Solids (Robert Krieger Publishing Co.) pp 101 Vol I

[3] Einstein, A 1916 The Foundation of the General Relativity (Annalen der Physik) 354 (7) 790

IOP Conf. Series: Journal of Physics: Conf. Series **1075** (2018) 012041 doi:10.1088/1742-6596/1075/1/012041

# Real-time impulse response estimation based on mirrored virtual sound sources

**Sadjad Siddiq**

Audio Researcher, Advanced Technology Division, Square Enix Co. LTD., 6-27-30 Shinjuku, Tokyo, Japan

E-mail: siddsadj@square-enix.com

**Abstract.** A fast method to estimate the impulse response of a fixed sound source to a listener at a movable location in a room of arbitrary shape in two dimensions is proposed. The method can be used in video games to calculate the reverberation of sounds placed in a level depending on the position of the listener and can also be used in 3d environments in which the listener mainly moves in two dimensions. After simplifying the geometry of the room using clustering and converting it to a 2d map, early reflections are modelled in detail using an enhanced image source method that takes occlusion into account: the position of the sound source is mirrored on all walls that are hit directly by its sound and virtual sound sources are created at the mirrored locations. This step is repeated recursively to calculate higher order reflections. Based on this pre-computation and the position of the listener, an impulse response of the early reflections in the room is generated in real time. The reflection data is enhanced stochastically using noise bursts to yield a more convincing acoustic effect. The late reflections in the tail of the impulse response are estimated based on the earlier part of the impulse response.

## 1. Introduction

Typical levels of role playing games in which players can move through three dimensional space usually feature a lot of different acoustic environments like small rooms, big halls or open space. Rendering the sound in such environments in a realistic way increases the immersiveness of the player in the game. Additionally sound that is spatialized in a realistic way can also serve as an additional cue for the player, carrying information about the location of the sound source. Especially in virtual reality applications the realistic spatialization of sound is a very important illusion to trick the player's brain into believing it is actually inside the virtual environment.

Methods used to calculate the propagation of sound can be classified into two categories: methods based on solutions of the wave equation and methods based on geometrical acoustics. A notable example of a method based on solutions of the wave equation by Raghuvanshi and Snyder [2] allows to render sound with high quality at a high memory cost (around 100 Mb of compressed data needed for large game maps) and long computation time, which is typical for such solutions.

Savioja et al. [1] give a detailed overview over methods based on geometrical acoustics. Geometrical methods are in general much faster than methods that are based on solutions of the wave equation but they have the disadvantage that diffraction and scattering of waves, which play an important role in sound propagation, are neglected since they are not described by the underlying models that are based on specular reflections. To take diffraction and scattering into account, they have to be implemented on top of the system based on geometrical acoustics. Methods of geometrical acoustics include ray tracing ([3], [4]), beam tracing ([5]), image sources ([8], [10], [11], [12]), surface based techniques like radiosity ([6], [7]) or hybrids, for example of image sources and beam tracing ([9], [13], [14]). The method presented in this paper is also a combination of the image source method and beam tracing.

In the image source method a sound source is mirrored at reflecting surfaces by placing a virtual sound source at a mirrored location at the other side of the surface. This process is repeated for all reflecting surfaces. After mirroring the sound source at all surfaces, the virtual sound sources created in this process are again mirrored at reflecting surfaces. This process can be repeated until an appropriate number of virtual sound sources are found. The virtual sound sources are usually calculated offline. During runtime the time of the incoming reflections at any point in space can be calculated based on the distance between this point and the virtual sound sources [8].

The basic algorithm does not include any mechanism to check if sound from a virtual sound source actually reaches the point of the listener or if the reflection path is obstructed. In an early implementation based on the image source method and beam tracing that can be applied to arbitrarily shaped rooms by Borish reflections paths are evaluated at runtime to check if they reach the listener [9].

The method proposed by Funkhouser et al. also checks if reflections paths lead to the listener without obstruction and they can be used for complex environments, because looking up beams is implemented in an efficient way by subdividing the environment into cells. [14]

There are also some solutions that take diffraction into account. Mechel added additional virtual sources in corners to account for diffraction and reduced the environment to two dimensions for some evaluations for efficient implementation. [10] Stephenson proposed an implementation in which new beams are inserted into the simulation at certain points to take the diffraction of sound waves around corners into account. Beams are grouped together for efficient computation. [13]

In this paper we describe a method that uses image sources and beam tracing to calculate the early reflections of the impulse response of a fixed sound source position to a movable listener based on simplified geometry of the environment. Diffraction and diffusion are not taken into account during these simulations but the obtained impulse response is enriched stochastically. Informal listening tests have confirmed that the enriched impulse response leads to more realistic results than the plain impulse response.

## 2. Overview

The goal of the method presented in this paper is to find the impulse response $h$ that captures the acoustic effects impacted on sound travelling from an immobile sound source at point $S$ to the position of the player $L$. Direct sound and early reflections are calculated based on the image source method, while late reflections are estimated based on the calculated early reflections. The sound source is reflected at the walls of the level and virtual sound sources are created for all reflections. Occlusion is taken into account by limiting the area in which each virtual sound source is valid. These calculations are preformed before the execution of the game.

During game execution the player's position is compared to the areas in which the sound sources are valid. This is done for each sound source. If the player is in an area where a virtual sound source is audible, the reflection corresponding to that sound source is added to the impulse response $h$. After calculating all early reflections, the late reflections are estimated based on the amplitude and decay of the early reflections. The impulse response $h$ is then applied to the sound played at location $S$ to give the acoustic impression that the sound was actually played at position $S$ in the level.

IOP Conf. Series: Journal of Physics: Conf. Series **1075** (2018) 012041 doi:10.1088/1742-6596/1075/1/012041

## 3. Simplification of Geometry

In modern games meshes describing the geometry of a level tend to be very complex. This complexity is a disadvantage for our method, because the sound source is mirrored at every wall in the level. To make this process faster, we simplify the geometry. The level is first converted to a 2d map and then the 2d map is simplified. This is done by taking a slice through the geometry at the height of the head of the player. This is done by first calculating a surface parallel to the ground the player walks on at the height of the head of the player. We then only consider those triangles or rectangles of the mesh that touch or cross this surface. These triangles or rectangles are simplified to two-dimensional lines by discarding the 3rd dimension of their vertices. Lines with zero length are also discarded.

In the following step the geometry of the 2d map is simplified. If the distance between a point and its closest neighbour is under a certain threshold the mean of the point and the neighbour is calculated and this new point then replaces the point and its neighbour. This algorithm is iterated repeatedly over all points until no pairs of points exist whose distance is under the threshold. See figure 1 for an example. Note that the lines in the simplified geometry are not necessarily parallel to the lines in the original geometry, but this discrepancy is negligible because the exact direction of the reflections calculated in the next step is not important.

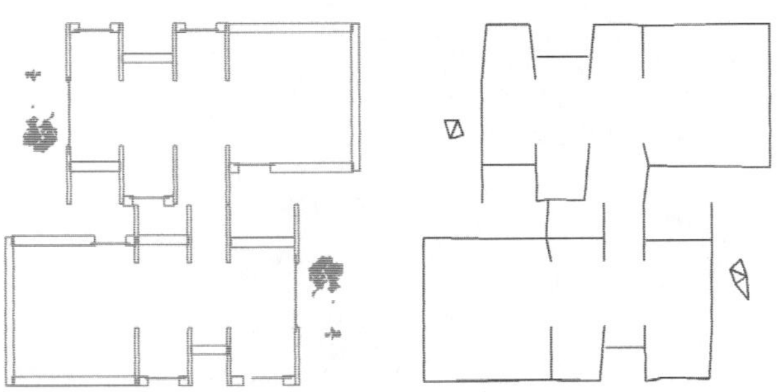

**Figure 1.** Slice through original complex geometry (left side; 5646 points, 18760 lines) and simplified geometry (right side; 64 points, 123 lines)

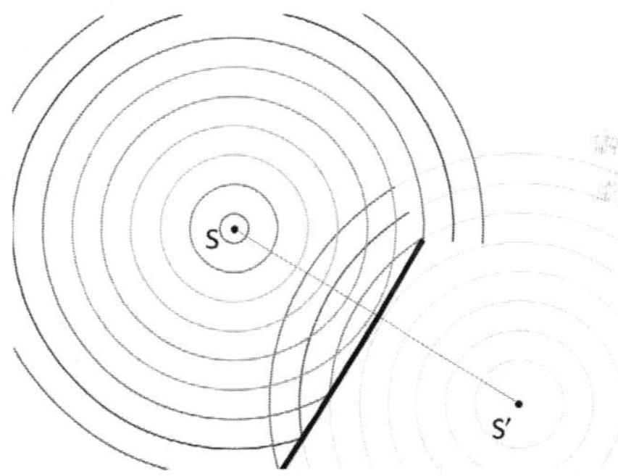


**Figure 2.** The sound propagating as spherical waves from sound source S is reflected at an obstacle (thick line). The reflected sound can be thought of as coming from point S', which is a mirrored version of S

## 4. Virtual sound sources

The sound emitted by a sound source at point $S$ propagates through space as a spherical wave until the wave hits an obstacle and is reflected by this obstacle. The part of the wave being reflected at the obstacle can be represented as a wave coming from point $S'$, which is point $S$ mirrored at the surface of the obstacle as can be seen in figure 2. The reflected waves now continue to travel though space until they in turn hit an obstacle and are reflected once again. This reflection can also be represented by calculating spherical waves emitted from a virtual sound source $S''$ that is a mirrored version of $S'$, using the surface of the reflecting obstacle as mirror axis. This process continues until the energy of the sound wave is completely absorbed by the reflecting surfaces or by the medium the wave is travelling in. Note that at this stage we do not consider wave diffraction and scattering.

In two dimensional space, the wave propagates as a circular wave around $S$ and obstacles are modelled by line segments. The propagation of circular waves around point $S$ is calculated in

IOP Conf. Series: Journal of Physics: Conf. Series **1075** (2018) 012041 doi:10.1088/1742-6596/1075/1/012041

beams represented by circular sectors. Cutting the circle around $S$ in circular sectors facilitates the task of detecting which obstacles are hit by the circular wave, because the wave front is cut into small portions that can be examined independently by looking only into the direction of the beam. Each beam is thus represented by a circular sector defined by the center point $Q$ of its underlying circle, the angles $\omega_1$ and $\omega_2$ and their minimum and maximum radius $r_1$ and $r_2$. The radii represent the distance from the virtual sound source $Q$ between which the beam is valid.

Assuming that a beam is reflected by the line segment between the points $P_1$ and $P_2$, a new circular sector from point $Q'$, which is point $Q$ mirrored at the line $P_1P_2$, is created to represent the reflected beam. The beam is only valid from the point where it hits the obstacle and is reflected. This information is stored in $r_1'$, which is set to the distance between the new center point $Q'$ and line segment $P_1P_2$, calculated as $r_1' = \min(\overline{QP_1}, \overline{QP_2})$, where min returns the smallest value of its arguments. The beam that was reflected is terminated at this distance, so $r_2$ is set to the same value. We then check if the new beam hits any obstacle to find its value for $r_2'$.

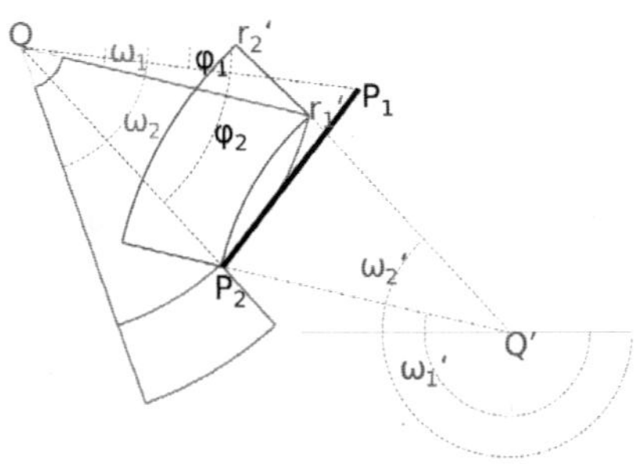


**Figure 3.** A circular sector hits an obstacle represented by a line segment and is reflected partially

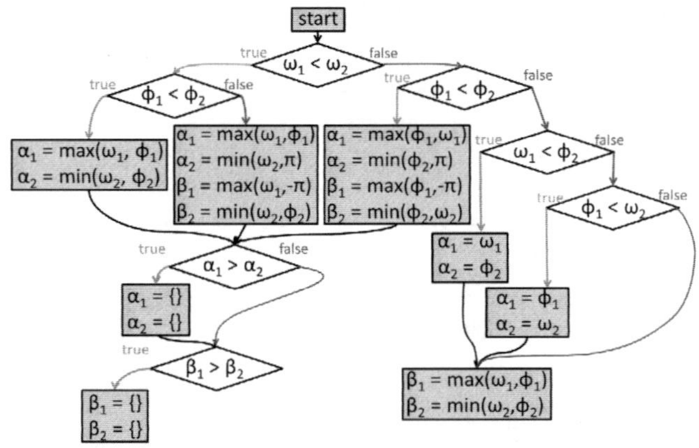


**Figure 4.** Flow chart of the algorithm used to find the intersection(s) of two beams. There can be 0 to 2 intersections. The intersections are represented by $\alpha_1$, $\alpha_2$ and $\beta_1$, $\beta_2$. If values are not set no intersection is found. {} means that a possible solution previously considered is discarded. The function min and max return the smallest or greatest value of their arguments.

To check if a beam starting at radius $r_1$ hits an obstacle represented by the line segment $P_1$ and $P_2$, first the distances between $Q$ and $P_1$ and $P_2$ are calculated. If the distance to one of the points is greater than $r_1$ and the line segment lies in the direction of propagation of the beam, it is hit. To find out if the beam propagates towards the line segment, we first calculate the angles between the x-axis and the vectors $\vec{QP_1}$ and $\vec{QP_2}$ as $\phi_1$ and $\phi_1$ respectively, which can also be interpreted as a beam from $Q$ to the line segment. We then compare the angles of the actual beam representing the wave and the beam to the line segment according to the flow chart in figure 4 to find out if the wave beam hits the line segment. There are three cases: (1) the line segment is not hit, (2) the beam hits the line segment and is completely reflected, (3) the beam hits the line segment and is partially reflected. In case (3) only the portion of the beam hitting the obstacle is reflected. Such a case us shown in figure 3. Only the beam between angles $\phi_1$ and $\phi_2$ is reflected. The beam is then split into two sectors. One sector in whose direction the wave continues to travel and one sector which terminates at radius $r$, which is the distance to line segment $P_1P_2$. The beam is then reflected at the obstacle as described above.

Note that a beam being partially reflected can also be split into three parts, if the line segment reflects only the center portion of the beam, leaving both edges intact. This is why the algorithm in the flow chart in figure 4 returns up to two angle pairs: $\alpha_1$, $\alpha_2$ and $\beta_1$, $\beta_2$.

After examining all beams, the newly created beams are examined and it is checked if they hit an obstacle. If so they are again reflected after splitting them into a reflected and a non-reflected part if necessary. This algorithm is repeated until the sound has travelled for a period longer than the time for which we want to calculate early reflections.

Every beam is thus represented by a circular sector and stands for a reflection of the sound. In order to synthesize an impulse response we also store the number of reflections $N_R$ the wave has undergone until it reaches the area described by the circular sector for each beam.

## 5. Calculation of early reflections

Based on the pre-calculated circular sectors, the early reflections at any point $L$ in the level can be calculated. For every circular sector, we check if the point $L$ lies between the angles $\phi_1$ and $\phi_2$ and if its distance from the center point of the circular vector is between $r_1$ and $r_2$. If all of these conditions are true, the reflection represented by the circular vector is added to the impulse response containing the early reflections.

A simple way to model reflections is by adding an impulse at time $r/c$, where $c$ is the speed of sound in air and $r$ the distance between the center point of the circular sector $P$ and $L$. To generate a multichannel output signal, panning can be applied to the impulse based on the angle between the orientation of the player and the direction of the virtual sound source. The amplitude of the impulse can be calculated in proportion to $r$, which is the distance travelled by the sound wave until it reaches $L$. The drop in sound pressure is proportional to $1/r$. The number of reflections should also be taken into account, since obstacles usually do not reflect the whole energy of the incoming wave. The loss of energy due to a reflection can be calculated based on the reflective property of the material of the geometric objects. In our implementation we use a reflection coefficient specifying how much of the energy is reflected.

Reflections also alter the frequency content of the reflected wave. Rough materials do not reflect the energy in the high frequency components of a signal as good as the low frequency components, leading to a low pass filtering effect. This effect can be modelled by placing low-pass filtered versions of an impulse into the frequency response instead of using impulses. Gauss-curves of a specific width can be used for this purpose.

Using a negative reflection coefficient for each reflection lets us calculate an impulse response whose amplitude is centered around zero, which is usually the case in recorded room impulse responses, as was also remarked by Lehmann and Johansson [11]. When placing the impulses or Gauss curves into the impulse response waveform, the sign on the placed impulse or curve is calculated as $(-1)^{N_R}$, where $N_R$ is the number of times the wave has been reflected until it reaches point $L$.

## 6. Impulse response based on early reflections

Calculating the impulse response representing the early reflections this way has a major disadvantage. When the listener moves from point $L$ to a point $L'$, the distances between the listener's position and the virtual sound sources change and so do the sample numbers at which the impulses or curves for each reflection are placed. When the player moves around slowly, the moving impulses in the impulse response inflict an easily noticeable comb filter like quality onto the signal. This artefact can be seen in figure 5, which shows the spectrogram of white noise to which impulse responses based on the different listener positions have been applied. The changing resonant frequencies can be seen clearly. This effect is much stronger than

IOP Conf. Series: Journal of Physics: Conf. Series **1075** (2018) 012041 doi:10.1088/1742-6596/1075/1/012041

in reality, because the number of reflections considered in our simulation is only a fraction of the number of reflections found in reality. This discrepancy is due to the neglecting of important effects such as diffraction and dispersion and the simulation in two dimensional space instead of three dimensional space.

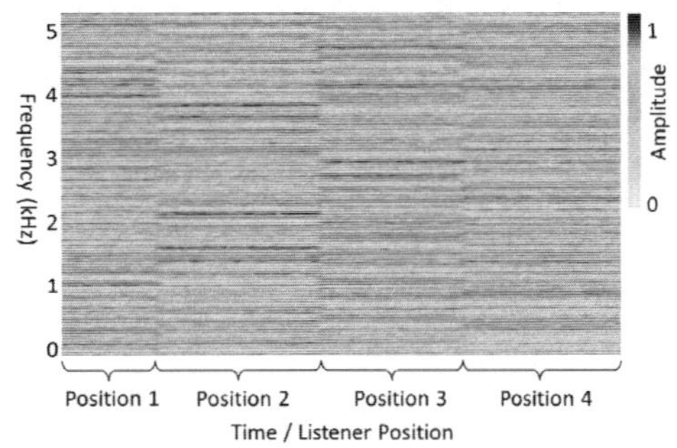


**Figure 5.** Impulse responses of early reflections calculated for 4 different positions have been calculated using only impulses to model the reflections. The resonance in thin frequency bands caused by these impulse responses is clearly visible on the white noise the impulse responses have been applied to.

To mitigate this effect, short bursts of white noise can be placed into the signal instead of using impulses or gauss curves. We use noise bursts whose length is proportional to the number of reflections the wave has undergone until it reaches point $L$. An envelope in the shape of a gauss curve is applied to the noise bursts. These noise bursts can be interpreted as dispersed versions of reflections. To simulate the low-pass filtering characteristics of the reflecting material, a low-pass filter whose characteristics depend on the number of reflections can be applied to the noise used in the bursts.

Whenever the position of the listener is updated, the impulse response needs to be recalculated. If the listener's position changes only slightly or gradually over time, which is often the case when a player walks from one point to another in a video game, the impulse response should also change gradually. To fulfill this condition also when using noise bursts for all reflections, the same noise samples need to be used for the same reflection every time. This can be achieved by reading noise from a noise buffer at a random position that is constant for each reflection or by providing a seed specific to each reflection to the random number generator before drawing the noise samples.

During the calculation of the circular sectors, the circle was divided into a certain number of sectors right from the beginning. When neighbouring beams hit the same obstacle, they are reflected in the same way, i.e. their underlying circles will have the same center point. Such neighbouring sectors need to be treated as the same reflection using the same noise samples, otherwise the player will hear a transition when moving from the area of one sector to the area of another sector. Two sectors $i$ and $j$ are considered as being neighbours when $P_i = P_j$, $N_{R_i} = N_{R_j}$ and either $\phi_{1_i} = \phi_{2_j}$ or $\phi_{2_i} = \phi_{1_j}$, where the subscripts mark the properties of the respective sectors.

## 7. Calculation of late reflections

The late reflections are estimated based on the early reflections and on parameters that can be controlled by the sound designer, since the desired amount of reverberation is often a question of game design rather than fidelity to physics.

Late reflections are modelled by white noise with an exponentially decaying amplitude. The amplitude of the noise is based on the amplitude of the later part of the impulse response of the early reflections. The mean amplitude is calculated as follows:

$$A = \frac{\sum_{n=N_h-k}^{N_h} |h[n]|}{k \cdot M} \tag{1}$$

IOP Conf. Series: Journal of Physics: Conf. Series **1075** (2018) 012041 doi:10.1088/1742-6596/1075/1/012041

where $h[n]$ is the impulse response containing the early reflections, $N_h$ the number of samples in $h[n]$, $k$ a constant that we define as $k = \lfloor N_h/4 \rceil$ and $M$, used to normalize the average, is the mean of the absolute samples of white noise with amplitude 1.

The exponential decay $\lambda$ can be calculated by calculating as

$$\lambda = -\frac{\log(A/A_0)}{3k} \tag{2}$$

where $A_0$ is the amplitude of the first $k$ samples of the impulse response, calculated in a similar way as the amplitude $A$ as

$$A_0 = \frac{\sum_{n=0}^{k} |h[n]|}{k \cdot M}. \tag{3}$$

The decay value can then be used to calculate the amplitude at sample $t$ as

$$A[t] = A_0 * e^{-\lambda * t}. \tag{4}$$

In our implementation the decay is often set to a custom value by a sound designer overriding the calculated value in favour of sound design aspects over fidelity to physical relations.

The samples of the white noise used to create the late reflections are the same each time the impulse response of the late reflections is recreated. Also here the reason is that this helps us to create a consistent sound that won't change unnaturally whenever the late reflections are re-estimated. A low pass filter can be applied to the noise being used to account for the fact that higher frequency components are often dampened in reflected sounds waves.

The decaying noise is added to $h$ to yield a complete impulse response, representing the direct sound, the early reflections and the late reflections.

## 8. Convolution and implementation details

The reverberation represented by the impulse response $h$ is applied to the sound signal of the sound source by convolving the sound with the impulse response $h$. Executing the convolution in the frequency domain is very efficient. If the convolution and the FFT necessary to transform the signal to the frequency domain are implemented on a GPU, considerable gains in speed can be expected.

Convolving part of a signal of $M$ samples length with an impulse response of $N$ samples length will yield a convolution product of $L = M + N - 1$ samples length and hence requires a Fourier transform of at least $L$ samples length to accommodate the whole product. Since typical impulse responses for large spaces are a few seconds long, this would mean that blocks of $L$ samples, where $L$ is a number corresponding to a few seconds, need to be processed. This would also increase the latency of the system to a few seconds. One alternative is to fill buffers of $L$ samples with only $M$ samples of the signal and transform them to the frequency domain for convolution, keeping the latency caused by the convolution as low as $M$ samples, but increasing the computational cost significantly by using FFTs with $L$ coefficients while only processing $M$ samples.

This problem can be solved by using partitioned fast convolution [16], in which the impulse response is cut into short frames of $M$ samples length and each frame is transformed into the frequency domain with a $2M$ order FFT. These frames are stored in the array $H[t]$, where $t = 0..T-1$ is the index of the frame and $T = \lceil L/M \rceil$. To process a frame of $M$ samples of the input signal at frame number $n$, it is also transformed to the frequency domain with a $2M$ order FFT, yielding frame $S[n]$. The output frame at frame number $n$ is calculated as

$$X[n] = \sum_{t=0}^{T-1} H[t] \cdot S[n-t], \tag{5}$$

where the dot stands for the sample-wise complex multiplication of two frames. To implement this formula, $T-1$ frames of the past signal have to be kept in memory but for every frame of $M$ samples, only one FFT and one iFFT of order $2M$ are necessary.

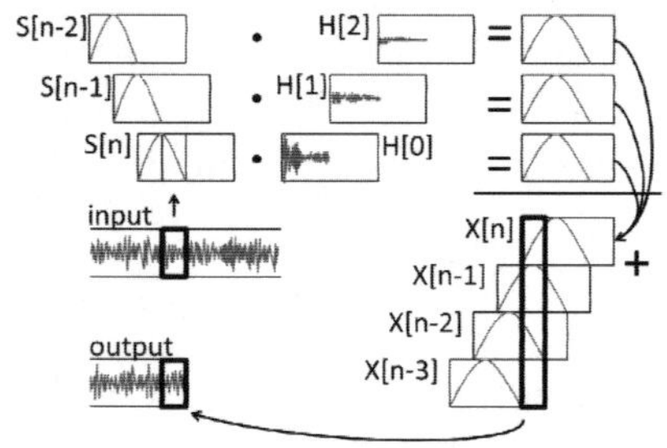


**Figure 6.** Summary of the partitioned convolution algorithm with overlapping input and output frames.

When the impulse response $h$ is updated frequently, as is the case when the player moves through the level, the timings where the early reflections are placed in the impulse response change frequently. This can cause artefacts in the output because succeeding frames of the output signal are convolved with different impulse responses. An easy solution to this problem is to process overlapping buffers to which an appropriate windowing function is applied before transforming them into the frequency domain. We use a periodic Hann window and a 50 % overlap. The relation between the frame number $n$ and its first sample $i$ in the time domain is $i = n * M/2$, so the frame $S[n]$ is constructed by taking $M$ samples starting from sample $i$, then multiplying it with the Hann window, padding it with zeros to $2M$ samples and transforming it to the frequency domain. Note that the output $X[n]$ of the convolution needs to be overlapped with the same overlap, which means that the relation between frame number and sample number in the time domain is the same. Figure 6 summarizes the convolution algorithm graphically.

To generate multichannel output the impulse response must be constructed as a multi channel signal. Each channel of the impulse response is then individually applied to the mono input signal.

## 9. Summary and outlook

The proposed method is a fast way to implement reverberation for virtual environments. Even though diffraction and dispersion are neglected during pre-computation of the early reflections, stochastic enhancement of the calculated impulse response lets us create convincingly sounding acoustic effects. The method works well for environments that can be easily converted to 2d maps but needs to be extended for environments with a more complicated geometry. Further optimization of the image source pre-calculations would enable us to also implement this part of the algorithm in real-time and thus allow us to calculate the sound for moving sound sources.

## References

[1] Lauri Savioja and U Peter Svensson, "Overview of geometrical room acoustic modeling techniques," *The Journal of the Acoustical Society of America*, vol. 138, no. 2, pp. 708–730, 2015.

[2] Nikunj Raghuvanshi and John Snyder, "Parametric wave field coding for precomputed sound propagation," *ACM Transactions on Graphics (TOG)*, vol. 33, no. 4, pp. 38, 2014.

[3] Carl Schissler and Dinesh Manocha, "Interactive sound propagation and rendering for large multi-source scenes," *ACM Transactions on Graphics (TOG)*, vol. 36, no. 1, pp. 2, 2016.

[4] Dirk Schröder, Philipp Dross, and Michael Vorländer, "A fast reverberation estimator for virtual environments," in *Audio Engineering Society Conference: 30th International Conference: Intelligent Audio Environments.* Audio Engineering Society, 2007.

[5] Fabio Antonacci, Marco Foco, Augusto Sarti, and Stefano Tubaro, "Fast tracing of acoustic beams and paths through visibility lookup," *IEEE Transactions on Audio, Speech, and Language Processing*, vol. 16, no. 4, pp. 812–824, 2008.

[6] Eva-Marie Nosal, Murray Hodgson, and Ian Ashdown, "Improved algorithms and methods for room sound-field prediction by acoustical radiosity in arbitrary polyhedral rooms," *The Journal of the Acoustical Society of America*, vol. 116, no. 2, pp. 970–980, 2004.
[7] Lakulish Antani, Anish Chandak, Micah Taylor, and Dinesh Manocha, "Direct-to-indirect acoustic radiance transfer," *IEEE Transactions on Visualization and Computer Graphics*, vol. 18, no. 2, pp. 261–269, 2012.
[8] Jont B Allen and David A Berkley, "Image method for efficiently simulating small-room acoustics," *The Journal of the Acoustical Society of America*, vol. 65, no. 4, pp. 943–950, 1979.
[9] Jeffrey Borish, "Extension of the image model to arbitrary polyhedra," *The Journal of the Acoustical Society of America*, vol. 75, no. 6, pp. 1827–1836, 1984.
[10] FP Mechel, "Improved mirror source method in roomacoustics," *Journal of sound and vibration*, vol. 256, no. 5, pp. 873–940, 2002.
[11] Eric A Lehmann and Anders M Johansson, "Prediction of energy decay in room impulse responses simulated with an image-source model," *The Journal of the Acoustical Society of America*, vol. 124, no. 1, pp. 269–277, 2008.
[12] Hansung Kim, Richard J Hughes, Luca Remaggi, Philip JB Jackson, Adrian Hilton, Trevor J Cox, and Ben Shirley, "Acoustic room modelling using a spherical camera for reverberant spatial audio objects," in *Audio Engineering Society Convention 142*. Audio Engineering Society, 2017.
[13] Uwe M Stephenson, "Quantized pyramidal beam tracing-a new algorithm for room acoustics and noise immission prognosis," *Acta Acustica united with Acustica*, vol. 82, no. 3, pp. 517–525, 1996.
[14] Thomas Funkhouser, Ingrid Carlbom, Gary Elko, Gopal Pingali, Mohan Sondhi, and Jim West, "A beam tracing approach to acoustic modeling for interactive virtual environments," in *Proceedings of the 25th annual conference on Computer graphics and interactive techniques*. ACM, 1998, pp. 21–32.
[15] Vesa Valimaki, Julian D Parker, Lauri Savioja, Julius O Smith, and Jonathan S Abel, "Fifty years of artificial reverberation," *IEEE Transactions on Audio, Speech, and Language Processing*, vol. 20, no. 5, pp. 1421–1448, 2012.
[16] Barry D Kulp, "Digital equalization using fourier transform techniques," in *Audio Engineering Society Convention 85*. Audio Engineering Society, 1988.

IOP Conf. Series: Journal of Physics: Conf. Series **1075** (2018) 012042 doi:10.1088/1742-6596/1075/1/012042

# Measurement of Acoustic Performance of an Auditorium

**C. M. F. Pangau, M. S. Prawirasasra, Abrar Ismardi**
Telkom University, Indonesia
E-mail: bibinprawirasasra@telkomuniversity.ac.id

**Abstract.** Objective of this paper is to obtain speech-related acoustic parameters in auditorium. The object of this research is a room that has T-shape geometry with amphitheater seating. Evaluation was carried out by measurement in empty room with two different sound source position. The following impulse response objective parameters were analyzed: Early Decay Time, Reverberation Time and Definition. Convergence of each configurations was analyzed through standard deviation and just noticeable difference (JND). Additionally, noise criteria considered as evaluation since there are active ACs inside the object.

## 1. Introduction

Currently, most of academic activities are held at enclosed space. Every physical boundary contributes to draw phenomenon of reflection, absorption, diffusion and transmission whenever sound wave interacts with it. The peculiar behavior of sound field might be perceived as pleasant or irritating sound. Acoustic performance can be analyzed from impulse response parameters such as Reverberation Time ($T_{30}$), Early Decay Time (EDT) and Definition ($D_{50}$). Those parameters are dependent to room geometry, position of sound source and receivers, sound level of the source and absorption [1]. In addition, background noise contributes to sound masking hence it needs to be rated as parameters that represent noise quality inside room in noise criteria (NC).

Measurement procedure complies to ISO 3382-1:2009 with previously stated parameters are analyzed. Since parameters potentially have spatial variation across floor plan [3],[4], arrival sound might be perceived differently. The analysis of data variation is through least difference value that human ear could detect. It is expressed in Just Noticeable Difference (JND). According to ISO:3382-1:2009, consideration of JND is only at 500 Hz and 1000 Hz. JND value is presented at Table 1. JND analysis leads to identify impact of sound source position to perceived sound.

**Table 1.** Acoustic Parameters and JND value

| Acoustic Parameter | JND | Typical Range |
|---|---|---|
| EDT | rel. 5 % | 1.0 s;3.0 s |
| $D_{50}$ | 5 % | 30 %; 70% |
| $T_{30}$ | 5 % | |

## 2. Description of an object

The object of this research is one of academic facilities which is use for speech purposes. It has unique geometry featured, the floor plan is T-shaped with stage location is in intersection between line of word T. However, room is designed to accommodate clear and comfort view for audience from every section of seating area to the stage. So, seat direction is set into stage and ground floor for seating area is inclined. Illustration of floor plan and view from last row to stage is at Figure 1 and Figure 2, respectively. Contrarily to inclined seating, the distance between floor and ceiling does not have constant distance. Furthermore, air temperature is controlled by ACs that installed at several strategic spot on ceiling.

Total area of this room is about 135.68 $m^2$ and height dimension from lowest floor to ceiling is around 3.16 m. Seating capacity of this room is around 95 persons. Each seating was made of fabric and no surfaces of floor is exposed since it is entirely covered by carpet (Figure 3). Differently, there are some section of wall surfaces is exposed while the rest is already covered by fabric. Location of uncovered part is around stage area.

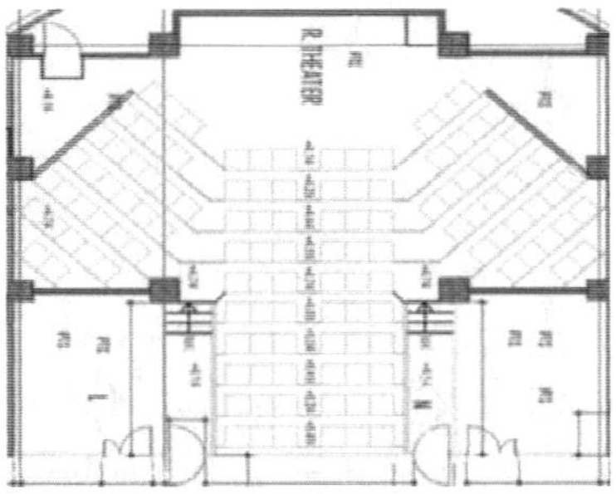


**Figure 1.** Floor plan

**Figure 2.** Inclined seating

**Figure 3.** Material surfaces

## 3. Acoustic Parameters

Considering purpose of the object is for speech activity, impulse response parameters are chosen so it may illustrate room performance relate with it. In this research, those parameters are Early Decay Time (EDT), Reverberation Time ($T_{30}$) and Definition ($D_{50}$). Noise Criteria is assessed too since ACs may cause intolerable noise for specific activities inside closed space.

Early Decay Time (EDT) is defined as time necessary for the sound level to decrease by first 10 dB. EDT has chosen because it shows better correlation to geometry shape and strongly influenced by early reflection rather than reverberation time [1]. Reverberation time ($T_{30}$) is required time for room to decrease the energy by 30 dB after source is turned off.

Speech intelligibility is represented by definition $D_{50}$. Theoretically, it defines as ratio of useful energy within first 50 ms to total energy. It is expected that arrival energy within first 50 ms is having large energy. Mathematical expression to calculate $D_{50}$ as follow

$$D_{50} = \frac{\int_0^{0.05} h^2(t)dt}{\int_0^{\infty} h^2(t)dt} \qquad (1)$$

The satisfaction condition for speech inteligibillity and general living environments is depend on energy level of background noise, hence, it needs to be rated. Noise criteria or NC is a single number or rating used to define the maximum allowable background noise in enclosed space. t is also used to establish satisfaction conditions for speech intelligibility and general living environments [5].

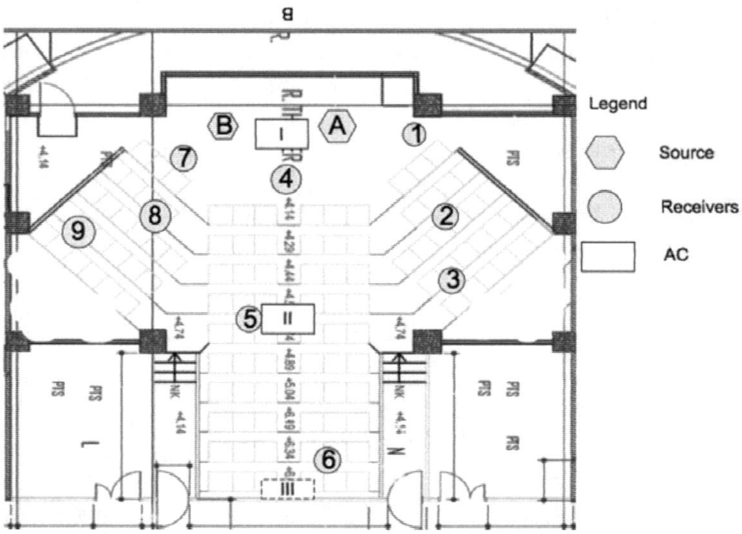


**Figure 4.** Receivers Position

## 4. Measurement method

Acoustic parameters can be carried out through measurement with impulse response technique that comply to ISO 3382-1:2009. Acoustic parameters are taken to analyze decaying sound in audience area. Measurement points are used to represent it. In this research, every section is divided into 3 measurement points that spread symmetrically from rear to front seat. Measurement is conducted while room in unoccupied condition with one active AC (number II). Location of AC is at Figure 4.

The sound source is dodecahedron that located in front area of auditorium. Signal MLS is emitted to source then sensors capture the decaying signal as room impulse response (RIR). Measurement is conducted with two different excitation positions of sound source. Position of sound source and receivers is plot at Figure 4. Also, data is taken about 5 times at each receiver position for every position of source. The purpose of this configuration is to find the measurement accuracy.

## 5. Experimental Result and Analysis

### *5.1. Noise Criteria Analysis*

According to experiment and calculation, NC has range from 30-33. This paper presents only the highest level of background noise from each section. It is showed by point 1, 4 and 7 with value of NC value is about 33 for all points.

### *5.2. Acoustic Parameters Analysis*

For impulse response parameters, data is presented in two type: (1) standard deviation (STDEV) and (2) Average (Ave.). Both analysis are tabulated on two frequency (500 Hz and 1000 Hz) for all measurement positions. JND is analyzed by two way. First is by comparing deviation in particular measurement point and second to find the least difference between measurement points. Analyzed frequency is only at 500 Hz and 1000 Hz.

The result of repeated measurements of acoustic parameters is presented at Table 2, Table 3, and Table 4 for EDT, $D_{50}$ and $T_{30}$, respectively. Standard deviation for EDT (Table 2) is vary for source position at A. Some point has standard deviation less than JND and the other is large than 2 JND for both analysis frequencies. It occurs at point 7, 8 and 9. However, source position at B have different pattern to position at A. Standard deviation value for each receivers point is less than 1 JND. Moreover, JND of average value at both frequencies between receiver positions is more than 1 JND.

**Table 2.** Standard deviation and average of EDT (s)

| | Receivers position (Sound source position: A) | | | | | | | | |
|---|---|---|---|---|---|---|---|---|---|
| f (Hz) | 1 | 2 | 3 | 4 | 5 | 6 | 7 | 8 | 9 |
| 500 (STDEV) | 0.00 | 0.00 | 0.01 | 0.01 | 0.02 | 0.00 | 0.46 | 4.92 | 1.62 |
| 500 (Ave.) | 0.37 | 0.43 | 0.45 | 0.45 | 0.42 | 0.36 | 0.53 | 2.54 | 1.17 |
| 1000 (STDEV) | 0.00 | 0.00 | 0.01 | 0.01 | 0.03 | 0.00 | 0.96 | 4.14 | 0.10 |
| 1000 (Ave.) | 0.39 | 0.51 | 0.57 | 0.35 | 0.50 | 0.58 | 0.89 | 2.27 | 0.50 |
| | Receivers position (Sound source position: B) | | | | | | | | |
| f (Hz) | 1 | 2 | 3 | 4 | 5 | 6 | 7 | 8 | 9 |
| 500 (STDEV) | 0.02 | 0.00 | 0.01 | 0.01 | 0.02 | 0.11 | 0.01 | 0.02 | 0.00 |
| 500 (Ave.) | 0.44 | 0.38 | 0.40 | 0.37 | 0.43 | 0.58 | 0.27 | 0.34 | 0.51 |
| 1000 (STDEV) | 0.02 | 0.00 | 0.01 | 0.01 | 0.00 | 0.01 | 0.01 | 0.01 | 0.00 |
| 1000 (Ave.) | 0.44 | 0.43 | 0.39 | 0.40 | 0.65 | 0.50 | 0.35 | 0.45 | 0.37 |

$D_{50}$ has quite similar pattern with EDT. Point 7, 8 and 9 has standard deviation larger than 1 JND for source position at A. Contrarily, sound source position at B has no impact to repeated measurement since it has standard deviation less than 1 JND. However, the different between the range of average value (maximum value - minimum value) among receivers point can be divided into more than 3 JND for both frequencies.

**Table 3.** Standard deviation and average of $D_{50}$ (%)

| | Receivers position (Sound source position: A) | | | | | | | | |
|---|---|---|---|---|---|---|---|---|---|
| f (Hz) | 1 | 2 | 3 | 4 | 5 | 6 | 7 | 8 | 9 |
| 500 (STDEV) | 0.59 | 0.29 | 0.75 | 0.96 | 1.02 | 0.71 | 5.69 | 10.30 | 9.50 |
| 500 (Ave.) | 89.02 | 79.60 | 84.01 | 82.61 | 85.13 | 85.11 | 83.93 | 85.30 | 76.88 |
| 1000 (STDEV) | 0.27 | 0.25 | 1.62 | 0.40 | 1.04 | 0.89 | 10.77 | 11.37 | 3.87 |
| 1000 (Ave.) | 85.42 | 81.28 | 77.85 | 87.72 | 82.16 | 72.50 | 78.20 | 76.48 | 75.73 |
| | Receivers position (Sound source position: B) | | | | | | | | |
| f (Hz) | 1 | 2 | 3 | 4 | 5 | 6 | 7 | 8 | 9 |
| 500 (STDEV) | 1.66 | 0.43 | 0.52 | 1.53 | 6.68 | 2.57 | 0.90 | 1.23 | 1.19 |
| 500 (Ave.) | 79.58 | 85.83 | 83.87 | 91.38 | 74.75 | 75.71 | 91.56 | 86.44 | 78.47 |
| 1000 (STDEV) | 1.08 | 0.17 | 0.39 | 0.59 | 3.32 | 3.50 | 0.88 | 0.67 | 0.26 |
| 1000 (Ave.) | 81.35 | 82.96 | 84.04 | 86.62 | 74.96 | 70.14 | 87.55 | 81.73 | 84.21 |

The range of standard deviation of $T_{30}$ (Table 4) is quite different than previous parameters. JND for both configuration (position A & B) has vary value from less than 1 JND to large than 1 JND. By comparing average value at particular point to the others, the difference is also large than 1 JND.

**Table 4.** Standard deviation and average of $T_{30}$ (s)

| Receivers position (Sound source position: A) | | | | | | | | | |
|---|---|---|---|---|---|---|---|---|---|
| f (Hz) | 1 | 2 | 3 | 4 | 5 | 6 | 7 | 8 | 9 |
| 500 (STDEV) | 0.48 | 0.10 | 0.14 | 0.29 | 0.83 | 0.83 | 1.55 | 0.26 | 1.41 |
| 500 (Ave.) | 1.10 | 1.26 | 1.37 | 1.28 | 1.48 | 1.62 | 1.90 | 1.38 | 1.94 |
| 1000 (STDEV) | 0.01 | 0.00 | 0.41 | 0.66 | 1.06 | 0.27 | 0.19 | 0.44 | 3.05 |
| 1000 (Ave.) | 0.43 | 0.43 | 0.91 | 0.93 | 1.27 | 0.91 | 0.74 | 0.69 | 2.07 |
| All (Ave.) | 1.24 | 1.29 | 1.29 | 1.30 | 1.44 | 1.41 | 1.45 | 1.23 | 1.58 |
| Receivers position (Sound source position: B) | | | | | | | | | |
| f (Hz) | 1 | 2 | 3 | 4 | 5 | 6 | 7 | 8 | 9 |
| 500 (STDEV) | 0.58 | 0.10 | 0.37 | 1.03 | 0.56 | 0.55 | 0.79 | 0.56 | 1.22 |
| 500 (Ave.) | 2.06 | 1.00 | 0.85 | 1.41 | 1.28 | 1.60 | 1.35 | 1.32 | 1.22 |
| 1000 (STDEV) | 0.46 | 0.00 | 0.04 | 0.90 | 0.62 | 0.72 | 0.99 | 0.36 | 0.52 |
| 1000 (Ave.) | 1.76 | 0.46 | 0.49 | 0.90 | 0.90 | 1.54 | 1.12 | 0.71 | 0.52 |
| All (Ave.) | 1.75 | 1.24 | 1.13 | 1.34 | 1.44 | 1.72 | 1.36 | 1.27 | 1.31 |

## 6. Conclusion

The paper presents acoustical performance of T-shaped auditorium by using two position of sound source. The result indicates that the accuracy of EDT and $D_{50}$ measurement depend on sound source position since there are variation of standard deviation value from less to larger than 1 JND. But $T_{30}$ is independent to sound source position since the standard deviation is larger than JND. Comparison of different average value between receivers point to JND value shows that the impression of arrival sound is perceived differently for most audience area.

Energy of background noise reaches the peak at area around noise source (AC). It is shown by three receivers position at point 1,4 and 7 which is represent maximum background noise for each section. As it is getting further from noise source, background noise level is decaying.

## References

[1] Topa, M. D., Toma, N., Kirei, B. S., Srcu, I., & Farina, A. 2012. Experimental acoustic evaluation of an auditorium. *Advances in Acoustics and Vibration.*

[2] Stojanovi, V., Milivojevi, Z. 2014. The Evaluation of Certain Acoustic Parameters of the Orthodox Church. *International Journal of Innovative Research in Advanced Engineering (IJIRAE)*, **1** (10).

[3] Ettore, C.,Francesco, M. 2007 On the spatial variation of acoustical parameters in churches. *Int.Congress on Acoustics* (Madrid)

[4] Ruvalcaba Esquivel, D. 2005. Variability of acoustic parameters over audience areas in auditoria.

[5] Kanev, N. G., Livshits, A. Y.,Mller, H. 2013. Acoustics of the Great Hall of the Moscow State Conservatory after reconstruction in 20102011. *Acoustical Physics*, **59** (3), 361-368.

[6] Mller, H., Vehvilinen, S., Tishko, D., Wulfrank, T., Rozanov, S. I. 2010. Acoustic description of the Great hall of the Moscow PI Tchaikovsky Conservatory. *Proc. of the 20th International Congress on Acoustics* (Sydney)

[7] Patania, F., Gagliano, A., Nocera, F., Cicero, A. 2014 Evaluation of the acoustic performance of a theatrical space set up in a restored Latomia in Ragusa Iblea *Internoise*, (Melbourne).

# Identification disorders of vocal cords through the public communications VoIP network-based IVR with dysphonia severity index (DSI)

A B Ariyanto[1], F Y Supratman[1] and H Bethaningtyas[1]
[1]Telkom University, 40257 Bandung, Indonesia

E-mail: ahmad.99bagus@gmail.com

**Abstract.** The sounds that have been recorded by the system telephone networks Voice over Internet Protocol (VoIP) will be done the noise reduction using the filter process FastICA and Biorthogonal Wavelet. Voice recording process done with a vowel/a/ in one breath for 10 seconds. The result of the process reduction noise sound will be analyzed using the method of Dysphonia Severity Index (DSI) and Harmonic Noise to Ratio (HNR). The DSI method and HNR is used to determine data quality sound healthy and disorder by looking at the value of the both parameter.

## 1. Introduction

Disorders of the vocal cords causing trouble in talking. The disruption on the phonation organ caused by human vocal cords nodules that are experiencing swelling, irritation or infection [1]. So it takes a Medical Check-Up as early as possible to find out the various possibilities of disease that occur in the vocal cords [2]. Medical check up the vocal cords by doctors in general is using minimally invasive methods, that is to place the laringoskopi and stroboskopi into the mouth of the patient, this would make patients uncomfortable. Limitations equipment in the hospital became the primary factor for patients not to do a check-up. On the other hand, check-up the vocal cords to the hospital requires more time and cost. Based on a lot of problems that arose in the detection of vocal cords through a minimally invasive method, needed another alternative to help the medical world in detecting disease at the vocal cords, one of which is a non-invasive method. This method is done by searching for the value of earmarks in the sound produced by the patient to be a benchmark in determining disease experienced. The technique on the a non-invasive method is to ask the patient to pronounce the nodules of vocal/a/ continuously in one breath for a few seconds [3].

## 2. Methods and Methodology

### *2.1. VoIP*

Voice over Internet Protocol (VoIP) is a technology that is able to transmit the voice data in realtime with the network Internet Protocol (IP) [4]. VoIP technology works by changing the sound analog signal into a digital signal that can be transmitted over a network that utilizes IP. Once converted to digital signals, the noise will be translation into IP packets are then transmitted over a network. VoIP is used in an organization with the aim to reduce the cost of communication and makes it easy to communicate with. In the scale of LAN, VoIP is used to connect one room to another room, then a Metropolitan Area

Network (MAN), VoIP is used to establish lines of communication between the location of one with another location, while a Wide Area Network (WAN), VoIP is typically used to create lines of communication between countries or even continents [5].

### *2.2. IVR*

Interactive Voice Response (IVR) is a computer system that interact with the caller, where the callers will be give the input to the system by pressing the keypad on the phone (DTMF dualtone multifrequency) or by saying a something (a natural language to be known) [6]. The basic principle of IVR system is how the caller read menu and choose an option from the menu to perform an action, or as an alternative to entering information (in the format number via the keypad emphasis).

### *2.3. FastICA*

FastICA is one of the ICA methods. This is allows for a step whitening prelimenary for mixed signal zero mean, to increase the speed of convergence of the ICA. Reduction of noise by using FastICA method will give you good results and fast in terms of computation in comparison with other algorithms in a way doing the extraction processused in Independent Component Analysis [7].

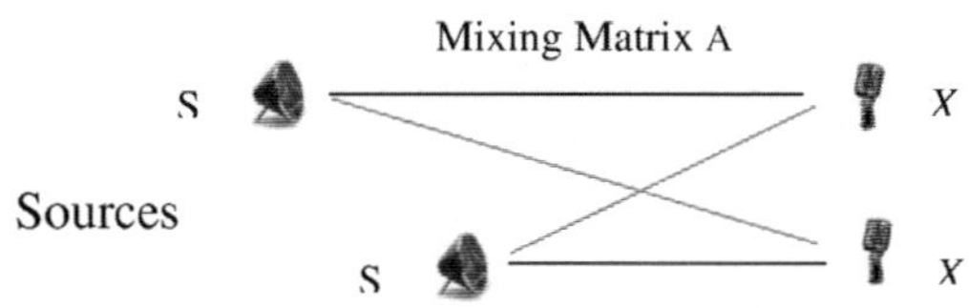


**Figure 1.** Mixing of two sources.

Figure 1 shows the mixing of two sources of sound, *S* is assumed as the source of sound, distance or intensity of the source Matrix mixer is A mixed signal as X, and t is the time. Thus, can be written equations 1 and 2.

$$x1(t) = a_{11}s_1 + a_{12}s_2 \quad (1)$$
$$x2(t) = a_{21}s_1 + a_{22}s_2 \quad (2)$$

In equations 1 and 2, no longer use time dimension of *t* in the equation, because ICA components at each observation that presume that they one another ( x1 and x2) are components that are free, as well as the components of the sn. So the equation can be simplified into the equation 3.

$$xj = a_{j1}s_1 + a_{j2}s_2 + \cdots + a_{jn}s_n \quad (3)$$

ICA uses the following equations:

$$\mathrm{X} = \mathrm{AS} \quad (4)$$

Where S is the signal source, parameter A matrix describing the mechanism of incorporation of signal source and signal mixture from the merger describes by *x*. The basic principle of the ICA is to find the source of the original signal *S,* and assuming the original signal mutually independent of each other tatistically. FastICA iteration scheme is based on the calculation of fixed-point to find the maximum value of WTx nongaussianty [7]. Steps to find a variable value FastICA is as follows:

- Select the value vector of *w*, can be taken random.
- Find $w+ = E\{xg\{WTx\}\} - E\{g'\{WTx\}\}W$
- Then locate the variables $w = \frac{w^+}{\|w^+\|}$

  with $W$ = invers matrix $x$, $E$ = eigen value, and $g$ = approximations of negentropy

IOP Conf. Series: Journal of Physics: Conf. Series **1075** (2018) 012043 doi:10.1088/1742-6596/1075/1/012043

### 2.4. Wavelet Biorthogonal

Wavelet transform is a function generated from one basic function translation and dilatation. Wavelet is used as a tool of analysis of data from a filter, to reduce pollutant factors (noise). Wavelet have many types based on the application, one of which is the Biorthogonal Wavelet [9]. It is Orthogonal Wavelet expansion, which allows for more degrees of freedom. With the addition of the degrees freedom to allow for compose a more symmetric wavelet functions [16]. In the orthogonal wavelet filter analysis and synthesis filter is the inverse of time with one another (for example, h (n) = h (n), g (n) = g (n)). When h = hˇ then brings about the relationship between the coeffetien scale and coeffetien wavelet for wavelet orthogonal [10].

## 3. Vocal Parameter Acoustic

### *3.1. Dysphonia Severity Index*

Dysphonia Severity Index (DSI) was designed as an objective correlation and quantitative of sound quality [11]. It is based on the results of a test run by a Belgium study group analyzing multivariate normal and sound database of pain. Parameters based on the combination of the DSI influence the measurement of the sound basis of the formant frequencies, namely vowel sounds (F1) then low intensity level (in dB), the maximum time the blast (MPT in units of seconds) and Jitter (%)18. DSI is used to analyze the patient's sound quality for subsequent use as parameters in the determination of the value of votes [12].

$$DSI = 0.13MPT + 0.0053F1 - 0.26I - 1.18Jitter + 12.4 \quad (5)$$

where

MPT = Aerodynamic Measurements of Variables (s)
F1 = Formant Frequency Maximum Base (Hz)
L = Sound Intensity Level (dB)
Jitter = Time difference from the Two Periods (%)

### *3.2. HNR*

Harmonic to Noise Ratio (HNR) is the acoustic parameters measured as the overall characteristics of the signal, and not as a function of the frequency [13]. The overall value of the signal HNR vary due to the different vocal tract resulting in a different amplitude. The greater the value of the resulting HNR then semkain harmonious voice signal. HNR equation shown in equation (6).

$$HNR = 10 * log_{10} \frac{AC_v(T)}{AC_v(0) - AC_v(T)} \quad (6)$$

ACv (0) = Coeffetien initial autocorrelation
ACv (T) = Coeffetien autocorrelation in accordance with fundamental period

## 4. Result and Discussion

Voice data source used there are two kinds, primary data and secondary data. The primary voice data is the sound that recorded with the procedures directly by patients. While secondary data is the sound that is already available in the database. This secondary data obtained from *Computational Physics Laboratory Databases, Telkom University and International Saarbrucken Voice Database (SVD)* [14]. The voice data will be recorded through the telephone network to find out the influence of noise caused by communication network. Voice data used in testing as many as 175 final research data, consisting of 115 and 60 healthy data pain.

### *4.1. The Difference in Healthy and Diseased Sound Characteristics*

The sound quality has healthy enough significant differences compared to the sound of misbehaving on the vocal cords. Based on previous research and reference the test results, the value of the parameter of the DSI and the sound of disorders HNR has lower values compared to sound healthy. This is due to the sound of disorders harmony voice signal has a value which is less good compared to sound healthy. Here are the differences parameter DSI and HNR on sound healthy and disorders.

IOP Conf. Series: Journal of Physics: Conf. Series **1075** (2018) 012043 doi:10.1088/1742-6596/1075/1/012043

**Table 1.** Differences of sound the sound of disorders and healthy

| Parameters | Healthy | | Sick | |
|---|---|---|---|---|
| | DSI | HNR | DSI | HNR |
| Average | 1.169 | 15.804 | 0.139 | 13.789 |
| The Maximum Value | 3.263 | 21.289 | 2.617 | 16.108 |
| The Minimum Value | -0.996 | 10.484 | -4.267 | 11.980 |

Table 1 shows the difference between DSI and HNR sound healthy on the recordings directly. At the sound of the DSI value healthy average of 1.169 HNR value and an average of 15.804. While the sound of disorders DSI value of 0139 and HNR value average of 13,789.

*4.2. The Distinction of Direct Recording and Telephone Network*

Sound recording through the telephone network are experiencing a decrease in sound quality due to the presence of noise which causes the addition of foreign signals that may damage the information contained in the data is sound. So, will experience the error detection on sound data.

**Table 2.** The difference value healthy of DSI and HNR on real-time recording and through the phone

| Parameters | Directly | | Phone | | | |
|---|---|---|---|---|---|---|
| | | | Before Reduction | | After Reduction | |
| | DSI | HNR | DSI | HNR | DSI | HNR |
| Average | 10.253 | 19.001 | 3.935 | 8.522 | 6.707 | 9.513 |
| The Maximum Value | 16.595 | 27.443 | 8.278 | 16.494 | 12.248 | 18.410 |
| The Minimum Value | 5.442 | 10.526 | -4.405 | 0.421 | -0.747 | 3.524 |

In table 2 indicates that the sound after recording through the telephone network will result in a decreased in the value. This proves that by doing the recording via the telephone network will provide the noise which caused the value of DSI declined by 61% and HNR of 55.15%. However, after a process of separation noise the sound data back to increase. On the parameters of the DSI increase by 41.3% and the HNR of 10.42%.

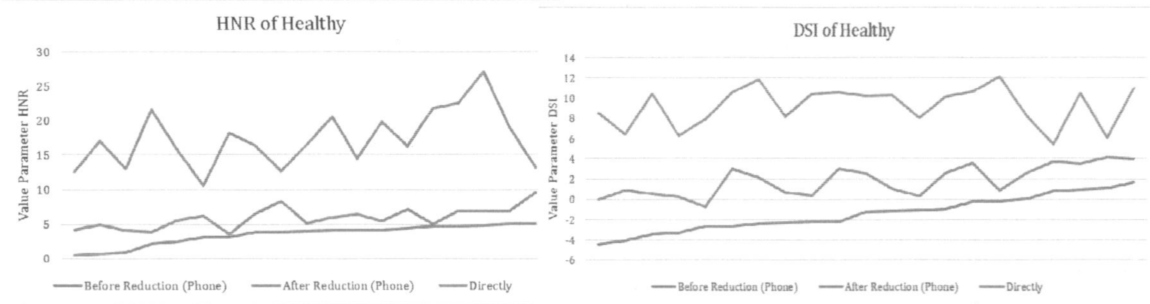


(a) (b)

**Figure 2.** Graph Difference HNR and DSI on sound healthy. (a) the HNR. (b) the DSI

**Table 3.** Difference value of DSI and HNR on real-time recording and telephone at the sound of disorders

| Parameters | Directly | | Phone | | | |
|---|---|---|---|---|---|---|
| | | | Before Reduction | | After Reduction | |
| | DSI | HNR | DSI | HNR | DSI | HNR |
| Average | 7.953 | 13.638 | 4.137 | 2.820 | 6.220 | 4.147 |
| The Maximum Value | 12.272 | 16.425 | 6.387 | 0.322 | 9.123 | 8.250 |
| The Minimum Value | -1.136 | 10.278 | -5.018 | 9.531 | -0.588 | 1.583 |

Table 3 shows the difference between the DSI and HNR the sound of disorders. In general the same as healthy sound, there was a decrease after recording is done through the telephone network. The decline on DSI of 47.98%, whereas in HNR amounting to 79.32%. Then the value of the parameter of the DSI and HNR back rise, after reduction of noise. The parameters of the DSI went up by 33.49% and on the parameters of the HNR of 32%. Thus proving that with the reduction of noise data the sound back to the actual information based on recording directly.

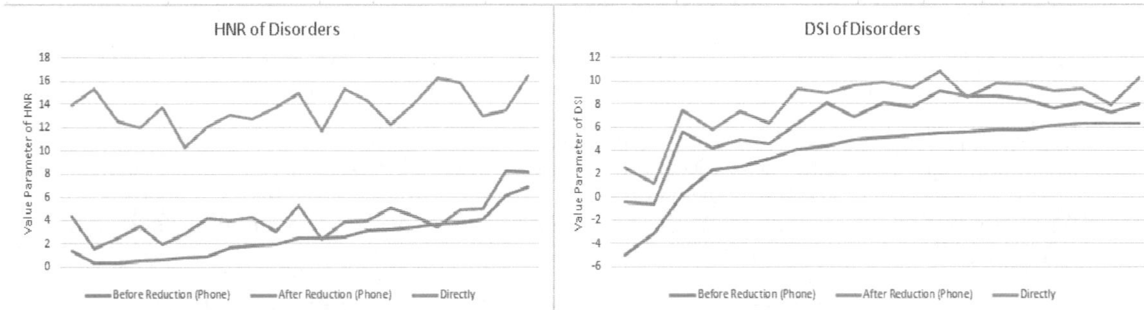


(a) (b)

**Figure 3.** Graph difference of HNR and DSI on sound of disorders. (a) the HNR. (b) the DSI

*4.3. Analysis Of The Results Of Detection System*

The addition of Biorthoghonal wavelet filters on the system led to increased sound quality approaching the data at the recording directly. In previous research, has done testing the system noise reduction using FastICA algorithm, difference Table addition of wavelet filter can be seen in table 4 below [15], [16].

**Table 4** Wavelet Filter Addition Difference

| FastICA | | FastICA+Wavelet | |
|---|---|---|---|
| **HNR** | **DSI** | **HNR** | **DSI** |
| 5.114 | 4.728 | 6.285 | 7.690 |

Based on table 4 above, there is an increase in the quality of the existence DSI and HNR addition of Biorthoghonal Wavelet filters. Thus, the data will be more sound approach to data recording with sound directly which is the original data without any noise due to communication network.

## 5. Conclusion

Calculation analysis of the data on value DSI and HNR from 175 data through the phone networks recording obtained difference values between the healthy and disorders voice data for system identification. After the recording process via the telephone network, the value of DSI and HNR decline due to noise that causes the change information from the voice signal. DSI value on healthy voice has the range (-4,405-8,278), whereas in HNR is at range (0421-16,494). The disorders sound, data on the value of the DSI sound data lower than healthy. The value of the DSI sound healthy is at the range (5,018-6,387), whereas in HNR is at range (0322-9,531). Then, after the reduction of the value of the DSI on healthy voices are on the range (-0747-12,248) and HNR is at range (3,524-18,410). DSI Value next to the sound of disorders had lower values compared to the sound of disorders, DSI value range to the sound of disorders is (-0588-9,123). While the HNR is at range (1,583-8,250).

## 6. References

[1] N Yuwono and S Novita 2014 Nodul Pita Suara vol 41 no 6 pp 428–431 (Kalimantan Barat)

[2] Department of Health ,"Attitude and Practice of Medical Checkup,", Republic of China, 2008.

[3] Arifinto D and Sekartedjo 2004 Speech Disorder Analysis using Time-Varying Autoregressive *Proc. IEEE-MWSCAS* pp 191-94 (Hiroshima, Japan)

[4] Barry W J and Putzer M 2016 Saarbrucken Voice Database, *Institute of Phonitics, Univ. Of Saarland* http://www.stimmdatenbank.coli.unisaarland.de/

[5] Wahyuddin M I 2009 Implementasi VOIP Computer to Computer Berbasis Freeware Menggunakan Session Initiation Protocol *Jurnal Artificial* (ICT Research Center UNAS)

[6] M Imam 2015 Aplikasi Reduksi Noise dalam Perbaikan Kualitas Suara untuk Deteksi Gangguan Pita Suara (Universitas Telkom)

[7] H Aapo and E Oja 2000 Independet Component Analysis: Algorithm and Application (Helsinky Univesity of Techonolgy)

[8] A Niknejad 2005 Noise in Analog Comunication System (UC Barkeley: University of California)

[9] N Firda 2016 Analisa Deteksi Jenis Kelainan Pita Suara dengan Menggunakan Transformasi Wavelet Biorthogonal (Universitas Telkom)

[10] Keinert Fritz 1996 Biorthogonal Wavelets For Fast Matrix Computations I(owa State University)

[11] M S De Bodt and F L Wuyts. 2010 The Dysphonia Severity Index (DSI) (University of Antwerp)

[12] M. Farrús, J. Hernando, and P. Ejarque 2007 Jitter and Shimmer Measurements for Speaker Recognition (Spain: Technical University of Catalonia)

[13] T. Wahyu and R. Ningsih 2015 Nilai Forman dan Karakteristik Bunyi Ujaran Pada Autism Spektrum Disoders (ASD) dam Anak Normal vol 6 no 1972 pp 125–133 (Universitas Gunadarma)

[14] Mifflin 2002 American Heritage Dictionary of the English Language, 4th Edition (Houghton Mifflin Company)

[15] A. Cohen, Daubechies I and Feauveu 1990 Biorthogonal Bases of Compactly Supported Wavelets (Universite Paris)

[16] Wuyts F L, Geert M and Marc R 2010 The Dysphonia Severity Index: An Objective Measure of Vocal Quality Based on a Multiparameter Approach (Belgium: University of Antwerp)

IOP Conf. Series: Journal of Physics: Conf. Series **1075** (2018) 012044 doi:10.1088/1742-6596/1075/1/012044

# A multidimensional valuation of the human perception of construction vibration

**A A Noviyanti[1] and J S Woodcock[2]**

[1] School of Applied Science, Technology, Engineering and Mathematics, Prasetiya Mulya University, Tangerang, Indonesia
[2] Acoustics Research Centre, University of Salford, Greater Manchester, United Kingdom

E-mail: anastasia.noviyanti@prasetiyamulya.ac.id

**Abstract.** Previous studies have shown that annoyance due to vibration does not simply depend on the magnitude of vibration exposure. However, only little is known about the important features that contribute the human perception of vibration. This paper presents a study investigating seated subjects' perception of construction induced groundborne vibration in the vertical direction with the method of multidimensional scaling. The investigation was conducted using a paired comparison subjective test. Sixteen subjects were asked about their perception of the annoyance and similarity of sixty-six pairs of vibration stimuli. Through non-metric multidimensional scaling analysis, it is presented that the perception of construction vibration activities are dependent on up to three perceptual dimensions: duration of the signal, vibration energy in the 31.5 Hz of ⅓ octave band, and the centre of mass of the frequency spectrum of the signals. A regression model was built from the relationship between the annoyance ratings and the objective descriptors revealed through the multidimensional scaling analysis. This model was found to describe 79% of the variance in the measured perceived annoyance ratings data with *p-level* < 0.01.

## 1. Introduction

Vibration can be sensed by human through the auditory, visual and nervous system. The potential effects of vibration exposure depend on many factors, including the characteristics of the motion, the characteristics of the exposed person, activities of the exposed person and other aspects of the environment. These factors can come together to influence how the vibration is perceived and may cause annoyance, discomfort, activity disturbance, impaired health or motion sickness [1]. Exposure-response relationships are an important guidance for policy makers and planners to give permission of the area where the population is growing in an environment.

Field surveys of human response in residential environment has been held in Norway, Sweden, and North America [2,3,4] which each studies has led to exposure response relationships for annoyance due to vibration in residential environments. However, the fundamental question of which vibration exposure descriptor correlates with annoyance is not fully investigated in these studies. Afterwards, study about a number of objective features of vibration exposure which can influence human response including the amplitude, direction, frequency content, duration, and envelope characteristics of the vibration was developed [5].

The investigation of subjective response to combined noise and vibration exposure from railways [6][7] exposed that the magnitude of noise exposure had a significant effect on the judgment of annoyance caused by vibration. On the other hand, there was no significant effect of vibration exposure on the judgment of annoyance caused by exposure to noise. Another study of combined effects of railway induced vibration and noise [8], was found that vibration did not affect noise annoyance, but the total annoyance caused by combined noise and vibration was extremely greater than the annoyance caused by noise alone. Exposure to vibration from environmental sources such as railways and construction activities is considered giving effects such as annoyance and sleep disturbance. Another studies of the human perception of noise and vibration also investigated about multidimensional phenomenon that can be explained by multiple objective parameters that establish a perceptual space, using a technique known as multidimensional scaling. Recently studies focused on the perception of vibration of railway activities were conducted [9] and combined noise and vibration of railway activities [10] has given good results of perceptual dimensions and its relationship with the objective features of the stimuli. This study presented in this paper explores the dimensions of the construction activities generated groundborne vibration and how these perceptual dimensions relate to annoyance.

## 2. Subjective Test Design and Method

### *2.1 Stimuli*

The University of Salford in the United Kingdom conducted a research project into the human response to environmental vibration funded by Department for Environment, Food, and Rural Affairs (DEFRA) [11]. In this project, measurements of vibration from construction activities were made close to the construction of a new light rail systems; the measured activities included digging, piling, breaking, moving, starting, stopping, loading, and unloading [12]. Twelve representative vibration events were selected from these data to be used as stimuli for the test described in this section.

### *2.2 Test Setup*

The test set up used a transducer (Buttkicker LFE) to reproduce vibration, rigidly attached to the underside of a chair with four bolts. The transducer is an electro-dynamic shaker powered by a 1000 W amplifier. The reproduced vibration was measured by seat pad accelerometer Svantek SV38 with a sensitivity of 100 mV/(m/s2) connected to Svantek 957 sound and vibration analyser. The accelerometer was placed above the chair when the stimuli was played. The vibration were reproduced at vertical direction and at levels that comfortable for the subjects and to produce VDV less than number stated in BS 7085:1989, with total daily exposure of whole body vibration for a human subjects should not exceeds a VDV of 15 $m/s^{1.75}$[13].

### *2.3 Subjective Test*

The subjective test took place in Semi-Anechoic Chamber in University of Salford. Sixteen subjects participated in the paired comparison test using twelve stimuli. Subjects were asked to maintain a comfortable upright posture throughout the test. One paired comparison of the vibration was played before the test in order to allow the subjects familiarize to the stimuli and the interface.

Participants were played every possible unique pairing of the twelve stimuli. The ordering of the stimuli used in the subjective test was randomised according to a Ross series [14] that ensures the greatest separation of pairs with common stimuli [15], for each pair they were asked to make a judgement on:

1. Which of the vibrations would bother, disturb or annoy you most if you felt them in your own home?
2. How similar do you perceive the pair of vibrations to be?

Each subject marked their response to both of these questions via continuous sliders on a GUI developed in Matlab, which were coded between -0.5 (stimulus 1) to 0.5 (stimulus 2) for question 1

IOP Conf. Series: Journal of Physics: Conf. Series **1075** (2018) 012044 doi:10.1088/1742-6596/1075/1/012044

and 0 (very similar) to 1 (very different) for question 2. The option of "Neither" was included as it has been shown in paired comparison tests to assess sound quality that allowing a tie minimises circular errors [16].

## 3. Subjective Test Results

### *3.1. Single figure perceived annoyance ratings*
Single figure annoyance ratings were calculated from the paired comparison of annoyance data using the following formula:

$$A_{av} = \frac{1}{N}\sum_{j\neq 1} A_{j,i} \tag{1}$$

where $A_{av}$ is the average perceived annoyance ratings for subject with stimulus *i*, *N* is the number of subject and $A_{j,i,}$ is the summation of paired comparison of annoyance rating for stimuli *i* and *j* across the subject group. Calculation of single figure annoyance scores are using this method assumes that perceived annoyance scores are based on an interval level psychological scale. This method was utilised previously in the estimation of single figure merit scores in sound quality tests of car door closures [17] and to estimate a single figure annoyance for comparing annoyance due to railway vibration [9].

Figure 1 shows the single figure annoyance scores calculated using this method for each of the twelve construction vibration stimuli. An ANOVA of the single figure perceived annoyance ratings for each subject showed that there are no significant differences between the subject's ratings (*p-level* = 0.29). A high score indicates a high degree of annoyance. The overall magnitude between the models of annoyance cannot be known. However, it is clear that stimulus 1 has been judged to be most annoying stimuli.

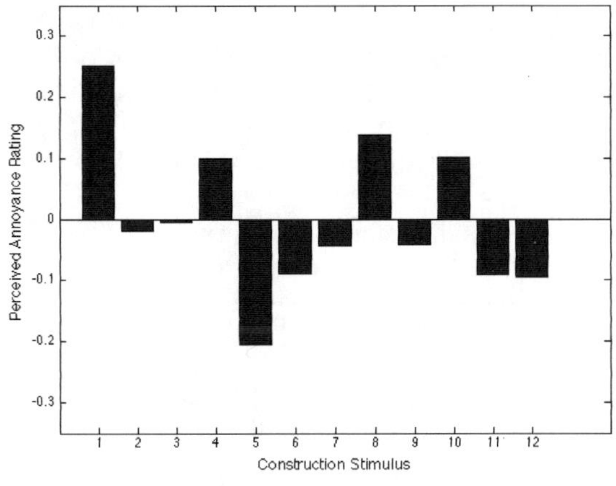


**Figure 1.** Perceived annoyance ratings of twelve construction stimuli.

### *3.2. Multidimensional perception of dissimilarity judgments*
Multidimensional scaling (MDS) is a method that represents measurements of similarity (or dissimilarity) among pairs of object [18]. It develops a configuration of a set of stimuli in an multidimensional space to provide a representation of pairwise (dis)similarities and also psychological distances between the stimuli.

An important step when building an MDS model is to consider the number of dimensions that will make up the MDS configuration. The goodness of fit of the multidimensional scaling solution to the original pairwise dissimilarities can be measured by the stress of the configuration. A reasonable choice of dimensionality is one such that the stress is acceptably small and for which a further increase in dimensionality does not significantly reduce stress [19]. Stress was calculated for configuration in a number different dimensions; this suggested that a solution with 2-3 dimensions was sufficient to describe the data.

The three dimensional non-metric MDS solution is shown in figure 2. Each point in the group space represents one of the twelve construction vibration stimuli. It can be seen in the figure that the stimuli are fairly distributed across all dimensions, even tough stimuli number 11 and 12 are separate in dimension 1. However, both stimuli are not separate from other stimuli in other dimensions. It can be concluded all subjects made their judgments based upon the full continuum of each perceptual dimension rather than simply categorise the stimuli.

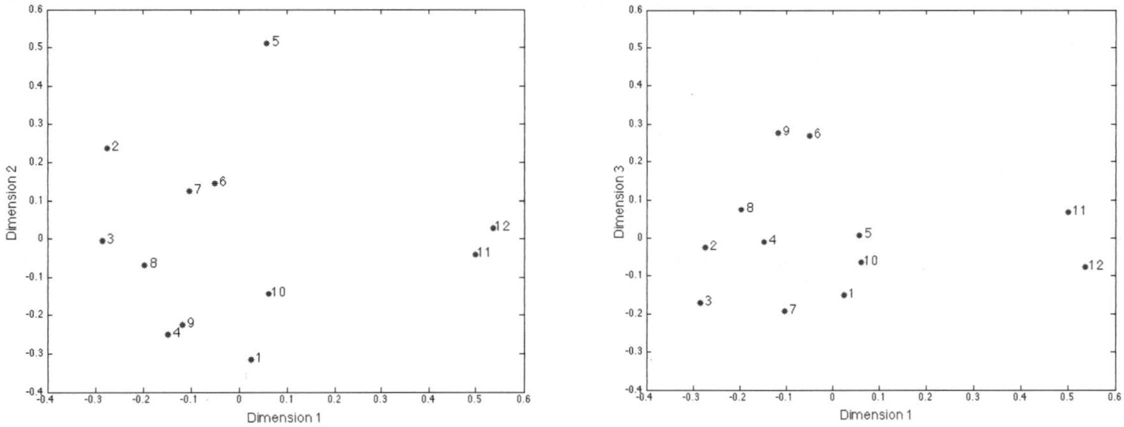


**Figure 2.** Three dimensional non-metric multidimensional scaling configuration.

*3.3. Interpretation of the perceptual space*
All the objective parameters analysed for the subjective test were including the crest factor and duration defined by the 10 dB down points of the signal envelope, spectral centroid and the maximum frequency, vibration dose value, RMS acceleration, and peak acceleration. These energy descriptors were also calculated with the application of the Wb frequency weighting, discussed in BS 6472-1:2008 [20]. The vibration dose value was calculated based on standard described in BS 2631-1:1997 [21].

**Table 1.** Pearson's correlation coefficient between parameter of construction signals of three perceptual dimensions. * *p-level < 0.05.* ** *p-level < 0.01.*

| Parameter | Dimension 1 | Dimension 2 | Dimension 3 |
|---|---|---|---|
| **Spectral centroid** | -- | -- | 0.505 |
| **10 dB down points duration** | - 0.843** | -- | -- |
| $f_{max}$ | -- | - 0.614* | -- |
| $f_{max}$ $W_b$ | 0.841** | -- | -- |
| **RMS acceleration** | -- | 0.845** | -- |
| **RMS acceleration $W_b$** | -- | 0.882** | -- |

Correlation is the commonly used approach when there are many measures of association of variables. Evans [22] suggests the strength of correlations for absolute Pearson's correlation coefficient, usually used for interval or ratio level scales. Table 1 shows the Pearson's correlation coefficient between objective parameters of the vibration stimuli and the positions of the stimuli in a three dimensional MDS solution.

## 4. Model for predicting annoyance due to construction vibration
In the previous section, the correlations of each perceptual dimension and the objective parameters were presented. Recently studies of human perception due of annoyance of railway vibration [9] and

combined railway noise and vibration [10] found that the effects of the stimuli could be investigated by multiple linear regression analysis. Multiple linear regression is a technique which can derive a model whereby several predictor variables are used to model a single response variable [23].

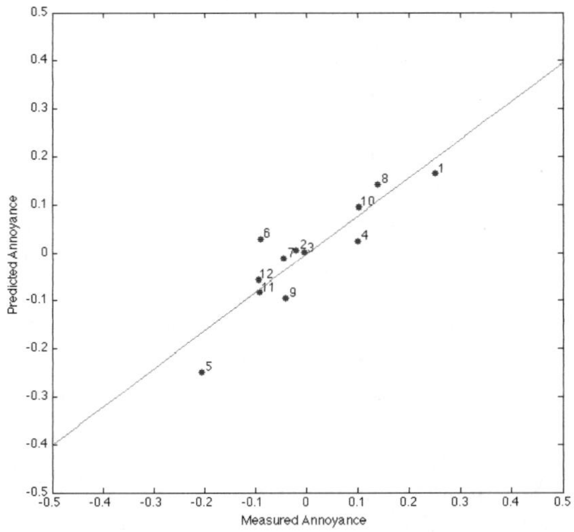


**Figure 3.** Comparison of measured single figure annoyance ratings from the subjective test and those predicted using equation 2.

Based on these consideration, dimension 1 is influenced by the temporal aspect of the signals, which is 10 dB downpoints duration; dimension 2 is influenced by the energy of the signals, which is 31.5 Hz 1/3 octave band $W_b$ weighted RMS acceleration; dimension 3 is influenced by the frequency content of the signal, which is spectral centroid. The multiple linear regressions were computed using the perceived annoyance ratings as the dependent variable and three possible parameters that have a highest correlation with the perceptual dimensions as independent variables is presented in equation 2.

$$A = 0.9583 + 0.0064T_{10dB} - 0.8171RMS_{Wb31.5} - 0.0201Sc \quad (2)$$

where $A$ is predicted annoyance, $T_{10dB}$ is duration of 10 dB down points, $RMS_{Wb31.5}$ is the $W_b$ weighted RMS acceleration 31.5 Hz, $Sc$ is spectral centroid. Figure 3 shows an excellent agreement between measured single figure annoyance scores and those predicted using equation 2 ($R^2 = 0.798$, $p\text{-}level < 0.01$). This calculation suggests that these perceived annoyance rating model shows 79% variance of the data of annoyance. The Pearson's correlation coefficient between the annoyance ratings measured in the subjective test and the annoyance ratings predicted by the model was 0.89. The parameter the $T_{10dB}$ is failed to reach the significant level ($p\text{-}level = 0.12$), the $RMS_{Wb31.5}$ is significant to 0.01 level, and $Sc$ is significant to 0.05 level.

## 5. Summary

This paper presents analysis of the results of paired comparison subjective test design to investigate the perception of construction vibration. Using the method of multidimensional scaling, the perception of construction vibration is influenced by three perceptual dimensions. The first perceptual dimension was influenced by temporal characteristic and described by the duration of the signal, which is 10 dB down points of the signal. The second perceptual dimension was found to be related to vibration energy in the 31.5 Hz third octave band. The third perceptual dimension was influenced by frequency content described by centre mass of the frequency spectrum of the signals, which is spectral centroid.

The objective descriptors revealed through the multidimensional scaling analysis were related to these annoyance ratings via a multiple linear regression model. The Pearson's correlation coefficient between the annoyance ratings measured in the subjective test and the annoyance ratings predicted by the model has very strong correlation. This model was used to predict annoyance ratings measured for the set of vibration stimuli used in the paired comparison test.

## 6. References

[1] Griffin M J 1996 Handbook of human vibration *London: Academic Press Limited*

[2] Klæboe R, Turunen-Rise I, Hårvik L and Madshus C 2003 Vibration in dwellings from road and rail traffic — part II: exposure–effect relationships based on ordinal logit and logistic regression models *Applied Acoustics*. **64** 89 - 109

[3] Gidlöf-Gunnarsson A, Ögren M, Jerson T and Ohrström E 2012 Railway noise annoyance and the importance of number of trains, ground vibration, and building situational factors. *Noise & Health*. **14** 190 - 201

[4] Zapfe J, Saurenman H and Fidell S 2009 *Ground-borne noise and vibration in buildings caused by rail transit* Contractor's final report for TCRP Project D-12

[5] Sica G, Peris E, Woodcock J, Moorhouse A and Waddington D 2013 Design of measurement methodology for the evaluation of human exposure to vibration in residential environments *Science of the Total Environment In Press*. 482-3 461 - 71

[6] Howarth H V C and Griffin M J 1990 Subjective response to combined noise and vibration: summation and interaction effects *Journal of Sound and Vibration* **143** 443 - 54

[7] Howarth H V C and M J Griffin: 1990 The relative importance of noise and vibration from railways *Applied Ergonomics* **21** 129 - 34.

[8] Lee P J and Griffin M J 2013 Combined effect of noise and vibration produced by high-speed trains on annoyance in buildings *Journal of the Acoustical Society of America* **133** 2126 - 35

[9] Woodcock J, Moorhouse A and Waddington D 2014 A Multidimensional Evaluation of the Perception and Annoyance Caused by Railway Induced Ground-borne Vibration *Acta Acustica United with Acustica* **100** 614 - 27

[10] Sharp C, Woodcock J and Waddington D 2014 Considering the perception of combined railway noise and vibration as a multidimensional phenomenon *Inter noise* **249** 888 - 97

[11] Waddington D, Moorhouse A, Steele A, Woodcock J, et al 2011 *Human response to vibration in residential environments (NANR209) Final Project Report* London : DEFRA

[12] Waddington D C, Woodcock, J, Peris E, Condie J, Sica G, Moorhouse A T and Steele A 2014 Human response to vibration in residential environments *The Journal of the Acoustical Society of America*. **135**(1) 182 - 93

[13] British Standards Institution BS 7085:1989 *Guide to safety aspects of experiments in which people are exposed to mechanical vibration and shock*

[14] Ross R T 1934 Optimum orders for the presentation of pairs in the method of paired comparisons *Journal of Educational Psychology* **25** 375 - 82

[15] David, H A 1988 *The method of paired comparisons* 2nd ed (London, UK : C. Griffin)

[16] Parizet E 2002 Paired comparison listening tests and circular error rates *Acta Acustica United with Acustica* **88** 594 - 98

[17] Parizet E, Guyader E and Nosulenko V 2008 Analysis of car door closing sound quality *Applied Acoustics*. **69**(1) 12–22

[18] Groenen P and Borg I 2005 *Modern multidimensional scaling : Theory and application* 2nd ed (New York : Springer)

[19] Kruskal J B 1964 Multidimensional scaling by optimizing goodness of fit to a nonmetric hypothesis *Psychometrika* **29**(1) 1–27

[20] British Standards Institution BS 6472-1:2008 *Guide to evaluation of human exposure to vibration in buildings - part 1: vibration sources other than blasting*

[21] British Standards Institution BS ISO 2631-1:1997 *Mechanical vibration and shock — evaluation of human exposure to whole-body vibration - part 1: general requirements*

[22] Evans J D 1996 *Straightforward statistics for the behavioral sciences* (Pacific Grove, CA: Brooks/Cole Publishing)

[23] Weisberg S 2005 *Applied linear regression* 3rd ed (United States of America : John Wiley & Sons, Inc)

# The development of acoustic environment simulator for passenger's train soundscape

**K W Zakri[1], A S Sudarsono[1], J Sarwono[1], S S Utami[2], N Hidayah[1] and N N Hamdani[1]**

[1]Laboratory of Building Physics and Acoustics, Engineering Physics, Institut Teknologi Bandung, Indonesia
[2]Acoustics and Visual Studio, Engineering Physics, Universitas Gadjah Mada, Indonesia

E-mail: keyshawz@gmail.com

**Abstract.** The study of soundscape in laboratory usually conducted by reproducing acoustic environment recording or simulated acoustic environment. Different method has been developed to understand the soundscape in the laboratory using acoustic environment composition. The soundscape composition has been validated and implemented for outdoor urban soundscape but not the indoor soundscape. The previous soundscape simulator only considers the behaviour of sound sources (movement, and position in space) without considering the sound reflection. This study tries to develop an acoustic environment simulator for passenger train soundscape study. In the passenger train, there are some annoying sound source that cannot be removed such as the sound of bogie or the sound of buffers and chain coupler. The soundscape composition concept using acoustic environment simulator can be implemented to understand the existing soundscape, to fix the soundscape (by selecting different sound source for masking and adjust the sound level), and to evaluate the effect of the acoustic treatment inside the passenger train. The simulator can reproduce realistic acoustic environment using ambisonic reproduction concept. Furthermore, the reflection inside the passenger train simulated by using the concept of auralization.

## 1. Introduction

The soundscape study usually conducted in situ and laboratory experiment [1]. The in-situ experiment usually conducted using soundwalk which include site survey to several locations [2] [3]. The soundwalk approach has been widely used especially for analysing urban soundscape. The other method includes the reproduction and simulation of outdoor soundscape in laboratory. Soundscape reproduction system using ambisonic reproduction system has been validated using Semantic Differential Analysis [4] [5] and Semantic Categorization [6]. Simulated soundscape has been validated by Sudarsono et al using Semantic Differential Analysis [7]. Both system has been validated for urban soundscape analysis. Another approach to analysing soundscape has been developed using composition approach [8]. This system has been developed and validated also for urban soundscape. In general, soundscape analysis method has been developed but mainly for outdoor soundscape and urban area.

Soundscape concept can be implemented not only for outdoor area but also indoor area such as hospital [9], library [10], office [11], restaurant [12] or passenger transportation [13]. The passenger transportation become an interesting object because the passenger must be in the specific acoustic

IOP Conf. Series: Journal of Physics: Conf. Series **1075** (2018) 012045 doi:10.1088/1742-6596/1075/1/012045

environment for hours without the opportunity to move to different environment. The study for passenger transportation mostly conducted using energy based approach [14] without considering the perception of the passenger.

The aim of this study is to develop a system of acoustic environment simulator for analysing the soundscape in a passenger transportation with the focus on passenger train. The simulator is able to reproduce the acoustic environment and also able to include/exclude sound source to design the better soundscape according to the concept of soundscape composition.

## 2. Soundscape in passenger train

Rapid developments in transportation forced railway companies to provide excellent servants. After managing safety in passenger train, noise has become one of issue to passenger railways. Objective measurement and subjective measurement have been done to get the sound in passenger train. Sound pressure level usually used as the parameter for acoustic environment, without considering the passenger's perception. The sound source that considered as noise in passenger train cannot be removed from acoustic environment directly, since it will lose a sensation of being in passenger train and the sound occurred into a unity during the trip.

**Figure 1.** The Picture of Executive Passenger Train Class in Indonesia.

**Figure 2.** The Picture of Economy Passenger Train Class in Indonesia.

In Indonesia, passenger train consist of three different class such as executive (K1), business (K2), and economy (K3) class. Executive class passenger train in Indonesia have 50 seats and economy class passenger have 80 seats with maximum velocity at 100 km/hour, temperature range at $24^0C$ – $26^0C$ and relative humidity in range 60% -70% [15]. The picture of the passenger train is shown in figure 1 and figure 2.

Using soundscape approach, the result of objective measurement is supported by subjective measurement result. Background noise is produced by mechanical activity during the train movement.

Although the objective result is not exceeded the standard for passenger train, passengers stated that the acoustic condition of passenger train is less satisfying.

The dominant sound, annoying sound, and important sound have been found for passenger train, especially for Indonesia's train [13]. The noticeable sound which giving higher background noise in passenger train were bogie, announcement, buffers and chain coupler. Although, announcement have become one of the dominant sound, it did not perceive as the annoying sound. The noise sources that could affect the privacy and sense of comfort in the passenger train were bogie, buffers and chain coupler, and rail connection, which is formed by mechanical activity during train movement. Meanwhile, announcement is considered as the important sound, following by television as the second important sound.

## 3. Soundscape composition as an approach for passenger train soundscape treatment

Soundscape composition has been developed for urban soundscape analysis [8]. This method has been validated using semantic differential analysis resulting the same soundscape dimensions with the in-situ study [16]. This method has been implemented to understand the relationship between sound objects in urban area with the soundscape dimensions. Furthermore, the model has been implemented in the acoustic environment simulator so the simulator can predict the perception of urban acoustic environment [7].

Soundscape in passenger train is unique because the acoustic environment is dominated by the constant background noise especially the mechanical noise. Some study tried to improve the quality of the soundscape by reducing the mechanical noise. A different approach can be implemented by using the concept of masking by adding sound objects.

The masking concept has been widely used in the urban soundscape. Some studies focus on the effect of masking urban noise using water feature [17] [18]. In addition, the sound of bird chirping is also considered as suitable masking sound [19]. This concept also can be implemented for passenger train soundscape. The study for identifying the suitable sound source and how loud the sound source should be heard need to be done for passenger train soundscape. One of the method to identify this matter is by using soundscape composition in an acoustic environment simulator.

## 4. Acoustic environment simulator development

Acoustic environment simulator has been developed especially for urban acoustic environment simulation. Two systems of acoustic environment have been developed: using headphone using multi-speaker reproduction, and using augmented reality.

The acoustic environment simulator mainly developed for urban area. An acoustic environment simulator has been developed by Lunden et al, to simulate the movement in the urban traffic using headphone [20]. Further development also includes the visual stimulus using projector as developed by Hong and Jeon [21]. This simulator does not consider the movement of the sound source in the acoustic environment.

Bruce et al developed urban acoustic environment simulator with the focus on sound source addition of existing acoustic environment [4]. The background sound is made using soundfield microphone and reproduce using multi-speaker system. This acoustic environment simulator has been develop further using object oriented approach allowing the soundscape designer to compose an acoustic environment only using the sound source [7]. The object-oriented concept was implemented in this simulator by considering three parameters: the position of a sound object (for static sound objects), the sound level of a sound object, and the movement of a sound object.

The concept of augmented reality also has been adapted to make an acoustic environment simulator. The Listen project tries to implement immersive audio-augmented environment using virtual reality [22].

Most of the simulator developed for urban area and rarely used for simulating indoor acoustic environment, hence the reflection factor is neglected in the simulator. Further development can be done by including the sound reflection in closed space in the simulation.

## 5. System concept and design

The acoustic environment simulator is developed using several concepts: Reproduction-Simulator, and object oriented [23]. The reproduction-simulator concept is the same concept with Bruce's simulator. This concept includes the acoustic environment reproduction using ambisonic system and the addition of sound source. The sound source in this simulator does not has behaviour such as movement or position.

The object-oriented concept was developed in Sudarsono's simulator [16]. This simulator can be used to compose an acoustic environment only using the sound source. The sound source is considered as an object with several behaviours: sound level, movement, and position. Although the simulator has succeeded in simulating urban area, this simulator cannot be used to design indoor soundscape because the simulator does not consider the reflection in a room. Further development is implemented in acoustic environment simulator by considering the reflection phenomenon.

The system is shown in Figure 3. Acoustic environment recording is taken using ambisonic recording (B-Format Signal). The sound sources are recorded in mono signal using unidirectional microphone. The position and the movement of the sound source is simulated using ambisonic panner and the automation of the ambisonic panner. This simulator implements Wigware ambiPan x-y 1-3D (Developed by Bruce Wiggins) ambisonic panner to simulate the position and the movement of sound source. The output of this panner is a B-Format. Further signal processing is the addition of room reflection using auralization with the B-Format impuls from the actual train. The auralization is implemented using Reverberate LE VST Plugin.

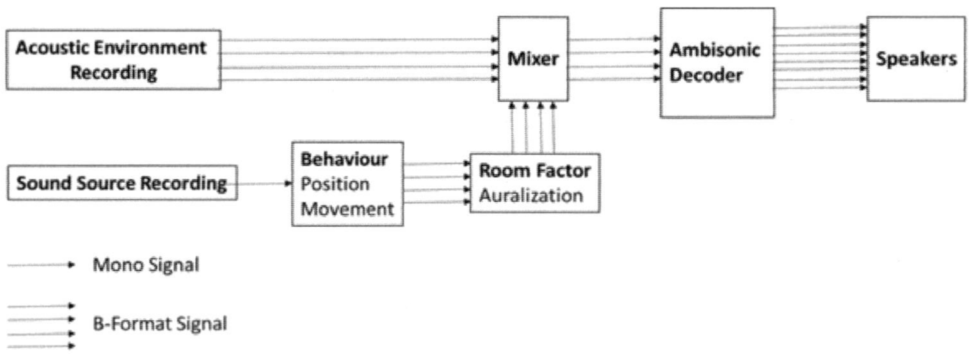


**Figure 3.** Acoustic Environment Simulator for Passenger's Train Concept.

The signal from both acoustic environment recording and sound source recording are mixed together and decoded into eight channel outputs. This simulator is using WigWare Regular Shaped 1$^{st}$ Order Ambisonic Decoder developed by Bruce Wiggins to decode the B-Format signals to the signals sent to the speakers.

The simulator is developed using several software: PureData, LoopMIDI and Reaper. PureData is used to make the interface act as a MIDI controller. LoopMIDI is a software developed by Tobias Erichsen which allows the MIDI connection between PureData and Reaper. Reaper is a Digital Audio Workstation (DAW) software which usually used for music mixing.

The interface of the simulator is shown in Figure 4. The participant will listen to the recording of passenger train and requested to make the soundscape better by selecting sound sources and adjust the sound level. In this simulator, it is assumed that the sound addition will be reproduced from the Public-Address system which is located in a static position so the variation of sound source position is not needed.

The development of this simulator enables us to understand several aspects in passenger train:

- To determine of soundscape dimension for passenger train
- To understand the perception of the existing passenger train
- To determine the sound level of the suitable sound to make the soundscape better
- To determine the effect sound source addition to the perception of acoustic environment in the passenger train

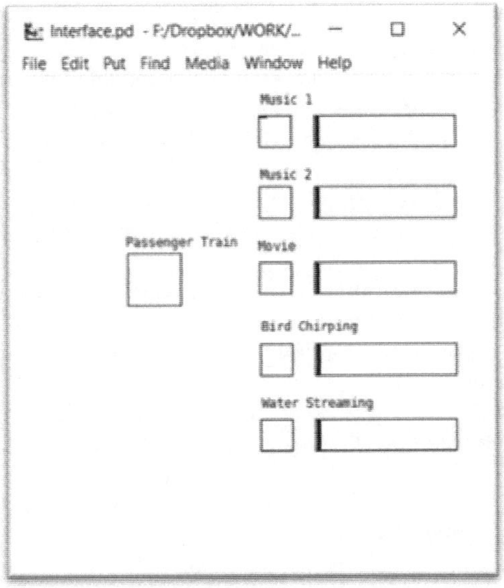


**Figure 4.** The Interface of Acoustic Environment Simulator for Passenger Train.

## 6. Measurement using simulator

The experiment was conducted by using the acoustic environment simulator for passenger train. The participant was requested to listen to two stimulus and rated the stimulus based on three perceptions: safety, relaxation, and dynamic. The first stimulus was the ambisonic recording of passenger's train and the second stimulus was the previous recording with the addition of music. The type of music and the sound level reproduction was taken based on the previous experiment using the acoustic environment simulator. In those experiment, the participant was requested to add the sound of music and adjusted the sound level. Twenty participants (ten males and ten females) joined the experiment voluntarily and the entire participants were master and undergraduate student of Engineering Physics Department, Institut Teknologi Bandung.

Analysis using T-Test indicates that the addition of music significantly ($sig<0.05$) affect the perception of safety, relaxation, and dynamic. The mean difference of the rating is 1.381-2.381 as shown in Table 1.

**Table 1**. The result of T-Test and Mean Difference (MD)

| | T-Test | MD |
|---|---|---|
| Safety | 0.000 | 1.714 |
| Relaxation | 0.002 | 1.381 |
| Dynamic | 0.000 | 2.381 |

This result indicates that the acoustic environment simulator is suitable for soundscape analysis. The simulator has been implemented for soundscape intervention and the result of the intervention also has been tested with a positive result.

## 7. Conclusion

An acoustic environment simulator has been developed for passenger train. The simulator is developed by combining the ambisonic recording, sound source addition, and room reflection aspect. This simulator can be implemented for soundscape research in passenger train such as: to determine of soundscape dimension for passenger train, to understand the perception of the existing passenger train, to determine the sound level of the suitable sound to make the soundscape better, and to determine the effect sound source addition to the perception of acoustic environment in the passenger train.

## 8. References

[1] Aletta F, Kang J and Axelsson Ö 2016 Landscape and Urban Planning Soundscape descriptors and a conceptual framework for developing predictive soundscape models Landsc. Urban Plan. 149 65– 74
[2] Adams, Bruce, Davies, Cain, Jennings, Carlyle, Cusack, Hume and Plack 2008 Soundwalking s Methodology for Understanding Soundscapes Institute of Acoustics Spring Conference 2008 vol 30(Reading: the Institute of Acoustics)
[3] Semidor C 2006 Listening to a city with soundwalk method Acta Acoust. United with Acoust. 92 959– 64
[4] Davies W J, Bruce N S and Murphy J E 2014 Soundscape Reproduction and Synthesis Acta Acust. united with Acust. 100 285–92
[5] Sudarsono A S, Lam Y W and Davies W J 2016 The effect of sound level on perception of reproduced soundscapes Appl. Acoust. 110 53–60
[6] Guastavino C, Katz B F G, Polack J and Levitin D J 2005 Ecological Validity of Soundscape Reproduction Acta Acoust. United with Acoust. 91 333–41
[7] Sudarsono A S, Lam Y W and Davies W J 2017 The Validation of Soundscape Environment Simulator to Determine the Relationship between Sound Objects and the Perception of Soundscape Acta Acust. united with Acust.
[8] Sudarsono A S, Lam Y W and Davies W J 2016 Soundscape Perception Analysis Using Soundscape Simulator Inter Noise 2016 (Hamburg) pp 6868–75
[9] Azzahra I, Sarwono J, Prasetiyo I and Utami S Measurement Analysis of the Soundscape of Care Units in a Private Hospital in Bandung
[10] Ikhwanuddin R, Sarwono J, Sudarsono A S and Utami S S Library Soundscape : Higher Education Students ' Perception 73–9
[11] Ma H and Yu B 2016 Is a good soundscape element a restorative environment factor in open office? INTER-NOISE NOISE-CON Congr. Conf. Proc. 253 4930--4936
[12] Zhang S and Kang J 2016 The Influence of Crowd Density on Evaluation of Soundscape in Typical Chinese Restaurant INTER-NOISE NOISE-CON Congr. Conf. Proc. 253 155–60
[13] Zakri K W, Hidayah N, Hamdani N N, Sarwono J, Utami S S and Sudarsono A S Evaluating Perceived Acoustic Environment in Executive Class Passenger Train using Soundscape Approach 3633–8
[14] Hardy A 2000 Measurement and assessment of noise within passenger trains J. Sound Vib. 231 819–29
[15] Anon Album Kereta
[16] Sudarsono A S, Lam Y W and Davies W J 2017 The validation of acoustic environment simulator to determine the relationship between sound objects and soundscape Acta Acust. united with Acust.103
[17] Jeon J Y, Lee P J, You J and Kang J 2012 Acoustical characteristics of water sounds for soundscape enhancement in urban open spaces. J. Acoust. Soc. Am. 131 2101–9
[18] Axelsson Ö, Nilsson M E, Hellström B and Lundén P 2014 A field experiment on the impact of sounds from a jet-and-basin fountain on soundscape quality in an urban park Landsc. Urban Plan. 123 49–60
[19] Hong J Y and Jeon J Y 2017 Exploring spatial relationships among soundscape variables in urban areas: A spatial statistical modelling approach Landsc. Urban Plan. 157 352–64
[20] Lundén P, Gustin M, Nilsson M E, Forssén J and Hellström B 2010 Psychoacoustic evaluation as a tool for optimization in the development of an urban soundscape simulator Proc. 5th Audio Most. Conf. A Conf. Interact. with Sound 1–6
[21] Hong J Y and Jeon J Y 2013 Designing sound and visual components for enhancement of urban soundscapes. J. Acoust. Soc. Am. 134 2026–36
[22] Eckel G 2001 Immersive Audio-Augmented Environments The LISTEN Project Proc. Fifth Int. Conf. Inf. Vis. 571–3
[23] Bruce N S, Davies W J and Adams M D 2009 Development of a soundscape simulator tool Inter-Noise 2009 (Ottawa)For *Journal of Physics: Conference Series*, please use the Vancouver numerical system where references are numbered sequentially throughout the text. The numbers occur within square brackets, like this [2], and one number can be used to designate several references. The reference list gives the references in numerical, not alphabetical, order.

# Evaluation of Time-Difference-of-Arrival Error of Acoustic Beacons Caused by Velocity of Microphone Array

**Satoki Ogiso[1], Koichi Mizutani*[1,2], Naoto Wakatsuki[1,2] and Tadashi Ebihara[1,2]**

[1]Ph. D. Program in Empowerment Informatics, School of Integrative and Global Majors, University of Tsukuba, 1-1-1 Tennodai, Tsukuba, Ibaraki, 305-8573 Japan
[2]Faculty of Engineering, Information and Systems, University of Tsukuba, 1-1-1 Tennodai, Tsukuba, Ibaraki, 305-8573 Japan

E-mail: mizutani@iit.tsukuba.ac.jp

**Abstract.** In this paper, we described a criterion of estimating the quality of time-difference-of-arrival (TDOA) estimation with a moving microphone array. The microphone array was supposed to receive a known signal from an acoustic beacon which transmitted a binary code with binary phase shift keying (BPSK). The movement of the microphone array may cause two kinds of error on the TDOA estimation; physical change of TDOA caused by the movement, and apparent change caused by the lower correlation values of Doppler-shifted signals. For evaluating the latter error, we employed the peak to average power ratio (PAPR) of the cross correlation function. The Doppler effect on TDOA error were evaluated with a computer simulation. The microphone array was moved at -2 to 2 m/s, covering the velocity range of indoor mobile robot. Besides the velocity, following three parameters of the sound were investigated. The types of the code for evaluation were chosen as M-, Gold and Kasami sequences, carrier frequencies were 0-40 kHz, and chip rates of the binary code were 0.5 - 20 kHz. As the results, it was confirmed that if PAPR is lower than a certain value, 40 in this simulation, the peak detection of the cross correlation is difficult and it results in high and unpredictable TDOA errors. Higher chip rates and lower carrier frequencies showed better resilience against the Doppler effect. The proposed criterion may infer the maximum velocities at which TDOA can be estimated without any measure against the Doppler effect for realistic parameters. This criterion would contribute to the on-line evaluation of the quality of TDOA, which may be beneficial for estimating sound direction of arrival.

## 1. Introduction

Self-localization of mobile robots are one of the essential functions for intelligent tasks. Especially in indoor environments, there is no chance to use GNSS/GPS (Global Navigation Satellite Systems / Global Positioning Systems) because the satellites are not observable. Under these conditions, there are some possible methods for estimating locations of the mobile robots. One of the conventional methods is wheel-based or IMU(inertial measurement unit)-based odometry, which accumulates velocity and angular velocity of a mobile robot[1, 2]. Although it is quite easy to implement, the accuracy cannot be guaranteed owing to accumulation of measurement errors. Contrarily, there are localization methods which use not only the sensors above, but also

IOP Conf. Series: Journal of Physics: Conf. Series **1075** (2018) 012046 doi:10.1088/1742-6596/1075/1/012046

sensors to acquire information about the surrounding environment. The typical methods are camera-based or LIDAR(Light Detection and Ranging)-based SLAM (simultaneous localization and mapping)[3]. This type of the localization method can acquire information of surrounding environment and correct localization error according to the information. However, it is usually a costly calculation that extraction of plenty of feature points (>500) to be tracked from measurement and tracking them over time robustly. Also, the LIDARs or cameras for practical use themselves are expensive.

To overcome these problems of calculation cost and expensive equipments, acoustic localization methods have been proposed. There are several types of acoustic localization. One is multi-transmitter single-receiver type. This type of the localization method uses TDOA for each sounds from different but known locations. Although it is possible to estimate the location, the pose of receiver cannot be obtained, as single receiver does not provide enough information. It is critical for applications such as mobile robot localizations, which the pose is quite important as well as a location for movement.

Our research group has proposed a self-localization method using acoustic beacons in surrounding environment and microphone array on mobile robot[4, 5, 6], as multi-transmitter and multi-receiver type. This method can estimate time-difference-of-arrival (TDOA) between the received signals of each microphone element of the microphone array, and conduct self-localization based on TDOA. This method does not require time synchronization, and can estimate not only position, but also the pose of mobile robots.

For these acoustic localization methods, the Doppler effect on the sounds used for localization is one of the major problems. In this problem, there are studies to evaluate the effect on cross-correlation values[7], localization[8], communication of sound emitters and receivers[9]. However, these studies are only for single receiver and not intended for array signal processing. Therefore, TDOA of the receiver microphones were not evaluated.

In this paper, we evaluate the effect of the Doppler effect on TDOA error for binary phase shift keying (BPSK) of acoustic beacons. There are mainly two ways that the Doppler effect affects TDOA. One is the apparent change of sound speed caused by the microphone movement, which causes actual TDOA changes on microphone signals. This change can be predicted by theoretical analysis. The other one is change of cross-correlation characteristics of received and transmitted signals. TDOA of the BPSK signals are usually detected as a time at which the maximum value of cross-correlation function of receiving signals and replica signal of transmitting signal was observed. The Doppler effect may affect to this characteristic, and deteriorates both the peak time and the peak height. This change is not easy to predict theoretically. Our aim of this paper is to evaluate the TDOA error from these two aspects focusing on TDOA error. This observation may make a criteria to predict a velocity range that TDOA estimation can be done for realistic parameters of BPSK signal.

## 2. Computer simulation of localization

### *2.1. Conditions of simulation*

Computer simulations were performed assuming movement of a mobile robot along a straight line in certain velocities. Figure 1 shows a description of simulation condition. The mobile robot had two microphones which were placed in distance of 1 m, and the microphones were unidirectional and synchronized. A loudspeaker was placed on 5 m apart from the microphones in order to assume a far sound field. The loudspeaker emitted BPSK-modulated signal repeatedly. The generation polynomials, the initial conditions of each sequence and other common conditions of the emitted signals are described in Table 1. Different BPSK-modulated signals for various carrier frequencies, $f_c$, and chip rates, $f_b$, were examined. Signals of the microphones were recorded for a period of BPSK-modulated signal, namely $n/f_b$ (s), at a sampling rate of 400 kHz. The TDOA of recorded signals were calculated by cross-correlation method, which estimates

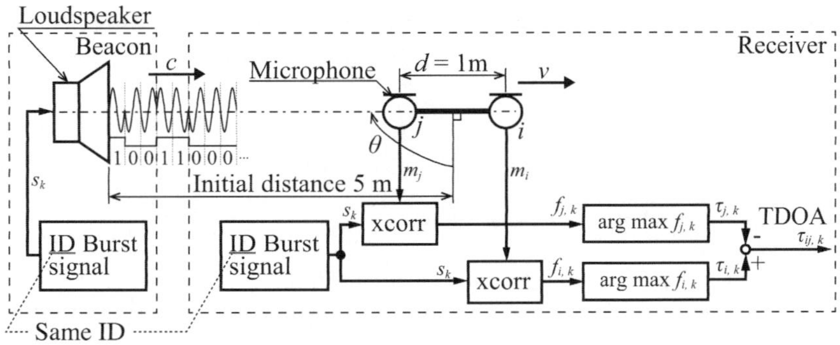


**Figure 1.** Simulation condition and description of TDOA estimation method

**Table 1.** Common parameters of simulation

| | |
|---|---|
| Sequence type | M, Gold, and Kasami |
| Sequence length $n$ | 255 (8th order) |
| Generation polynomials | |
| M sequence | [8, 7, 6, 5, 2, 1, 0] |
| Gold sequence | [8, 7, 6, 5, 2, 1, 0] |
| | [8, 7, 6, 1, 0] |
| Kasami sequence | [8, 4, 3, 2, 0] |
| | $m = 1$ |
| Initial conditions | [0, 0, 0, 0, 0, 0, 0, 1] for all sequences |
| Sampling rate | 400 kHz |

TDOA as the time at which the cross correlation function of the recorded signals takes the maximum value. The window length of the cross correlation $w$ was set to be $n/f_b$ (s) in order to calculate for 1 period of a signal. Following two variables were evaluated to evaluate the effect of BPSK signal parameters on resiliency against the Doppler effect

(i) Different carrier frequencies, $f_c$, under condition of a constant chip rate $f_b = 5$ kHz. Higher carrier frequency would result in a larger frequency shift by the Doppler effect. To evaluate the effect, practical carrier frequencies were chosen from 0 Hz (base band signal) to 40 kHz (common in ultrasonic localization systems).

(ii) Different chip rate, $f_b$, under condition of a constant carrier frequency $f_c = 20$ kHz. Higher chip rate widely spread spectrum and would have resilience against the Doppler effect. To evaluate the effect practically, a chip rate range of 0.5–20 kHz was simulated.

For each condition, the theoretical TDOA error $\Delta\bar{\tau}$ caused by the apparent sound speed change was calculated as following equation using variables shown in Fig. 1:

$$\Delta\bar{\tau} = \frac{d\sin(\theta)}{c-v} - \frac{d\sin(\theta)}{c}. \tag{1}$$

The first term represents TDOA with movement at a velocity of $v$, and the second term represents TDOA without movement.

Also, it is important to evaluate characteristics of cross correlation functions that how easily the maximum peak can be found. This characteristic was evaluated with peak to average power ratio (PAPR) with variables shown in Fig. 2. PAPR was calculated as following equation:

$$p = \frac{\max {f_{i,k}}^2}{\mu}, \tag{2}$$

where $\mu$ denotes the mean of power value except values around the maximum peak. $\mu$ is calculated as:

$$\mu = \frac{1}{\arg_t \max(f_{i,k}) - g/2 + w} \int_{t=-w}^{\arg_t \max(f_{i,k})-g/2} f_{i,k}^2 dt + \frac{1}{w - (\arg_t \max(f_{i,k}) + g/2)} \int_{t=\arg_t \max(f_{i,k})+g/2}^{w} f_{i,k}^2 dt. \quad (3)$$

The variable $g$ denotes a guard bandwidth for ignoring the maximum peak and values around it of the cross correlation function. Eq. 2 holds high value if the peak value is higher than values

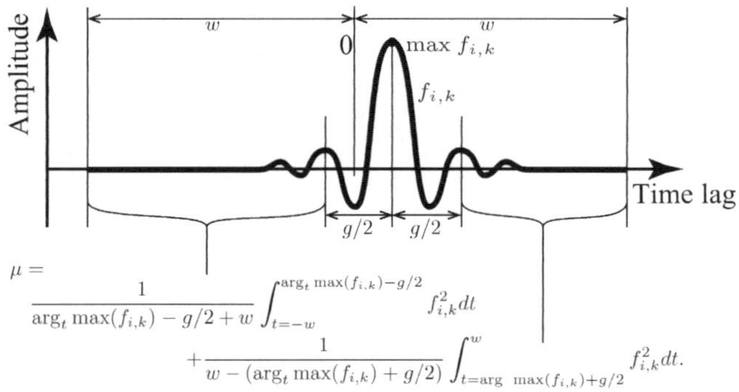


**Figure 2.** Definition of variable used for calculation of PAPR

of other region. Higher $p$ indicates easier to detect the maximum peak, and lower $p$ indicates more difficult to find out the correct peak of the cross correlation function. In the simulation, $g$ was set to 10% of window length $w$.

### *2.2. Results and Discussions*

The results of computer simulations are shown in Fig. 3. Figures 3(a)–(j) show the results under different $f_c$. Figures 3(a)–3(e) and Fig. 3(k)–3(o) show results of TDOA estimation error, and Fig. 3(f)–3(j) and Fig. 3(p)–3(t) show corresponding PAPR, $p$. From the figures, a trend can be observed that TDOA error depends on $f_c$ even at the same velocity. Also, shown as the shaded area, the TDOA error was drastically increased if the velocity exceeded a certain amount. This drastic increase from theoretical TDOA error can be explained by PAPR changes. For example, the cross correlation function at P1, $v$ = -1 m/s, and P2, $v$ = 0 m/s, in Fig. 3(e) are shown in Fig. 4(a) and Fig. 4(b). As shown in Fig. 4, the peak of the cross correlation function is obvious if $p$ takes higher value. These results may indicate that when PAPR take lower values, corresponding TDOA may be quite inaccurate. The threshold of $p$ for accurate or inaccurate TDOA was approximately 40 in this case as seen in Fig. 3. One exception is Fig. 3(f), which takes exceptionally low values without drastic change on TDOA. This is because its condition is $f_c = 0$, namely there is no carrier energy on the PAPR where all other conditions have.

Figures 3(k)–(t) show the results under different $f_b$. From the figures, a trend can be observed that $f_b$ affected TDOA error and lower $f_b$ resulted in higher TDOA error at the same velocity. Also, this drastic change from theoretical TDOA error can be explained by PAPR changes, as a range at which PAPR is less than approximately 40 is corresponding to inaccurate TDOA ranges. The lower $f_b$ means narrower spectrum spreading in frequency and shorter signal period. Lower chip rate $f_b$ cause lower cross correlation peak at higher velocity because of the longer signal period and narrower spectrum, and affects the TDOA error. With higher $f_b$, the period of the signal was shortened, and the frequency was spread widely. The shorter signal period and

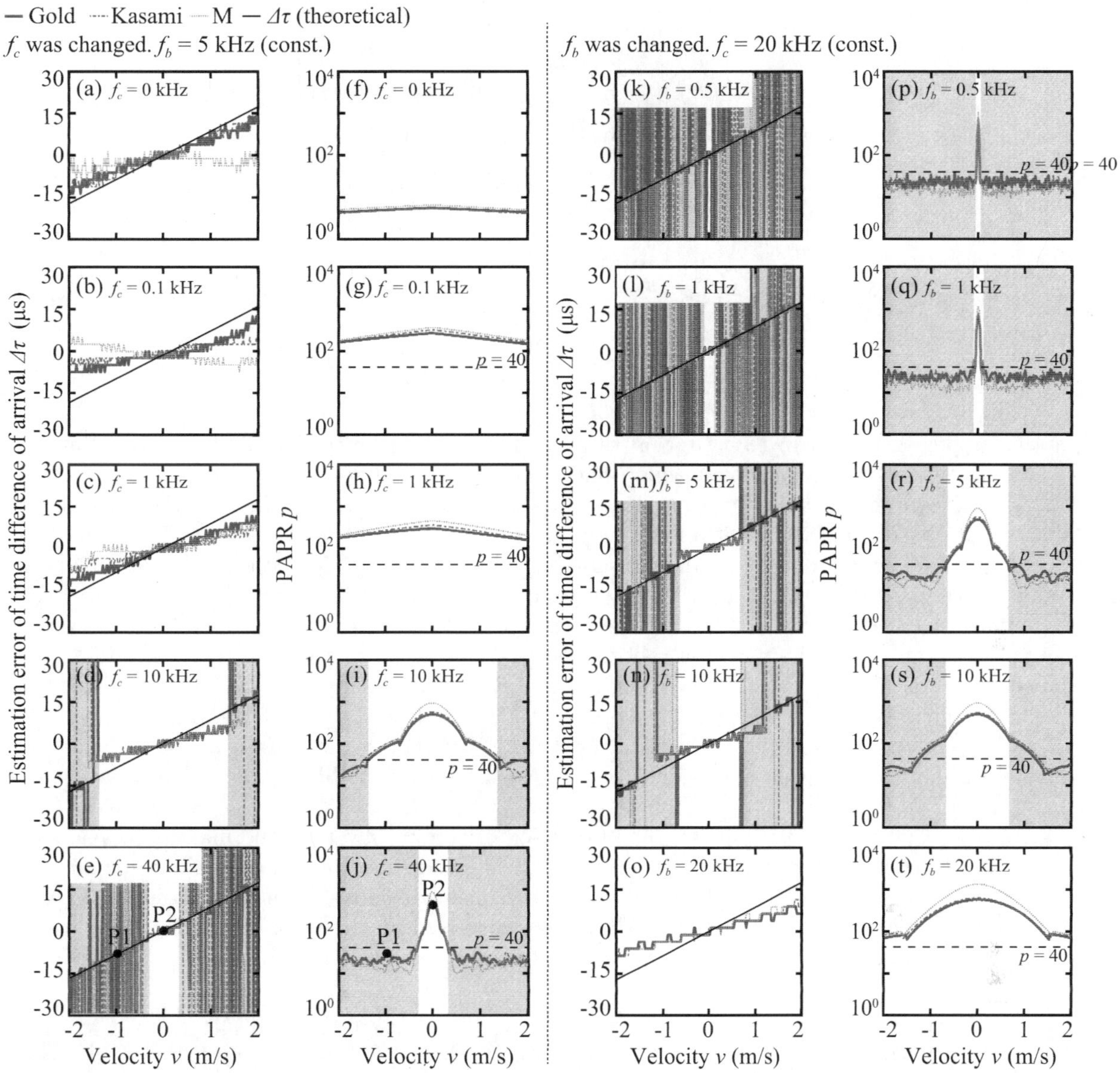


**Figure 3.** Velocity $v$ vs. error of time difference of arrival $\Delta\bar{\tau}$ and velocity $v$ vs. PAPR $p$ for different $f_c$ and $f_b$. Shaded areas represent inaccurate TDOA.

widely spread spectrum makes resilience to the Doppler shift, and results in good correlation characteristics as $p$ indicates. From the simulation, there are velocity ranges for each $f_b$ where TDOA can be estimated correctly. Under condition of $f_c$ = 20 kHz, the velocity range of -0.1–0.1 m/s for $f_b$ = 0.5 kHz, -0.2–0.2 m/s for $f_b$ = 1 kHz, -0.6–0.6 m/s for $f_b$ = 5 kHz and 10 kHz, and more than -2–2 m/s for $f_b$ = 20 kHz. Especially under the condition of $f_b$ = 20 kHz, TDOA would not become inaccurate without any measure against the Doppler shift. Also, it can be confirmed with PAPR shown in Fig. 3(t), as the PAPR is enough high to estimate the peak of the cross correlation function. This can be a criteria to determine $f_b$ and whether measure against the Doppler shift is needed.

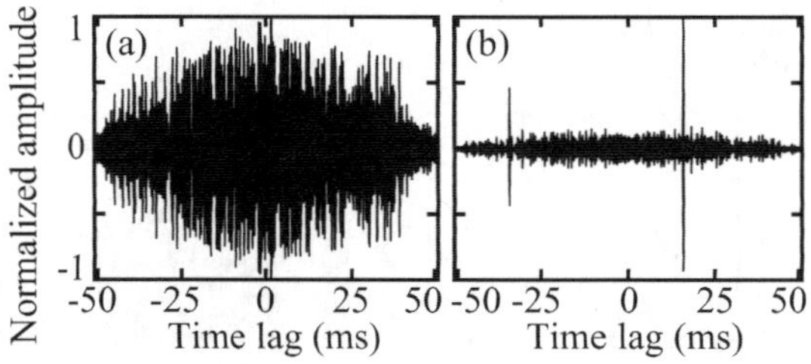


**Figure 4.** Normalized cross correlation functions $f_{i,k}$ of Kasami sequence for each highlighted points in Fig. 3. (a) P1, $f_c$ = 40 kHz, $f_b$ = 5 kHz, $v$ = -1 m/s, (b) P2, same condition to (a), $v$ = 0 m/s.

## 3. Conclusion

In this paper, we described a criterion of estimating the quality of time-difference-of-arrival (TDOA) estimation with a moving microphone array. The microphone array was supposed to receive a known signal from an acoustic beacon which transmitted a binary code with binary phase shift keying (BPSK). Within the Doppler effect, we focused on the apparent change of TDOA by the lower correlation values of Doppler-shifted signals. For evaluating this error, we employed the PAPR of the cross correlation function. The Doppler effect on TDOA error were evaluated with a computer simulation. The microphone array was moved at -2 to 2 m/s, covering the velocity range of indoor mobile robot. Besides the velocity, following three parameters of the sound were investigated. The types of the code for evaluation were chosen as M-, Gold and Kasami sequences, carrier frequencies were 0-40 kHz, and chip rates of the binary code were 0.5 - 20 kHz. As the results, it was confirmed that if PAPR is lower than a certain value, 40 in this simulation, the peak detection of the cross correlation is difficult and it results in high and unpredictable TDOA errors. Higher chip rates and lower carrier frequencies showed better resilience against the Doppler effect. The proposed criterion may infer the maximum velocities at which TDOA can be estimated without any measure against the Doppler effect for realistic parameters. This criterion would contribute to the on-line evaluation of the quality of TDOA, which may be beneficial for minimizing TDOA error caused by the Doppler effect by processing the received signals to maximize the PAPR. As future work, experimental results are needed.

## Acknowledgment

This work was supported by the Japan Society for the Promotion of Science (JSPS) KAKENHI Grant Number JP16J03531

## References

[1] Dudek G and Jenkin M 2008 *Springer Handbook of Robotics* (Springer) pp 477–490 ISBN 978-3-540-23957-4
[2] Weston J L and Titterton D H 2000 *Electron. Commun. Eng. J.* **12** 49–64 ISSN 09540695
[3] Thrun S, Fox D, Burgard W and Dellaert F 2001 *Artificial Intelligence* **128** 99–141 ISSN 00043702
[4] Ogiso S, Kawagishi T, Mizutani K, Wakatsuki N and Zempo K 2015 *ROBOMECH Journal* **2** 1–12 ISSN 2197-4225
[5] Ogiso S, Kawagishi T, Mizutani K, Wakatsuki N and Zempo K 2015 *Proceedings of the 22th International Congress on Sound & Vibration (ICSV22)* July (Florence) pp T01.RS01–0626(R)
[6] Ogiso S, Kawagishi T, Koichi M, Wakatsuki N and Zempo K 2016 *The 7th International Conference on Indoor Positioning and Indoor Navigation, IPIN2016* p 141_RP (6 pages) ISBN 9781509024254
[7] Perez-Lorenzo J, Viciana-Abad R, Reche-Lopez P, Rivas F and Escolano J 2012 *Applied Acoustics* **73** 698–712 ISSN 0003682X
[8] Albuquerque D F, Vieira J M N, Lopes S I and Aguilera T 2016 *The 7th International Conference on Indoor Positioning and Indoor Navigation, IPIN2016* October pp 140–RP(8 pages) ISBN 9781509024254
[9] Ebihara T and Leus G 2016 *IEEE Journal of Oceanic Engineering* **41** 408–427 ISSN 03649059

IOP Conf. Series: Journal of Physics: Conf. Series **1075** (2018) 012047 doi:10.1088/1742-6596/1075/1/012047

# Frequency Measurement of Bonang Barung and Peking in Javanese Gamelan using Audacity

**Y Pramudya[1], L Widayanti[1] and F Melliagrina[1]**

[1]Program Studi Magister Pendidikan Fisika, Universitas Ahmad Dahlan, Yogyakarta, Indonesia

E-mail: yudhiakto.pramudya@pfis.uad.ac.id

**Abstract.** Javanese Gamelan has been used as acoustical research to investigate the tune in terms of frequency and intensity. In this research, the bonang barung and peking frequencies were measured using Audacity. It is low cost software to record and analyze the sound. It can process the sounds to generate the FFT of them. The bonang barung and peking are part of Gamelan Gending Bahana Yogyakarta ensemble. The sounds were measured in the hall where the gamelan is located. The lowest and the highest frequencies of the first row of Laras Pelog of Bonang Barung are 609.6 ± 0.1265 Hz and 1050 ± 0.09487 Hz, respectively. For the second row, the lowest and the highest frequencies are 300.1 Hz and 512.8 Hz, respectively. The lowest and the highest frequencies of Laras Pelog of Peking are 1176.1±1.1 Hz and 2101.0±0.0 Hz, respectively. The lowest and the highest frequencies of Laras Slendro of Peking frequencies are 928.1±2.2 Hz and 2118.1±1.1 Hz, respectively. Hence, by using low cost software and non-acoustical room, the gamelan frequency measurement has been performed with high accuracy and precision.

## 1. Introduction

Javanese Gamelan has unique set of instruments that has their own characteristic in term of the way of playing and their pitch. Hence, the intensity and frequency of the instruments are different between each other's. There are two types of tone set in Gamelan. There are Laras Pelog and Laras Slendro. The instruments also have different shapes and material. The shape of Bonang Barung kettle is similar to inverted bowl. The Saron wilah shape is a bar with rounded on the top side. However, the Bonang Barung and Saron have the same material [1]. The frequency has been measured using oscilloscope on Gamelan Swastigitha and Gamelan Kyai Kaduk Manis [2]. The other measurement using sonogram to measure frequency of Gong [3].

The freeware software is important to introduce the sound produced by musical instrument to the students and the player. The Audacity is the example of freeware software. The software is useful tools to introduce the student to understand the signal of voice and noise [4]. The Audacity software has been used to measure frequency of copper flute instrument [5]. Even, the software is useful to be

IOP Conf. Series: Journal of Physics: Conf. Series **1075** (2018) 012047 doi:10.1088/1742-6596/1075/1/012047

employed to assist the students to understand the gravitational wave [6]. Hence, the gamelan frequency especially the Laras Pelog of Bonang Barung is possible to be measured using Audacity [7]. In this paper, we investigate the comparison the frequency measurement between the Bonang Barung and Peking using Audacity.

## 2. Methodology

We measured frequency of Bonang Barung and Peking at the Hall of Universitas Ahmad Dahlan. It is a regular meeting room. Hence, we need to wait until night to reduce the noise. The Bonang Barung and Peking are part of gamelan ensemble in Universitas Ahmad Dahlan. The name of the gamelan is the Gamelan Gending Bahana. The scheme of Bonang Barung of Laras Pelog is shown in figure 1. There are two rows in Bonang Barung. The top row has higher pitch than the lower one. The player need to play the kettles in the first row and the second row, simultaneously. The kettels or small gong is hit by padded stick or tabuh.

| | | | | | | |
|---|---|---|---|---|---|---|
| 4 | 6 | 5 | 3 | 2 | 1 | 7 |
| 7 | 1 | 2 | 3 | 5 | 6 | 4 |

**Figure 1.** The Laras Pelog of Bonang Barung have two rows of kettle or small gong with different pitch.

The Peking has similar arrangement of note as shown in figure 2. There are two types of peking that we used for frequency measurement. There are Laras Pelog and Laras Slendro. They have different wilah or bar arrangement. The peking is also known as Saron Panerus.

| | | | | | | |
|---|---|---|---|---|---|---|
| 1 | 2 | 3 | 4 | 5 | 6 | 7 |

Laras pelog

| | | | | | | |
|---|---|---|---|---|---|---|
| 1 | 2 | 3 | 5 | 6 | 7 | 1' |

Laras slendro

**Figure 2.** The Laras Pelog and Laras Slendro of Peking have different wilah or bar arrangement.

The Audacity software was employed to measure the the Intensity and Time of the signal. The Fast Fourier Transform (FFT) is utilized to analyze the dominant frequency for each tone. The measurement was repeated to obtain the average and standard of deviation of frequencies. The Laras Pelog of Bonang Barung frequency measurement can be shown in figure 3(a). Similar arrangement of microphone and laptop for frequency measurement of Peking can be shown in figure 3(b). The microphone is placed right under the wilah or bar.

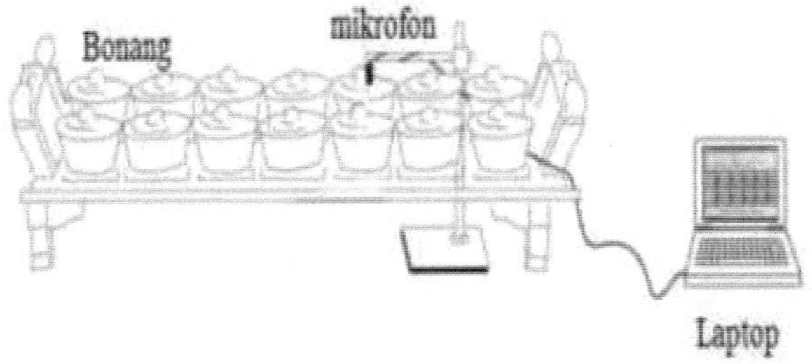


(a)

(b)

**Figure 3.** The frequency measurement scheme of (a) Bonang Barung using Audacity and Laptop and (b) Peking using Audacity and Laptop.

## 3. Result and discussion

### *3.1. Result*

The result of frequency measurement of Bonang Barung and Peking can be discussed in this section. The sound spectrum of Laras Pelog of Bonang Barung and Peking can be shown in figure 4. Each kettle of laras Pelog of Bonang Barung was hit 10 times for statistical purpose. The number of repetition for hitting each wilah or bar was 7 times.

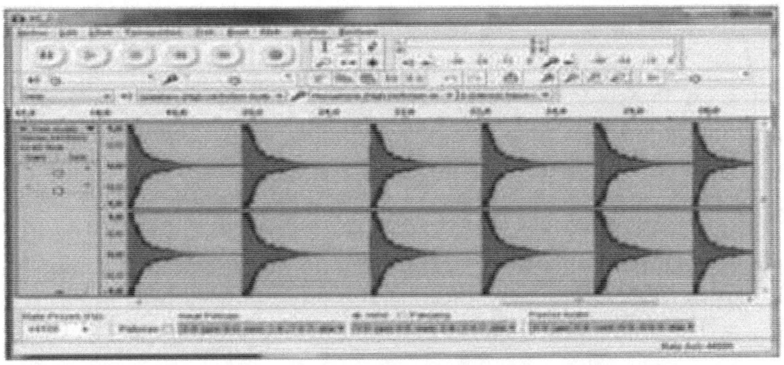

(a)

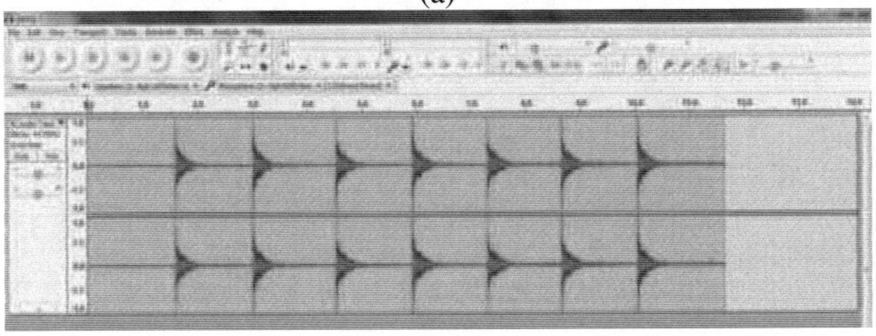

(b)

**Figure 4.** (a) The sound spectrum of Laras Pelog of Bonang Barung in Audacity. (b) The sound spectrum of Laras Pelog of Peking in Audacity.

After the data sound was captured and recorded by using Audacity, the FFT menu of the software was employed to get dominant frequency for each tone as shown in figure 5.

IOP Conf. Series: Journal of Physics: Conf. Series **1075** (2018) 012047 doi:10.1088/1742-6596/1075/1/012047

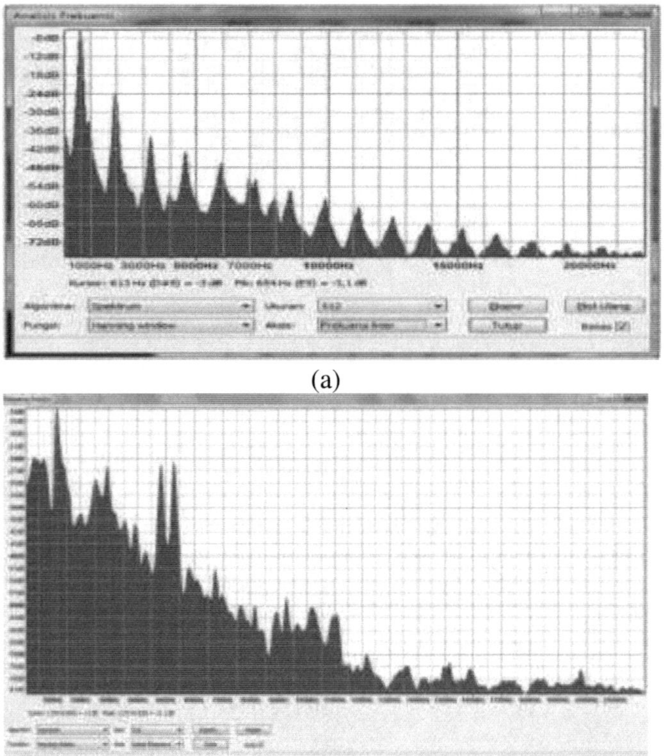

(a)

(b)

**Figure 5.** (a) The FFT analysis of Laras Pelog of Bonang Barung in Audacity. (b) The FFT analysis of Laras Pelog of Peking in Audacity.

From the FFT analysis, we can obtain the dominant frequency of Laras Pelog of Bonang Barung and Peking. The frequency is chosen that has the highest intensity. The cursor on the software is the right tool to pick up the dominant frequency. The frequency of Laras Pelog of Bonang Barung for the higher pitch and lower pitch can be shown in table 1 and table 2, respectively. The table also shows the comparison to the other frequency measurement on different gamelan ensembles and different methods.

**Table 1.** The frequencies of each kettle on the first row (higher pitch) of Laras Pelog of Bonang Barung and their comparison to the other measurements.

| Tone | Frequency (Hz) | Gamelan Swastigitha Frequency (Hz) [2] | Gamelan Kyai Kaduk Manis Frequency (Hz) [2] |
|---|---|---|---|
| 1 | $609.6 \pm 0.1265$ | 602 | 604 |
| 2 | $653.9 \pm 0.03162$ | 643 | 682 |
| 3 | $705.8 \pm 0.06325$ | 708 | 732 |
| 4 | $822.3 \pm 0.03162$ | 828 | 840 |
| 5 | $887.1 \pm 0.03162$ | 887 | 892 |
| 6 | $954.5 \pm 0.09487$ | 950 | 976 |
| 7 | $1050 \pm 0.09487$ | 1052 | 1077 |

IOP Conf. Series: Journal of Physics: Conf. Series **1075** (2018) 012047 doi:10.1088/1742-6596/1075/1/012047

**Table 2.** The frequencies of each kettle on the second row (lower pitch) of Laras Pelog of Bonang Barung and their comparison to the other measurements.

| Tone | Frequency (Hz) | Gamelan Swastigitha Frequency (Hz) [2] | Gamelan Kyai Kaduk Manis Frequency (Hz) [2] |
|---|---|---|---|
| 1 | 300.1 ± 0.03162 | 300 | 310 |
| 2 | 322.3 ± 0.06352 | 324 | 336 |
| 3 | 345.8 ± 0.09487 | 353 | 362 |
| 4 | 411.0 ± 0.06325 | 415 | 424 |
| 5 | 459.5 ± 0.06325 | 444 | 445 |
| 6 | 476.1 ± 0.03162 | 472 | 482 |
| 7 | 512.8 ± 0.1265 | 525 | 538 |

The frequency of Laras Pelog and Laras Slendro of Peking can be shown in table 3. The table also show the comparison to the other frequency measurement on different gamelan ensembles and different methods.

**Table 3.** The frequencies of each wilah of Peking their comparison to the other measurements.

| Tone of Laras Pelog | Frequency (Hz) | Gamelan ITB Frequency using Sound Analyzer (Hz) [8] | Tone of Laras Slendro | Frequency (Hz) | Gamelan ITB Frequency using Sound Analyzer (Hz) [8] |
|---|---|---|---|---|---|
| 1 | 1176 ± 0.4158 | 1208 | 6 | 928.1 ± 0.8315 | 928 |
| 2 | 1272 ± 0.7181 | 1300 | 1 | 1075 ± 0.4158 | 1073 |
| 3 | 1409 ± 0.1890 | 1391 | 2 | 1234 ± 0.3024 | 1246 |
| 4 | 1643 ± 1.398 | 1639 | 3 | 1423 ± 0.1890 | 1418 |
| 5 | 1765 ± 0.2268 | 1757 | 5 | 1636 ± 0.4914 | 1639 |
| 6 | 1862 ± 0.2646 | 1854 | 6 | 1870 ± 1.550 | 1854 |
| 7 | 2101 ± 0 | 2050 | 1 | 2118 ± 0.4158 | 2167 |

### *3.2. Discussion*

We have done measurement of frequency of Laras Pelog of Bonang Barung and Laras Pelog and Laras Slendro of Peking using Audacity software. The results are compared to the other measurements using different gamelan ensembles and methods. The different ensembles and methods are used as references. From the results, we can see that the frequency measurements are having small differences compared to the references. These differences due to different gamelan ensembles and methods. Besides that, the rooms for taking the data are also different.

The frequency measurement of Laras Pelog and Laras Slendro of Peking are having similar deviation from the references compared to the frequency measurement of Laras Pelog of Bonang Barung. This fact is interesting since the way of playing the instrument is different. The kettle of Bonang Barung is hit gently by using the padded stick. In the other hand, usually the wilah or bar of Peking is hit loudly. It is important to investigate the influence of hitting force to wilah to the frequency of wilah. The mechanism for hitting the kettle and wilah need also to be improved in such a way the force is similar between number of hit to repeat the experiment. The shape of kettle and wilah will be interesting part to be studied. Since, they have completely different shape.

The ability of Audacity software to capture the sound and generating analysis data have potential to be used for diverse purposes. The gamelan players and gamelan maker can use this software to assist the tuning process. Since there is no fee for installment, the software is useful for the educator to introduce the concept of wave and oscillation and their application using simple and inexpensive software.

## 4. Conclusion

The Audacity has been employed to measure the frequency of kettle of Laras Pelog of Bonang Barung and wilah of Laras Pelog and Laras Slendro of Peking. The inexpensive and easy method was able to generate the result that has small differences from the references. The results quality need to be improved by investigating the factor of hitting force to the instruments. Hence, the method and software can be also used for measuring the other instrument in the gamelan ensemble. Moreover, the Audacity is powerful software to get fast measurement to assist the player and the gamelan maker to tune the instruments.

## References

[1] Sethares W A 2005 *Tuning, Timbre, Spectrum, Scale* (Berlin: Springer-Verlag) p 202 - 207
[2] Surjodiningrat W, Sudarjana, P J, Susanto 1993 *Tone Measurements of Outstanding Javanese Gamelan in Yogyakarta and Surakarta* (Yogyakarta: Gadjah Mada University Press)
[3] Kuswanto H 2012 *Prosiding Seminar Nasional Penelitian, Pendidikan dan Penerapan MIPA* (Yogyakarta) p F33
[4] Gailey A 2015 *The Physics Teacher* vol 53 p 239
[5] Brazzle B 2011 *The Physics Teacher* vol 49 p 228
[6] Farr B, Schelbert G, Trouille L 2012 *American Journal of Physics* vol 80 p 898
[7] Widayanti L, Pramudya Y 2014 *Prosiding Pertemuan Ilmiah XXVIII HFI Jateng & DIY* (Yogyakarta) p 24
[8] Suyatno, Tjokronegoro H A, Merthayasa IGN, Supanggah R 2013 *Prosiding Seminar Fisika dan Aplikasinya* (Surabaya) p 261

IOP Conf. Series: Journal of Physics: Conf. Series **1075** (2018) 012048 doi:10.1088/1742-6596/1075/1/012048

# The usage of natural materials for the green acoustic panels based on the coconut fibers and the citric acid solutions

**Z A Rachman[1], S S Utami[1], J Sarwono[2], R Widyorini[3] and H R Hapsari[1]**

[1]Engineering Physics, Faculty of Engineering, Universitas Gadjah Mada, Yogyakarta, Indonesia
[2]Engineering Physics, Faculty of Industrial Technology, Intitut Teknologi Bandung, Indonesia
[3]Forest Product Technology Department, Faculty of Forestry, Universitas Gadjah Mada, Yogyakarta Indonesia

E-mail: Zulfi.arachman@gmail.com

**Abstract.** Acoustic panels of sound absorbers are used to overcome noise in an environment. Expensive panels using fabricated materials, which contains a lot of chemicals are mostly used in practice for noise reduction. Natural fibers have been extensively used to produce environmentally friendly composite materials and therefore, known as green materials. Coconut fibers hold great potential for substituting the expensive synthetic fibers in manufacturing acoustic absorption boards. This study aims to determine the acoustical performance of a green particleboard made from coconut fiber and citric acid solution. The acoustical performance is represented by sound absorption coefficient. The sound absorption coefficient of these panels were tested based on ASTM E-1050-98 standard using B&K impedance tube. The small test samples were tested in the frequency range of 50 Hz to 6,400 Hz. The results show that the tested panels are insufficient in absorbing sound of low frequencies. The panels absorption ability is improving in the range of 1,000 to 6,400 Hz. Furthermore, a morphological analysis has been done by using Phenom Pro X Scanning Electron Microscope to observe the effectiveness of citric acid solution in gluing the fiber. The variations of citric acid solution are 20%, 10% and 5%. It is shown that the lowest concentration has more pores distribution.

## 1. Introduction

Noise is an unwanted sound. It can be produced by many sources such as human's voice, running engine, operating machine tool, etc. Noise has become one of the major pollution types in the world and continues to get worse due to the rapid developments of modern industries and transportation[1][2]. Constant exposure to noises can cause all kinds of health problems. According to the World Health Organization's Guidelines for the community noise, the significant health-related effects of noise pollution are hearing impairment, interference with speech communication, degradation of speech performance, annoyance, and sleep disturbance[3][4].

Noise in a building can be eliminated by building materials with acoustical characteristic of a good sound energy absorption. Acoustic is one of the major parameters that determine a comfortable environment in the building. Acoustic comfort refers to the limits in decibels that must be respected in order to preserve the auditory health, making the environment acoustically appropriate and agreeable to the individual [4][5][6].

Currently, the acoustic panels (that are able to absorb the sound energy) available in market for acoustic treatment are made from synthetic materials such glass wool and foam glass. These synthetic materials are found to be not only harmful to human health, but may also cause pollutions and global warming [7].

These issues explore an opportunity to look for alternative materials from natural fibers to be developed as "green" acoustic panel as a new solution to the harmful materials. It is important to produce less-cost and green (eco-friendly) materials that can reduce noise in the buildings. Natural fibers are

IOP Conf. Series: Journal of Physics: Conf. Series **1075** (2018) 012048 doi:10.1088/1742-6596/1075/1/012048

extensively used to produce eco-friendly composite materials. Natural fibers, such as wood, hemp, and coconut fibers, possess great potential to replace the expensive synthetic fibers in manufacturing acoustic absorption boards. It is because natural fibers are easy to find (available in the nature) and do not require high cost to process. Natural fibers also have natural cellular structure that can effectively absorb acoustic energy [8][9].

Porous absorbers are the most commonly used sound absorbing materials. To be an excellent sound absorber, the sound waves must penetrate into the absorbing material where the sound is dissipated in the form of friction and heat. The amount of sound which enters the porous material depends on the wavelength of the sound and its angle of incidence on the absorbing material. The ability of a material to absorb sound of a particular frequency is often described by a sound-absorption coefficient ($\alpha$). [4] [9].

In 2014 according to Badan Pusat Statistik (Indonesia's National Statistics Institution), Indonesia produced 35,000 tons of coconut fiber per year. Unfortunately, its utilization is still insufficient. Therefore, the development of fiber-based sound absorber in Indonesia is really promising. This research aims to study eco-friendly and low cost acoustic panel (sound absorber) based on the coconut fiber and citric acid (natural materials) to reduce noise in the building. [10]

## 2. Materials and methods

This research aims to create/ or study the eco-friendly and low cost acoustic panel as sound absorber based on the coconut fiber and citric acid as the natural materials. The quality of the acoustic panel has been successfully analysed using impedance tube method and Scanning Electron Microscope (SEM).

### *2.1. Materials preparation*

Coconut fibers as the main materials were collected from Gunung Kidul, Yogyakarta. The fibers were air-dried to reduce the moisture content and were cut into 2.5 cm $\pm$ 1 cm. Citric acid as the main material of the adhesive solutions was dissolved in water to a solution concentration of 59 – 60%. The solution was used as adhesive and distributed/mixed manually onto fibers at 5%, 10% and 20% based on the weight of the air-dried fibers.

### *2.2. Acoustic panel manufacturing process*

The mixed fibers were then oven-dried over night at 80°C to reduce the moisture content. The fibers were hand formed into a mat using a forming box of 150 mm x 150 mm, followed by hot pressing into acoustic panel (board) with a distance bar of 10 mm to control the board thickness. The boards were pressed at 180 °C for 10 minutes under a pressure of 2.5 MPa. The target densities of the board were 0.5 g/cm$^3$, 0.25 g/cm$^3$ and 0.125 g/cm$^3$. Prior to the evaluation, all boards were conditioned at ambient conditions for approximately 7 days.

**Figure 1.** Raw materials

**Figure 2.** Forming box

**Figure 3.** Hot pressing process

**Figure 4.** Final product

IOP Conf. Series: Journal of Physics: Conf. Series **1075** (2018) 012048 doi:10.1088/1742-6596/1075/1/012048

### *2.3. Laboratory testing*

The quality of the acoustic panels is represented by sound absorption coefficient and morphological analysis of the board's surface. In total, there are 7 samples that have been used in this research. The test was done by varying the density and citric acid concentration (Table 1).

**Table 1.** List of the samples

| Sample | Density | Concentration of adhesive solution |
|---|---|---|
| 1 | 0.5 g/cm$^3$ | 20% |
| 2 | 0.5 g/cm$^3$ | 10% |
| 3 | 0.5 g/cm$^3$ | 5% |
| 4 | 0.25 g/cm$^3$ | 10% |
| 5 | 0.25 g/cm$^3$ | 5% |
| 6 | 0.125 g/cm$^3$ | 10% |
| 7 | 0.125 g/cm$^3$ | 5% |

*2.3.1. Sound absorption evaluation.* The sound absorption coefficient was evaluated by using the impedance tube with two microphones. The diameter of the sample of this test was 3 cm with 1 cm of thickness.

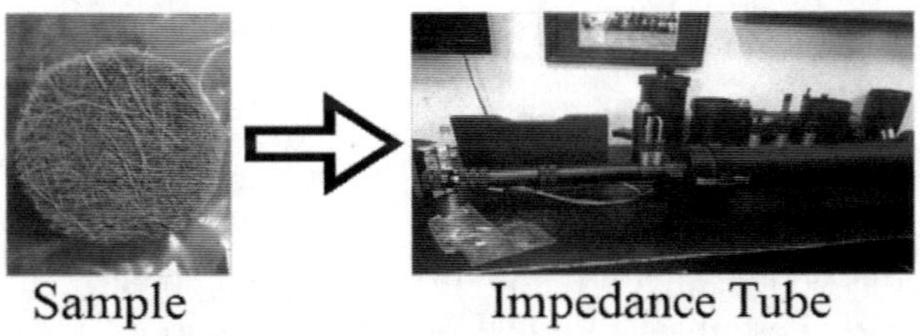


**Figure 5. The** sample and equipment

The sample was tested is in the frequency range of 50 Hz – 6,400 Hz. The sound absorption coefficient is derived by measuring the incident and reflected components of noise, which are generated inside the impedance tube by the sound source. The reflected component is affected by the acoustic properties of the tested sample (sample placed in a holder at one end of a tube).

*2.3.2. Morphological evaluation.* The surface morphology of the boards was observed using SEM. An accelerating voltage of 10 – 15 kV was used to collect the micrographs on the surface of the samples. The size of the samples was 5 mm x 5mm. The morphology of the boards was observed at the room temperature. Before testing, the samples were sliced and mounted onto SEM stubs using double sided adhesive tape.

**Figure 6.** Sample for SEM testing

**Figure 7.** Micrographs recording process

## 3. Results and discussions

### *3.1. Morphological Evaluation*

The analysis was performed with two samples which had 20% and 10% of citric acid solution. Unfortunately, during the analysis, sample 3 which has 5% of citric acid solution was unable to meet the standard of the sample to be analyzed in the SEM. This analysis aimed to see the effectiveness of citric acid in creating the pores.

It is observed that both of the samples had irregular porous shape. The porous distribution represented by the red line in the figures. The captured micrographs of the porous of samples show that the porous distribution is decreased when the concentration of citric acid solution is increased from 10% to 20%.

**Figure 8.** Porous distribution of 10% citric acid solution sample

**Figure 9.** Porous distribution of 20% citric acid solution sample

These porous distributions will be affecting the sound absorbing quality. In Figure 8 the network of interlocking porous are clearly described. In a real field this network of interlocking porous plays an important part, because such a structure decreases air-flow resistivity and thus creates dissipates the sound energy in the pores.

### *3.2. Sound absorption evaluation*

The sound absorption coefficients are graphically illustrated in Figure 10. The zero value in the y-axis indicates perfect sound reflection and the value of one implies a perfect sound absorption. The results showed that tested boards are insufficient in absorbing sound energy of low frequencies.

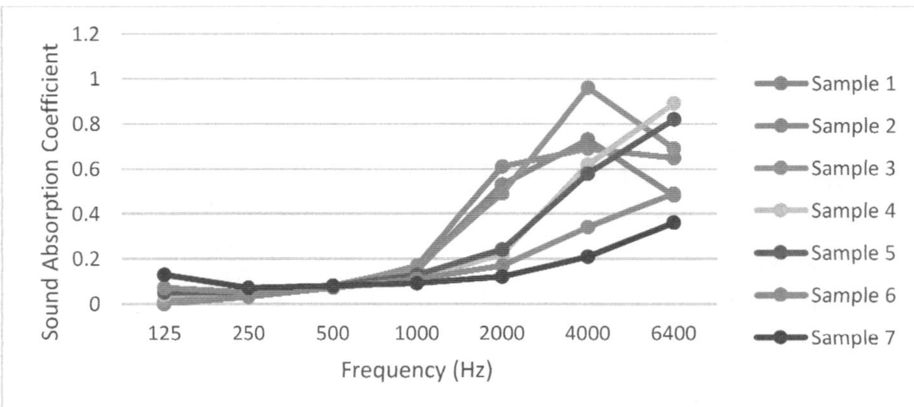


**Figure 10.** Sound absorption coefficient for all samples

The value varies between 0 – 0.13 in the range of 125 Hz – 500 Hz. It occurred because the boards do not have a sufficient thickness. It claims the theory that in order for the sound absorber to have at least sound absorption of 80%, the thickness must be larger than 1/10 of the desired wavelength. Various studies that deals with sound absorption in porous materials have concluded that sound absorption at low frequencies has direct relationship with thickness.

The panel's absorption ability is improving in the range of 1000 Hz to 4000 Hz. The values vary between 0.09 – 0.96. Sample 2, which contains of 10% of citric acid solution and has 0.5 gr/cm$^3$ of density, indicates the best performance in the frequency of 1000 Hz and 2000 Hz, while the best performance in the frequency of 4000 Hz is Sample 3, which contains 5% of citric acid solution and has 0.5 gr/cm$^3$ of density. The increased absorption coefficient value in the range of 1000 Hz – 4000 Hz occurred because the locked thickness of the boards (0.01 m) nearly met the minimum thickness required as shown in Table 2.

**Table 2.** The minimum thickness required

| Sound velocity | Desired frequency | Wave length | Thickness |
|---|---|---|---|
| 343.5 m/s | 125 Hz | 2.750 m | 0.275 m |
| | 250 Hz | 1.370 m | 0.137 m |
| | 500 Hz | 0.690 m | 0.069 m |
| | 1,000 Hz | 0.340 m | 0.034 m |
| | 2,000 Hz | 0.170 m | 0.017 m |
| | 4,000 Hz | 0.080 m | 0.008 m |
| | 6,400 Hz | 0.050 m | 0.005 m |

At the frequencies above 4,000 Hz the absorption capability of Sample 1 – Sample 3 are decreasing. But Sample 1 – 5 still have a great ability in absorbing the sound. In the highest frequency, the best absorption capability is Sample 4 which has 0.25 gr/cm$^3$ of density and contents 10% of citric acid solution. This indicates that at higher frequencies, thickness has insignificant effect on sound absorption.

Moreover, all the panels experienced increasing sound absorption coefficient because of the structure of the coconut fiber, which results in high absorption coefficient.

**Figure 11.** Micrographs of cavity on the coconut fiber

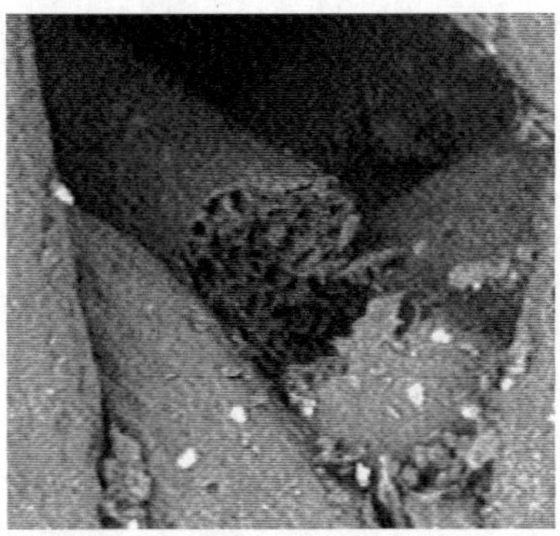

**Figure 12.** Micrograph of lumen inside the fiber

The white dots in Figure 5.9 are indicated as the cavity of the coconut fiber. The cavity on the surface of fiber is a great feature to improve the sound absorbing quality because additional thermal energy was dissipated more rapidly due to the increased frictional surface area. The sound absorption coefficient was therefore correspondingly increasing. The cavity inside the fiber (lumen) had been captured during the morphological analysis test. The sound waves propagated vibration energy through the air spaces in the fiber. Part of this sound energy was converted into heat.

IOP Conf. Series: Journal of Physics: Conf. Series **1075** (2018) 012048 doi:10.1088/1742-6596/1075/1/012048

## 4. Conclusions

Coconut fibers and citric acid solutions are found to be suitable materials to create acoustic panels. It has been proofed by micrographs on the surface of the samples. It is observed that coconut fiber has natural cellular structure (cavity and lumen) structure, which can effectively absorb the sound energy. Citric acid plays an important role in the bonding process and it is also observed that porous distribution is decreased when the concentration of citric acid is increased. Moreover, The acoustic panels are insufficient in absorbing sound in the low frequencies. The acoustic panel's ability is improving in the frequency 1,000 Hz – 4,000 Hz. The best panel in those frequencies is sample 3, which can absorb more than 90% of sound energy in frequency of 4,000 Hz.

## References

[1] Y. Wang, C. Zhang, L. Ren, M. Ichchou, M.-A. Galland, and O. Bareille, "Influences of rice hull in polyurethane foam on its sound absorption characteristics," *Polym. Compos.*, vol. 34, no. 11, pp. 1847–1855, Nov. 2013.

[2] J. G. Gwon, S. K. Kim, and J. H. Kim, "Sound absorption behavior of flexible polyurethane foams with distinct cellular structures," *Mater. Des.*, vol. 89, pp. 448–454, Jan. 2016.

[3] WHO, "Guidelines dor Comumunity Noise." World Health Organization, 2000.

[4] A. Gupta, *Industrial safety and environment*. New Delhi: Laxmi Publications (P) LTD, 2006.

[5] C. Mediastika, "Potensi Jerami Padi Sebagai Bahan Baku Panel Akustik," Yogyakarta, vol. 35, N0. 2, Desember 2017, pp. 183–189.

[6] J. F. dos Santos, L. Seligman, and T. M. Tochetto, "Conforto acústico na percepção de escolares alfabetizados," *Rev. Soc. Bras. Fonoaudiol.*, vol. 17, no. 3, pp. 254–259, 2012.

[7] A. Putra, Y. Abdullah, H. Efendy, W. M. F. W. Mohamad, and N. L. Salleh, "Biomass from Paddy Waste Fibers as Sustainable Acoustic Material," *Adv. Acoust. Vib.*, vol. 2013, pp. 1–7, 2013.

[8] L. Peng, B. Song, J. Wang, and D. Wang, "Mechanic and Acoustic Properties of the Sound-Absorbing Material Made from Natural Fiber and Polyester," *Adv. Mater. Sci. Eng.*, vol. 2015, pp. 1–5, 2015.

[9] J. Sataloff, R. T. Sataloff, and R. T. Sataloff, Eds., *Occupational hearing loss*, 3rd ed. Boca Raton, FL: CRC Taylor & Francis, 2006.

[10] Badan Pusat Statistik, "Statistik Indonesia Tahun 2014," 2015. .

[11] J. E. G. van Dam *et al.*, "Process for production of high density/high performance binderless boards from whole coconut husk," *Ind. Crops Prod.*, vol. 24, no. 2, pp. 96–104, Sep. 2006.

[12] J. Khedari, N. Nankongnab, J. Hirunlabh, and S. Teekasap, "New low-cost insulation particleboards from mixture of durian peel and coconut coir," *Build. Environ.*, vol. 39, no. 1, pp. 59–65, Jan. 2004.

IOP Conf. Series: Journal of Physics: Conf. Series **1075** (2018) 012049 doi:10.1088/1742-6596/1075/1/012049

# Study of the Effect of Splaying Wall to Modify Acoustic Modes Distribution in Small Room

**H Gunawan[1] and T Aditanoyo[1]**

[1]ALTA Integra (Jakarta, Indonesia)

Email: info@vokuz.com

**Abstract**. Modes are known as ones of the important acoustical issues in small room. These phenomena could drastically change the spectral and temporal characteristic of sound. Studies have been done in optimizing rectangular room dimension, optimizing source and listener position, and optimizing low frequency with acoustic panel. To follow the best room ratio, a built room has to be cut a lot. This paper is studied the effect of splaying wall in attempt to shift the modes without trimming too much space. Axial room modes from different room ratios – including random, splay wall, Bolt, Louden, Cox – were calculated and compared. Low frequency characteristic of small rooms were observed aided by FEM simulation so that SPL result could be graph and statistically analyzed. It was found that splaying some walls from the random room dimension could distribute the modes evenly, similar with Cox', though the SPL standard deviation was raised. Further research will be continued through psychoacoustic evaluation.

**Keywords**: Room Modes, Room Ratio, Splay Wall, Modes Distribution, FEM

## 1.Introduction

Room modes add acoustic distortion that could drastically change the spectral and temporal characteristic of sound in critical listening space. Studies have been done in optimizing rectangular room dimension, with the essence of method is to avoid degenerate modes by evenly distributing modal frequencies. Design criteria suggested is (1) to avoid multiple modal in one third octave (2) to avoid the absent of modes at every one third octave.

Bolt [1] using design charts method that determine the average modal spacing and evenly spaced modes in a room. Ratios of 2:3:5 and 1: 2.33:4.33 and 1:1.26:1.59 are suggested consecutively.

Louden [2] calculated the modal distribution for a large number of room ratios and published a list of preferred dimensions based on a single figure of merit with the lowest the standard deviation of the intermodal spacing. One of the room ratio is 1:1.4:1.9.

Bonello [3] developed a criterion that the modal density in one third octave frequency band should never decrease when going up from a band to the next higher band. Modes with coincidental frequency are only tolerated in one third octave band with five or more modes.

Cox et al. [4] continue Louden research by developed computer algorithm using image source modelling to find the best room ratio. The computer algorithm can go through thousands of variations with more precise calculation. The calculated result was compared to Bolt, Louden, IEC room ratio and showing that Cox' room ratio has the lowest standard deviation.

In practice, it is difficult to have the room comply with ideal room ratio suggested above since the listening room usually has to be designed in a room that already built. To follow the best room ratio, a room's one or more dimensions have to be shortened and it will reduce a lot of space. To save room space and cost, this paper suggests an alternative method to evenly distribute modal frequency by splaying wall or ceiling instead of preserving the shoebox shape room. Bolt [5] said boundary irregularities may produce interactions and consequent spreading of energy among modes, resulting in greater sound diffusion throughout the room. Thus the fewer shape symmetries a room possesses, the greater will be the energy mixing and sound diffusion. In contrast, D'Antonio [6] said that making the room non-rectangular does not make modes disappear and does not make the modal behavior less pronounced than in an equivalent rectangular room, just different.

This paper is to study the effectiveness of splaying wall in attempt to evenly distribute the modal frequency and reduce peaks and dips due to the presence of modes. The contradiction of Bolt's and D'Antonio's statement is also discussed.

## 2. Comparison of Established Shoebox Room Ratios

Assumed we have random room with dimension 6.583-meter length, 4.888 meters width and 2.732-meter height. This random room does not comply with any ratio recommendation according to Bolt, Louden, or Cox. **Table 1** show that if the random room wants to follow Cox ratio for small listening room [7], the space will be reduced from about 87 $m^3$ to about 60 $m^3$. Space loss due to reducing dimension is expensive considering expensive land cost at big city.

**Table 1**. Established room ratios, dimension, and volume

| Room Ratio | Dimension | | | Volume ($m^3$) |
|---|---|---|---|---|
| | Length (m) | Width (m) | Height (m) | |
| Random (1:1.79:2.41) | 6.583 | 4.888 | 2.732 | 87.909 |
| Bolt (2:3:5) | 6.583 | 3.950 | 2.633 | 69.466 |
| Louden (1:1.4:1.9) | 5.191 | 3.825 | 2.732 | 54.245 |
| Cox (1:1.56:1.86) | 5.082 | 4.262 | 2.732 | 59.174 |

The formula for determining the frequency of each mode inside a room, $f$, is

$$f = \frac{c}{2}\sqrt{\left(\frac{n_x}{L}\right)^2 + \left(\frac{n_y}{W}\right)^2 + \left(\frac{n_z}{H}\right)^2}, n = 1,2,3, \ldots$$

with c is the speed of sound; L, W, H are consecutively length, width, and height of the room; n is the index of the room modes. Primary axial modes frequency of length (1,0,0) is 26 Hz; primary axial modes frequency of width (0,1,0) is 35 Hz; and primary axial modes frequency of height (0,0,1) is 63 Hz. **Table 2** shows the distribution of axial modes based on one third frequency propose by ISO 266 axial modes of experiments room. Primary axial modes of Random room is absent on the 3rd and 4th one third octave. In our experience, the absent of modes for 2 of one third octave bandwidth give the feeling of lack of bass in this room. Bolt, Louden, and Cox show more evenly spaced modes compared to the random room.

FEM (Finite Element Method) was utilized to predict the SPL (Sound Pressure Level) in a point inside a room. Each room was drawn based on the dimension in **Table 1**. The air inside the room had density 1.2 $kg/m^3$ and speed of sound 343 m/s. All of the surfaces were defined as totally reflective boundaries to emphasize the effect of standing wave inside the room. The room then was broken down into elements of tetrahedral. The maximum size of each element was 0.5 m so the calculation duration did not take too much time while the observable valid frequency still maintained below 343 Hz. Then, mode shapes could be calculated.

**Table 2**. Distribution of primary axial modes of various room ratios

| ISO 266 | | Axial Primary Modes (Hz) | | | |
|---|---|---|---|---|---|
| 1/3 Octave | Freq (Hz) | Random | Bolt | Louden | Cox |
| 1st | 22.1 – 27.8 | (1,0,0) 26 | (1,0,0) 26 | | |
| 2nd | 27.8 – 35.1 | (0,1,0) 35 | | (1,0,0) 33 | (1,0,0) 34 |
| 3rd | 35.1 – 44.2 | | (0,1,0) 43 | | (0,1,0) 40 |
| 4th | 44.2 – 55.7 | | | (0,1,0) 45 | |
| 5th | 55.7 – 70.2 | (0,0,1) 63 | (0,0,1) 65 | (0,0,1) 63 | (0,0,1) 63 |

The source was defined as monopole point source at a corner of room. Monopole point represents omnidirectional behavior of speaker when exciting low frequency sound. The corner was chosen to nullify the speaker-boundary interference effect. The acoustic power excited by the source was $1\times10^{-6}$ Watt and flat throughout the whole frequency band. The SPL probe was located at the 3D diagonally opposite corner of the source. At a mode frequency, the dimensional corner is the location with highest pressure fluctuation hence resulting highest SPL among other locations. Each simulation was run from 20 Hz to 200 Hz with 1 Hz increment.

**Figure 1** shows SPL calculation results of 4 different room ratios. The SPL standard deviation and average is provided in **Table 3**. It is shown that Cox room ratio has the lowest standard deviation with high average. Considering that the room's SPL response with low standard deviation is preferable [4], it indicates Cox room ratio may be the best room ratio with less coloration and more powerful bass sound. Meanwhile, the random room comes out as the second best despite of not complying with any established room ratio.

**Table 3**. SPL standard deviation and average of various room ratios

| | random | bolt | louden | cox |
|---|---|---|---|---|
| Stdev (dB) | 11.66612 | 13.70007 | 15.21464 | 11.32772 |
| Average (dB) | 74.47975 | 72.78577 | 75.24404 | 74.25697 |

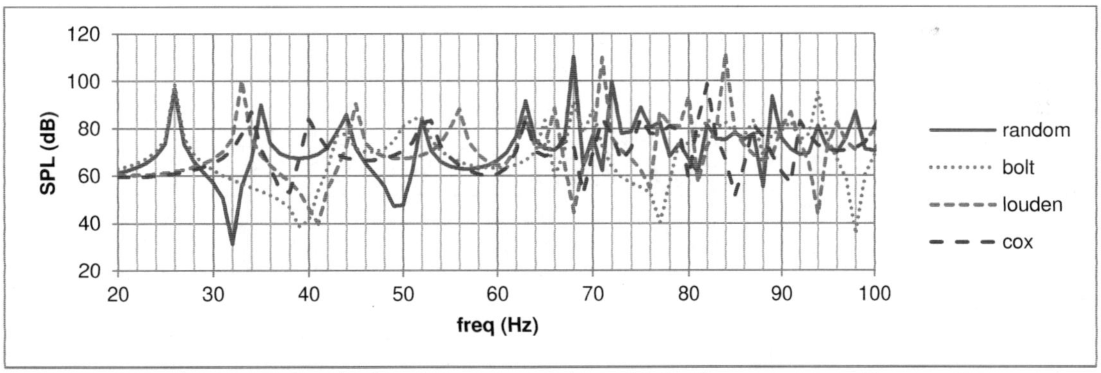


**Figure 1**. Sound pressure level of various room ratios with both source as well as receiver at corner and 3D diagonally opposing each other

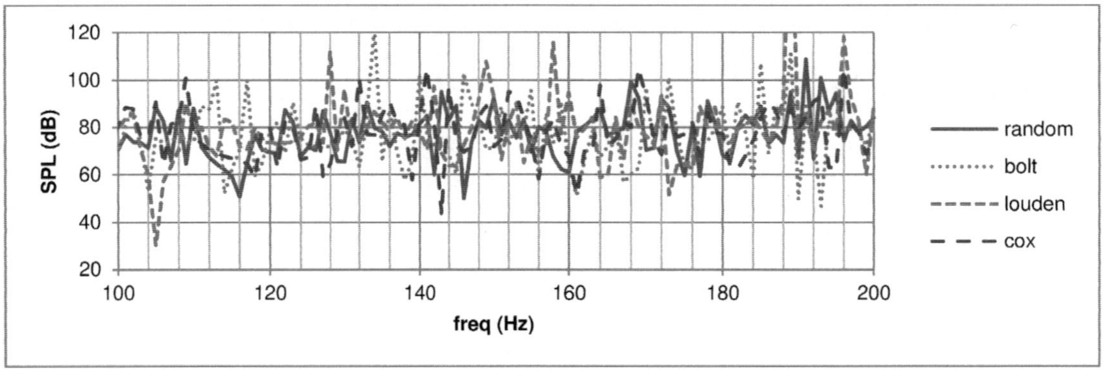


**Figure 1**. Sound pressure level of various room ratios with both source as well as receiver at corner and 3D diagonally opposing each other (continued)

## 3. Splay Wall Room Experiment

Splaying walls may be the solution to balance between minimizing reduced room volume and achieving room with less modes degenerate problems. The splay room illustration can be found at **Figure 2**. According to **Table 4**, cutting volume from random room to splay room is a half compared to the cox room ratio..

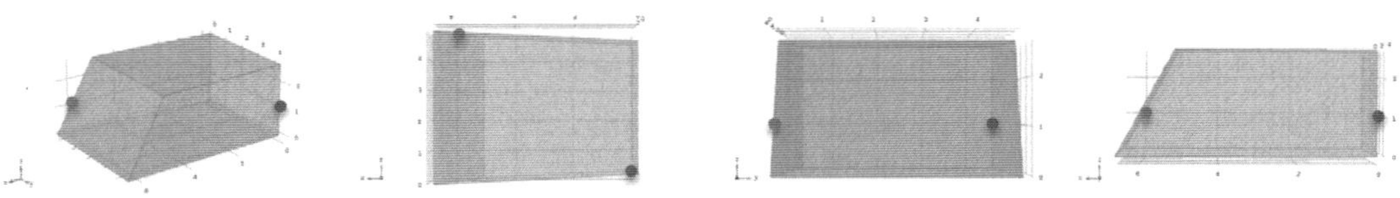

(a) (b) (c) (d)

**Figure 2**. Splay room (a) 3D view, (b) top view, (c) rear view, (d) left-side view; blue dot is the souce, red dot is the receiver

**Table 4**. Random, cox, and splay wall room ratios, dimension, and volume

| Room Ratio | Dimension | | | Volume ($m^3$) |
|---|---|---|---|---|
| | Length (m) | Width (m) | Height (m) | |
| Random (1:1.79:2.41) | 6.583 | 4.888 | 2.732 | 87.909 |
| Cox (1:1.56:1.86) | 5.082 | 4.262 | 2.732 | 59.174 |
| Splayed (1:1.56~1.79:1.86~2.41) | 5.082~6.583 | 4.262~4.888 | 2.732 | 72.274 |

**Table 5** shows better distribution of axial modes in splay wall room with absent of axial modal only at 4$^{th}$ one third octave; compared to the random room with absent of axial modal at 3$^{rd}$ and 4$^{th}$ one third octave. The splay wall room has the same axial modes pattern with the cox room ratio. The axial modes shape of random, cox, and splay wall room are depicted in **Figure 3** in isosurfaces plot. Surfaces with different color indicates different pressure with blue color means the lowest sound pressure that lay below zero and red color means the highest sound pressure that lay above zero. As the sound pressure oscillates periodically, both blue and red surfaces imply the area with the highest sound pressure level and known as antinodes. Green yellowish surfaces indicate zero sound pressure which means no sound at the area and known as nodes. Irregularity in boundary surface like splayed wall can undermine the modes shape in a shoebox room. This disruption is preferable as the sound field is more diffuse.

To analyze the sound pressure level of splay wall room in frequency response, FEM with similar configuration was applied and the result can be found at **Figure 4**. The differences with previous FEM depicted in **Figure 1** are position of both monopole sound source and receiver probe that were set at 1

IOP Conf. Series: Journal of Physics: Conf. Series **1075** (2018) 012049 doi:10.1088/1742-6596/1075/1/012049

m height above floor while were remained opposing each other 2D diagonally (see **Figure 2**). This configuration is implemented to make the result practically relevant for ear level of a sitting person is around 1 m. Splayed room showing highest peak at 55 Hz at 107 dB and lowest dip at 106 Hz at 0.3 dB. Statistical calculation of SPL from these three different rooms can be seen at **Table 6**. Curiously, SPL standard deviation in splay wall room comes as the highest with its average is marked as the lowest among random room and cox ratio room.

**Table 5**. Distribution of primary axial modes of random, cox, and splay wall room

| ISO 266 | | Axial Primary Modes (n,n,n) Hz | | |
|---|---|---|---|---|
| 1/3 Octave | Freq (Hz) | Random | Cox | Splayed |
| 1st | 22.1 – 27.8 | (1,0,0) 26 | | |
| 2nd | 27.8 – 35.1 | (0,1,0) 35 | (1,0,0) 34 | (1,0,0) 29 |
| 3rd | 35.1 – 44.2 | | (0,1,0) 40 | (0,1,0) 38 |
| 4th | 44.2 – 55.7 | | | |
| 5th | 55.7 – 70.2 | (0,0,1) 63 | (0,0,1) | (0,0,1) 64 |

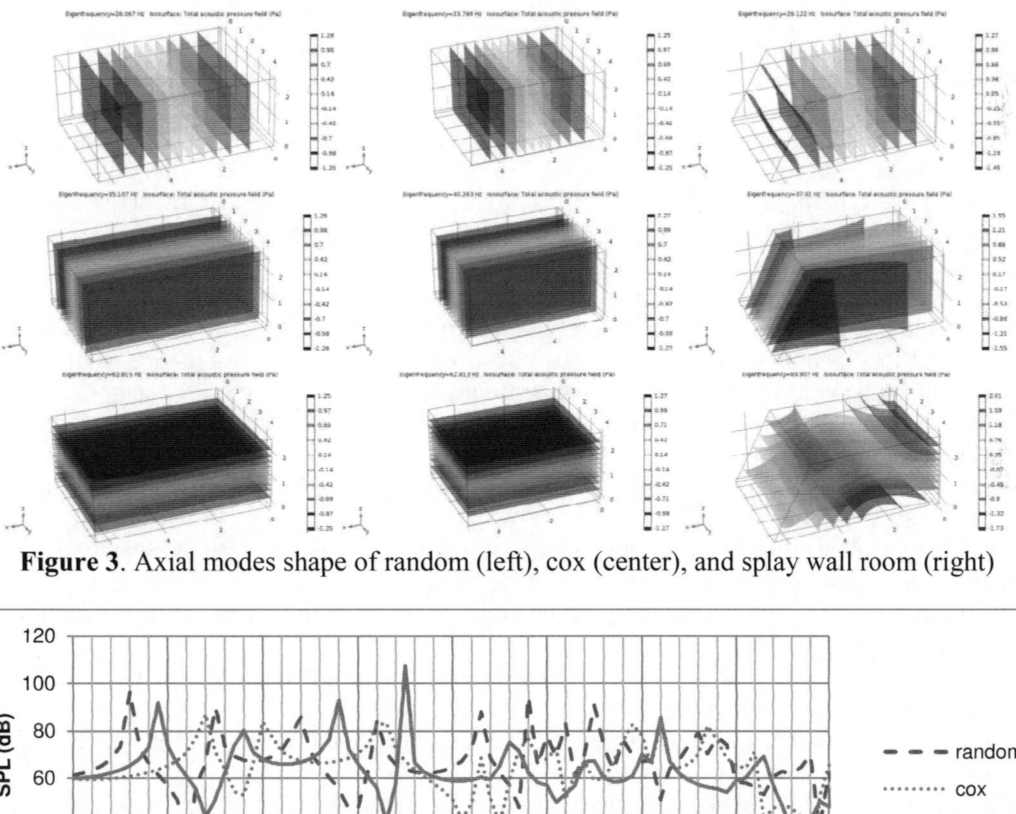

**Figure 3**. Axial modes shape of random (left), cox (center), and splay wall room (right)

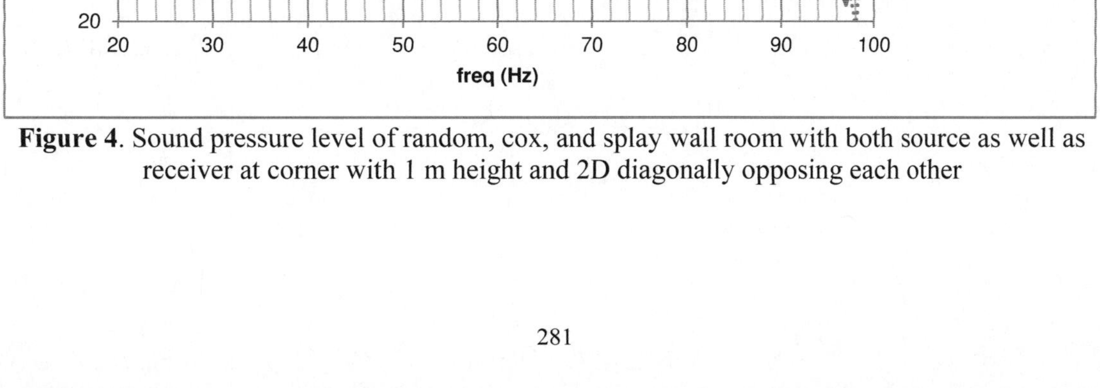


**Figure 4**. Sound pressure level of random, cox, and splay wall room with both source as well as receiver at corner with 1 m height and 2D diagonally opposing each other

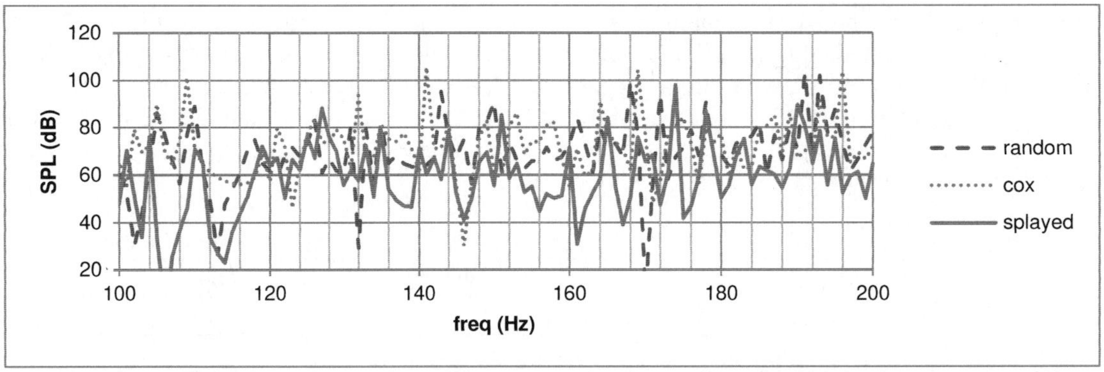


**Figure 4**. Sound pressure level of random, cox, and splay wall room with both source as well as receiver at corner with 1 m height and 2D diagonally opposing each other (continued)

**Table 6**. SPL standard deviation and average of random, cox, and splay wall room

| | random | cox | splay |
|---|---|---|---|
| Stdev (dB) | 13.21111 | 12.56185 | 14.21583 |
| Average (dB) | 68.00068 | 68.34318 | 60.41051 |

## 4.Conclusion

Splaying the wall of small room can shift the axial modes and make more even distribution without cutting too much space compared to resizing room while maintaining the shoebox room shape. The splay wall room also shows more diffuse sound field inside the room. Some of the peaks and dips in SPL in the splay wall room are become more extreme. Hence, this room is recommended to take further treatment such as considering the source and listener position or applying sound absorption panel that tuned at heavily boosted peak frequencies.

After all of the applied treatment to the small room, it is still unknown which kind of character is preferred; whether it is with evenly distributed modes rather than modes with coincidence frequency, or with lowest SPL standard deviation rather than ignoring this parameter. In practice, some people go for the room that a bit colorized with peak in some frequencies.

## References

[1] R. H. Bolt, 'Note on normal frequency statistics for rectangular rooms', *J. Acoust. Soc. Am.*, vol. 18, no. 1, pp. 130–133, 1946.
[2] M. Louden, 'Dimension-ratios of rectangular rooms with good distribution of eigentones', *Acta Acust. United Acust.*, vol. 24, no. 2, pp. 101–104, 1971.
[3] O. J. Bonello, 'A new criterion for the distribution of normal room modes', *J. Audio Eng. Soc.*, vol. 29, no. 9, pp. 597–606, 1981.
[4] T. J. Cox, P. D'Antonio, and M. R. Avis, 'Room sizing and optimization at low frequencies', *J. Audio Eng. Soc.*, vol. 52, no. 6, pp. 640–651, 2004.
[5] R. H. Bolt, H. Feshbach, and A. Clogston, 'Perturbation of sound waves in irregular rooms', *J. Acoust. Soc. Am.*, vol. 14, no. 1, pp. 65–73, 1942.
[6] P. D'Antonio, 'Minimizing acoustic distortion in project studios', *Disponıvel Em Httpwww Rpginc Comcgi-Binbyteserver PlnewslibraryPS AcD Pdf Acesso Em*, vol. 5, 2004.
[7] 'University of Salford | A Greater Manchester University'. [Online]. Available: http://www.acoustics.salford.ac.uk/acoustics_info/room_sizing/?content=best. [Accessed: 12-Sep-2017].

IOP Conf. Series: Journal of Physics: Conf. Series **1075** (2018) 012050 doi:10.1088/1742-6596/1075/1/012050

# Evaluation of acoustic performance open-plan office at Jawa Pos building based on ISO 3382-3

**Thabrani H[1], Susilo I[1], Suyatno[1]**

[1]Acoustics laboratory, Departement of Physics, Institut Teknologi Sepuluh Nopember, Surabaya, Indonesia

E-mail: kangyatno@physics.its.ac.id

**Abstract.** The acoustic performance of open-plan office has become an important topic for discussion along with the development of office centers in major cities in Indonesia, including in Surabaya. Acoustic problems at open plan office often arise, such as background noise and speech privacy. Based on ISO 3382-3 2012, an open plan office must meet certain standards to keep productivity and concentration. These parameters are Average A-Weighted Background Noise (Lp, A, B), A-Weighted SPL of Speech at 4 m (Lp, A, s, 4m), Spatial Decay Rate of A-Weighted SPL of Speech (D2,s), Distraction Distance (rD), Privacy Distance (rP). From measurement and calculation did to an office building in Surabaya obtained a value of Lp, A, B equal to 52 dBA, Lp, A, s, 4m 52 dBA, D2, S 4,19 dBA, rD 11,75 m, rP 23, 8 m, and STI at the nearest table 0.78. These results indicate that further acoustic treatment is required to meet the standards specified by ISO.

## 1. Introduction

Acoustics is a rapidly growing topic around the world, especially in big city cities. Along with the proliferation of buildings, buildings, housing with all its uniqueness, it takes a good plan, including in terms of a sound system. Not only concert halls, auditoriums, classrooms, office space also require convenience for increased productivity. In concept and space efficiency, open office space will require less space.

Open plan office was introduced in 1960 by Quickborner Team in Germany. There are many advantages to this office concept, making the office more efficient, efficient place, communication, coordination and easy supervision, and can increase morale.

Open-plan office is currently used for all types of offices, including offices that require high concentration. From here, the concept of open-plan office raises many problems, especially in terms of speech privacy and noise control. Various regulations and standardization were created to address this issue. ISO (International Standard Organization) has published a special standard for open-plan office to be applicable to various types of work with an emphasis on optimal performance. In this study, the evaluation of open office space in Surabaya along with the solution recommendation to improve performance and acoustic comfort for employees can be achieved.

ISO 3382-3: 2012 describes the method of measuring acoustic space in an open-plan office. In it, there are measurement procedures, required equipment, provisions, data retrieval methods, and presentations on reports. This standard is used in medium and large office space. The parameters measured are the distraction distance *rD*, the decay of *D2, S* sound, the sound at a distance of 4 meters, and the STI on the nearest work table *rP*. all these measurements use A (weighting) weighting. Each point will do 4 values, that is, SPL pink noise *Lp, Ls, STI*, background noise *Lp, B*, and distance from the source.

Outside and indoor sounds that appear permanently and stable at some level without the presence of a prominent noise source are called background noise. Measurement of background noise is done first to find out NC (Noise Criteria) a room for each frequency.

Speech Transmission Index (STI) is an objective parameter to determine the clarity of the conversations transmitted from the speaker to the listener through the transmission line. The STI method has continued to be developed and improved to date since its introduction in the 70s. Major developments in revising the STI have been made with the International Electro technical Commission (IEC 60268-16) generated in 2011. The conversations are based on how to communicate with people. In many situations, the conversation signal will decrease in the transmission line between the speaker and the listener, which means less clarity of the conversation captured by the listener's ears. To calculate the decrease in sound clarity, a rapid and objective method has been developed, Speech Transmission Index (STI). The STI method applies to certain signals being recovered from the receiver. The quality of STI is derived and expressed at a value between 0 to 1.

**Table 1.** Speech Transmission Index (STI) Quality (IEC 60268-16)

| Label STI | Nilai STI |
|---|---|
| Bad | 0 - 0.3 |
| Poor | 0.3 - 0.45 |
| Fair | 0.45 - 0.6 |
| Good | 0.6 - 0.75 |
| Excellent | 0.75 - 1 |

Based on Table 1, the higher the STI value indicates the better the voice clarity when the acoustic privacy speech is required, the STI value should be low. (Stout, 2015).

In this paper is reported about the acoustic condition of the space of an open office in Surabaya. The acoustic parameters used in the evaluation are background noise, distraction distance, privacy distance, spatial distance, spatial decay rate, and Lp, A, s, 4m.

**Table 2.** Standard value of open office space based on ISO 3382-3

| No | **Parameter** | **Annotation** | **Good** | **Poor** |
|---|---|---|---|---|
| 1 | *Background Noise* | | < 50 dB | > 50 dB |
| 2 | *Distraction distance* | STI 0,5 | < 5 m | > 10 m |
| 3 | *Privacy distance* | STI 0,2 | | |
| 4 | *Spatial decay rate* | Sound Atennuation | > 7 dB | < 5 dB |
| 5 | Lp,A,s,4m | SPL speech 4 m | < 48 dB | > 50 dB |

In part II discussed the methodology, part III contains the results of the evaluation and analysis of parameter values. At the end, the conclusions from the research are presented.

## 2. Methodology

The room that became the object of research is one open office in Surabaya with the area (34x36 meters.) Figure 1 shows a sketch of the space in the perusal.

**Figure 1**. Room Condition

In the existing condition, inside the room there are various office supplies, Work Desk (234), Meeting (5), Chair (278), Computer (234), Center stage (1), Lamp 34, (4), Speaker (4), Television (5), scanner (3), Photocopy machine (1), cleaning equipment, Billiard Table (1), Ducting AC (4), Hydrant (4), Plant Pot (4), Hanging Banner (8), Trash Can, Kardus, ATK and Books. Figure 2 shows one corner of the space.

The space evaluated appears to be a relatively symmetrical space. Therefore data retrieval can be taken only by a quarter of the office. In accordance with ISO 3382-3 standards, measurements are carried out in length as in line 1 and 2. Figure 3 shows the measurement plan.

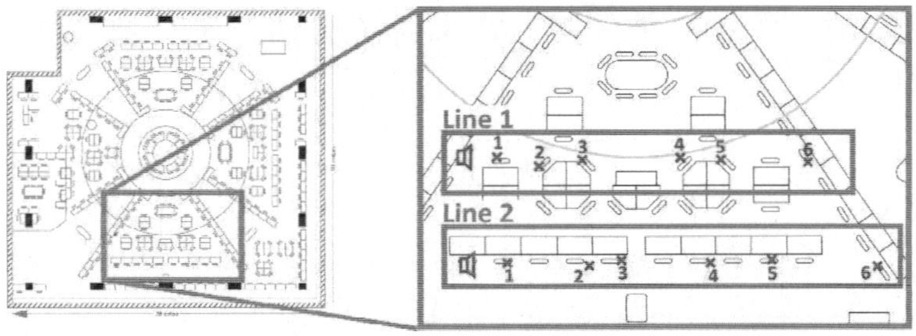


**Figure 2.** Measurement plan

Based on Fig. 3, the acoustic parameters are measured, including background noise, distraction distance, privacy distance, spatial distance, spatial decay rate, and Lp, A, s, 4m. Measurements are performed by impulse response as the source of sound and microphone as a listener.

The measurement result is a .wav signal from every microphone position. The measurement signal (recording) is then analyzed using the sound analyzer to obtain the desired acoustic parameters.

## 3. Result and Discussion

As mentioned in section I, open plan office is a concept of office layout without large divider barrier between employees. Inside a large space lined a desk table is very much placed coincide. In addition to forming a good communication process in the work, there are some negative effects such as noise. Noise can interfere with employee productivity. In this case, a single source of noise will affect many people around it. Fixed noise is background noise mostly caused by AC, computer CPU. Based on ISO 3382-3, the maximum default value for Noise Criteria in open plan office space is 40 dB. If the value exceeds 40 dB, it can disrupt the performance of its employees.

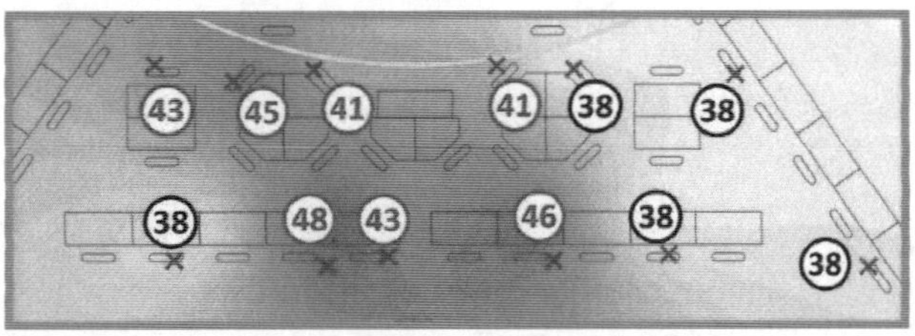


**Figure 3.** Mapping *Background noise* at each measurement position

IOP Conf. Series: Journal of Physics: Conf. Series **1075** (2018) 012050 doi:10.1088/1742-6596/1075/1/012050

In addition to the above noise, there is also an impulsive or temporal noise, which is caused by the voice of the person speaking. In addition to NC In ISO, the parameters used for evaluation are SPL within 4m (Lp, A, s 4m), and decibel decrease value during multiplication distance (D2, s).

Based on the measurement result, the value of noise criteria as in Figure 4 is obtained.

The height of the background noise in the middle of the picture is caused by the cooling of the room which is flowed through ducting enclosed by aluminum foil. The holes of the ducting are split into two. The possible value of 48 dBA is obtained because it is located just below the ducting, and the right part is not. In the middle, there is also a printing press that sounds quite perceivable. This automatically affects the background noise value.

In the existing condition, in addition to the element of violence of two-way communication voice should also note the value of clarity. If the sound of one of the employees is not audible, then the sound can be categorized as background noise. But if the sound is clear, then the concentration of employees around it will be disrupted, the focus on the work will be distracted by the clear voice. Therefore, please note the distance where the clarity of the sound started badly.

Referring to table 1, the open plan office parameters are based on ISO 3382-3, there are parameters that need to be obtained is distraction distance (rD) where STI = 0.5 and privacy distance (rP) where STI = 0.2. In addition, it is also necessary to measure the STI on the nearest workbench.

A decreasing SPL against the increase in distance. From here it can be done plotting the graph decrease the voice level of the conversation to the distance. Then obtained Figure 4 as follows.

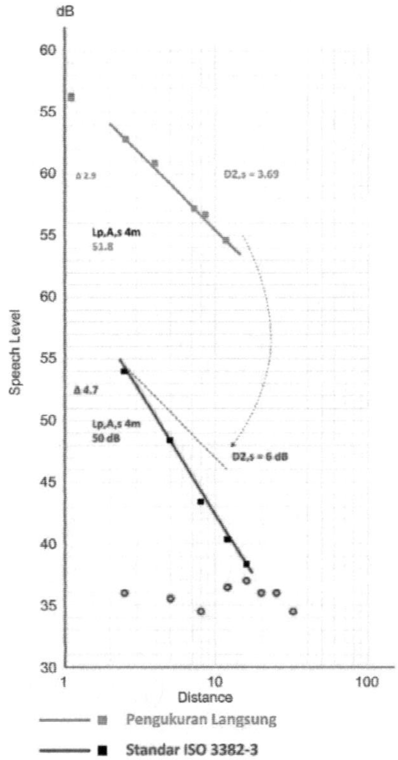


**Figure 4.** SPL graph of distance on Line 1

For the open-plan office, the recommended STI value is where it is located less than 10 meters away for STI = 0.5. The image also shows the trend line formed from the point of test point on line 1 and 2 which is done separately.

Based on Figure 5, for the distance distraction (distraction distance) line 1 is at a value of 9.68 meters. While on line 2 is at a distance of 13.69 meters. From here it can be analyzed that line 1 still meets the standards and line 2 does not meet. Viewed from the side of the slope line is also known in line 2 is very gentle with the equation of the line y = -0.0223x + 0.8061. When the STI on the nearest table is 0.8 it is necessary to have a value b below 0.028 to achieve STI 0.5 before the distance of 10 meters. As in line 1 which has a slope of the limit, but still meet the standards of open plan office requirements.

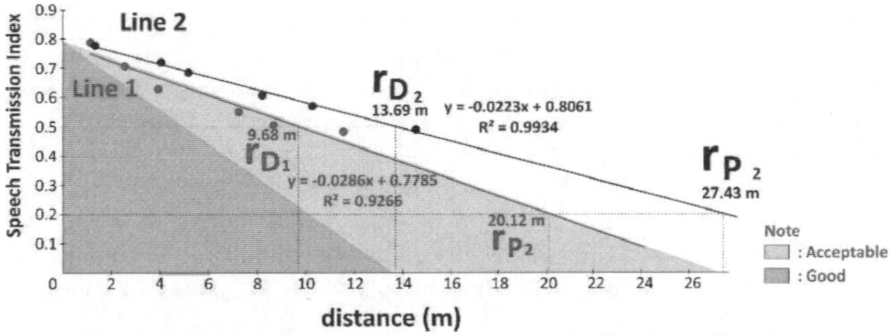


**Figure 5.** STI graphs at each measurement point line 1 and line 2

The uniqueness of open plan office design has a difference of roof height. At the edge has a height of 4 meters while in the middle has a height of 8 meters. To the lower edges Because there are another office space and balcony. With the condition of the curved balcony as in Figure 6 below.

**Figure 6.** Microphone position and speaker at *open-plan office*

Viewed from the side position as in Figure 6, line 1 on the left and line 2 on the right. Line 1 is farther from the right wall, and line 2 is closer to the right wall. Line 1 has a border roof between 8 meters and 4 meters, while Line 2 has a roof of 4 meters. Automatically the reflection of sound received by line 2 will be bigger and faster from the direction of the right wall and roof. This large reflection will make the SPL decline run very slowly and the clarity of the sound is higher. So the slope difference of STI graph to distance on line 2 is more sloping than line 1.

From Figure 6, the SPL graph of the above distance, on line 1 obtained the value at a distance of 4 meters of 51.8 dB and doubling the distance of 3.69 dB. In line 2 we get a value of 4 meters at 52 dB and a doubling of 4.02 dB. For this SPL measurement ISO cleans good standard of 48 dB for a distance of 4 meters. The value is obtained Because expected normal human voice has decayed at that distance. And below the standard background noise that is worth 50 dBA on All Frequency. With these results then there is still a little voice perceived.

Direct measurement results show that there are still many deficiencies in the value of predefined parameters. Therefore it is necessary to improve the acoustic condition of the room in the simulation.

To know the value of improvement, to be a more precise target and efficient then conditioned the first result of simulation which refers to the result of measurement directly.

After the two are equated, some improvement plans are made possible. Briefly, the high background noise value is mainly due to the central air ducts. Then the repair will be tried with the installation of the absorber on the ducting AC. Speech levels that have low attenuation are caused by a long drone. The buzz is due to the number of reflections in the room. Then the number of reflections resolved by the absorber laid on the ceiling of the office. The STI value is also problematic will be better after adding the absorber on the work table. The addition of sound masking on each work table will also worsen STI in the place.

## 4. Conclusion

Along with the development of acoustics in developed and developing countries, the acoustic object of open-plan office space needs to be applied also especially in big cities like Surabaya. The conclusions obtained from this research include:

1. Background Noise (background noise) is still not meet the standards, especially those under ducting AC. Therefore it is necessary to install a foam-based noise absorbers on the ducting.
2. The voice level of the conversation is at a distance of 4 meters of 51.8 dBA. The number is still not meet the standards because it can still be felt the difference with the background noise limit. While the attenuation of the sound of 3.69 dBA on line 1, and 4.02 dBA on Line 2. also less meet the 5 dBA limit standard. Therefore it is necessary to do the treatment in the form of the addition of absorber layer made of blanket in the sky of the room sky.
3. Distraction distance is at a distance of 9.7 on line 1. While on line 2 is 13.7 dBA. For Line 2 less meet. In addition to the addition of blankets in the sky, also need to be added silencer on the workbench.
4. Sound masking also needs to be prepared and activated conditionally to disguise the clarity of other voice conversations while interrupting the work concentration.

## References

[1] F A Everest. 2001. *The Master Handbook of Acoustics.* New York: Mc Graw Hill.
[2] ANSI (2008). S12.2-2008: Criteria for Evaluating Room Noise (ANSI)
[3] C E Mediastika 2005 Akustika Bangura: Prinsip-prinsip dan Penerapannya di Indonesia (Yogyakarta: Penerbit Erlangga)
[4] D M Jones and W J Macken 1995. Auditory Babble and Cognitive Efficiency: Role of Number of Voices and Their Location. *Journal Experimental Psychology*
[5] Howard, David M and Angus, J. 1996. Acoustics & Psychoacoustics. Oxford: Focal Press.
[6] IEC 60268-16: 2011. Sound System Equipment – Part 16: Objective rating of speech intelligibility by speech transmission index. BSI Standards Publication
[7] ISO 3382-3, Acoustics -- Measurement of Room Acoustic Parameters -- Part 3 : Open plan office
[8] Long, Marshall. 2006. "Architectural Acoustics". San Diego, California: Elsevier Academic Press.
[9] Stout, J. 2015. Speech Privacy Standard. Cambridge Sound Management, Inc.
[10] T. Vervoort & M. Vercammen. 2015. Background Noise Level to Determine the Speech Privacy in Open Plan Offces. Maastricht: EuroNoise 2015

## Acknowledgement

This research was supported/partially supported by Physics Department, Institut Teknologi Sepuluh Nopember Surabaya. We are thankful to our colleagues in the Acoustics and Building Physics Laboratory who provided expertise that greatly assisted the research, although they may not agree with all the interpretations provided in this paper.

We are also grateful to Dr. Suyatno for their comments on an earlier version of the manuscript, although any errors are our own and should not tarnish the reputations of these esteemed professionals, and Susilo Indrawati M,Si., and Gontjang Prajitno, M.Si. who moderated this paper and, in that line, improved the manuscript significantly. We must express our appreciation to the M. Wildan Habibie, Mudito Tejo B. for sharing their pearls of wisdom with us during this research.

# Picture of Hearing Health with Exposure to Environmental Noise on Boat Ketek Riders in Region of Ancol Tanggo Rajo Jambi 2017

**A Akbar[1], I Fadlan[1] and I Sukandar[2]**

[1]Department of Otorhinolaryngology-Head and Neck Surgery, Faculty of Medicine and Health Sciences, Universitas Jambi, Jambi 36361, Indonesia
[2]Public Health Department of Jambi City, Jambi 36128, Indonesia

E-mail: ismeliaent@gmail.com

**Abstract.** Hearing loss is a term used to describe hearing loss in one or both ears. Normally the human ear can hear a 20 - 20000 Hz frequency with an intensity below 80 dB. If a person continually hears a sound above the normal threshold, it will damage the function of hair cells resulting in hearing loss. Hearing loss is a problem most commonly encountered in workers who work in environments with high levels of noise. The diagnosis can be confirmed with audiometry test. This study aims to see a picture of hearing health of ketek driver in Ancol Tanggo Rajo Jambi City. This research is a descriptive cross sectional study. The sample was taken by random sampling. This research was carried out at harbor boat ketek Ancol Tanggo Rajo Jambi City on October 2017. The samples were taken from the population of ketek driver as many as 30 people. The characteristics of ketek driver are in the >35 years age category were 29 people (96.7%), driver ketek who works >10 years are 26 people (86.7%). The result of audiometry test were 30 patients (100%) with Hearing Loss. The most type of hearing loss is Noise Induce Hearing Loss with 15 people (50%), followed by Conductive Hearing Loss with 9 people (30%), Sensorineural Hearing Loss with 3 people (10%), and Mix Hearing Loss with 3 people (10%).

## 1. Introduction

Hearing loss is the most common problem in workers working in high-noise environments, such as workers in industrial estates including mining, exploration (e.g. drilling), shipping, aviation, textile machinery and testing of machines jet. This will be very detrimental to the workers because it can cause permanent threshold shift. So, it is necessary to do early detection of hearing loss to prevent temporary threshold shift to permanent threshold shift [1]. According to the World Health Organization (WHO) recorded the number of hearing loss in 2000 there are 250 million (4.2%) of the world population who suffer from hearing loss and approximately half (75-140 million) are in Southeast Asia. From the results of the WHO MultiClinic Study in 1998, Indonesia included four countries in Southeast Asia with a high prevalence of hearing loss (4.6%), the other three countries were Sri Lanka (8.8%), Myanmar (8.4%), and India (6.3%) [2].

Indonesia data based on Health Survey of Hearing and Vision Sensory in 1994-1996 also showed high morbidity, ear diseases (18.5%), prevalence of hearing loss (16.8%), while deafness was found at (0.4%) population [3]. From the data obtained from the Public Health Service of Jambi City in 2016

IOP Conf. Series: Journal of Physics: Conf. Series **1075** (2018) 012051 doi:10.1088/1742-6596/1075/1/012051

also recorded the result of hearing impairment occurring in more than half the number of Community Health Clinic located within the area of Jambi City. With the highest number of hearing loss patients, Simpang IV Sipin Public Health Clinic (10 patients), Simpang Kawat Public Health Clinic (13 patients), Kebun Handil Community Health Clinic (73 patients), Pakuan Baru Public Health Clinic (85 patients), and the most are occupied by Community Health Clinic of Olak Kemang (261 patients) which have significant numbers among other Community Health Clinic. Where the working area of Community Health Clinic Olak Kemang is the domicile of all boat ketek riders who work on the port of boat ketek in Ancol Tanggo Rajo Jambi.

Noise is a disturbing or unwelcome noise or sound. From this definition indicates that the actual noise is very subjective, depending on each individual, time and place of the occurrence of noise. While in audiologically, noise is a mixture of pure tone sounds with various frequencies [4].

Noise generated by the engine of boat ketek has exceeded the threshold tolerated by the hearing in accordance with the Decree of the Minister of Manpower Number: KEP-51/men/1999, on the Threshold Level of noise in the workplace, set at 85 dB with the longest duration of exposure for 8 hours. But the fact is in the engine of boat ketek located in the port of Ancol Tanggo Rajo Jambi City, the average noise of the machine is 110 dB with average exposure time for more than 60 minutes per day. That should only be exposed for only about 90 seconds [5] [6].

## 2. Research purposes

The purpose of this study was to look at the picture of hearing health with exposure to continuous noise for long periods of time.

## 3. Methods and sampling

This research was conducted on boat ketek riders in Ancol Tanggo Rajo Jambi. This research is descriptive research with cross sectional study using sampling technique of sampling by random sampling. This research was carried out at port boat of Ancol Tanggo Rajo Jambi on October 29, 2017. The sample was taken from the population of the boat ketek riders in the port of boat Ancol Tanggo Rajo Jambi City as many as 30 people. The variables studied were exposure of noise intensity, length of service, hearing loss and degree of hearing loss. How to collect data are through direct interviews, otoscopy checks and audiometric tests. The procedure of measuring noise at work with a sound level meter device held by a distance of 1.5 meters from a noisy source, is performed 5 times at each noise source [7].

## 4. Result and discussion

The result of this research shows the characteristic of the boat ketek rider Ancol Tanggo Rajo Jambi, found that all of the responden in the age more than 30. Most ketek boat riders are in the age category> 35 years of 29 subjects (96.7%) while the least are in the 18-34 year age category amounting to 1 person (3.3%).

The audiometric picture obtained, shows all subjects experiencing hearing loss, with the spread of each type of hearing loss, a conductive hearing loss of 9 people (30%), sensorineural hearing loss of 3 people (10%), mixed hearing loss 3 people (10%), and the most common type of hearing loss is Noise Induce Hearing Loss of 15 people (50%). An illustration of the frequency distribution of the subject based on the type of hearing loss described in table 1 below.

IOP Conf. Series: Journal of Physics: Conf. Series **1075** (2018) 012051 doi:10.1088/1742-6596/1075/1/012051

**Table 1.** Frequency distribution of the subject based on the type of hearing loss.

| **Hearing Status** | Frequency (n) | Percentage (%) |
|---|---|---|
| Normal | 0 | 0 |
| Conductive | 9 | 30 |
| Hearing Loss Sensorineural | 3 | 10 |
| Noise Induce Hearing Loss | 15 | 50 |
| Mix Hearing Loss | 3 | 10 |
| Total | 30 | 100.0 |

From the condition of hearing as can be seen in table 1, there is also a difference in the degree of hearing loss in either the left ear or right ear, with the spread of the degree of hearing loss each, the normal left ear 5 people (16.7%), the normal right ear 0 people ( 0%), left ear mild hearing loss 16 people (53.3%), right ear mild hearing loss 18 people (60%), ear hearing loss moderate 7 people (23.3%), right ear hearing loss moderate 10 people (33.3% ), left ear hearing loss was moderate to heavy 2 people (6.7%), right ear hearing loss was moderate to heavy 2 people (6.7%). as can be seen in tables 2 and 3 below.

**Table 2.** Frequency distribution of the subject based on the degree of hearing (left-ear).

| **Degree of Hearing Loss** | Frequency (n) | Percentage (%) |
|---|---|---|
| Normal | 5 | 16.7 |
| Mild | 16 | 53.3 |
| Moderate | 7 | 23.3 |
| Moderate to Heavy | 2 | 6.7 |
| Total | 30 | 100.0 |

**Table 3.** Frequency distribution of the subject based on the degree of hearing (right-ear).

| **Degree of Hearing Loss** | Frequency (n) | Percentage (%) |
|---|---|---|
| Mild | 18 | 60 |
| Moderate | 10 | 33.3 |
| Moderate to Heavy | 2 | 6.7 |
| Total | 30 | 100.0 |

The noise intensity obtained in the respondent workspace ranges from 110-112 dB by using sound level meter equipment, depending on the speed of boat ketek.

The intensity of noisy exposure can cause hearing loss of conductive or sensorineural hearing loss. Exposure to noisy intensity over long periods and exceeding tolerable thresholds may cause damage to the corti organs which will result in sensorineural hearing loss [8]. Conductive hearing loss occurs when there is damage to the outer ear or middle ear so that sound waves cannot be transmitted to vibrate the fluid in the ear in [9]. Conductive hearing loss may be caused by interference with the outer ear or middle ear. Disorders of the external ear which may cause conductive hearing loss such as ear canal atresia, blockage by the serumen, external otitis sirkumskripta, osteoma of the ear canal. Meanwhile interference with the middle ear that can cause conductive hearing loss are blockage of tuba eustachius, otitis media, otosklerosis, timpanosklerosis, hemotimpanum, and dislocation of hearing bone [10].

Sensorineural hearing loss is divided in sensorineural hearing loss of cochlea and retrochlea. Sensorineural cochlear hearing loss is caused by aplasia (congenital), labirintitis (bacteria or virus) and drug intoxication (streptomycin, kanamycin, garamycin, neomycin, quinine, acetosal, or alcohol). In addition, it can also be caused by sudden deafness, capitis trauma, acoustic trauma and noisy exposure. Sensorineural cetrocochlear hearing loss is caused by acoustic neuroma, cerebellar pons corner tumor, multiple myeloma, brain injury, cerebral hemorrhage, and other brain abnormities [11].

In research conducted by Mutiarani Z proves that the working period is a risk factor of hearing loss in workers [12].

The degree of hearing loss in the right ear has a higher degree compared to the left ear. The riding position of the ketek boat is one of the factors, since the distance between the ears from the sound source should also be considered [13]. Where the ketek boat engine is placed behind the ketek boat in the right side, beside the riders of the ketek boat. And this position applies to every boat ketek that exist in the area of Ancol, Tanggorajo, Jambi City.

## 5. Conclusion

The boat riders who suffer from hearing loss due to noise at the port of Ancol Tanggo Rajo Jambi as many as 30 respondents (100%). Of the 30 respondents who experienced hearing loss, 9 respondents had conductive hearing loss, 3 respondents had sensorineural hearing loss, 15 respondents had hearing loss due to noise, and 3 respondents had mixed hearing loss.

The degree of right ear hearing loss is greater than the left ear, due to ear distance with a sound source closer to the right ear. The working time of the boat riders makes the likelihood of hearing loss greater (> 10 years). Environmental noise that occurs in ketek boat engine ranges between 110-112 dB, depending on the speed of the boat ketek. And is the intensity of noise that is above the normal tolerated ear threshold.

## 6. References

[1] World Health Organization (WHO) 2007 State of Hearing & Ear Care in South East Asia Region *WHO Regional Office SEARO* (http://who.int/pbd/deafness/activities/hearing_care/advance.pdf)

[2] Komite Nasional Penanggulangan Gangguan Pendengaran dan Ketulian (Komnas PGPKT) 2006 Rencana Strategi Nasional Penanggulangan GangguanPendengaran dan Ketulian untuk Mencapai Sound Hearing 2030. (http://www.ketulian.com/v1/web/index.php?to=home&show=detail)

[3] Depkes Kementerian Kesehatan Republik Indonesia 2010 Telinga Sehat Pendengaran Baik (http://www.depkes.go.id/index. php/berita/press-release/840-telinga-sehat-pendengaran-baik.html.)

[4] Soetirto I 1990 Tuli Akibat Bising (Noise Induced Hearing Loss) *In Soepardi EA, Iskandar N, Ed. Book of Otorhinolaryngology-Head and Neck Surgery* Ed 3 (Jakarta: Balai Penerbit FK UI) pp 37-39

[5] Keputusan Menteri Tenaga Kerja Nomor: KEP-51/MEN/1999 1999 Tentang. Nilai Ambang Batas Faktor Fisika di Tempat Kerja

[6] Zachreini I and Khairunisa C 2013 Gambaran Tingkat Pengetahuan dan Kepatuhan Penggunaan Alat Pelindung Diri (APD) Telinga Terhadap Paparan Kebisingan Pada Siswa SMK Jurusan Teknik Mesin di Kota Lhokseumawe, Pertemuan Ilmiah Tahunan Otologi (PITO) KE-8, Jakarta.

[7] Komnas Penanggulangan Gangguan Pendengaran dan Ketulian 2013 Gangguan Pendengaran (Jakarta: Komite Nasional Penanggulangan Gangguan Pendengaran dan Ketulian)

[8] Anizar 2012 Teknik Keselamatan dan Kesehatan Kerja di Industri (Yogyakarta: Graha Ilmu)

[9] Bhargava K B, Bhargava S K and Shah T M 2002 Deafness & Examination of the Ear *In A Short Textbook of E.N.T. Diseases* 5th ed (Mumbai: Usha Publications) pp 119-125 & 21-40

[10] Bashiruddin J and Soetirto I 2007 Gangguan pendengaran akibat bising (*noise-induced hearing loss*) *In Soepardi, E.A., Iskandar, N., Bashiruddin, J. dan Restuti R.D. ed VI. Book of Otorhinolaryngology-Head and Neck Surgery* (Jakarta: Balai Penerbit FK UI) pp 20-22 49-52

[11] Eryani Y 2016 Hubungan Intensitas Kebisingan, Durasi Paparan dan Penggunaan Alat Pelindung Diri dengan Gangguan Pendengaran Akibat Bising pada Karyawan PT. BUKIT ASAM (PERSERO) TBK Bandar Lampung. [Undergraduate Thesis] (Bandar Lampung: Universitas Lampung) (http://digilib.unila.ac.id)

[12] Mutiara Z 2010 Faktor risiko terjadinya gangguan pendengaran pada operator mesin Shuttle bagian Weaving PT. X

[13] Meyer S F 1997 Pemaparan Bising Industri dan Kurang Pendengaran *In Ballengger JJ, editor of Otorhinolaryngology-Head and Neck Surgery Ed 13* (Jakarta: Bina Rupa Aksara) pp 305-331

IOP Conf. Series: Journal of Physics: Conf. Series **1075** (2018) 012052 doi:10.1088/1742-6596/1075/1/012052

# Measurement traceability of acoustics and vibration instruments in Indonesia

**D Hermawanto[1], M R Palupi[1], D Rusjadi[1], N R Prasasti[1], B Dwisetyo[1], C C Putri[1] and H A Akil[1]**

[1]Acoustics & Vibration Metrology Laboratory, Research Center for Metrology (RCM-LIPI), Kompleks PUSPIPTEK, Gedung 420, Setu, Tangerang Selatan, 15314 – Banten

E-mail: denny.hermawanto@lipi.go.id

**Abstract**. This paper presents the chain of calibration for acoustics and vibration measurement instruments to national standard which must be maintained by NMI (National Metrology Institute). At the present time, NMI in Indonesia is Research Center for Metrology - Indonesian Institute of Sciences (RCM-LIPI). For acoustics measurement instruments are microphone, pistonphone and sound level meter, and for vibration measurement instruments are accelerometer and vibration meter. The implementation of calibration method based on IEC standard is described along with the uncertainty components that contribute to the calibration results. International recognition of calibration and measurement capabilities of RCM-LIPI for acoustics and vibration are shown.

## 1. Introduction

Acoustics & vibration measurement plays important role for supporting industrial development in many areas. For example, noise emission assessment is usually used by the automotive manufacture and construction. Effective noise control plan could not be applied without proper noise measurement technique and accurate measurement result. In the field of vibration, vibration meter is used for machine health monitoring and product testing. The health of employees related to acoustics & vibration is also protected by government regulation which is issued by Ministry of Manpower of the Republic of Indonesia Number: PER.13/MEN/X/2011, about threshold limits of physical and chemical factors in the workplace is 85dBA and $4m/s^2$ respectively [1]. Measurement technique and instruments plays important rule to obtain reliable measurement result and correct interpretation for the judgment.

Instrument measurement result reliability can be assessed by calibration process where the instrument under test is compared with the reference value and this process is conducted by the calibration laboratory. In order to assure the reliability of reference value used by calibration laboratories, the reference value should be calibrated to standard value as the realization of unit definition and has the highest accuracy. Unit standard in each country is realized and maintained by national metrology institute (NMI). Research Center for Metrology LIPI (RCM-LIPI), previously named KIM-LIPI, is Indonesian NMI which has been registered in International Bureau of Weights and Measures (BIPM) which is the international standard organization for maintaining International System of Units (SI). RCM-LIPI also signs the CIPM-MRA which is show the international equivalence of measurement standards and the certificate among the CIPM member. RCM-LIPI has been accredited ISO 17025 by KAN, Indonesian accreditation body, as calibration laboratory with the identifier LK 070 IDN [2].

This paper presents the calibration capability provided by acoustics & vibration metrology laboratory of RCM-LIPI. The implementation of traceability from the measurement instrument to the primary standard will be described. Inter-comparison result with other NMIs will also be presented.

## 2. Calibration Methods

### *2.1. Acoustics Calibration*

Instruments used for acoustics measurement are sound level meter (SLM), acoustics calibrator or pistonphone and microphone the method for calibration of sound level meter is explained in IEC 61672-3:2013, Electroacoustics – Sound Level Meters – Part 3: Periodic Test [3]. This standard gives standard procedure to assess conformity of sound level meter to the standard SLM described in the IEC 61672-1 Electroacoustics – Sound Level Meters – Part 1: Specification [4]. The method use for calibration of SLM is substitution method. Sound generated by speaker is measured by reference microphone at first and then this microphone is substituted by the sound level meter under test. By this process, the sound level meter is traceable to the reference microphone. For the calibration of acoustics calibrator or pistonphone, the method used is insert voltage referring to the IEC 60942:2003, Electro acoustics–Sound Calibrator [5]. In this method, the sound pressure generated by acoustics calibrator is measured by the reference microphone and its voltage output ($V_1$) is measured by the voltmeter. This output voltage corresponds to the SPL produce by acoustics calibrator which includes environmental condition effect to the sound pressure like temperature and air pressure. An equivalent electrical signal is then inserted ($V_{ins}$) directly through the microphone backplate pin bypassing the diaphragm so that the output voltage of microphone equal to $V_1$. The insert voltage is equal to sound pressure of acoustics calibrator in the ideal condition without environmental effect. The ideal sound pressure generated by acoustics calibrator is then calculated from the reference microphone sensitivity so that the sound pressure produce by acoustics calibrator is traceable to standard unit through the reference microphone sensitivity.

Reference microphone can be calibrated either by secondary and primary method. Secondary method of microphone calibration employs comparison method which is described in IEC 61094-5:2016. In this method, microphone under test and reference microphone are mounted face-to-face with small distance to ensure that both microphones receive the same sound pressure. The sound pressure is generated by pistonphone and output voltage of both unknown microphone and reference microphone are measured. Then, the sensitivity of unknown microphone can be calculated from reference microphone sensitivity [6]. Most precise microphone calibration can be implemented by primary method called reciprocity which is described in IEC 61094-2:2009. In this calibration method, microphone is utilized as sound transmitter and receiver so that ratio of electrical transfer impedance ($Z_e$) and acoustics transfer impedance ($Z_a$) between the microphones can be measured and calculated [7]. Three microphones ($Mic_1$, $Mic_2$, $Mic_3$) are used in the calibration and configured as transmitter-receiver in three combination ($Mic_1$-$Mic_2$, $Mic_1$-$Mic_3$, $Mic_2$-$Mic_3$) namely configuration A, B, C respectively. Sensitivity of microphones is then obtained by solving these equations:

$$M_1 M_2 = \left(\frac{Z_e}{Z_a}\right)_A \qquad M_1 M_3 = \left(\frac{Z_e}{Z_a}\right)_B \qquad M_2 M_3 = \left(\frac{Z_e}{Z_a}\right)_C \tag{1}$$

### *2.2. Vibration Calibration*

In the vibration field, accelerometer is employed by vibration meter to measure displacement, velocity and acceleration of the measurement object. Sensitivity of accelerometer can be expressed in charge sensitivity or voltage sensitivity with unit of (pC/m/s$^2$) and (mV/m/s$^2$) respectively. The vibration meter is calibrated using the secondary method named comparison method or back-to-back calibration method. It is called back-to-back because vibration meter's accelerometer and reference accelerometer are mounted inline and direct contact on the top vibration exciter. Measurement of displacement, velocity and acceleration by the vibration meter is compared with the reference accelerometer result to obtain correction value for vibration meter under test. By this method, vibration meter is traceable through the reference accelerometer sensitivity. Explanation of calibration technique and uncertainty estimation for this secondary method can be found in the ISO 16063-21:2003 [8].

The primary method for calibration of accelerometer which has the most accurate result is implemented by laser interferometry and briefly described in standard ISO 16063-11:1999. Basis of the method is application of fringe which is constructive and destructive interference of two coherent laser waves. According to the standard, there are 3 methods which can be implemented namely Fringe Counting Method (FCM), Minimum Point Method (MPM) and Sine Approximation Method (SAM) [9]. The differences between the methods are frequency range of calibration and complexity of the method. FCM method determines the sensitivity of accelerometer by measuring fringe frequency with the counter and suitable for calibration of accelerometer in the frequency range of 1 Hz to 800 Hz. In the MPM methods, sensitivity of accelerometer is calculated from the displacement which is measured by using the arguments to the zero crossing of the Bessel function of the first kind and first order. The limitation of this method is the difficulty to detect zero crossing in low frequency therefore frequency range for this method is started from 800 Hz to 10 kHz. Comparing with FCM and MPM, SAM method has the widest calibration frequency range. In the SAM method, quadrature laser interferometer is used and the output which has 2 output signals sampled as $\{u_1(t_i)\}$ and $\{u_2(t_i)\}$. From these output signals, the series of modulation of phase will be:

$$\varphi_{Mod}(t_i) = \arctan\left(\frac{u_2(t_i)}{u_1(t_i)}\right) + n\pi \tag{2}$$

## 3. Calibration Uncertainty Components

### 3.1. Acoustics

The calibration of sound level meter is conducted in the RCM-LIPI's full anechoic chamber which has the dimension 10 m x 10 m x10 m, frequency cut-off 63 Hz and background noise about 17 dB. As the reference, microphone type B&K 4180 serial number 1395438 is used for calibration. Uncertainty components of sound level meter calibration are calibration instruments reading resolution, reference microphone sensitivity, digital voltmeter, sound attenuator, acoustics calibrator, drift of reference microphone sensitivity, stability of calibration system and other factor such as the anechoic chamber characteristic and sound level meter placement. These components yield the uncertainty of 0.38 dB within frequency range 63 Hz to 10000 Hz in the sound level meter calibration system. For acoustics calibrator calibration using insert voltage method, uncertainty components contributed by digital voltmeter, reference microphone sensitivity, microphone sensitivity drift, barometer, pressure change, thermometer, temperature change, system stability and rounding error. Total uncertainty for acoustics calibrator calibration is 0.1 dB for the SPL 94 dB, 114 dB, and 124 dB. In the microphone calibration using reciprocity method, calibration of 1 inch microphone and ½ inch use the same system but the uncertainty values for each frequency are different because calibration parameters between 2 microphones are different. Since RCM-LIPI use reciprocity method which is the most accurate method for microphone calibration, many components are considered in the calculation of calibration uncertainty. These components are described as follow:

- Microphone parameter: front depth, front volume, equivalent volume, resonance frequency, loss factor
- Electrical parameter: voltage ratio, capacitance, crosstalk, Polarization voltage, frequency, receiver ground shield, transmitter ground shield, distortion
- Coupler properties: length, diameter, volume, surface area, air leakage

IOP Conf. Series: Journal of Physics: Conf. Series **1075** (2018) 012052 doi:10.1088/1742-6596/1075/1/012052

- Ambient Conditions: static pressure, pressure variation, pressure coefficient, temperature coefficient, microphone temperature, temperature variation, relative humidity, microphone, relative humidity
- Physical Correction: radial wave motion, viscosity losses, heat conduction, adding excess value
- Repeatability and rounding error

Minimum uncertainty for microphone sensitivity calibration for ½ inch microphone is 0.06dB and the maximum uncertainty is 0.12dB in the frequency range 20 Hz to 10000 Hz. For 1 inch microphone, the frequency range of calibration is 20 Hz to 25000 Hz with the range of uncertainty from 0.06 dB to 0.17 dB.

*3.2. Vibration*

The implementation of vibration meter calibration using back to back calibration method is according to IEC 16063-21:2003. Instrument used for calibration are reference accelerometer type B&K 8305 serial 1499872, conditioning amplifier, vibration exciter, amplifier and FFT analyzer. These instruments contribute to calibration uncertainty together with other parameters such as reference accelerometer drift, measurement repeatability, reading resolution of device under test, stability of the system, mounting influence, transducer non-linearity, gravity influence and influence of exciter's magnetic field. These produce uncertainties ranging from 1.1 % to 1.3 % in the frequency range 20 Hz to 5000 Hz. For the primary calibration of accelerometer, RCM-LIPI implements IEC 16063-11:1999 method 3: sine approximation method. Instruments used for this calibration are Polytech laser Doppler vibrometer, conditioning amplifier and calibration software B&K 3629. In the uncertainty calculation, this primary calibration system considers following components:

- Acceleration amplitude: signal, generator frequency, interferometer signal filtering effect, laser wavelength stability, motion disturbance effect
- Voltage amplitude: voltage measurement error, transverse motion effect
- Conditioning amplifier: capacitance uncertainty, input voltage uncertainty, output voltage uncertainty, conditioning amplifier uncertainty
- Repeatability

Uncertainties obtained by this calibration system are within the range of 0.7 % to 0.9 % in the frequency range of 10 Hz to 5000 Hz.

## 4. International Comparisons Result

*4.1. APMP.AUV.A-K1.1*

This registered bilateral comparison conducted between RCM-LIPI with Korea Research Institute of Standard (KRISS) for pressure sensitivity of 1-inch microphone LS1p using reciprocity method. The frequency range of comparison is from 63 Hz to 8000 Hz in octave frequencies and the artefact is microphone B&K 4160 serial number 1792662. The initial calibration was conducted at KRISS and then continued by RCM-LIPI and finally recalibrated by KRISS.

Maximum deviation of microphone sensitivity calibration result between RCM-LIPI and KRISS is 0.05 dB which occurs at frequency 8000 Hz. Compared to the expanded uncertainties, all the deviation still within the range of measurement uncertainties therefore show the equality between RCM-LIPI and KRISS. The final comparison report has been published in the BIPM key comparison database (KCDB) website as the supplementary report [10].

IOP Conf. Series: Journal of Physics: Conf. Series **1075** (2018) 012052 doi:10.1088/1742-6596/1075/1/012052

#### *4.2. APMP.AUV.A-K3.1*

Bilateral comparison was conducted between RCM-LIPI and KRISS with the artefact of ½ inch LS2p microphone type B&K 4180 serial number 2341431. Calibration frequency range were 31.5, 63, 125, 250, 500, 1000, 2000, 4000, 6300, 8000, 10000, 12500, 16000, 20000 and 25000 Hz.

Maximum deviation of sensitivity result is 0.1 dB at frequency 25000 Hz and this value is still in the range of stated measurement uncertainty which shows the degree of equivalence between KRISS and RCM-LIPI. The final comparison report has been published in KCDB websites [11].

#### *4.3. APMP.AUV.A-S1*

The subject of this comparison was the measurement of sound pressure level, frequency and total harmonic distortion of multi-frequency sound calibrator. The comparison was joined by 13 countries in Asia which were NMIA Australia, SCL Hongkong, RCM-LIPI Indonesia, CMS/ITRI Taiwan, NML/SIRIM Malaysia, NIM China, KRISS Korea, A*Star Singapore, NMIJ Japan, NPLI India, NMISA South Africa, NIS Egypt and NIMT Thailand. The protocol said that calibration method used for comparison follow IEC 60942:2003. The measured sound pressure were 94dB and 114dB with the frequency of measurement were 31.5, 63, 125, 250, 500, 1000, 2000, 4000, 8000, 12500 and 16000 Hz. Complete final comparison report which includes sound pressure level, frequency and THD measurement at each frequency can be seen in the comparison result published in the technical supplement of Metrologia journal [12].

#### *4.4. APMP.AUV.V-K1.2*

The participants of this comparison were NIM China, RCM-LIPI Indonesia and NPLI India with NIM act as the pilot laboratory. The artefacts of comparison were accelerometer single ended (SE) accelerometer type B&K 8305-001 serial 2519436 and back to back (BB) accelerometer type B&K 8305 serial 2440139. The complex charge sensitivity of both accelerometers should be measured with the method described in IEC 16063-11:1999 for the following frequencies: 40, 50, 63, 80, 100, 125, 160, 200, 250, 315, 400, 500, 630, 800, 1000, 1250, 1600, 2000, 2500, 3150, 4000, 5000 Hz. The sensitivity value and the uncertainty of RCM-LIPI and NIM China has shown the equivalence value. The comparison result for each frequency and each BB and SE accelerometer and its uncertainty was published in the KCDB website [13].

#### *4.5. EURAMET.AUV.V-P1*

This comparison was conducted between RCM-LIPI and LNE France with the artefact of accelerometer SE type 8305-001 serial 2481861 and BB type B&K 8305 serial 2679379. Sine approximation method as described in the IEC 16063-11:1999 should be used for the calibration of charge sensitivity of accelerometer over frequency range 10 Hz to 10 kHz. The sensitivity value measured by RCM-LIPI was close to the result measured by LNE for SE accelerometer. There is a difference about 0.0005 $pC/ms^2$ in low and medium frequency region for BB accelerometer, but these differences are still within the range of uncertainty of both participants which are around 0.001 $pC/ms^2$. These results show that accelerometer calibration using laser interferometry method by RCM-LIPI has degree of equivalence with the LNE. The report of this comparison has been approved for the publication in the KCDB website and technical supplementary of Metrologia journal[14].

## 5. Conclusion

The traceability of acoustics and vibration measurement has been realized by RCM-LIPI as the Indonesian national metrology institute from secondary to primary method. Participation in the international comparison with satisfactory result shows the validity of calibration result and technical capability to perform the calibration according to the standard procedure.

**References**

[1] Peraturan Menteri Tenaga Kerja dan Transmigrasi, Nomor PER.13/MEN/X/2011, Tentang Nilai Ambang Batas Faktor Fisika dan Faktor Kimia di Tempat Kerja.

[2] ISO Standard. ISO/IEC 17025:2008: General requirements for the competence of testing and calibration laboratories.

[3] International Electrotechnical Commission (2006). IEC 61672-3 Electroacoustics - Sound level meters - Part 3: Periodic tests.

[4] International Electrotechnical Commission (2013). IEC 61672-1 Electroacoustics – Sound Level Meters – Part 1: Specification.

[5] International Electrotechnical Commission (2003). IEC 60942:2003, Electro acoustics – Sound Calibrator.

[6] International Electrotechnical Commission (2016). IEC 61094-5:2016, Electroacoustics - Measurement microphones - Part 5: Methods for pressure calibration of working standard microphones by comparison.

[7] International Electrotechnical Commission (2009). IEC 61094-2:2009, Electroacoustics - Measurement microphones - Part 2: Primary method for pressure calibration of laboratory standard microphones by the reciprocity technique

[8] International Organization for Standardization (2003). ISO 16063-21:2003, Methods for the calibration of vibration and shock transducers — Part 21: Vibration calibration by comparison to a reference transducer

[9] International Organization for Standardization (1999). ISO 16063-11:1999, Methods for the calibration of vibration and shock transducers -- Part 11: Primary vibration calibration by laser interferometry

[10] Sang Joon Suh and Jae Gap Suh. (2009). Bilateral Comparison of LS1P Microphone Calibrations between KRISS and KIM-LIPI APMP.AUV.A-K1.1. http://kcdb.bipm.org/appendixB/appbresults/ccauv.a-k1/apmp.auv.a-k1.1_final_report.pdf

[11] Sang Joon Suh and Achmad Suwandi. (2010). A Bilateral Comparison of LS2P Microphone Pressure Sensitivity between KRISS and the KIM-LIPI APMP.AUV.A-K3.1. http://www.bipm.org/utils/common/pdf/final_reports/AUV/A-K3/APMP.AUV.A-K3.1.pdf

[12] Virat Plangsangmas and Surat Leeudomwong and Andrew Scott and Bo Zhong and Yuchung Huang. (2014). Final report on the regional supplementary comparison APMP.AUV.A-S1. Metrologi vol. 51. 09004

[13] Qiao Sun and Aidong Liu. (2010). Final report on the key comparison APMP.AUV.V-K1.2. Metrologia. vol. 47. 09004.

[14] Claire Bartoli and Denny Hermawanto. (2017). Pilot Study EURAMET.AUV.V-P1: Bilateral comparison on magnitude of the complex charge sensitivity of accelerometers from 10 Hz to 10 kHz. Metrologia. vol. 54. 09002

# Tank experiment and simulation of sunken hazardous and noxious substances detection using high frequency active sonar

**D G Han[1], J W Choi[1]* and S U Son[2]**

[1] Department of Marine Sciences and Convergence Engineering, Hanyang University, Ansan, 15588, Republic of Korea
[2] The 6th R&D Institute, Agency for Defense Development, Changwon, 51698, Republic of Korea choijw@hanyang.ac.kr

**Abstract**. Hazardous and Noxious Substances are defined as substances that are likely to create hazards to the marine ecosystem and human health when they are released into the sea. Because even a small amount of leakage may cause a fatal effect, it is important to develop a technology to handle effectively the HNS spill incidents. Especially, it is difficult to detect or monitor the HNS deposited on the seabed after HNS spill incident using a direct observation method. In this paper, the remote sensing technique using underwater acoustic method is suggested as an alternative method to observe effectively the HNS sunk to ocean bottom. First, to examine the detection possibility of underwater acoustic method, tank experiment was carried out using 200-kHz active sonar. The target HNS was chloroform, which is one of major sunken HNS. However, castor oil having a similar acoustic impedance with the chloroform was used as an alternative substance due to the toxicity of chloroform. Second, a computer simulation to investigate the availability of side scan sonar to detect the sunken HNS was carried out. It is assumed that the chloroform and tetrachloride, which is another sunken HNS, are randomly distributed on the several types of seabed. The simulation algorithm is based on the difference of backscattering strengths from the ocean bottom interface and the sunken HNS.

## 1. Introduction

As the volume of Hazardous and Noxious Substances (HNS) transported by ships increases, the risk of leakage accident is increasing. HNS are likely to cause critical damage to marine ecosystems and the human health even with a small amount of spillage. Internationally, the OPRC-HNS Protocol was adopted by the IMO (International Maritime Organization) in March 2000 in order to prepare a countermeasure against HNS spills at sea, and it became effective internationally in June 2007. Also studies on the accident response system have been actively carried out to minimize marine pollution caused by hazardous substances.

The SEBC (European Behavior Classification System) categorizes HNS into 12 groups on physical and chemical behavior as gas (G), gas/ dissolver (GD), evaporator (E), evaporator/ dissolver (ED), floater/evaporator (FE), floater/ evaporator/ dissolver (FED), floater (F), floater/ dissolver (FD), dissolver/ evaporator (DE), dissolver (D), sinker/ dissolver (SD), sinker (S). It is important to classify the HNS based on the behavior characteristics to establish effectively the accident response system when it flows out to the ocean. In a case of SD or S, optical observation using the methods such as satellites or SAR (synthetic aperture radar) is not possible. In this case, remote sensing technique using underwater acoustics could be a good alternative method to detect the HNS belonging to SD or S category (hereafter called HNS_S). However, there are little previous researches for acoustic detection of HNS_S. In this paper, the measurement results of water tank experiment are presented, which was conducted to investigate the possibility of acoustic detection of the HNS_S. For this, castor oil was used as an alternative substance due to the toxicity of HNS. The castor oil has the almost same impedance with the Chloroform, which is one of representative HNS_S. In addition, image simulation of side scan sonar was made on condition that HNS_S lies on the seabed interface to investigate the possibility of HNS_S detection using side scan sonar.

## 2. Measurements of reflection loss for castor oil layer

The Chloroform was chosen as a target HNS_S because its huge volume is transported by ship worldwide. It was assumed that Chloroform lay on the sediment. However, because the Chloroform has a very high toxicity, castor oil was used in the tank experiment. The experiment was carried out in a mini-water tank having 0.25 × 0.25 × 0.68 m in dimension. Although castor has a similar acoustic property to the Chloroform, the specific gravity of castor oil is smaller than that of water, whereas that of Chloroform is greater than that of water. For this reason, castor oil floats on water. Since the goal of this study is to investigate the detection possibility of HNS_S, the geometry was considered to be upside down. A directional transducer (Simrad, 200-7G) was stalled on the bottom of water tank and transmitted the 0.025-ms 200 kHz continuous wave toward the surface. First, measurements were made without castor oil layer [Fig. 1(a)], and it was repeated with 0.1-m thick castor oil layer on top [Fig 1(b)].

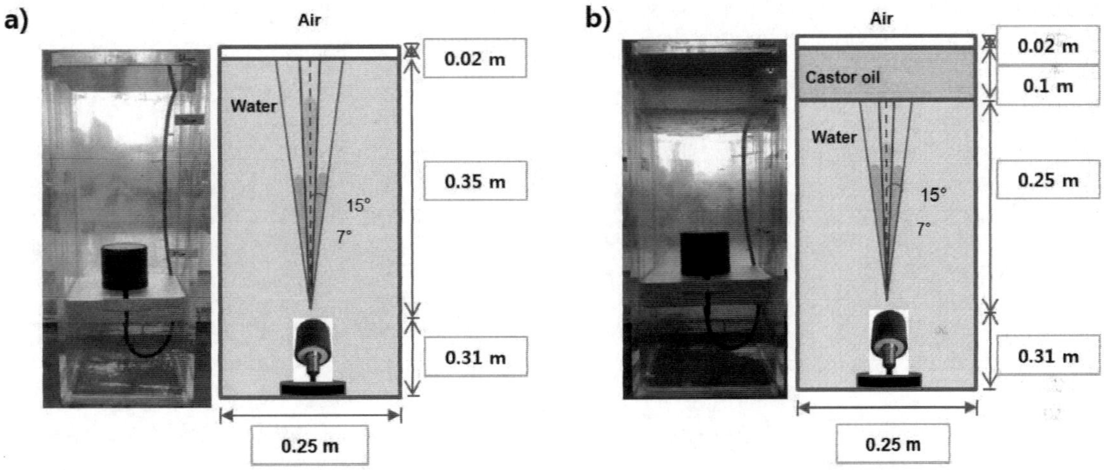


**Figure 1.** Geometry of water tank experiment for (a) water only and (b) 0.1-m thick castor oil on top.

### *2.1. Results*

Figure 2(a) shows the example of the received signals for water only (blue) and castor oil on water (red). The echo reflected from the water-castor oil interface is received about 0.35 ms earlier than that of air-water interface. Figure 2(b) shows the comparison of received levels (in dB re 1$\mu$Pa) between water only and castor oil on water. The measurement results imply that HNS_S may be detectable by acoustic method.

IOP Conf. Series: Journal of Physics: Conf. Series **1075** (2018) 012054 doi:10.1088/1742-6596/1075/1/012054

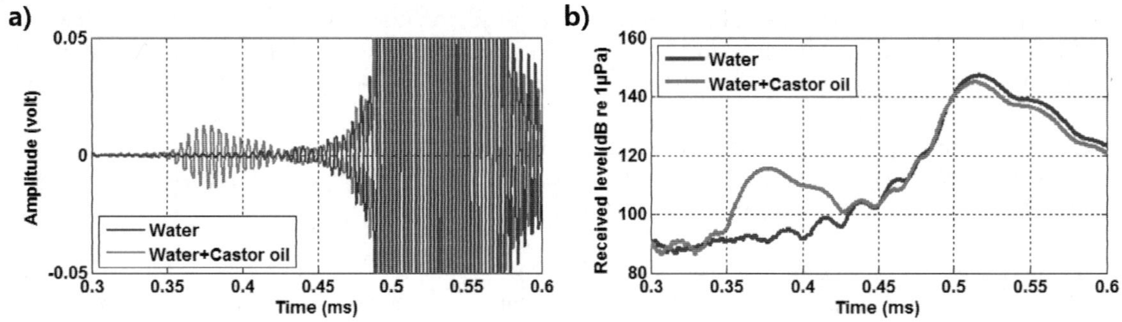


**Figure 2.** Measurement comparison between water only and castor oil on top. (a) time series of received signals and (b) received levels in dB re 1 $\mu$Pa.

### *2.2. Reflection loss estimate of Castor oil layer*
Reflection loss from castor oil was estimated from the difference between the received levels measured for water only and castor oil on water and it was compared to the theoretical value estimated by Eq. (1). Figure 3 (a) shows the reflection loss predicted as a function of grazing angle, in which the sound speed and density of castor oil was estimated using the empirical equations in Ref. 1 and 2. Figure 3(b) shows a comparison of reflection loss between the modelling and measurements at normal incidence angle. The measured reflection loss is in good agreement with the prediction with a small difference less than 3 dB.

$$R = \frac{\rho_o c_o/\sin\theta_2 - \rho_w c_w/\sin\theta_1}{\rho_o c_o/\sin\theta_2 + \rho_w c_w/\sin\theta_1} \qquad (1)$$

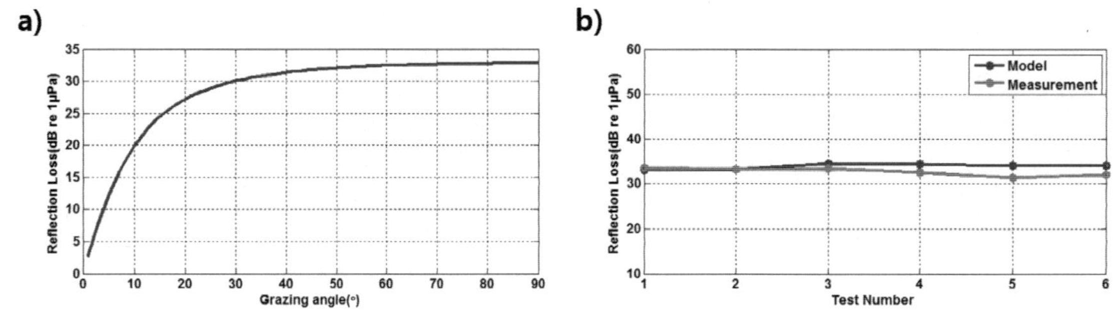


**Figure 3.** (a) Castor oil reflection loss estimated as a function of grazing angle and (b) comparison of theoretical reflection loss of castor oil at normal incident angle and measurements.

## 3. Simulation of side scan sonar image
Side scan sonar has been widely operated for ocean bottom survey. It uses high frequency signals to obtain high resolution image. In this study, the side scan image was simulated on condition that HNS_S lies on the seabed interface to investigate the detection possibility.

### *3.1. Simulation process*
It was assumed that 100-kHz side scan sonar was operated at a depth of 50 m in waters 100 m in depth, and was moved along the y-axis direction, and could cover 500 m wide on each side. Figure 4 shows the geometry of side scan sonar survey and its simulation result. A red dot and arrow in Fig. 4(a) indicate the starting point and towing direction of the side scan sonar. The received acoustic intensities were simulated using a sonar equation, in which the transmission loss was estimated from the two-way distance from the side scan sonar to the seabed. Backscattering strengths were predicted using APL-UW

backscattering model [3]. The surficial sediment was assumed to be sandy sediment with a mean grain size of 1 ϕ. Black parts in Fig. 4 (a) indicate the locations where two types of HNS_S lie on the sediment.

### *3.2. Simulation results*

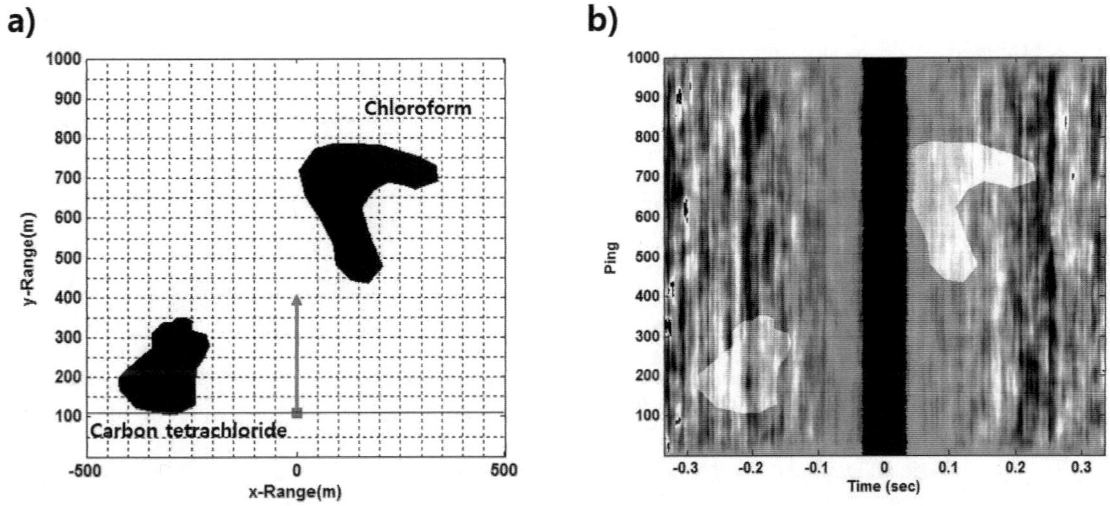


**Figure 4.** Simulation image of side scan sonar. (a) geometry of targeted area where HNS_S lie on the sediment and (b) simulation results.

Figure 4(b) shows the simulation image of side scan sonar. The results imply that if the acoustic impedance of HNS differs from those of water medium and sediment, there is a possibility of detection when HNS_S lies sunken on the ocean bottom.

## 4. Conclusion

In order to investigate the possibility of acoustic detection of HNS_S, water tank experiment was conducted using castor oil which has similar acoustic impedance with the Chloroform, and we confirmed the possibility of acoustic detection. Second, the side scan sonar image was simulated on condition that HNS_S lies on the seabed interface. A sonar equation was used to simulate the sonar image and the results showed that the side scan sonar may be a good tool to monitor the HNS_S lying on the sediment interface.

## References

[1] Timme R W 1972 *J. Acoust. Soc. Am.* **52**, 989
[2] Del Grosso V A and Mader C W 1972 *J. Acoust. Soc. Am.* **52**, 1442
[3] *APL - UW High Frequency Ocean Environmental Acoustic Models Handbook: Bottom,* APL-UW TR-9407, 1994.

## Acknowledgments

This research was a part of the project titled 'Development of Management Technology for HNS Accident', funded by the Ministry of Oceans and Fisheries, Korea.

IOP Conf. Series: Journal of Physics: Conf. Series **1075** (2018) 012055 doi:10.1088/1742-6596/1075/1/012055

# The study of resonator-diffuser acoustics performance using PVC pipe

**Suyatno[1], Andy A[1], Indrawati S[1] and Prajitno G[1]**

[1]Acoustic Laboratory, Department of Physics, Institut Teknologi Sepuluh Nopember Surabaya, Indonesia.

E-mail: kangyatno@physics.its.ac.id

**Abstract.** A preliminary study of a resonator panel and diffuser using polyvinyl carbonate (PVC). The resonator and diffuser parameters of the panels are tested in the chamber reverberation chamber and anechoic chambers in the laboratory of acoustic physics, the physics department of FIA ITS. The resonator parameter tested is in the anechoic chamber by looking spectral for SPL differences when without and the panel existed. As a diffuser, scattering patterns are tested in the anechoic chamber, while scattering coefficients (S) at a reverberation chamber. In this paper, the panel of resonators in the test made from pipes of 0.5 ", 1" and 3/2 "diameters with panel sizes are 60 cm x 60 cm. base on measurement, the panel has a resonance frequency at 540 Hz. While as a diffuser, the scattering coefficient (S) value for a small pipe is 0.98 at a frequency of 125 Hz with an angle of 0o, a medium pipe is 0.99 at a frequency of 500 Hz with an angle of 30°, a large pipe is 0.93 at a frequency of 500 Hz with Angle 0°.

## 1. Introduction

Resonance and diffusion are an important part at room acoustic, as both events can affect the acoustic quality in a room. Resonance is the process that an object vibrates because another object was vibrated, while diffusion is the event where sound spreading or scattering. In acoustic, resonance and diffusion are controlled by using acoustic panels, like a resonator and diffuser panels.

### *1.1. Resonator*

In acoustics, the resonator is used to adjust the sound frequency by eliminating or amplifying a sound at a certain frequency depending on the resonance frequency of materials. Figure 1 shows the acoustic resonator that called Helmholtz resonator.

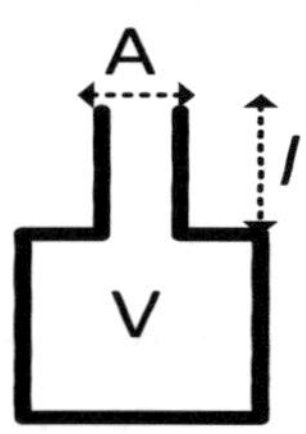


**Figure 1.** Helmhotz resonator.

Based on the fig. 1, the resonator with open space area A and length of neck l, then the volume of the cavity is v will produce the resonance frequency like equation (1).

$$f = \frac{c}{2\pi}\sqrt{\frac{A}{lV}} \quad (1)$$

Which $f$ is resonance frequency (Hz), c is velocity of sound in air (m/s), A is area of the holes ($m^2$), l is long of neck (m), and V is volume of cavity ($m^3$).

While based on research Mitchell et al (2016) for the resonator with an elastic material, the resonance frequency value can be calculated using the equation (2).

$$f = \frac{1}{2\pi}\sqrt{\frac{3\,Ec}{2\,Rl\,t\,\rho l}} \quad (2)$$

Which $E_C$ is elasticity modulus of skin ($N/m^2$), R is radius of core (m), ρ is density of core ($Kg/m^3$), t is thinness of skin (m), and l is line of organ pipe (m).

In other cases, the resonant frequency value for tube material (sonic crystal) can also be calculated using the principle of open and closed organ pipe used with equation (3).

$$f = \frac{c}{4l} \quad (3)$$

Which $l$ is leng of pipe and c is velocity of sound.

### *1.2. Diffuser*

Based on Trevor [1], the performance of the diffuser can be obtained by using the scattering coefficient. Scattering coefficient is the ratio between secularly reflected energy and scattered reflected energy. Specular reflected is the amount of energy reflected in the same way. if the energy is on a flat surface, while scattered is the amount of energy that is not specular reflected or that called is randomly reflected. Figure 1 describes this Phenom of reflected of sound at the surface.

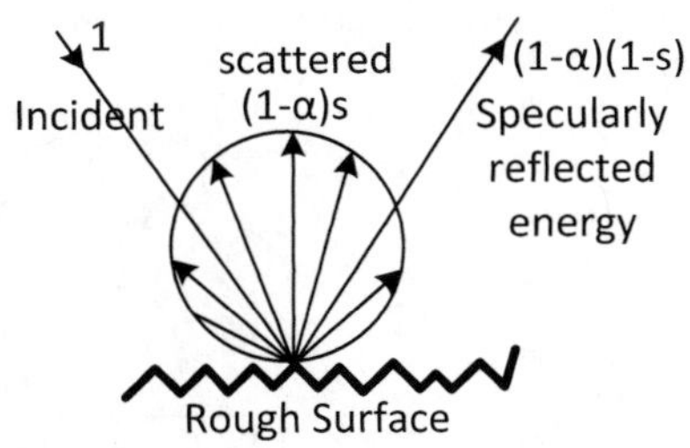


**Figure 2.** Scattering of sound at surface material.

Based on fig 1, to find the value of scattering coefficient or S diffuser can use equation as follows [1] using the reverberation time of the room.

$$s = \frac{\alpha_{spec}-\alpha}{1-\alpha} = 1 - \frac{E_{spec}}{E_{total}} \quad (4)$$

$$\alpha = 0{,}16\frac{V}{A}\left(\frac{1}{T_2} - \frac{1}{T_1}\right) \quad (5)$$

$$\alpha_{spec} = 0{,}16\frac{V}{A}\left(\frac{1}{T_4} - \frac{1}{T_3}\right) \quad (6)$$

Which, S is scattering coefficient, $\alpha_{spec}$ is specular absorption coefficient, A is absorption coefficient, $E_{Spec}$ is specular energy, $E_{ref\ total}$ is total energy, V is volume of chamber ($m^3$), A is area of panel ($m^2$), $T_1$ is reverberation time when the room empty and turn table not rotate (s), T2 is reverberation time when turn table and panel are not rotated (s), T3 is reverberation time when the room empty and turntable are rotated (second), and T4 is reverberation time when turn table and panel are rotated (second).

Currently, in the laboratory of Physics Acoustics, studies of several forms and designs of resonators and diffusers made of some natural materials is being conducted, like shell (Alfianti et al, 2016), coconut shell (Indrawati et al, 2016). This paper discusses the preliminary study of the resonator and

IOP Conf. Series: Journal of Physics: Conf. Series **1075** (2018) 012055 doi:10.1088/1742-6596/1075/1/012055

diffuser design in the form of the cylinder using the PVC pipe material. The panels made has 3 variations of diameter 0.5 ", 1" and 3/2 ". While the acoustic parameter that measure is the resonant frequency and the scattering coefficient of the refuser panel was made. In part II, explained about the measurement procedure of the acoustic parameters. Section III discusses the results of measurement and analysis, and the last section is the conclusion.

## 2. Methods

As mentioned in Part I, the performance of the material designed include the resonance properties and diffusivity of the panel. In this paper, the resonator panels have 3 variations diameter of the pipes, there are 0.5 ", 1", and 3/2 ", with panel sizes are 60 cm x 60 cm. The panels are made as shown in Figure 3.

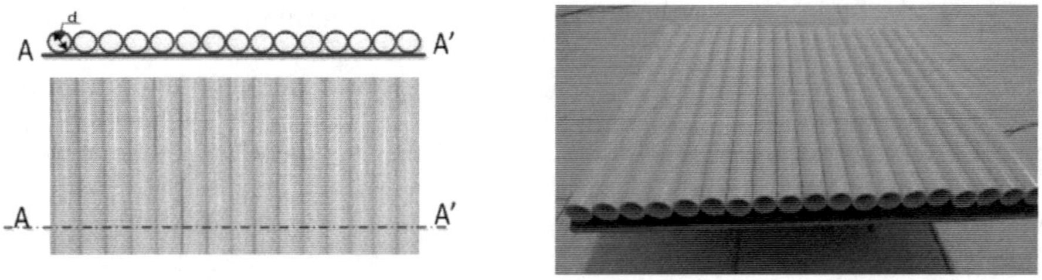


**Figure 3.** Refuser panel.

As a resonator, the test is performed at an anechoic chamber by analyzing the sound signals received by the microphone when the panel is installed and empty. By using the method, we can find the difference amplitude of sound energy at some frequency. While the nature of diffusivity (scattering), the measurement of scattering coefficients is done in the room reverberation room, and the measurement of scattering pattern is done in an anechoic room. For measurement of scattering pattern, based on Trevor (2004), the position of the panel, microphone and speaker is in position 0.2, 0.4 and 0.6 from the size of space length. Figure 4 shows the measurement setup.

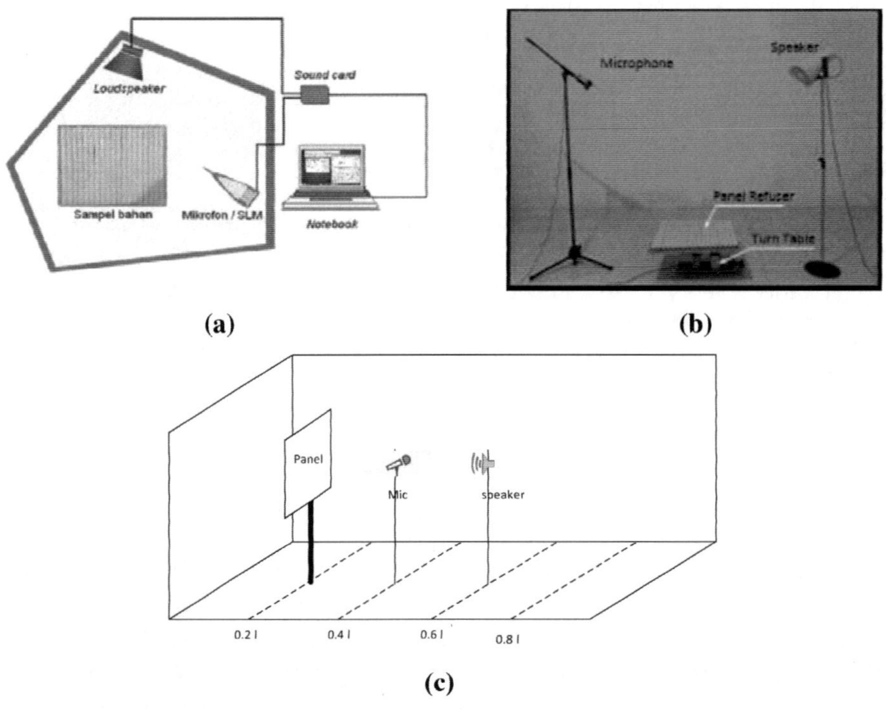


**Figure 4.** Measurement set up. (a) absorbtion coeffisient, (b) scattering coeficient, and (c) dispertion patern

IOP Conf. Series: Journal of Physics: Conf. Series **1075** (2018) 012055 doi:10.1088/1742-6596/1075/1/012055

## 3. Result and Discussion

### *3.1. Resonance*

Refers to equations (2), (3), and (4), the resonant frequency of the created panels can be obtained. Table 1 shows the magnitude of the resonance frequency.

**Table 1.** A slightly more complex table with a narrow caption.

| Sample | Diameter | Kjaer (Hz) | Mitchel (Hz) |
|---|---|---|---|
| | 0.5" | 1082 | 71.53 |
| Resonator | 1" | 1082 | 50.38 |
| | 3/2" | 1082 | 58.25 |

Theoretically, based on equation 3, the resonance frequency of pipe resonator is not dependent on the dimension, especially area of the hole.

For the result of resonance frequency by measurement, analyze was doing with compare of SPL that received by the microphone when the panel is installed and not installed. Figure 5 is showing the response frequency when panel installs and not installed.

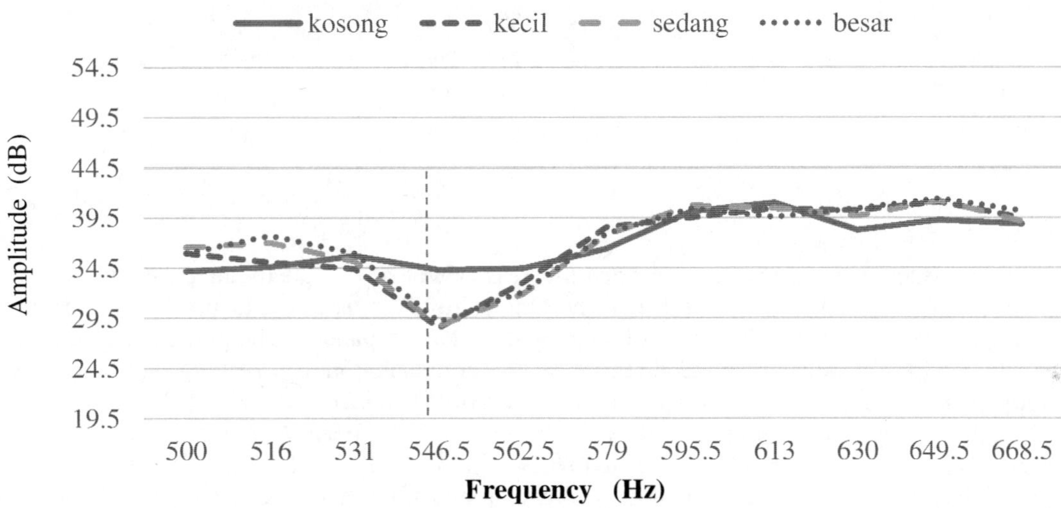


**Figure 5.** Graphic of resonance frequency of panel.

### *3.2. Scattering*

Based on measurement and calculation that refers to equation (1), to get the scattering coefficient, the measurement of RT parameter was done when the room empty and the panels exist. Table 2 shows the RT value of the chamber. Refer to equation (4), (5) and (6), the value of scattering coeficient is follow the table 3.

Based on the data shown in Table 2, the performance of the panel with diameter 0.5" is the best performance, while for a diameter 1.5" has the worst performance. This can be seen from the many scattering coefficient values between 0 and 1. This can happen because a small pipe it can be arranged more when compared to a larger diameter. This means that in small diameter of the resonator, there will be a more surface curve. It can more scatter the sound that comes than the larger diameter of pipes.

**Table 2.** The value of reverberation time of the chamber and scattering coefficients

| Frequency (Hz) | T1 (s) | T2 (s) | | | Scattering Coefficients | | |
|---|---|---|---|---|---|---|---|
| | | 0.5" | 1" | 1.5" | 0.5" | 1" | 1.5" |
| **125** | 0.398 | 0.900 | 0.896 | 0.363 | 0.98 | -0.84 | 0.74 |
| **160** | 1.058 | 0.571 | 1.046 | 1.054 | 0.39 | 1.56 | -6.12 |
| **200** | 1.142 | 1.344 | 1.300 | 1.107 | 1.60 | -0.70 | -0.34 |
| **250** | 0.813 | 0.834 | 0.847 | 0.650 | 0.95 | 3.00 | 0.93 |
| **315** | 1.367 | 1.777 | 1.891 | 2.375 | -43.51 | -5.09 | 0.93 |
| **400** | 2.048 | 2.258 | 2.325 | 2.364 | 0.56 | 0.16 | 0.10 |
| **500** | 1.425 | 1.478 | 1.484 | 1.529 | 0.25 | 0.25 | 0.93 |
| **630** | 1.686 | 1.725 | 1.878 | 1.960 | 0.07 | -0.06 | 0.75 |
| **800** | 2.118 | 2.138 | 2.126 | 2.113 | -1.92 | -3.44 | -5.60 |
| **1000** | 2.236 | 2.224 | 2.233 | 2.045 | 0.81 | -2.13 | 0.07 |
| **1250** | 2.112 | 2.295 | 2.368 | 2.335 | 0.52 | -0.02 | -0.48 |
| **1600** | 2.083 | 2.118 | 2.109 | 2.145 | 0.89 | 0.33 | -0.55 |
| **2000** | 2.364 | 2.100 | 2.252 | 2.206 | 0.93 | 4.02 | -0.62 |
| **2500** | 2.111 | 2.061 | 2.147 | 2.109 | -1.88 | -0.27 | -0.46 |
| **3150** | 2.168 | 2.188 | 2.204 | 2.263 | 0.51 | 0.33 | 0.63 |
| **4000** | 2.324 | 2.257 | 2.234 | 2.441 | 7.56 | 25.77 | 0.28 |
| **5000** | 2.226 | 2.214 | 2.217 | 2.331 | 0.30 | 0.05 | -0.70 |
| **6300** | 2.219 | 2.413 | 2.351 | 2.350 | 0.69 | 0.56 | -0.07 |
| **8000** | 2.296 | 2.348 | 2.454 | 2.363 | 1.14 | 0.25 | -0.65 |

### *3.3. Dispersion patern of scattering*

The distribution pattern is obtained by comparing the SPL value that reflects by the panel and received by the microphone at an angle of 0° with the other angle. As a diffuser, the dispersion of scattering pattern is important to know the directivity of sound will reflect by panels. The purpose of measuring the distribution of sound is to know the direction of the dominant sound reflection when the sound strikes to the panel. Figure 6 shows the distribution pattern of the created diffuser panel.

Based on the scattering diagram that is shown in fig. 6 and 7. It appears that the dispersion pattern of the three diffusers, seen uniformly at low frequency or 125 to 500 Hz, but, at high frequencies, ie, 1000 to 4000 Hz, there are several anomalies. For example, diffuser with diameter pipe 1" for angle 70° and 80° at 1000 Hz, for diameter 0.5" for an angle of 60 ° at 2000 Hz. That same also diffuser with diameter 0.5" for an angle 30 at 4000 Hz.

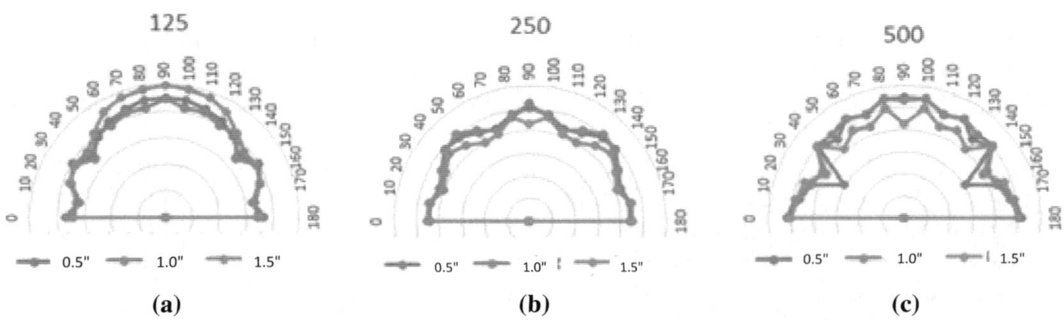


**Figure 6.** Dispersion pattern of panel in frequency (a) 125 Hz, (b) 250 Hz, and 500 Hz.

IOP Conf. Series: Journal of Physics: Conf. Series **1075** (2018) 012055 doi:10.1088/1742-6596/1075/1/012055

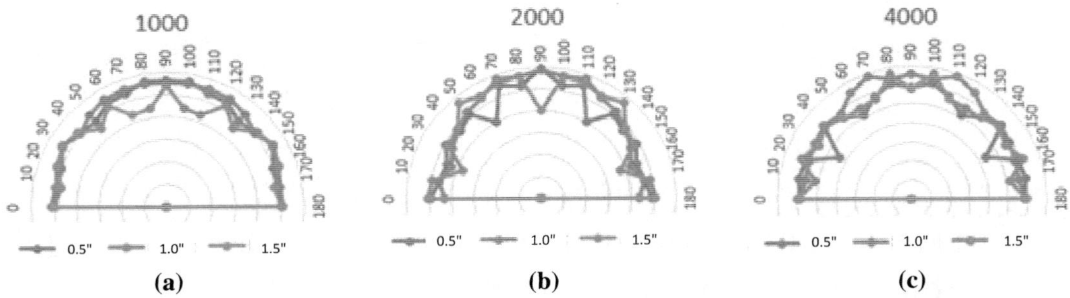


**Figure 7.** Dispersion pattern of panel in frequency (a) 1000 Hz, (b) 2000 Hz, and 4000 Hz.

## 4. Conclusions

Based on the measurement, the resonance frequency of the refuser panel is 546.5 Hz. The value of the scattering coefficient for the diffuser with diameter 0.5" has a good performance, while 1.5" have a poor performance. And at 125 Hz frequency, the performance of the three panels gives the best response that is uniform, this indicates that this panel will be able to distribute the sound well in all directions. But at a frequency of 500 Hz, the panel will propagate to the front only.

For further research is needed regarding the number of pipes and pipe size combinations a panel. Note that as a general principle, for large tables font sizes can be reduced to make the table fit on a page or fit to the width of the text.

## 5. References

[1]. D'Antonio P, TJ Cox. 2004. "Acoustic absorbers and diffuser: theory, design and application". Spoon Press: London.
[2]. K. B. Ginn. 1978. "Architectural Acoustics". Bruel & Kjær : Denmark
[3]. Stephanie J. Mitchell. 2016. "Metaconcrete: Engineered aggregates for enhanced dynamic performance". California Institute of Technology.
[4]. Dita A. A, Suyatno Susilo I, 2017, "Utilizing the Uniqueness of Blood-seashell (Anadara granosa) as an Alternative Acoustics Material for a Diffuser Panels", Procedia Engineering 170 , page 261 – 265
[5]. Susilo I, Suyatno, 2017, "Innovative Coco Shell Resonator (CSR) Panels for Acoustic Performance", Procedia Engineering 170, page 293 – 298

IOP Conf. Series: Journal of Physics: Conf. Series **1075** (2018) 012056 doi:10.1088/1742-6596/1075/1/012056

# The mother's womb acoustic environment: study of the original sounds and replication for pre-term infants

*

**G Dealessandri[1] and M Vivalda[2]**

[1]Bioacoustics department of ICOSTECH srl – Pinerolo (Italy)
[2]M.D. NICU S. Anna Hospital Città della Salute e della Scienza - Torino (Italy)

E-mail: germano.dealessandri@icostech.com

**Abstract**. An experimental study of the physical phenomena that regulated the sound propagation inside the human body and the amniotic fluid was carried out, by analysing the acoustic climate inside the uterus of pregnant women. The aim of this work was to reproduce the original sounds that a fetus can hear inside the mother's womb, as faithfully as possible, in order to provide preterm neonates the rich acoustic stimuli they would otherwise be deprived of. This work describes a natural and new method in the field of new-born care programs which assists early new-born development. The method involves the replication of the intrauterine acoustical environment of the mother which may play an important role in new-born development, especially those born pre-term.

## 1. Introduction

Several studies have shown that intrauterine environment during pregnancy is characterised by the presence of noises and sounds. These acoustic signals can propagate reaching the foetus's ears [1]. The foetus is able to perceive acoustic stimuli starting from the sixth month of pregnancy [2] [3]. Noises and sounds inside the mother's womb can influence the human development. Previous researches have shown the possibility of the foetus to react to external sounds [2] [3] and his capability to recognise voices or sounds during the intrauterine life [4] [5] [6].

Sounds from the external environment to the foetus must propagate through the air and the body with different transfer functions. Indeed exogenous sounds are modified, both in frequency and intensity, when they propagate through the mother's abdominal tissue, while endogenous sounds are modified by all the mediums they find in their path (human organs, bones, etc.). However, notwithstanding the nature and acoustic stimuli propagation modality, before reaching the foetus's ears they must propagate through the amniotic liquid.

The target of the research was to discover the transfer functions of the several acoustic stimuli to allow the artificial recreation of the womb's sounds lost from the pre term infant.

## 2. Study methodology

It is not possible to measure directly the individual intrauterine acoustic sounds during the pregnancy for obvious difficult and ethical reasons. Therefore the study focused on measuring the overall acoustic sounds and then decomposing the sounds into their different fundamental components. Subsequently and successively, analytical mathematical models were created to describe the sounds transformation during the propagation through the physical mediums of woman's tissue and amniotic liquid (transfer functions) and eventually the transformed sounds are recomposed. Thus, the methodology has been focused on three different steps:

- Signal acquisition;
- Post processing and transfer function analysing
- Rebuilt of the womb's acoustic environment

## 3. Signal acquisition

The fundamental actions of this acquisition stage have been the following:

- Endogenous sounds registration by means of an intra-vaginal microphone in some of the pregnant women volunteers;
- Study of the modification of sounds during the propagation through maternal tissues, recorded by means of an intra-vaginal microphone and a series of sounds generated outside;
- Analysis of the sound modification during the propagation in liquid (amniotic fluid). Taking into account the scientific literature on underwater acoustics, this aspect has been measured in laboratory by simulations able to investigate the variation of the sound transmission with the liquid deepness (distance of the foetus's ear from the amniotic sac). The influence of the foetus oscillation in the fluid was also studied.

The women voluntarily co-operated in this measurement study. They agreed, using previous informed consent, with written declarations that were conformed to the Helsinki Convention regarding experiments on human subjects.

The first recorded sounds were those present inside the abdomen cavity of the women (endogenous, or background sounds) when the external environment was silent.

The second series of sounds recorded, were generated by a couple of loudspeakers (see figure 1). These sounds included: pink noise, pop music, classical music and mother's voice reading a lullaby. Both the external and the internal signals were recorded simultaneously. The measured sound levels were recorded and the endogenous signals contribution removed by logarithmic difference between the measured signals and the background noise.

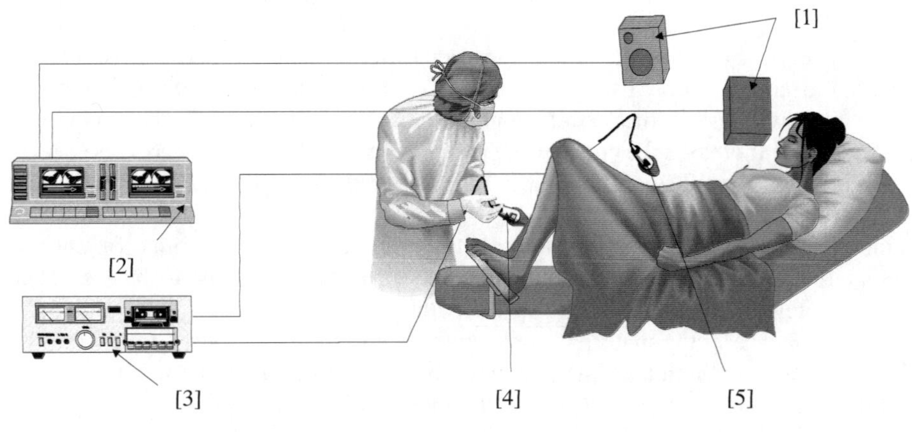


[1] Loudspeakers; [2] Sound source; [3] Dual channels analyser; [4] Internal microphone; [5] External microphone.

**Figure 1**. Description of experimental phase for live measurements.

IOP Conf. Series: Journal of Physics: Conf. Series **1075** (2018) 012056 doi:10.1088/1742-6596/1075/1/012056

The relative transfer function of the mother's voice was determined by comparing the external recording of their voice with the internal signal recorded by the probe.

The measurements for the amniotic fluid transfer function determination were carried out by recording on digital recording equipment the signal received from a condenser microphone (protected by a lattice membrane) merged into a recipient filled with a saline liquid, with characteristics similar to the amniotic fluid. Simultaneously the digital equipment recorded the sound source, which was pink noise diffused by a loudspeaker outside the recipient. The measurements were taken at variable depths of 1 cm, between 1 cm and 8 cm of fluid, to simulate the various possible foetus ear positions. Inside the water was placed also a small doll for interference simulation of the foetus.

## 4. Post processing and transfer function analysing

The background noise inside the abdominal cavity is composed by a time-fluctuating signal, due greatly to the maternal heart-beat and to the intestinal gurgling. The equivalent sound pressure level of this noise is generally about 30 dB(A). Figure 2 shows the equivalent levels for each frequency (averaged over the values obtained with several woman). The major part of the acoustic energy is concentrated on the middle-low frequencies.

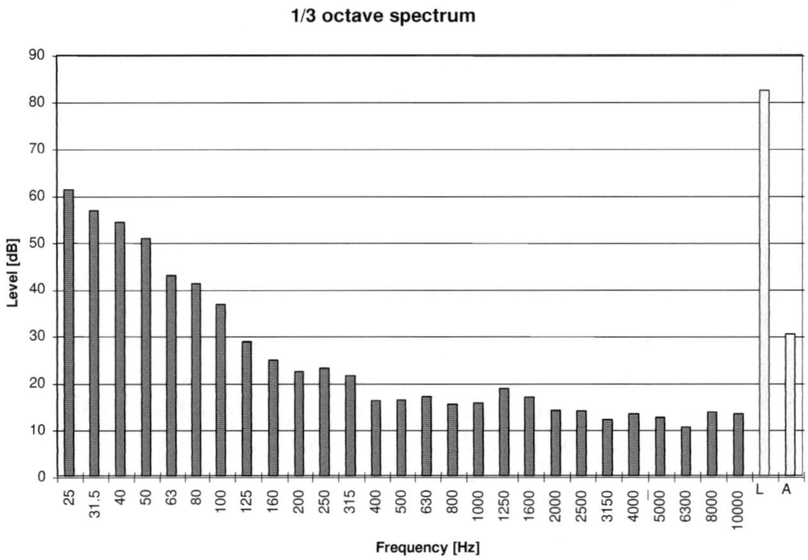


**Figure 2.** Averaged background noise level over the tested women in 1:3 octave frequency span.

The abdominal tissue transfer function is the transformation of the acoustic signals propagating through the human tissues. Figure 3 shows an example of a transfer function related to a pink noise source. From this it is evident that the sound attenuation due to the abdominal tissues is increasing with frequency. As frequency increases the column referring to the 'inside sound' decreases, while the outside sound remains fairly constant. The abdominal tissue attenuates more the medium-high frequencies, of which the human ear is more sensitive.

It was found during this study that the maternal voice signal propagates more efficiently with respect to other sound sources. It reaches the foetus through two different mechanisms: by the outside airborne path and by the preferential structural (body) inside path.

Figure 4 shows one case of the maternal voice analysis and its alterations. Comparing this data with Figure 3, it can be seen that the maternal voice undergoes amplification at the low frequencies, from 80 to 200Hz, and the attenuation is less sensitive at the medium-high frequencies causing a complete and audible signal transmission.

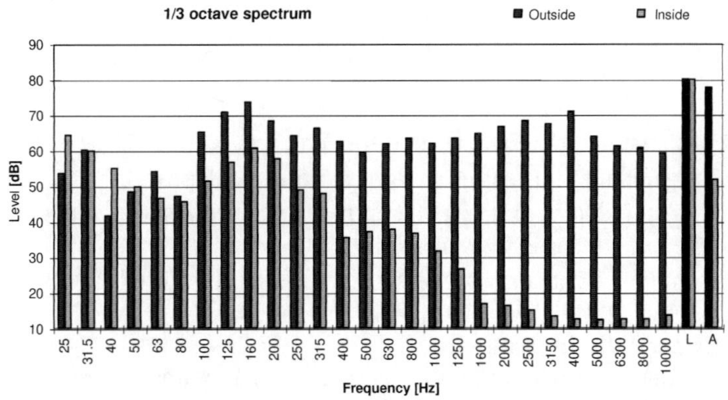


**Figure 3**. Example of sounds alteration from abdominal tissues in 1:3 octave bands analysis.

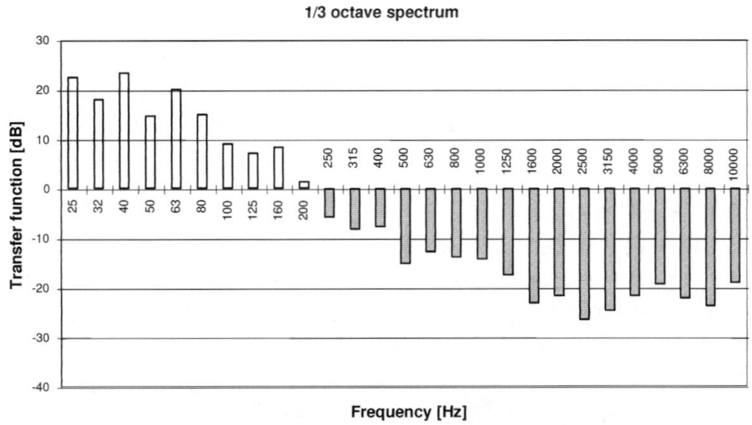


**Figure 4**. Example of mother's voice transfer function.

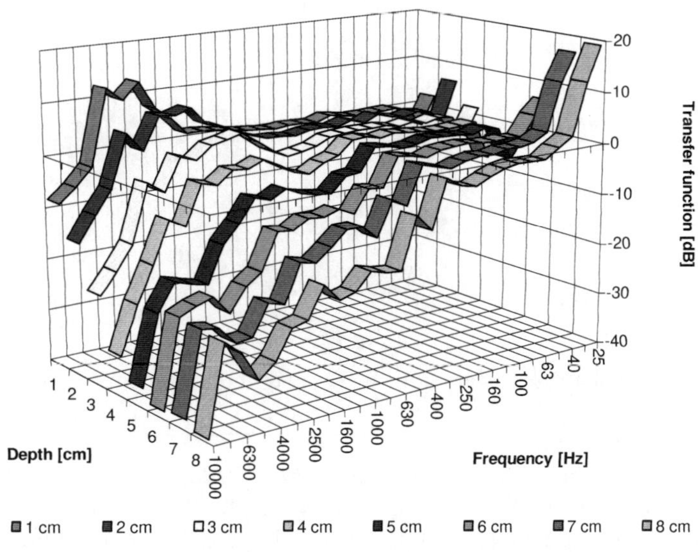


**Figure 5.** Example of amniotic fluid transfer function.

IOP Conf. Series: Journal of Physics: Conf. Series **1075** (2018) 012056 doi:10.1088/1742-6596/1075/1/012056

The amniotic fluid has another transfer function on the sound transmission. This effect assumes a different behaviour depending on the different depth at which the signal arrive. From the experimental data (see figure 5) it can be observed that:

- the sound wave is not attenuated between the frequency bands from 50 to 315 Hz;
- above 315 Hz the transfer function varies according with the depth;
- between 1 cm and 3 cm depth of fluid the transfer function shows an amplification at the frequencies where the human ear is more sensitive;
- For depths above 3 cm the attenuation of sound increases proportionally with the frequency.

## 5. Rebuilt of the womb's acoustic environment

A device able to acquire, sample and elaborate the acoustic signals with the aim to rebuild the proper intrauterine sounds was built. The procedure is based on:

- Signal sampling
- Decomposing the digitalised signals into their fundamental components by mean of FFT
- Transfer functions application (modification of the signals to simulate the propagation through the barriers represented by the human tissue and the amniotic fluid);
- The superposition to the transformed signals of the underlying acoustic endogenous sounds like mother's heart-beat and intestinal mumbles (figure 6).

This process is able not only to recreate the womb sounds environment that the foetus perceives, but also to 'personalise' it. Indeed, by the voice acquisition of the actual mother and of her typical heart-beat whilst relaxed, it is possible to reply to pre-term not a generic acoustic environment, but the sounds effectively 'lost' from the womb in which the foetus grows during the pregnancy. The quality of this process strongly depends on the optimal transfer function determination and the accuracy of the sounds acquisition.

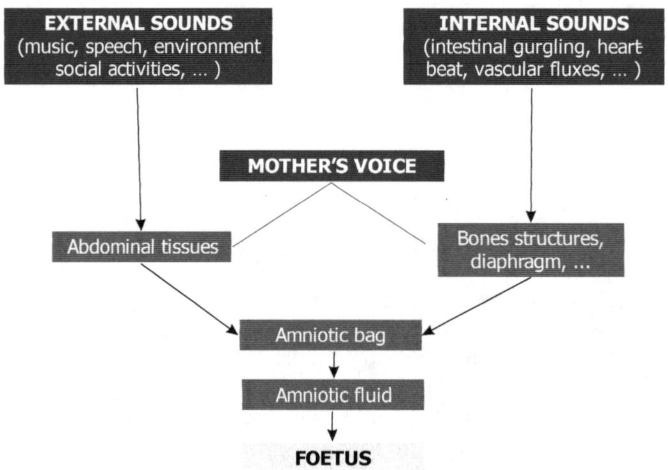


**Figure 6.** Typical flow chart of different sound components composition.

## 6. Experimental tests in Hospital

Based on the results of the research a proper device has been developed to be used in the hospitals (8). The developed device can playback the customized sounds from the mother of the pre term infant. That means there is a way to follow the daily sounds surrounding the mother, making them in relationship with her heartbeat and her voice. The device is studied to surround the pre term in the same monaural hearing environment as he's used to listen inside the womb.

The environment inside the incubator and the external world is separated by mean of an interchangeable membrane useful both for playback the sound, keep hygiene level and climate barrier. The device can be connected to all the kinds of existing incubators so it can be supplied ad an add-on. There is only the need to prepare an adaptor for one of the windows of the NICU itself.

The design and dimension of the playback device can be customized in order to be hosted below the incubator and move with it.

## 7. Discussion and prospects

The aim of this work is not to discuss about the characteristic of the information that reach the foetus, as they appear in the recreation process but it is important to underline the following acoustic aspects:

- High frequency sounds are attenuated more by the system abdomen-amniotic fluid in respect to low frequency sounds which leads to the question of just how important are low frequency sounds to the foetus?;
- The dynamic aspect of the foetus of listening to exogenous sounds. The foetus may decide to hear more distinctly some external signals by moving closer to the mother's abdominal wall, or to acoustically isolate itself from the outside world by moving to the centre of the amniotic sac. It has been shown that by increasing positional depth the attenuation effects of the abdominal wall and the amniotic fluid become synergetic;
- The fundamental role that the mother's play. Their sounds are transmitted with all their fundamental characteristics, distinctly reaching the foetus and stimulating him or her for several hours in a day (the problems in the immature-borne infants linguistic acquisitions may have in this deprivation a cause?);
- The direct link that the foetus has with the maternal heart-beat. This sensation is consistently fed to the foetus listening to the endogenous intrauterine background sounds (This raises several factors: could this represent a psychological enforce; a synchronisation with the circadian mother's rhythms and with the mother's emotive status and an education to the recognition of external sounds as dangerous or relaxing?);
- The fact that the male voice is sufficiently transmitted through the tissues while outside (externally) the female's voice is more difficult to be perceived. With practise the foetus will be able to recognise the male voices of the community in which the mother lives (so also but not uniquely the father's voice). During this period a unique conscience of the mother's voice is developed.

## 8. References

[1] Richards D S, Frentzen B, Gerhardt K J, McCann M E and Abrams R.M 1992 Sound levels in the human uterus *Obstet. Gynecol* 80(2) pp 186-190
[2] 9. Querleu D, Renard X and Versyp F 1988 Foetal Hearing *Eur. J. Obstet. Gynecol. Reprod. Biol.* 28 (3) pp 191-212
[3] 10.Querleu D, Renard X and Boutteville C 1989 Hearing by the human foetus? *Semin. Perinatol* 13 (5) pp 409-420
[4] 11.Walsh E J and McGee J 1990 Development of auditory coding in the central nervous system: implications for in uterus hearing *Semin. Perinatol* 14 (4) pp 281-293
[5] 12.Perris E E, Myers N A and Clifton R K 1990 Long term memory for a single infancy experience *Child. Dev.* 61 pp 1796-1807
[6] 13.DeCrasper A J and Fifer W P 1980 Of human bounding. Newborns prefer their mothers' voices *Science* 208 pp 1174-1176
[7] 14.Standley J 2012 Music therapy research in the NICU: an updated meta-analysis *Neonatal Netw.*
[8] 15.Mauro V, Germano D, Marco F, Giovanni B B, Paolo M and Daniele F 2012 A pilot study to assess the effects of biological sounds on preterm new-borns *3rd International Conference on Clinical Neonatology* (Turin)

IOP Conf. Series: Journal of Physics: Conf. Series **1075** (2018) 012057 doi:10.1088/1742-6596/1075/1/012057

# Enclosure design of train of eating and generator on production PT. INKA Madiun using computational fluid dynamics

**H R Amrullah*, P K Deo and W A Asmoro**

Department of Engineering Physics, Institut Teknologi Sepuluh Nopember, Sukolilo, Surabaya 60111, Indonesia

*E-mail: hafizhamrull@gmail.com

**Abstract.** Trains are a means of transportation is much in demand by the community because it is considered effective and efficient. The railway has several sections, one of which is on the railway carriage and the power plant. In this study, train carriage and power plant with the loading of one railway carriage, designed to be quieter, with reference not exceeding the standard limit based on Kepmen LH Number 48 of 1996. Design used by altering the enclosure insulation in the generator chamber. In the existing condition, the train carriage has not met the standard at station noise. This study obtained the noise insulation results from enclosure design on railway carriages and generators made with various variations, so obtained enclosure with 25% grill ratio and ventilation on the floor, that is with the value of sound pressure level at a distance of 6 meters of 65.77 and 65.56 dBA.

## 1. Introduction

Trains generally consisted of locomotives and several carriages to transport passengers and goods in bulk either between cities, provinces and interstate. This means of transportation was much in demand by society because it was considered effective and efficient. However, in addition to having a positive impact, the railway also gave a negative impact that causes noise to the environment, both internal and external environment of the train. Some sources of rail noise included noise from wheel and rail area, generator system on railway carriage, other than noise from inverter, compressor and noise from AC system (air conditioner) [1].

The railway has several parts, one of them on the carriage section of the carriage. The generator train was part of a series of trains that had the function of supplying electrical energy into the circuit, in which there was a generator to supply electricity during travel, usually placed at the end of the circuit. The carriage included a noise source inside the railway circuit. According to Regulation of the Minister of Transportation No. 15 of 2011, that the generator train must had a maximum noise test result of 85 dBA on 6 meters [2]. Noise caused by generating machinery would affect the hearing health of both passengers and operators on duty on the carriage. One of the most effective and widely used forms of noise reduction today is the passive noise control which can be achieved with the use of an acoustic barrier, such as an enclosure [3].

From the previous research, an optimal acoustic enclosure was developed for portable generators using passive noise control techniques. In an effort to develop a commercially attractive enclosure to reduce the noise was emitted from the generator, this investigation would determine the correct design method intended to optimize the enclosure performance for the generator [4]. Based on information about the generator train which was the largest noise source in the railway circuit, therefore it was necessary to conduct research on the design of the power plant with the enclosure at PT. INKA Madiun which was then analyzed using CFD (Computational Fluid Dynamic) whose goal was to reduce the noise level generated by the carriage, so it does not negatively affect the health of passengers and operators.

IOP Conf. Series: Journal of Physics: Conf. Series **1075** (2018) 012057 doi:10.1088/1742-6596/1075/1/012057

## 2. Methods

This paper described an objective measurement and a computer simulation method using acoustics diffusion equation. Acoustics data of the existing room was needing to ensure the validity of the simulation model.

### *2.1. Measurement*

2.2. The acoustical condition of train of eating and generator in PT. INKA was measured. The train had dimension with a length of 20 m, width of 2.9 m and height of 3.63 m. Noise object was a generator train with loading of one railway carriage, had dimension with a length 3.6 m, width 1.6 m and height 2.1 m. The first thing to do was to collect the generator data, to knew the characteristics of the noise source with the measuring point of the noise. Data was 1 meter away from the generator and it was at 2 points (Fig. 1). Two sensors of SLM DEKKO Model: SL-130, were used to get a noise value.

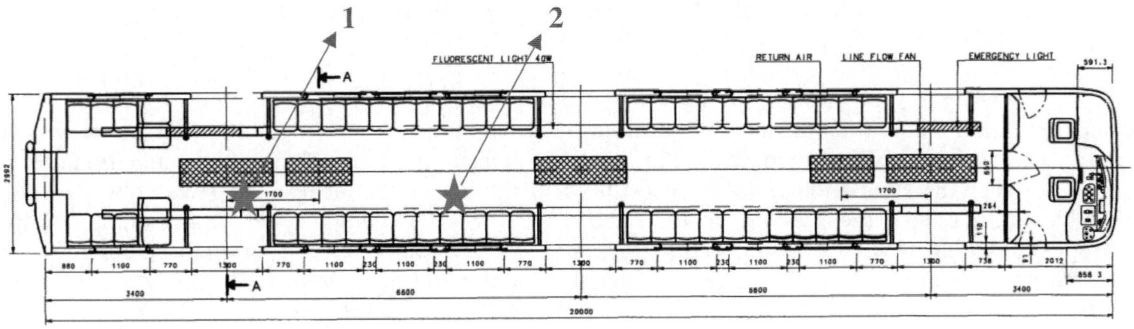


**Figure 1.** Point of Measurement Noise Generator

Method of measurement refers to the method described in ISO 3381-2005 *Railway Applications – Acoustics – Measurement of Noise Inside Railbound Vehicles*, which in principles of the source and microphones. Microphone were positioned inside train, in the dining room there are 10 points of measurement with the height of the sensor at each point of 1.2 (sitting position, point 1) and 1.6 meters (standing position, point 2) (Fig. 2).

The second method of noise measurement were described in Ministry of Transport Regulation No. 15 of 2011, noise data measurements were measured on the outside of the carriages within 6 meters of the carriage.

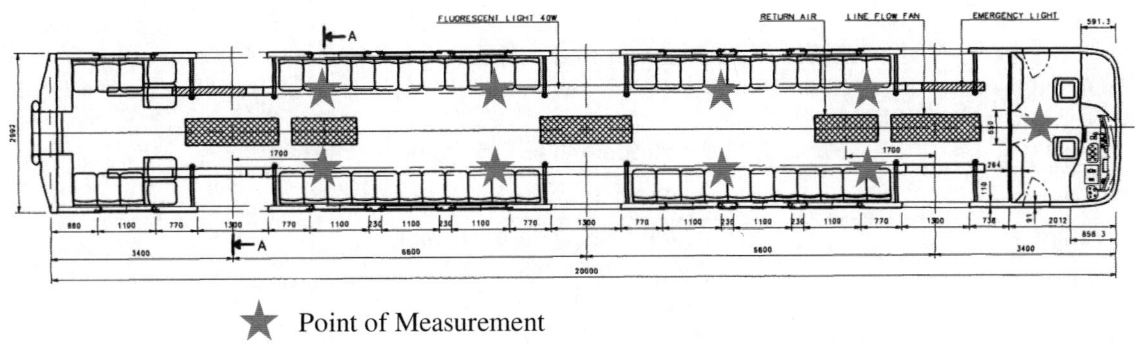


**Figure 2.** Point of Measurement Inside Railbound Vehicles

Next experiment, method of measurement refers to the method described in ASTM:E1332(10), "Determination of Outdoor-Indoor Transmission Class" to know ability of insulation of generator chamber. Microphone are positioned at 8 position (Fig. 3), inside and outside generator chamber, and outside train.

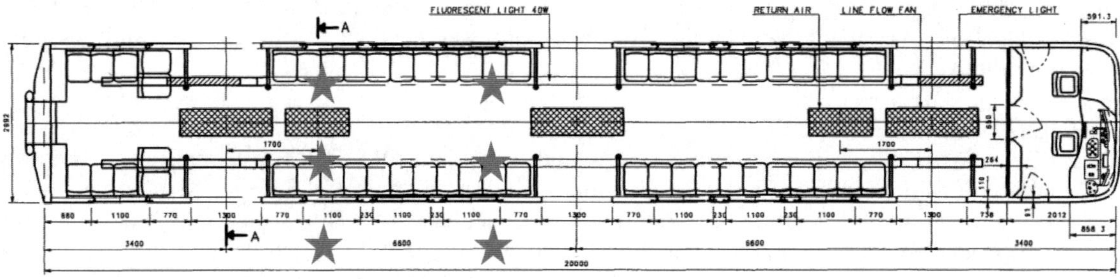


★ Point of Measurement

**Figure 3.** Point of Measurement Transmission Class

*2.3. Simulation Design*

There were several simulation steps by using CFD analysis, drawing geometry, determining the boundary, determining the material, determining the variation on the enclosure, determining the boundary condition, meshing, then plotting the simulation result used in analyzing the simulation (see Figure 4). When the simulation results were similar to the data retrieval results, it was redesigned at the enclosure to reduce the pressure level. Firstly, we drew the geometry to representation shape of train simulation. On the drawing geometry, obtained 2 variation on the simulation. First simulation was figured enclosure (generator chamber, shown in Fig. 5) with a 75 grill horizontal and second simulation was figured enclosure with a 25% grill and ventilation under the floor. Meaning of 75% and 25% grill was comparison between grill and outside of wall generator chamber (Fig. 6). The constituent material on partition C was used in arranging the walls, roof and floor in the enclosure, while the B partition was still used to arrange the B partitions in the enclosure.

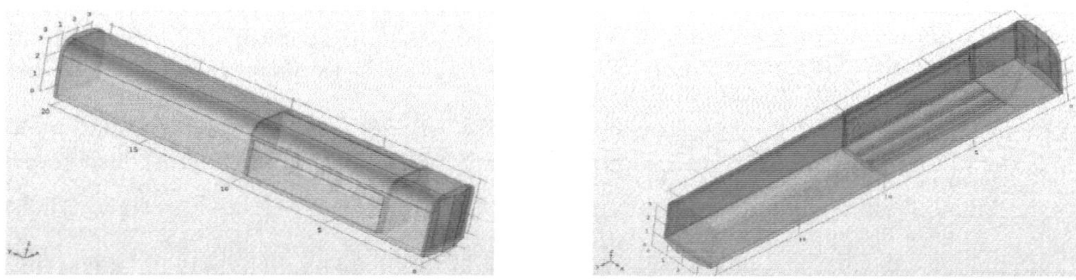

**Figure 4.** Geometry of Body Train Appear 3D Axis

On the partition of generator chamber, had 2 partition with symbols B and C. The specification of the partition, on Table 1.

**Table 1.** The Partition of The Generator Chamber

| Partition | Material and Size |
|---|---|
| Partition B | *Aluminium perforated 2mm* |
| | *Rockwool 50 mm* |
| | *Air Space 50 mm* |
| | *Rockwool 50 mm* |
| | *SS400 3 2mm* |
| Partition C | *Aluminium perforated 2mm* |
| | *Rockwool 50 mm* |
| | *Air Space 50 mm* |
| | *Rockwool 50 mm* |
| | *SS400 3 2mm* |

IOP Conf. Series: Journal of Physics: Conf. Series **1075** (2018) 012057 doi:10.1088/1742-6596/1075/1/012057

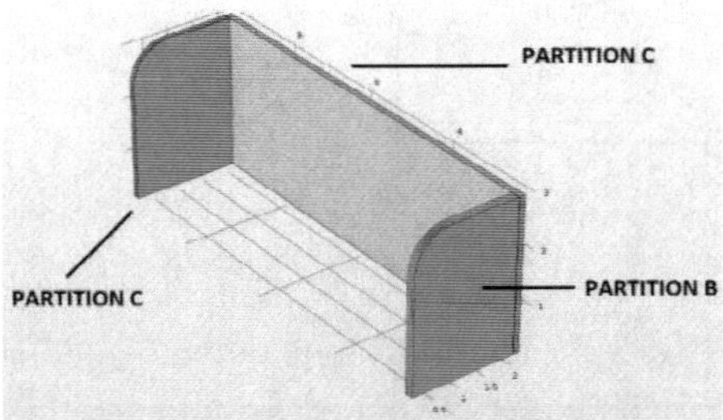


**Figure 5.** The Partition of Generator Chamber

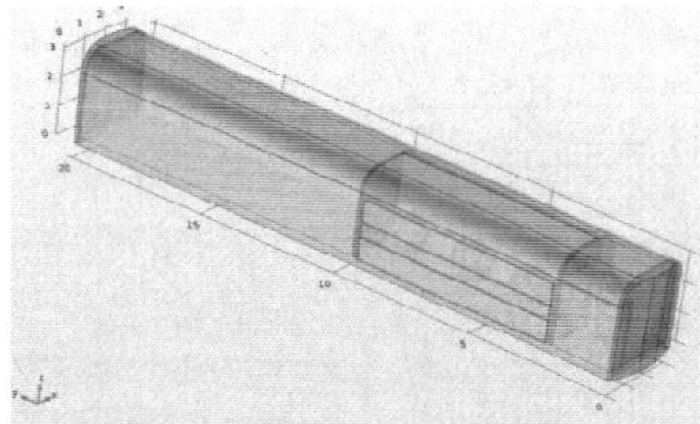

**Figure 6.** Geometry of Body Train Appear 3D Axis with 75% Grill Horizontal

## 3. Result

### 3.1. *Result of Measurement*

The measurement results obtained the characteristic of noise generator, the noise characteristic in the train interior and the transmission loss of enclosure.

*3.1.1. The characteristic of noise generator.* The measurement at the generator site at two measurement points obtained a spectrum of noise at the generator. As shown in Fig. 7, the higher spectrum amplitude occurs from 630 - 8000 Hz, therefore it consequently most of the noise are high frequency. This may also imply that the noise is in high frequency range. Hence, acoustical insulation was necessary.

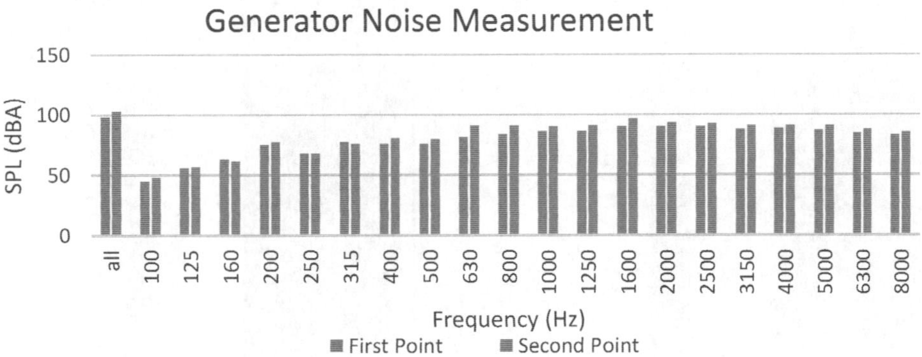


**Figure 7.** Noise Characteristics at the Generator train

*3.1.2. Result measurement of noise inside train.* Based on ISO 3381-2005 *Railway Applications – Acoustics – Measurement of Noise Inside Railbound Vehicles*, the sound contour map is shown in Fig. 8 by placing the microphone 1.2m above the floor to simulate the height of normal passenger sitting whilst the Fig. 9 shows the sound contour of people standing 1.6m. These two figures suggest that the sound pressure level at the far end from the generator was lower about 10 dB than that of close to the Generator. However, the loudness level was more acceptable to the passengers at the far end.

IOP Conf. Series: Journal of Physics: Conf. Series **1075** (2018) 012057 doi:10.1088/1742-6596/1075/1/012057

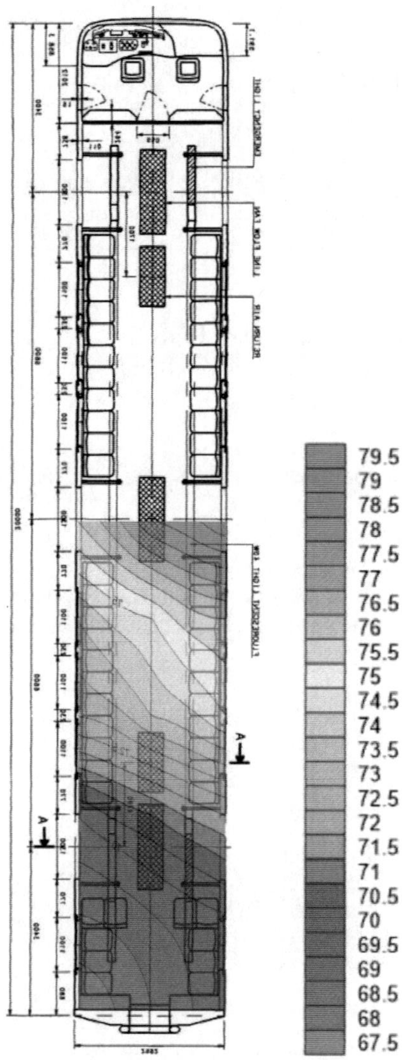


**Figure 8.** Results of noise data taking in the dining room, for the height of the measuring instrument 1.2 meters

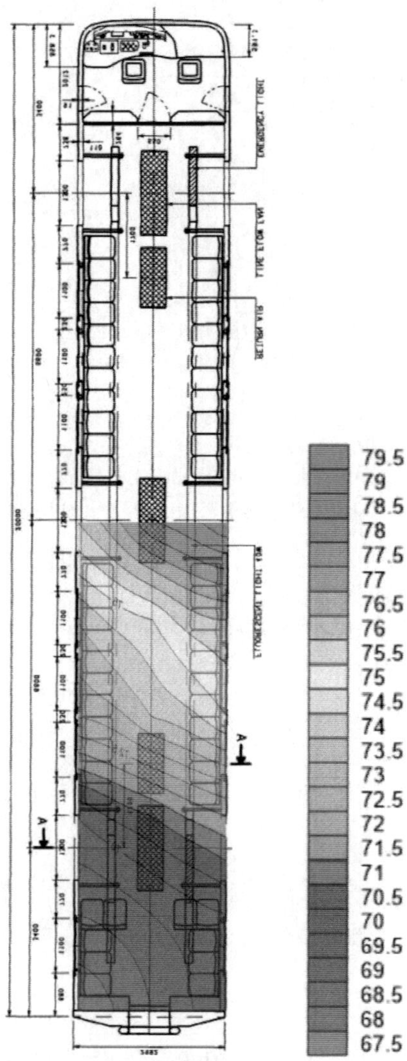


**Figure 9.** Results of noise data taking in the dining room, for the height of the measuring instrument 1.6 meters

Ministry of Transport Regulation No. 15 of 2011, data on the measurement of noise at a distance of 6 meters was obtained TTB (all) value of 74.9 dBA with loading of one railway carriage.

*3.1.3. Result measurement transmission loss of enclosure*. In the first data acquisition area obtained average transmission loss value of 8.2 dBA. While on the second area of data acquisition obtained average transmission loss value of 12.1 dBA. While on the third area data retrieval obtained the average transmission loss of 18.6 dBA.

3.2. *Result of Simulation*

On the result of simulation, obtained simulation with CFD analysis. After designed existing condition of train (Fig. 4), obtained result of simulation body train and enclosure with a grill 75% horizontal (Fig. 10 and 11), and with a 25% grill and ventilation under the floor (Fig. 12 and 13).

IOP Conf. Series: Journal of Physics: Conf. Series **1075** (2018) 012057 doi:10.1088/1742-6596/1075/1/012057

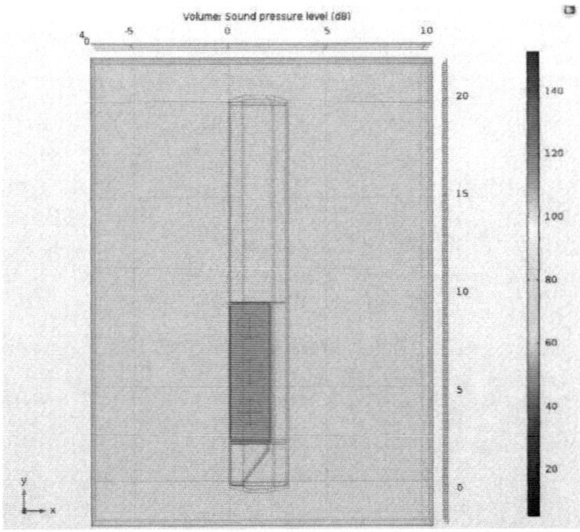


**Figure 10.** The First Simulation Appear on Top Side

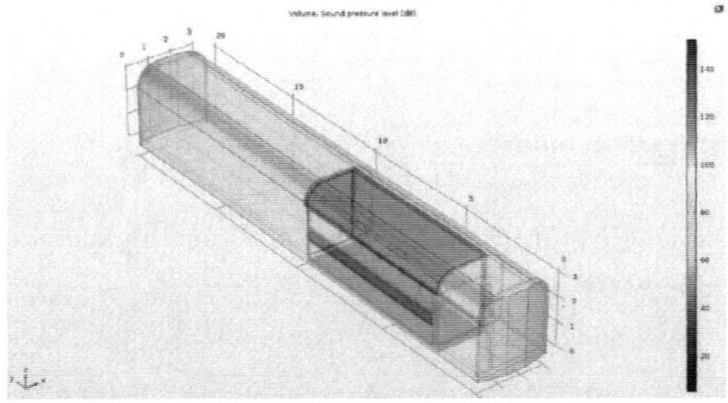

**Figure 11.** The First Simulation Appear on 3D Axis

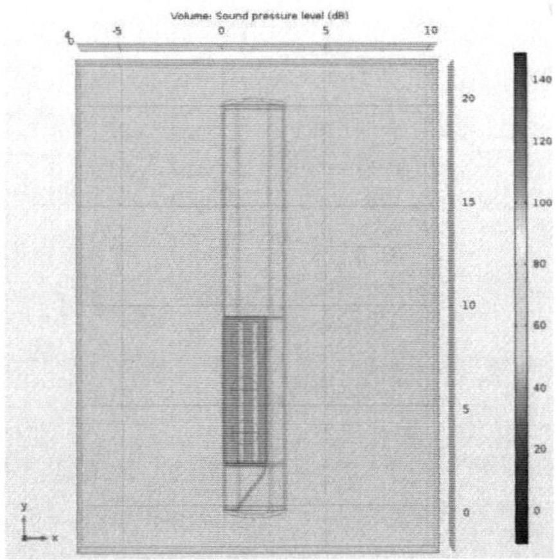


**Figure 12.** The Second Simulation Appear on Top Side

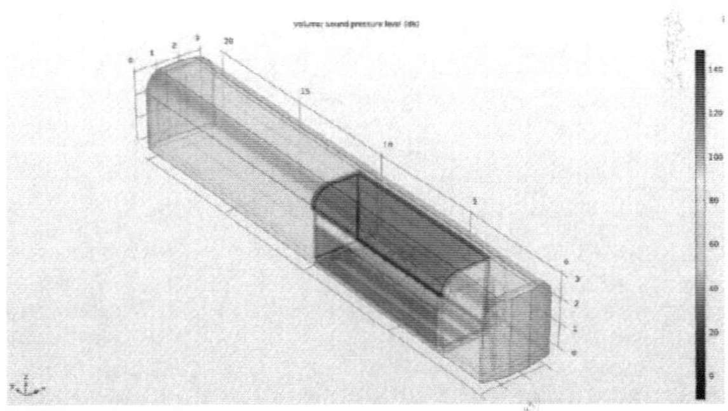

**Figure 13.** The Second Simulation Appear on 3D Axis

In Figure 10 and 11, we obtained the sound pressure level at a distance of 6 meters was 71.81 dBA, whilst in Figure 12 and 13, the sound pressure level at 6 meters was 65.88 dBA, respectively. The loudness level predictions were close to the measurement.

## 4. Conclusion

Simulation of several enclosure variations in a body train has been conducted. When compared between result of measurement and simulation that show variation of enclosure, simulation can lower sound pressure level value by 4.2% – 12.2%. the biggest noise reduction occurs on enclosure with 25% grill and ventilation under the floor, because it can reduce noise from the source due to the noise into the environment only through 25% of the hole from the wall enclosure and ventilation on the floor.

## References

[1] D. Thompson, Railway Noise and Vibration, Southampton: Elsevier, 2008.
[2] Minister of Transportation, Regulation of the Minister of Transportation No. 15 of 2011 on Standards, Procedures for Testing and Certification of Train Draw for Locomotives, Jakarta: Ministry for Transportation, 2011.

[3] J. E. Blanks, "Optimal Design of an Enclosure for a Portable Generator," Virginia Polytechnic Institute and State University, Blacksburg, Virginia, 1997.
[4] Aaberg, D. (2007). *Generator Set Noise Solutions: Controlling Unwanted Noise From On-site Systems.* Cummins Power.

**Acknowledgements**
Department of Technology in PT. INKA, Madiun for providing chance to survey and measure the measurement object is K3 carriage, economic class of Indonesia's train.

IOP Conf. Series: Journal of Physics: Conf. Series **1075** (2018) 012058 doi:10.1088/1742-6596/1075/1/012058

# Designing ocean acoustic waveguide remote sensing for target detecting and tracking

**H M Manik[1], Susilohadi[2], B R Kusumah[3], A Dwinovantyo[3] and S Solikin[3]**

[1]Laboratory of Ocean Acoustic and Sonar System, Department of Marine Science and Technology, Faculty of Fisheries and Marine Sciences Bogor Agricultural University IPB Dramaga Campus, Bogor 16680, Indonesia.
[2]Marine Geological Institute, Ministry of Energy and Mineral Resources Republic of Indonesia, Bandung 40174, Indonesia
[3] Graduate School of Marine Technology, Bogor Agricultural University, Bogor 16680, Indonesia

E-mail: henrymanik@apps.ipb.ac.id

**Abstract**. Acoustic sensor used sound wave to transmit pulses into the underwater in order to gain underwater environment information. We developed Ocean Acoustics Waveguide Remote Sensing (OAWRS) instrument. The acoustic signal acquisition system has been designed which used to record the direct, reflected, scattered or echoed signals originated from transmitter transducer, targets or underwater objects. The acquisition system compose of interconnected hardware and software with a simple working principles. The software, which is written under Windows™ and Matlab, controlled the hardware by sending commands to the Transmitter Module located somewhere to generate a sound pulse and the Receiver Module to be ready to receive sound signals. The received sound data may be recorded for further analysis to developed OAWRS algorithm.

## 1. Introduction

Sound waves provide the efficient method to explore ocean environments since these waves can propagate to ranges orders of magnitude beyond the attenuation limits of electromagnetic waves [1]. This makes the application of active sonar systems useful for imaging, detecting, and classifying targets of interest, such as seafloor morphology, man-made objects, and marine life in an underwater environment [2]. Standard acoustic used for surveying and monitoring fish populations often produce abundance and behavioral records that are ambiguous, incomplete and have high uncertainty [3]. This is because these methods rely on highly-localized measurements that are restricted to the immediate vicinity. The limited temporal and spatial coverage of conventional systems often lead to a gross undersampling, in time and space, of the vast ocean environments that fish typically occupy [4].

To overcome these problems, we designed Ocean Acoustics Waveguide Remote Sensing (OAWRS) instrument to be capable of instantaneous wide-area sensing of marine fish. Continuous monitoring with OAWRS enables the production of wide-area coverage that detail the spatial and temporal distributions of fish population. These wide area can reveal behavioral patterns that may enable better modeling and prediction of fisheries condition.

IOP Conf. Series: Journal of Physics: Conf. Series **1075** (2018) 012058 doi:10.1088/1742-6596/1075/1/012058

## 2. Method

*2.1. Sonar Equation for OAWRS*

Projector source level, SLprojector is defined for acoustic pressure at reference range of 1 m from acoustic centre. Sound intensity is calculated by:

$$I = Ptx\ /\ Area\ (W/m^2) \tag{1}$$

and measure in dB re 1μPa. For omni directional projector, the surface area of the sphere ($4\pi\ r^2 = 12.6\ m^2$). Then,

$$SL_{\text{ projector}} = 10\ \log((P_{tx}/12.6)/I_{ref})\ \text{dB} \tag{2}$$

Ptx is total acoustic power used by projector and sound reference intensity become:

$$I_{ref} = (P\ a_{ref})^2\ /\ \rho\, c\ (Wm^{-2})$$

where reference of level pressure $P\ a_{ref}$ is 1μPa, $\rho$ is medium density and $c$ is sound speed. Equating of transmitter acoustic signal level ($SL_{\text{ projector}}$) at 1 m for omni-directional projector is:

$$SL_{\text{ projector}}\ (P) = 170.8 + 10\ \log P_{tx} \tag{3}$$

Projector for directional type with directivity index:

$$DI_{tx} = 10\ \log\ (I_{\text{dir}}/I_{\text{omni}}) \tag{4}$$

$I_{\text{omni}}$ is intensity and distributed spherically and $I_{\text{dir}}$ is intensity along the acoustic axis of beam pattern. $SL_{\text{ projector}}$ become:

$$SL_{\text{ projector}}\ (P,\ \eta,\ DI) = 170.8 + 10\ \log P_{\text{tx}} + 10\ \log \eta_{\text{tx}} + DI_{\text{tx}} \tag{5}$$

where $\eta_{tx}$ is efficiency of projector.

*2.2. Designing of OAWRS transmitting and receiving system*

The Signal to Noise Ratio (*SNR*) for OAWRS is expressed by [5]:

$$SNR = P_v^2\ /\ P_N^2 \tag{6}$$

where $P_v$ is the volume scattered echo pressure amplitude, $P_N$ is the noise pressure amplitude. Influencing the absorption and spreading attenuation, the square echo pressure is calculated by:

$$P_v^2 = P_o^2\ (c\tau/2)\ \Psi r^{-2}\ \exp(-4\alpha\ r)\ Sv \tag{7}$$

$$Po^2 = \rho\, c\, a^2\ Ic\ D_I\ /\ 2 \tag{8}$$

where $Po$ is the source pressure, $c$ is sound speed in the seawater, $\tau$ is pulse width, $c\tau/2$ is range resolution, $\Psi$ is the equivalent beam width, $r$ is range, α is absorption coefficient (neper/m), $Sv$ is the linear variable of $SV$, $\rho$ is density of seawater, $a$ is radius of the transducer, $Ic$ is cavitation threshold, and $DI$ is the directivity index. The equivalent beam width $\Psi$ is calculated by [6]:

$$\Psi = (2/\pi)\ (\lambda/2a)^2 \tag{9}$$

where $\lambda$ is the wave length. The directivity index $DI$ is shown by:

$$DI = (ka)^2 \tag{10}$$

For a circular piston transducer, where $k = 2\pi / \lambda$ is the wave number. In order to gain high $SNR$, the full capability of transducer was obtained. Therefore, the source pressure should be determined by $Ic$, which is calculated by [7]:

$$Ic = 1.7\ \tau^{0.22}\ 10^{\ b+3}(1+r_t/10+T_a)^2 \tag{11}$$

$$b = 0.116 f^{0.52} \tag{12}$$

where $rt$ is the transducer depth from sea surface in m, $Ta$ is tensile strength of seawater in atm, $f$ is frequency in kHz. The noise spectrum level $Np$ is calculated by [7]:

$$N_P = N_{po}\ f^{-1.8} \tag{13}$$

$$NP_o = 10 \log N_{po} = 145\ \text{dB}\ (0\ \text{dB} = 1\ \mu\text{Pa/Hz}^{1/2}) \tag{14}$$

The received noise pressure is calculated by:

$$P_N^2 = NP^2\ \Delta f / DI \tag{15}$$

where $\Delta f$ is the bandwidth of the receiving system. The final expression of SNR [8] :

$$SNR = 2\ \pi^3\ \rho\ \tau\ a^4\ f^{5.6}\ I_c \exp(-4\ \alpha\ r)\ n\ Ts/r^2\ Npo^2\ \Delta f \tag{16}$$

*2.3. Data Acquisition*

Designing of acoustic transmitter and receiving system is shown in Figs. 1 and 2.

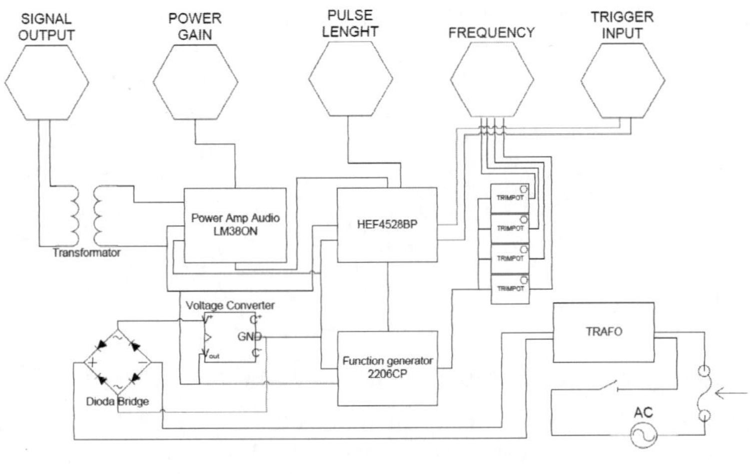


**Figure 1**. Designing of acoustic transmitter system.

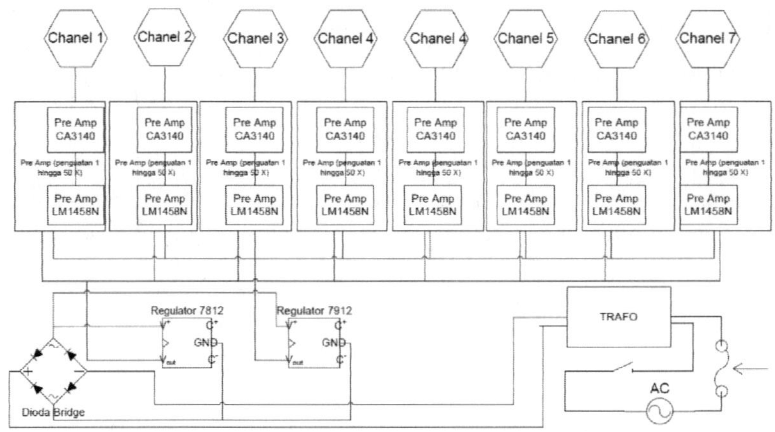


**Figure 2**. Designing of acoustic receiver system.

## 3. Results and Discussion

Detail designed of transmitter was shown in Figure 3. The transmitter consist of 4 *Integrated Circuit* (IC), *Trigger Generator*, *Signal Generator*, and *Amplifier, Matching Impedance Transformer*.

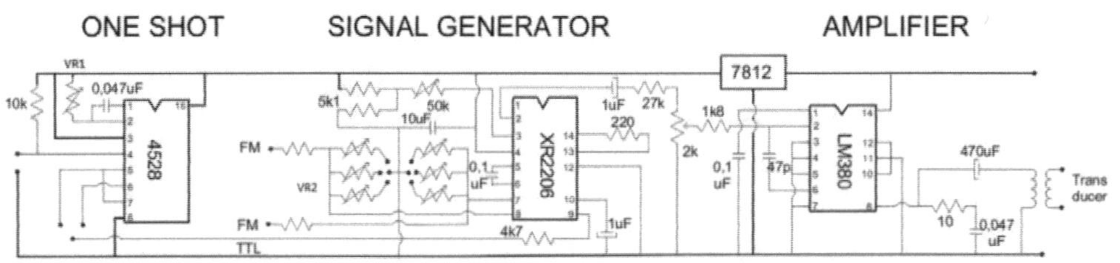


**Figure 3.** Schematic of acoustic transmitter OAWRS.

Electric circuit of receiver OAWRS was shown in Figure 4.

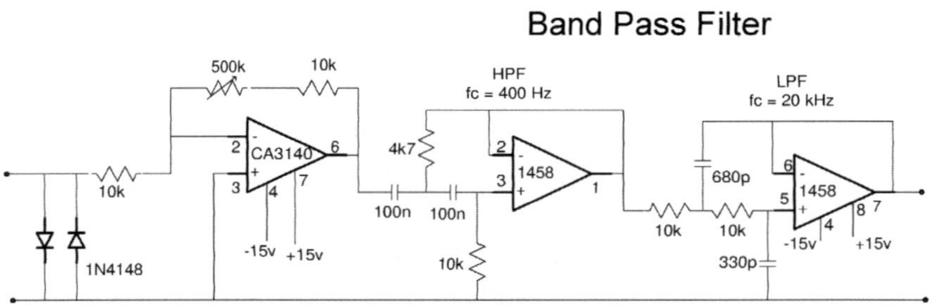


**Figure 4**. Schematic diagram of received signal of OAWRS/

The signal from the receiver transducer is inputted to the Linear Amplifier through a voltage divider circuit in the form of a ring diode and a 10 kohm resistor. The diode serves to limit the signal voltage to no more than 0.7 volts. IC CA3140 is configured as a linear amplifier through a 10 kohm feedback back resistor and a 500 kohm potentiometer. The gain value is determined by the ratio of the feedback resistor and the input resistor, in this design the gain can be set from 1 to 500 times. Windows Console of acquisition system was shown in Fig. 5. Figure 6 show the receive echo signal obtained from ocean acoustic tank.

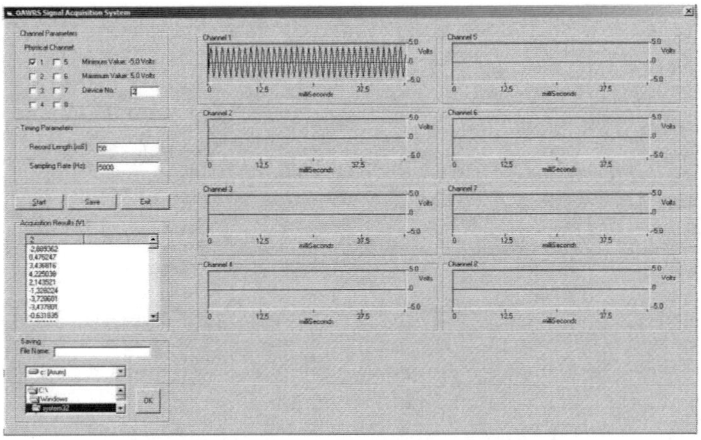

**Figure 5**. Windows Console of acquisition system,

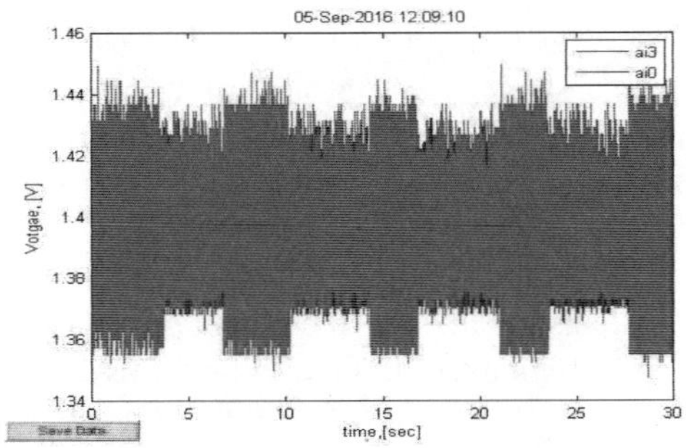


**Figure 6**. The received echo signal,

Sound propagation in the ocean is influenced by the physical and chemical properties of seawater and the geometry of the channel. Transmission Loss of OAWRS was shown in Figure 7. Underwater acoustic signal experiments are affected by attenuation caused by spreading and absorption. Propagation of acoustic ray was shown in Fig. 8. This figure show the characteristic of ray is multipath propagation. The sound intensity level was decreased in long ranges. This condition is agreed with the previous research [9] [10].

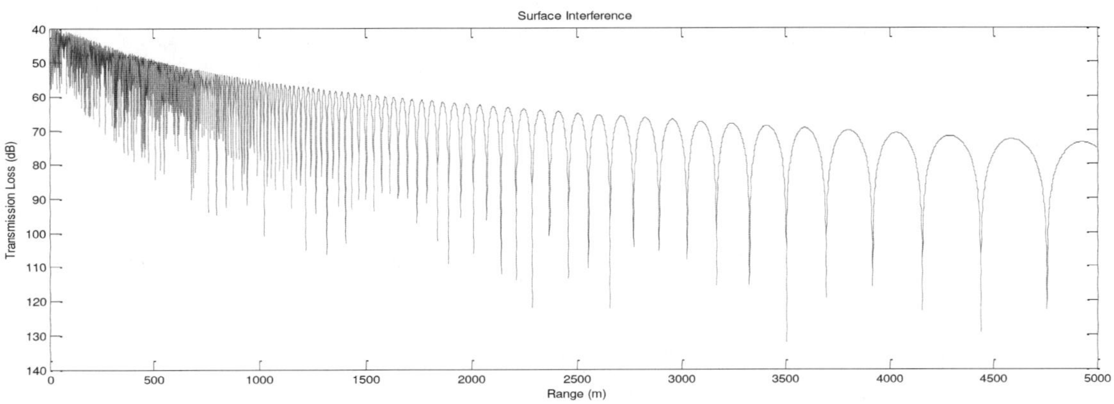


**Figure 7**. Transmission Loss of OAWRS.

IOP Conf. Series: Journal of Physics: Conf. Series **1075** (2018) 012058 doi:10.1088/1742-6596/1075/1/012058

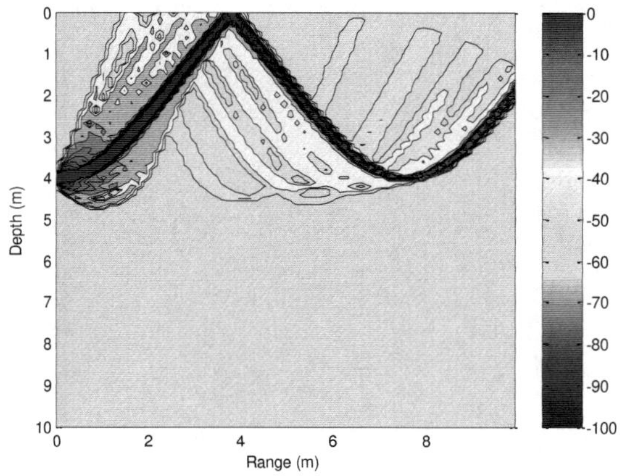


**Figure 8**. Acoustic ray propagation

## 4. Conclusions

In this research has successfully designed transmitter system to transmit acoustic signal and receiver system (hydrophone). Measurement of transmitting power and transmission loss parameters has been done. Sonar equation for ocean acoustic waveguide propagation has been successfully conducted and applied to OAWRS instrument. Acoustic ray computations are based on the wave propagation properties of different propagating mediums.

## References

[1] Gong Z, Andrews M, Jagannathan S, Patel R, Jech J M, Makris N C and Ratilal P 2010 Low-frequency target strength and abundance of shoaling atlantic herring (Clupea Harengus) in the Gulf of Maine during the ocean acoustic waveguide remote sensing 2006 experiment *The Journal of the Acoustical Society of America* **127** 104

[2] Chen T, Ratilal P and Makris N C 2005 Mean and variance of the forward field propagated through three-dimensional random internal waves in a continental-shelf waveguide *The Journal of the Acoustical Society of America* **118** 3560–3574

[3] Galinde A, Donabed N, Andrews M, Lee S, Makris N C and Ratilal P 2008 Range-dependent waveguide scattering model calibrated for bottom reverberation in a continental shelf environment. *The Journal of the Acoustical Society of America* **123** 1270–1281

[4] Burdic W S 1991 *Underwater Acoustic System Analysis*. New York (US): Springer-Verlag

[5] Liu X, Furusawa M, Hamada E, Aoyama C 1999 Designing acoustic transmitting-receiving system for volume backscattering measurement of zooplankton *Fisheries Science* **65** 410–419

[6] Furusawa M 1991 Designing quantitative echo sounders *The Journal of the Acoustical Society of America* **90** 26–36

[7] Nishimura M 1969 Study on the optimum frequency of echo sounders [Ph.D. thesis], Tohoku University, Sendai, Japan (in Japanese).

[8] Urick R 1983 Principles of Underwater Sound; New York (US): McGraw Hill

[9] Huang W, Wang D, Garcia H, Godø OR, Ratilal P 2017 Continental shelf-scale passive acoustic detection and characterization of diesel-electric ships using a coherent hydrophone array. *Remote Sensing* **9** 772

[10] Jagannathan S, Küsel E T, Ratilal P, Makris N C 2012 Scattering from extended targets in range-dependent fluctuating ocean-waveguides with clutter from theory and experiments. *The Journal of the Acoustical Society of America* **132** 680–693

## Acknowledgments

This research was supported by the INSINAS Program of Indonesia Ministry of Research, Technology, and Higher Education.

IOP Conf. Series: Journal of Physics: Conf. Series **1075** (2018) 012059 doi:10.1088/1742-6596/1075/1/012059

# Nonlinear stochastic drillstrings vibrations: parametric study for stick-slip suppression

**M Y Alqaradawi[1], M Kanzari[1] and B Balachandran[2]**

[1]Department of Mechanical and Industrial Engineering, College of Engineering, Qatar University, Doha 2713, Qatar
[2]Department of Mechanical Engineering, University of Maryland, College Park, Maryland, USA

E-mail: myq@qu.edu.qa

**Abstract**. Unstable vibrations caused by the transition between static and kinetic frictions produces stick- slip motions which is one of the primary causes of poor drilling efficiency and mechanism failure in the drilling process. Coupled lateral and torsional drill string dynamics are explored numerically and experimentally within this article. A reduced order model able to capture the non-linear responses, observed during dry-friction whirling, is presented along with scaled experimental apparatus tests. To reduce stick-slip oscillations, the control strategies are mostly focused on the angular velocity of the upper part of the drill string, the torque on bit and the torque of friction. These parameters are investigated, and the simulation results are seen to be in good agreement with experimental observations.

## 1. Introduction

Rotary drilling systems employs rotational drill bit, which is one of the most important component impacting, directly, on the rock surface. Drill bit is driven from the surface by a slender structure of pipes, called the drill string, which transports the energy from the motor to the drill bit [1-3]. Structurally, the drill-string components are housed inside a drill pipe and includes the drill string, drill collar, and drill bit. Based on the functions of the drill collar and bit, the combination of these two parts is referred to as the Bottom-Hole-Assembly (BHA).

Most structural failures occurring on the BHA are due to bending and torsion motions experienced by the drill string. Downhole/ surface monitoring have indicated that drill string exhibits severe stick-slip vibrations causing violent torsional motion and whirling leading to lateral vibrations with large amplitudes [3-7].

Stick-slip phenomenon has been subjected to many prior efforts, including theoretical analysis and field measurement. Several models have been investigated to explain this type of dysfunction, such as the model of mass-spring on a moving belt [8] and the torsional pendulum model [9,10].

Drilling wells production have opened new challenges for experts to investigate the parameters and phenomenon enhancing drilling vibration and affecting drilling costs. Before the 1960's, research efforts were focused on the drillstring' s material strength components, the challenges have since changed to emphasize on its dynamic behavior [11]. As a result, various control strategies appeared to compensate the drill string stick-slip phenomenon, such as manipulating the top rotary system in soft manner, and thus, helping the absorption of torsional harmful waves at the surface. This solution has been used in different drilling wells and it has significantly reduced the bit stick-slip oscillations [12]. In order to understand the reasons of this type of vibrations, laboratory drilling experiments have been carried out [13] and theoretical models have been discussed. Kyllingstad and Halsey [14] used single Degree of Freedom 1DOF pendulum model in which the drillstring is assumed to be suspended at a rotary table. Constant coulomb friction torque is applied to the drill bit to analyze its stick-slip motion. Classic Coulomb torque model has provided a good estimation of torsional vibrations in many cases: Jansen and van den Steen [15] employed this model and divided the torque-on-bit (TOB) into static and sliding. Stick slip vibrations were observed for different conditions. Leine et al. used two-degree-of-freedom model to study stick-slip and whirl vibrations of the drill string. The BHA was modeled as a rigid disk at the end of a mass less flexible drill pipe.

Based on drilling parameters supervision, many control methodologies have appeared to compensate drill string non-linear complex vibration. A workshop sponsored by the International Association of Drilling Contractors (IADC) regarding stick-slip mitigation was held in 2010 to help the petroleum industry.

In this work, stick-slip control and suppression is focused on. The organization of the article is as follow: in sec 2, a reduced order dry friction model of the drilling system using closed loop control system is proposed, numerical studies conducted with this model are presented. Simulation results are compared for different input parameters and presented in Sec 3. In Sec 4, qualitative comparison of torsional behavior between the experimental observations and numerical results is carried out, then, concluding remarks are collected in Sec.5

## 2. Dynamic model development

The drill string consists of the BHA and drill pipes screwed end to end to each other to form a long rod. As shown in Figure 1 (a), the BHA englobe the bit which is the cutting tool, the stabilizers, which prevent the drill string from unbalancing, and a connected pipe sections known as drill collars. The length of the BHA is almost constant, however, the length of drill pipes depends on the borehole depth. The drillstring is persistently driven by a rotating table from which the energy is transferred to the drill bit.

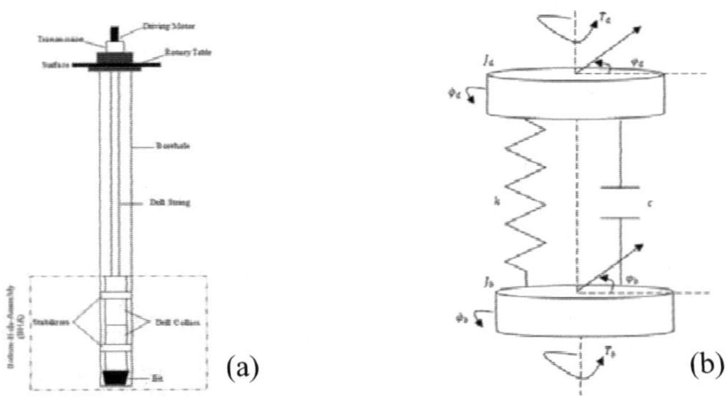


**Figure 1.** (a) Schematic of Typical Drill String (b) Mechanical model of a drill string

In this section, a discrete system is developed to model the qualitative dynamics causing stick slip motions. Same assumptions are those used in previous studies [8] are used. A representative mechanical model which can capture axial and torsional model in presented in Figure 1(b). The drilling system is

IOP Conf. Series: Journal of Physics: Conf. Series **1075** (2018) 012059 doi:10.1088/1742-6596/1075/1/012059

assumed to be a torsional pendulum system driven by a top motor described by a lumped parameter model with two-degree-of-freedom with an equivalent torsional stiffness $K$ and an equivalent rotational mass moment of inertia $J_b$. The interaction Bit-rock is modeled by a dry friction model. The presented model is similar to the ones given in the literature with the difference that, here, the TOB and the torque of drill string are considered, and their effects is analyzed. The equation of motion of the two-degree-of-freedom model of drill string can be written as:

$$J_d \ddot{\varphi}_d + k(\dot{\varphi}_d - \dot{\varphi}_b) + c(\varphi_d - \varphi_b) = T_M(t) - T_d(\dot{\varphi}_d) \quad (1)$$

$$J_b \ddot{\varphi}_b + k(\dot{\varphi}_d - \dot{\varphi}_b) - c(\varphi_d - \varphi_b) = -T_b(\dot{\varphi}_b) \quad (2)$$

Such as $\varphi_d$ and $\varphi_b$ are, respectively, the angular displacement of the rotary top driving system and the angular displacement of the BHA. $J_d$ is the drive moment inertia and $J_b$ is the inertia of the drill string associated with the bottom hole assembly described by:

$$J_b = \rho \frac{\pi}{96}[L_p(D_p^4 - d_p^4) + L_c(D_c^4 - d_c^4)] \quad (3)$$

The driving motor torque is presented by $T_M$ , the dry friction torque by $T_b$ and the viscous damping torque by $T_d$ , where:

$$T_b(\dot{\varphi}_b) = c_b\dot{\varphi}_b + T_{fb} sgn(\dot{\varphi}_b) \quad (4)$$

$$T_d(\dot{\varphi}_d) = c_d\dot{\varphi}_d + T_{cd} sgn(\dot{\varphi}_d) \quad (5)$$

The damping viscous coefficient associated with top driving system are given by the coefficients $c_d$ and $c_b$ , $T_{cd}$ is the coulomb friction torque and $T_{fb}$ is the friction torque which is the variation of the stribeck torque with the static dry friction described as:

$$T_{fb}(X) = \begin{cases} Te\,(X) & if\ \dot{\varphi}_b = 0\ and\ |T_e| \le T_{sf} \\ T_{sf} sgn\,(Te(\gamma)) & if\ \dot{\varphi}_b = 0\ and\ |T_e| > T_{sf} \\ T_{sf}\ \frac{\mu_b(\dot{\varphi}_b)}{\mu_{sf}} & if\ \dot{\varphi}_b \ne 0 \end{cases} \quad (6)$$

Such as $X$ is the system state vector, $Te$ is the external torque facing the static friction torque $T_{sf}$ to move the bit:

$$Te\,(X) = c(\dot{\varphi}_d - \dot{\varphi}_b) + k(\varphi_d - \varphi_b) - c_b\dot{\varphi}_b \quad (7)$$

To describe the zero-speed friction, an exponential decaying law at the sliding phase is expressed by the dry friction coefficient:

$$\mu_b(\dot{\varphi}_b) = [\mu_{cf} + (\mu_{sf} - \mu_{cf})e^{-\gamma_b|\dot{\varphi}_b|} sgn\,(\dot{\varphi}_b) \quad (8)$$

Where, $\gamma_b$ defines the velocity decrease rate $(0 < \gamma_b < 1)$ and $\mu_{cf}$ and $\mu_{sf}$ are the coulomb and static friction coefficients:

$$T_{sf} = r_b W_{ob} \mu_{sf} \qquad T_{cf} = T_{sf} \frac{\mu_{cf}}{\mu_{sf}}, \tag{9}$$

where $r_b$ is the bit radius and $W_{ob}$ is the Work On Bit (WOB). $\dot{\varphi}_s$ is the stribeck speed and $\dot{\varphi}_b \epsilon [-\dot{\varphi}_s, \dot{\varphi}_s]$.
From the damping spring model, the drill string torque and the TOB are obtained

$$\boldsymbol{T_{DS}} = \boldsymbol{K} \int (\dot{\boldsymbol{\varphi}}_{\boldsymbol{d}} - \dot{\boldsymbol{\varphi}}_{\boldsymbol{b}}) \boldsymbol{dt} + \mathrm{c}\, \dot{\boldsymbol{\varphi}}_{\boldsymbol{b}} \tag{10}$$

$$\boldsymbol{T_{OB}} = \boldsymbol{T_{DS}} - \boldsymbol{\mu_b}(\dot{\boldsymbol{\varphi}}_{\mathbf{b}}) \tag{11}$$

## 3. Parametric studies and results

To understand the phenomenon of stick-slip, also known as micro-stalling, a close loop system has been developed. Interactive graphical editor blocks of MATALB/ SIMULINK have been used to the numerical simulation of the proposed model given by equation (1) – (11).

**Table 1.** System parameter values used to generate numerical results

| Parameter | Value | Units |
|---|---|---|
| $D_r$ | *0.15* | $m$ |
| $D_s$ | *0.18* | $m$ |
| $L$ | *2×10^3* | $m$ |
| $J_d$ | *500* | $Kg\, m^2$ |
| $J_b$ | *300* | $Kg\, m^2$ |
| $K$ | $500\text{-}10^5$ | $Nm^{-1}$ |
| $c$ | *0.1* | $N.ms.rad\text{-}^1$ |
| $\mu_{sf}$ | *80* | - |
| $\mu_{cf}$ | *50* | - |
| $\Omega$ | *60-200* | *RPM* |
| $G$ | $25.10^9$ | *Pa* |
| $E$ | $70.10^9$ | *Pa* |

Using linear stability analysis of the two-degree-of-freedom model, varied stability charts are obtained. Different torque profiles showing the stick slip vibration are presented below:

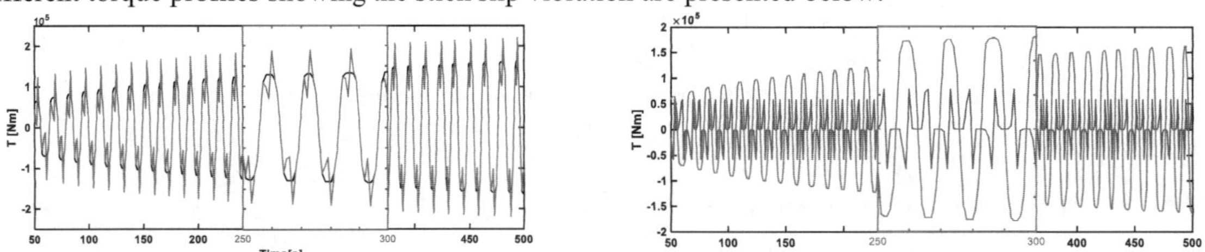


(a) (b)

**Figure 2.** Torque variation during stick-slip at Ω = 100 RPM (a) Top drive motor torque (black line) and TOB (red line) (b) Toque of friction (blue line) and drill string torque (brown line).

Figure 2 shows the variation of the different system torques for stick-slip vibration. For the maximum of driving torque, an increased demand from the bit to achieve penetration, that cannot be met by the drilling motor, causing a complete standstill during stick-slip, then, the bit alternately rotates faster than the driving speed, and the torque on bit increase rapidly which can damage the drill string.

During stick slip vibration, the dry friction curve, presented in Figure 2(b) shows a stribeck effect resulting on an instable steady rotation and uneven loading across the bit rock interface creating an increased friction forces.

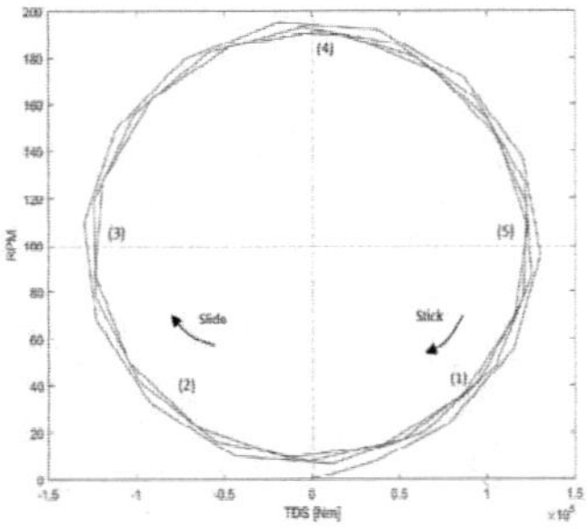

**Figure 3**. Drill string stick-slip vibration phase plane simulation

Figure 3 illustrate phase plan simulations of the drill string stick-slip vibrations. The horizontal axis presents the twisting force of the drill string and the vertical axis presents the angular velocity of the bit. Initially, the bit is sticking, and the friction torque balances the drilling torque [1]. When the friction torque exceeds the maximum static friction, the bit starts sliding [2-3]. The friction torque decreases, and the bit speed reaches its maximum [4]. The behavior of the drill string dynamics under different driving conditions have been studied. The torsional drill string model discussed in the previous part has been used to investigate the influencing parameters on stick slip behavior. A range of selected parameters have been selected for different simulation runs.

### *3.1 The effect of driving speed on stick slip*

Low and high motor/ upper-hole driving speed values has been selected. The simulation shows the same trend for the different speeds, however, more the driving speed is high, smoother the drill string torque is. The stick-slip vibration are highly observed for low velocities. For low RPM's, the bit motion become stationary for certain period (long sticking period). For high RPM's the stick-slip vibration are transformed to regular torsional oscillations, the bit speed decreases without stoppage.

### *3.2 The Effect of stiffness on stick slip*

The effect of torsional stiffness of drill string on bit-rock interaction was rarely investigated in the literature.

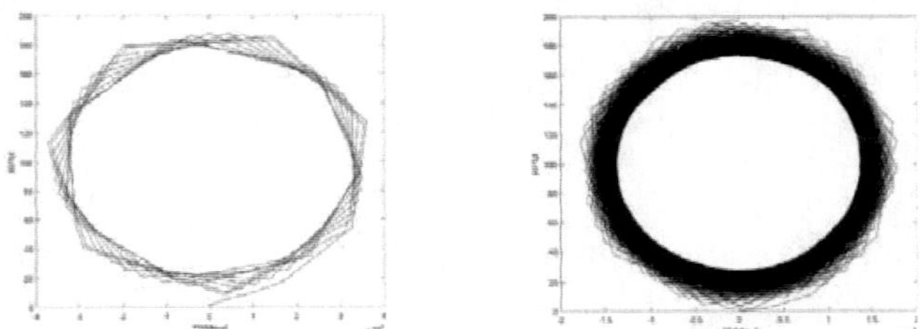

(a) (b)

**Figure 4.** Effect of Torsional drill string stiffness on stick slip vibration $\Omega$= 100 RPM (a) K=500Nm$^{-1}$ (b) K=10$^3$ Nm$^{-1}$

As shown in Figure 4, the stiffness effect was tested for three different values of K using the base case parameters. The drill string torque and the TOB increase with high stiffness and stick-slip vibration are transferred to torsional rolling with more stability regions. From the simulation results it can be also noted that for high stiffness, the amplitude (peak to peak) of the drill string torque decreases which can reduce the severity of sticking.

## 4. Experimental Design and Arrangement

Laboratory scaled drilling apparatus has been constructed to simulate the working conditions experienced by drill string. The BHA was modeled by a rigid disk /rotor, the borehole by a rigid cylinder/ stator and the drill string by a long aluminum shaft. The upper end of the shaft, was driven by a servo motor with controlled rotation speed.

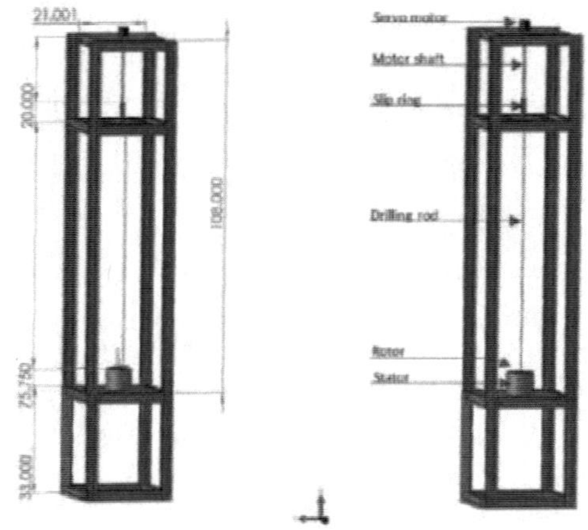


**Figure 5.** 3D Design of the experimental apparatus

### *4.1 Numerical /experimental comparative study*

The intent of this part is to present the experimental observations and measurement to make model predictions for qualitative comparison with the numerical findings. Experimental data of the torsional strain during dry friction whirl for different drive speed has been recorded assuming that only the first torsional mode of vibration is excited, so that the strain measurement is directly promotional to the torsional displacement. For low driving speed, torsional vibrations are irregular with high fluctuations leading to sticking phases. However, when the table speed increase, these vibrations are converted to regular torsional oscillations over time.

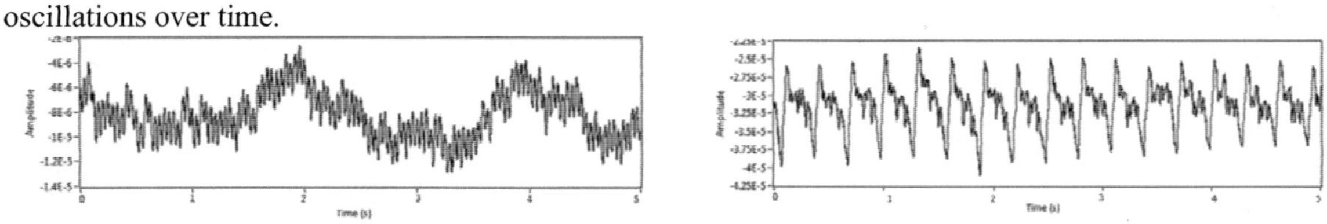

(a) (b)

**Figure 6.** Experimentally observed torsional strain at Ω= 60 RPM (red line) and Ω= 100 RPM (black line).

In fact, an Increase in RPM leads to torsional strain reduction and so reduced torsional displacement/ vibration. These results compare well with numerical simulation (sec. 3.1).

## 5. Concluding Remarks

In this paper, the authors have presented a dynamic two-degree-of-freedom model for torsional drill string focusing on bit-rock stick-slip interactions. Simulations were carried based on nonlinear differential equations considering nonlinear dry friction forces. Resulting periodic self-excited stick-slip vibration was observed in the bit, drill string and top torques. The influence of torsional stiffness and upper driving speed was studied. In order to explain the observed phenomena, experimental tests was carried out and qualitative comparison with the model predictions was done.

## References

[1] Teale, R., 1965, The concept of specific energy in rock drilling, *Int. J. Rock Mech. Min. Sci & Geom.,* **2**, 57–73.

[2] Batako, A. D., Babitsky, V.I. and Halliwell, N.A., 2003, A self-excited system for percussive-rotary drilling, *J. Sound Vib.* **259**, 97–118.

[3] Bailey, J.J. and Finnie, I., 1960, An analytical study of drillstring vibration. *Trans. J. Eng. Ind.*, **82**, 122-28.

[4] Denny, M., 2004, Stick-slip motion: an important example of self-excited oscillation, *Eur. J. Phys.* **25** 311-22.

[5] Berlioz, A., Der Hagopian D., Dufour, J. R. and Draoui, E., 1996, Dynamic behavior of a drill-string: experimental investigation of lateral instabilities, J. Vib. Acoust. acoustics **118**, 292–98.

[6] Macdonald, K.A. and Bjune, J. V., 2007, Failure analysis of drill strings, *Engineering Failure Analysis* **14**, 1641–1666.

[7] Navarro-Lopez, E. and Cortes, D., 2007, Sliding-mode of a multi-DOF oil well drillstring with stick-slip oscillations. *Proceedings of American Control Conference*. New York City, USA, 3837-42.

[8] Liao, C.-M., Balachandran, B. and Karkoub, M., 2007, Drill String Dynamics: Reduced-Order Models, *J. Vib. Acouct.***2009**-10339-349.

[9] Mihajlovic, N., Van de Wouw, N., Rosielle, P.C., and Nijmeijer, H., 2007, Interaction Between Torsional and Lateral Vibrations in Flexible Rotor Systems with Discontinuous Friction, *Nonlinear Dyn.*, **50**, 679-699.

[10] Leine, R. I, Nijmeijer, H., 2004, Dynamics and Bifurcations of Non-Smooth Mechanical Systems, Lecture *Notes in Applied and Computational Mechanics*, **18,** Springer-Verlag, Berlin.

[11] Whittington, T. De Weegh, A. O. Penetration rates optimized by Aligning Torque and Rotary speed, *J. Oil& Gas*, **97**, 55-57.

[12] Sauoikone, P. Kmorhlma, O., and white, D.B., 1992, A Field Method for Conuolling Drillsuing Torsional Vibrations, *IADCLSPE Drilling Conference*, New Orleans, Louisiana,43-45.

[13] Westermann, H., Gorelik, I. Rudat, J., Moritz, C. Neubaur, M., Wallaschrek, J., and Hon, O. 2010, A new test rig of experimental studies of drillstring vibrations, *SPE Drilling &completion*, **30**,119-128.

[14] Kyllingstad Edwards S., Lees A., and Friswell M.,1999, The influence of torsion on rotor/stator contact in rotating machinery. *J Sound Vib*,767–78.

[15] Jansen, J.D., Van den Steen, L., 1995, Active Damping of Self-Excited Torsional Vibrations in Oil Well Drillstrings, *J. Sound. Vib.,* **179**, 647–668.

### Acknowledgement

This work was made possible by NPRP grant # [7-083-2-041] from the Qatar National Research Fund (a member of Qatar Foundation). The statements made herein are solely the responsibility of the authors.

IOP Conf. Series: Journal of Physics: Conf. Series **1075** (2018) 012060 doi:10.1088/1742-6596/1075/1/012060

# Sound insulation design improvement for reducing noise of paper factory office

**M S Kanteyan[1], S S Utami[1], I Prasetiyo[2] and R J Yanti[1]**

[1] Department of Nuclear Engineering and Engineering Physics, Universitas Gadjah Mada, Yogyakarta, Jl. Grafika No. 2, Yogyakarta, INA
[2] Department of Engineering Physics, Institut Teknologi Bandung, INA

E-mail: maya.satih.k@mail.ugm.ac.id

**Abstract**. Noise exposure in industrial building such us a paper production factory can be harmful for the employees whom are working in the outdoor area as well as in the offices. Some employees that are working in the open-plan office area in this factory layout are complaining about the noise produced by machines. The noise produced from the machines makes them hard to focus on their jobs. It is found that the closest noise source position from the office produces 79 dBA while the noise level of 54 dBA is present in the office. To deal with this situation, a noise mapping was done using Sound Level Meter (SLM) for measurement. The noise mapping can help for noise assessment to know the noise distribution in the factory that give an impact to office. The Transmission Loss (TL) of the office walls was evaluated by measuring the noise level spectrum inside and outside of the office, while T30 (reverberation time) values were also measured to obtain for room absorption assessment as it affects the TL values. These two parameters used to obtain the Sound Transmission Class (STC). It also enables to systematically predict the sound insulation improvement of the offices particularly as a basis to choose suitable material and wall construction. In this research, INSUL software is utilized to calculate the TL of the proposed wall design. The result shows that improvement of existing partition insulation using double glazing laminated glass and gypsum can meet the preferable noise level inside of the office that is 38 dBA using the new STC value of 23.

## 1. Introduction

Acoustics condition in the office affect the workers productivity. The noise in the workplace is indeed a serious problem and may decrease the workers productivity [1]. Increased of the noise levels in the office, require the employees to raise their voice during speaking, a phenomenon known as Lombard Effect phenomenon [2]. Noise control in office space is necessary in order to achieve acoustical or noise criteria from an open site tread of 35 to 40 (or equivalent impact level "$L_{eq}$" of 45-50 dBA) specified by ASHRAE 1987: Interior Noise Design Goals [3].

Similar researchers were done by many researchers in the past. Samir N.Y. Greges, conducted research on industrial noise sources [4]. The study said that a sound field from a workplace is a combination of several noise sources such as air-borne noise, structural borne noise, diffraction in engine boundaries, reflection of the floor, walls, roof and surface of the machine, absorption of the surface, etc. Further research by Tan Kia Tang (1995) shows the result of a paper manufacturing and paper factory in Singapore has a noise range of 85-102 dBA and an average of 92 dBA. Luis Conde Santos (2008) and W J P Casas (2014) conducted a noise mapping in industrial area as a noise impact

IOP Conf. Series: Journal of Physics: Conf. Series **1075** (2018) 012060 doi:10.1088/1742-6596/1075/1/012060

assessment and noise control decision maker [5], [6]. Meanwhile, Faramarz Majidi (2016) used Geographic Information System (GIS) software for noise mapping of food industry in Iran as a noise assessment. Majidi set the elevation of measurement 1.5 meters from the floor and 1 meter of distance from reflective material. Classification of the colour scale divided into green for allowable area, yellow for dangerous area, and red for not allowable area [7].

James P. Conroy and John S. Roland conducted a study of Sound Transmission Class (STC) measurement according to ASTM E366-96 standard for office which has rectangular geometry. Transmission Loss (TL) should be conducted to get STC rating for the partition to know the insulation material characteristic. According to the standard, loudspeaker is not required for the measurement but it should be diffused. Microphones are placed in the source room and the receiver room with the source height at 3.1 feet above the floor [8].

Besides doing a noise mapping as a noise assessment in industrial area, improvement of insulation material should be done to get better sound insulation. A paper factory office has a SPL of 54 dBA that was caused by source of *single face* production machine. It is predicted that the higher noise level in office is an impact of worst material insulation of the single glazing window glass. Thus, sound insulation design improvement should be done to get preferable noise criteria.

## 2. Methods

### 2.1. Noise Mapping Measurement

The noise mapping measurement was conducted in machinery room behind the office wall partition. Ten SPL measurement points were taken using Sound Level Meter (SLM) with A-weighting (dBA) positioned at the height of 1.5 meters from the floor. The data is plotted in to Surfer Geographic Information System software with a colour scale red as the highest level and blue as the lowest level.

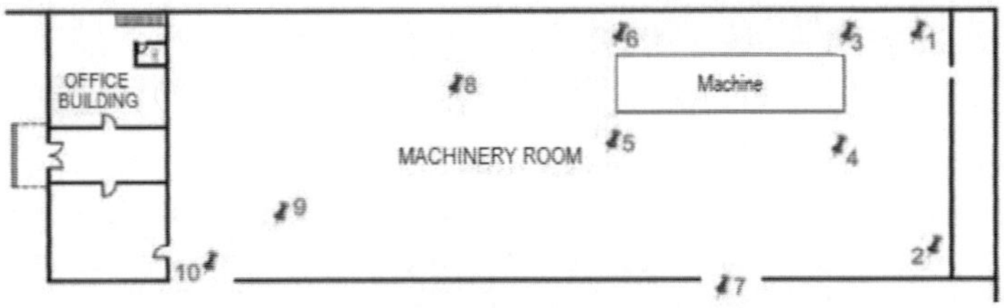


**Figure 1.** Measurement Points in the Machinery Room

### 2.2. Transmission Loss Measurement and STC Evaluation

Transmission loss measurement for partition was conducted in 3 points. One is in the source room, another in the receiver room in front of wall partition, and 1 point in receiver room in front of window glass partition. Recorder Zoom H6 and Mid Side microphone is used to record the sound and set sampling frequency 44.100 Hz/16 bit with a height of 1.5 meters and 1.5 meters away from the partition.

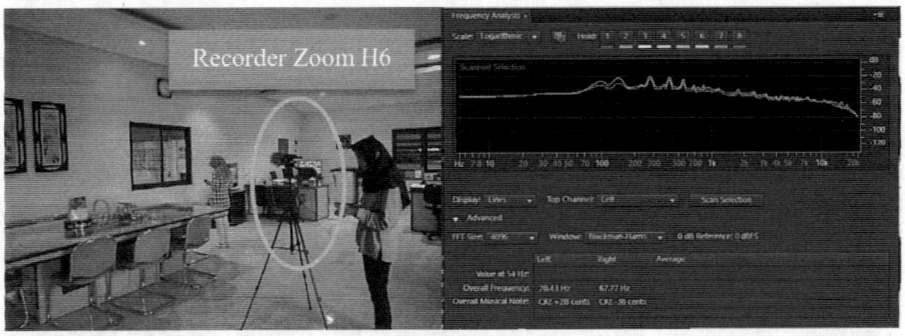


**Figure 2.** Transmission Loss Measurement and Signal Processing in Adobe Audition

For 5 minutes recording at each point, a *single face* machine was operated and assumed as a diffused source in the 9145 $m^3$ room volume. Signal processing was done in Adobe Audition using Blackman-Harris Window and produced a frequency spectrum in third octave band from 125 Hz to 4000 Hz. Reverberation time (T30) measurement in the office was also conducted to obtain absorption coefficient equivalent of materials (A). Impulse signal was generated and recorded by BSWA microphone. The data was analysed in Real Time Analyzer (RTA) software to obtain reverberation time in each third octave band frequency. STC rating of partition is obtained from transmission loss and STC references. To obtain a composite wall (overall partition material) STC, calculation of composite wall transmission loss should be conducted from TL of the wall and TL of the window glass. Thus, after obtaining the SPL of source room, receiver room, absorption coefficient, and partition surface area (S), TL is calculated based on the formula below:

$$TL = (SPLsource - SPLreceiver) + 10\log\left(\frac{S}{A}\right) \quad (1)$$

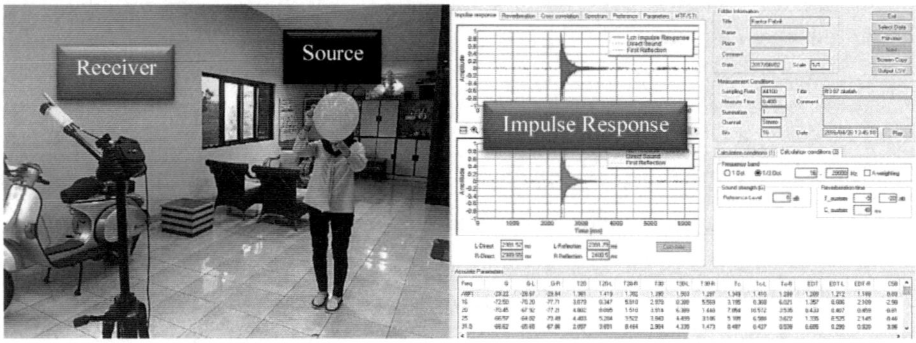


**Figure 3.** Reverberation Time (T30) Measurement in the Office and Signal Processing using RTA

## 2.3. *Insulation Design Improvement Recommendation*

INSUL 9.0 software is used to design appropriate partition material to achieve a preferable noise criteria for the office. The scenario of design improvement was giving a treatment to the lowest TL material, and adding a material to obtain the higher of STC rating value.

## 3. Result and Discussion

### 3.1. *Noise Mapping*

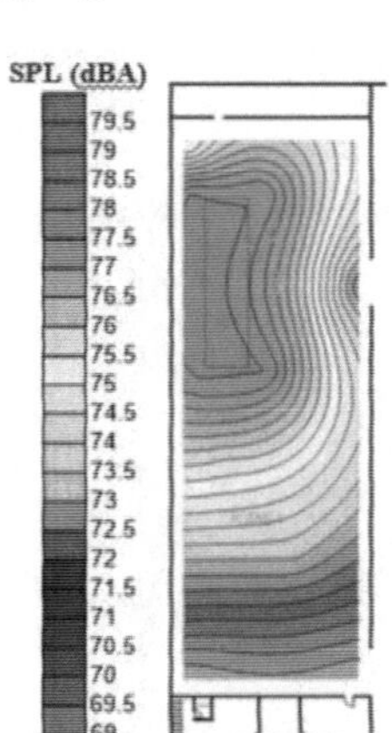


**Figure 4.** Noise Mapping of Machine

**Figure 5.** Paper Stocks Material

IOP Conf. Series: Journal of Physics: Conf. Series **1075** (2018) 012060 doi:10.1088/1742-6596/1075/1/012060

Noise mapping of the machine room is shown in Figure 1. The highest SPL is 79.5 dBA and the lowest SPL is 68.5 dBA. The red color represent the highest level of the noise which is showed by single face machine (rectangular box) and the blue color represent the lowest level that caused by the distance and paper stocks material that absorbs the acoustics energy. The lower SPL, near the office partition, shows an estimation of the SPL that will be isolated.

### *3.2. Transmission Loss (TL) and Sound Transmission Class (STC)*

Figure 6 describes the transmission loss of the composite wall (overall partition material) that shows a lower value due to the window glass TL. The low value of TL, that is the window glass, gives an impact to the composite wall TL value that is lower than the wall TL value. So, it needs an improvement to achieve a higher insulation of the composite wall material. The graph shows that the lowest TL value lies on frequency 250 Hz. It is proved by Table 1 that the noise source has the harder SPL on frequency 250 Hz as well as 125 Hz. This result implies that at low frequency, between the window glass and the wall have poor insulation. Meanwhile, a single glazing of window glass is a poor insulation material because it is not stiff, heavy, and thick enough to isolate the air-borne noise and also prevent the vibration that cause an structure-borne noise.

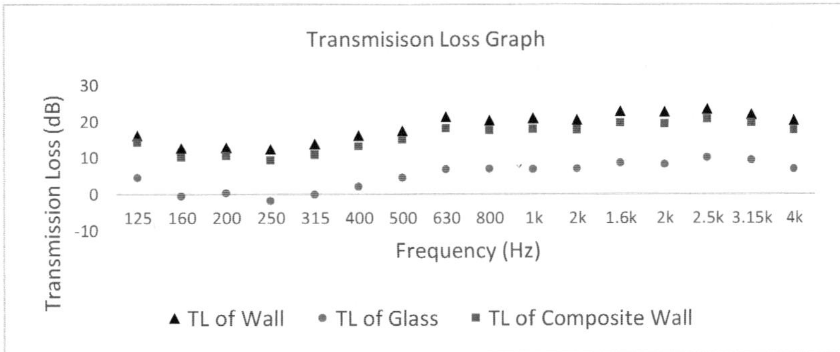


**Figure 6.** Transmission Loss Graph

**Table 1** Frequency Response of *Single Face Macine*

| Frequency (Hz) | SPL (dB) | Frequency (Hz) | SPL (dB) |
|---|---|---|---|
| 125 | 67.85 | 800 | 56.02 |
| 160 | 60.58 | 1000 | 54.63 |
| 200 | 59.54 | 1250 | 50.79 |
| 250 | 66.83 | 1600 | 49.91 |
| 315 | 58.95 | 2000 | 48.53 |
| 400 | 64.14 | 2500 | 46.48 |
| 500 | 61.29 | 3150 | 44.19 |
| 630 | 56.69 | 4000 | 42.08 |

STC of the composite wall result is obtained from STC of the window and the wall. By following the ASTM E413 to know the characteristic of acoustics insulation on material, Table 2 shows the STC result of the composite wall. STC of material is showed in frequency 500 Hz and achieved only when the deficiency of STC trial and the composite wall TL is no more than 8 dB, also the total deficiency is no more than 32 dB. The result achieved STC rating of 17 for the composite wall on third trial.

The result of field measurement will not be the same as measurement in a laboratory field which the noise is controlled. We cannot avoid the structure borne, room resonance, and flanking noise effects that are coming from the ceiling, the door-floor gap, electrical circuit, and the window facing the main road. During the measurement, it is very hard to control the noise since the building is in the main road of an industrial area. The same difficulties occurred during night time measurement. The

IOP Conf. Series: Journal of Physics: Conf. Series **1075** (2018) 012060 doi:10.1088/1742-6596/1075/1/012060

effect of flanking noise during the measurement created a lower value of TL and STC as compared to the simulation in INSUL 9.0 software using a quite similar material. The STC rating becomes 43 for composite wall. To obtain a specific Tl and STC result of materials, an absorption coefficient of each materials should be obtained than using the absorption coefficient equivalent of materials (A) from reverberation time values.

**Table 2** STC of Composite Wall

| **Frequency (Hz)** | **TL CW (dB)** | **STC References** | **Trial 1** | **Trial 2** | **Trial 3** | **Def** | **Trial 4** | **Def** |
|---|---|---|---|---|---|---|---|---|
| 125 | 14.30 | -16 | -1 | 0 | 1 | - 13.30 | 2 | - 12.30 |
| 160 | 10.16 | -13 | 2 | 3 | 4 | -6.16 | 5 | -5.16 |
| 200 | 10.57 | -10 | 5 | 6 | 7 | -3.57 | 8 | -2.57 |
| 250 | 9.32 | -7 | 8 | 9 | 10 | 0.68 | 11 | 1.68 |
| 315 | 10.89 | -4 | 11 | 12 | 13 | 2.11 | 14 | 3.11 |
| 400 | 13.18 | -1 | 14 | 15 | 16 | 2.82 | 17 | 3.82 |
| **Frequency (Hz)** | **TL CW (dB)** | **STC References** | **Trial 1** | **Trial 2** | **Trial 3** | **Def** | **Trial 4** | **Def** |
| 500 | 15.07 | 0 | 15 | 16 | 17 | 1.93 | 18 | 2.93 |
| 630 | 18.13 | 1 | 16 | 17 | 18 | -0.13 | 19 | 0.87 |
| 800 | 17.69 | 2 | 17 | 18 | 19 | 1.31 | 20 | 2.31 |
| 1k | 17.93 | 3 | 18 | 19 | 20 | 2.07 | 21 | 3.07 |
| 1.25k | 17.80 | 4 | 19 | 20 | 21 | 3.20 | 22 | 4.20 |
| 1.6k | 19.70 | 4 | 19 | 20 | 21 | 1.30 | 22 | 2.30 |
| 2k | 19.41 | 4 | 19 | 20 | 21 | 1.59 | 22 | 2.59 |
| 2.5k | 20.75 | 4 | 19 | 20 | 21 | 0.25 | 22 | 1.25 |
| 3.15k | 19.68 | 4 | 19 | 20 | 21 | 1.32 | 22 | 2.32 |
| 4k | 17.72 | 4 | 19 | 20 | 21 | 3.28 | 22 | 4.28 |
| **Total of Deficiency** | | | | | | 21.88 | | 34.75 |

*3.3. Recommendation*

Recommendation for improvement of insulation partition material to achieve a preferable noise criterion in the office room is done by replacing the lowest TL value of window glass with a better glass material and design. A double glazing and laminated glass are chosen. Double glazing design of glass has an advantage that the vacuum gap between the glasses has a role to avoid the acoustics wave transferred to the receiver room. A laminated glass has a role to block the acoustics wave because it has a hard structure that also makes the glass stronger and decrease the low frequency caused by the vibration. The specific material of glass is a Laminated Glass Trosifol (SC Interlayer) that has a lamination 0.380 mm each glass, and set the gap 86 mm. To support the better insulation of the glass, the frame should be very tight to avoid the flanking noise.

Besides replacing the glass, a single layer of gypsum is used as a finishing element of the wall partition. A gypsum with thickness of 20 mm has specification of 1/2" Type C Gypsum Board. The gypsum material is a hard material to block the acoustics that will be transferred into the office. From this recommendation, Figure 4 and 5 shows that the composite wall STC achieved 23, which increased the existing STC value being 4 rating higher. To meet the required SPL value in the office, the conversion process from dB to dBA is conducted. We achieved a SPL value of 38 dBA that meets the preferable noise criteria for office.

Figure 7 and Figure 8 shows that although there is an improvement of transmission loss value in each frequency, the lower frequency has lower transmission loss. It is very hard to control lower frequency of noise. Thus, a noise control should be done not only for the insulation partition material that facing the source room but also controlling the machine noise since it causes a structure borne noise. This can be done by adding a vibration damper. Another approach is install an absorber material in the machinery room to reduce the low frequency noise. Replacing a window glasses that facing the main road and adding a noise shield on the window frame are some ideas to improve better acoustics condition in the office room. Finally, flanking noise can be reduced by providing insulation treatment on door-floor gap and repairing the ceiling.

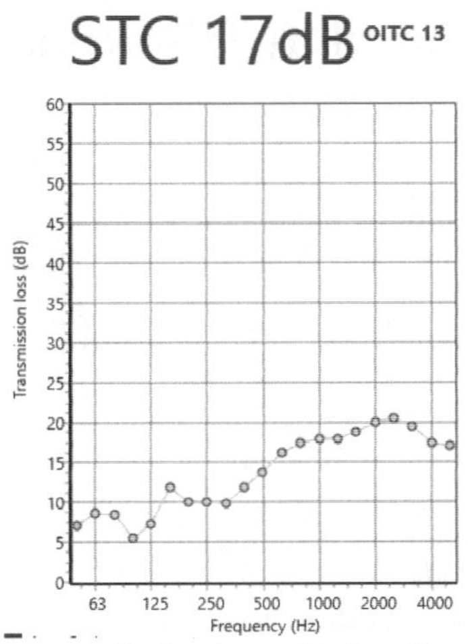


**Figure 7.** Transmission Loss Before Improvement

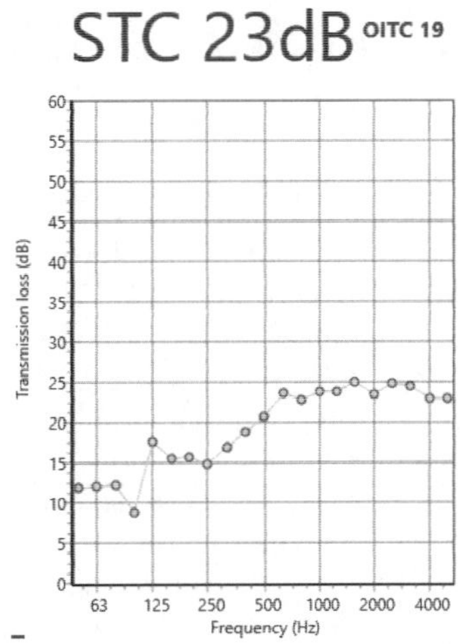


**Figure 8.** Transmission Loss After Improvement

## 4. Conclusion

The SPL of factory office exceeded the maximum value suggested by ASHRAE 1987: Interior Noise Design Goals with the range of 35 to 40 (or equivalent impact level "$L_{eq}$" of 45-50 dBA). Noise mapping shows SPL of the nearest from partition reached 68.5 dBA to 79.5 dBA and represent the noise level that will be transmitted to the office. Sound Transmission Class rating of composite wall partition has been obtained that is 17 based on field measurement of Transmission Loss. By following the ASTM E413 recommendation design of insulation material improvement is obtained by replacing single glazing window with double glazing window and cover the outside wall using gypsum. The improvement reaches 4 STC rating higher than before that is 23 and meet a preferable noise criterion for office that is 38 dBA. Another approach installs an absorber material in the machinery room to reduce the low frequency noise. Replacing a window glasses that facing the main road and adding a noise shield on the window frame are some ideas to improve better acoustics condition in the office room.

## References

[1] D M Sykes 1972 Productivity: How Acoustics Affect Workers' Performance in Offices & Open Areas
[2] H L Pick, G M Siegel, P W Fox, S R Garber and J K Kearney 1989 Inhibiting the Lombard Effect *J. Acoust. Soc. Am.* vol 85 no 2 pp 894–900
[3] M. Long 2000 Architectural Acoustics (Elsevier Ltd)
[4] S N Y Gerges 1992 Noise Sources
[5] L C Santos, C Matias and F Vieira 2008 Noise mapping of Industrial Sources *Acustica* pp 1–12
[6] W. J. P. Casas, E. P. Cordeiro, T. C. Mello, and P. H. T. Zannin, "Noise Mapping as a Tool for Controlling Industrial Noise Pollution," J. Sci. Ind. Res. (India)., vol. 73, no. 4, pp. 262–266, 2014.
[7] F. Majidi and N. Rezai, "Study of Noise Map and its Features in an Indoor Work Environment through GIS-Based Software," *J. Human, Environ. Heal. Promot.*, vol. 1, no. 3, pp. 138–142, 2016.
[8] J P Conroy and J S Roland 2003 Sound Transmission Class–Field Testing and Results pp 10–14

IOP Conf. Series: Journal of Physics: Conf. Series **1075** (2018) 012061 doi:10.1088/1742-6596/1075/1/012061

# Outdoor traffic noise effect in indoor sound distribution

**Ridho Muhammad Hasan[1], Sentagi S. Utami[1], Heri Sutanta[2]**

[1]Department of Nuclear Engineering and Engineering Physics, Universitas Gadjah Mada, Indonesia
[2]Department of Geodetic Engineering, Universitas Gadjah Mada, Indonesia

ridho.muhammad.h@mail.ugm.ac.id

**Abstract**. Noise mapping is one of the common acoustic project done, especially in urban environment. Researches have used different location, studied the usability area, and used simulation with numerical analysis as well as variety of data collecting techniques. However, information about noise distribution in an area that considers the sound distribution inside a building is still limited, since the indoor acoustics are mostly discussed in the field of room acoustics. Meanwhile, people tend to spend more time indoor instead of outdoor and therefore, such information above is considered important. This research propose a specific study about traffic noise effect to the indoor sound level distributions. Locations of this study are Faculty of Law (FH UGM) and Department of Electrical And Informatics Engineering Vocational College Universitas Gadjah Mada (DTEDI SV UGM). Selected based on interview results with the building management. The discussions are zoning of data collection points, identification of traffic noise from inside the building, analysis of sound wave attenuation, then the data are visualized via software that based on Geographic Information System or GIS. The identified traffic noise using simultaneous recording is categorized as continuous noise and intermittent noise such as police's whistle, modified exhaust, and honking. Police siren is categorized as an outlierwith the highest sound pressure. The study proved that the external sound sources do produce noise that affect the distribution of sound level inside the building. Detailed results and possible recommendations for this issue have been submitted to FH UGM and DTEDI SV UGM.

## 1. Background

Previous research about noise mapping in some area of Universitas Gadjah Mada (UGM) by Dian et al., found that there are high noise sources in several street intersections such as at the shopping center, main road, and food court intersections[1]. This condition is estimated to effect the indoor noise distribution inside the academic buildingsthat are Faculty of Law building (FH) in food court area, Department of Electrical And Informatics Engineering Vocational College building (DTEDI SV) in shopping center area, and Magister of Management building (MM) in main road area inside Universitas Gadjah Mada (UGM). Discussion with the building management for each location was also held, where itcan be concluded that the indoor space of MM building is not affected by external noise because of the structure of the building itself is designed for isolating external noise But the different result is found in FH and DTEDI, that in some parts of FH and DTEDI building there are external noises that are affecting the indoor sound distribution.

Given the result of the preliminary study above, the purpose of this research is to observe issues related with the external noise in these two locations and to provide some possible recommendation to reduce the external noise.

Many researches discussed about traffic noise but only few that studied on how the traffic noise affected indoor sound level distribution. Study on traffic noise is the domain area of environmental acoustic, focuses on outdoor noise. However, people tend to spend their time inside buildings where the acoustics quality is observed through room acoustics area of study[2]. How human perceive the differences of sound pressure level (SPL) is also related with the study on noise. The lowest SPL differences that a human can sense is around 1 dBA, although they are more confident with 3 dBA differences [3].

Noise level is using the unit of dBA since the frequency weighted of A-weighting function is representing the human's ear sensitivity. The noise distribution data is generated using Geographic Information System (GIS) as what have been done in the previous research about noise mapping[1], [4]–[6].

## 2. Method

### *2.1. Data collection*

Acoustics data was collected using recorder since information about the dynamic condition in each point can be re-played by having this recorded sound. Other information about identification type of noise and noise source can be provided using notes that were taken during recording.

The location used for this research is in the UGM area specifically FH UGM and DTEDI SV UGM. Selected based on the result from discussion with the building management. Point data assessment is determined based on adaptive sampling that considers the difference of location for each point, inside and outside building. By knowing the different acoustics conditions of the indoor and outdoor locations, thus the data collection are grouped into three zones. First is the outdoor zone near street area, indoor zone is area inside the building, then transition zone is zone between street and building. Total points for FH UGM area are 27 points as shown in Figure 1. Then for DTEDI SV UGM are 25 points as shown in Figure 2.

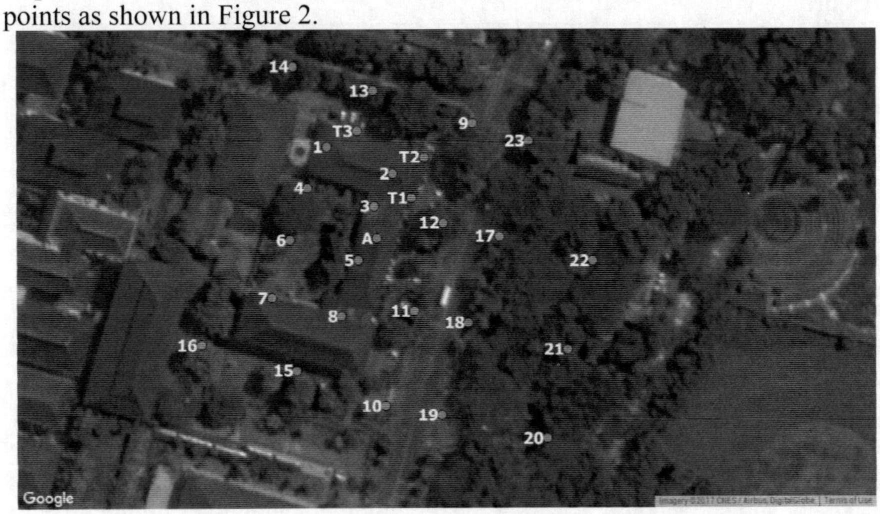


**Figure 1.** Location and point data assessment at FH UGM.

The locations of each point inside the building is only in the first floor for this research with the detail information shown in Table 1. Point locations of the data collection in building area are located in the first floor. Because in these two building, the first floor is already used for academic activities.

**Figure 2**. Location and point data assessment at DTEDI SV UGM.

Table 1. Detailed indoor point data locations

| Detailed point locations inside building (FH UGM) | Point Assessment |
|---|---|
| Hallway | 1 |
| Office lobby | 2 |
| Porch | 3, 5, 7, 8, A |
| Court | 4, 6 |
| **Detailed point locations inside building (DTEDI SV UGM)** | **Point Assessment** |
| Lobby | 1 |
| Student's waiting room | RT |
| Porch | 2, 3, 4U, 4S, 5, 6, 12 |

Before performing the actual data collection, preliminary field review in the area was performed. Results from the preliminary survey has identified several noise sources coming from street, which are police's whistle, honk, and modified exhaust. The dominant noise sources inside the building are chatter and steps. Considering the transition zone, the dominant noise source is moving vehicles that are heading to the parking lot. After knowing the condition in this area, the recorders were prepared. At least two recorders were for simultaneously recording.

To get the information needed, recording time is limited to two conditions. First is when the building was empty and second is when there were activities inside the building. Therefore, the two conditions were divided into two recording-time, 06.00 – 07.15 AM (morning time) and 09.30 – 11.00 AM (daily activity time). These two conditions are expected to cover the information needed.

*2.2. Result analysis*

The first analysis to see whether the traffic noise is affecting the indoor noise distribution or not, with the absolute value of difference between two times condition. Sound wave attenuation from outdoor zone and transition zone is then performed to show the actual condition of the parking lot route. After that, the identification of specific noise to prove the statement of each building management about the part of the building where there is external noise.

## 3. Result and discussion

The first result is the difference noise level in each points within two measurement time. This result will shows the effect of traffic noise to the changes in the noise level distribution inside the building. Each locations show that there is no correlation between the changes of noise level inside building with the traffic noise, as shown in Figure 3. In the road area the blue colour shows that there is only a little difference between two measurement times, but inside the building area the colour began turn into yellowish and reddish to show that there are many changes of noise level. This yellow and red colour indicate that the noise level has different value in the two measurement times.

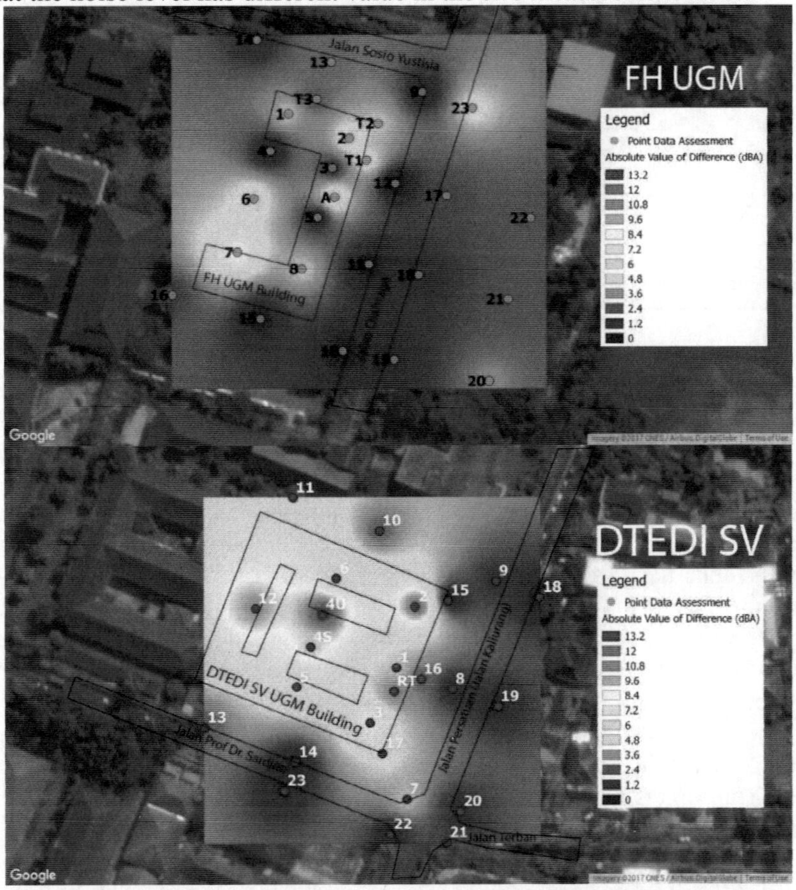


**Figure 3.** Result of absolute value of difference.

Even though there is no correlation between the changes of noise level inside building from the effect of traffic noise, it is still depicted that there is still contribution from the external noise such as moving vehicle. This condition is showed by the outcome of comparison between calculated sound pressure level and recorded sound pressure level with the outdoor zone as *L1* and transition zone as *L2* as in Table 2.

From Table 2 with the tolerance difference ±3 dBA, it can be concluded that the actual sound pressure level is higher than the calculated ones. This condition shows that the transition zone is contributing external noise as the vehicles are moving into parking lot in each area of study.

But there is a problem with the occurrence of police siren when point DTEDI 7 daily activity time was being recorded, because of the limitation of this research, the DTEDI 7 is considered as an outlier and it makes the calculated noise level higher than the actual noise level. Therefore, it can be neglected.

Table 2. Comparison between calculated sound pressure level and recorded sound pressure level

| Outdoor Zone Points | *L1* (dBA) | *r1* (meter) | *r2* (meter) | Calculated *L2* (dBA) | Transition Zone Points | Recorded *L2* (dBA) | Difference (dBA) |
|---|---|---|---|---|---|---|---|
| FH UGM morning time | | | | | | | |
| FH 9 | 73.7 | 1.5 | 23.3 | 61.8 | FH T2 | 67.8 | 6.0 |
| FH 13 | 74.7 | 1.5 | 17.7 | 64.0 | FH T3 | 65.5 | 1.5 |
| FH 12 | 71.1 | 1.5 | 16.9 | 60.6 | FH T1 | 64.8 | 4.2 |
| FH UGM daily activity | | | | | | | |
| FH 9 | 72.7 | 1.5 | 23.3 | 60.8 | FH T2 | 63.1 | 2.3 |
| FH 13 | 71.7 | 1.5 | 17.7 | 61.0 | FH T3 | 62.8 | 1.8 |
| FH 12 | 71.1 | 1.5 | 16.9 | 60.6 | FH T1 | 59.4 | -1.2 |
| DTEDI SV UGM morning time | | | | | | | |
| DTEDI 9 | 71.1 | 1.5 | 26.9 | 58.6 | DTEDI 15 | 62.4 | 3.8 |
| DTEDI 8 | 72.9 | 1.5 | 14.9 | 62.9 | DTEDI 16 | 64.7 | 1.8 |
| DTEDI 7 | 75.3 | 1.5 | 22.8 | 63.5 | DTEDI 17 | 66.8 | 3.3 |
| DTEDI SV UGM daily activity | | | | | | | |
| DTEDI 9 | 72.8 | 1.5 | 26.3 | 60.4 | DTEDI 15 | 61.5 | 1.1 |
| DTEDI 8 | 71.9 | 1.5 | 14.9 | 61.9 | DTEDI 16 | 62.1 | 0.2 |
| DTEDI 7 | 79.9 | 1.5 | 22.8 | 68.1 | DTEDI 17 | 61.8 | -6.3 |

With the result as in Table 2, transition zone is contributing the increase of noise level outside the building, and the dominance of specific noise in this zone are moving vehicle and its components. The identified noise that is moving vehicle, being generated to be noise mapping as shown in Figure 4. Figure 4 shows the specific noise in each area when the daily activity time is being recorded. Each area shows that there is actually identified vehicle noise in the building area. Even though the dominance of occurrence and level is low to mid, it is indicated that the external noise (vehicle and its components) are affecting the indoor sound distribution, especially from transition zone because of the parking lot route being used by vehicles. Then this issue needs to be controlled.

Specifically speaking, the condition in FH building is better than DTEDI SV, but in some points inside FH building area indicated that there is another sound source that dominating in these points, which is people chattering as part of the human activity (shown in white color Figure 4 (i)). Because this area is used as people to get together and have discussion.

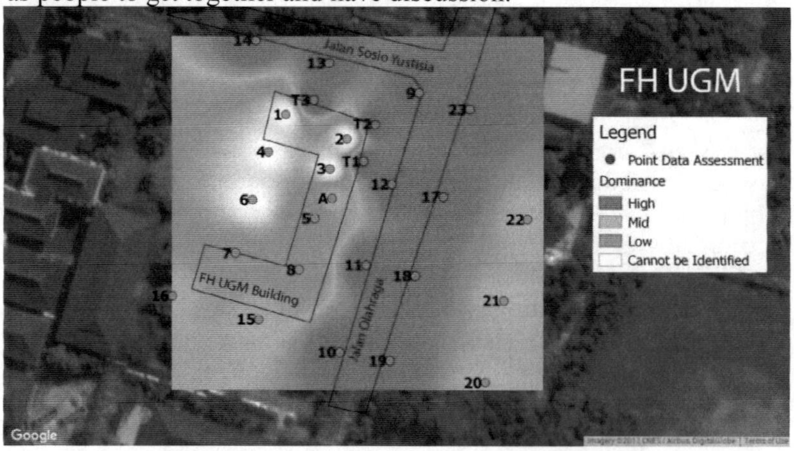


**Figure 4.** Identified distribution vehicle (i)

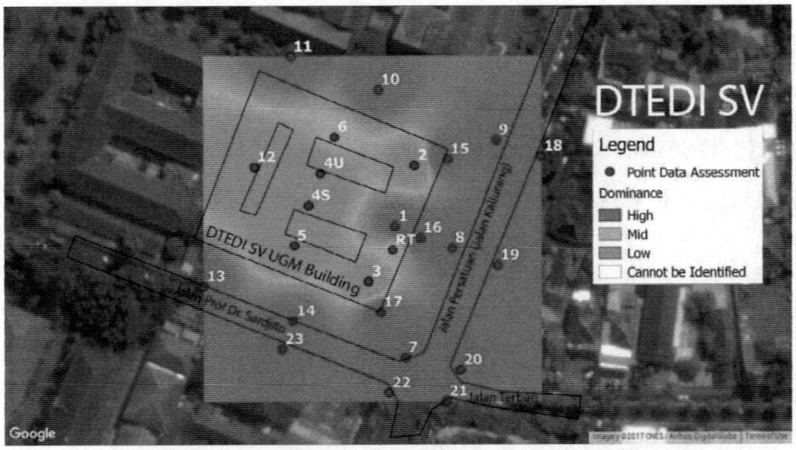


**Figure 4.** Identified distribution vehicle (ii)

It cannot be said whether this condition is bothering or not for people who are actually inside the building. Because it needs next study about sound perception in this area. The recommendation is to improve the acoustics quality in these two areas by changing the parking lot route and regulation about the modified exhaust. Both were easily identified through analyzing the recorded audio during the play-back process.

## 4. Conclusion

This research found that there is no correlation between the effects of traffic noise with the changes of indoor noise level. But there are still contribution from traffic noise as specific noise like intermittent noise and continuous noise that is identified as moving vehicle in indoor area along with noise caused by human activities inside building. The noise issue is caused by modified exhaust vehicle and parking route in the surrounding building area. Thus, the recommendation is to reduce this noise issue by changing the parking lot route and regulation about modified exhaust, with the consideration being returned to building management of each locations.

## References

[1] D. Đ. Avoressi, S. S. Utami, K. Zakiya, dan H. Sutanta, "SOUND LEVEL MAPPING USING GEOGRAPHIC INFORMATION SYSTEM ( GIS ) TO OPTIMIZE A GREEN CAMPUS ENVIRONMENT QUALITY," vol. 11, no. 6, hal. 4058–4064, 2016.

[2] N. Maisonneuve, M. Stevens, M. E. Niessen, dan L. Steels, "NoiseTube: Measuring and mapping noise pollution with mobile phones," *Environ. Sci. Eng. (Subseries Environ. Sci.*, hal. 215–228, 2009.

[3] Brüel&Kjær Sound & Vibration Measurement A/S, "Environmental Noise," 2000.

[4] P. Tao, Y. Chen, dan L. Tong, "GIS-based city noise mapping research and development," *Int. Geosci. Remote Sens. Symp.*, hal. 6729–6732, 2012.

[5] M. Cai, J. Zou, J. Xie, dan X. Ma, "Road traffic noise mapping in Guangzhou using GIS and GPS," *Appl. Acoust.*, vol. 87, hal. 94–102, 2015.

[6] A. Schindler, G. Information, dan S. Gis, "Soundscape Mapping : Spatial Variability of Sound at Furman University," 2016.

# Study report: association between pure tone average and ossicular status in chronic suppurative otitis media

**N Shinta[1], N Purnami[1] and T H Ahadiah[1]**

[1]Department of Otorhinolaryngology Head and Neck Surgery
Medical Faculty Airlangga University – Dr. Soetomo Hospital Surabaya

E-mail: nyilo@fk.unair.ac.id

**Abstract.** Prediction of ossicular status in chronic suppurative otitis media (CSOM) before tympanomastoidectomy was required for selection of surgical techniques. Some studies about prediction of ossicular status withpure tone average had revealed different results. This study aimed to prove correlation between pure tone average and ossicular status in CSOM patients whom underwent timpanomastoidectomy at Department of Otorhinolaryngology Head and Neck Surgery Dr. Soetomo Hospital Surabaya. This was an observational analytic study with retrospective cross sectional design using secondary data of medical record. Simple randomized sampling was derived from CSOM patients whom underwent tympanomastoidectomy and appropriated with inclusion and exclusion criterias. Ossicular status during the surgery was obtained from operation report based on Austin-Kartush criterias. Pure tone averageis the average threshold of hearing in speech frequencies that were 500, 1000, 2000, and 4000 Hz from audigram. About 65 CSOM patients who underwent tympanomastoidectomy were included to this. The ossicular erosion in the study was 63,1% while the normal ossicles (ossicular status O) was 36,9%. The most common ossicular erosion all ossicles (ossicular status D) 36,9%; followed by incus and stapes (ossicular status B) as 9,2%; incus (ossicular status A) as 10,7%; maleus and incus (ossicular status C) as 6,2%; and fixation of stapes (ossicular status F) as 3,1%. Two-way Anova test resulted p value = 0,00 ($p < 0,05$). There was correlation between pure tone average and ossicular statusin chronic suppurative otitis media patients. Increased pure tone average could estimate presence of ossicular erosion before surgery obtained in chronic suppurative otitis media patients.

## 1. Introduction

Hearing loss in CSOM patients is primarily conductive type but may also be mixed type. Degree of hearing loss depends on the presence of ossicular damage. Ossicular erosion and fixation result in hearing loss more than 30 dB (Browning, et al., 2008; Mittal, et al., 2015). Ossicular erosion is a frequent CSOM complication of about one third of all cases of CSOM. Ossicular erosion is assested during surgery. Ossicular status is an assessment of the integrity of each ossicles [1].

Prediction of ossicular status before tympanomastoidectomy is required as a basic for selection of surgical techniques. Estimates of preoperative ossicular status can be through by pure tone audiometry and high resolution computed tomography (HRCT) mastoid scan. Pure tone audiometry is a routine examination prior to ear surgery to assess hearing function and quantitatively assess the pure-tone average (PTA) of CSOM patients [2].

Some studies of PTA as a preoperative parameter of ossicular status in chronic suppurative otitis media had gived different results [3], [4]. Study by Feng found that PTA in CSOM patients with ossicular erosion was higher than patients without ossicular erosion and could be used to estimate ossicular status [4]. Another study for predictive value of the audiogram on ossicular status showed that PTA was higher in the group with ossicular erosion but did not provide significant predictive value for ossicular status [3]. The objective of the study was to analyze the association between PTA and ossicular status in patients with CSOM in Dr. Soetomo Hospital Surabaya.

## 2. Methods

This is an analytic observational study with cross sectional retrospective design using secondary data. The data were taken from medical record of Otology Division ORL-HNS Dr. Soetomo Hospital Surabaya for two years period from January 1st, 2015 to December 31st, 2016. Subjects of this study were CSOM patients whom had performed tympanomastoidectomy and according to the criterias of inclusion and exclusion by simple random sampling. Inclusion criterias were patients with complete medical record data, preoperative audiogram, and has an operation report of ossicular status assessment. Exclusion criteria are otosclerosis, middle tissue polyp, ossicular malformation, ossicular agenesis, sensorineural type hearing loss, and previous history of tympanomastoidectomy.

The pure tone average was calculated as the mean of sum of air conduction at frequency at 500, 1000, 2000, and 4000 Hz from audiogram. Ossicular status is an assessment of the ossicular chain when tympanomastoidectomy by ORL-HNS Specialist Consultant of Otology and then classified by Austin and Kartush into seven types.

**Table 1.** Ossicular status based on Austin and Kartush [5].

| No | Type | Ossicular Status |
|---|---|---|
| 1 | O | M+I+S+ |
| 2 | A | M+S+ |
| 3 | B | M+S- |
| 4 | C | M-S+ |
| 5 | D | M-S- |
| 6 | E | Maleus fixation |
| 7 | F | Stapes fixation |

Statistical test to analyze the relationship between pure tone average and ossicular status using two-way Anova test with significance level (α) of 0,05.

## 3. Result

Hearing loss moderate to severe degree was found in this study of 29,3%, followed by moderate degree as much as 24,6%, severe degree as much as 20,0%, profound degree as much as 13,8%, and mild degree as much as 12,3%.

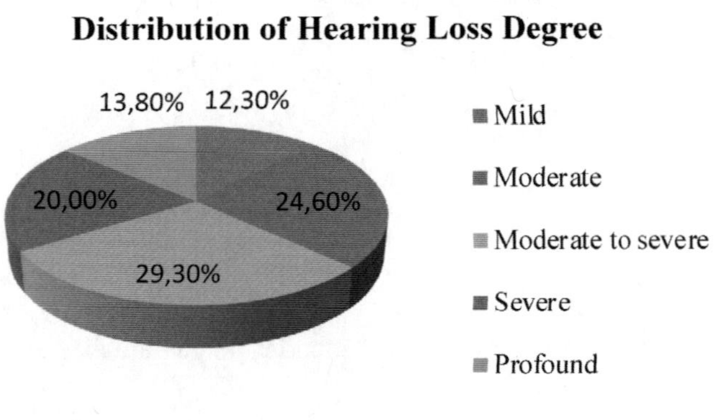


**Figure 1.** Distribution of degree of hearing loss.

Maleus was obtained most intact condition as much as 56,9% while destruction asmuch as 43,1%.

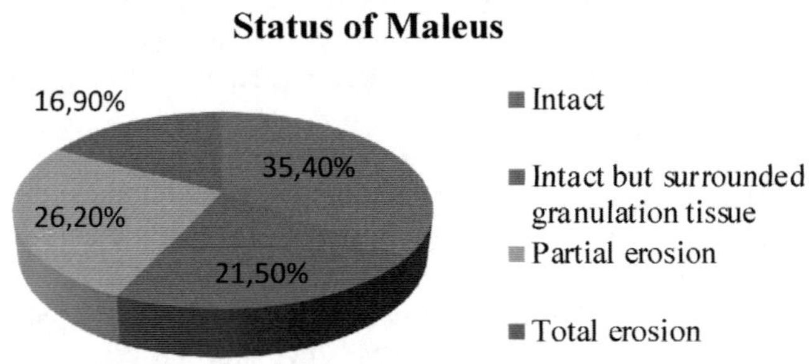


**Figure 2.** Distribution status of maleus.

Incus was found most destructed condition as much as 60,0% while intact was only 40,0%.

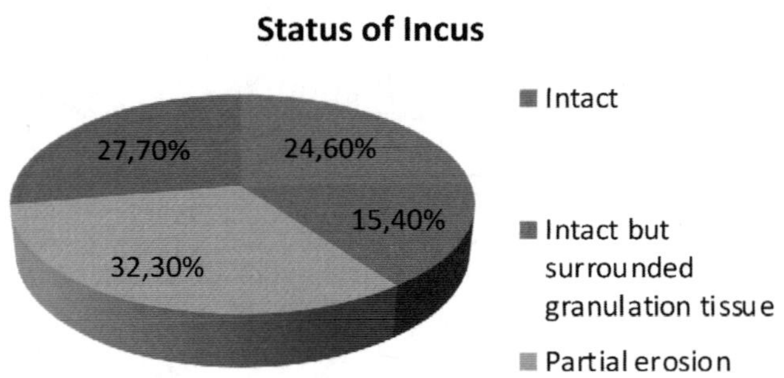


**Figure 3.** Distribution status of incus.

Stapes was obtained most intact condition as much as 50,8%, followed destruction as much as 46,2%, and fixation as much as 3,1%.

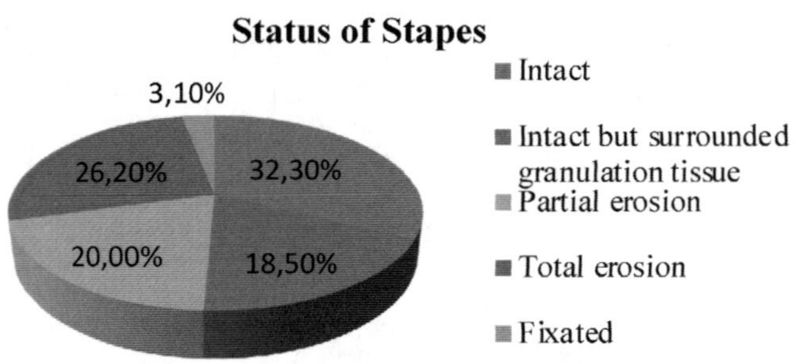


**Figure 4.** Distribution status of stapes.

The most common ossicular status was type O and D as much as 36,9%, type B as much as 9,2%, type A as much as 10,7%, type C as much as 6,2%, and type Fas much as 3,1%. There was no type E ossicular status in this study.

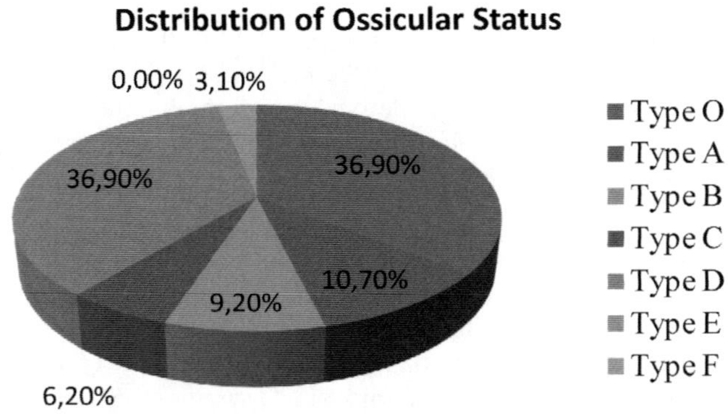


**Figure 5.** Distribution of ossicullar status type based on Austin Kartush.

A two-way anova test between the groups revealed p = 0.000 (p <0,05). This result showed a significant relationship between pure tone average and ossicular status.

## 4. Discussion

The highest degree of hearing loss in this study was moderate weight of 29.3%, followed by moderate degree as much as 24,6%, severe degree as much as 20,0%, very severe degree as much as 13,8%, and mild degree as much as 12,3 %. Chronic suppurative otitis media causes mild to moderate degree conductive hearing loss in over 50% of cases. Degree of hearing loss in CSOM is influenced by size and location of tympanic membrane perforation, granulation tissue, cholesteatoma, and ossicular erosion.

The most common intact ossicle in this study was maleus (55,4%), followed by stapes (50,8%), and incus (38,5%). The most destructed ossicle was incus (61,5%), followed by stapes (49,2%), and maleus (44,6%). The long process incus was an ossicle that has a long and hanging structure, a location in mesotympanum, as well as rare vascularization so susceptible to destruction [6].

Ossicular status in this study found that the ossicular status O (intact) as much as 36,9% and ossicular destruction as much as 63,1%. Most ossicular destruction was found in three ossicles (ossicular status D), which was 36,9%, followed by incus and stapes (ossicular status B) as much as 9,2%, incus destruction (ossicular status A) as much as 10,7%, destruction of maleus and incus (ocsicular status C) as much as 6,2%, and stapes fixation (ossicular status F) as much as 3,1%.

This study showed that there was a relationship between pure tone average and ossicular status in CSOM patients with p value between groups = 0.000 (p <0,05). This was appropiate with other studies that showed the value of AC and air bone gap on ossicular erosion higher than without ossicular erosion. That study concluded that AC and air bone gap values were reliable indicators for ossicular status in the CSOM [7]. Ear-canal collapse due to pressure of the earphone on the pinna during audiometry can display erroneous air bone gap at high frequencies. High frequency air bone gap have also been implicated to be the result of instrument calibration errors.

Ossicular destruction may cause conductive hearing loss, which is also affected by perforation of the tympanic membrane, granulation tissue, and cholesteatoma. Large perforations of the tympanic membrane generally cause hearing loss of 40-50 dB, whereas total perforation is up to 60 dB. Differences in degree of hearing loss at the same perforated size can be due to differences in the volume of the middle ear and mastoid cavity of each ear. The degree of hearing loss due to colesteatoma depends on the size and ability of cholesteatoma to erode the ossicullar chain. The ossicular destruction interferes the aerodynamic acoustic transmission process, thus lowering air conduction not followed by bone conduction in the audiogram. Decreased of air conduction increases the pure tone average of audiogram.

## 5. References

[1] Haidar H, Sheikh R, Larem A, Abdulkarim H, Ashkanani S and Alqahtani A 2015 Ossicular chain erosion in chronic suppurative otitis media *Otolaryngol* 5 (4) pp 1-4

[2] Sheikh A, Haidar H, Abdulkarim H, Aslam W, Larem A and Alsaadi A 2016 Preoperative predictors in chronic suppurative otitis media for ossicular chain discontinuity: across-sectional study *Audiol Neurotol* 21: pp 231–236.

[3] Jeng F, Tsai M and Brown C 2003 Relationship of preoperative findings and ossicular discontinuity in chronic otitis media *Otol Neurotol* 24 (29) pp 32-40

[4] Feng H, Chen Y and Ding Y 2005 Analysis of preoperative findings and ossicular condition in chronic suppurative otitis media. *Lin Chuang Er Bi Yan Hou Ke Za Zhi* 2005 19 (1) pp 7-10

[5] Anglitoiu A, Balica N, Lupescu S, Vintila R and Cotulbea S 2011 Ossicular chain status in the otological pathology of the ENT clinic Timisoara *Medicine in Evolution* 17 (4) pp 344-352

[6] Nayak GK, Barhma D, Chatterjee P and Sharma P 2016 Ossicular chain status in chronic suppurative otitis media *J. of Dental and Medical Sci* 15 (3) pp 20-23.

[7] Yazdi AK, Saedi B, Fayezizadah and Seifmanes H 2011 Association between audiometric profile and intraoperative findings in patients with chronic suppurative otitis media. *Iranian J of Otorhinolaryngol* 2 (23) pp 37-43

IOP Conf. Series: Journal of Physics: Conf. Series **1075** (2018) 012063 doi:10.1088/1742-6596/1075/1/012063

# Improving the airborne sound insulation properties of vacuum insulating glazing using dissimilar pane thickness and a laminated pane: experimental results

**A C Nugroho[1], D Cabrera[1] and C Kocer[2]**

[1]School of Architecture, Design, and Planning. University of Sydney, NSW 2006, Australia.
[2]School of Physics, University of Sydney, NSW 2006, Australia

E-mail: anug8272@uni.sydney.edu.au

**Abstract**. Vacuum insulating glazing (VIG) is a relatively new glazing technology that was first developed primarily for its thermal insulation properties. This research aims to continue a preceding study about the acoustic performance of VIG and to analyse the effects of dissimilar thickness and additional damping to the sound insulation performance of VIG. In this research, the airborne sound insulation performance of VIG is examined with theoretical and experimental approaches. Sound insulation performance was measured using the field sound intensity method for 5 configurations of VIG. Measurements were well-predicted by theory at low frequencies, where it is mostly governed by the mass-law with additional corrections. Sound insulation performance was heavily affected by the coincidence dip, and using dissimilar thickness in the VIG does not alleviate the problem. Additional damping affects the coincidence behaviour of the VIG, shifting the critical frequency and increases the sound insulation performance at and above the critical frequency with improvements up to 9 dB.

## 1. Introduction

As sustainability efforts in the built environment are rising, more novel ways are currently being developed to reduce energy consumption. Generally, windows are the weakest insulating element in a building's façade due to its relatively high thermal transmittance (U-value) compared to other building elements (1). Even though windows contribute to 30%-60% of a building's energy loss (2) (3), they also provide a huge potential to provide large amounts of energy savings by using glazing systems that have a high thermal insulation. This could be achieved by using state-of-the-art technology such as vacuum glazing, which can provide minimum heat loss in a slimmer construction.

Vacuum insulating glazing (VIG) is a relatively new glazing technology that is first developed primarily for its thermal insulation properties (4). It is made of two parallel pieces of glass separated by an array of support pillars (or spacers) with an evacuated space between which is hermetically sealed at the edges using solder glass.

The sparse literature available about the sound insulation performance of vacuum glazing means that this is still an ongoing field of research. A prior study by Cabrera, Ashmore and Kocer (5) investigated the acoustic performance of several commercial VIG products and comparing them to monolithic pieces of glass. Measurements showed that the VIG behaves like a single glazing construction in the low frequencies, following the mass-law principle of an equivalent monolithic piece of glass with equal surface density to the VIG. Measurement results also correspond well with Asano et al's (2) findings about the VIG's behaviour in the coincidence frequency, i.e., that the coincidence dip occurs at the same frequency as for a single constituent pane from the VIG assembly. It is also found that the coincidence dip can be greatly reduced by clamping another piece of glass into one of the glass panes, creating an asymmetric configuration with one side considerably thicker than the other. The additional glass pane introduces additional damping into the VIG, reducing the depth of the notch in the coincidence frequency which subsequently improves its single number ratings. It also shifts the coincidence frequency to that of the clamped pane pair, suggesting that behaviour near the coincidence frequency is influenced more by the thicker of the two layers of glass (as has been observed in double glazing).

This research aims to continue Cabrera, Ashmore, and Kocer's work and to investigate their hypothesis by measuring several VIG configurations with different thicknesses and damping, and observing their sound insulation performance as well as their single number quantities. This study would also lead to a more general understanding of the acoustic properties of VIG, and to propose improvements in the design to provide better sound insulation.

## 2. Airborne Sound Insulation in Vacuum Insulating Glazing

The sound insulation properties of vacuum glazing may be approximated as that of a single isotropic panel as pointed out in previous studies (2) (5). Sound insulation is dependent on a panel's mass, stiffness, damping, and dimensions – with each component affecting different regions across the whole frequency range (6). In the very low frequencies, sound reduction would be mainly determined by the stiffness as well as the physical dimension of the panel. Resonances can occur in a panel with finite size in the so called fundamental frequency $f_{1,1}$ In this frequency, the sound reduction will be at its lowest point, although usually this is outside the frequency range of interest.

Above the fundamental resonance frequency, sound reduction would generally increase as a function of its surface density. The sound reduction performance in this region can be estimated using several approaches, and for this study the field incidence sound reduction is shown in Equation 1 (7), where $f$ is frequency (Hz) and $\rho_s$ is the surface density of the panel (kg/m$^2$).

$$R = 20\ \log_{10}(f\rho_s) - 48\ dB \tag{1}$$

The sound reduction continues to increase until it approaches the critical frequency $f_c$, where the wavelength of the incident sound on the panel is equal to the wavelength of the bending waves within the panel (7). Since the maximum and minimum peaks of the wavelengths are spatially matched, the resulting sound energy is easily transmitted through the panel (8), resulting in a significant decrease in sound reduction performance and can be estimated using Equation 2,

$$f_c = \frac{{c_0}^2}{2\pi h}\sqrt{\frac{12\ (1-\sigma^2)\ \rho_p}{E}}\ Hz \tag{2}$$

Here, $c_0$ is the speed of sound (m/s), $h$ is the panel thickness (m), $\sigma$ is Poisson's ratio, $\rho_p$ is the panel density (kg/m$^3$), and $E$ is the Young's modulus (Pa). Substituting the relevant constants of glass into Equation 3 results in a simple approximation (7):

$$f_c \cong \frac{12}{h}\ Hz \tag{3}$$

After determining the critical frequency, the sound reduction index at that point can be estimated using Equation 4 (8):

$$R_{f=f_c} = 20\log_{10}\left(\frac{\omega\ \rho_s}{2\rho_0 c_0}\right) + 10\ \log_{10}\left[\frac{2\,\eta}{\pi}\left(\frac{\Delta f}{f_c}\right)\right]\ dB \tag{4}$$

Where $\eta$ is the panel's damping coefficient / loss factor, and $\Delta f$ is the bandwidth (Hz). Just below the critical frequency, the sound reduction index is estimated by drawing a straight line between the mass-law curve at $f_c/2$ and the estimated sound reduction index calculated by Equation 5. At and above

IOP Conf. Series: Journal of Physics: Conf. Series **1075** (2018) 012063 doi:10.1088/1742-6596/1075/1/012063

the critical frequency, sound reduction is mostly governed by the damping properties of the panel as indicated by the $\eta$ variable. The sound reduction above the critical frequency is given by Equation 5 (9):

$$R = 20\log_{10}\left(\frac{\omega\,\rho_s}{2\rho_0 c_0}\right) + 10\ \log_{10}\left(\frac{\eta\,f}{f_c}\right)\ + 10\ \log_{10}\left(1-\frac{f_c}{f}\right) - 44.5\ dB \qquad (5)$$

## 3. Measurement Method

### *3.1. Quantitative assessment of airborne sound insulation*

Sound reduction index, or $R$, is defined as the ratio of sound power incident on the panel to the sound power that is transmitted through the specimen. This study uses the intensity method in accordance with ISO 15186-2 (10). The sound reduction index for the intensity method ($R_I$) is calculated according to Equation 6, where $L_{p1}$ is the average sound pressure of the source room, $L_I$ is the measured sound intensity that is transmitted into the receiving room, $S_m$ is the measurement area and $S$ is the specimen area.

$$R_I = L_{p1} - L_I - 6 + 10\log\left(\frac{S_m}{S}\right) \qquad (6)$$

In the ISO standard, the Waterhouse correction is applied to the calculation to avoid an underestimation of the low-frequency sound reduction performance. It is calculated by Equation 7, where $S_{b2}$ is the area of the boundary surfaces in the receiving room (m$^2$), $V_2$ is the receiving room volume (m$^3$), and $\lambda$ is the wavelength of the frequency that is calculated (m).

$$K_c = 10\log\left(1+\frac{S_{b2}\,\lambda}{8V_2}\right)\text{dB} \qquad (7)$$

Thus, the complete modified intensity sound reduction index $R_{IM}$ would be:

$$R_{IM} = L_{p1} - L_I - 6 + 10\ log\left(\frac{S_m}{S}\right) + K_c \qquad (8)$$

### *3.2. Measurement Setup*

#### *3.2.1. Measured Samples*

In total, 5 samples of VIG were prepared with various combinations for the thickness of the glass pane. Two samples had identical pane thickness, with 3 mm (3V3) and 5 mm (5V5) for each pane. One VIG sample was made from a combination of 3mm and 5mm panes (3V5) to test the hypothesis regarding the dissimilar thickness. The last two samples were 3V3 VIG samples combined with a single sheet of monolithic glass with a PVB interlayer in-between (3V3L3, 3V3L5) to investigate how damping affects the sound insulation performance of VIG. All VIG samples were provided by a Japanese manufacturer.

#### *3.2.2. Laboratory Arrangement*

The VIG samples were mounted in the wall between the reverberation room and the general laboratory area (Figure 1). The samples were mounted in the test aperture measuring 1200 mm x 640 mm. The samples were sealed around the edges with closed-cell foam tape, and were tightly clamped into the aperture using metal strips with additional closed-cell foam tape. Each sample was clamped at the source room side, creating a constant shallow niche with a depth of 150 mm from the receiving room.

The six microphones were used to measure the sound pressure level in the source room and were distributed in the room adhering to ISO 15186-2. Since the rooms were in reality not perfectly diffuse, the sound pressure level at the source must be measured at a number of locations and then averaged (11), to obtain a spatially averaged sound pressure level used to determine the sound reduction index of the VIG samples. The microphones were also set in different heights to obtain a better spatial averaging of the room.

**Figure 1.** General arrangement of the measurement setup.

**Figure 2.** Absorptive panels in front of the niche.

The two Turbosound TA-500 loudspeakers were used as the primary sound source. The loudspeakers were placed in the corners of the source room as shown in Figure 2, and each of the sound sources produced steady state mutually incoherent pink noise. Since the two channels are not correlated with each other, destructive interference between the two sources could be ignored, rendering the sound field to be homogenous and isotropic and the average squared pressures within the rooms to be additive (12). The spatially averaged sound pressure level in the source room was kept around 112 dB for all measurements. The 6 microphones and two sources that were used are equivalent of 12 source-receiver combinations for each measurement. Acrylic diffusers measuring at 1.22 m x 0.92 m were distributed pseudo-randomly within the room to improve the diffusivity of the sound field by redirecting the incident sound energy, which increases the sound field isotropy.

*3.2.3. Measurement Procedure*

The transmitted sound intensity was measured in the receiving room using a pair of ½-inch B&K 4197 microphones that are connected to a handheld B&K 2260 Investigator. A 12-mm spacer was chosen for its suitable range from 100 Hz – 5 kHz, which is of particular interest for this research.

The average sound intensity level, $L_I$, was measured on the receiving side of the VIG sample by scanning the intensity probe continuously over the measurement surface at a fixed distance. Measurements were made with the intensity probe's orientation normal to the measurement surface. Absorptive panels were also placed around the opening on the receiving room side (Figure 2). This is mainly to optimise the condition of the sound field from background noise as well as to avoid the intensity probe registering negative intensity from sound that is reflecting back into the probe.

Background noise measurements were made before each measurement, and a signal-to-noise ratio of at least 15 dB was always maintained at every frequency band. In total, each VIG sample was measured a minimum of 4 times with different source-receiver combinations. This was done to reduce the uncertainty that can be caused by measurement errors from various factors. Results were then averaged to obtain 1/3-octave band sound insulation performance and to determine the single number quantities.

## 4. Measurement Results

*4.1. Single VIG*

Measurement results showed that the VIG is in good agreement with the mass-law theoretical predictions following the combined panel surface density as per equation 2. A significant dip in the critical frequency is still present in all samples. For the 3V3 and 5V5 sample, the critical frequency closely follows the theoretical prediction for 3 mm glass and 5 mm glass respectively.

The 3V5 sample produces an interesting result; the measured critical frequency does not match the predicted critical frequency of the thicker glass (5 mm), as per the initial hypothesis. By observing Figure 3, it is clearly seen that the measured coincidence dip frequency lies between the predicted values for 3 mm & 5 mm glass, which means that the 3V5 sample has a similar effective critical frequency as 4 mm glass. In addition, using a dissimilar glass thickness does not significantly reduce the depth of the dip.

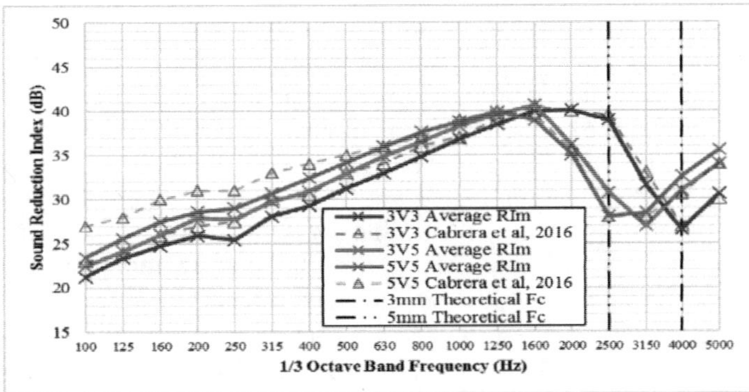


**Figure 3.** Comparison between the measured results and theoretical predictions for single VIG.

### *4.2. Laminated VIG*

Likewise, with the measurement results for the single pane VIG, the laminated composite VIG samples are also in close agreement with the mass-law prediction calculated from equation (4) as observed from Figure 4. It is also noted that since the laminated VIG are thicker than the single VIG, higher sound reduction indexes are expected because of the increased surface density.

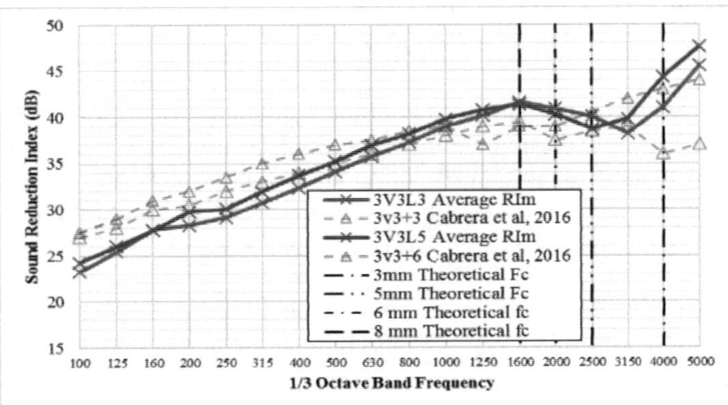


**Figure 4.** Comparison between the measured results and theoretical predictions for laminated VIG

Prediction of the critical frequency for the laminated composite VIG is much more difficult compared to the individual VIG samples. For the 3V3L3 sample, the predicted critical frequency is off by two 1/3 octave bands as shown in Figure 4. The same pattern is also observed in the 3V3L5 sample, with the measured critical frequency corresponds to that of a 5 mm glass.

he additional effects of laminated glass can also be observed by comparing it with Cabrera, Ashmore, and Kocer's results (5). From Figure 4, we can observe that the 3V3+3 has two dips that are visible, one at 2000 Hz and another one at 4000 Hz. Those frequencies correspond to the critical frequencies of 6 mm and 3 mm glass respectively. Also, there is a small dip present in 1250 Hz for the 3V3+6 sample, which correspond to the theoretical critical frequency of a 9 mm glass. It appears that combining the glasses together creates an additional coincidence dip that corresponds to the thickness of the clamped pane. By comparing it to the sound reduction indexes of the laminated composite VIG from this research, it can be said that adding an intermediate layer would improve the sound reduction performance as well as shifting the coincidence dip to higher frequencies.

IOP Conf. Series: Journal of Physics: Conf. Series **1075** (2018) 012063 doi:10.1088/1742-6596/1075/1/012063

The added intermediate layer affects the depth of the coincidence dip considerably. Improvements of 8-9 dB are achieved when the laminated VIG is compared with the 5V5 sample, which has a similar thickness (9 mm for the 3V3L3 and 11 mm for 3V3L5, compared to the combined 5V5 thickness of 10 mm). These findings agree with the reported 8-11 dB improvement in laminated glass by Marsh (13).

### *4.3. Single Number Quantities*

Table 1 shows the single number quantities (SNQ) that are obtained from the measured sound reduction indexes. Interestingly, the single VIGs more or less produce the same numbers even though the samples are different in thickness. The thicker sample should have a better sound insulation performance, but this added value is offset by the large dip at the critical frequency that lowers the performance significantly. This is also proven by the STC ratings that are 3-4 dB lower than their respective $R_w$ counterparts for these samples, as the 8 dB maximum deviation rule for STC plays a prominent part in lowering the STC values. Even so, the SNQ of these samples are comparable to that of an insulating glass unit that consists of 10 mm & 6 mm float glass with an air gap between 6 mm - 16 mm (14).

**Table 1.** Summary of the single number quantities from the measured VIG samples

| VIG Sample | $R_w$ $(C, C_{tr})$ | $R_w + C_{tr}$ | STC |
|---|---|---|---|
| 3v3 | 35 (-1, -3) | 32 | 31 |
| 3v5 | 35 (-3, -3) | 32 | 31 |
| 5v5 | 35 (-3, -2) | 33 | 32 |
| 3v3L3 | 38 (-1, -4) | 34 | 38 |
| 3v3L5 | 38 (-1, -3) | 35 | 38 |

The laminated VIG produces respectably high single number values considering its thickness. Values are close to those for 19 mm monolithic glazing (15) which is almost double the thickness. Values are also comparable to an insulating glazing configuration of 6 mm and 8 mm laminated glass with a 16 mm air gap filled with argon gas (14). For laminated VIG, the STC values match the $R_w$ values due to the absence of a large coincidence dip as opposed to the single VIG.

## 5. Discussion

The general sound insulation performance of VIG has been reasonably predicted using an approximation from an isotropic thin panel. Experimental results have shown that the VIG will follow the mass-law depending on its surface density in the low frequencies, and exhibiting a reduced performance in the higher frequencies which is affected by bending waves.

Cabrera, Ashmore, and Kocer's hypothesis of improving the sound insulation performance of VIG by using a dissimilar thickness was tested with the 3V5 VIG sample, and improvements in the $f_c$ are not as expected by the hypothesis. The original thought of using a dissimilar thickness is based on reducing mutually-reinforced bending waves due to the matched thickness of the individual panes. Since one of the panes is only 2 mm thicker, the difference in bending wave speed in this configuration might be insufficient to remove the constructive interference. Assessing the SNQ also showed that there was no improvement, with the 3V5 VIG showing a lower $C$ correction factor than the 3V3 sample.

Results shown that adding an intermediate resilient layer into a VIG construction is beneficial in increasing the sound reduction index at high frequencies. The coincidence dip is still present, but it has been greatly reduced due to the added damping by the intermediate layer. Direct improvements to the sound insulation performance are evident, especially above the critical frequency.

An interesting observation from the measurement results is that the $f_c$ for the laminated VIG samples does not follow the prediction of the $f_c$ of a combined pane thickness. The measured $f_c$ is two 1/3-octave bands higher than the predicted values, which may be affected by the additional damping that is introduced into the glazing. Further investigation is required to predict the coincidence behaviour in laminated VIG.

VIG offers a reasonable sound insulation performance even though it was developed primarily for its excellent thermal properties. Although the sound insulation performance in the lower frequencies and at the critical frequency is still a concern, incorporating the VIG into a double glazing or even triple glazing construction should improve both of its acoustic and thermal performance significantly, which would make VIG an excellent choice for a high-performing façade in a sustainable environment.

## 6. Conclusion

Methods for improving the airborne sound insulation performance of VIG has been explored to a certain degree of success. Five VIG samples with different configurations have been analysed with theoretical and experimental approaches. The hypothesis that using a dissimilar pane thickness would improve the sound insulation performance of VIG has not been thoroughly proven, and further testing is required to fully understand its potential benefits. The addition of an intermediate resilient layer in the VIG construction significantly improves the sound insulation performance in the higher frequencies. The coincidence behaviour of laminated VIG is affected by the additional damping, reducing the coincidence dip and shifting its critical frequency. The overall conclusion is that further studies must be made to obtain a more thorough understanding of the sound insulation performance of VIG, especially concerning the damping properties.

## 7. References

[1] Cuce E and Cuce P M 2015 Vacuum glazing for highly insulating windows: Recent developments and future prospects. *Renewable and Sustainable Energy Reviews* 54 pp 1345–1357

[2] Asano O, Misonou M, Kato H and Nagasaka S 1999 Advanced window incorporating vacuum glazing *In SPIE Conference on Solar Optical Materials XVI* (Denver: Colorado)

[3] Jelle BP, Hynd A, Gustavsen A, Arasteh D, Goudey H and Hart R 2012 Fenestration of today and tomorrow: A state-of-the-art review and future research opportunities *Solar Energy Materials & Solar Cells 96* pp 1-28

[4] Collins RE, Turner GM, Fischer-Cripps AC, Tang JZ, Simko TM, Dey CJ, et al 1995 Vacuum Glazing - A New Component for Insulating Windows *Building and Environment* Vol 30 No 4

[5] Cabrera D, Ashmore N and Kocer C. 2016 Airborne sound insulation of vacuum insulating glazing: General observations from measurements. *Building Acoustics* Vol 23 (3-4) pp 193-206

[6] BSI. BS EN ISO 15186-2:2010 Acoustics — Measurement of sound insulation in buildings and of building elements using sound intensity Part 2: Field measurements. 2010.

[7] Dijckmans A and Vermeir G 2013 Numerical Investigation of the Repeatability and Reproducibility of Laboratory Sound Insulation Measurements *Acta Acustica United with Acustica* Vol 99

[8] Jacobsen F 2011 The sound field in a reverberation room Lyngby (Denmark: Technical University of Denmark)

[9] Marsh J 1971 The airborne sound insulation of glass: Part 2 *Applied Acoustics* 4

[10] Pilkington UK Ltd. Pilkington Optiphon™ Laminated Glass for noise control. 2014.

[11] Viridian Glass. Viridian Glass. [Online]. 2017. Available from: http://www.viridianglass.com/~/media/viridian-glass/files/downloads/tech-direct/sound-and-noise.pdf.

**Acknowledgement**

The authors would like to thank Jonothan Holmes, Luis Miranda Jofre, Manuj Yadav, and David Spargo for their advice and assistance throughout this research. The lead author would also like to thank the Indonesian Endowment Fund for Education (LPDP) for its support during the author's studies.

IOP Conf. Series: Journal of Physics: Conf. Series **1075** (2018) 012064 doi:10.1088/1742-6596/1075/1/012064

# Reactive oxygen species levels are high risk worker of noise induced hearing loss in hospitals

**N Purnami[1] and S P P Manyakori[2]**

[1]Department of Otolaryngology Head and Neck Surgery, Dr. Soetomo General Hospital, Faculty of Medicine Universitas Airlangga, Jalan Mayjen Prof. Dr. Moestopo No. 47 Surabaya, Surabaya 60132, Indonesia
[2]Faculty of Medicine Universitas Airlangga, Jalan Mayjen Prof. Dr. Moestopo No. 47 Surabaya, 60132, Indonesia

E-mail: nyilo@fk.unair.ac.id

**Abstract.** Excessive noise exposure could increase the production of reactive oxygen species in the cochlea, thus causing the risk of noise-induced hearing loss (NIHL). Environment noise are commonly found in industrial sites but its possibility of hospitals also include noisy locations. Objectives: To analyze the correlation of reactive oxygen species and hearing impairment to employees at risk in the Hospital. Method: Participants was obtained by identifying the employees in hospital from 3,813. They were examined for baseline characteristics, hearing loss and reactive oxygen species. Hearing loss was defined as audiometry and tympanometry level. Result: The proportion of participant was 42.43 ± 10.72 years old in women (58.33%) and noise levels at Dr. Soetomo General Hospital was 98.15 ± 8.16 dB in range 85.39 to 112.90 dB. The prevalence of NIHL was 47.92% (audiometry) and 70.83% (otoacoustic emission). Reactive oxygen species estimated 5.55 ± 4.39 ng/ml. Statistical analysis of reactive oxygen species to audiometry ($p = 0.993$) and reactive oxygen species to otoacoustic emission ($p = 0.647$). Conclusion: increased production of reactive oxygen species that cause hearing loss, but there was no correlation between reactive oxygen species and hearing loss in risk worker at the hospital.

## 1. Introduction

Noise possibly cause health problems in the form of noise-induced hearing loss (NIHL). In Nepal, it was reported that 31% of carpenters and 44% sawing experience NIHL [1]. In the Tanzanian mining area reported by 12% have poor hearing and 35% have a mild hearing loss. Every year the NIHL population increases due to the hearing exposure [2].

Noisy exposure could increase the production of reactive oxygen species (ROS) in the cochlea. ROS is a mediator of cochlear cell damage, in addition biochemically ROS causes peroxidation of cochlear lipids and produces toxic substances. ROS itself was reported to persist cochlear for 7 - 10 days by post-exposure to the noise. If individuals are exposed to noise every day it will be very risky to experience noise-induced hearing loss and other hearing loss [3].

Enforcement of a diagnosis of hearing loss requires a tool for strengthening the diagnosis including using audiometry and optoacoustic emission (OAE) [4]. Audiometry is used to measure the subjective

IOP Conf. Series: Journal of Physics: Conf. Series **1075** (2018) 012064 doi:10.1088/1742-6596/1075/1/012064

hearing impairment of the patient while OAE is used to evaluate the patient's hearing loss objectively [4] [5]. Thus, both examinations are important as the basis for effective therapy.

Indonesia is a country that has a large area and has various tribes [6]. Characteristics of workers in Indonesia is very unique, often found its workers tend not to comply for in the use of Personal Protective Equipment. Government regulations related to occupational health especially noise thresholds in the workplace [7] [8]. Hospitals are one of the places to work and where to seek medical help, but some places in the hospital are a source of noise pollution. Dr. Soetomo General Hospital, Surabaya Indonesia is one of the places of work that has several noisy locations such as medical facility maintenance instantiation, environmental sanitation installation, and nutrient installation. Based on the description above it need to be performed the correlation analysis of ROS content with audiometric and OAE measurements and at employees at risk in hospital.

## 2. Materials and methods

The subjects of this study were employees of Dr. Soetomo General Hospital, Surabaya, Indonesia. The subjects must meet inclusion criteria and exclusion criteria. The number of employees was 3,813 people and after being identified based on inclusion and exclusion criteria it got 56 subjects. Inclusion criteria include employees at risk of exposure to noise pollution with noise <85 dB, in hospitals (medical facility maintenance instantiation, environmental sanitation installation, and nutrient installation) and willing to participate the research by filling out informed consent sheets. Exclusion criteria include having ear abnormalities, results of type B tympanometric screening, and moderate or ever taking muscle-drugs drugs (figure 1).

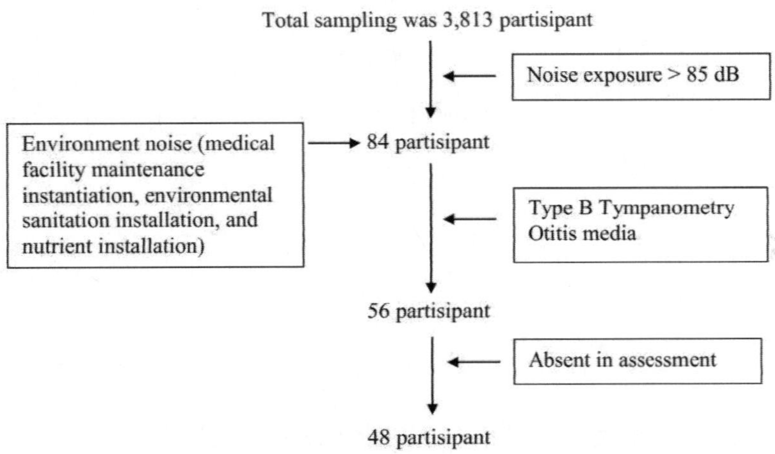


**Figure 1.** Sampling of participants.

The study used observational analytic study. Then, prior to conducting the data retrieval the researchers first conducted a test of ethics with the ethics committee of Dr. Soetomo General Hospital, Surabaya Indonesia (219/Panke.KKE/III/2017). Data collection was done in some potential locations to be a source of noise in Dr. Soetomo General Hospital Surabaya Indonesia. Some area was value noise pollution levels. Its worker was do collection data (baseline characteristics, level of plasma ROS, audiometry, and OAE).

Baseline characteristics involved sex, age, education, environment noise, and hearing exposure. To determine the plasma ROS levels, we sample 5 ml fasting venous blood, then isolated its by room of temperature 10-15$^0$ C. The blood was analyzed using Enzyme Linked Immune Sorbent Assay (ELISA) reader (Bio-Rad laboratories, Inc., Hercules, USA) and used Ferric Reducing Ability of Plasma (FRAP) reagent kit. Furthermore, a pure tone audiometry (GSI Arrow, Eden Prairie MN, USA).

Assessment AOE used distortion product otoacoustic emissions (AuDX Biologic, Pleasanton CA, USA). If employers were an absent in assessment schedule, so they were drop out in research.

The research data were presented as mean ± standard error and frequency. All data of inspection result was processed using SPSS version 23.0 (SPSS, Inc., Chicago, IL). Statistical evaluation of plasma ROS levels and OAE were using chi-square test ($p < 0.05$). Plasma ROS levels associated with audiometry used chi-square test ($p < 0.05$).

## 3. Result

### *3.1. Demographic data*

Demographic data was divided into two base on frequency and average data. Frequency data were sex, education, and work area in dr. Soetomo General Hospital. Average data were age, noise exposure, and duration of work. Most of the respondents were female (58.33%), education of senior high school (81.25%), and work in nutrition room (68.75%). Frequency data in this study can be seen detail in table 1. The age of participant were 42.43 ± 10.72 years old and range 22 to 58 years old. The some area in dr. Soetomo General Hospital have average of noise exposure (>85 dB) of 98.58 ± 4.54 from range 85.39 to 112.90 dB. They had duration of work in its of 19.54 ± 9.56 years and range 1 to 36 years (table 2).

**Table 1.** Frequency of domegraphic participant.

| Variables | | % |
|---|---|---|
| Sex | Male | 41.67 |
| | Female | 58.33 |
| Education | Elementary School | 2.08 |
| | Junior High School | 0.00 |
| | Senior High School | 81.25 |
| | Bachelor Degree | 16.67 |
| Noise | Sanitation | 14.58 |
| Pollution | Maintenance medic | 16.67 |
| Areas | Nutrition | 68.75 |

Value of ROS was 5.55 ± 4.38 ng/dl with range 3.99 to 37.60 ng/dl. Value of OAE was mostly in category refer (70.83%) and mostly audiometry was category normal (52.08%). Detail about of data can be seen in table 3. So, based on of the study was most participant have noise inducted hearing loss.

**Table 2.** Average of respondents characteristic.

| Variables | Mean ± SD | *p* |
|---|---|---|
| UAge | 42.43 ± 10.72 | 0.023 |
| KNoise | 98.58 ± 4.54 | 0.167 |
| Duration of work | 19.54 ± 9.56 | 0.042 |

### *3.2. ROS associated with hearing loss*

The participant was mostly in OAE category of refer and ROS category of low (54.17%). They was not mostly in audiology category, difference on OAE category. They was most audiology category of normal and ROS category of low (35.42). The result of the statistical test with chi square was found no correlation between plasma ROS level and audiometry value ($p = 0.993$). The result of the statistical test with chi square was found no significant correlation between serum ROS level and audiometry ($p = 0.647$). The result of analyzed was shown in table 3.

IOP Conf. Series: Journal of Physics: Conf. Series **1075** (2018) 012064 doi:10.1088/1742-6596/1075/1/012064

**Table 3.** ROS associated with hearing loss.

| characteristic | Category | ROS | | *p* |
|---|---|---|---|---|
| | | Low | High | |
| OAE | Refer | 54.17 | 16.67 | 0.993 |
| | Pass | 14.58 | 14.58 | |
| audiometry | hearing loss | 25.00 | 22.92 | 0.647 |
| | normal | 35.42 | 16.66 | |

## 4. Discussion

NIHL is a hearing loss that experienced by many workers. Noisy exposure over 85 dB for 8 hours is believed to interfere auditory organ function [9]. Some recent studies suggest that NIHL hearing loss is due to an increase in the production of toxic ROCs of cochlea [3] [10]. In vivo studies also show that ROS causes mutation of the apoptotic-inducing factor (AIF) gene. AIF regulation reduction in neuronal cells makes it more sensitive and easily damaged [11].

ROS is considered to be a toxic cellular metabolic product but it also serves as a molecule that regulates many physiological processes. ROS plays an important role in the induction of apoptosis under physiological and pathological conditions. Increased ROS formation and subsequent apoptotic induction have been implicated in the development of some hearing loss pathologies. Furthermore, mitochondrial dysfunction plays an important role in some types of hearing loss [10].

Based on this research there was no correlation between ROS and hearing loss. It was proved by no correlation between ROS content with audiometry and OAE. The audiometric examination was a subjective examination of the patient about the individual's hearing threshold [12]. Meanwhile, OAE was an examination of the inner ear objectively, especially the function of outer cochlear cell [5]. Both examinations were often used as indicators for early detection of hearing loss [4] [13].

Currently, ROS was alleged to be one of the main causes of impaired hearing function. ROS has characteristics as an unpaired electron, activating toxic chemical reactions to cellular and subcellular structures [14]. ROS levels in the cochlea cause decreased hearing function by damaging the cochlear hair cells [15]. However, some research suggests the levels of ROS in cochlea have increased for the first time that exposed to the noise and decreased subsequent exposure [3] [10]. The progression of ROS to a decrease in hearing function takes a long time. The condition was supported by several studies that suggest an age increase to stimulate the number of ROS in the cochlea that causes hearing loss [11].

The main sources of ROS production within the cochlea appear to be the hair cells' mitochondria, or enzymes such as xanthine oxidase and NADPH oxidase. Once generated, ROS are responsible for direct cellular damage to lipids, proteins, and DNA, triggering apoptosis or necrosis and may also diffuse among the inner ear scale [16]. ROS was produced by mitochondria, cytochrome p450 metabolism, microsomes, nitric oxide synthase and other inflammation-related processes. The overproduction of free radicals is highly toxic and triggers a complex mechanism of damage due to the peroxidation of membrane lipids, denaturation of cellular proteins, damage of DNA and cell death in several pathological conditions such as mutagenesis, carcinogenesis, aging, neurodegenerative and inflammatory diseases [15] [17].

There is increasing evidence that ROS-mediated damage in the cochlear may be a common factor for hearing loss caused by many factor. Based on the results obtained no significant of plasma ROS level on hearing loss. It is needed to find specific reseach influence ROS related by NIHL.

## 5. Conclusion

The number of workers who experience hearing loss in some of the noisy environment in the hospital was quite a lot. ROS production was influenced by many factors, one of age and noise exposure. Based on evidence was an increased production of ROS causing of hearing loss, but no correlation of reactive oxygen species with hearing impairment in high risk employees at the Hospital. ROS levels in blood not specific toward hearing loss but research in vivo and evidence was significant correlation of ROS and hearing impairment.

## 6. References

[1] Robinson T, Whittaker J, Acharya A, Singh D and Smith M 2015 Prevalence of noise-induced hearing loss among woodworkers in Nepal: a pilot study *Int J Occup Environ Health* **21** pp 14-22

[2] Musiba Z 2015 The prevalence of noise-induced hearing loss among Tanzanian miners. *Occup Med (Lond)* **65** pp 386-90

[3] Kurabi A, Keithley E M, Housley G D, Ryan A F and Wong A C 2017 Cellular mechanisms of noise-induced hearing loss *Hear Res* **349** pp 129-137

[4] Sulaiman A H, Husain R and Seluakumaran K 2014 Evaluation of early hearing damage in personal listening device users using extended high-frequency audiometry and otoacoustic emissions. *Eur Arch Otorhinolaryngol* **271** pp 1463-70

[5] Mokrian H, Shaibanizadeh A, Farahani S, Jalaie S, Mahdi P, Amali A and Arian N H 2014 Evaluation of distortion and transient evoked otoacoustic emission in tinnitus patients with normal hearing *Iran J Otorhinolaryngol* **26** pp 19-24

[6] Efendi F 2012 Health worker recruitment and deployment in remote areas of Indonesia *Rural Remote Health* **12**

[7] Peraturan Menteri Kesehatan Republik Indonesia Nomor 70 Tahun 2016 tentang Standar dan Persyaratan Kesehatan Lingkungan Kerja Industri

[8] Peraturan Menteri Tenaga Kerja dan Transmigrasi Nomor Per.13/Men/X/2011 Tahun 2011 Tentang Nilai Ambang Batas Faktor Fisika dan Faktor Kimia di Tempat Kerja

[9] Yost W A 2013 *Fundamentals of hearing: an introduction* (Lieden; Boston, Brill)

[10] Kamogashira T, Fujimoto C and Yamasoba T 2015 Reactive oxygen species, apoptosis, and mitochondrial dysfunction in hearing loss *Biomed Res Int* **617207**

[11] Seidman M D, Ahmad N, Joshi D, Seidman J, Thawani S and Quirk W S 2004 Age-related hearing loss and its association with reactive oxygen species and mitochondrial DNA damage *Acta Otolaryngol Suppl.* pp 16-24

[12] Govender S M and Mars M 2018 Validity of automated threshold audiometry in school aged children *Int J Pediatr Otorhinolaryngol* **105 pp** 97-102

[13] Keppler H, Degeest S and Dhooge I 2017 The relationship between tinnitus pitch and parameters of audiometry and distortion product 5ptoacoustic emissions *J Laryngol Otol* **131 pp** 1017-1025

[14] Hwang J H, Chen J C, Hsu C J, Yang W S and Liu T C 2012 Plasma reactive oxygen species levels are correlated with severity of age-related hearing impairment in humans *Neurobiol Aging* **33** 1920-6

[15] Henderson D, Bielefeld E C, Harris K C and Hu B H 2006 The role of oxidative stress in noise-induced hearing loss *Ear Hear* **27 pp** 1-19

[16] Ciorba A, Gasparini P, Chicca M, Pinamonti S and Martini A 2010. Reactive oxygen species in human inner ear perilymph *Acta Oto-Laryngologica* **130 pp** 240-246

[17] Falasca V, Greco A, Ralli M 2017 Noise induced hearing loss: The role of oxidative stress *Otolaryngol Open J*

**Acknowledgments**

This project was supported by a research grant form Dr. Soetomo General Hospital Surabaya, Indonesia.

IOP Conf. Series: Journal of Physics: Conf. Series **1075** (2018) 012065 doi:10.1088/1742-6596/1075/1/012065

# Low-frequency impact sound pressure level characteristics with three standard impact sources in Korean reinforced concrete structured apartment buildings

**J H Jeong**[1]

[1]1030 Gyeongghung-daero, Ganam-eup, Yeoju-si, Gyeonggi-do, Korea 12661

Email: jhjeong92@gmail.com

**Abstract**. Floor impact sound is one of the major low frequency sound which are generated in residential buildings. In ISO standard, rubber ball standardized as heavy/soft impact source and modified tapping machine was proposed to simulate frequency characteristics of real impact sound. Rubber ball is 2.5 kg silicone rubber hollow ball and free drop from 1 m height. However, low frequency impact sound pressure level of tapping machine in a residential buildings with high floor impact isolation performance is relatively lower than real impact sound. It is need to be checked that measurement of floor impact sound isolation performance in residential buildings with resilient materials using tapping machine is suitable method. In this study, floor impact sound isolation performances of reinforced concrete structured residential buildings with resilient buildings were measured using tapping machine, bang machine and rubber ball. The low frequency measurement results were compared. It was found from measurement results that tapping machine impact sound pressure level below 100 Hz bands were lower than rubber ball and bang machine impact sound in Korean residential building where 20 mm thick resilient material was installed. Single number quantity of tapping machine impact sound pressure level in Korean residential building was 50 dB in $L'_{n,w}$. From these results, it can be concluded that measurement of low frequency impact sound isolation performance using tapping machine in reinforced concrete structured residential buildings with resilient material is not suitable.

## 1. Introduction

Floor impact sound is one of the major low frequency sound which are generated in residential buildings. In Korea, a lot of disputes and complains on low-frequency impact sound have been occurring. In order to measure and evaluation low-frequency impact sound isolation performance, bang machine and rubber ball have been used as standard impact sources are used in Korea and Japan [1, 2]. Rubber ball and bang machine was developed to simulate children's running and jumping in residential buildings. It was reported that physical and subjective characteristics of rubber ball was similar with children's running and jumping in Korean reinforced concrete structured multi story residential buildings[3, 4]. Rubber ball was standardized in ISO from 2010[5, 6, 7] as the only heavy/soft impact source. Rubber ball is 2.5 kg silicone rubber hollow ball. The detailed chemical composition and physical properties were standardized in ISO 10140-5 and ISO 16283-2. In order to generate low frequency impact sound, the operator have to make a free drop from 1 m height.

IOP Conf. Series: Journal of Physics: Conf. Series **1075** (2018) 012065 doi:10.1088/1742-6596/1075/1/012065

Rubber ball and bang machine directly impact floor system on the upper floor and measure transmitted impact sound in lower unit of the residential buildings. Impact force and impact sound pressure level of rubber ball and bang machine is usually concentrated in low-frequency band below 100 Hz.
Tapping machine have been widely used in ISO standard. Tapping machine simulate high heel drop and dropping hard material on the floor and usually generate mid and high frequency impact sound. In order to change frequency characteristics of tapping machine impact sound, resilient material or pad was used. By using a resilient material or pad, mid and high frequency tapping machine impact sound reduced and relatively high level of low frequency impact sound obtained; this method is called modified tapping machine. In ISO 10140-5 modified tapping machine was proposed.
Floor impact isolation performance of wooden structured residential buildings in low-frequency band is relatively low, therefore modified tapping machine can generate enough impact sound pressure level in low frequency band. However, the floor impact isolation performance of reinforced concrete structured residential building with resilient material is better than wooden structured buildings. It is need to be checked that measurement of floor impact sound isolation performance in residential buildings with resilient materials using tapping machine is suitable method.
In this study, floor impact sound isolation performances of reinforced concrete structured residential buildings with resilient buildings were measured using tapping machine, bang machine and rubber ball. The low frequency measurement results were compared.

## 2. Measurement of floor impact sound pressure level

Floor impact isolation performance of reinforced concrete residential building in Korea was measured according to KS F 2810-1 and 2, using fixed microphone method. Three kinds of standard impact sources were used. One is tapping machine which was standardized in ISO 10140-5[2] and KS F 2810-1[8], another is bang machine which regulated in KS F 2810-2[6] and the other is rubber ball which is standardized in ISO 10140-5[2] in 2010 as a heavy/soft impact source. Five positions on upper floor were impacted and impact sound five positions on lower floor including center position. Measurement setup was shown in Figure 2 (a).

(a) Tapping machine (b) Bang machine (c) Rubber ball

**Figure 1.** Three kinds of standard impact sources

Floor impact sound isolation performance measurements were conducted in 40 typical Korean multi story residential units. Floating floor system to isolate transmission of floor impact energy into lower unit, shall be used all of the Korean residential buildings. Figure 2 (b) shows floor system of typical Korean reinforced concrete structured residential buildings. The thickness of reinforced concrete slabs was 210 mm.
On the reinforced concrete slab, 20 mm or 30 mm thick resilient material was installed to reduce floor impact sound. Then aerated light-weight concrete was installed thermal insulation. Floor heating system and cement mortar usually installed on the aerated light-weight concrete. Finally, finishing material such as wood floor or linoleum installed on the surface of cement mortar. Floor impact sound pressure level was measured in the living room of each units. The area of the living rooms was from 19.5 $m^2$ to 29.3 $m^2$. Frequency range of tapping machine impact sound was from 50 Hz to 3 150 Hz in 1/3 octave band.

IOP Conf. Series: Journal of Physics: Conf. Series **1075** (2018) 012065 doi:10.1088/1742-6596/1075/1/012065

Bang machine and rubber ball impact was measured from 25 Hz to 630 Hz.

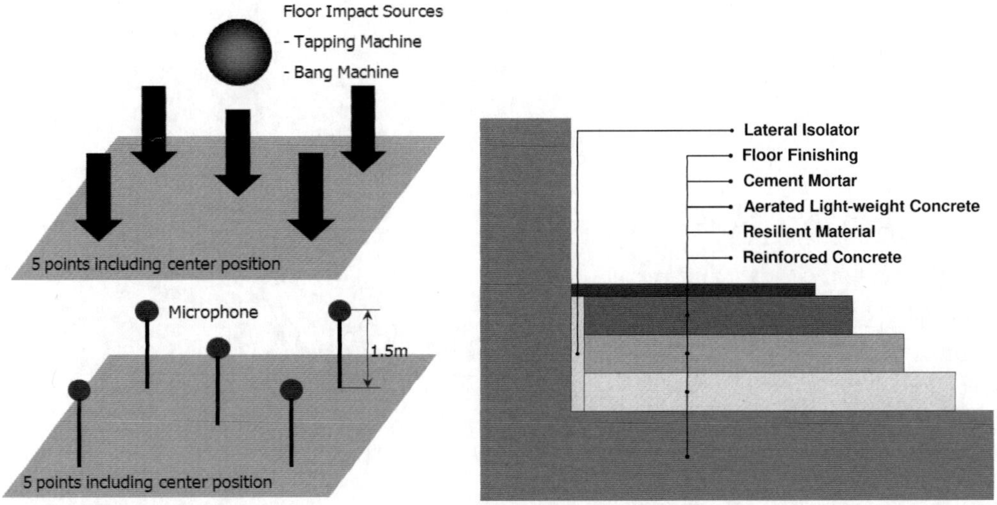


(a) Measurement setup (b) Floor heating system with resilient material

**Figure 2.** Floor impact sound pressure level measurement setup and floor system of typical Korean reinforced concrete residential buildings

## 3. Results of floor impact sound pressure level measurements

Measurement results of impact sound pressure level from 40 reinforced concrete residential buildings using three kinds of standard impact sources were shown in Figure 3. Figure 3 (a) shows measurement results of tapping machine impact sound. Tapping machine impact sound shows relatively low impact sound pressure level above 200 Hz bands. Mid and high frequency tapping machine impact was reduced my resilient material. Below 100 Hz bands, tapping machine impact sound pressure level decreased.

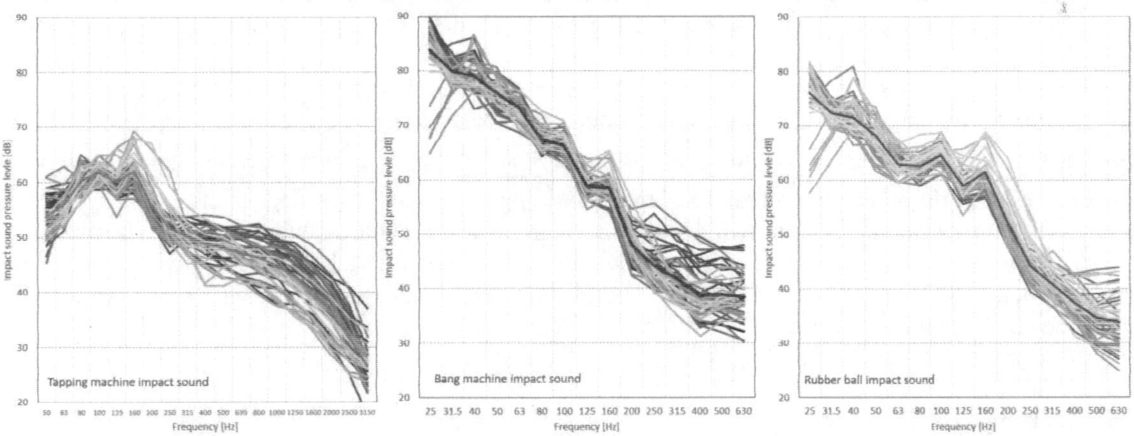


(a) Tapping machine (b) Bang machine (c) Rubber ball

**Figure 3.** Measurement results of three kinds of standard impact sources in Korean reinforced concrete structured residential buildings

Figure 3 (b) shows measured results of bang machine impact sound. Rubber ball results was shown in Figure 3 (c). Bang machine and rubber ball impact have high relative high impact sound pressure level in low frequency bands. Both of them have similar frequency characteristics. However, bang machine impact sound pressure level is higher than rubber ball impact sound in the frequency range below 100 Hz bands. Because, impact force of the bang machine is about three times higher than impact force of rubber ball impact.

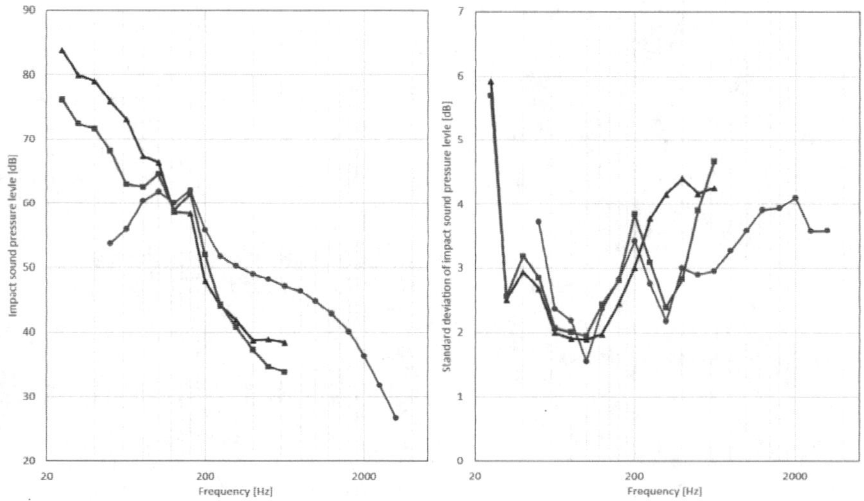


(a) Averaged impact sound pressure level (b) Standard deviation

**Figure 4.** Averaged impact sound pressure levels and standard deviations of three kinds of standard impact sources

Figure 4 (a) shows average frequency spectra of three kinds of impact sounds. Above 125 Hz bands, impact sound pressure level of tapping machine was higher than bang machine and rubber ball impact sound. However, below 100 Hz bands bang machine and rubber ball impact sound pressure levels were higher than tapping machine impact sound. In the case 50 Hz band, rubber ball impact sound pressure level was 14.3 dB higher than tapping machine impact sound. Even 63 Hz band, rubber ball impact sound was higher than 6.9 dB comparing with tapping machine impact sound. When comparing bang machine impact sound with tapping machine impact sound in 50 Hz and 63 Hz bands, bang machine impact sound pressure levels were higher 22.1 dB and 17 dB than tapping machine impact sound respectively.

Single number quantity of averaged tapping machine impact sound pressure level; $L'_{n,W}$ ; was 50 dB. It was reported that average impact sound isolation performance of European dwellings was Class D; $L'_{n,W}$ ≤56 dB[9]. When classification scheme from EU COST action applied on the tapping machine impact sound isolation performance of Korean residential buildings, it can be classified as Class C.

Figure 4 (b) shows standard deviation of 40 measurement results. Standard deviation of tapping machine impact sound above 100 Hz bands were smaller than bang machine and rubber ball impact sound. However, below 80 Hz bands, tapping impact sound show the highest standard deviation values among three kinds of floor impact sound.

## 4. Conclusions

Low frequency impact sound is one of the major noise sources in residential buildings. In order to reduce and control the low frequency impact sound, proper measurement and evaluation method is needed. In ISO standard, rubber ball standardized as heavy/soft impact source and modified tapping machine was proposed to simulate frequency characteristics of real impact sound. However, low frequency impact sound pressure level of tapping machine in a residential buildings with high floor impact isolation performance is relatively lower than real impact sound.

In this study floor impact sound isolation performance of 40 Korean reinforced concrete structured residential buildings with resilient material was measured using three kinds of standard impact sources for the comparison of low frequency impact sound pressure level characteristics. It was found from measurement results that tapping machine impact sound pressure level below 100 Hz bands were lower than rubber ball and bang machine impact sound in Korean residential building where 20 mm thick resilient material was installed. Single number quantity of tapping machine impact sound pressure level in Korean residential building was 50 dB in $L'_{n,w}$.
From these results, it can be concluded that measurement of low frequency impact sound isolation performance using tapping machine in reinforced concrete structured residential buildings with resilient material is not suitable.

## References

[1] ISO 10140-3 Acoustics - Laboratory measurement of sound insulation of building elements - Part 3: Measurement of impact sound insulation.
[2] ISO 10140-5 Acoustics - Laboratory measurement of sound insulation of building elements - Part 5: Requirements for test facilities and equipment.
[3] J. Y. Jeon, J. H. Jeong, M. Voländer and R. Thaden: Evaluation of Floor Impact Sound Insulation in Reinforced Concrete Buildings. Acustica 90 (2004) 313-318.
[4] J. Y. Jeon, J. H. Jeong and Y. Ando: Objective and subjective evaluation of floor impact noise. Journal of Temporal Design in Architectural and the Environment, 2(1), pp.20-28.
[5] ISO 16283-2 Acoustics - Field measurement of sound insulation in buildings and of building elements - Part 2: Impact sound insulation.
[6] KS F 2810-2, Field measurement of floor impact sound insulation of buildings – Part 2: Method using standard heavy impact sources.
[7] JIS A 1418-2, Acoustics – Measurement of floor impact sound insulation of buildings – Part 2: Method using standard heavy impact sources.
[8] KS F 2810-1, Field measurement of floor impact sound insulation of buildings – Part 1: Method using standard light impact sources.
[9] COST Action YU0901 : Integration and Harminizing Sound Insulation Aspects in Sustaibable Urban Housing Constructions, (2014).

# The modified whispered test for screening of hearing impairment in children at the elementary school

**N Purnami[1]**

[1]Department of Otorhinolaryngology Head and Neck Surgery
Faculty of Medicine, Universitas Airlangga-Dr. Soetomo General Hospital
Surabaya 60231 Indonesia

E-mail: nyilo@fk.unair.ac.id

**Abstract.** Hearing impairment in children could be identified by several methods of hearing test. The conventional whispered test is designed with the distance of 6 meters measurement. There is a need to get smaller examination room, then there is a method to modify the distance. this method is Modified Whispered Test, that is quite practical, simple and can be performed in smaller room and for the large groups. The study was designed to know whether the modified whispered test was sensitive or not to screen the hearing impairment among the elementary school students. This study will improve the program of early detection of the hearing impairment among the elementary school students in Surabaya Indonesia. A cross sectional study of 100 elementary school students has been conducted at Sekolah Dasar Negeri Kali Rungkut I, II and III Surabaya. The study was analyzed the validity of whispered test compare to the gold standard, Audiometry. The population-based sample of 100 children (aged 6 to 13 years), was taken randomly and proportionally from the 1st to 6th grade students. It was concluded that the modified whispered test applicable for screening of hearing impairment among elementary school children, with 80% sensitivity, 95,2% specificity and 0,752 Youden Index on 41-55 dB degree of hearing loss (Moderate Hearing Loss). Further Studies of hearing impaired children are necessary to plan preventive and curative program.

## 1. Introduction

Hearing impairment of children could be identified by several methods of hearing test [1]. Whispered test was one of those tests that was quite practical, simple and available for large groups [2] [3]. The modified Whispered test has been used to reduce the distance between the physician and the patient of the conventional method. Each of these manoeuvres, the physician is positioned face to face to the patient, then turning the head on the opposite site to the patient's tested ear, as well as the patient's head on the contrary sides. It may have resulted reducing the distance of about one-third from the conventional method which is 6 meters distance [4]. The modified Whispered test has been used as screening method of hearing impairment among candidate students and employees of Dr. Soetomo general hospital in Surabaya for years [5]. So far there was no experience to examine the students on the elementary school, therefore this study would be applicable to children.

The problem in this study to find the validity, whether the modified Whispered test sensitive and specific results in finding children with hearing loss. The benefits to be achieved from this research was to get the method of modified whispered test as one of the hearing screening methods in school children. This study would find the advantages and disadvantages in the implementation and also for the reference in preparing the children to perform the hearing impairment detection program.

## 2. Theory

### *2.1. Hearing mechanism*

The ear acts as a hearing organ, in which the incoming sound waves are captured by the ear and transmitted to the cerebral cortex, so that the sound sensation can be heard. This mechanism occurs through two stages, conduction and perception. At this stage sound waves through the outer ear canal, the tympanic membrane, the hearing bone chain and ends in the foramen ovale.

At this stage the sound waves received by the foramen ovale will be continued by the perilymph on the vestibule scale. Through these endolymph sound waves will move the Reisner membrane and basal membrane on a media scale, then it will vibrate the hair cells in the Corti organon and result in a change of electrical potential. The potential of electricity was channelled as a stimulus of hearing in the cerebral cortex [6].

### *2.2. Hearing screening*

Screening is a method to identify early hearing loss and a part of program planning to address it. So, this activity is not merely that stands alone, but more a component of a wider strategy. Screening is very useful in national program planning for early detection, intervention and rehabilitation of hearing loss and is a potential step undertaken in the primary health care strategy [7].

### *2.3. Hearing examination*

A frequent and routine hearing examination with existing facilities and infrastructure will improve the skill of the examiner. The following are described various types of hearing tests that are often performed, and the results of such examinations may be complementary for the proper diagnosis [8]. The hearing test that are commonly performed as Tuning Fork Test, Speech Test which consists of conversation test and whispered test, Pure Tone Audiometry, Speech Audiometry, Tympanometry.

### *2.4. Hearing screening with a modified whispered test*

The modified whispered test was as follows [9]: 1) the child sits face to face to the examiner, the examiner's head turning to the opposite of the tested ear , 2) the finger index of the examiner push on the tragus of the untested ear, then it was continues to move to make a sound as a masking, 3) whispered words was tested by turning the examiner head to the untested ear, 4) when the test done with the right side of tested ear, then change the direction with the other side of ear, 5) the whispered words were taken from the Phonetical Balanced list (PB list), whispered 5 to10 words, then the student was told to imitate loudly when it had been heard. 6) if the student can correctly repeat of 80% the whispered words, then the ears were considered as good or pass and expressed similarly with the conventional whisper test of 6 meters. On the contrary the result was considered fail when the student could not imitate 80% of the words whispered.

### *2.5. Hearing test*

Hearing examination at elementary school age was an important step for catching hearing disorders within the school environment. Test screening ever done the sound test sound 3 m distance from the ear by repeating ten words whispered correctly. This test has a sensitivity of 100% and a diagnostic specificity of 74.4%. Children in the age group of 6-13 years (elementary school age) are considered to have reliable hearing examination results, either by audiometry or a whispering test. At the age ranging from 6 years, the child has passed a critical period of speech and language speech so that it was considered quite skilled in undergoing the auditory examination. This factor was important to be considered in the examination of the whispered test, because the child must be able to capture the spoken word and repeat what it hears. Generally, children older than 5 years of normal intelligence may follow a standard pure audiometric examination intended for adults [10].

Speech was an oral or verbal language that consists of a series of words or sentences. The word consists of syllables and finally the smallest sound unit that can distinguish the meaning was called phoneme. Seen from the lexical aspect of the smallest idea unit called morpheme. In Indonesia there were two forms of morphemes. The bound morpheme such as prefix, insertion and suffix. The free

IOP Conf. Series: Journal of Physics: Conf. Series **1075** (2018) 012066 doi:10.1088/1742-6596/1075/1/012066

and independent morpheme was called the word. When viewed from the physical aspect, speech was a form of sound that consists of complex vibrations that always change both the tone, the tempo and the volume of the voice. The smallest sound unit of speech or phoneme can be divided into two types: vowel sound and consonant sound. Each phoneme has a different frequency with each other. This was important for the recognition of individual sounds. Consonant frequencies were generally higher than vowels, but consonants were more sensitive to distortion than vowels. In general, perceptual deafness was damaged in high tones, most of which are consonants.

## 3. Methods

### *3.1. Type of research*

The type of research was using comparison of Cross Sectional Study.

### *3.2. Procedure*

The procedure for this research was:

*3.2.1. Preparation of examination room.*

*3.2.2. Finding the clean and quite class room that located at the Elementary School of Kali Rungkut I.*

*3.2.3. Noise Measurement.* Noise was measured with a calibrated sound level meter, 1/2 meter from a 1.2-meter-high window from the ground. Battery in good condition. Measurements on the "A" frequency weights and "Fast" time weights.

*3.2.4. Measurement of the intensity of the auditor's whisper.* When the examiner whisper voice, and was listened to by the child's ears, the intensity was measured by a sound meter level device placed next to the child's ear with the direction of the microphone facing in accordance with the child's ears.

*3.2.5. Sampling was selected randomly to determine the sample of study participants was taken tests, the modified Whispered test and audiometry.* The examiner of the whisper test was 1 qualified person from audiology clinic. This test was conducted in the selected examination room located at SDN Kali Rungkut I. While the audiometric examination was done in the Audiology clinic room of RSUD Dr. Soetomo, with examiners from qualified audiometrician other than examiners of whisper tests. Routine physical examination of the ears, nose and throat was also performed on all students performed by the researcher.

*3.2.6. Performing the modified whispered test.* Examiners sit face to face with the participants in the examination room. When the right ear was examined, the examiner looked away right. the index finger of the examiner closed the left ear canal by depressing the tragus and moving repeatedly to create an ear-masking sound, thus the sound to be heard only from the right ear. Sound whispered out with air spare in the lungs after ordinary expiration. Whispered 10 words from PB list of available words. The child was asked to repeat what has been heard. When spoken correctly 80% of the words above, was considered normal hearing and passed. If not, then it was recorded as abnormal or failed which means not pass in the test.

*3.2.7. Audiometric Test.* It was in the audiology room of Dr. Soetomo. The examiner was an audiologist. The examination room used was soundproof. The tool used was an electroacoustic device that can produce pure tunes with measurable frequency and intensity. The sound was received by the child through headphones placed over the outer ear canal for air delivery and through the vibrator for bone conduction. The audiometer used was a type of "Madsen".

*3.2.8. Pure Tone Audiometry Test.* It will be given an instruction to the children that he would hear a sound in one ear for each examination. When the children hear the sound, noted with raising the hand fingers according to which ear hears. The right ear heard with the right hand, and vice versa. After the sound was not heard again immediately hand down. The right ear was checked first then left, except when the left ear was better hearing from the right, then the left ear was checked first. Audiometry was performed first checks start at 1000 Hz frequency. Then, checked it again for 2000, 3000, 4000, 6000, 8000 Hz. After that check again 1000 Hz tone followed by the examination of low tones: 500, 250 and 125 Hz. Weber tests on all frequencies with vibrators placed on the median line. Conducted hearing threshold examination through bone delivery from the frequency of 250 Hz continue to raise up to 8000 Hz. Result from the examination was determined the value of the child's hearing threshold.

## 4. Data analysis

The study was designed to know whether the modified Whispered test was sensitive or not to screen the hearing impairment among children in the elementary school students. This study will improve the program of early detection of the hearing impairment among the elementary school students. It was expected that the modified Whispered test would be sensitive in screening the hearing impairment.

A cross sectional study of 100 elementary school students has been conducted at SDN Kali Rungkut I, II and III Surabaya, from October to November 1997. The study would compare the first group (which was tested by modified whispered voice test) and the second group (which was tested by audiogram as the gold Standard). The sensitivity, the specificity and Youden Index were counted. The Youden Index should be closer to 1.

The population-based sample of 100 children (aged 6 to 13 years), was taken randomly and proportionally from the $1^{st}$ to $6^{th}$ grade students. The ratio between male and female was 1:1. The study revealed the following results. There were 86 parents (86%) without any complain about their children and 14 parents (14%) complained that their children had hearing problem. Otoscope examinations, it was detected 130 normal ears (65%), 64 ears (32%) with retracted tympanic membrane, 3 ears (1,5%) with perforated tympanic membrane and 3 ears (1,5%) with cicatrix. Ear, Nose and Throat examination revealed 36 normal children (36%) and 54 children (54%) with several diseases. The common disorders were Chronic adeno tonsilitis in 25 children (25%) and Chronic pharyngitis in 23 children (23%). The noise rate of school ranged from 50 dB to 80 dB and 45-55 dB of the examination room. There were 185 ears (92,5%) who passed the screening modified Whispered test (could retell 80% or more of the whispered words) and 15 ears (7,5%) who failed. From the pure tone audiometry of which the frequency ranged from 250 to 8000 Hz were found 166 normal ears (83%), 29 ears (14,5%) with mild hearing loss (26-40 dB) and 5 ears (2,5%) with moderate hearing loss. From the audiogram in 3 frequencies (500 – 1000 - 2000 Hz) were found 170 normal ears (85%), 25 ears (12,5%) with mild hearing loss (26-40 dB) and 5 ears (2,5%) with moderate hearing loss. The modified whispered test of mild hearing loss had 10,3 % of sensitivity, 95,2 % of specificity and 0,055 of Youden Index. For moderate hearing loss, it had 80 % of sensitivity, 95,2% of specificity and 0,752 of Youden Index. The mean hearing level of 166 normal ears (83%) had air conduction ranged from 14,8 to 20,5 dB and unnormal ears were 26,1 to 33,5 dB. There were 8 ears (4%) of sensorineural hearing loss with 30-50 dB hearing loss at frequency of 4-6 kHz. Air conduction hearing level of normal otoscope ears ranged from 16 dB to 21,2 dB and unnormal ears ranged from 18,1 dB to 25 dB.

## 5. Conclusion

It was concluded that the modified Whispered test applicable for screening of hearing impairment among elementary school children, with 80% sensitivity, 95,2% specificity and 0,752 Youden Index on 41-55 dB degree of hearing loss.

Further Studies may suggest planning preventive and curative program, based on the results of the prevalent hearing impaired among the children.

## 6. References

[1] C. B. Me, Screening for Hearing Impairment in Young Children, London: Crom Helm Ltd., 1988.

[2] S. S. Djoko, S. and A. H. Soetarno, "Pembakuan tes bisik kata bisilabik Bahasa Indonesia untuk

"screening" pendengaran anak sekolah dasar," in *Kongres Nasional VII*, Surabaya, 1983.

[3] R. Sedjawidada and A. Manukbua, "Tes Bisik," in *Kongres Nasional V*, Semarang, 1977.

[4] S. B. Hulley and S. R. Cummings, Designing clinical research an epidemiologic research, Baltimore: Williams & Wilkins, 1988.

[5] M. S. Wiyadi and H. Amatpoero, "Pemeriksaan Pendengaran Karyawan RSUD Dr. Soetomo," in *Kongres Nasional XI*, Yogyakarta, 1995.

[6] G. E. Shambough and M. E. Glascock, Surgery of the ear 3rd edition, Philadelphia: WB Saunders Co., 1980.

[7] P. A. Braveman and E. Tarimo, Screening in Primary Health Care, Geneva: WHO, 1994.

[8] G. L. Adams, L. R. Boles and P. A. Hilger, Boies fundamentals of otolaryngology 6th edition, Philadelphia: WB Saunders Co, 1989.

[9] M. S. Wiyadi, Tes Pendengaran Suara Bisik, Media Perhati, 1997.

[10] J. L. Northern and M. P. Downs, Hearing in Children 4th edition, Baltimore: Williams & Wilkins, 1991.

# Noise exposure in vocational school

**V Melyani[1] and N Purnami[1]**

[1]Department of Otorhinolaryngology Head and Neck Surgery
Faculty of Medicine, Universitas Airlangga-Dr. Soetomo General Hospital
Surabaya 60231 Indonesia

E-mail: nyilo@fk.unair.ac.id

**Abstract.** Hearing Conservation Program is a series of activities designed to prevent hearing loss in Vocational High School students. The objectives of the research were to know the difference between 10th grade and 12th grade student's questionnaires result which includes the presence of hearing loss, personal protective equipment, the student's knowledge of hearing loss, and both ears Pure Tone Average (PTA) value of the 10th grade and 12th grade students that exposed to noise. The research was analytic observational by assessing the result of questionnaire and audiometric examination using cross sectional study. The questionnaires charging, audiometric examination, and the measurement of equipment's noise with Sound Level Meter (SLM) have been conducted for 3 months. The samples were 74 students taken from each 10th grade and 12th grade classes at light vehicle engineering majors. The data were processed, analyzed statistically. Assessment of questionnaire results, there was no difference between 10th grade and 12th grade students (Mann-Whitney test, $p = 1$). There were no differences in right-ear PTA value between 10th grade and 12th grade students (Two-Sample t-Test, $p = 0.10$), whereas left ear PTA indicate differences between 10th grade and 12th grade students (Mann-Whitney test, $p = 0.001$). The questionnaires data indicate that schools haven't done yet a hearing conservation program for 10th grade and 12th grade students. The differences in the student's right and left ear PTA indicate the possibility that noise exposure on students of light vehicle engineering majors have some effect.

## 1. Introduction

Noise Induced Hearing Loss (NIHL) is often occurs due to high level of noise exposure over long period of time. This type of hearing loss, in general, caused by prolonged occupational exposure to noise. Noise induced hearing loss is sensorineural type of hearing loss. Factors which affect hearing loss that is high intensity noise, high frequency, duration of noise exposure, individual sensitivity and other factors that can lead to hearing loss [1], [2].

Educational process in vocation school are not only done in class but requires learning in the form of a practicum that uses machine in long period of time. Lack of information that the students receive about the danger of noises may influence their hearing loss so that a hearing conservation program is necessary [3].

This research aimed to know the implementation of hearing conservation program in vocational school and to know the hearing loss in group of students exposed to noise.

## 2. Method

The research was analytic observational using cross sectional study design. The research population is all students of vocational school majoring in light vehicle engineering that using machine, the sample in this research 10th grade students as control group which have not exposed yet to noise and 12th grade students as case group in light vehicle engineering majors which fulfill the inclusion and exclusion research criteria.

Questionnaires were filled in by answering the options deemed correct by the respondent and the ears are sure to be clean of dirt prior to the audiometric examination. In this research the audiometric examination (GSI-Arrow brand) was conducted on both ears. The examination results obtained values on four frequencies of 500Hz, 1000Hz, 2000Hz, 4000Hz and the PTA calculated on all four frequencies in both ears. Measurement of noise level of practicum tool was done with SLM tool (Tenmars brand), the data obtained was processed and analyzed using statistical method. Place of research is SMK Negeri 5 Surabaya, research conducted in March until May 2016.

## 3. Results

The study was conducted in two groups of students. 10th grade students as a control group of 74 students with age ranged between 15 years to 17 years and 12th grade students as case group of 74 students, with the age ranged between 17 years to 19 years. All the students picked as a sample is male. The result of statistical analysis of the questionnaire results showed that there was no difference in all three categories includes hearing loss, the use of personal protective equipment, and the students knowledge (Mann-Whitney test, $p = 1$), this may indicate the school have not done yet the hearing conservation program for the students, so that there are no distinguishable results of the three categories of questionnaire data assessment. Whereas the audiometry examination of both groups showed difference PTA value of right ear and left ear in the amount of 1,71 dB (Table 1).

**Table 1.** PTA Data of 10th and 12th grade

| | **PTA (dB)** |
|---|---|
| Right Ear | 25,70 |
| Left Ear | 23,99 |

Normality test (Kolmogorov-Smirnov test) performed on statistical analysis results of PTA right and left ear 10th and 12th grade. The result of PTA of right ear obtained normally distributed data ($p = 0{,}18$), then Two-Sample t-Test was performed and there was no difference on right ear PTA ($p = 0{,}10$). Whereas the result of PTA left ear normality test obtained abnormally distributed data ($p = 0{,}004$), then Mann-Whitney test was performed and obtained difference of PTA of left ear ($p = 0{,}001$) (Table 2).

**Table 2.** Statistical Analysis Results of 10th and 12th grade PTA

| **Statistical Analysis** | **PTA Right Ear** | **PTA Left Ear** |
|---|---|---|
| Kolmogorov-Smirnov Test | p=0,18 | p=0,004 |
| Two-Sample t-Test | p=0,10 | Not performed |
| Mann-Whitney Test | Not performed | p=0,001 |

At the measurement of engine noise level, the students were exposed to the machine noisy for ± 9 hours (08.00 am-05.00 pm). There were 3 types of machines that caused noise above the limits of

IOP Conf. Series: Journal of Physics: Conf. Series **1075** (2018) 012067 doi:10.1088/1742-6596/1075/1/012067

harmful noise level that is the compressor engine with noise level of 84.8 dB-102.8 dB, iron cutting machine (grinders) of 81.9 dB-93.5 dB and car lab of 88 dB (Table 3).

**Table 3.** Practicum Machine Noise Level

| Type of Machine | Noise Level (dB) |
|---|---|
| Compressor | 84.8 – 102.8 |
| Iron cutting machine (grinders) | 81.9 – 93.5 |
| Car lab | 88 |

## 4. Discussion

Modern technology can induce few problems on hearing, which also has wide effect towards concentration and communication impairments. It leads toward permanent hearing loss. The main effect of noise exposure is impairment in hearing organs, causing temporary or permanent hearing impairments, depends on the intensity, time of exposure, and host's sensitivity [4].

Environmental noises are the main health problem in various countries, including Indonesia. Based on World Health Organization (WHO), noise is classified as a pollutant. A structured program is needed to prevent hearing impairments and reduce noise in school environment. Hearing conservation program is a chain of activities, which aiming to prevent Noise Induced Hearing Loss [5].

In this experiment, the result is gained through questionnaire in tenth and twelve graders. There is not any significant difference on hearing impairment, utilization of personal protection equipment and knowledge about the danger of noise problem, since the school has not applying the hearing conservation program in students. Lack of knowledge about the danger of noise problem and no utilization of personal protection equipment while doing the experiment causing danger, which raise the risk of hearing impairments in students.

Based on WHO (2000), there are significant elevations in hearing impairments and hearing loss prevalence. Indonesia is one of four South East Asia countries with a quite high hearing impairment prevalence (4,6%), 328 million of the population (about 91%) is adult, meanwhile the other 32 million is children [6].

Based on *Survei Kesehatan Indera Penglihatan dan Pendengaran* in year 1993 until 1996, it is showed that the morbidity rate data is high: disease of the ear is 18,5%, hearing impairments prevalence is 16,8%, meanwhile hearing loss is 0,4% and occurs mostly in school age (about 7-18 years old) [7].

In this experiment, measurement is done in two groups. Tenth grader is the control group and twelve graders are the case group. There is difference of Pure Tone Average (PTA) in left ear, because located near the source of the noise. (PTA > 25 dB) It is classified as mild hearing impairment.

Noise induced by the practicum machine exceeds the save margin (85 dB). Protecting the students from noise exposure is important, such as utilizing personal protection equipment for ear. It is also important to educate the students as soon as possible so they can avoid noise exposure. For the future, hearing conservation program needs to be applied in school.

## 5. Conclusion

There is not any significant difference in the questionnaire result, showing that the school has not applying hearing conservation program in tenth and twelve graders. There is not any PTA difference in right ear, means the students are in high risk of the danger of experiment machine (>85 dB). There is difference between right and left ear. Could be caused by the location of the noise source. The left ear is at the higher risk of experiment machine's noise.

## 6. References

[1] Stach B A 2012 Clinical audiology an introduction (*San Diego: Singular Publishing Group Inc*) pp 137-41

[2] Rabinowitz P M 2009 Noise-induced hearing loss (*Available from: www.ec.europa.eu/health/phrisk/risk.en.htm* ) Accessed on November 8, 2015

[3] Heggins II J 2013 The effects of industrial noise on hearing (*Available from: www.bcm.edu/cmb/index.cfm?pmid=497* ) Accessed on October 10, 2015

[4] Dobie R A 2011 Noise induced hearing loss *in Bailey B J Ed. Head and neck surgery otolaryngology* Vol 2 (Philadelphia: JB Lippincott Company) pp 82-91 **19**

[5] Marisol C B, Diarmid C L and Kyle S 2004 Occupational noise *Environmental Burden of Disease Series* No 9 (*Geneva: World Health Organization Protection of the Human Environment*)

[6] Mathers C, Smith A and Concha M 2000 Global burden of hearing loss in the year 2000 (*Geneva: World Health Organization*)

[7] Departemen Kesehatan Direktorat Jenderal Bina Kesehatan Masyarakat 1998 Hasil survei kesehatan indera penglihatan dan pendengaran 1993-1996 *Available from: http://kink.onesearch.id/Record/IOS8.DEPKES-OAI:78* (Jakarta: Departemen Kesehatan Republik Indonesia)

IOP Conf. Series: Journal of Physics: Conf. Series **1075** (2018) 012068 doi:10.1088/1742-6596/1075/1/012068

# Cochlear dysfunction with acoustic trauma in fire shooting training

**N Purnami[1], F Helmi[1], B Utomo[1], D F Anissa[1] and D Arifianto[2]**

[1]Department of Otorhinolaryngology-Head and Neck Surgery, Faculty of medicine Universitas Airlangga – Dr. Soetomo General Hospital
[2]Department of Engineering Physics, Institut Teknologi Sepuluh Nopember, Surabaya, Indonesia

E-mail: nyilo@fk.unair.ac.id

**Abstract.** Acoustic trauma risk for athlete shooter is huge. Continuous exposure to firearms may further increase the risk of acoustic trauma. Continuous exposure causes damage to hair cells in the cochlea that cause cochlear dysfunction and will increase the hearing threshold. The aim of this study was to determine the difference of hair cell damage in cochlea which can cause cochlear dysfunction due to acoustic trauma after shooting practice based on duration and frequency of training. There was no statistically significant difference in audiogram results between the first group and the second group (Mann-Whitney U Test, $p<0.05$). The distortion product of otoacoustic emission (DPOAE) showed a significant difference between the first group and the second group (Mann-Whitney U Test, $p<0.05$). Different test results on audiogram and DPOAE between right and left in each group showed no significant difference (Wilcoxon Signed Ranks Test, $p<0,05$). There was a significant difference in the incidence of hair cell damage in cochlea with the duration and frequency of training.

## 1. Introduction

Large acoustic energy (explosions, gunfire, bombs) can cause acoustic trauma. This can cause mechanical damage in the form of tympanic membrane rupture, ossicular damage, hydrolimfatic rupture, basilar cell membrane damage, and hair cell injury can also cause metabolic damage in the form of oxidative stress in Corti organs [1], [2]. Oxidative stress is a condition of stress in the cells due to an unbalanced amount of increased reactive oxygen species (ROS) production with an inexpensive intracellular system of superoxide dismutase (SOD) and glutathione peroxidase (GPx) [3]. Oxidative stress on cochlea that accumulates in the long term causes apoptosis and ends with cochlear dysfunction [4].

The hair cells of Corti's organs are essential for converting acoustic energy into electrical stimuli that are channeled into the hearing center of the brain. Sensorineural hearing loss (SNHL) is the most common type of hearing loss due to hair cell damage in the Corti organ. It is a disability affecting one person and over 48 million Americans. The causes of SNHL are numerous, one of which is acoustic trauma [5]. A pure tone audiometric examination (ANM) is a subjective examination that can assess the auditory function at the overall level of the auditory system from the outer ear to the auditory cortex. Examination of distortion product of otoacoustic emission (DPOAE) is an objective examination useful for assessing hearing function at the level of outside hair cells in the cochle [6].

World Health Organization (WHO) data in 2005 showed that there were 278 million (4.2%) of the world's population suffering from hearing loss. According to the Southeast Asia Multi Center Study survey, Indonesia is one of 4 countries with a high prevalence of deafness that is 4.6% of the total population of Indonesia [7]. This prevalence can cause social problems in the community. This prevalence is high enough to create social problems in the community. Hearing disorders experienced by veterans in the United States in third ranks after mental disorders and musculoskeletal disorders [8]. Study in Magelang, shows the incidence rate of acoustic trauma at the police academy was 12.4%. Study in Makassar police department found an acoustic trauma event of 16.67% [9]. A study of young soldiers in France who were observed for 15 days using DPOAE showed a high frequency hearing loss with persistent tinnitus, which occurred 24 hours after exposure to an automatic machine-gun blast. Research in brazil with DPOAE showed significant hearing loss at 3-6 kHz at 24-72 hours after noise trauma, accompanied by persistent tinnitus symptoms [10].

The more exposed to the sound of firearms, will increasing the risk of acoustic trauma. The aim of this study was to determine the difference of hair cell damage of the hearing organ (cochlea) due to acoustic trauma after shooting practice based on duration and frequency of exercise.

## 2. Methods

This study compared two groups that performed once shooting practice with hair cell damage after doing 5 shooting exercises. Samples include a student of the Sekolah Polisi Negara (SPN) of East Java which has undergone shooting practice. It was random sampling. Both was taken 100 students from 700 students of the academic year 2017-2018. Students who are unwilling to attend research and students with disorders in the middle ear are not included in the study. Intensity of short guns and the long-barreled guns was measured in December 2016, at 1 meter and 10 meters.

Definitions of the hearing loss are permanent hearing loss due to the strong acoustic exposure and abrupt energy, which by examination of pure tone audiometry has obtained a threshold increase of 15 dB or more at 500, 1000, or 2000 Hz; or 20 dB or more at a frequency of 3000, 4000 or 6000 Hz, or acquired an acoustic notch at a frequency of 4000 Hz without or accompanied by an increase in the 500 Hz, 1000 Hz, 2000 Hz frequency threshold that occurs in SPN students. On DPOAE examination obtained referral results that show hair cells damage at a frequency of 3000-6000 Hz, especially the frequency of 4000 Hz. Sound level meters are placed 10 meters and 1 meter from the shooter.

Materials and tools consist of letter of agreement, stationery, ENT examination tool. Sound level meter, pure tone audiometry, DPOAE examination tool. The results were descriptive frequency distribution data. The audiogram results are rated at 4000 Hz. DPOAE results are also rated at 4000 Hz.

## 3. Results

The sound intensity of guns recorded on the sound meter level was 91 to 106 dB, compare the long barrel of 97 to 106 dB. In 1 meter, gun intensity level was 111 to 128 dB, while the long barrel was 117 to 126 dB. Conclusion from the results explain that all sound intensity levels reach more than 90 dB could damage to hair cells. Audiogram result shows that the first group that got increase of hearing threshold was 7 students (7%), left ear was 15 students (15%). The second group got increase of hearing threshold was 5 students (5%), the left ear was 7 students (7%). Refer DPOAE results to first group; right ear 17 students (17%) compare to 30 students (30%). The second group of references on the right ear 61 students (61%) compare to 59 students (59%).

There was no statistically significant difference in audiogram results between the first group and the second group (Mann-Whitney U Test, $p < 0.05$). The DPOAE results showed a significant difference between the first group and the second group (Mann-Whitney U Test, $p < 0.05$) (Table 1). Different test results on audiogram and DPOAE results between right ear and left ear in each group showed without significant difference (Wilcoxon Signed Ranks Test, $p < 0.05$) (Table 2).

**Table 1.** Test result for the first and second group.

| | OAE_KA | OAE_KI | AUDIO_KA | AUDIO_KI |
|---|---|---|---|---|
| **Mann-Whitney U** | 2800.000 | 3550.000 | 4900.000 | 4550.000 |
| **Wilcoxon W** | 7850.000 | 8600.000 | 9950.000 | 9600.000 |
| **sZ** | -6.363 | -4.116 | -.594 | -1.990 |
| **Asymp. Sig. (2-tailed)** | .000 | .000 | .553 | .047 |

**Table 2.** Formatting sections, subsections and subsubsections.

| | OAE_KI - OAE_KA | AUDIO_KI - AUDIO_KA |
|---|---|---|
| Z | $-2.600^{b}$ | $-2.324^{b}$ |
| Asymp. Sig. (2-tailed) | .009 | .020 |

## 4. Discussion

This study shows that the intensity level generated by the gun, measured at 10 meters with 100 dB, and 1 meter with 120 dB. The average intensity level of the long-barrelled gun at 10 meters was 102 dB, and at a f 1 meter was 122 dB. This is contrast due to another reference which is the average intensity level of long-barrelled gun is 162 dB and for a 158 dB [10]. This difference is due to the intensity measured at 10 meters and 1 meter from the source of the eruption, not on the tip of the gun. the measurement is also done in the open airfield so that the results are influenced by wind, and there was less objects that reflect sounds like trees, building walls, and other hard objects. Some references showed that the intensity of gun lies at the tip of the gun and the noise exposure depends on the intensity, direction and the presence of objects that reflect or muffle the sound [11], [12], [13].

From this research, the result of audiogram in the first group as many as 19 students (19%) experienced an increase of hearing threshold. The right ear 7 students (7%) and the left ear 15 students (15%), and both ears as many as 4 students. There were 10 (10%) students who experienced an increase in the threshold. right ear 5 students (5%), left ear 7 students (7%), and both ears 1 student (1%). From the results of this study showed that the left ear has more hearing threshold increase, but there is no significant difference. This is consistent with Budiyanto's research at Akpol Semarang reporting 11% left ear acoustic trauma and right ear 6.2%, while both ear (1.4%). Acoustic trauma events in the left ear are more than the right ear, this can be caused by the position of the body when firing, where the gun is placed right with the gun body attached to the right shoulder, allowing ear distance closer to the source of the explosion than the right ear. so that the intensity of gun eruptions that affect the ear is larger than the right ear and also the result of a head-shadow effect, in which the head obstructs or protects the eruption of guns on the right and left [9].

From the results of this study there was no significant difference in audiogram results between two groups. There is a significant difference of DPOAE results between the first group and the second group. Audiometry examination is a subjective examination that assesses hearing function at the overall level of the auditory system from the outer ear to the auditory cortex. The DPOAE examination is an objective examination that is useful for assessing hearing function at the level of the hair cell in the cochlea [14]. The results of this study indicate that DPOAE is more sensitive than audiometry in assessing cochlear dysfunction. Exposure to continuous noise causes a shift in basilar membrane and damages stereo cilia, hair cells, and organ of Corti. This explains why in the second group more damaged hair cells, especially at the frequency of 4000 Hz. Cochlear damage due to frequency and high intensity centred on the frequency of 4000 Hz. Approximately 10 mm from the oval window is the area with the weakest anatomical structure and 4000 Hz receptor is the largest amplitude hair cell and receives the greatest energy from noise exposure. This place is the weakest point on the organ of Corti.

IOP Conf. Series: Journal of Physics: Conf. Series **1075** (2018) 012068 doi:10.1088/1742-6596/1075/1/012068

## 5. Conclusion

Exposure to gunshot explosion sound over a long period of time can damage the hair cells in the auditory organs, especially at 4000 Hz frequency that can be detected by DPOAE examination. There were significant differences in hair cells damage in cochlea caused by acoustic trauma with duration and frequency of exercise. SPN students are advised to use ear protection while undergoing shooting exercises. There was time limitations of this study. Further research is needed to investigate more about these case.

## 6. References

[1] Gates G A and Clark W W 2012 Occupational hearing loss *In Lalwani AK, ed Current diagnosis & treatment 3th edition* (New York: The McGraw-Hill Companies, Inc.,) pp 747 59

[2] Alberty P W 2003 Occupational hearing loss *In Snow JB, Ballenger JJ, eds Ballenger's Otorhinolaryngology Head and Neck Surgery, 16th edition* (Philadelphia: Lea and Febiger) pp 357-73

[3] Sohal R S and Orr W C 2012 The redox stress hypothesis af aging *Free Radical Biology and Medicine* **52** : 539-55

[4] Venditti P, Stefano L D and Meo S D 2013 Mitochondrial metabolism of reactive oxygen species *Mitochondrion* **13** : 71-82

[5] Dinh C T, Goncalves S and Bas E 2015 Molocular regulation of auditory hair cell death and approaches to protect sensory receptor cell and/or stimulate repair following acoustic trauma. *Available from : http://www.researchgate.net/publication/275047084* (Acessed August 9, 2017)

[6] Bess F H and Humes L E 2008 Audiology as profession *In Katz J, Medwetsky L, Burkard R, Hood LJ, eds. Audiology, the fundamentals, 4th edition* (Philadelphia: Lippincott Williams & Wilkins) pp 4-9

[7] WHO Regional Office SEARO 2004 State of hearing and ear care in South East Asia Region

[8] Brian K 2004 Drug may help troops battle combat-realated hearing loss *Avalaible from: http://www.drug-may-help-troops-battle-combat-realated-hearing-loss* (Acessed August 9, 2017)

[9] Budiyanto A 2003 Trauma akustik akibat latihan menembak pada taruna akademi kepolisian semarang *Karya akhir: SMF Ilmu penyakit Telinga Hidung dan Tenggorok Bedah Kepala dan Leher Kedokteran UNDIP* (RS Dr. Kariadi Semarang)

[10] Campanella L, Bonani A, Finnoti E and Tomasseti M 2004 Biosensors for determination of total and natural antioxidant capacily of red and white wine: Comparison with other spectrophonic and fluorimetric methods *Biosense Bioelectron* pp 19-641

[11] Dobie R A 2014 Noise induced hearing *In Johnson JT, Rosen CA, eds. Bailey's head and neck surgery otolar yngology, 5th edition* Vol 2 (Philadelphia: Lippincot Williams and Wilkins) pp 2530-40

[12] Fox M S 1996 Noise-induce hearing loss *In Ballenger JJ, ed. Disease of the nose, throat, ear, head and neck. 15th ed* (Philadelphia: Lea and Febiger) pp 967-8

[13] Alberty P W 2003 Noise and ear *In Stephen D, editor. Adult audiology. Scott-Brown's Otorhinolaryngology head & neck surgery 16th ed.* (Philadelphia: Lea and Febiger) pp 357-73

[14] Soetirto I, Hendarmin H and Bashiruddin J 2010 Gangguan pendengaran *In Soepardi EA, Iskandar N, Bashiruddin J, Restuti RD, ed. Buku ajar ilmu kesehatan telinga hidung tenggorok kepala dan leher, 6th ed.* (Jakarta: Balai Penerbit FKUI) pp 43-5

IOP Conf. Series: Journal of Physics: Conf. Series **1075** (2018) 012069 doi:10.1088/1742-6596/1075/1/012069

# A general description of measurement method for sound absorption and surface impedance of materials using ensemble averaging

**N Sakamoto[1], T Otsuru[2], R Tomiku[2] and Saki Yamauchi[1]**

[1]Graduate school of Engineering, Oita University, Japan
[2]Department of Architecture and Mechatronics, Oita University, Japan

E-mail: v17f2001@oita-u.ac.jp

**Abstract.** Sound absorption characteristics of materials are important for acoustical researches as well as for practical designs. Various measurement methods have been proposed in this research work for sound absorption characteristics. Some of them are standardized in the International Organization for Standardization(ISO) and in the Japanese Industrial Standards (JIS). There are two well-known methods of laboratory measurement of absorption, one is the "reverberation room method" and the other is the "tube method". However, these two methods have several drawbacks and disadvantages respectively, as they are not suitable for "in-*situ*" measurement. Furthermore, sufficient amount of impedance database has not yet been accumulated to meet the necessity for constructing boundary conditions of wave-based room acoustical simulations. To overcome the situation, the authors have presented a method utilizing "ensemble averaging", namely the "EA method". There are two types of EA methods. One uses two-microphones and the other uses a pressure-velocity probe (Microflown PU-sensor). Herein, an overview of the EA method is described.

## 1. Introduction
Sound absorption coefficients of acoustical materials are measurable both by means of the reverberation room method and by the tube method according to ISO standards [1], [2]. The results of the sound absorption coefficient of the former in a diffused sound field are of importance for building acoustics as well as for noise control. However, this method cannot measure the surface impedance of a material which is indispensable for wave-based-room-acoustical simulations like finite element method (FEM) in constructing sound fields' absorptive boundary conditions. The tube method can measure the surface impedance of a material and at the same time it can measure the sound absorption coefficient. Nevertheless, there are not enough amount of impedance database so far, and various alternative methods have been proposed. (Allard [3], [4], [5], Garai [6], Mommertz [7], Nocke [8]) Some of the authors also have been proposed a method using an ensemble averaging technique (EA method) [9 – 14]. In this paper, the outline of EA method is summarized in comparison with conventional methods.

## 2. Conventional Method
There are two well-known conventional methods. One is the "reverberation room method" and the other is the "impedance tube method". They are standardized in the International Organization for Standardization (ISO) [1], [2] and in the Japanese Industrial Standards (JIS) also.

### *2.1. Reverberation room method*
The Reverberation room method is widely used as a practical absorption index. However, the creation of diffuse sound field requires a large and costly reverberation room. A completely diffuse sound field can rarely be achieved. The edge effect and dependence on characteristics of reverberation room itself are not negligible [9]. Size of specimen requires approximately $10m^2$, which is disadvantageous regarding cost and practicability. Occasionally absorption coefficient exceeds 1.0 because of edge effect. Moreover, an accurate value of complex impedance cannot be derived, which is essential for wave-based room acoustical simulations.

### *2.2. Impedance tube method*
The Impedance Tube method can obtain both normal surface impedance and absorption coefficient at the same time. The setup system consists of two microphones, sound tube, FFT (Fast Fourier Transform) frequency analyzer. It is a rather compact and simple apparatus in comparison with the reverberation room method, and the specimen size is large enough to set it up in the sound tube ends and is somewhat smaller than the one used in the reverberation room method. On the other hand, the measurable frequency range depends on the diameter of sound tube. Consequently, one needs several sound tubes corresponding to the purpose. Furthermore, the conditions of the specimen's cutting edge and installation status affects the results considerably.

### *2.3. Other method*
There are some other conventional methods proposed in recent years:

- Allard anechoic room two-microphone [3], [4], [5].
- Garai MLS reflection method [6].
- Nocke Free field transfer function method [8].

These methods are categorized in reflection method. The following are disadvantages [9].

- low-frequency limitation
- inaccuracy with weakly absorbing materials
- the need for complicated equipment when used *in-situ*
- only one sound source can be employed
- sometimes absorption coefficient exceeds 1.0s because of edge effect and so on

## 3. EA method
EA signifies "Environmental Anonymous" at first, and later altered "Ensemble Average". Consistently, these naming means *in-situ* measurement by using ensemble average of random incidence sound source (environmental anonymous noise). Figure1 shows the schematic diagram of EA method measurement.

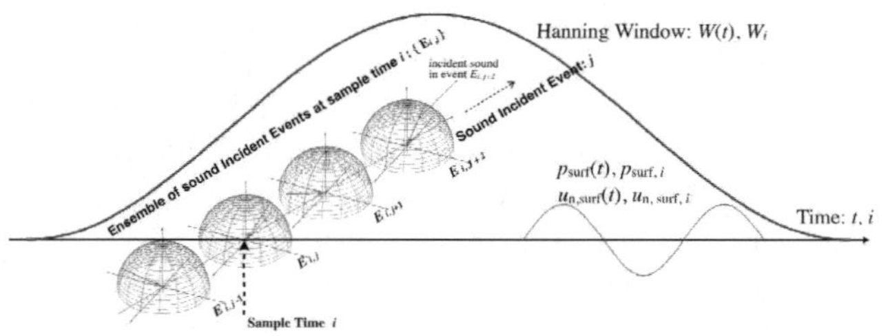


**Figure 1.** Schematic diagram of EA method measurement mechanism [13].

### *3.1. Advantages and disadvantages*
The main purpose of EA method is to obtain the absorption coefficient and normalized surface impedance for using the boundary condition of numerical analysis by *in-situ* measurement. Environmental anonymous noise works well effectively as a random incidence sound source, (but it is desirable to add sound sources (pink noise, white noise) in case signal-to-noise ratio is not enough to

measure at the measurement place.) The process of ensemble averaging is easy to be performed on an ordinary 2-ch fast-Fourier-transform instrument and it is also helpful to decrease the "edge-effects (the effect of diffracted waves from the specimen's edge)" that is one of the most disturbing obstructions in various sound absorption measurement method [18]. The mechanism of cancelling the edge-effect is explained [10] and sample-size-independency of EA method is examined in a literature [11]. A condition of sound field is not restricted practically, if only it has relevant capacity of sound field space. Random incidence value is measurable, which nearly coincides with a normal surface incidence value. A typical disadvantage is that EA method results are not equal to other conventional values like Sabine absorption coefficient, normal impedance and so on, Nonetheless, incidence angle averaged surface normal impedance of the material can be obtained. The EA method consists of two types [Figure 2], one is the "EApp" and the other is the "EApu". The main difference between these two methods is in its setup system, the number of microphones and its characteristics. The EApp method requires that two microphones be used, and the EApu method requires that one pressure-velocity probe (Microflown PU-sensor) [Figure 3] be used.

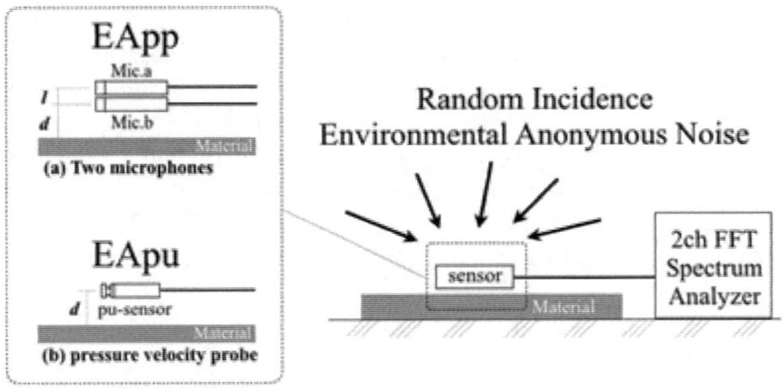


**Figure 2.** A Block diagram of the measurement system EApp (a) and EApu (b).

*3.2. EApp method*

This method uses two microphones to measure transfer function between them for obtaining the values. In comparison with later-mentioned EApu, two microphones are much less costly than the PU-sensor, Furthermore, there is no need for sensitive calibrations like PU-sensor. On the other hand, the assumption of sound receiving point is geometrically not clear (distance between two microphones cannot be ignored). It takes more time and effort than EApu, because the replacement of the two microphones is indispensable in each measurement procedure.

*3.3. EApu method*

This method uses pressure-velocity probe. In comparison with the above-mentioned EApp method, sound receiving point is geometrically clear, because it is measured by only one PU-sensor. However, the PU-sensor is considerably expensive than the usual microphone used in EApp. Furthermore, each sensor has different characteristics, and their characteristics are fluctuated by humidity and temperature individually [14]. Therefore, calibration is necessary for each PU-sensor and in each measurement procedure.

## 4. Calibration of PU - sensor

This section refers to the calibration of PU-sensor (pu-sensor) [Figure 3] for EApu. PU-sensor is very sensitive and is susceptible to the surrounding environment. Calibration is one of the most important and critical procedures in EApu as mentioned earlier in this paper. Accurate and precise calibration can derive the values worthwhile. The following three techniques are adopted as our bench mark through investigation.

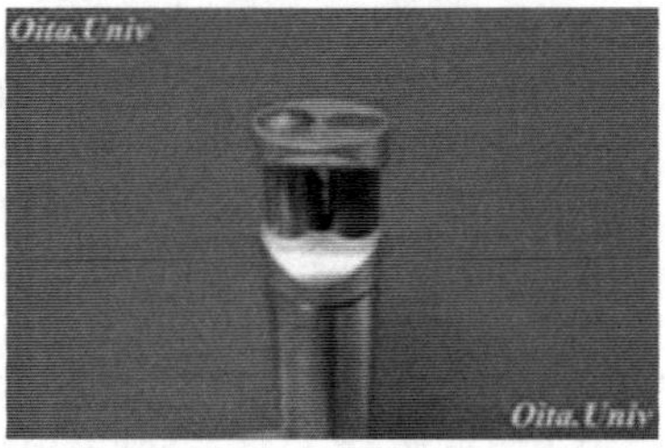


**Figure 3.** Pressure-velocity probe (Microflown PU-sensor).

*4.1. Standing wave tube method*

The standing wave tube method is less likely to be affected by the surrounding environment because of its setup system specifications. This method consists of two types: The first one is the long tube technique [16] and the second one is the short tube technique [Figure 4] [16]. One of our purpose is *in-situ* measurement, so we prefer a short tube technique for easy handling. Measurable frequency range of the calibration depends on the sound tube diameter. The bandwidth is limited for high frequencies. The smaller the tube diameter is, the higher the frequency can be measured. The PU-sensor setting also significantly affects the values. At lower frequencies (f<100Hz) small leakages in the probe mountings influence the measurements. To prevent these influences, the probes must be sealed with e.g. a rubber ring [16]. Our recent research improves the calibration values for the lower frequencies by applying curve-fitting approximation [18]. In other words, the calibration of PU-sensor is the most important and difficult step to obtain the relevant values throughout the measurement procedure of EApu.

*4.2. POS (piston on a sphere)*

POS signifies piston on a sphere. PU-sensor calibration with POS setup [Figure 4] needs to be performed in a quiet environment with low reflections but anechoic conditions are not required [19]. The calibration process and procedure are susceptible to errors because of: probe and reference sensor allocations, reference sensor response variation, difficult assessment of the proper noise level, and lastly complicated model fitting criteria. Therefore, it is recommended that the POS calibrator be used as a checkup tool and not as an actual calibrator [19].

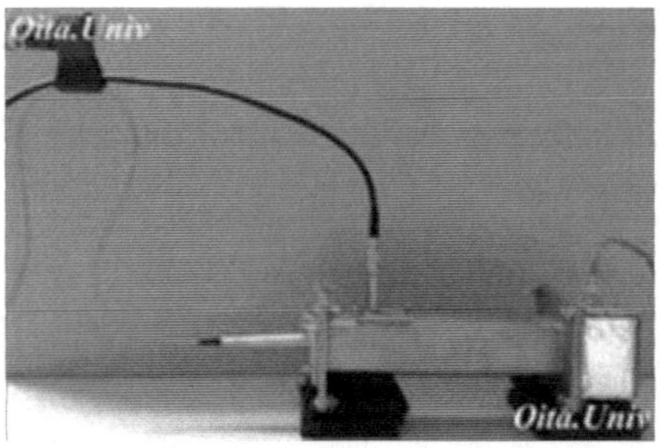


(a)

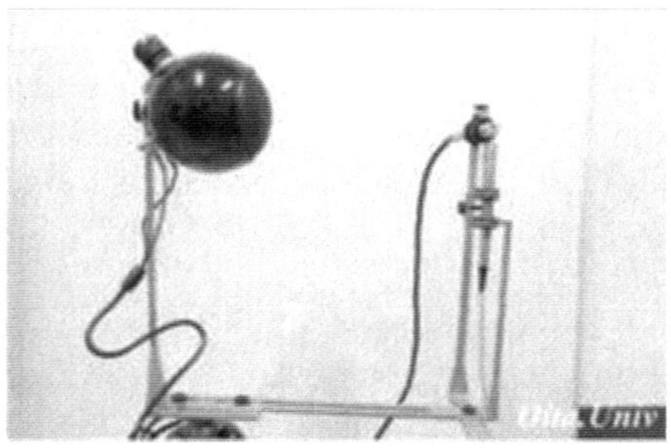

(b)

**Figure 4.** Setup image of (a) standing wave tube and (b) piston on a sphere

*4.3. Manufacture's shipping data*

Microflown provides manufacture's calibration value for each PU-sensor, but our investigation reached the conclusion that it is unacceptable, since values are fluctuated by humidity [14], [15]. Therefore, we decided to undertake research for making further refinements to the calibration by means of standing wave tube techniques was begun.

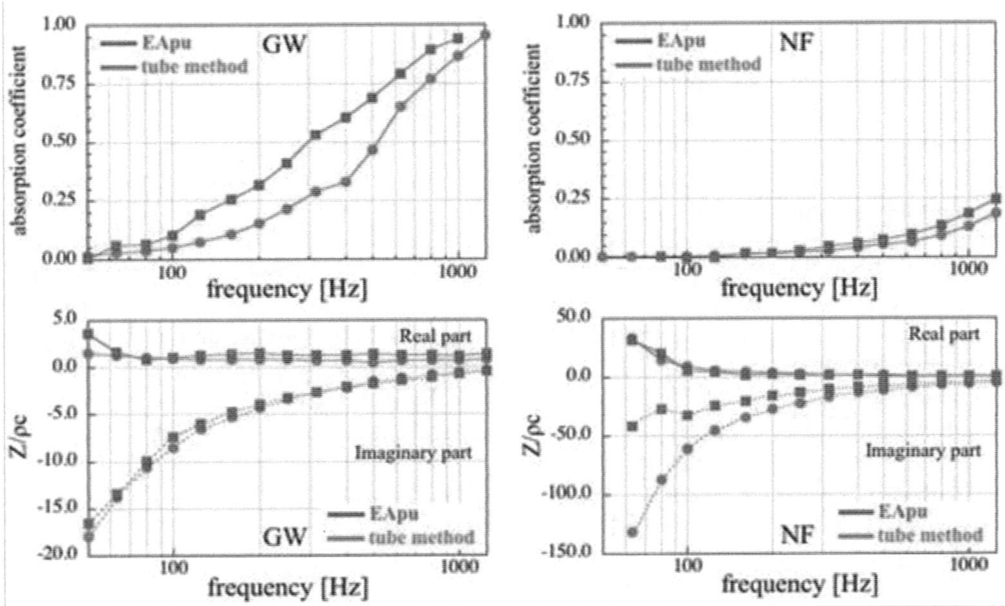


**Figure 5.** Comparison of sound absorption coefficients and surface normalized impedance of glass wool and needle felt measured by EApu method and tube method.

## 5. Example results

Figure 5 shows comparisons of sound absorption coefficients and surface normal impedance of glass wool (GW:32kg/m$^3$) and needle felt (NF) measured by EApu method and tube method. The blue lines show the values by EApu method, and the red lines show the values by tube method. The values of the NF absorption coefficient used by these two methods have similar tendencies. The values of the GW absorption coefficient by two methods also show a similar tendency, though the certain discrepancies are observed around 200 - 400 Hz area. It can be inferred that they were caused by the noise problem from the conditions of specimen's installation. This is one of the most serious matter of the tube method. For this reason, it demands further investigation. The values of the GW surface normalized impedance used by two methods agree well each other. On the other hand, the imaginary part of the NF surface normalized impedance deviate from each other in the lower frequency area.

## 6. Conclusion

This paper provides a general description of EA method by comparing with the other conventional methods. EA method can obtain the rather robust values of normal surface impedance and absorption co-efficient at the same time in *in-situ* measurements. EApp has advantages because it is not costly, and it has reliable setup system, and there is no need to calibrate. Nonetheless, it takes more time and effort than EApu because of its measurement procedure. On the other hand, the EApu is costly and needs calibration in each measurement procedure set. However, it is still valuable regarding its theoretical and geometrical straightforwardness. Further experiments and investigations are necessary to improve the EA method itself. Impedance database for wave-based-room-acoustical simulations is still insufficient. EA method makes a substantial contribution to the construction of materials' impedance database.

## 7. References

[1] ISO 354:2003 Acoustics – Measurement of sound absorption in a reverberation room

[2] ISO 10534-2:1998 (en) Acoustics — Determination of sound absorption coefficient and impedance in impedance tubes — Part 2: Transfer-function method

[3] Allard J F, Champoux Y and Nicolas J 1989 Pressure variation above a layer of absorbing material and impedance measurement at oblique incidence and low frequencies *J. Acoust. Soc. Am.* **86** 766-770

[4] Allard J F and Champoux Y 1989 In situ two-microphone technique for the measurement of

the acoustic surface impedance of materials *Noise Control Eng. J.* **32** 15-23

[5] Allard J F, Depollier C and Guignouard P 1989 Free field surface impedance measurements of sound-absorbing materials with surface coatings *Appl. Acoust.* **26** 199-207

[6] Garai M 1993 Measurement of the sound-absorption coefficient In situ: The reflection method using periodic pseudo-random sequences of maximum length *Appl. Acoust.* **39** 119-139

[7] Mommertz E 1995 Angle-dependent in-situ measurements of reflection coefficients using a subtraction technique *Appl. Acoust.* **46** 251-263

[8] Nocke C 2000 In-situ acoustic impedance measurement using a free-field transfer function method *Appl. Acoust.* **59** 253-264

[9] Takahashi Y, Otsuru T and Tomiku R 2005 In situ measurements of surface impedance and absorption coefficients of porous materials using two microphones and ambient noise *Appl. Acoust.* **66** 845-865

[10] T. Otsuru, R. Tomiku , NBC. Din, N. Okamoto, M. Murakami. Ensemble averaged surface normal impedance of material using an in-situ technique: preliminary study using boundary element method. J Acoust Soc Am 2009 125:3784-91

[11] NBC. Din, T. Otsuru, R. Tomiku, N. Okamoto, K. Asniawaty. Measurement method with a pressure-velocity sensor for measuring surface normal impedance of materials using ensemble averaging: com-parison with other methods and its geometrical configuration. Acoust Sci Tech 2012 33:86-95.

[12] NBC. Din, T. Otsuru, R. Tomiku, N. Okamoto, K. Asniawaty. Reproducibility and applicability of ensemble averaged surface normal impedance of materials using an in-situ technique. Acoust Aust 2013 41:207-12.

[13] T. Otsuru, R. Tomiku, N. Okamoto, E. Ueda, A. Nakamura. Uncertainty examination of ensemble averag-ing method for sound absorption of materials in a reverberation room. Acoust Sci Tech 2016 37:129-32.

[14] K. Asniawaty, T. Otsuru, R. Tomiku, N. Okamoto. The effect of humidity on the stability of pressure-velocity sensors - a measurement method of absorption characteristics of building materials by using ensemble averaging -. AIJ Technol Des 2013 19(41):179-84

[15] T. Otsuru, R. Tomiku, N. Okamoto, S. Yamauchi, Sound absorption deviation measured using pressure-velocity sensors by ensemble averaging technique under different relative humidity conditions, Appl. Acoust. 122 (2017) 121-127

[16] Microflown e-Book, Chapter4: Calibration pp 4 – 42 **43**

[17] T. Otsuru, R. Tomiku, S. Noriaki, S. Yamauchi, An overview of measurement method for sound absorp-tion and surface impedance of materials using ensemble averaging. ASJ 2017(in print, in Japanese)

[18] S. Noriaki, T. Otsuru, R. Tomiku, S. Yamauchi, An in-situ sound absorption measurement method by en-semble averaging technique - expansion of measurable frequency range. ASJ 2017(in print, in Japanese)

[19] Manual, Piston on a sphere, Microflown Technologies pp 3-10

**Acknowledgments**

This work was partly supported by JSPS, a Grant in Aid for Scientific Research (B) 16H04465, 2016-18.

IOP Conf. Series: Journal of Physics: Conf. Series **1075** (2018) 012070 doi:10.1088/1742-6596/1075/1/012070

# Configuration and welding characteristics of ultrasonic complex vibration welding systems using two-dimensional vibration stress

**J Tsujino**[1]

[1]Kanagawa Unversity, Rokkakubashi, Yokohama, Japan
Asahi EMS Co. Ltd, LINK-US Co. Ltd, Shinyokohama, Yokohama, Japan

E-mail: tsujino@kanagawa-u.ac.jp

**Abstract.** Various complex vibration systems with two-dimensional vibration locus were developed for improving welding characteristics of ultrasonic metal welding. Ultrasonic complex vibration welding system using two-dimensional vibration stress could be applied effectively for joining same and dissimilar metal and ceramics, and has superior quality compared with conventional welding with linear vibration locus. Welding of dissimilar metal specimens are required for electronic devices and fuel cell, battery or EDLC electrodes and the other various industry fields.

## 1. Introduction

Ultrasonic complex vibration systems could be configured using multiple vibration systems using same and different frequencies and multiple power amplifiers, and various complex vibration systems up to 1 MHz were developed. Complex vibration converter using diagonal slits driven by a longitudinal vibration source was proposed. The longitudinal to torsional complex vibration converter vibrates in circular to elliptical vibration locus at a free edge of the converter in the case where vibration phase difference between longitudinal and torsional vibrations is about 90 degrees. The converter was designed using equivalent electrical transmission line method and FEM. Various welding tips could be installed in the free edge part with a connecting bolt. Using complex vibration, required vibration velocity becomes one-third to quarter compared with conventional welding using linear vibration and weld strength near to material strength was obtained independent of specimen position and direction, and multiple spot or continuous seam welding becomes possible using complex vibration equipment. Various dissimilar metal specimens including many metal foils such as aluminum and copper were successfully welded using ultrasonic complex vibration welding systems. These welding are essential for electronic devices and multi-layer fuel cell, battery or EDLC electrodes for electric or hybrid automobile and other various industry fields.

## 2. Configuration of ultrasonic complex vibration welding equipment

### *2.1. Ultrasonic complex welding system using a complex vibration converter with diagonal slit*

Figure 1 (a) shows 19.5 kHz, 2 kW ultrasonic complex vibration welding equipment using a complex vibration converter with a half-wavelength complex transverse vibration welding tip. The ultrasonic welding system consists of a 19.5 kHz bolt-clamped Langevin type longitudinal transducer (BLT, 50

IOP Conf. Series: Journal of Physics: Conf. Series **1075** (2018) 012070 doi:10.1088/1742-6596/1075/1/012070

mm in diameter), a stepped horn for enlarging vibration velocity with a supporting flange, a complex vibration converter with diagonal slits. The complex vibration converter is designed using equivalent electrical transmission line method and FEM analysis. The converter is 40 mm in diameter. Twelve diagonal slits were cut directly along the circumference of the converter rod using a spark or milling machine. Longitudinal vibration is partially converted by diagonal slits to torsional vibration. The complex vibration system is driven using a 2-kW frequency auto-tracking and constant vibration velocity control driving system.

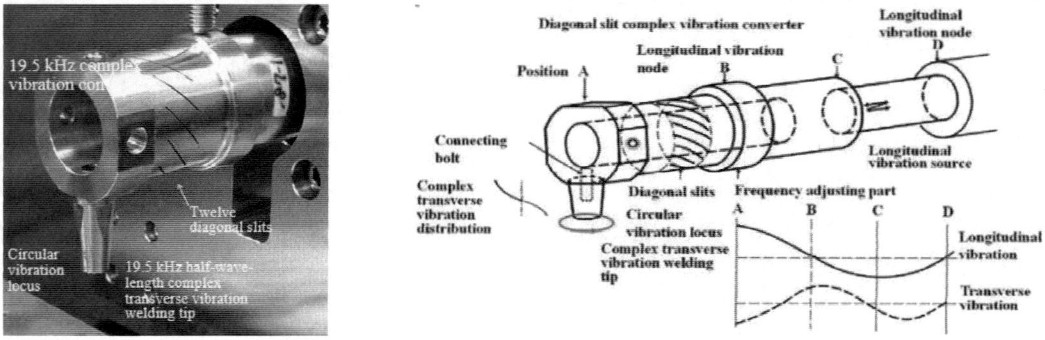


(a) (b)

**Figure 1.** (a) 19.5 kHz ultrasonic complex vibration welding equipment using a half-wave-length longitudinal mode complex vibration converter with a slit part installed a half-wave-length complex transverse vibration welding tip at a free edge (25-mm-length), (b) Schematic diagram of the longitudinal half wave-length mode converter with a slit part and longitudinal and torsional vibration distributions along the converter and a driving longitudinal vibration source.

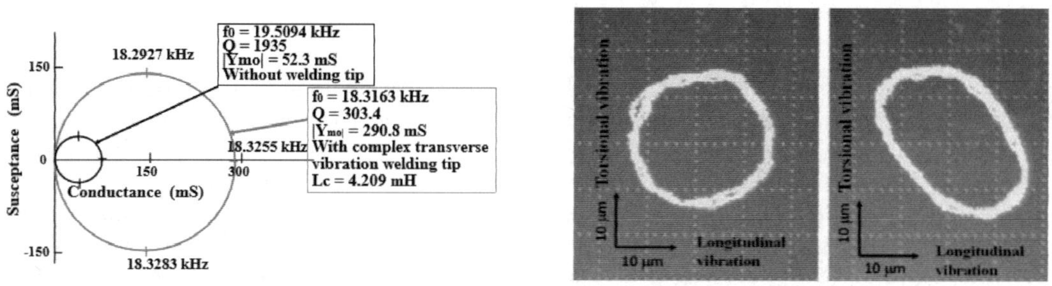


(a) (b)

**Figure 2.** (a) Free admittance loops of the vibration system without welding tip and with a half wave-length complex transverse complex vibration welding tip of the half wavelength complex transverse vibration welding tip using a power factor compensating inductance $L_c$, (2) Vibration loci of the half wavelength complex transverse vibration welding tip under welding condition.

Figure 1 (b) shows schematic diagram of converter and longitudinal and torsional vibration distribution along a complex vibration converter and driving stepped horn. The converter and stepped horn vibrate 3/4 wave-length longitudinal mode and 5/4 wave-length torsional mode between free edge of converter A and nodal position of stepped horn C.

Figure 2 (a) shows free admittance loops of the vibration system without welding tip and with a half wave-length complex transverse complex vibration welding tip using a power factor compensating inductance $L_c$. These loops are single loop due to near longitudinal and torsional resonance frequencies. Inserting inductance $L_c$, quality factor decreases from 1935 to 303, but

IOP Conf. Series: Journal of Physics: Conf. Series **1075** (2018) 012070 doi:10.1088/1742-6596/1075/1/012070

motional admittance $|Y_{mo}|$ increases from 52.3 mS to 290.8 mS and the complex vibration system with a complex vibration welding tip could be driven effectively.

Figure 2 (b) shows vibration loci of the half-wave length complex transverse vibration welding tip. The welding tip vibrates in almost circular vibration locus during welding process.

### *2.2. Ultrasonic complex vibration system with 20-mm-wide complex vibration welding tip*

Figure 3 (a) shows 19.5 kHz complex vibration system with 20-mm-wide, 15-mm-depth, 9-mm-height complex vibration welding tip (22 gram in weight) for welding of stacked many thin electrode foils installed at free end of complex vibration converter. This welding tip affects resonance frequency of the complex vibration system. The complex vibration converter was redesigned considering the welding tip weight.

This non-resonant welding tip vibrates nearly as a mass but vibrates slightly transversally. Transverse resonance frequencies of the welding tip are 35.62 kHz and 44.15 kHz in the longitudinal and torsional vibration directions. The 20-mm-wide complex transverse welding tip is effective for welding of many stacked aluminum and copper foils and terminals.

Figure 3 (b) shows free admittance loops of the converter with 20-mm-wide welding tip under no load condition and loaded by static clamping force of 1,000 N. Motional admittance $|Y_{mo}|$ were improved to 516 mS and 499.7 mS using inductance $L_c$ = 2.288 mS. Using the complex vibration converter considered welding tip weight, $|Y_{mo}|$ under static pressure 1,000 N decreases only slightly.

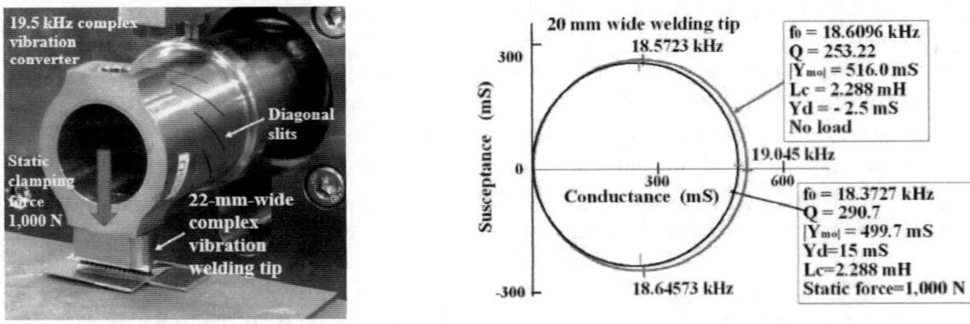


(a) (b)

**Figure 3.** (a) 19.5 kHz complex vibration system with 20-mm-wide complex transverse vibration welding tip using the converter designed considering welding tip weight of 22 g, (2) Free admittance loops of the converter with 22-mm-wide welding tip under no load condition and loaded by static clamping force of 1,000 N.

### *2.3. Complex vibration welding systems with long and narrow welding tip*

Figure 4 (a) shows 19.5 kHz hard metal 3.0-mm-diameter, 79-mm-long transverse vibration welding tip for welding of deep and narrow area. The 3.0-mm-diameter hard metal complex transverse vibration welding tip is fixed to half wave-length transverse vibration holder by shrinkage fit. The welding tip is installed in free edge of complex vibration converter using a connecting bolt. Transverse vibration distributions measured by a laser Doppler vibrometer along half-wave transverse vibration holder and 3.0-mm-diameter, 79-mm-long hard metal rod is shown in the figure. The hard metal rod vibrates transversally in complex vibration mode with four nodal points. Free end of welding tip vibrates in circular to elliptical locus.

Figure 4 (b) shows free admittance loops of 19.5 kHz complex vibration welding system with 3.0-mm-diameter, 78.70-mm and 78.95-mm-long hard metal complex transverse vibration welding tip without power factor compensating inductance $L_c$.

Resonance frequencies of complex vibration system with 78.70-mm and 78.95-mm-long hard metal welding tip change from 19.0039 kHz to 18.9770 kHz, quality factors are 1703.9 to 2065.91 and $|Y_{mo}|$ are 36.655 mS to 23.311 mS by 0.25-mm-length difference. But, motional admittances could be improved by inserting adequate power factor compensating inductance $L_C$, and the complex vibration system can be driven effectively. The complex vibration system could be driven effectively using frequency auto tracking amplifier system but calibration and monitoring vibration amplitude of welding tip is required during welding process.

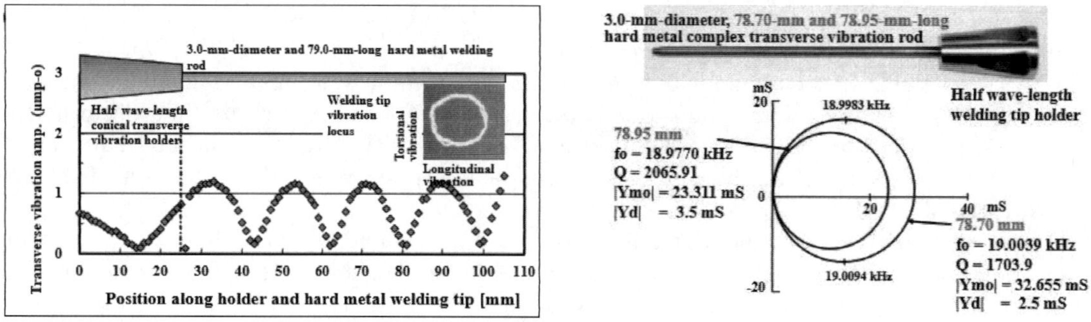


(a) (b)

**Figure 4.** (a) Transverse vibration distribution along 19.5 kHz half-wave length transverse vibration holder and 3.0-mm-diameter, 79-mm-long hard meal complex transverse vibration welding tip, (b) Free admittance loops of 19.5 kHz complex vibration welding system with 3.0-mm-diameter, 78.70-mm and 78.95-mm-long hard metal complex transverse vibration welding tip without power factor compensating inductance $L_c$.

## 3. Welding conditions of ultrasonic complex vibration welding systems

Figure 5 (a) shows welding condition of 1.0-mm-thick aluminium and copper plate using a 27 kHz 10-mm-diameter complex vibration welding tip of circular vibration locus. 1.0-mm-thick aluminium and copper plates are welded stably from only 1.5 mmp-o (peak-to-zero value) to 2.1 mmp-o with specimen material strength. Stable weld range is wider compared that using conventional linear vibration equipment with linear vibration locus.

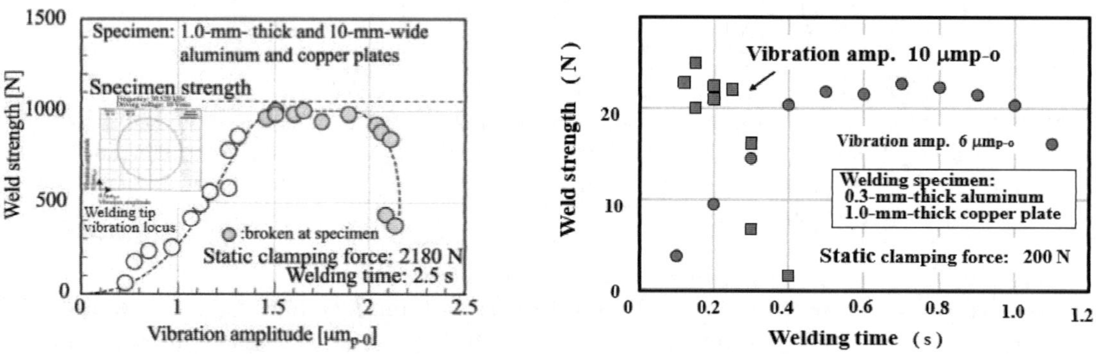


(a) (b)

**Figure 5.** (a) Welding condition of 1.0-mm-thick aluminum and copper plate using a 27 kHz 10-mm-diameter complex vibration welding tip of circular vibration locus, (b) Welding condition of 0.3-mm-thick aluminum and copper plate using a 19.5 kHz 3.0-mm-diameter hard metal welding tip.

Figure 5 (b) shows welding condition of 0.3-mm-thick aluminium and 1.0-mm-thick copper plate using a 19.5 kHz 3.0-mm-diameter, 79-mm-length hard metal complex vibration welding tip of circular vibration locus. Maximum welding strength over 20 N is obtained from 0.1 s to 0.25 s welding time under 10 mmp-o vibration amplitude and maximum strength is obtained between 0.4 s and .0 s. Maximum strength was obtained at wider welding condition compared with a conventional linear vibration welding system.

Using a complex vibration welding tip, required vibration amplitude and static pressure become smaller and stable welding range becomes wider. Furthermore, large and uniform welding strength is obtained independent of welding specimen direction.

## 4. Welded conditions of ultrasonic complex vibration welding systems

Figure 6 shows conditions of 0.3-mm-thick aluminium and 1.0-mm-thick copper plate welded using a 19.5 kHz 3.0-mm-diameter, 79-mm-length hard metal complex vibration welding tip of circular vibration locus. Aluminum specimens were broken at welding surface or circumference of welded area with weld strength almost martial strength. Center part of welded area were welded on copper plate completely.

Figure 7 (a) shows conditions of 30 lapped 0.02-mm-thick copper, electrode foils and Ni coated copper terminal welded using 19.5 kHz 20-mm-wide complex transverse vibration welding tip. Electrode copper foils and Ni coated copper terminal were welded uniformly with small damage and sufficient weld strength.

Figure 7 (b) shows conditions of 30 lapped 0.02-mm-thick aluminum foils and Ni coated copper terminal welded using 20-mm-wide complex transverse vibration welding tip. These specimens were welded successfully with small deformation, no damage and sufficient weld strength.

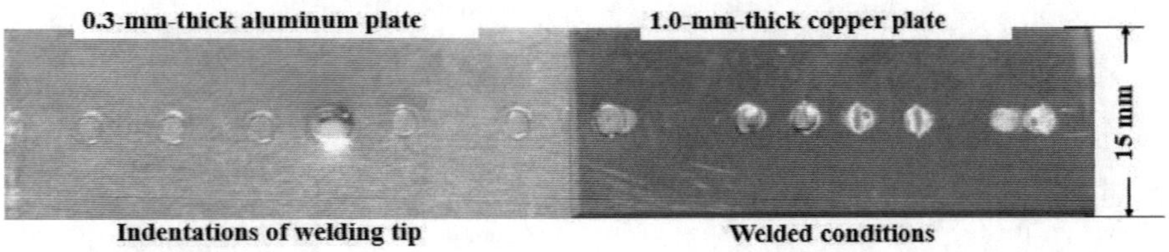


**Figure 6.** Indentations and broken conditions of 0.3-mm-thick aluminum and 1.0-mm-thick copper plate welded using 0.3-mm-diameter hard metal complex transverse vibration welding tip after tensile tests.

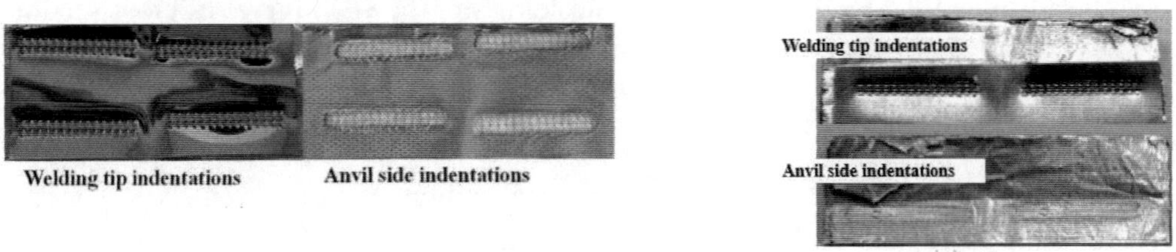


(a) (b)

**Figure 7.** (a) Conditions of 30 lapped 0.02-mm-thick copper, electrode foils and Ni coated copper terminal, (b) Conditions of 30 lapped 0.02-mm-thick aluminum foils and Ni coated copper terminal welded using 20-mm-wide complex transverse vibration welding tip.

**Figure 8.** Conditions of two 0.16-mm-thick, 3.0-mm-wide nickel clad copper terminals and deep nickel coated steel case bottom welded using a 19.5 kHz 3.0-mm-diameter, 79-mm-long hard metal complex transverse vibration welding tip.

Figure 8 shows conditions of two 0.16-mm-thick, 3.0-mm-wide nickel clad copper terminals and deep nickel coated steel case bottom welded using a 19.5 kHz 3.0-mm-diameter, 79-mm-long hard

metal complex transverse vibration welding tip. These two lapped terminals were successfully welded with their material strength. Required welding time is under 0.1 s.
Three or more lapped terminals can be welded with sufficient weld strength. Right side specimens were welded higher welding conditions and welded area was partially colored due to temperature rise at weldment by higher welding condition.

## 5. Cross sections of welded specimens

Figure 9 (a) shows scanning of electron microscope (TEM) images of cross sections of aluminium copper and aluminium-nickel plate specimens welded using 19.5 kHz 10-mm-diameter welding tip under sufficient welding conditions. These specimens were successfully welded under sufficient welding conditions. No diffusion area and no intermetallic structure and no specific structure are observed in welded interfaces.

Figure 9 (b) shows SEM image of cross section of thirty lapped 0.015-mm-thick aluminium electrode foils and 0.8-mm-thick aluminium terminals welded using 19.5 kHz 30-mm-wide complex transverse vibration welding tip. These lapped aluminium foils and terminals were successfully welded with sufficient weld strength, small deformation and no damage.

(a) (b)

**Figure 9.** (a) TEM images of cross sections of aluminum-copper and aluminum-nickel plate specimens welded using 19.5 kHz 10-mm-diameter welding tip, (b) SEM image of cross section of thirty lapped 0.015-mm-thick aluminum electrode foils and 0.8-mm-thick aluminum terminals welded using 19.5 kHz 30-mm-wide complex transverse vibration welding tip.

## 6. Conclusions

Various complex vibration systems with two-dimensional vibration locus were developed for improving welding characteristics of ultrasonic metal welding. Ultrasonic complex vibration welding system using two-dimensional vibration stress has superior welding characteristics and required vibration velocity and clamping force were decreased significantly compared with conventional linear vibration system. Welding of dissimilar metal specimens including many metal foils such as aluminium and copper which are essential for electronic devices and multi-layer fuel cell, battery or EDLC electrodes for electric or hybrid automobile and other various industry fields. Complex vibration using two-dimensional and three-dimensional vibration is effectively applied various applications of higher power ultrasonic.

## 7. References

[1] Tsujino J and Sugimoto E 2015 Ultrasonic welding using a long and thin complex transverse vibration welding tip with vibration detector and static pressure controller *Proc. 2015 IEEE International Ultrasonics Symposium* (IEEE Xplore)

[2] Tsujino J and Sugimoto E 2015 Ultrasonic welding using a long and thin complex transverse vibration welding tip with vibration detector and static pressure controller *Proc. 2015 IEEE*

[3] Tsujino J 2016 Configurations of Ultrasonic Complex Vibration Welding Systems and Welding Characteristics Using Two-dimensional Vibration Stress *Proc. 10th International Conference Trends in Welding Research & 9th International Welding Symposium of JWS* **pp** 254-257

IOP Conf. Series: Journal of Physics: Conf. Series **1075** (2018) 012071 doi:10.1088/1742-6596/1075/1/012071

# Identification vocal cord condition during treatment process with dysphonia severity index

**M H Trihatmojo[1], Suprayogi[1] and H Bethaningtyas[1]**

[1]Engineering Physics, Faculty of Electrical Engineering, Universitas Telkom

E-mail: hilmantrihatmojo@gmail.com

**Abstract**. The purpose of this study was to identify vocal cord condition during treatment process with Dysphonia Severity Index (DSI). The DSI is based on combination of acoustic parameter of MPT(s), F1, Sound Pressure Level (SPL) and jitter that represent vocal quality. Direct measurements of 24 data voice based on 8 patients with vocal cord disorder was taken. The subjects were measured on 3 different day with an interval of 7-9 days. The difference in DSI between each measurement was observed. The mean difference DSI on patients with vocal cord disorder on 3 different day measurements was 0.54, 0.88 and 1.54 respectively. The more positive the patients index, the better vocal quality of the patients.

## 1. Introduction

The vocal cords are one of the human organs that have an important role in communication process. A person who have a disorder in their vocal cord will have difficulty pronouncing vowel sound clearly in a sustained time [1]. Patients with severe vocal cord disorder will require outpatient care to monitor their health condition or so called monitoring. The purpose of monitoring is to carefully observe a certain condition or behavior so the information obtained could be the basis for making further action decisions. The patient's medical record information during the outpatient process is required to obtain the results of the physical examination, diagnosis, treatment measures and other services that have been provided to the patient [2]. The invasive method will make the patients uncomfortable to check their health condition [3]. Therefore, a diagnosis based on sound analysis will aid the health worker to detect a disorder of the vocal cords and make the patient more comfortable to monitor their health condition.

Signal speech processing allow the extraction of a set of sound parameters that can be used to diagnose disorder in vocal cords. Currently, the acoustic parameters used in the application of acoustic analysis are the fundamental frequency (F0), jitter, shimmer and HNR[4]. However the measurements of jitter, shimmer, and HNR could not indicate the disorder severity of vocal cord patients [5]. Based on previous research, the research in acoustic analysis of jitter, shimmer, HNR and detection the type of vocal cord disorder using wavelet biorthogonal trasnformation[1] has been done, but the studies to monitor the patient's cure rate during treatment process with parameter acoustic have not been performed.

Rate of the vocal cord disorder severity could be measured with Dysphonia Severity Index(DSI). DSI is a weighted combination acoustic parameter of Maximum Phonation Time (MPT), Formant 1 (F1), Sound Pressure Level (SPL) and jitter. In DSI index, a positive index value shown a better vocal

IOP Conf. Series: Journal of Physics: Conf. Series **1075** (2018) 012071 doi:10.1088/1742-6596/1075/1/012071

quality and negative index value shown the worst vocal quality [5]. The measurements of patients voice quality based on DSI will aid medical personnel to assess the progress of patients health.

## 2. Methodology

### *2.1. Human Voice Production Process*

The process of sound production in humans can be divided into three physiological processes, i.e. the formation of airflow from the lungs, the change of air flow from the lungs into voiced and unvoiced sound known as phonation and articulation which is the process of modulation or sound settings be a specific sound. Figure 2.1 shows the body organs involved in the production process of human voice [6]. The air flow generated by the lung muscle drive is constant. When the vocal cords are in contraction, the passing airflow makes it vibrate [6]. The air flow will be cut by the movement of the vocal cords into a quasi-periodic pulse signal [7]. Voice signals have characteristics that are formed from a combination of frequency, amplitude and phase. The speaker's voice characteristics have two categories. The first is through the subjective perception of the listener. For example pitch, where one person has a low / heavy base sound while the other is higher or both have the same gender. The second criterion is objective physical characteristics of the sound acoustic emission of sound waves, for example the fundamental frequency with units of Hertz [1].

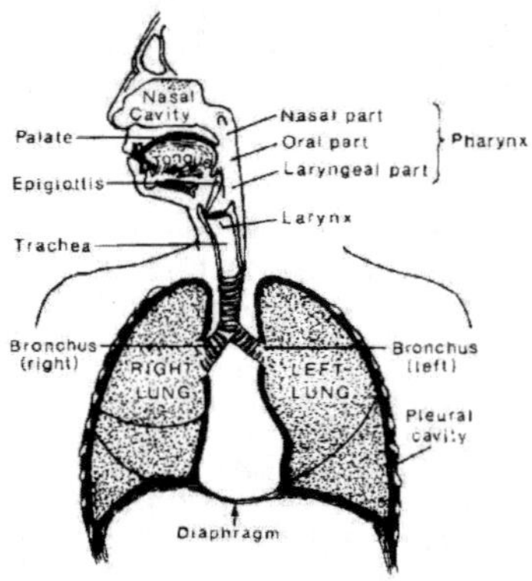


**Figure 2.1** Human Speech Organs

### *2.2. Vocal Cord Disorder*

The vocal cords are place of vibration in human voice. Uncontrolled open and closure processes in the vocal cords are called vocal cord disorder or dysphonia. The various vocal cords disorder include vocal cord Nodules which are bilateral swell on vocal cords of varying size found on the part with the vocal cord membrane, Laryngopharyngeal Reflux (LPR), a state in which gastric acid moves towards the upper esophagus of the larynx, and Laryngitis which is an inflammation that occurs in the larynx, and others [1].

### *2.3. Accousic Parameter*

The acoustic parameters used in the application of acoustic analysis are the Maximum Phonation Time (MPT), Formant 1 (F1), Sound Pressure Level (dB), and jitter. The process begins with sound input from a microphone connected to the matlab program and stored in wav format. Furthermore, the calculation of acoustic parameters and DSI values are measured. The calculation of DSI through a combination of acoustic parameters is performed to see the value index. If the value obtained is still below the normal sound standard then the voice in the next week will be re-insert to obtain a comparison of DSI values during the recovery process.

IOP Conf. Series: Journal of Physics: Conf. Series **1075** (2018) 012071 doi:10.1088/1742-6596/1075/1/012071

*2.3.1. Maximum Phonation Time (MPT).* Maximum Phonation Time is the maximum time (in seconds) for which a person can sustain a vowel sound when produced on one deep breath at a relatively comfortable sound[5].

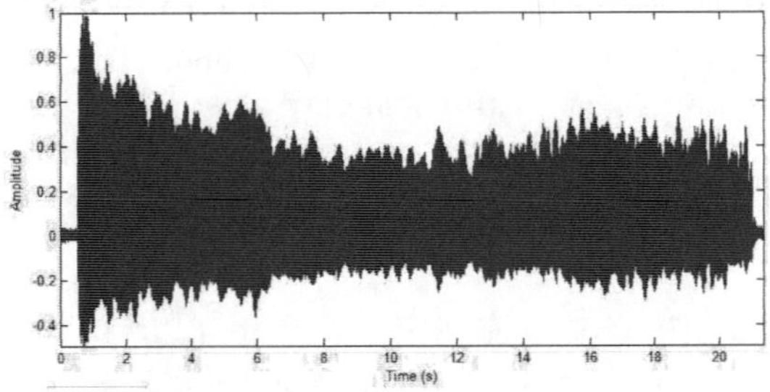


**Figure 2.2** Time Domain Signal

Figure 2.2 shows the time domain signal with the x-axis showing the time (s) and the y-axis as Amplitude. The maximum value on the x axis is the MPT value to be measured.

*2.3.2. Formant Frequencies.* Formant frequency is the peak of the spectral sound which is wholly influenced by the articulator. In the processing of speech signals, the formant is an acoustic resonance of the human vocal tract [8]. Formant frequencies obtained from the determination at the peak of resonance on the voice signal segment. Formant parameters are determined by "peak picking" on the filter response curve or by solving the square root equation [9]. The formant frequency equation can be seen in equation 1, where fs is the sample frequency and is the first formant position.

$$F = \frac{F\theta}{2\pi}\theta_0 \text{ Hz} \tag{1}$$

*2.3.3. Sound Pressure Level.* SPL is a logarithmic measure of the root mean square of sound pressure compared to the reference sound pressure measured in decibels (dB) [10]. The sound pressure level equation can be seen in equation 2 where P0 is the reference pressure in air (P0 = 20 μPa) and P represents the rms of the measured sound pressure [11].

$$\text{SPL} = 20\ \log(\frac{P}{Po})\ \text{Db} \tag{2}$$

*2.3.4. Jitter.* Jitter represents the difference in the absolute average value between successive periods divided by the average period. Jitter representation can be seen in equation 3 [3].Ti representing the length of the obtained base frequency period, N represents the number of basic frequency periods obtained.

$$\text{Jitter (lokal)} = 1 + \frac{\frac{1}{N-1}\sum_{i=1}^{N-i}|Ti - T_{i+2}|}{\frac{1}{N}\sum_{i=1}^{N-i}Ti} x\ 100\ \% \tag{3}$$

*2.3.5. Dysphonia Severity Index.* DSI is a multiparameter measurement consisting of a combination of acoustic parameter variables to represent one's vocal quality [4]. DSI mathematical equations can be seen in equation 4 with parameters used in DSI is the maximum pronunciation time (MPT) Formant 1, Sound Pressure Level (SPL), and jitter.

$$\text{DSI} = [0.13 \text{ x MPT} + 0.0053 \text{ x } F0(high) - 0.26 \text{ x } I\,(low) - 1.18 \text{ x } jitter + 12.4\ ] \tag{4}$$

## 3. Result

*3.1. Data Collecting*
The data were collected by recording the patient's voice at Dustira Cimahi Hospital using Roland Quad-Capture UA-55 soundcard and Shure SM-58 microphone. Direct recording and retrieval of 24 voice data were performed on 8 patients with vocal cords disorder. The subjects were measured three times in a different day, with a time interval of approximately 1 week to monitor the condition of the patient's vocal cord. Then the DSI value is measured to see the difference on each data measured.

*3.2. Data Processing*
The data that has been taken is processed using Matlab R2013b to get the value of Maximum Phonation Time (s), Formant 1 (Hz), Sound Pressure Level (dB), jitter, and DSI. The mean values of parameters and DSi on each measurement of control patients can be seen in Table 3.1

**Table 3.1** Mean Value of DSI and Acoustic Parameter in each measurement

| No | Measurement | MPT(s) | F1 (Hz) | SPL (dB) | Jitter (%) | DSI |
|---|---|---|---|---|---|---|
| 1 | Week 1 | 6.909 | 718.912 | 59.761 | 0.874 | 0.539 |
| 2 | Week 2 | 8.509 | 758.203 | 60.275 | 0.8213 | 0.884 |
| 3 | Week 3 | 12.2 | 822.661 | 60.987 | 0.8037 | 1.541 |

Table 3.1 shows the mean values of the parameters and the DSI value of patients on each measurement. In the 1st week the average DSI is 0.5391, the 2nd week is 0.884 and the 3rd week is 1.5412. The DSI values in the 1st and 3rd measurements show significant differences.

In this study 24 voice data were taken from 8 patients in three different days. In 8 patients with vocal cord disease there are 3 different types of diseases: 1 laryngitis, 5 Laryngopharyngeal Reflux (LPR), 2 thyroids. The development of each acoustic parameter and the DSI value was observed.

*3.2.1. MPT Measurement Result.* MPT is obtained by taking the maximum phonation time during the recording duration in a sustained voice / a /. The MPT value of each patient during the measurement process had an average value of weeks 1, 2 and 3 of 11,649,12,34,21,38 (s) for laryngitis patients, 6,954, 8,762 and 11.63 (s) for LPR patients as well as 4,425, 5,94 and 9,041 for patients thyroid. For each patient in each measurement there is an increase in MPT value. The increasing trend of value indicates that the sound quality of patients is getting better. MPT view in each patient can be seen in Figure 3.1

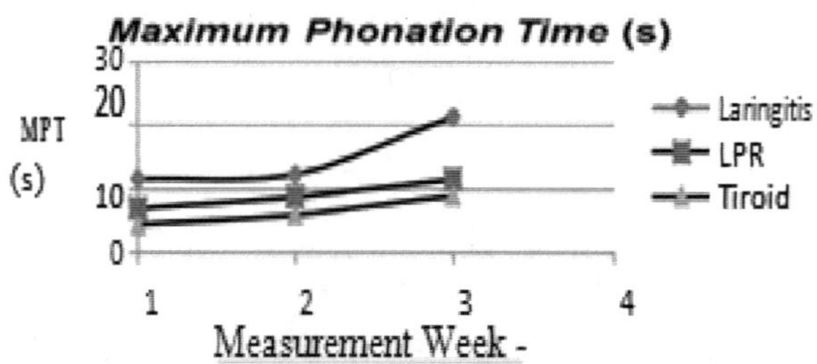


**Figure 3.1** MPT(s) Parameter in Each Measurement

*3.2.2. Formant 1 Measurement Result.* Formant 1 measurements were performed by taking voice data in patients with vocal cord disorder. Formant frequency is an acoustic resonance of human vocal tract [8]. The patient's Formant 1 values had mean values of weeks 1,2 and 3 of 557,352,469,14 and 574,847 (Hz) for laryngitis patients, 753,535, 824,237 and 838,863 (Hz) for LPR patients as well as 713.136, 737.648 and 906,065 (Hz ) for thyroid patients.

Figure 3.3 shows the value of Formant 1 in each patient in each measurement. The F1 value of each patient shows the variation in the value of each measurement, indicating that the F1 parameter still has different values for each type of dissorder at each measurement.

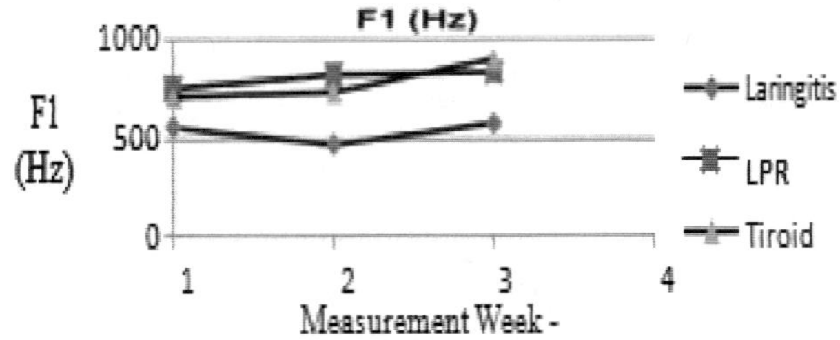


**Figure 3.2** F1 Parameter (Hz) in Each Measurement

*3.2.3. Sound Pressure Level Measurement.* The SPL value at each measurement show varying values in each patient with different disorder types. In some patients there was an increasing tendency in SPL values during routine control but in some patients there was an unstable SPL value. The SPL scores for each patient varied with the mean values of measurements at weeks 1, 2 and 3 of 60,496, 57,476, and 58,401 (dB) for laryngitis patients, 60,698, 61,964 and 61,834 (dB) for LPR patients and 57,052, 57,451 and 60,163 (dB) for thyroid patients. Display values of patient SPL parameters during the measurement process can be seen in Figure 3.4.

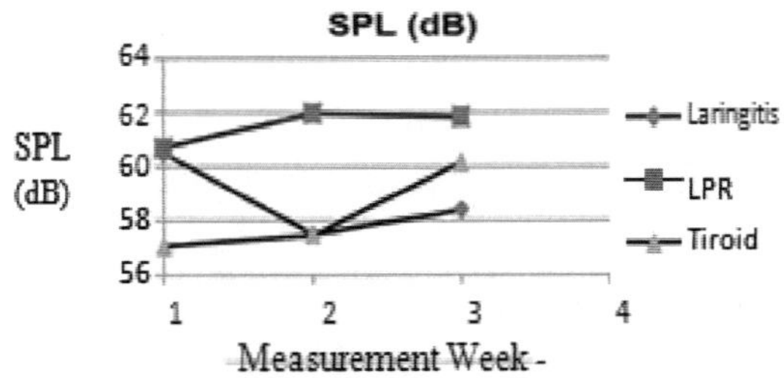


**Figure 3.3** SPL(dB) Parameter in Each Measurement

*3.2.4. Jitter Measurement Result.* Jitter measurements are obtained in the same way as other parameters, and the jitter value of patient's during the measurement process was stored. The mean values of the jitter parameters at measurements weeks 1, 2 and 3 were 0.98, 0.941 and 0.92 (%) for laryngitis patients and LPR patients were 0.833, 0.816, 0.767 (%) and 0.924, 0.775 and 0.838 for thyroid patients.

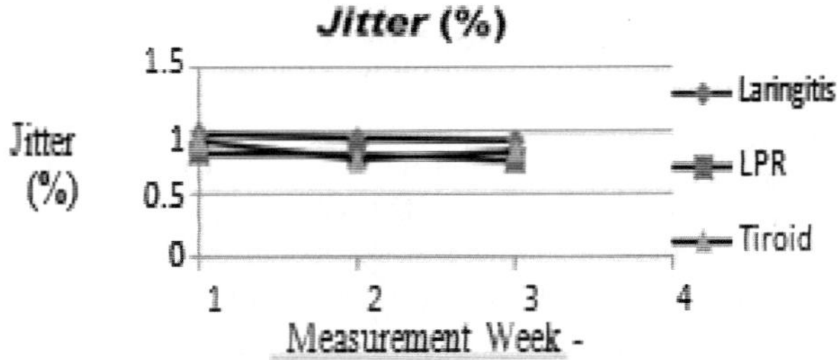


**Figure 3.4** Jitter Parameter in Each Measurement

*3.2.5. DSI Measurement Result.* DSI measurements were obtained by calculating equation 4 which consisted of combination of Maximum Phonation Time (s), Formant 1 (Hz), Sound Pressure Level (dB) and jitter. The DSI value of each patient showed an increasing trend in each measurement on the different types of disorder with the mean values of weeks 1, 2 and 3 measurements of -0.017, 0.437

and 1.956 for laryngitis patients and in LPR patients of 0.534, 0.834, and 1.376 and DSI of 0.831, 1.232, and 1.747 for thyroid patients.

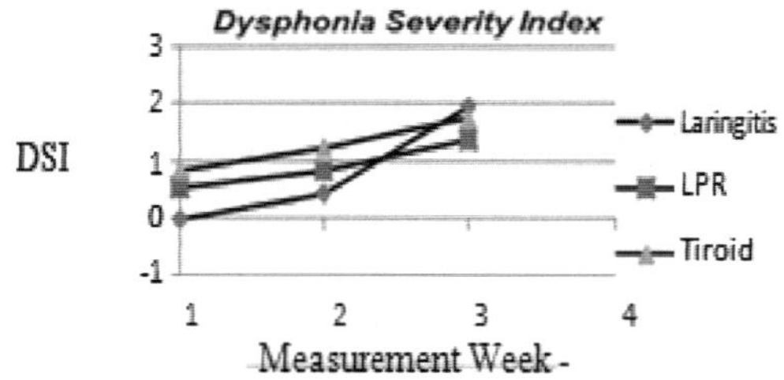


**Figure 3.5** DSI Parameter in Each Measurement

DSI values can represent the quality of a person's vocal, an increasingly positive DSI score indicating good sound quality [5]. In Figure 3.6 we can see the results of measurements on the DSI values of patients with different types of diseases.

## 4. Conclusion

Acoustic parameter analysis results show an increase in I-low, F0-High and MPT and decrease in jitter value. With the values of each parameter at weeks 1, 2 and 3 for I-low (SPL) of 59,761, 60,275 and 60,987 (dB), F0-High of 718,912, 758,203 and 822,661 (Hz), MPT of 6,909,8,509 and 12,201 ( s) as well as jitter of 0.874, 0.821 and 0.804 (%).

The DSI score of each patient with vocal cord disease shows an increasing trend in each measurement with the lowest DSI value to the highest range (-0.9133 - 2.5212) and the mean values of weeks 1, 2 and 3 are 0.5391, 0.884 and 1.5412.

## References

[1] Firda Nurmalida, "Analisa Deteksi Jenis Kelainan Pita Suara Dengan Menggunakan Transformasi Wavelet Biorthogonal",Tugas Akhir, Universitas Telkom, 2016
[2] Suryadi, Irfani dan Lingga Jaya Bermana Putra."Sistem Informasi Rekam Medis Rawat Jalan Pada Rumah Sakit Umum Daerah Palembang BARI".
[3] Achmad, Tri dan Luqman Hakim. "Identifikasi Suara Serak Berbasis Tranformasi Wavelet dan ALgoritma Jaringan Syaraf Tiruan" Seminar on Intelligent Technology and its Application, ITS, 2012. Hal : 119
[4] Joao Paulo Teixeira dan Paula Odete Fernandes "Acoustic Analysis of Vocal Dysphonia" Polytechnic Institute of Braganaça – Campus Sta. Apolónia 5301 857 Bragança – Portugal,(2015).
[5] Florts L. Wuyts. "The Dysphonia Severity Index: An Objective Measure of Vocal Quality Based on Multiparameter Approach", Journal of Speech Language and Hearing Research.2000
[6] Borden, "Speech Science Primer:Physiology, Accoustic, and Perception of Speech",1994.
[7] ArryA.A,"ProsesPembentukandanKarakteristikSinyalUcapan",Teknik Elektro,ITB,Bandung.
[8] Ashish, Bageshree V. Sathe-Pathak. "Extraction of Pitch and Formants and its Analysis to identify 3 different emotional states of a person" Pune University,International Journal of Computer Science Issues, 2012
[9] Fausto,Roy C. Sneel, "Formant Location From LPC Analysis Data",IEEE Transactions On Speech and Audio Processing,1993.
[10] Errede Steven,"The Human Ear-Hearing, Sound Intensity and Loudness Levels",UIUC Physics 406 Acoustical Physics of Musics,University of Illinois.
[11] Joseph,Marc Schoenwiesner,"Implementing Loudness Models in Matlab", Int.Conference on Digital Audio Effects,2004.

# Acoustic properties of Temen and Wulung bamboo as a material for Gambang: Sundanese traditional musical instrument

**K Leonardo[1], J Sarwono[1] and I Prasetiyo[1]**

[1]Acoustic Laboratory, Engineering Physics, Institut Teknologi Bandung, Ganesa 10, Bandung 40132, Indonesia

Email: kevin.leonardo273@gmail.com

**Abstract.** Wulung bamboo (*Gigantochloa atroviolacea Widjaja)* and Temen bamboo (*Gigantochloa atter*) are common materials for making Sundanese traditional music like Gambang in West java. From subjective judgment of the maker like Mr. Handiman Diratmasasmita, it is perceived that Temen bamboo is more musical and softer than Wulung bamboo when the Gambang are played. Hence, it is instructive to investigate further in order to find the acoustic parameters of both the bamboos that associate with the differences. To deal with the problem, a series of measurements were carried out in ITB anechoic chamber. Two similar-tuned rattle tube from each bamboo are considered in this work. To capture all required acoustical data, a microphone was positioned behind the opening of the rattle tubes. The sounds of two bamboos are excited by impact hammer and the Frequency Response Function (FRF) obtained by spectrum analyzer system. The results are then analyzed and discussed by considering two frequency ranges namely low-mid frequencies and high frequencies. For low-mid frequency range, it is found that the overtones ratio are not much different for example f, 2.4 f, 3.4 f, 4.9 f, 6.9 f are present for the Wulung bamboo while f, 2.2 f, 2.76 f, 4.82 f, 7.02 f exist in the Temen bamboo where f of 527.3 Hz is fundamental frequency or pitch. However, the overall fundamental and harmonics amplitude of Temen bamboo are higher than Wulung bamboo. At high frequency range the amplitudes of the Wulung bamboo are significantly higher than the Temen bamboo particularly for frequencies above 10 kHz. These results can be attributed to a more musical softer sound of the Temen bamboo.

## 1. Introduction

Gambang is a Sundanese musical instrument that made from bamboo. This instrument consists of bamboo tubes, which tuned in diatonic scale and arranged in parallel from low note to high note. The bamboo tube has similar shape compared to angklung, but this instrument played by hitting the tube with rubber mallet. According to Handiman Diratmasasmita, a Sundanese musical instrument maker, there are two materials that used to build the bamboo tubes. The first one is Wulung bamboo (*Gigantochloa Atroviolacea Widjaja*) and the second one is called Temen bamboo (*Gigantochloa Atter*). The later, according to the maker, are more musical and softer than the first.

There are some researches about the bamboo tube and bamboo musical instrument. Zainal et al.

presented analytical solution to determine pitch of angklung compared to the sound measurement. Siswanto et al. [2] presented the sound characteristic of three-rattle angklung. Sudarsono et al. [3] presented the acoustic analysis of pentatonic angklung. Dwiartama et al. [4] presented early observation of temen bamboo from Surade region in Indonesia. Budi et al. [5] presented acoustical analysis of temen bambu taken from Kuningan and Sukabumi district in Indonesia. Unfortunately, there's still no research about the sound quality of these two bamboos compared to the subjective preference of the maker.

To find the correlation between sound quality of two bamboos and subjective preference of the maker, then acoustic measurement is required. The obtained frequency spectrum data of both bamboos will be compared to analyse their differences objectively.

(a) (b)

**Figure 1.** (a) Sundanese Gambang (b) Gambang played by Handiman Diratmasasmita.

## 2. Methods and material

### *2.1. Bamboo Tubes*

Two tubes from each kind of bamboo are similarly tuned to C5 note (527.25 Hz). These two bamboos are taken from Surade district, Bandung. The tube length, diameter, and age of each bamboo are adjusted to Handiman's preference. In table 1 the physical properties of the finished two bamboo tubes are shown.

**Table 1.** Physical properties of each bamboo tube.

| Name | Length (cm) | Diameter (cm) | Weight (g) |
|---|---|---|---|
| Wulung bamboo (*Gigantochloa Atroviolacea Widjaja*) | 48.1 | 4.41 | 183 |
| Temen bamboo (*Gigantochloa Atter*) | 47.9 | 4.29 | 133 |

### *2.2. Experimental Setup*

The acoustical properties from two tuned bamboo tubes were measured on the anechoic chamber in Bandung Institute of Technology. First, the bamboo was hanged using the rope. Then, three microphones were placed 50 centimeters behind of the tube opening, the side and the top of the tube as shown in Figure 3. Another three microphones were placed in the same sides, but they were placed 100 centimeters from the tube. The tubes were struck five times by impact hammer and the sound pressure data obtained using the microphones and Reaper software. The obtained data from each microphone were averaged. In this paper, data from microphones placed 50 centimeters behind the tube opening of each bamboo are analyzed.

IOP Conf. Series: Journal of Physics: Conf. Series **1075** (2018) 012072 doi:10.1088/1742-6596/1075/1/012072

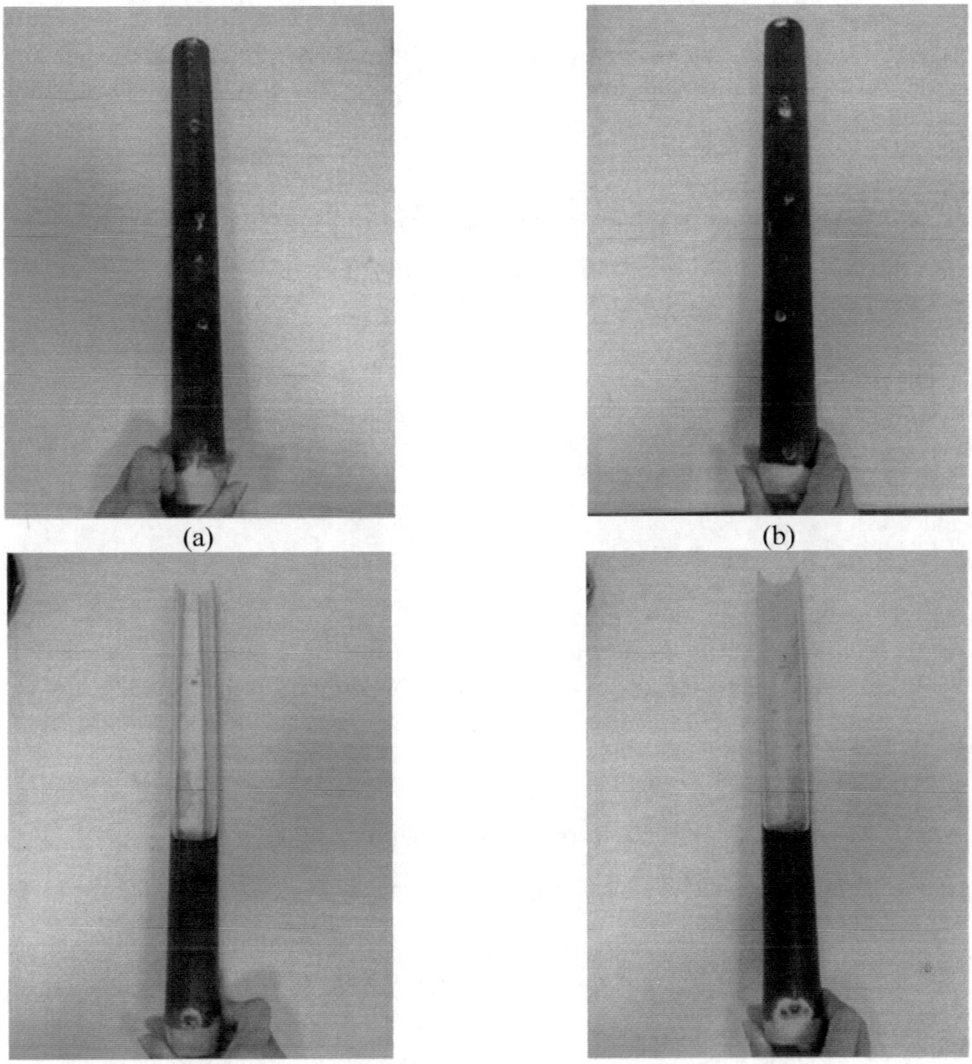

(a) (b) (c) (d)

**Figure 2.** Wulung bamboo (a) back (c) top, and Temen bamboo (b) back (d) top.

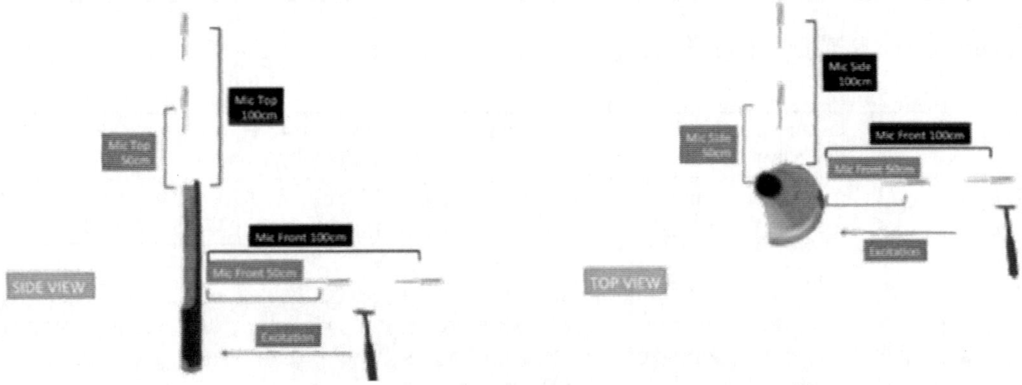


(a) (b)

**Figure 3.** Microphone placement (a) side view (b) top view.

IOP Conf. Series: Journal of Physics: Conf. Series **1075** (2018) 012072 doi:10.1088/1742-6596/1075/1/012072

## 3. Results and discussion

The results show the comparison of averaged spectral frequency between Wulung bamboo and Temen bamboo as can be seen in Figure 4 for low-mid frequencies between 100 Hz – 6000 Hz and Figure 5 for high frequencies.

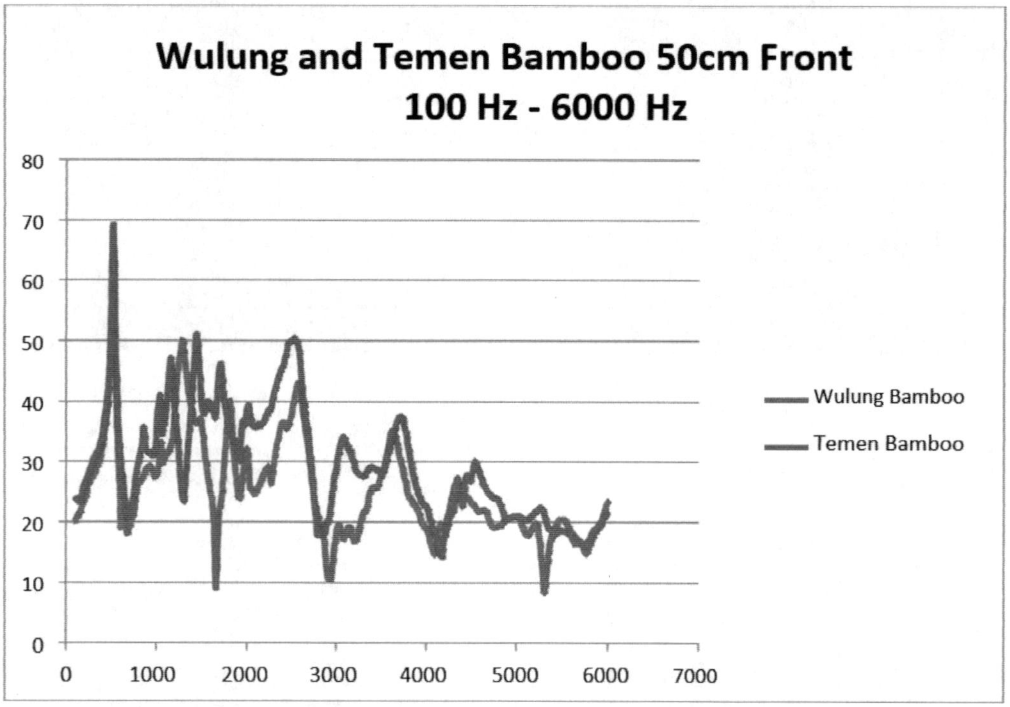


**Figure 4.** Comparison between Wulung and Temen bamboo for low-mid frequencies between 100 Hz – 6000 Hz.

**Table 2.** The fundamental and harmonics comparison of Wulung and Temen bamboo (1st and 2nd harmonics).

| Name | Fundamental (Hz) | dB | 1st Harmonics (Hz) | dB | Ratio ($f_1/f_0$) | 2nd Harmonics (Hz) | dB | Ratio ($f_2/f_0$) |
|---|---|---|---|---|---|---|---|---|
| **Wulung** | 527.3 | 66.45 | 1289.060 | 50.00 | 2.40 f0 | 1816.400 | 39.9 | 3.40 f0 |
| **Temen** | 527.3 | 69.10 | 1160.150 | 46.96 | 2.20 f0 | 1453.125 | 51.1 | 2.75 f0 |

**Table 3.** The comparison of Wulung and Temen bamboo for 100 Hz – 6000 Hz (3rd and 4th harmonics).

| Name | Fundamental (Hz) | dB | 3rd Harmonics (Hz) | dB | Ratio ($f_3/f_0$) | 4th Harmonics (Hz) | dB | Ratio ($f_4/f_0$) |
|---|---|---|---|---|---|---|---|---|
| **Wulung** | 527.3 | 66.45 | 2589.80 | 42.90 | 4.90 f0 | 3644.500 | 34.73 | 6.90 f0 |
| **Temen** | 527.3 | 69.07 | 2542.97 | 50.43 | 4.82 f0 | 3703.125 | 37.43 | 7.02 f0 |

In low-mid frequency range, the fundamental and harmonic frequency has been analyzed. As we can see in Figure 4, each harmonic from two bamboos are related. However, the overall amplitude of fundamental and harmonic frequency of Temen bamboo is higher than Wulung bamboo; only 1st harmonic of Wulung bamboo is higher than Temen bamboo.

In high frequency range, the amplitude above 10 kHz of Wulung bamboo is much higher than Temen bamboo. Moreover, there's no amplitude in frequency 12 kHz to 14 kHz of Temen bamboo.

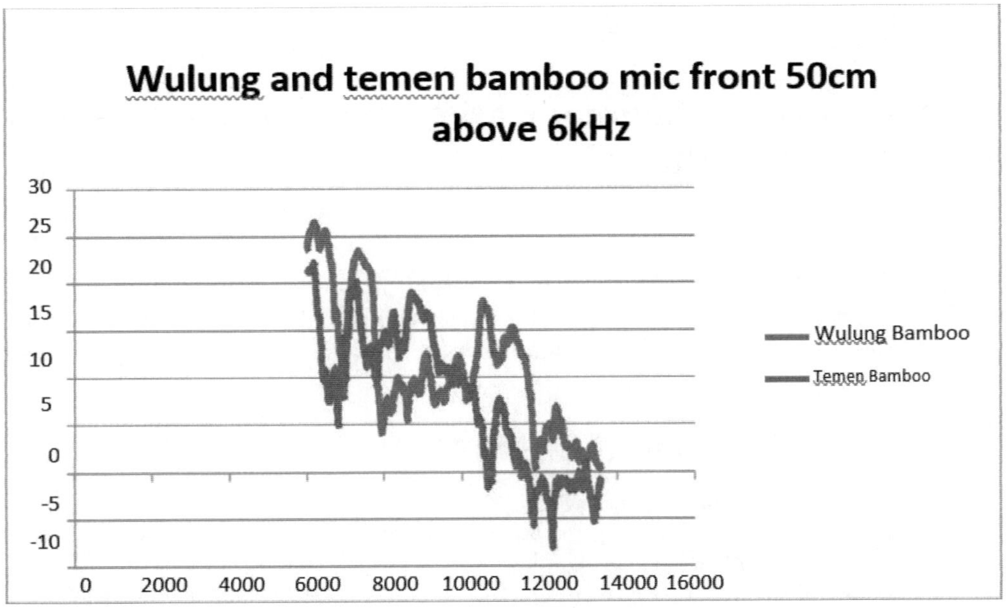


**Figure 5.** Comparison between Wulung and Temen bamboo for frequency above 6kHz.

## 4. Conclusion

Based on this measurement results, Temen bamboo has higher overall fundamental and harmonic amplitude and Wulung bamboo has higher amplitude in high frequency range especially above 10 kHz. These results can be contributed to the more musical and softer sound of Temen bamboo related to subjective preference of the maker.

## 5. References

[1] M R Zainal, S A Hussain and C H Azhari 2009 Pitch and timbre determination of the angklung *Am. J. Appl. Sci.* vol 6 no. 1 p. 24

[2] W A Siswanto, L Tam and M Z Kasron 2012 Sound Characteristic and Sound Prediction of the Traditional Musical Instrument the Three-Rattle Angklung *J. Acoust. Vib.* vol. 17 no. 3 p. 120

[3] A S Sudarsono and I G N Merthayasa 2013 Acoustic analysis from pentatonic Angklung *Proceedings of Meetings on Acoustics* vol. 19 p. 035078

[4] A Dwiartama, E M Budi, J Sarwono, E Ekawati, M F Pratama, M Hersaputri, E Aprilia and Listyana 2016 Pemanfaatan Sumberdaya Hayati Berkelanjutan melalui Indikasi Geografis: Bambu Temen Surade sebagai Bahan Baku Angklung Padaeng *Prosiding Seminar Nasional Pendidikan Biologi* p. B-1

[5] E M Budi, *et al.* 2016 Pengujian Karakteristik dan Kualitas Bambu Temen Hitam (*Gigantochloa Atroviolacea Widjaja*) Kabupaten Sukabumi dan Kabupaten Kuningan sebagai Bahan Baku Angklung, Institut Teknologi Bandung

**Acknowledgments**

The authors would like to give thanks to Mr. Handiman Diratmasasmita and fellow Sundanese musical instrument maker in Bale Angklung Bandung for subjective preference, history of the instrument, and made bamboo tubes for this research.

# Correlation between earphone usage and visiting game centre with hearing threshold

**S N Wicaksono[1], L Dewanti[1], and N Purnami[1]**

[1]Department of Otorhinolaryngology Head and Neck Surgery, Faculty of Medicine, Universitas Airlangga-Dr. Soetomo General Hospital, Surabaya 60231 Indonesia

E-mail: simon_nathanael_95@yahoo.com

**Abstract.** Noise-induced hearing loss is affecting more people than in the past. For the young adult NIHL is related to noises produced during recreational and leisure activity. In Surabaya, there are still little to no data regarding hearing loss related to recreational and leisure activity. The objective of this study was to know the correlation between earphone usage and visiting game centre with hearing threshold. This observational analytic with cross sectional design study involved 16 students who visited game centre in November 2016-January 2017 and fulfilled the inclusive criteria and exclusive criteria. The data collected using questionnaire and hearing threshold examination and analysed using Spearman's rho test with p-value (p) <0.05 and confidence interval (CI) 95%. There were no significant correlation between volume of earphone usage with hearing threshold. There was no significant correlation between duration of earphone usage with hearing threshold. There was no significant correlation between frequency of earphone usage with hearing threshold. There was no significant correlation between duration of visiting game centre with hearing threshold. There was no significant correlation between frequency of visiting game centre with hearing threshold. There was no correlation between earphone usage and visiting game centre with hearing threshold.

## 1. Introduction

Hearing is the neural perception of sound [1]. Sound is produced by everything in our surrounding that is capable of producing a disturbance pattern in air molecules [1]. There is a special receptor inside our body that is capable to transduce fluid movement into neural signals. This special receptor is auditory hair cell in the organ of Corte inside the cochlea. There are approximately 15.000 hair cells within each cochlea [1]. Hearing loss, or deafness, may be temporary or permanent, partial or complete [1]. Hearing loss is classified into two types, conductive hearing loss and sensorineural hearing loss. In sensorineural hearing loss, sound waves are transmitted to the inner ear, but they are not translated into nerve signals that are interpreted by the brain as sound sensation. This condition can be caused by defect in organ of Corte, in the auditory nerves, or rarely in the ascending auditory. One of the common causes of partial hearing loss is a degenerative, age-related process that occurs to the hair cell called neural presbycusis. This process of partial hearing loss is unavoidable because exposure to even ordinary modern-day sounds eventually damages hair cells. The sensorineural hearing impairment that develops over years of exposure to noise at moderately high levels is called noise-induced hearing loss (NIHL) [2]. Unfortunately, partial hearing loss due to the excessive exposure to loud noises is affecting more people than in the past. This phenomenon happens because we live in an increasingly noisy environment. In Australia, 1 in 6 people have some degree of hearing loss [3]. Around 10 million people, about 1 in 6 people, in the United Kingdom suffer from some

IOP Conf. Series: Journal of Physics: Conf. Series **1075** (2018) 012073 doi:10.1088/1742-6596/1075/1/012073

degree of hearing loss [4]. And this number is expected to get higher. Not only that the number is increasing but partial hearing loss also affect younger people. There is audiological evidence of hearing loss in 40% of students between the ages of 16 and 25 years, and 12,5% of students between the ages of 6 to 19 years already suffer from noise- induced threshold shifts [5]. NIHL doesn't only affected the daily activities, but also reduce physical and social execution behavior status and may induce psychological symptoms of feeling of isolation and exclusion, anxiety, depression, and cognitive disorder [5]. So NIHL is a very disabling condition for younger people in studying period.
Fortunately, this process can be prevented, unlike the neural presbycusis, partial hearing loss due to excessive exposure to loud noise can be prevented by simply avoiding the source of the noise or cover the ear with protective devices, such as earplug and other devices. This fact leads to several studies to identify the source of noise. One of the well- known source being workplace with noisy machinery or technological processes in industrial settings [2]. But for the young people, teenagers and young adults who don't have to work, NIHL is related to noises produced during recreational and leisure activity [5]. One of the major sources of noise exposures in teenagers and young adults is music, whether it is from personal music player or other devices with music player function [6].
One of the common type of portable music player is the MP3 player and the most common type of headphone is the earphone [5]. The noise produced by personal music player can range widely from 60 to 120 dB, with weekly exposure time from <1h to 14h [6]. These indicate that up to 25% of this population is at risk of developing hearing loss when listening to music at this level for 8h, over a long period of time [6]. The other source of noise that can affect the teenagers and young adults is game center. One of the studies to determine the noise exposure level of game center shows that the background noise of a game center was recorded at 61 dBA. When the electronic games where performed, the level of noise reached to 88-90 dBA. This result indicates that game center is one of the possible sources of excessive noise [7].
In Surabaya, there are still little to no data regarding hearing loss related to recreational and leisure activity. After understanding the possible effect of recreational and leisure activity to hearing loss, we assessed the correlation of the use of earphone and playing in game center with hearing threshold.

## 2. Conceptual Framework

Hearing loss can be caused by either age, ototoxic drug, heredity, disease and infection, and noise. This study focusses on noise, which is one of the possible cause that can be prevented. There are two main source of noise that can cause hearing loss: workplace, leisure activity, and living environment. There are already several studies involving the workplace and living environment but very little to no data about the leisure activity. This study concentrates on leisure activity as the risk factor, and try to minimize the exposure to the other risk factor.
Two of the common leisure activity done by teenagers or young adults are earphone usage and visiting game center. Both can affect the hearing threshold depend on the volume or the loudness of the sound, duration of time spends per usage or visit, and the frequency. This study concentrates on collecting data for the volume, duration, and frequency of both earphone usage and visiting game center and the effect of those factors to hearing threshold.

## 3. Methods

This research was an observational analytic study, with the purpose of finding the correlation between earphone usage and visiting game center with hearing threshold. The research design used was cross sectional study design.

### *3.1. Population, Sample, and Sample Size*

The population that was reached through this research was a group of students that visited a game centre in Surabaya. Total sampling of a group of students that visits game centre in the last three months at least one times who were willing to participate in this research. There were 16 respondents. The age of the subjects ranges from 19 to 22 years old with a mean age of 20,44 and standard deviation of 1,031. The distribution of gender is 10 (62,5%) male and 6 (37,5%) female. All of the subjects live in Surabaya and do not exposed to noisy environment. The distribution of the occupation is 15 (93,75%) students and 1 (6,25%) worker. All of the subjects do not expose to noisy workplace.

*3.2. Location*
This research was conducted in Ear Nose and Throat Head and Neck Surgery (ENT-HNS) Department's outpatient clinic RSUD Dr. Soetomo Surabaya during November 2016 – January 2017.

*3.3. Procedure*
The data of this research was taken by:

*3.3.1. Measuring the sound level inside the game center.* Measure the sound intensity level inside the game center using the sound level meter.

*3.3.2. Giving informed consent.* introducing name, institution, and purpose of the research to the respondents should be done before doing the research. Moreover, approval to be respondents is the most important thing that should be obtained.

*3.3.3. Doing the interview.* Asking the respondents about identity, exclusion criteria, earphone usage, and visiting game center habit.

*3.3.4. Measuring the degree of hearing loss.* Measure the hearing threshold in anechoic chamber, RSUD Dr. Soetomo, data collection and record.

## 4. Results

*4.1. Earphone Usage*

Based on the data collected from the questionnaire, the most frequent data (25%) of the volume of earphone usage is 0%. The mean volume of earphone usage is 41,25% with a standard deviation of 27,958. The result of Saphiro-Wilk, test of normality, for the volume of earphone usage is 0,015 ($p<0,05$) which mean the distribution is not normal. The characteristics of respondent's duration of earphone usage can be observed on table 2. The most frequent data of the duration of earphone usage is 0 minutes (25%). The mean duration of earphone usage is 108,75 minutes/day with a standard deviation of 114,302. The result of Saphiro-Wilk, test of normality, for the duration of earphone usage is 0,016 ($p<0,05$) which mean the distribution is not normal.

The most frequent data of the frequency of earphone usage is 30 x/months (56,3%). The mean frequency of earphone usage is 19,88 x/month with a standard deviation of 13,594. The result of Saphiro-Wilk, test of normality, for the frequency of earphone usage is 0,000 ($p<0,05$) which mean the distribution is not normal.

*4.2. Visiting Game Center*
Based on the data collected from the questionnaire, the characteristics of respondent's duration of visiting game centre can be observed on table 4. The most frequent data of the duration of visiting game centre are 1 hour/visit (25%) and 2 hours/visit (25%). The mean duration of visiting game centre is 3 hours/visit with a standard deviation of 2. The result of Saphiro-Wilk, test of normality, for the duration of visiting game centre is 0,032 ($p<0,05$) which mean the distribution is not normal.

The most frequent data of the frequency of visiting game centre is 4 x/months (25%). The mean frequency of visiting game centre is 5,19 x/months with a standard deviation of 3,885. The result of Saphiro-Wilk, test of normality, for the frequency of visiting game centre is 0,024 ($p<0,05$) which mean the distribution is not normal. The mean maximum volume of the game centre is 87,7417 dBA with a standard deviation of 7,6914. The mean minimum volume of the game centre is 77,5167 dBA with a standard deviation of 3,56392.

*4.3. Correlation between Volume of Earphone Usage with Audiometry Result*
Based on the data collected from the audiometry test, the correlation between volume of earphone usage with audiometry result on 4000 Hz can be observed on table 1. Due to the distribution data of

the volume of earphone usage being not normal, the correlation test used for testing the correlation between volume of earphone usage and audiometry result on 4000 Hz is Spearman's rho. The result for both, right ear 0,766 and left ear 0,238 is not significant (p>0,05).

**Table 1.** Correlation between Volume of Earphone Usage with Audiometry Result on 4000 Hz

| Volume of Earphone Usage (%) | Right Ear (dBA) | | | | | Left Ear (dBA) | | | | |
|---|---|---|---|---|---|---|---|---|---|---|
| | 0 | 5 | 10 | 15 | 20 | 0 | 5 | 10 | 15 | 20 |
| **0** | 1 | 0 | 2 | 1 | 0 | 1 | 2 | 1 | 0 | 0 |
| **25** | 0 | 0 | 1 | 0 | 0 | 0 | 1 | 0 | 0 | 0 |
| **30** | 0 | 0 | 0 | 1 | 0 | 0 | 0 | 0 | 1 | 0 |
| **50** | 0 | 2 | 1 | 0 | 0 | 2 | 1 | 0 | 0 | 0 |
| **55** | 0 | 0 | 1 | 0 | 0 | 0 | 0 | 0 | 1 | 0 |
| **60** | 0 | 1 | 1 | 0 | 0 | 0 | 0 | 1 | 1 | 0 |
| **65** | 0 | 2 | 0 | 0 | 0 | 1 | 1 | 0 | 0 | 0 |
| **75** | 0 | 0 | 1 | 1 | 0 | 0 | 0 | 1 | 0 | 1 |
| **Total** | 1 | 5 | 7 | 3 | 0 | 4 | 5 | 3 | 3 | 1 |

### *4.4. Correlation between Duration of Earphone Usage with Audiometry Result*

Based on the data collected from the audiometry test, the correlation between duration of earphone usage with audiometry result on 4000 Hz can be observed on table 3. Due to the distribution data of the duration of earphone usage being not normal, the correlation test used for testing the correlation between duration of earphone usage and audiometry result on 4000 Hz is Spearman's rho. The result for both right ear 0,582 and left ear 0,802 is not significant (p>0,05).

**Table 2.** Correlation between Duration of Earphone Usage with Audiometry Result on 4000 Hz

| Duration of Earphone Usage (minutes/ day) | Right Ear (dBA) | | | | | Left Ear (dBA) | | | | |
|---|---|---|---|---|---|---|---|---|---|---|
| | 0 | 5 | 10 | 15 | 20 | 0 | 5 | 10 | 15 | 20 |
| **0** | 1 | 0 | 2 | 1 | 0 | 1 | 2 | 1 | 0 | 0 |
| **30** | 0 | 1 | 0 | 0 | 0 | 0 | 1 | 0 | 0 | 0 |
| **45** | 0 | 1 | 2 | 0 | 0 | 1 | 1 | 0 | 1 | 0 |
| **120** | 0 | 0 | 0 | 2 | 0 | 0 | 0 | 1 | 1 | 0 |
| **150** | 0 | 0 | 2 | 0 | 0 | 0 | 0 | 0 | 1 | 1 |
| **180** | 0 | 1 | 0 | 0 | 0 | 0 | 0 | 1 | 0 | 0 |
| **195** | 0 | 0 | 1 | 0 | 0 | 1 | 0 | 0 | 0 | 0 |
| **240** | 0 | 1 | 0 | 0 | 0 | 1 | 0 | 0 | 0 | 0 |
| **420** | 0 | 1 | 0 | 0 | 0 | 0 | 1 | 0 | 0 | 0 |
| **Total** | 1 | 5 | 7 | 3 | 0 | 4 | 5 | 3 | 3 | 1 |

### *4.5. Correlation between Frequency of Visiting Game Centre with Audiometry Result*

Based on the data collected from the audiometry test, the correlation between frequency of visiting game center with audiometry result on 4000 Hz can be observed on table 3. Due to the distribution data of the frequency of visiting game center being not normal, the correlation test used for testing the correlation between frequency of visiting game center and audiometry result on 4000 Hz is Spearman's rho. The result for right ear 0,028 is significant (p<0,05). The result for the left ear 0,908 is not significant (p>0,05).

**Table 3.** Correlation between Frequency of Visiting Game Center with Audiometry Result on 4000 Hz

| Frequency of Visiting Game Center (x/months) | Right Ear (dBA) | | | | | Left Ear (dBA) | | | | |
|---|---|---|---|---|---|---|---|---|---|---|
| | 0 | 5 | 10 | 15 | 20 | 0 | 5 | 10 | 15 | 20 |

| | | | | | | | | | | |
|---|---|---|---|---|---|---|---|---|---|---|
| **1** | 0 | 0 | 2 | 0 | 0 | 1 | 2 | 1 | 0 | 0 |
| **2** | 0 | 0 | 1 | 1 | 0 | 0 | 1 | 0 | 0 | 0 |
| **3** | 0 | 1 | 0 | 1 | 0 | 1 | 1 | 0 | 1 | 0 |
| **4** | 0 | 1 | 2 | 1 | 0 | 0 | 0 | 1 | 1 | 0 |
| **5** | 0 | 1 | 0 | 0 | 0 | 0 | 0 | 0 | 1 | 1 |
| **6** | 0 | 0 | 1 | 0 | 0 | 0 | 0 | 1 | 0 | 0 |
| **8** | 0 | 0 | 1 | 0 | 0 | 1 | 0 | 0 | 0 | 0 |
| **10** | 0 | 1 | 0 | 0 | 0 | 1 | 0 | 0 | 0 | 0 |
| **12** | 1 | 0 | 0 | 0 | | | | | | |
| **14** | 0 | 1 | 0 | 0 | 0 | 0 | 1 | 0 | 0 | 0 |
| **Total** | 1 | 5 | 7 | 3 | 0 | 4 | 5 | 3 | 3 | 1 |

## 5. Discussion

### *5.1. Correlation between Volume of Earphone Usage with Audiometry Result*
According to the result of the correlation test between volume of earphone usage with audiometry result, there is no correlation ($p>0,05$) between volume of earphone usage with audiometry result. Well know that the greater the intensity, or in this case volume, the greater is the risk of permanent damage [8]. The result of this study is in contradiction with the result of the study done by Jiang et al., on 2012 which mention that there is a correlation between the volume of earphone usage with audiometry result [9]. This contradiction might be cause by the different level of objectivity in measuring the volume. This study uses the percentage of volume in the personal music device, while the study done by Jiang et al. used the sound level of the device.

### *5.2. Correlation between Duration of Earphone Usage with Audiometry Result*
Based on the data collected in this study and the result of the correlation test, there is no correlation ($p>0,05$) between duration of earphone usage with audiometry result. Based on the theory noise can affect hearing in a variety way, one of them being the duration of the exposure. While a loud but short duration of noise can cause a temporary reduction in hearing ability, a moderate but long duration of noise can cause a permanent reduction in hearing ability [8]. The possible explanation of the result of this study is because duration is not a single factor for hearing ability. This possibility is also supported by another study done by Kim et al., on 2012, which found that student who had used these player for more than 5 years with cumulative usage time of 15 hours x years had significantly elevated hearing thresholds [5]. Hearing ability is affected by volume, frequency and duration at the same time. Unfortunately, in this study the volume, frequency and the duration cannot be analyze using statistic due to the abnormal data distribution.

### *5.3. Correlation between Frequency of Earphone Usage with Audiometry Result*
Based on the result of the correlation test between frequency of earphone usage with audiometry result, there is no correlation ($p>0,05$) between frequency of earphone usage with audiometry result. It is well known that hearing loss is associated with repeated daily exposure to noise [8]. The basic factor associated with hearing loss are intensity, duration, or in this study represent by the duration of earphone usage and frequency of earphone usage, and spectrum. These factors interact with each other in a complex way [8]. It is possible that the cause of the result is because in this study the correlation between frequency of earphone usage with audiometry is analyze separately with the other factor.

### *5.4. Correlation between Duration of Visiting Game Centre with Audiometry Result*
Based on the data collected in this study and the result of the correlation test, there is no correlation ($p>0,05$) between duration of visiting game center with audiometry result of left ear on 4000 Hz. But the duration of visiting game center, based on the correlation test, has correlation ($p<0,05$) with the audiometry result of the right ear on 4000 Hz. The study done by Pangemanan, et al., on 2012 stated that there is no correlation between noise and hearing threshold among game center worker in Manado [10]. Even though both study stated that there is no correlation between the duration of visiting game

center with audiometry result. It is well known that duration is one of the important factor of hearing loss [8]. The possible explanation is that in this study the duration only measures the duration spend on each visit. The study done by Kim et al., mention that student who had used these player for more than 5 years with cumulative usage time of 15 hours x years had significantly elevated hearing thresholds [5]. So, it is possible if there is also a data about how many years the exposure had taken place there will be a correlation between the duration and the hearing threshold.

### *5.5. Correlation between Frequency of Visiting Game Centre with Audiometry Result*

Based on the result of the correlation test between frequency of visiting game center with audiometry result, there is no correlation ($p>0,05$) between frequency of visiting game center with audiometry result. It is well known that the main factor for hearing loss is the intensity of the noise, and the next factor is the duration of the exposure [8]. In this study the duration of exposure is measure into how long do the subject stays in the game center every time the subject visits the game center (duration of visiting game center) and how many time the subjects visit the game center in a month (frequency of visiting game center). In this definition duration of visiting game center have more impact to the duration of exposure, which in turn the hearing loss, than frequency of visiting game center.

## 6. Conclusions

The conclusions of this study are there is no correlation between earphone usage with hearing threshold and There is no correlation between visiting the game center with hearing threshold.

## References

[1] S. L., Human Physiology: From Cells to Systems 7th edition, Cengage, 2010.

[2] Pawlacyzk-Luszczynska, Z. A., Z. K., Z. M. K. and M. Sliwinska-Kowalska, "Noise induced hearing loss: Research in central, eastern, and south-eastern Europe and newly independent states," *Noise Health,* vol. 15, no. 62, pp. 55-66, 2013.

[3] S. Bullock and M. Hales, Principles of Pathophysiology, Australia: Pearson Australia, 2013.

[4] NHS, "Hearing Loss," [Online]. Available: http://www.nhs.uk/conditions/Hearing-impairment/Pages/Introduction.aspx. [Accessed 17 January 2016].

[5] M. G. Kim, S. M. Hong, H. J. Shim, Y. D. Kim, C. I. Cha and S. G. Yeo, "Hearing threshold of Korean adolescent associated with the use of personal music players," *Yonsei Med. Journal,* vol. 50, no. 6, pp. 771-776, 2009.

[6] M. Sliwinska-Kowalska and A. Davis, "Noise-induced hearing loss," *Noise Health,* vol. 14, no. 61, pp. 274-280, 2012.

[7] S. M. Mirbod, R. Inaba, H. Yoshida, C. Nagata, Y. Komura and H. Iwata, "Noise Exposure Level While Operating Electronic Arcades Games as A Leisure Time Activity," *Industrial Health,* vol. 30, pp. 65-76, 1992.

[8] J. Katz, L. Medwetsky, R. Burkard and L. J. Hood, Handbook of Clinical Audiology 6th Edition, America: Lippincott Williams & Wilkins, 2009.

[9] W. Jiang, F. Zhao, N. Guderley and V. Manchaiah, "Daily music exposure dose and hearing problems using personal listening devices in adolescent and young adult: A systematic review," *International Journal of Audiology,* vol. 55, no. 4, pp. 197-205, 2016.

[10] D. H. Pangemanan, J. N. Engka and A. F. Kalesaran, "Pengaruh Pajanan Bising terhadap Pendengaran dan Tekanan Darah pada Pekerja Game Center," *Jurnal Biomedik,* vol. 4, no. 3, pp. 133-140, 2012.

IOP Conf. Series: Journal of Physics: Conf. Series **1075** (2018) 012074 doi:10.1088/1742-6596/1075/1/012074

# Characterization and evaluation the acoustics parameters of Graha Sepuluh Nopember ITS Surabaya

**Suyatno[1], B Kurnia[1], G Prajitno[1] and S Indrawati[1]**

[1]Department of Physics, Institut Teknologi Sepuluh Nopember, Surabaya, Indonesia

E-mail: kangyatno@physics.its.ac.id

**Abstract.** All articles *must* contain an abstract. The abstract text should be formatted using 10 point Times or Times New Roman and indented 25 mm from the left margin. Leave 10 mm space after the abstract before you begin the main text of your article, starting on the same page as the abstract. The abstract should give readers concise information about the content of the article and indicate the main results obtained and conclusions drawn. The abstract is not part of the text and should be complete in itself; no table numbers, figure numbers, references or displayed mathematical expressions should be included. It should be suitable for direct inclusion in abstracting services and should not normally exceed 200 words in a single paragraph. Since contemporary information-retrieval systems rely heavily on the content of titles and abstracts to identify relevant articles in literature searches, great care should be taken in constructing both.

## 1. Introduction

Acoustically, a space must be able to accommodate the sound quality needs of listeners and speakers according to their needs. There are several important acoustic parameters of a space according to the function of a space. These parameters are the distribution of SPL, Reverberation Time, clarity, and definitions [1,2]. Based on Ribeiro [3], a multipurpose room has an acoustic parameter with RT is 1.2 s to 1.8 s, a clarity (C80) value of more than +3dB and definition (D50) of 50%.

This paper discusses the characterization and evaluation of the acoustic parameters of Graha 10 November (GSN), ITS Surabaya. Later, from the parameters obtained can be used to improve the acoustic parameters of GSN, so that it will be suitable for the desired room function whether as a speech room and performance room. For example, distribute of the loudspeaker system, position, and type of speaker. GSN is a room which has 45000 m$^3$ of volume, and the area for listener is 2400 m$^2$ with capacity at 1st floor is 3000 audience. Figure 1 shows the situation of the GSN.

Currently, one of the functions this room is for graduation ceremonies and music performances, like as gamelan and traditional dance performances. As a graduation ceremony room that one of its functions is as a speech room (public lecture). Consequently, the room of GSN must be able to provide the clearly of sound (definition). Moreover, this space is also used as a concert hall, like gamelan and traditional dance also modern music. As a concert hall, this room should be able to provide clarity of sound from every played musical instrument to the audience. As a space for Javanese gamelan performances, the room must have an RT in the audience area is 1.2 s and clarity is C80 4 dB and D50 is 45% [4,5].

Part II explain the measurement procedure, and part III discusses the results of measurement and analysis, and in the last part is the conclusion.

**Figure 1.** Situation of graduation ceremony at GSN

## 2. Methodology

To obtain the acoustic parameters of the main room of the GSN, characterization is done by using two methods, IR and white noise. The IR method is used to find out the room response to the instantaneous sound from several positions in the audience area. While the white noise method used to know the distribution of SPL in the audience area. Both methods use the sound source at the front position (stage), while for the receiver microphones are installed scattered in the audience room.

From the IR method, the distribution of acoustic parameters such as RT, C80, and D50 is obtained, while by using white noise method, an acoustic parameter of SPL distribution is obtained. Measurements are made by using sound sources from a balloon burst for impulse response methods as well as synthesis sound from an audio generator. The position of the microphone is in the audience area that represents the front zone, rear, middle and side of the room. To obtain the desired acoustic parameters, the sound signal is obtained, and then analysed using a Yoshimasa electronic sound analyzer. Figure 2 shows the position of point measurement at GSN.

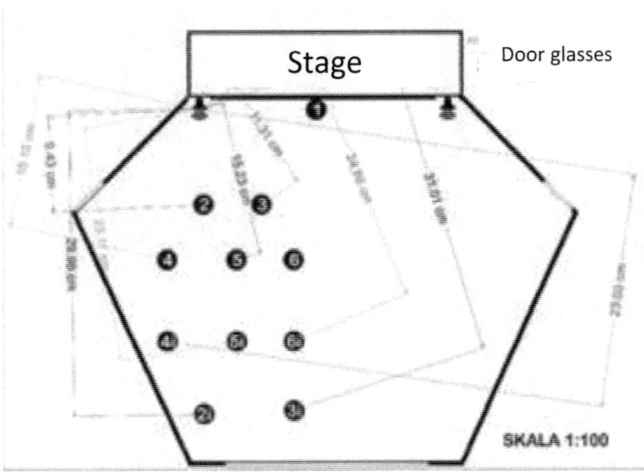


**Figure 2.** Position of measurement point.

## 3. Result and discussion

### *3.1. Distribution of SPL parameters*

The GSN position, which is close to the main road, has a contribute to the high a background noise. This is supported also by the semi-open building design on the 2nd and 3rd floors. As a result, the sound can get into the room through the 2nd and 3rd floor. In this paper, the measurement was

IOP Conf. Series: Journal of Physics: Conf. Series **1075** (2018) 012074 doi:10.1088/1742-6596/1075/1/012074

conducted on the 1st floor that the audience exists at every event. Figure 3 shows the position and the one of geometry of the GSN.

**Figure 3.** Location and architectural design of Graha Sepuluh Nopember.

Based on the measurement, the Background Noise of GSN at night when the measurement process is 60 dB and more at noon. The aim of measurement at night is to get the nature of the acoustic characteristic room with minimalizing noise. The value of this Background Noise will affect the minimum SPL value that should be generated in every activity inside the GSN building. In order to obtain the distribution of SPL parameters at the building, the sound source used must have a minimum value more 10 dB than Background Noise [1, 2]. Based on the measurement, distribution of BN and SPL as shown in Figure 4.

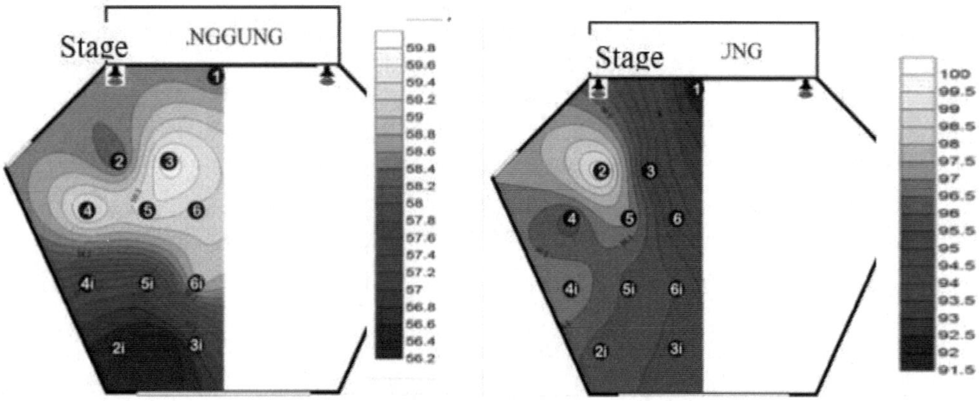


**Figure 4.** Distribution of SPL parameters.

Referring to the figure 4, the source with 100 dB of SPL on stage generates the SPL parameter of the distribution in the uneven GSN. The SPL difference between front and rear is more than 5 dB. This shows the energy that reaches the listener evenly. This is due to the dimensions of space large enough and high ceiling of about 35 m. So, the sound received by the listener is dominated by the direct sound. The geometry of room also makes the value will have an impact on the sound energy heard by the listener in the rear, especially if the resulting sound source has SPL of less than 70 dB.

### *3.2. Distribution of RT, C80, and D50*

As mentioned in the part II, the acoustic parameter of RT, C80, and D50 of space are obtained through IR measurements method. Figure 5 shows an image of the output signal generated by one of the measuring points in GSN.

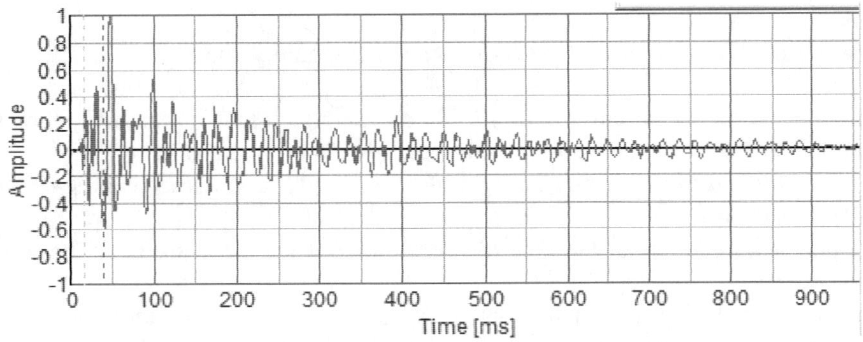


**Figure 5.** Impulse Signal at Graha Sepuluh Nopember.

Based on the signal IR in figure 5, parameter RT, C80, and D50 can be obtained. Figure 6 shows the distribution parameters RT, C80, and D50 at frequency 1 kHz.

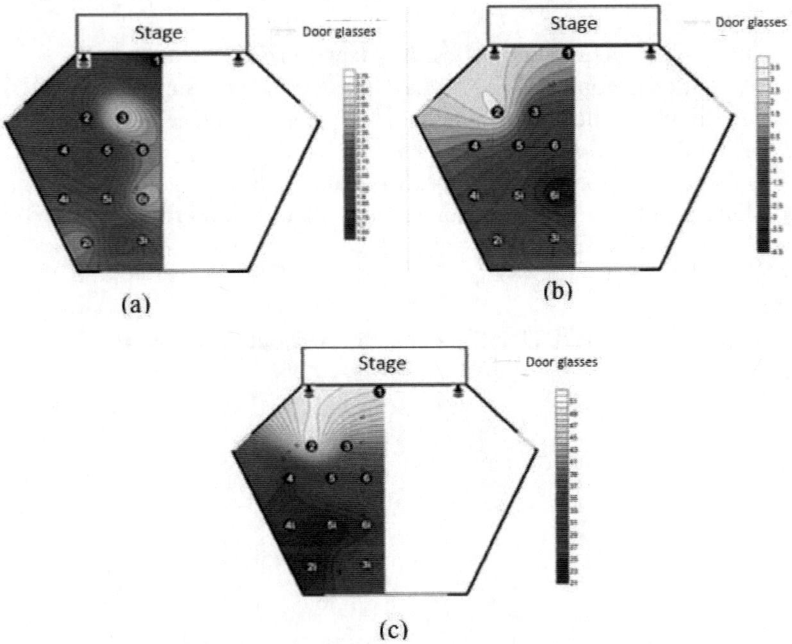


**Figure 6.** Distribution parameter of (a) Reverberation Time; (b) C80; (c) D50.

Based on Figure 6, the distribution of the clarity of music (C80) is less (-2dB) indicates that the clarity of instrument received by the listener is lacking. Similar with C80, the value of D50 less than 40% indicated the clarity of speech is poor. It is also supported by a fairly large RT value, which is between 1.6 to 2.75 seconds.

Refers to the parameter results, the acoustic parameters of the GSN main space should be increased so that it is suitable for speech room and performance room, especially for Javanese gamelan performances. Because in every big event there are always Javanese gamelan performances and traditional dances. According to Suyatno, et al (2016), RT parameters for Javanese gamelan performances are in the range of 1.2 s to 1.6 s, D50 is 66% to 74% and C80 is +4 dB to +6.2 dB. Improvements can be made by adding a sound system with intermediate frequency characteristics as well as positioning appropriate with characteristics of space.

## 4. Conclusions

Based on measurement, calculation, and analysis it can be concluded that:

- The distribution of sound energy in GSN is quite evenly, that is, the difference between the SPL value between the source and the farthest position around 5 dB.
- Valued of the acoustic parameter of RT is 1.6 - 2.75 s, sound clarity is less 40% for D50 and less (-2) dB for C80.
- This condition suggests that the acoustic conditions of the GSN are poor that for speech and concert activity.
- Improved the acoustic parameters can be done by adding an electronic sound system (loudspeaker) with characteristic sound at the mid-high frequency and directivity to the back position.

## 5. References

[1] Beranek L 2004 Concert Hall and Opera House (Springer, Verlag)
[2] Ribeirio 2008 Room Acoustic Quality of A Multipurpose Hall: A Case Study, Acoustics 2008, Paris.
[3] Marshall L., 2006, Architectural Acoustics, Elsevier Academic Press.
[4] Suyatno, H A Tjokronegoro, I G N Merthayasa, and R Supanggah 2016 Preference of reverberation time for musicians and audience of the Javanese traditional gamelan music *Journal of Physics: Conference Series* **776**
[5] Suyatno, H A Tjokronegoro, I G N Merthayasa and R Supanggah 2015 Acoustics Parameter Pendopo Mangkunegaran Surakarta for Javanese Gamelan Performance *Procedia Social and Behavioral Science* **184** 322-327

### Acknowledgments

The authors wish to thank to LPPM ITS Surabaya throughout BOPTN research 2017, and Graha Nopember ITS Surabaya.

IOP Conf. Series: Journal of Physics: Conf. Series **1075** (2018) 012075 doi:10.1088/1742-6596/1075/1/012075

# Study on measurement and habitability evaluation for floor impact vibration on prefabricated housing floor

**T. Matsuda[1], T. Shimizu[1]**

[1] Central Research Laboratory, Daiwa House Industry Co., Ltd., 6-6-2, Sakyo, Nara City 631-0801, Japan

toru.matsuda@daiwahouse.jp

**Abstract**. This paper reports an experimental study of a method used to select impact and measurement points based on the distribution properties of a floor to determine the habitability resistance performance in the case of floor vibration. The results confirm that it is possible to fully represent the vibration resistance of an entire floor area based on five impact points and measurement points distributed diagonally on the floor surface, excluding measurement points within 455 mm of the impact points.

## 1. Introduction

The diversification of housing environments and life styles that have occurred in recent years have been accompanied by the a wide range of demands for living environments. There has also been a rise in the level of housing performance related to residential floors, including sound-proofing, vibration-proofing, safety, walking sensation, and insulation performance required by clients and end users. Of these, blocking sound from the upper and lower stories is considered to be particularly important. Light-weight steel construction floors of detached houses and collective housing are often designed to enhance vibration control support. However, in some cases, as a result of excessive focus on improving sound-proof performance, the floor of a home blocks sound from the upper and lower floors, but cannot ensure vibration proofing, resulting in floor vibration problems and dissatisfaction with the functions of the floor. Present habitability evaluation standards for vibration of floors include the international standard ISO2631-2:2003 **[1]** and the Japanese standard AIJES-V0001-2004 **[2]** issued by the Architectural Institute of Japan. The concept of evaluation methods under these standards are indicated as the absolute performance of the floor under vibration; these standards do not stipulate the impact sources, impact methods, impact points, and measuring points. The result is that the evaluations include scattering of properties of impact force applied to the floor, impact points, and measuring points in addition to a floor's vibration habitability resistance performance (i.e., difficulty in vibrating the floor). The vibration field also requires the concept of evaluating a floor's vibration resistance performance itself, for example, the "floor impact sound insulation performance" from the field of architectural acoustics. A previous paper **[3]** reports on a study of an evaluation method using rubber balls with impact force properties of JIS A 1418-2:2000 **[4]** (ISO140-7:1998**[5]**) (below, "rubber balls"), which are a standard impact source of heavy floor impact sound insulation, as a method of regulating impact force. However, in order to obtain a floor habitability resistance performance, in addition to setting the impact source and impact force properties applied to the floor, it is also essential to clarify the impact points and measurement points of the floor that is being evaluated. Therefore, it is very useful, if possible, to determine the measuring method and the evaluation method specifying the impact point and the measurement point that can represent the vibration distribution of the entire floor. The "floor impact sound insulation performance," is an evaluation method that includes impact force properties, impact points, and measurement points, stipulated in JIS A 1418-2:2000 **[6]** (ISO140-7:1998 **[7]**). Another evaluation method has been specified in JIS A 1419-2:2000 **[8]** (ISO717-2:2013 **[9]**).

In this study, we conducted experiments on the method for selecting the impact point and measurement point from the vibration distribution characteristics of the floor. In addition, we examined the measurement method in order to evaluate the comfort of the entire floor rather than the evaluation of the occupancy of the representative points of the floor. The contents of this report are accepted in a previous report **[10]**, and the reanalysis of the measurement method is reported here.

## 2. Vibration distribution of floor

### *2-1. Experimental methods*

The experiment was conducted using five light-weight steel construction floors of prefabricated houses shown in **Table 1**. A rubber ball was used to cause impact vibration of the floor, and the vibration acceleration was measured. The experiment was conducted for the four patterns shown in a) and b) below. **Figures 1** to **2** show examples of plane figures, impact points, and measuring points.

(a) The rubber ball was dropped from 100 cm at five impact points shown in **Figure 1**, and the vibration acceleration was measured at intervals of 455 mm over the entire floor for each impact point.
(b) The rubber ball was dropped from 100 cm at the five impact points shown in **Figure 2**, and at the measuring points, the vibration acceleration at five points distributed diagonally within the floor plane was measured for each impact point.

### *2-2. Experimental results and discussion*

**Figures 3** and **4** show vibration distributions in the floor using formula (1). **Figures 3** and **4** show Floor A with the smallest floor area and Floor E with the largest floor area when impacted at five points distributed diagonally by experiment methods a) and b). The figures (i) show the following legends (A) and (C)and, the figures (ii) show the following legends (B) and (C). Furthermore, in the **Figures 3** and **4**, the ranges of 1σ and 2σ obtained from the measurement range of (A) are shown.

**Table 1** List of floor specifications used in the experiment. **[10]**

| Specimen | Structure | Structure of floor (mm) | Slab dimensions (mm) |
|---|---|---|---|
| Floor A | | Rubber vibration insulator: 19 (Hardness: 81) + Deck plate: 1.2+ Precast concrete slab: 80 + Asphalt mat: 12 + Wooden floor: 15 | 2730 × 3640 |
| Floor B | | Rubber vibration insulator: 19 (Hardness: 69) + Deck plate: 1.2+ Precast concrete slab: 80 + Asphalt mat: 12 + Wooden floor: 15 | 2730 × 3640 |
| Floor C | Lightweight Steel Structure | Rubber vibration insulator: 6.5 + Steel joist: 75 × 45 + Particle board: 25 + Plaster board: 9.5 + Particle board: 9 + Wooden floor: 12 | 3640 × 3640 |
| Floor D | | Rubber vibration insulator: 8 + Steel joist: 75 × 45 + Particle board: 20 + Asphalt mat: 6 + Particle board: 15 + Wooden floor: 13 | 3640 × 4550 |
| Floor E | | Autoclaved lightweight aerated concrete: 100 + Particle board: 15+ Wooden floor: 12 | 3640 × 8190 |

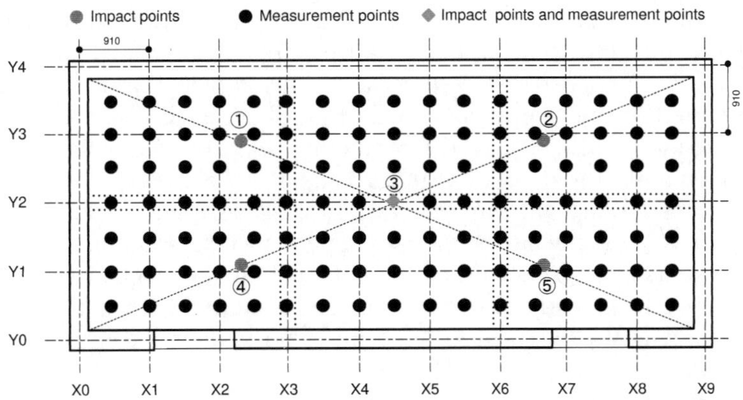


**Figure 1** Relationship between the impact points and the measurement points in the case of pattern (a). **[10]**

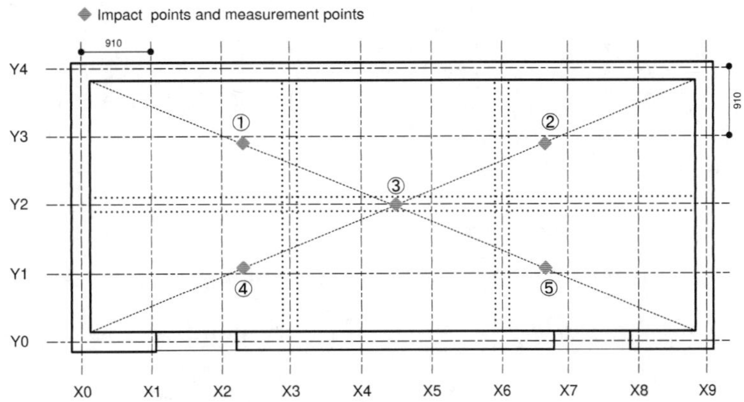


**Figure 2** Relationship between the impact points and the measurement points in the case of pattern (b). **[10]**

$$L_{Wm,max}(O.A) = 20log_{10}\left(\frac{a}{a_0}\right) \quad \text{(dB)} \qquad (1)$$

where $a$ denotes the vibration acceleration corrected 1–80 Hz by the vibration sensation as stipulated by ISO2631-2 **[1]** ($m/s^2$), and $a_0$ denotes the standard acceleration ($10^{-5}$ $m/s^2$).

Legend:
- (A): Entire floor measured on a grid at intervals of 455 mm for each impact point
- (B): Entire floor measured on a grid at intervals of 455 mm for each impact point, excluding points within 455 mm of the impact points
- (C): Five points distributed diagonally on the floor plane measured for each impact point
- (D): Five points distributed diagonally on the floor plane measured for each impact point, excluding points within 455 mm of the impact points

In the vibration distribution shown in the figure (i) of **Figures 3** and **4**, the acceleration response measured at five points distributed on the diagonal of (C) in the range measured on the grid at the entire floor of (A) at 455 mm intervals is generally within the range. However, there is an acceleration response far outside the measurement range of (A) for each impact point. In addition, there is also an acceleration

IOP Conf. Series: Journal of Physics: Conf. Series **1075** (2018) 012075 doi:10.1088/1742-6596/1075/1/012075

response outside the range of 2σ. The reason for this is considered to be that when the measurement point is in the vicinity of the impact point, there is a possibility that linearity cannot be maintained or it may be caused by local vibrations. Therefore, it is considered inappropriate to evaluate the vibration distribution of the entire floor by an evaluation that includes those measured values.

The vibration distributions of the (B) and (D) excluding measurement points within 455 mm from the impact point are shown in the figure (ii) of **Figures 3** and **4**.
(A) and (C) on the figure (i) are different, most of the measuring points are within the range of 2σ. Because of these reasons, by excluding the measuring points within 455 mm from the impact point, the measurement point excluding the singular point, it is thought that it can be selected. Subsequently, the difference between (A) and (C) and that between (B) and (D) are shown in **Figure 5** in order to check the influence of measurement points within 455 mm from the impact point. Similarly, the difference

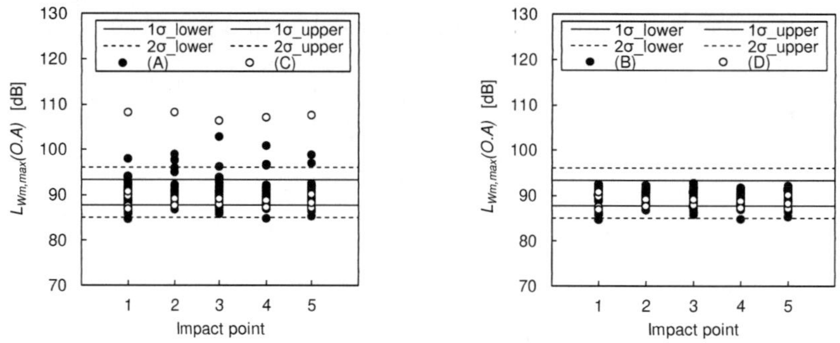


(i) (ii)

**Figure 3** Distribution of floor vibration in the case of impact at five points. (Floor A)

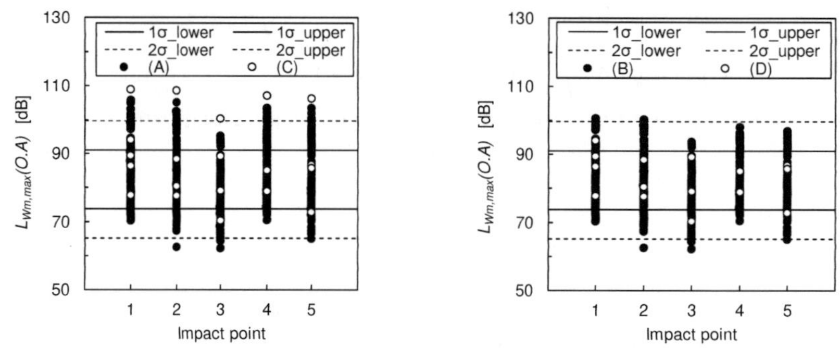


(i) (ii)

**Figure 4** Distribution of floor vibration in the case of impact at five points. (Floor E)

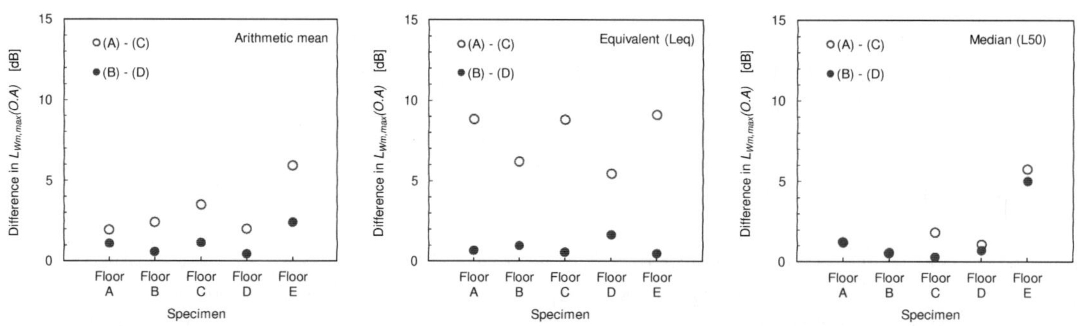


**Figure 5** (A)–(C) and (B)–(D) targeting arithmetic mean and equivalent (Leq) and median (L50).

between (A) and (C) and that between (B) and (D) are shown in **Figure 5** in order to check the influence of measurement points within 455 mm from the impact point. In the figure, the arithmetic mean, equivalent (Leq), and median (L50) are calculated and compared for each floor. In (A)–(C) including measurement points within 455 mm from the impact point, there is a difference of about 5 dB in Floor E in the case of the arithmetic mean value and median (L50). The differences in most of the cases other than Floor E are within 3 dB. However, with the equivalent (Leq), there is a difference of more than 5 dB in all floors, and the maximum difference is approximately 9 dB. Next, in the cases of (B)– (D), excluding measuring points within 455 mm from the impact point, no significant change is seen compared with (A)–(C) at the median (L50). However, the arithmetic mean and equivalent (Leq) are within 2 dB on most floors, which is obviously small.

For these reasons, excluding measurement points within 455 mm from the impact point is not affected by the singular point; it is considered to be appropriate as a measurement point that can represent the vibration distribution of the entire floor.

Judging from the above results, to represent the vibration distribution of an entire floor, it is unnecessary to measure many points on a grid; rather by measuring five points distributed diagonally in the floor plane excluding measuring points within 455 mm of impact points in show the **Figure 6**, these five points can be representative points that can clarify the vibration distribution of the entire floor.

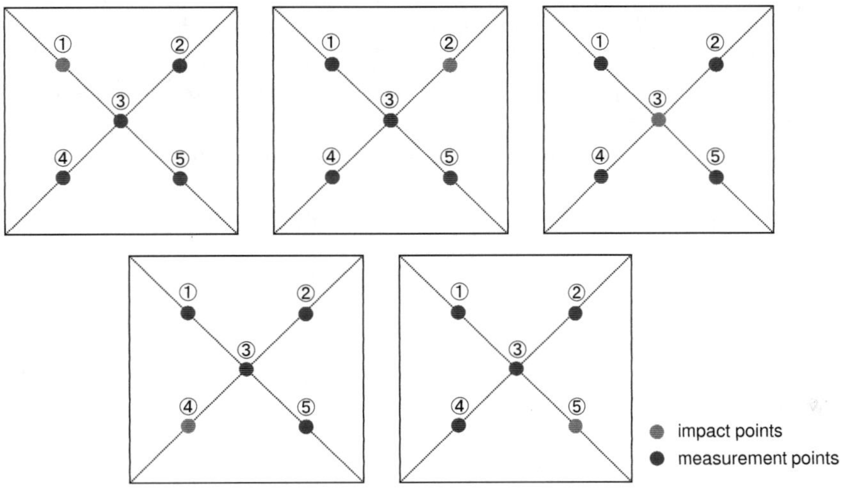


**Figure 6** Relationship between impact points and measurement points.[10]

## 3. Conclusion

This is a report on an experimental study of a method for selecting impact points and measuring points to calculate evaluation values based on distribution properties of an entire floor in order to calculate the habitability resistance performance to vibration of a floor (floor impact vibration habitability resistance performance). In this study, we confirmed that it is possible to fully represent the habitability vibration resistance of an entire floor area based on five impact points and measuring points distributed diagonally on the floor surface and excluding measuring points within 455 mm from the impact point.

In the previous report **[10]**, using the vibration acceleration maximum value measured by the relationship between the impact point and the measurement point described above, we examined the calculation method of the evaluation to evaluate the habitability resistance performance of the entire floor rather than the habitability evaluation at the representative point of the floor.

Furthermore, in order to confirm the adequacy of the evaluation, the relationship between the sensation scale and the evaluation quantity is also examined from the vibration sensation evaluation experiment. Future prospects of the study include further exploring the vibration control method from the viewpoint of living comfort of an end user.

**Reference**

[1] ISO 2631-2 (2003): Mechanical vibration and shock -- Evaluation of human exposure to whole-body vibration -- Part 2: Vibration in buildings (1 Hz to 80 Hz)
[2] AIJES-V0001 (2004): Guidelines for the evaluation of habitability to building vibration [in Japanese]
[3] R. Tomita, K. Inoue: Effect of Vibration Sense by Frequency Characteristics of Impact Vibration for Residential Floor, Journal of Environmental Engineering (Transactions of AIJ) Vol. 79 (2014) No. 705 p. 927-935 [in Japanese] http://doi.org/10.3130/aije.79.927
[4] JIS A 1418-2: 2000: Acoustics - Measurement of floor impact sound insulation of buildings – Part2: Method using standard heavy impact sources [in Japanese]
[5] ISO 140-7: 1998: Acoustics -- Measurement of sound insulation in buildings and of building elements -- Part 7: Field measurements of impact sound insulation of floors
[6] JIS A 1418-2, Acoustics — Measurement of floor impact sound insulation of buildings — Part 2: Method using standard heavy impact sources (2000) [in Japanese].
[7] ISO 140-7, Acoustics — Measurement of sound insulation in buildings and of building elements — Part 7: Field measurements of impact sound insulation of floors (1998).
[8] JIS A 1419-2, Acoustics — Rating of sound insulation in building and of building elements — Part 2: Floor impact sound insulation (2000) [in Japanese].
[9] ISO 717-2, Acoustics — Rating of sound insulation in buildings and of building elements — Part 2: Impact sound insulation (2013).
[10] T. Matsuda, T. Shimizu: Study on correspondence between evaluation of vibration sensation, average method, and measurement density of acceleration levels for impact vibration in prefabricated housing floor. Journal of Building Engineering 10(2017) 124-139.

IOP Conf. Series: Journal of Physics: Conf. Series **1075** (2018) 012076 doi:10.1088/1742-6596/1075/1/012076

# Reliability evaluation technique of compressor using pressure pulsation and vibration signals

**Y W Kim[1] and W B Jeong[1]**

[1]School of Mechanical Engineering, Pusan National University. Busan, South Korea

E-mail: wbjeong@pusan.ac.kr

**Abstract.** This paper dealt with the technique for determining the reliability of a compressor using pressure pulsation and vibration signals of the refrigerant compressor for a domestic refrigerator. Experiments were conducted according to the failure level and type of failure of the compressor. Experiments on the failure level of the compressor proceeded according to the degree of damage of the port of the valve. Experiments on the type of compressor failure were made according to the refrigerant leakage position. The pulsation of the refrigerant was measured in the suction and discharge valve of the compressor, and the vibration signal was measured on the surface of the compressor. Fisher's linear discriminant method was used to select the health index for reliability discrimination. The pressure pulsating signal of the suction valve was effective in most cases in the reliability discrimination of the domestic refrigerator compressor. The compressor surface acceleration was effective in some cases, and the pressure pulsating signal of the discharge valve was not effective in most cases.

## 1. Introduction

Many studies have been made on the fault diagnosis of compressors. Yang et al (2005) identified compressor failure using ANN (artificial neural networks) and SVM (supported vector machine) [1]. AlThobiani et al (2014) uses TKEO (Teager-Kaiser energy operator) and DBN (deep belief networks) to determine the compressor failure [2]. Ledesma et al (2015) determined the failure while controlling the speed of the compressor [3]. This study also uses ANN. Wang et al (2015) proposed a method for diagnosing the failure of a compressor valve by applying a combination of compressor experiment and computation [4]. Pichler et al (2016) proposed a method to detect the failure of a compressor valve by controlling the load applied to the compressor [5]. In this paper, we will comprehensively examine the fault detection with the pressure pulsation of the compressor valve and the acceleration of the compressor surface for the fault level and fault type. The compressor used in the experiment was a compact compressor for refrigerators using refrigerant. The failure level was selected as the magnitude of the breakage of the compressor valve port. The failure of the compressor according to the leakage position of the refrigerant was selected as the failure type. The health index to identify faults was selected in the time and frequency domains, respectively. The objective function for selecting the health index was Fisher's linear discriminant method.

## 2. Experiment

The failure of a small compressor for a refrigerator is largely due to the breakage of parts and the leakage of refrigerant due to poor assembly. Particulate damage occurs mainly in the valve port of the

IOP Conf. Series: Journal of Physics: Conf. Series **1075** (2018) 012076 doi:10.1088/1742-6596/1075/1/012076

compressor, and poor assembly occurs mainly in the suction and discharge piping of the refrigerant. Deep cracks in the valve port resulting from the piercing of the valve lead to fatal defects in the compressor drive unit and the suction valve and cause the compressor to break down. Also, leakage of refrigerant due to poor assembly can adversely affect compressor efficiency and stability.

### *2.1. Fault level*
The failure level of the compressor was divided into three stages of normal, crack and hole according to the degree of damage of the corner of the compressor valve port. The state change of the system according to the fault level is shown in table 1. Figure 1 shows the appearance of the valve port and the crack location. Figure 2 and 3 shows the shape of the crack and hole in the state of failure of the valve port, respectively.

**Table 1.** The change of the system according to the fault level.

| Fault level | Performance down | Breakdown |
|---|---|---|
| Normal | X | X |
| Crack | O | X |
| Hole | O | O |

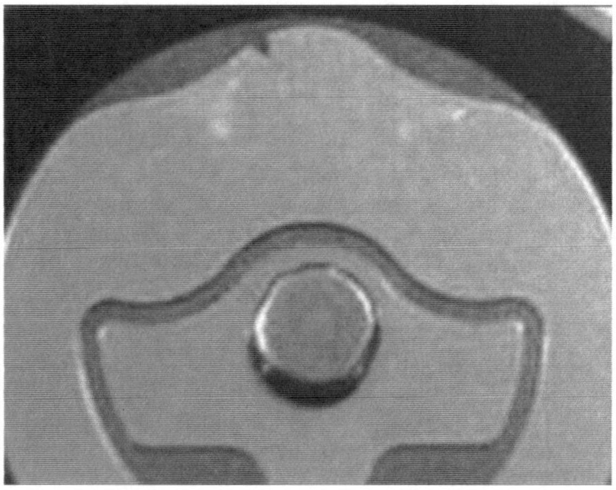

**Figure 1.** Suction valve port of compressor.

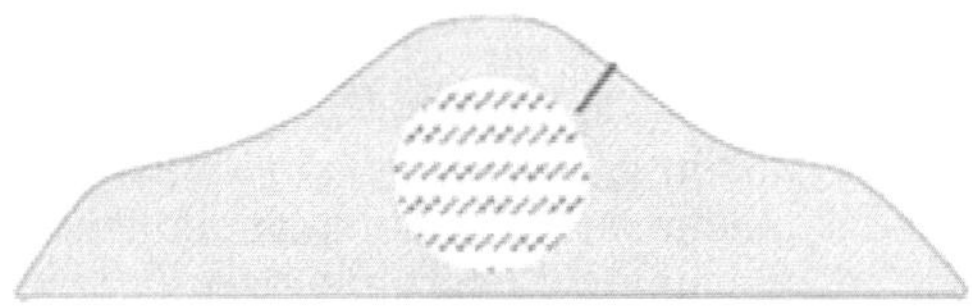

**Figure 2.** Fault type: Crack.

**Figure 3.** Fault type: Hole.

### *2.2. Fault type*
The types of faults were mainly classified as failure point, that is, discharge pipe, suction pipe, and suction valve (poor side clearance assembly).

### *2.3. Measured signal information*
Generally, a data signal used for fault diagnosis of a refrigerator compressor in an industrial field is a pressure pulsation signal of a refrigerant measured at an inlet and an outlet of a compressor refrigerant. The pressure pulsation signal of the refrigerant has the advantage of being able to detect the fault level intuitively without additional signal processing, but it has a disadvantage that it is difficult and time-consuming to embed the sensor in the piping to acquire the signal. On the other hand, the surface acceleration signal of the compressor tends to require additional signal processing, but the advantage is that the signal can be acquired relatively easily without additional work. In this study, the acceleration sensor is attached to the outer surface of the compressor to obtain the vibration signal of the compressor and compare it with the pressure pulse signal of the refrigerant.

IOP Conf. Series: Journal of Physics: Conf. Series **1075** (2018) 012076 doi:10.1088/1742-6596/1075/1/012076

## 3. Reliability evaluation

The signals obtained from the experiment pass through the filter and have a statistical representative value for each factor. And the factors 1st to 2nd that maximize the FDR are selected. The optimal reliability evaluation technique is then determined through linearization.

### *3.1. Fisher's linear discriminant ratio*

Fisher's linear discriminant ratio(FDR) is a useful discriminator that can simply classify each of the clusters with their variance and mean. The generalized FDR can be expressed as equation (1) [6].

$$FDR_{M\_class} = \sum_{i}^{M}\sum_{j\neq i}^{M} \frac{(\mu_i - \mu_j)^2}{\sigma_i^2 + \sigma_j^2}, \mathrm{M} > 2 \tag{1}$$

### *3.2. Health index candidates*

The procedure for selecting a suitable health index to perform a fault diagnosis analysis is divided into a signal processing section and a factor selection section. In this study, a total of 120 health index candidates were selected. The information for each filter is shown in the table 2.

**Table 2.** Information about filters.

| Data domain | Filter | Description |
|---|---|---|
| | Raw | Raw data |
| | Band (1-3) | Bandpass filter 60Hz, 120Hz, 180Hz+-10 |
| Time | High | Cut-off frequency 1000Hz |
| | Low | Cut-off frequency 1000Hz |
| | Envelope | Enveloped raw data |
| Frequency | FFT | Fast Fourier transformed data |
| | FFT with envelope | FFT with enveloped raw data |

The formula for each feature is shown in the table 3 [7]. Where $s(f_i)$ is the power spectrum function. Modified frequency domain features use $\sqrt{s(f_i)}$ instead of $s(f_i)$.

**Table 3.** Formula about features.

| Data Domain | Feature Name | Formula | Remarks |
|---|---|---|---|
| Time | Mean | $\bar{X} = \frac{1}{N}\sum_{i=1}^{N} X_i$ | |
| | rms | $X_{rms} = \sqrt{\frac{1}{N}\sum_{i=1}^{N} X_i^2}$ | Root mean square |
| | Skewness | $\frac{\frac{1}{N}\sum_{i=1}^{N}(X_i - \bar{X})^3}{\left(\sqrt{\frac{1}{N}\sum_{i=1}^{N}(X_i - \bar{X})^2}\right)^3}$ | |
| | Kurtosis | $\frac{\frac{1}{N}\sum_{i=1}^{N}(X_i - \bar{X})^4}{\left(\frac{1}{N}\sum_{i=1}^{N}(X_i - \bar{X})^2\right)^2}$ | |
| | FC | $FC = \frac{\int f_i \times s(f_i)\, df}{\int s(f_i)\, df}$ | Frequency center |
| Frequency | RMSF | $\frac{\int f_i^2 \times s(f_i)\, df}{\int s(f_i)\, df}$ | Root mean square frequency |
| | RVF | $\frac{\int (f_i - FC)^2 \times s(f_i)\, df}{\int s(f_i)\, df}$ | Root variance frequency |

### *3.3. Health index selection by FDR and linearize*

The top two factors are optimized by projecting on one straight line that maximizes the FDR. The results are shown in Figure 4 and 5. The red line is the line where the FDR is maximized when two indices are reduced to one dimension.

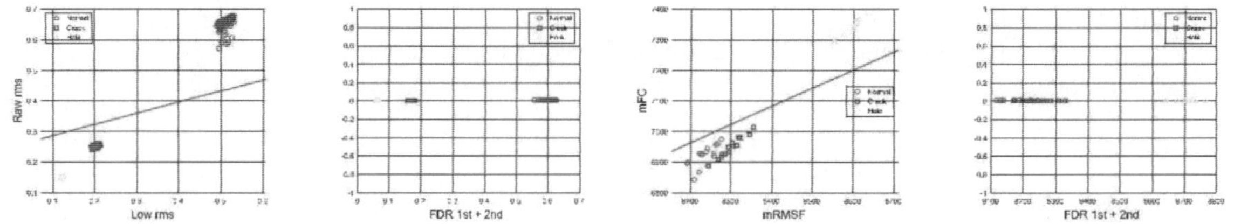

(a) Time domain, inlet pulsating pressure (b) Frequency domain, inlet pulsating pressure

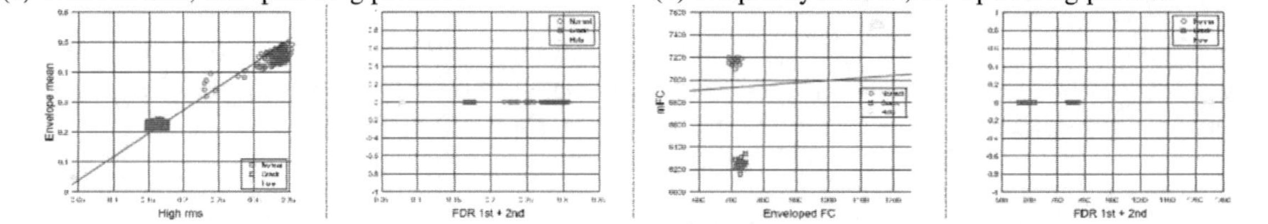

(c) Time domain, outlet pulsating pressure (d) Frequency domain, outlet pulsating pressure

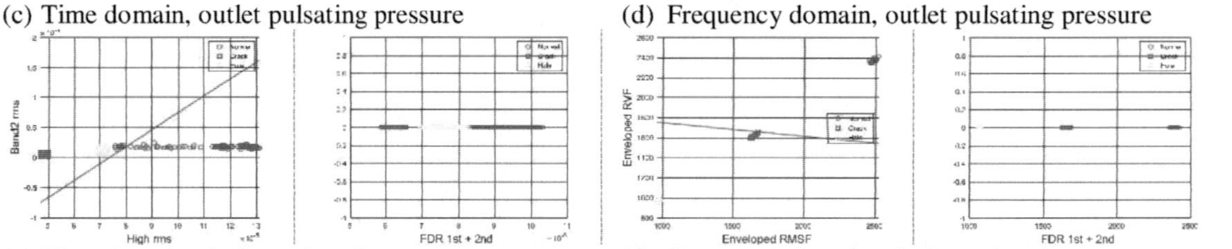

(e) Time domain, shell acceleration (f) Frequency domain, shell acceleration

**Figure 4.** Health index selection of compressor fault level.

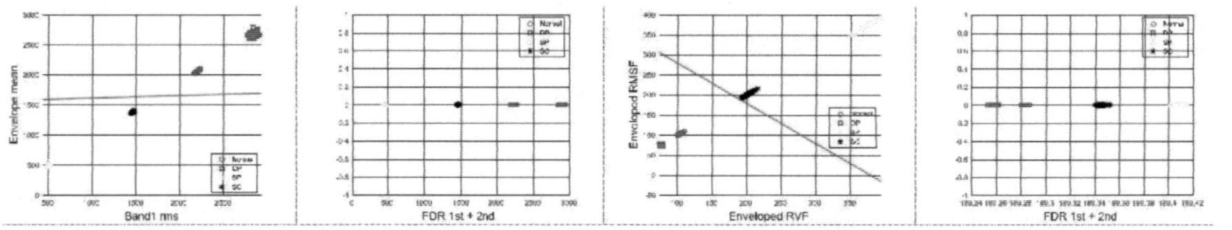

(a) Time domain, inlet pulsating pressure (b) Frequency domain, inlet pulsating pressure

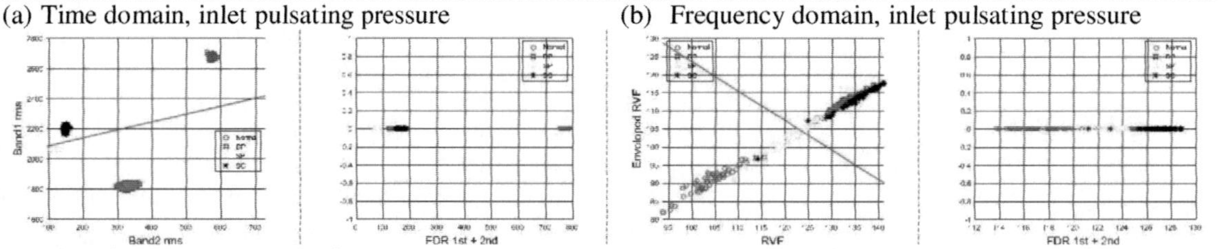

(c) Time domain, outlet pulsating pressure (d) Frequency domain, outlet pulsating pressure

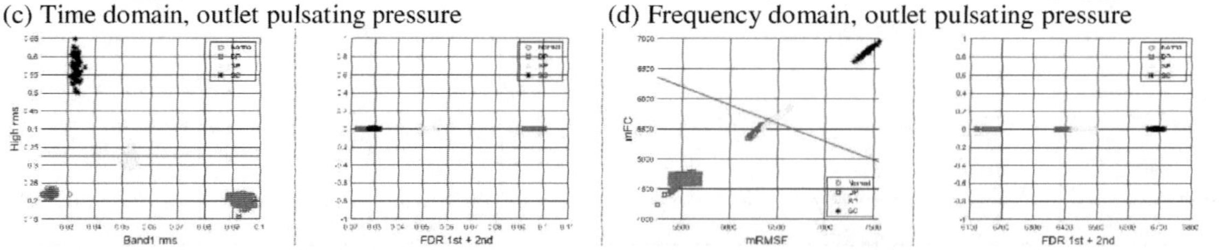

(e) Time domain, shell acceleration (f) Frequency domain, shell acceleration

**Figure 5.** Health index selection of compressor fault type

## 4. Conclusion

A reliability evaluation technique for the fault level and fault type of the compressor was established and proceeded using the pressure pulse signal and the surface acceleration signal of the compressor for the refrigerator. The reliability evaluation according to the type of failure should be clearly distinguished between each group. In the case of the reliability evaluation according to the failure level, the tendency should also be clearly shown according to the level. The results are summarized in table 4. The symbol ○ stands for a clearly classified signal. △ generally refers to signals that are separated. X indicates that the signal is indistinguishable at all.

**Table 4.** Reliability evaluation results

| Data domain | Data description | Fault level | Fault type |
|---|---|---|---|
| Time | Inlet pulsating pressure | ○ | ○ |
| | Outlet pulsating pressure | X | △ |
| | Shell acceleration | X | △ |
| Frequency | Inlet pulsating pressure | X | ○ |
| | Outlet pulsating pressure | X | X |
| | Shell acceleration | ○ | △ |

In this study, the inlet pulsating pressure was most suitable for determining the fault level and type. On the other hand, the outlet pulsating pressure was inadequate value for reliability determination in this study. The acceleration of the compressor surface showed the possibility of reliability discrimination.

## 5. References

[1] Yang B S, Hwang W W, Kim D J and Tan A C 2005 Condition classification of small reciprocating compressor for refrigerators using artificial neural networks and support vector machines *Mechanical Systems and Signal Processing* **19(2)** 371-390

[2] AlThobiani, F., & Ball, A. (2014). An approach to fault diagnosis of reciprocating compressor valves using Teager–Kaiser energy operator and deep belief networks. *Expert Systems with Applications,* **41(9)**, 4113-4122.

[3] Ledesma, S., Belman-Flores, J., & Barroso-Maldonado, J. (2015). Analysis and modeling of a variable speed reciprocating compressor using ANN. *International Journal of Refrigeration,* **59**, 190-197.

[4] Wang, Y., Xue, C., Jia, X., & Peng, X. (2015). Fault diagnosis of reciprocating compressor valve with the method integrating acoustic emission signal and simulated valve motion. *Mechanical Systems and Signal Processing,* **56**, 197-212.

[5] Pichler, K., Lughofer, E., Pichler, M., Buchegger, T., Klement, E. P., & Huschenbett, M. (2016). Fault detection in reciprocating compressor valves under varying load conditions. *Mechanical Systems and Signal Processing,* **70**, 104-119.

[6] Mika, S., Ratsch, G., Weston, J., Scholkopf, B., & Mullers, K.-R. (1999). *Fisher discriminant analysis with kernels.* Paper presented at the Neural Networks for Signal Processing IX, 1999. Proceedings of the 1999 IEEE Signal Processing Society Workshop.

[7] Jeon, B. C., Jung, J. H., Youn, B. D., Kim, Y.-W., & Bae, Y.-C. (2015). Datum unit optimization for robustness of a journal bearing diagnosis system. *International Journal of Precision Engineering and Manufacturing*, **16(11)**, 2411-2425.

IOP Conf. Series: Journal of Physics: Conf. Series **1075** (2018) 012077 doi:10.1088/1742-6596/1075/1/012077

# A case study of smartphone based platform for mapping and analyzing environmental noise in Singapore

**Saurabh Garg, Kian Meng Lim and Heow Pueh Lee**

Department of Mechanical Engineering, National University of Singapore, 1 Engineering Drive 2, Singapore 117576

E-mail: mpesg@nus.edu.sg

**Abstract.** In this paper, we present a smartphone based platform for mapping and analyzing environmental noise. Smartphones are used to record environmental noise (both audio and sound pressure level) as well as locations (latitude, longitude, and accuracy). The recorded data is uploaded to a server and noise map of a region is generated using the location information along with the sound pressure levels (SPL) at those locations. Noise maps can used to discover sources of noise in a region and also to compare SPL between different places. Thus, they can be an important tool to help implement mitigating steps for reducing noise or planning future developments. To illustrate the capabilities of the platform, we present a case study of noise mapping with five educational institutes and five residential neighborhoods in Singapore. We present noise maps for these places and discuss possible sources of noise. We also discuss potential reasons for observed differences in equivalent continuous sound pressure level between different educational institutes, and residential neighborhoods.

## 1. Introduction

Noise pollution has been on the rise due to increasing urbanization of cities. Increase in environmental noise levels has adverse physiological and psychological problems not only in humans [1, 2] but it also affects biodiversity in cities [3]. Thus, it is important to monitor environmental noise and take actions to provide suitable living and working environment. Traditionally, environmental noise monitoring has been done using few static measurement stations installed in the region to be monitored. Static stations can only provide noise data at the installed locations with no information in other places. However, with the advent of smartphones which are equipped with a microphone and GPS providing accurate location of the smartphone, it is possible to build a noise map of a region. A noise map is like a weather map but for noise. It uses color to show the sound pressure levels in a region (Figure 1). Noise maps can be used to discover sources of noise in a region and also to compare sound pressure levels between different places. Thus, they can be an important tool to help implement mitigating steps for reducing noise or planning future developments.

There are few reported works in literature for building environmental noise map. One of the first works on noise maps was NoiseTube [4]. The authors combined GPS information with noise data collected by smart phones to generate noise maps. Similar works were reported by Kanjo [5] and Schweizer et al. [6]. Rajib et al. took a different approach in [7] by building classifiers to

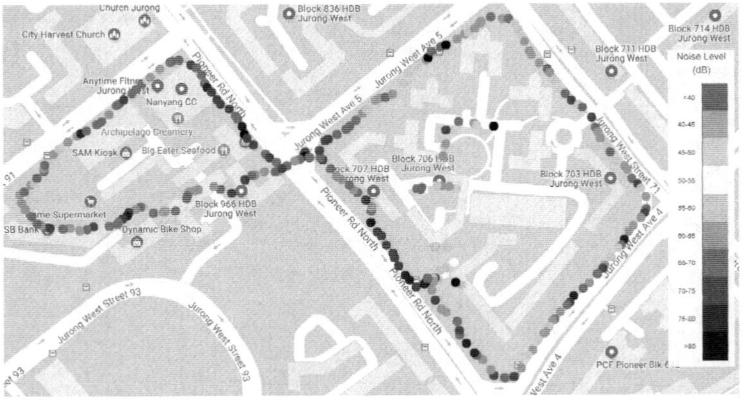

**Figure 1.** Noisemap of a residential neighborhood in Jurong West, Singapore.

detect the location (hand, pocket, or bag) of phone and automatically determine when to start measuring. This frees user from actively starting measurements and arguably more data can be collected. Shim et al. [8] conducted a study to determine feasibility of noise mapping using volunteers or citizen scientists. They were able to build a noise map of Seoul using 336 volunteers over a period of seven days. Data was recorded by each volunteer's phone automatically at 10 minutes interval. Zuo et al. [9] used an external microphone with a wind screen connected to smartphones to build noise maps. External microphone frees users from holding smartphone in their hands and pointing them correctly for a long period of time. Wind screen helps to reduce effects of movement of subjects and wind. They also proposed an interpolation method to build very detailed noise maps with the limited measured data.

In this paper, we present a system for recording environmental noise maps using Android based smartphones. We illustrate our system with two case studies in Singapore. The first case study compares noise levels in five educational institutes (polytechnics) and the second case study compares noise levels in five residential neighborhoods in Singapore.

## 2. Noise Mapping Platform

We have developed an Android app known as *NoiseExplorer* for measuring sound pressure level (SPL) using smartphones. NoiseExplorer gets the audio data from the built-in microphone of a smartphone and computes instantaneous dBA (A-weighted SPL). Additionally, NoiseExplorer also keeps track of the minimum, maximum and continuous equivalent (average) SPL for the current session. The user can choose to show either a live frequency spectrum or the current location of the smartphone on Google Maps using Google Maps Android API.

The locations and noise levels are recorded in a log file and are used to generate noise map. The data is logged to the file at a user specified frequency (default is five seconds because it is the fastest speed at which Google maps updates location). The user can optionally choose to record the audio as uncompressed WAV files. Although uncompressed WAV files occupy more space compared to compressed formats such as mp3 or aac, we chose WAV so that we can perform spectral and psychoacoustic analysis on the recordings. For most smartphones, the storage is not an issue these days, moreover, users can delete recorded WAV file or move them to an external storage device or micro SD cards once they have uploaded them to the server (or even without uploading). At present WAV and log files are stored on the smartphone itself and manually uploaded to a server after all measurements are finished. The server then generates noise maps from the log files. It is also possible to generate average spectrum and spectrogram from WAV files. Finally, the user can specify a calibration file containing a set of frequency and

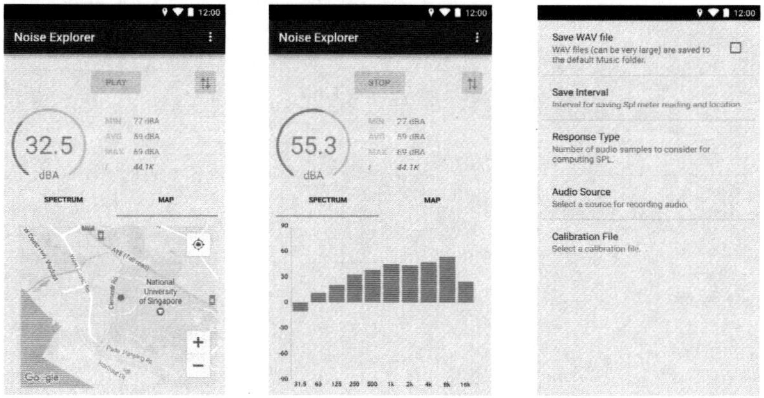


**Figure 2.** NoiseExplorer user interface.

correction factor pairs. The correction factors for frequencies not present in the calibration file are linearly interpolated. Figure 2 presents some screenshots of the app.

## 3. Case Studies

We have used NoiseExplorer app to record noise map for various places in Singapore and around the world. In this paper, we will present noise maps for five polytechnic institutes and five residential neighborhoods in Singapore. The noise maps were recorded by four smartphones: Samsung Galaxy S7 (relatively new), two different Samsung Galaxy S4 (both old and used), and a ZTE Blade Q Pro (low-cost about 60 USD). A team of four students walked with smartphone in the ten areas in Singapore to record noise levels. The data analysis and noise map generation was done offline on a computer.

The noise maps for five polytechnics are presented in Figure 3. The major source of noise in all five polytechnics are canteens and other areas where a lot of students gather around. Singapore and Ngee Ann polytechnic are next to major trunk roads and traffic noise is prominent for areas adjacent to roads. Table 1 shows the $L_{eq}$ (continuous equivalent SPL) and $L_{max}$ (maximum SPL) for all five polytechnics. Nanyang polytechnic is about 3dB noisier than other polytechnics because they have a child care center on campus and at the time of measurement there were couple of buses to pick children up from the center with many children running around. Additionally, they happened to have an event on campus with many people talking during measurement.

The noise maps for five residential neighborhoods are presented in Figure 4. The major source of noise in all five neighborhoods is traffic with occasional people talking and wind noise. Table 2 shows the $L_{eq}$ and $L_{max}$ for all five neighborhoods. Chai Chee is the quietest because most of the measurements are away from any major trunk road and Jurong West is noisiest because most of the measurements are next to a major trunk road with high traffic. Many measurements for Choa Chu Kang and Clementi were taken near train tracks and hence noise levels are high. For Woodlands, many measurements are near train tracks and major trunk roads, however, there are measurements down in park and other internal locations with less traffic thereby reducing the overall noise level.

## 4. Conclusions

In this paper, we presented an overview of the smartphone based platform for mapping and analyzing environmental noise. We discussed NoiseExplorer app for computing and recording sound pressure level along with location information via GPS. The location information with

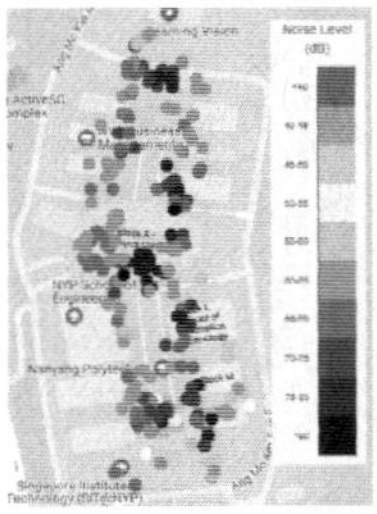

(a) Nanyang Polytechnic

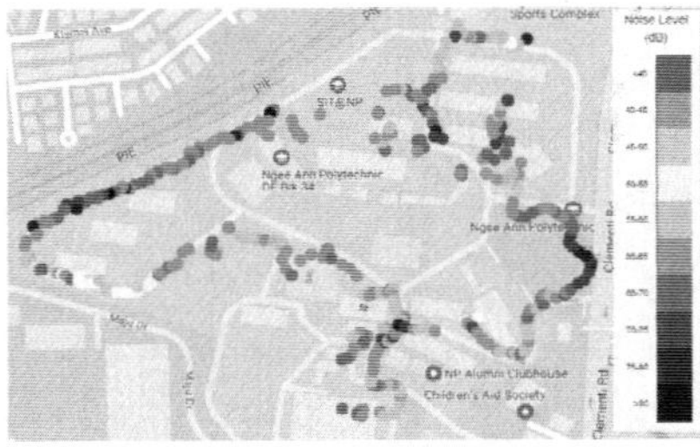

(b) Ngee Ann Polytechnic

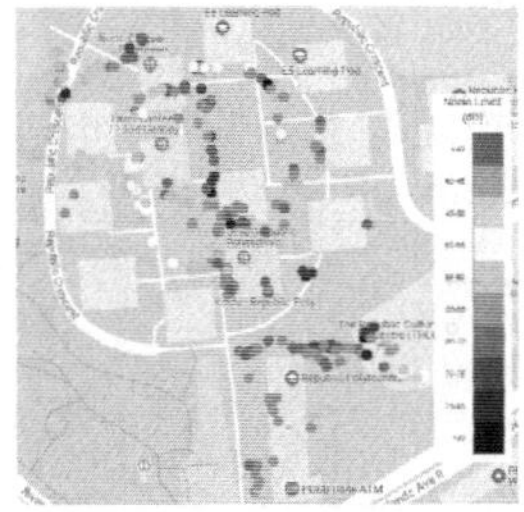

(c) Republic Polytechnic

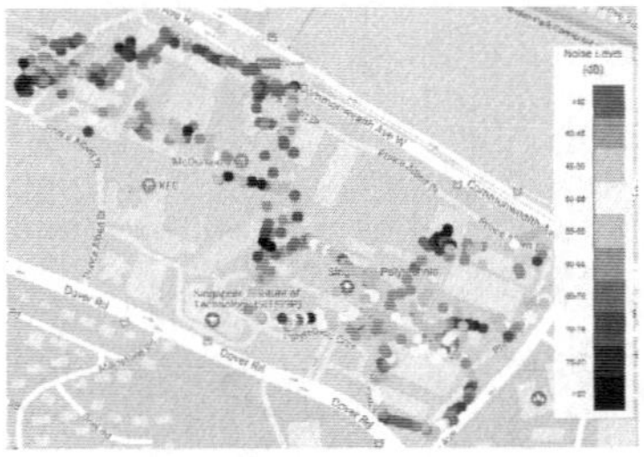

(d) Singapore Polytechnic

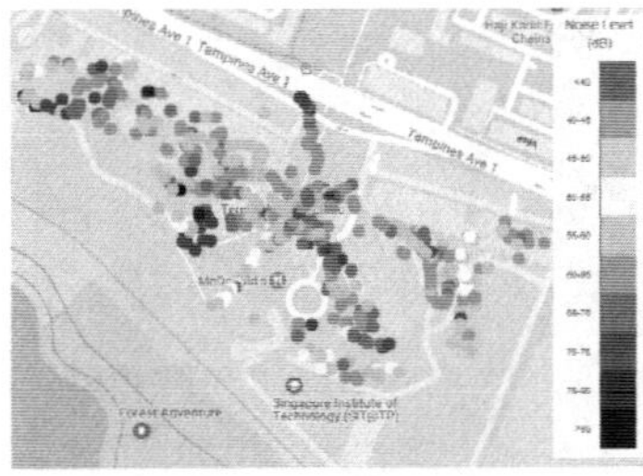

(e) Temasek Polytechnic

**Figure 3.** Noise maps for five polytechnics in Singapore.

**Table 1.** Noise levels for five polytechnics in Singapore.

| Polytechnic | LAFeq (dBA) | LAFmax (dBA) | Duration (hh:mm:ss) |
|---|---|---|---|
| Nanyang | 71.4 | 89.1 | 00:31:53 |
| Ngee Ann | 68.0 | 89.2 | 00:35:05 |
| Republic | 68.4 | 89.6 | 00:29:58 |
| Singapore | 68.8 | 93.1 | 01:04:26 |
| Temasek | 65.9 | 91.4 | 00:56:48 |

sound pressure levels is used to generate noise maps. Noise maps can be used to find sources of noise in a region and also allows to compare different regions in terms of sound pressure levels. We presented two case studies comparing noise levels among five polytechnic institutes and five residential neighborhoods in Singapore.

## 5. Acknowledgement

This material is based on research/work supported in part by the Singapore Ministry of National Development and National Research Foundation under L2NIC Award No. L2NICCFP1-2013-8. Any opinions, findings, and conclusions or recommendations expressed in this material are those of the author(s) and do not necessarily reflect the views of the Singapore Ministry of National Development and National Research Foundation.

## References

[1] Stansfeld S A and Matheson M P 2003 *British Medical Bulletin* **68** 243–257
[2] Kawada T 2004 *Journal of Nippon Medical School* **71** 5–10
[3] Newport J, Shorthouse D J and Manning A D 2014 *Ecological Management & Restoration* **15** 204–214 ISSN 1442-8903

IOP Conf. Series: Journal of Physics: Conf. Series **1075** (2018) 012077 doi:10.1088/1742-6596/1075/1/012077

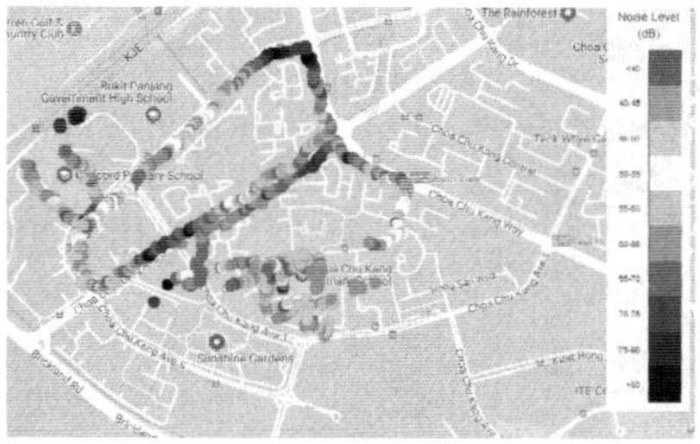

(a) Choa Chu Kang

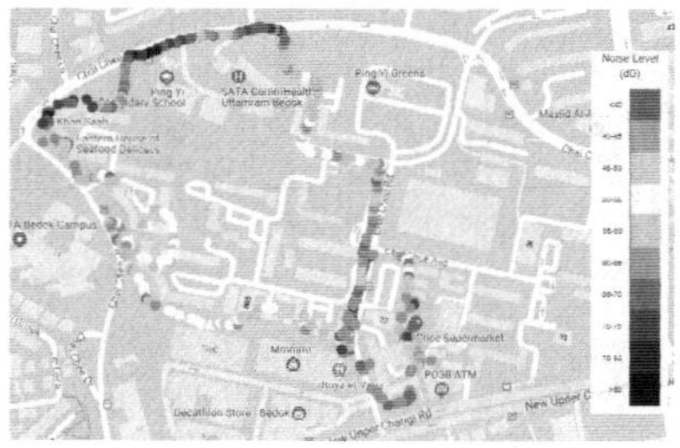

(b) Chai Chee

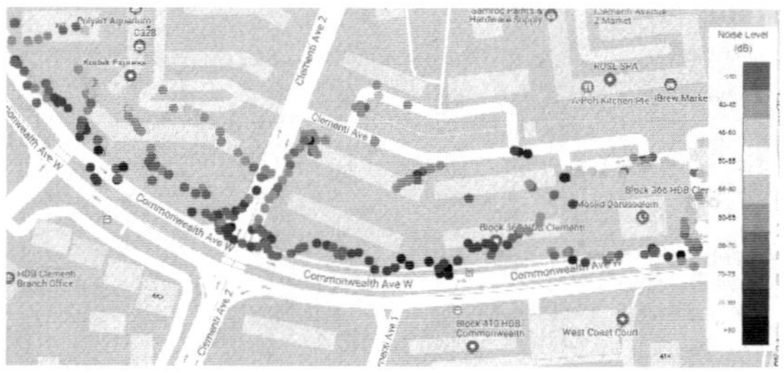

(c) Clementi

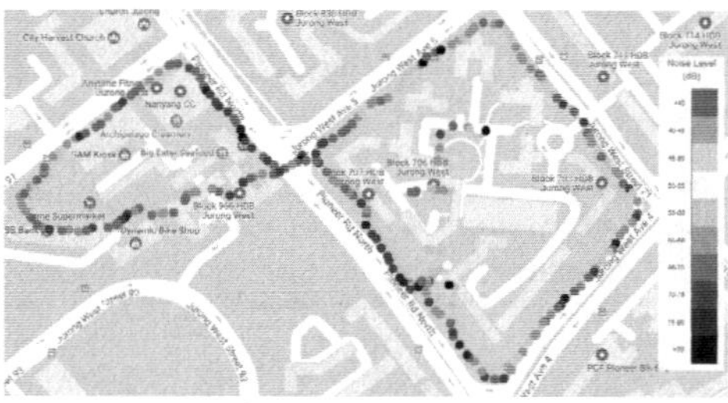

(d) Jurong West

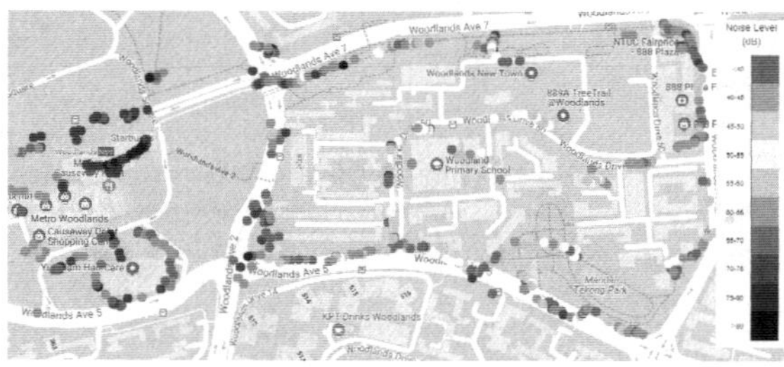

(e) Woodlands

**Figure 4.** Noise maps for five residential neighborhoods in Singapore.

**Table 2.** Noise levels for five residential neighborhoods in Singapore.

| Neighborhood | LAFeq (dBA) | LAFmax (dBA) | Duration (hh:mm:ss) |
|---|---|---|---|
| Chai Chee | 66.1 | 87.9 | 00:31:46 |
| Choa Chu Kang | 70.4 | 90.3 | 01:17:11 |
| Jurong West | 71.1 | 94.0 | 00:22:15 |
| Clementi | 69.8 | 86.9 | 00:25:04 |
| Woodlands | 68.8 | 90.4 | 01:02:40 |


[4] Maisonneuve N, Stevens M, Niessen M E and Steels L 2009 *Information Technologies in Environmental Engineering: Proceedings of the 4th International ICSC Symposium Thessaloniki*

[5] Kanjo E 2010 *Mobile Networks and Applications* **15** 562–574

[6] Schweizer I, Darmstadt T, Probst F, Bärtl R, Darmstadt T, Mühlhäuser M, Darmstadt T, Schulz A and Darmstadt T 2011 *In Second International Workshop on Sensing Applications on Mobile Phones*

[7] Rana R, Chou C T, Bulusu N, Kanhere S and Hu W 2015 *Pervasive and Mobile Computing* **17** 1–22

[8] Shim E, Kim D, Woo H and Cho Y 2016 *PLOS ONE* **11** 1–7

[9] Zuo J, Xia H, Liu S and Qiao Y 2016 *Sensors* **16**

IOP Conf. Series: Journal of Physics: Conf. Series **1075** (2018) 012078 doi:10.1088/1742-6596/1075/1/012078

# Insulation Performance of Classrooms in Green Building

**Ulvariandani, M. S. Prawirasasra, Suwandi**

Telkom University, Indonesia

E-mail: bibinprawirasasra@telkomuniversity.ac.id

**Abstract.** Physical boundary of object was designed specically to accommodate one requirement for green building. In general, composition of walls are two-panels gypsum combine with half and full-stickered glass of windows and louvered windows were placed in order to optimize daylighting usage and maintain air circulation. According to layout, object position is surrounded by 5 other classrooms. Whenever two or more class were active in the same time, airborne noise were potentially propagate from one classroom to another and vice versa hence it might creates excessive background noise inside it. The purpose of this research is to evaluate insulation performance of vertical structure by in-situ measurement.

## 1. Introduction

Education may be one of the most significant aspect to change the quality of live and most people are willing to spend around 12-16 years of their youngster time to get it. But good quality of education itself is not easily obtained, the successful rate of education is very dependent to four major factors such as teacher effectiveness, behavior management, time and instructional management or classroom environment [1] and also physical environment [2].

Acoustic as physical component is very important since excessive background noise and high reverberation time could be deterrent for both education achievement and education performance [3] and it has direct impact to verbal task [3], [6]. Background noise itself is might be originated from airborne noise and or structure-borne noise that enter into classroom from certain direction hence arriving energy is dependent to effectiveness of sound insulation from each boundary segment.

Determination of single number that represent quality of insulation can be carried out either by laboratory testing or field testing. The last stated method has been successfully implemented to determine insulation performance of boundary [5] and flanking noise can be identified too [6], good design of acoustic condition is frequently neglected by architect because they put more focused on another components of indoor air quality [7].

There are possibility acoustic factors did not consider as one of objective since philosophy design of object was to meet one concept of green building which is focus on energy consumption. Classrooms borders were designed to have large opening space in order to optimize daylight usage and attached several natural ventilation and fans to maintain indoor air quality. Though, in the 3rd floor has 6 nearly identic classrooms with composition two rows of rooms structure, each row consists of 3 classrooms. The width of corridor between classrooms was around 2 m. Whenever two or more classes were having teaching session in the same time, there were possibilities of

IOP Conf. Series: Journal of Physics: Conf. Series **1075** (2018) 012078 doi:10.1088/1742-6596/1075/1/012078

airborne noise to propagate into classes. The purpose of this research is to determine single number (DnTw) that represent the effectiveness of each segment to insulate airborne noise that propagate from outside to inside objects by in-situ measurement. Procedure of measurement comply with EN ISO 16283-1 Acoustics - Field measurement of sound insulation in buildings and of building elements - Part 1: Airborne sound insulation.

### *1.1. Description of Object*

The location of object was in 4th floor of the building. It consisted of 6 classrooms and two ubounded spaces dedicated for student's waiting room. Space layout of the floor is illustrated at Figure 1. The physical boundary facing East (E), South (S) and West (W) of classroom with number 402 has been selected as an object for this research. Classroom itself has dimension 7.23 x 7.175 x 3.53 m with E and W-facing wall was made of identic material: double panel gypsum while S-facing boundary composed of peculiar materials and specifically designed for airflow circulation. Top section of it were hollow louvered-window while the middle section were made of full-stickered glass window, double panel gypsum and metal-frame full-stickered glass door.

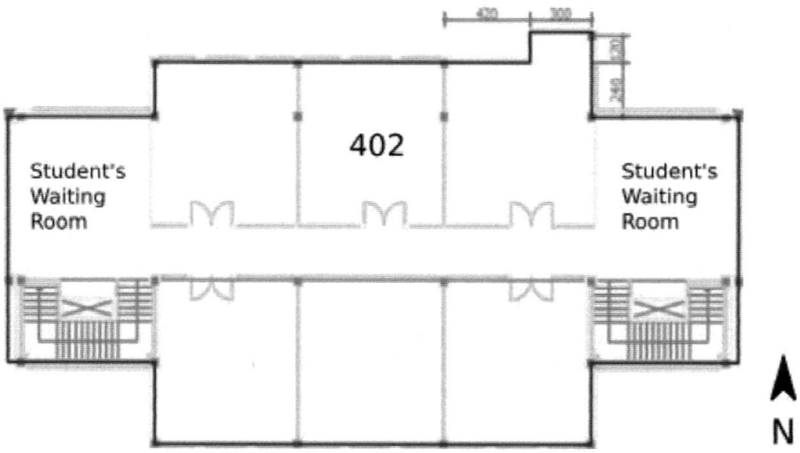


**Figure 1.** Building layout

## 2. Measurement Procedure

Determination of airborne insulation performance follows the procedure as stated in ISO 16283-1:2014. Principally, it requires two rooms , one is dedicated for source room where the loudspeaker was placed and the other has chosen as receiving room. Calculation required physical information obtained from measurement such as background noise, sound pressure level and reverberation time in receiving room.

In this research, measurement equipments were one omnidirectional loudspeaker and fixed microphone moved manually from one position to another. During measurement, loudspeaker was placed only in one position: in the center of source room or it has distance at least 1.0 m from common partition. Speaker was fed by broadband noise signal. Emitted signal was measured in both room, one for sound properties in source room while the other to extract information from transmitted signal through adjacent and flanking element. Acoustical information was captured by microphone in 7 different positions inside receiver room and no positions lied in the same plane relative to the boundaries at each room. Measurement configuration for each partition is illustrated at Figure 2. Mutual area between source room and receiver room was defined as object evaluation.

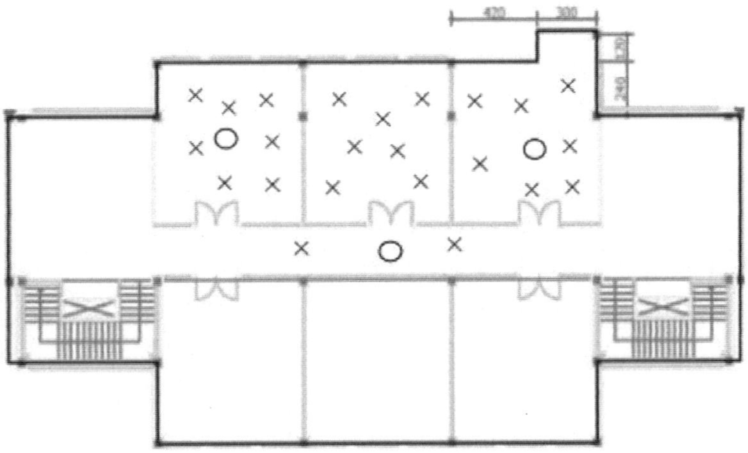


**Figure 2.** Measurement Configuration

Calculation was performed in 1/3 octave from 100 Hz to 3150 Hz. The difference of sound pressure level between source and receiving room is expressed in level difference which stated as follows

$$D = L_s - L_r \tag{1}$$

with $D$ is a level difference from energy-average from source room ($L_s$) and receiving room ($L_r$). Below is the equation to calculate average of level energy from measurement points

$$L_{ave} = 10log(\frac{1}{n}\sum_{i=1}^{n} 10^{\frac{L_i}{10}}) \tag{2}$$

where $L_{ave}$ is average of sound energy in dB and $L_i$ is pressure level in measurement number $i$ in dB.

Since classroom is closed space then reverberation occured and level difference required to be standardized into

$$D_{nT} = D + 10log\frac{T}{T_0} \tag{3}$$

where $D_nT$ is standardized level difference (dB), T is reverberation time in receiving room (s) and $T_0$ is reference reverberation time, 0.5 s. Intersection value between $D_nT$ at 500 Hz and shifted reference curve was value of $D_nT_w$.

IOP Conf. Series: Journal of Physics: Conf. Series **1075** (2018) 012078 doi:10.1088/1742-6596/1075/1/012078

## 3. Result and Discussion

Measured sound pressure level were tabulated at Table 1, 2 and 3 at source room and receiver room. The difference between those parameters described ability of partition to transmit sound energy. Since measured energy have similar magnitude at both room, it demonstrated almost all energy were transmitted from source room to receiver room.

**Table 1.** Sound Level at W facing wall

| f (Hz) | 100 | 125 | 160 | 200 | 250 | 315 | 400 | 500 |
|---|---|---|---|---|---|---|---|---|
| $L_s$ (dB) | 45 | 51 | 59 | 64 | 65 | 64 | 62 | 62 |
| $L_r$ (dB) | 57 | 57 | 60 | 61 | 62 | 62 | 60 | 58 |
| f (Hz) | 630 | 800 | 1000 | 1250 | 1600 | 2000 | 2500 | 3150 |
| $L_s$ (dB) | 61 | 61 | 63 | 63 | 63 | 62 | 64 | 62 |
| $L_r$ (dB) | 58 | 56 | 56 | 55 | 55 | 52 | 53 | 51 |

**Table 2.** Sound Level at E facing wall

| f (Hz) | 100 | 125 | 160 | 200 | 250 | 315 | 400 | 500 |
|---|---|---|---|---|---|---|---|---|
| $L_s$ (dB) | 49 | 55 | 64 | 69 | 69 | 68 | 67 | 65 |
| $L_r$ (dB) | 58 | 57 | 62 | 65 | 65 | 75 | 63 | 61 |
| f (Hz) | 630 | 800 | 1000 | 1250 | 1600 | 2000 | 2500 | 3150 |
| $L_s$ (dB) | 65 | 64 | 66 | 66 | 66 | 65 | 68 | 66 |
| $L_r$ (dB) | 59 | 57 | 57 | 57 | 57 | 55 | 56 | 53 |

**Table 3.** Sound Level at S facing wall

| f (Hz) | 100 | 125 | 160 | 200 | 250 | 315 | 400 | 500 |
|---|---|---|---|---|---|---|---|---|
| $L_s$ (dB) | 55 | 57 | 64 | 70 | 69 | 69 | 68 | 66 |
| $L_r$ (dB) | 49 | 52 | 57 | 66 | 65 | 64 | 63 | 61 |
| f (Hz) | 630 | 800 | 1000 | 1250 | 1600 | 2000 | 2500 | 3150 |
| $L_s$ (dB) | 66 | 68 | 69 | 69 | 68 | 68 | 67 | 67 |
| $L_r$ (dB) | 62 | 62 | 61 | 62 | 62 | 63 | 63 | 62 |

Evaluation of transmission performance for field measurement was ranked by single number named $D_nT_w$. It was determined at 500 Hz of $D_nT$ by shifted up or down the reference curve until total error from $D_nT$ at 1/3 octave did not exceed 32 dB. Performance ranked of these partitions required reference curve to be shifted down more than 37 dB. Value of $D_nT_w$ for each section of wall was plotted at Figure 3, 4 and 5 with adaptation to A-weighted pink noise.

Based on calculation, $D_nT_w$ + C for W, E and S facing wall were 13.7 + 0 dB, 11.8 -0.01 dB and 10-0.01 dB, respectively.

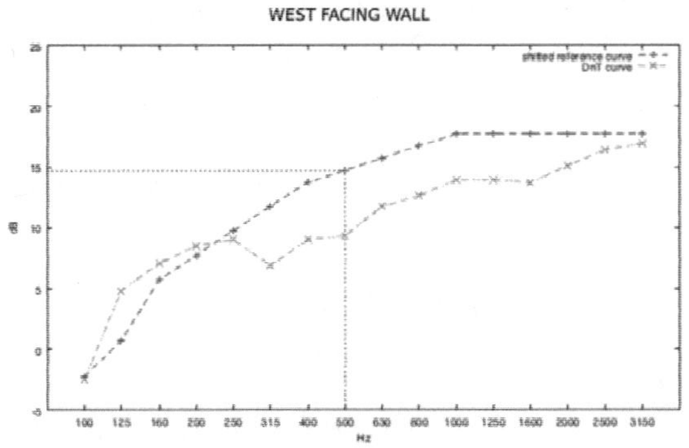


**Figure 3.** $D_nT_w$ of West facing wall.

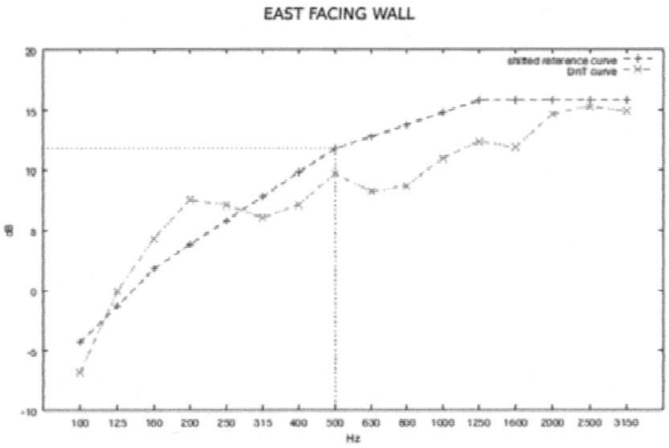


**Figure 4.** $D_nT_w$ of East facing wall.

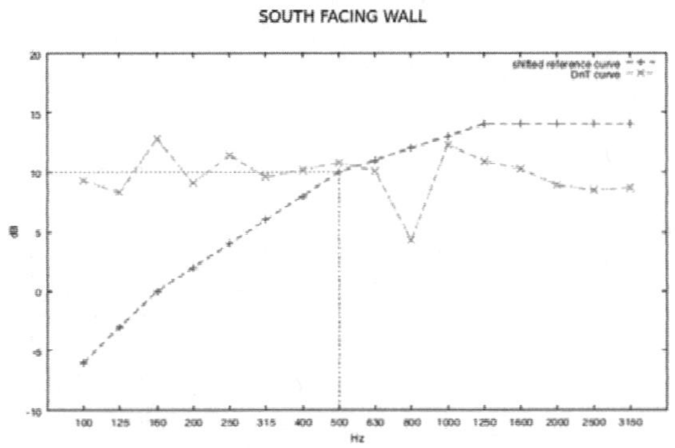


**Figure 5.** $D_nT_w$ of South facing wall.

## 4. Conclusion

The physical boundary's insulation performance of classroom in green building has been analyzed. The main feature of green building, especially natural air circulation, seems have significant impact to acoustic insulation performance. Since one segment of partition have hollow part, it provides no insulation effect whenever sound wave propagates through it. Consequently, either flanking noise or direct noise can easily be heared in classroom. In conclusion, these partitions do not have sufficient acoustical performance to insulate airborne noise for classroom application.

## References

[1] Hannah R. 2013. The Effect of Classroom Environment on Student Learning. Western Michigan University.

[2] Cheryan S and Ziegler S. A. and Plaut V.C. 2014 Designing classrooms to maximize student achievement *Behaviour and Brain Sciences* **I** (I): 412.

[3] Dockrell J. E. and Shield B. M. 2006 Acoustical barriers in classrooms: the impact of noise on performance in the classroom. *British Educational Research Journal* **32**: 509525.

[4] Fischer R, Boroditsky L and Dempsey R.(2006). Airborne noise flanking of shipboard vibration isolation systems. *Sound and Vibration* **40** (12).

[5] Jagniatinskis A, Fiks B and Girnius V. 2007 Airborne sound insulation performance of lightweight partitions for dwellings. VGTU Institute.

[6] Jelneka M, Rychtarikovab M and Vermeirb G. 2007 Distribution of background noise in the classroom and its influence to speech intelligibility. In: Inter-Noise.
[7] Jensen K, Arens E and Zagreus L. 2005 Acoustical quality in office workstations, as assessed by occupant surveys. In: Proceedings of 10th international conference on indoor air quality and climate, Indoor Environmental Quality (IEQ).

IOP Conf. Series: Journal of Physics: Conf. Series **1075** (2018) 012079 doi:10.1088/1742-6596/1075/1/012079

# Vibration field estimation of a plate in time domain using Kalman filter

**M J Sim[1], B K Jung[2] and W B Jung[1]**

[1]Department of Mechanical Engineering, Pusan National University
[2]Agency for Defence Development

E-mail : wbjeong@pusan.ac.kr

**Abstract**. The modal expansion method is a technique of visualizing the vibration field of the structure by estimating the MPF (modal participation factor) from the signals from a limited number of sensors. This paper will propose a method to obtain the reliable vibration field of the whole structure by using noise-contaminated signals. In this paper, the error of the predicted vibration field due to the noise signal is reduced by Kalman filter. The Kalman filter predicts more precise state variables from the noise-contaminated signal. With this characteristic of the Kalman Filter, a technique is introduced to predict the vibration field of the structure by using the acceleration vibration signals and the MPF of the structure, used as the state variables and the measured value, respectively. In this paper, the studied analytical model is a plate with 20 sensors evenly placed on the plate. The analytical model is excited by an aperiodic force, and MCS. Nastran was used for the vibration analysis. To confirm the accuracy of the predicted vibration field, the predicted vibration signals are compared with the exact vibration signals. The result shows that the proposed method of the vibration field visualization using Kalman Filter gives reliable results.

## 1. Introduction

One of the vibration field prediction methods for monitoring the vibration state of the structure is the modal expansion method. The modal expansion method is a technique of visualizing the vibration field of the structure by estimating the MPF (modal participation factor) from the signals from a limited number of sensors. As an example of a previous study, Jung [1]predicted the vibration field of the washing machine using the modal expansion method. Usually, the modal expansion method is applied in frequency domain. This paper, on the other hand, will propose a method to obtain the reliable vibration field of the whole structure by using noise-contaminated signals.

## 2. Theory of Kalman filter for vibration field estimation

To analyze the system in the time domain, a state space model representing the vibration system was designed. The state space is represented as equation (1) and (2).

$$X_{k+1} = AX_k + BF_k \tag{1}$$

$$Z_k = CX_k + DF_k \tag{2}$$

In the equation (1) and (2), X is the state variable vector, Z is a measurement vector, and F is a force vector in the physical coordinate system. The state variable vector consists of the velocity MPF of the structure and the displacement MPF. The Measurement Value Z consist of the noise-contaminated acceleration sensor signals. The subscript k represents time. In here, the matrix A, B, C, D is represented as equation (3), (4), (5) and (6).

$$A = I_{2m\times 2m} + \Delta t \begin{bmatrix} [0]_{m\times m} & [I]_{m\times m} \\ -[\omega]^2_{m\times m} & [0]_{m\times m} \end{bmatrix} \tag{3}$$

$$B = \begin{bmatrix} [0]_{m\times N} \\ \Delta t[\Phi]^T_{m\times N} \end{bmatrix} \tag{4}$$

$$C = [\tilde{\Phi}]_{n\times m}[-[\omega]^2_{m\times m} \quad 0]_{m\times 2m} \tag{5}$$

$$D = [\tilde{\Phi}]_{n\times m}[\Phi]^T_{m\times N} \tag{6}$$

In the equation (3), (4), (5) and (6), $\Phi$ is modal matrix, $\tilde{\Phi}$ is reconstructed modal matrix, and $[\omega]$ is the natural frequency matrix. We can estimate the MPF selected by the state variable with Kalman filter using the designed state space. With these estimated MPF, it is possible to predict not only the acceleration vibration field but also the displacement and velocity vibration field.

## 3. Analysis and Result

### *3.1. Analysis Model*

The analytical model covered in this paper is a finite element model in the form of a plate structure with six degrees of freedom constraint at the four vertexes. In this case, the thickness of the plate is 5 mm, and the material of the structure is steel.

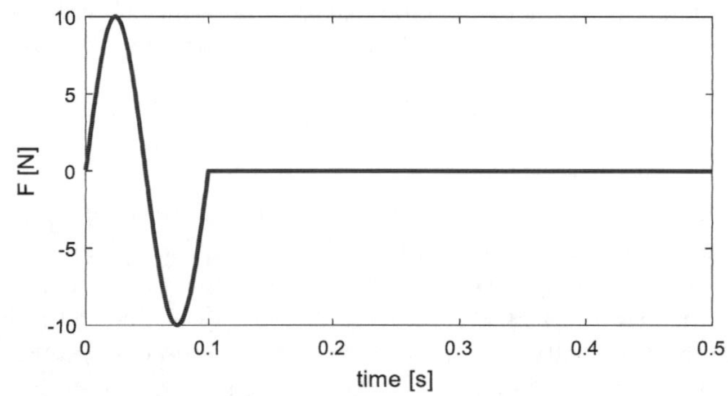


**Figure 1.** Aperiodic force.

IOP Conf. Series: Journal of Physics: Conf. Series **1075** (2018) 012079 doi:10.1088/1742-6596/1075/1/012079

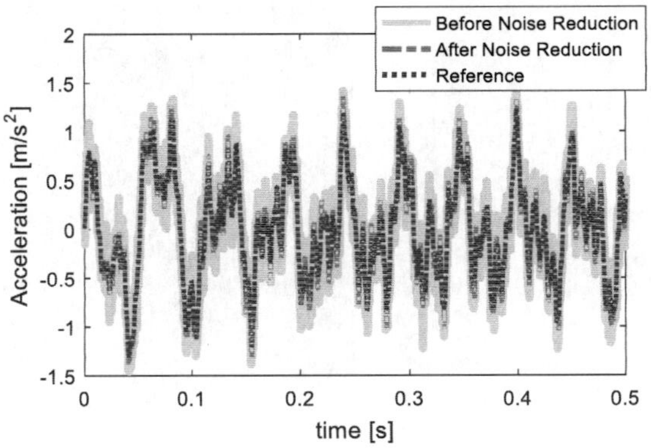


**a)** Acceleration

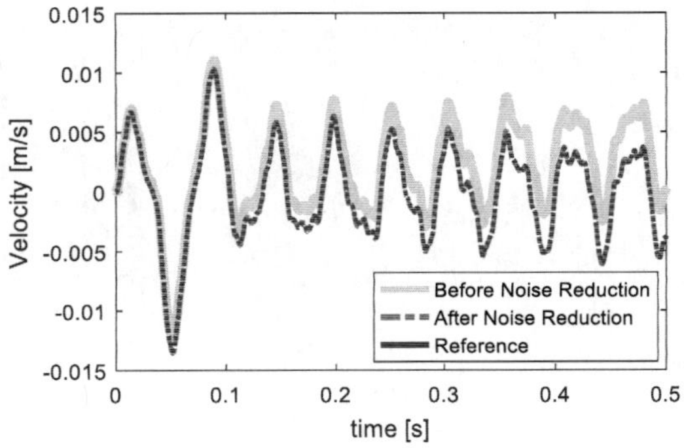


**b)** Velocity

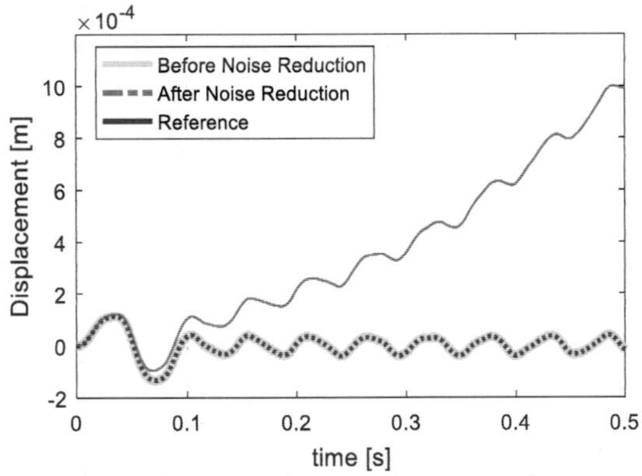


**c)** Displacement

**Figure 2.** Vibration signals estimation subjected to the Aperiodic force

*3.2. Vibration Field Prediction for Aperiodic Excitation.*

The vibration field was predicted when the structure is subjected to an aperiodic excitation force as shown in Figure 1. The measured signal from the sensors are assumed to be a signal obtained by mixing the Gaussian noise with the acceleration vibration response obtained from the transient response analysis. To verify the vibration signal estimation performance for the acceleration, velocity, and displacement of the Kalman filter, Figure 2 shows the vibration response graph at the non-measurement point. From the graph a) in Figure 2, the measurement noise of the acceleration sensor is reduced in the estimation signal from which the noise is removed. Next, From the graphs b) and c), 'the velocity signal and the displacement signal affected by the noise' show a phenomenon in which the error due to noise accumulates. On the other hand, the estimated velocity signal and the displacement signal are well estimated for the response vibration response signal.

To investigate the prediction performance of the Kalman filter, the acceleration vibration field graph at a randomly selected time of 0.101[s] is shown in Figure 3. From the Figure 3, when the vibration field is estimated using the Kalman filter, the noise effect of the sensor signal is reduced, and acceleration vibration field is estimated relatively accurately.

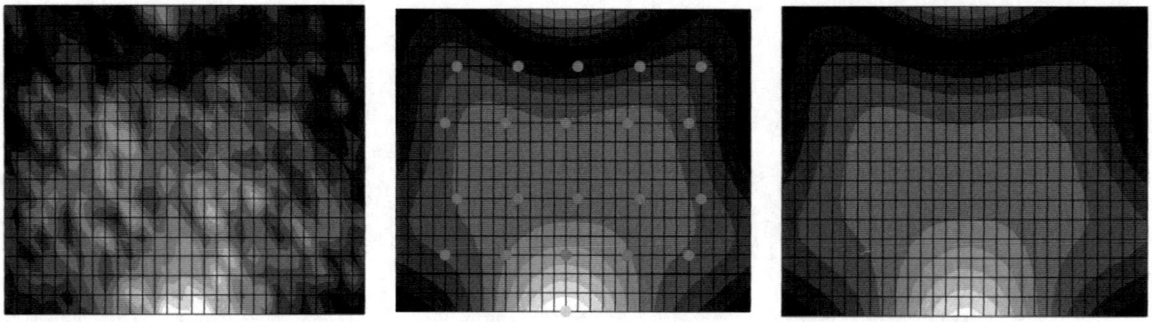

**a)** Before noise reduction **b)** After noise reduction **c)** Reference

**Figure 3.** Vibration patterns at 0.101[s] subjected to the Aperiodic force

## 4. Conclusion

This paper proposed a method of vibration field estimation in time domain using Kalman filter. By using Kalman filter, it was possible to obtain the reliable vibration field from the noise-contaminated signal. In this paper, a simple model, the plate structure, was studied, but it is expected that it will be useful for monitoring the vibration response of complicated shaped structure in the further studies.

## References

[1] Jung, B. K. and Jeong, W. B., 2015, Mode selection of modal expansion method estimating vibration field of washing machine, *Journal of Sound and Vibration*, Vol. 340, pp. 343 ~ 353.

[2] Jung, B. K., Jeong, W. B. and Kim, J. H., 2012, Estimation of Vibration and Radiated Noise of a Compressor by using Modal Expansion Method, *Proceeding of the KSNVE Annual Spring Conference 2012*, pp. 554~555.

[3] Jung, B. K., Cho, J. R. and Jeong, W. B., 2016, Reproduction of vibration patterns of elastic structures by block-wise modal expansion method, Smart Structures and Systems, Vol. 18, pp.819~838.

[4] Jung, B. K., Cho, J. R. and Jeong, W. B., 2015, Sensor placement optimization for structural modal identification of flexible structures using genetic algorithm, Journal of Mechanical Science and Technology, Vol. 29, No. 7, pp.2775~2783.

[5] Jung, B. K., Bae, K. W. and Jeong, W. B., 2015, Selection of Fitness Function of Genetic Algorithm for Optimal Sensor Placement for Estimation of Vibration Pattern of Structure, Transactions of Korean Society for Noise and Vibration Engineering, Vol. 25, No. 10, pp. 677~684.

[6] Sim, M. J., Jung, B. K. and Jeong, W. B., 2016, Estimation of Vibration Field Using Noisy Acceleration Signals Obtained from Sensors, Proceeding of the KSNVE Autumn Conference 2016, pp.69~69.

IOP Conf. Series: Journal of Physics: Conf. Series **1075** (2018) 012080 doi:10.1088/1742-6596/1075/1/012080

# Alternating Least Squares (ALS) blind source separation performance varied on dimension, salinity, and temperature water environment

**N P Yuwono[1], D Arifianto[1], Wirawan[2], W A Asmoro[1], E Widjiati[3] and M P Liana[1]**

[1*] Department of Physics Engineering, Institut Teknologi Sepuluh Nopember, Indonesia
[2*] Department of Electrical Engineering, Institut Teknologi Sepuluh Nopember, Indonesia
[3*] Indonesia Hydrodynamics Laboratory (BPPT, IHL), Indonesia

E-mail: arifianto@gmail.com

**Abstract.** This experiment conduct in mini semi-anechoic water test tank and towing tank at IHL. The purpose is to analyse performance from ALS algorithm in dimension, salinity and temperature water environment. For this purpose we varied water salinity and temperature in mini semi-anechoic water test tank, and for different dimension we took the experiment on two different size of test tank. The results showed that ALS have consistent result at salinity and temperature variety but at large dimension test tank $200 \times 10 \times 5.5$ m without variation on the water medium indicates an anomaly in the results of the ALS BSS technique performance.

## 1. Introduction

Effort to get information from underwater where human have limited in using their eyes makes them using ears. Leonardo Da Vinci do some experiment to hear the sound come from underwater using a tube which one side he put on his outer ear and another in underwater, so he can hear sound of ship from the distance [1]. In modern day, extracting information from underwater sound have many challenges, both from technic and source. Crowded activities on the surface and underwater produce many sound and extracting information to distinguish which is useful information and not by computer have many ways from researcher.

One of the alternative way is called blind sources separation (BSS). Blind sources separation (BSS) is a technique used for source classification by simply observing the mixed signal (observation) with the medium or mixing system (matrix mixing). When researcher want to know what sounds are recorded by the microphone with simply exploring the observation signal, then the math will do the operation of the decomposition of sound (demixing). The matrix notation of the sound mixing process and its decomposition will appear as below [2].

$$X(t) = AS(t) \tag{1}$$

with $X(t) = (x_1(t), \dots, x_J(t))$ is observation signal, $A$ is mixing matrix and $S(t)$ is source signal. Then to get source signal expectation, the equation can be assumed by reversing the mathematical model from (1) to,

$$\hat{S}(t) = X(t)A^{-1} \tag{2}$$

with $\hat{S}(t)$ represent source signal expectation, $A^{-1}$ is demixing matrix and $X(t)$ is observation signal.

## 2. Problem Address

*2.1. Alternating Least Squares (ALS).*
In order to apply equation (1) and (2) there are two problems: first the system matrix (mixing matrix) must be known, and the second place where mixing noise occur must noise free or ideal. This condition is impossible to achieve in real world case because ideal noise less environment without interruptions are none. Then adjustment for the equation above need to be made.

$$X(t) = \sum_{k=-\infty}^{\infty} AS(t-k) \tag{3}$$

Equation (3) occur because in un-ideal environment mixing sound process followed by echo effect. Delay echo effect in mathematic have same meaning with equation (4).

$$X(t) = \sum_{k=-\infty}^{\infty} A * S(t) \tag{4}$$

So equation (4) called convolutive mixture. Another problem from real world case is the only one information we have are observation signal $X(t)$ without information about mixing matrix $A$ and source signal $S(t)$. Then we choose mathematic solution called Alternating Least Squares (ALS).
[3] have scheme for demixing matrix like **Figure 1**.
Terms apply for this method are:

- Number of sensors ≥ number of sources ≥ 2.
- Source $s(t)$ have *zero-mean, non-stationary. Cross spectral density* of source $P_S(\omega, m)$ is diagonal for all $\omega$ and $m$ with $\omega$ is frequency bin and $m$ is epoch. Note: epoch is time duration when signal assumed to be stationer.
- $H(\omega)$ is *Discrete Fourier Transform* from $H(z)$ is *full column rank* matrix. *Full-rank* is matrix with all *linearly independent* vector inside.

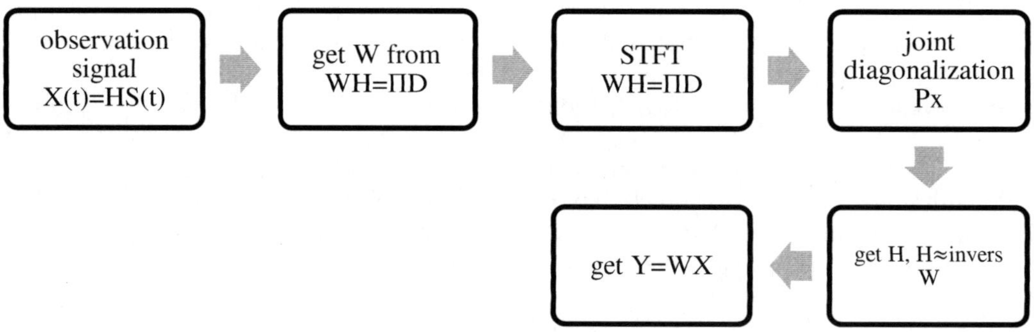


**Figure 1**. Flowchart getting a $W$ matrix.

with $H(z)$ is *z-transform* matrix from $J \times N$ *transfer function* mixing matrix. $J$ get from how many matrix observation signals while $N$ get from how many matrix source signals. Then $\Pi$ is permutation matrix and $D$ is diagonal matrix. For $W \approx A^{-1} \approx H^{-1}$ [4].

*2.2. Adjustment for water environment.*

Retrieving data by bringing equipment that consist of transmitter and receiver to real sea then perform scanning process have disadvantages such as cost, transportation, time and energy big consume. This situation then encourages researchers to conduct laboratory experiments to reduce the inefficiencies caused by classic way data retrieval.

In line with laboratory-scale data retrieval solution, the pre-research conducted on this research has developed a semi anechoic mini test tank has dimension $2 \times 1 \times 1$ m with thickness 12 mm shows in **Figure 2**, as representation of water environment but with physical variables like temperature and salinity adapted to Indonesia water characteristics [5]. For dimension variation, we conduct another data retrieval at towing tank Indonesia Hydrodynamics Laboratory (BPPT, IHL) with dimension $200 \times 10 \times 5.5$ m.

## 3. Experiment Method [5]

First scenario we vary water environment at semi anechoic mini test tank with temperature 13°C, 17°C, 21°C, 25°C and 29°C. Same with first scenario, for the second we change the water environment with room temperature but vary in salinity are: 3.1%, 3.2%, 3.3%, 3.4% and 3.5%. Data retrieval using sources generated from 2 underwater speakers and 3 hydrophones. The sound sources consist single tone and multi tone. The third scenario, we conduct data retrieval at towing tank Indonesia Hydrodynamics Laboratory (BPPT, IHL).

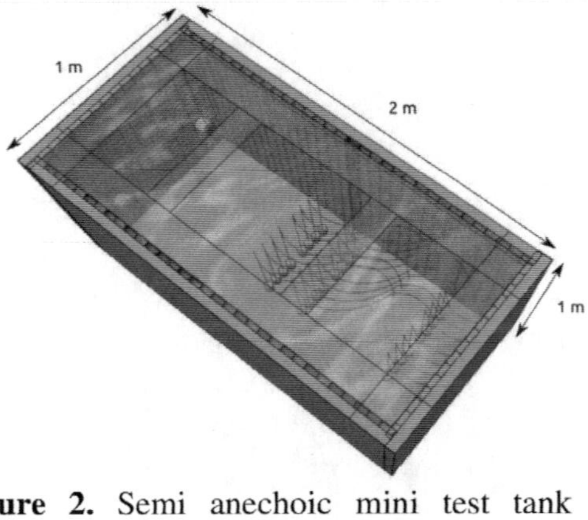


**Figure 2.** Semi anechoic mini test tank has dimension $2 \times 1 \times 1$ m with thickness 12 mm. There is sponge inside the test tank with high of bulge 2 cm for reduce the echo.

**Figure 3.** Towing tank Indonesia Hydrodynamics Laboratory (BPPT, IHL) with dimension $200 \times 10 \times 5.5$ m.

IOP Conf. Series: Journal of Physics: Conf. Series **1075** (2018) 012080 doi:10.1088/1742-6596/1075/1/012080

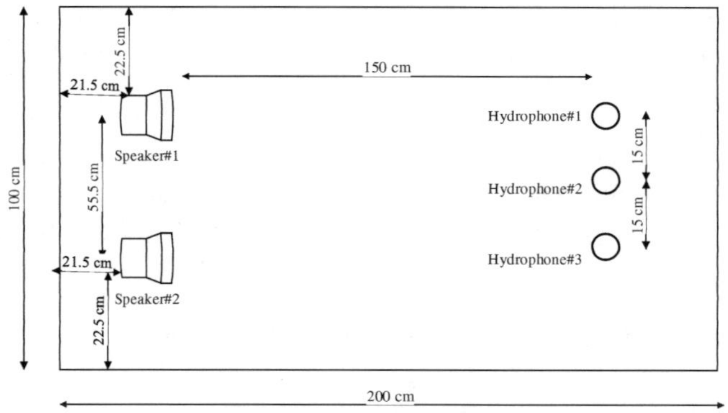


**Figure 4.** Setting underwater speaker and hydrophone top view.

## 4. Results

Based on mean squared error (MSE) measurement [4] from **Figure 5** and **Figure 6** shown that separation results (source expectation) at signal with kurtosis smaller give better approximation than another. This condition is consistent for all variation experiment, include salinity, temperature and dimension variation. Kurtosis is measure for how "peaky" probability density function (PDF) of an observation signal $\boldsymbol{X(t)}$. More "peaky" the signal shown by greater score for their kurtosis, better separation achieved on all variation in this research. Kurtosis score for observation signal we used shown at **Figure 7** and **Figure 8**.

## 5. Conclusion

Water environment variety we conduct in this research seems like have not significant effect to the results of BSS process algorithm using ALS. It shown that statistic properties have more roles than water environment variety.

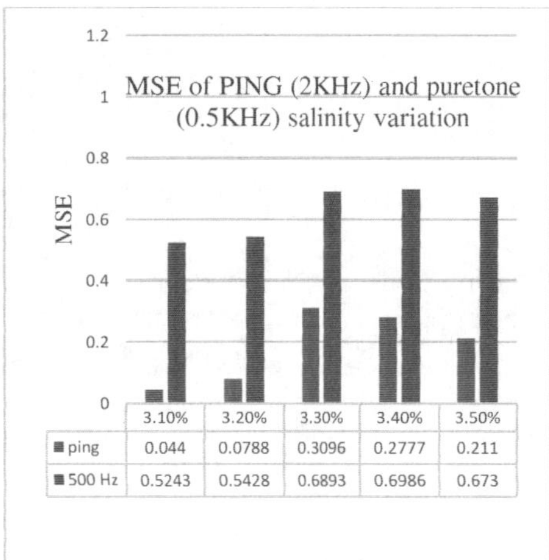


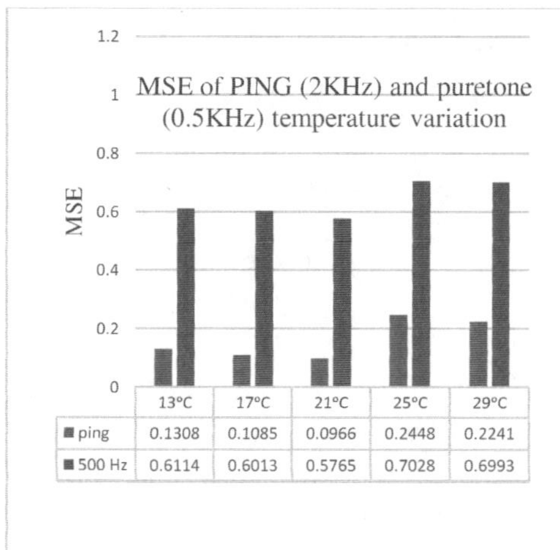

IOP Conf. Series: Journal of Physics: Conf. Series **1075** (2018) 012080 doi:10.1088/1742-6596/1075/1/012080

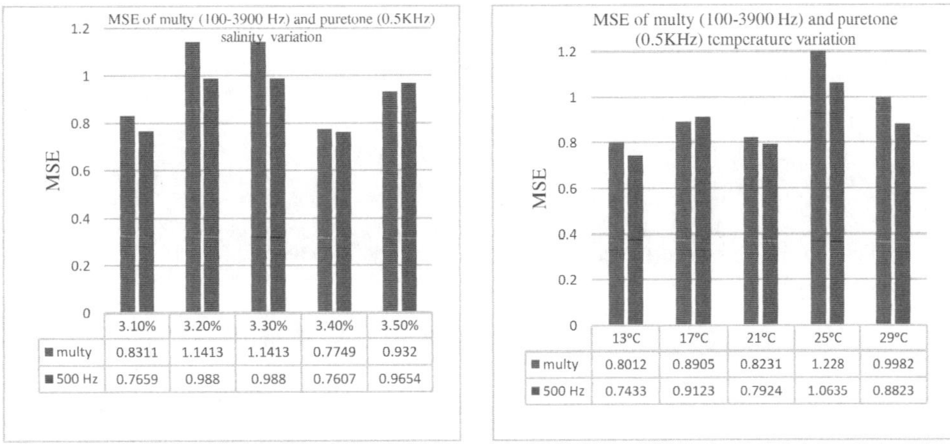


**Figure 5.** MSE separation results comparison on experiment conduct at semi anechoic mini test tank

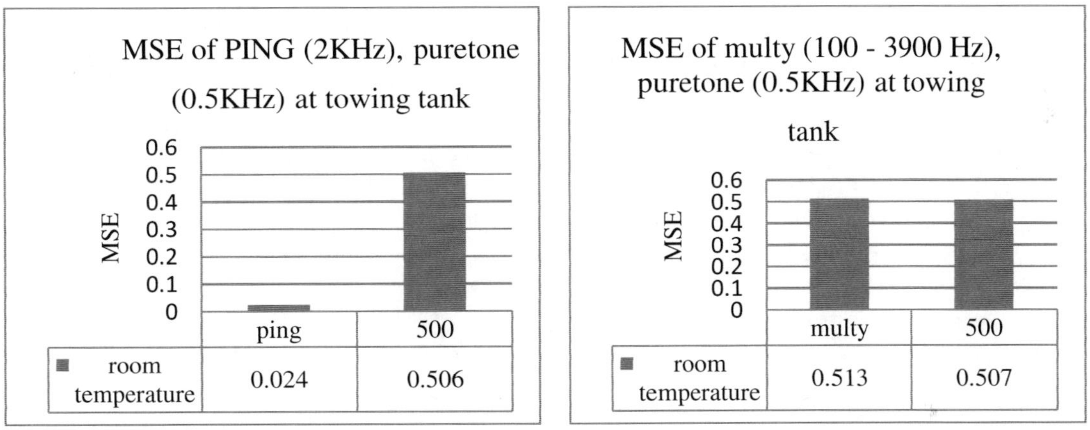


**Figure 6.** MSE separation results comparison on experiment conduct at towing tank

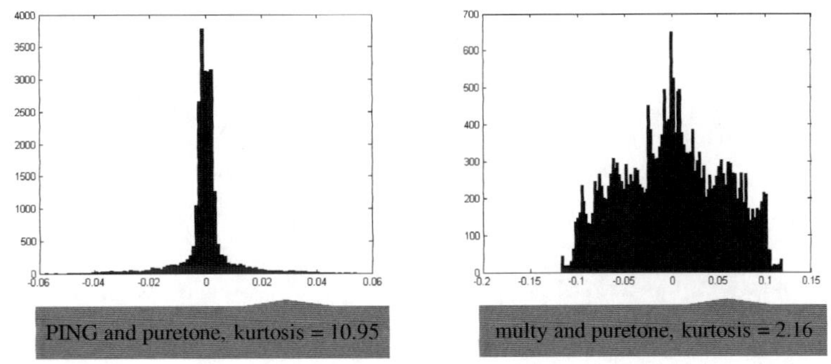


**Figure 7.** Joint distribution PDF from observation signal at semi anechoic mini test tank.

IOP Conf. Series: Journal of Physics: Conf. Series **1075** (2018) 012080 doi:10.1088/1742-6596/1075/1/012080

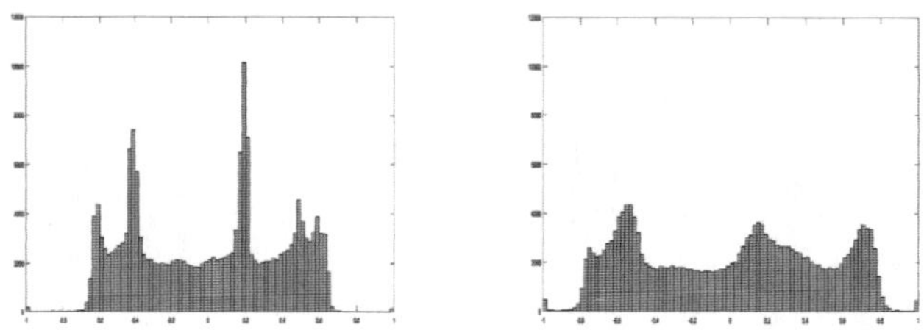

PING and puretone, kurtosis = 1.75 multy and puretone, kurtosis = 1.74

**Figure 8.** Joint distribution PDF from observation signal at towing tank.

## 6. References


[1] Urick, R.J, 1983. Principle of Underwater Sound 3rd Edition. McGraw-Hill

[2] Stone. (2004). "Independent Component Analysis. A Tutorial Introduction"

[3] Kamran. Rahbar, and Reilly, J. P., (2001). "Blind Source Separation of Convolved Sources by Joint Approximate Di Diagonalization of Cross-Spectral Density Matrices". IEEE Transaction on Audio, Speech, and Language Processing".

[4] Vincent, Gribonval dan Fevotte. (2006). "Performance Measurement in Blind Audio Source Separation" IEEE Transactions on Audio, Speech, And Language Processing, Vol. 14, No. 4

[5] Yuwono, Niken P, Arifianto, Dhany, Widjiati, Endang dan Wirawan (2014). "Underwater Sound Propagation Charateristics at Mini Underwater Test Tank with Varied Salinity and Temperature". IEEE

# A pilot study: the impact of continuous and intermittent noise on human blood pressure changes

**N Khoirunnisa[1], A Nadiroh[1], D Arifianto[1] and T Dhanardono[1]**

[1]Department of Engineering Physics, Institut Teknologi Sepuluh Nopember, Indonesia

*Email: najihah.nasa@yahoo.co.id

**Abstract**. Noise is one of factor that is often encountered as physical hazards in the workplace. Moreover, it can allow health problems in the circulatory system, especially blood pressure which is hypertension. Exposure of occupational noise in industries rise the heart rate, systolic, and diastolic blood pressure, but not generally in field studies what kind of noise characteristics that can increase the blood pressure changes. A pilot studies of the impact of continuous and intermittent noise on human blood pressure changes was conducted. The results show that variation of continuous and intermittent noise increase the blood pressure changes not significantly. Increased blood pressure changes by the intermittent noise is more give impact than continuous noise. Future research is needed adding more training time and prolong intermittent noise exposure.

## 1. Introduction

Noise is one of the physical factors that can give harmful effects in residential and industrial environment [1]. It is the major preventable cause of hearing loss. Noise-induced hearing loss can be caused by a one-time exposure to an intense impulse sound such as gunfire, or by steady state long-term exposure with sound pressure levels higher than LA 75–85 dB—e.g., in industrial settings. Moreover, the others impact of noise exposure are interference in conversation, annoyance, disturbs sleep and causes daytime sleepiness, affects patient outcomes and staff performance in hospitals, increases the occurrence of hypertension and cardiovascular disease, and impairs cognitive performance in school children [2] [3]. Noise exposure is also considered as the third cause of heart attack among the residents after smoking and air pollution [1].

The previous study was obtained the effect industrial noise to workers [4] [5]. The studies proved that occupational noise rise the heart rate, systolic, and diastolic blood pressure [4] also concluded that exposure to noise significantly increased systolic blood pressure but had no significant increase in the diastolic pressure of the workers [5]. The previous study also conducted to analyze the impact of noise exposure in climatic chamber [6]. Both systolic and diastolic blood pressures decreased in exposure to heat, while exposure to different levels of noise elevates systolic and diastolic blood pressures. This paper discusses the influence of continuous and intermittent noise on human systolic blood pressure changes in experimentally controlled conditions. Also, to know the noise characteristics that increase the blood pressure and differences impact between continuous and intermittent noise exposure.

IOP Conf. Series: Journal of Physics: Conf. Series **1075** (2018) 012081 doi:10.1088/1742-6596/1075/1/012081

## 2. Materials and Methods

### *2.1. Participants*
A total of 12 students (6 female, 6 male) aged 20-25 years took part in the experiment. All participants reported normal hearing. The 2 (1 female, 1 male) of participants were chosen to be control objects. The control objects are the participants that is not given any treatment as a control variable.

### *2.2. Measurement*
The tests were conducted in a semi-anechoic chamber with a length of 3.31 m, width of 2.74 m, and height of 2.37 m in which the background noise level was about 36.5 dB(A). The noise is generated from a SRS-X55 speaker. Sphygmomanometer OMRON HEM-7203 is used to measure the blood pressure.

*2.2.1. Noise equipment.* The noises designed to study the effect of noise variety to the blood pressure changes. Two types of noise exposed consist of continuous noise and intermittent noise. White noise and pink noise are used for continuous noise. Each noise was conditioned to adjust the ear response, in which the sound density forms a ramp curve for 20 ms as the active ear. Four types of sounds i.e. sound of airplanes, trains, ambulance sirens, and exploding bombs are used for intermittent noise.

*2.2.2. Test conditions.* The continuous noise was exposed for 10 minutes with three of sound pressure level that is 74 dB(A), 80 dB(A), and 86 dB(A) for three times for each exposure. The details of measurement procedure are shown in the steps below:

- Participants enter the test chamber and rest for 10 minutes
- The blood pressure of participants is measured to determine blood pressure before exposure to noises
- Noise at the 74 dB(A) sound pressure level was exposed to the participants for 10 minutes
- The blood pressure of participants is measured to determine blood pressure after exposure to noises
- Participants rest for 10 minutes
- The measurement procedure is also performed for the 80 dB(A) and 84 dB(A) sound pressure level

The measurement procedure for intermittent noise are shown in steps below:

- Participants enter the test chamber and rest for 10 minutes
- The blood pressure of participants is measured to determine blood pressure before exposure to noises
- Airplane sound was exposed to the participants for 1 minutes
- The blood pressure of participants is measured to determine blood pressure after exposure to noises
- Participants rest for 10 minutes
- The measurement procedure is also performed for trains, ambulance sirens, and exploding bomb sounds

## 3. Results

### *3.1. The Result of Blood Pressure Changes to Continuous Noise*

The test of blood pressure changes to continuous noise has been evaluated to 2 kinds of noises, i.e. white noise and pink noise. It can be shown by the graphic of systole value (Fig 1).

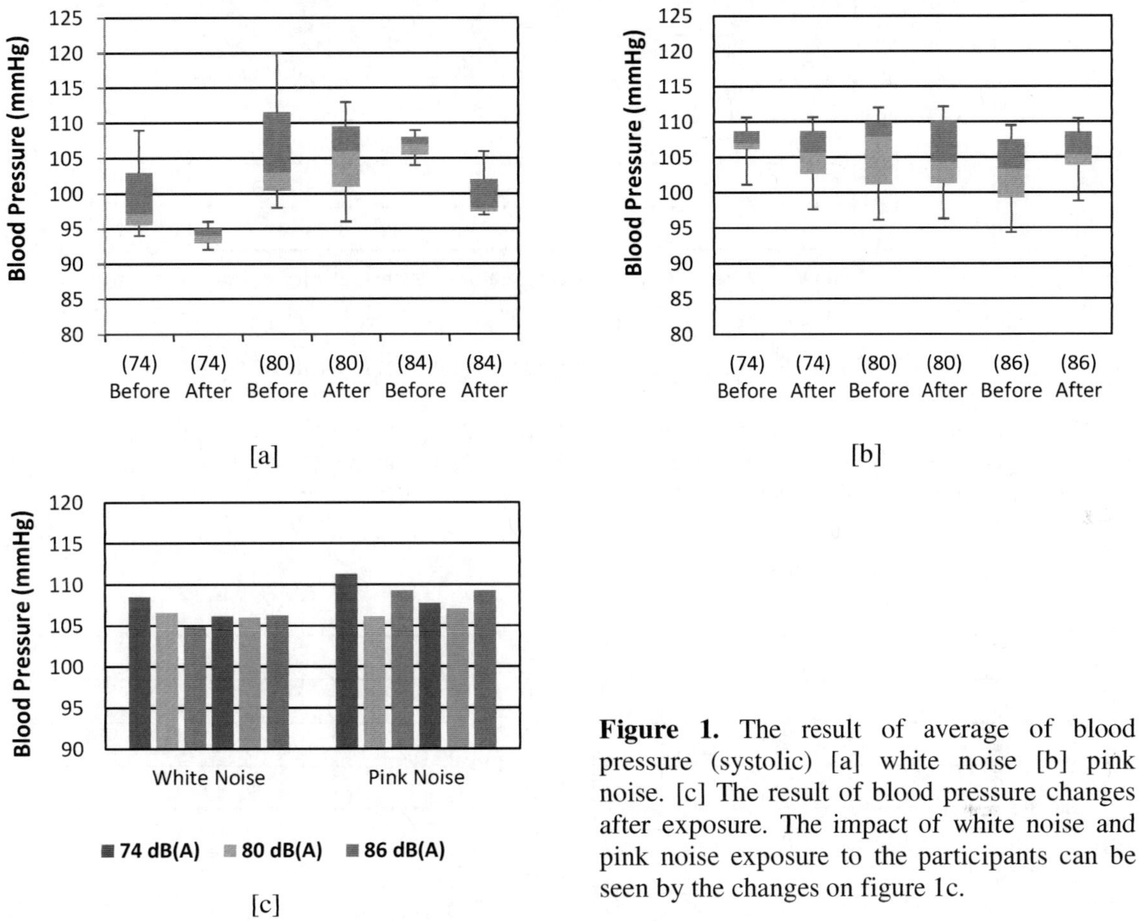


**Figure 1.** The result of average of blood pressure (systolic) [a] white noise [b] pink noise. [c] The result of blood pressure changes after exposure. The impact of white noise and pink noise exposure to the participants can be seen by the changes on figure 1c.

### *3.2. The Result of Blood Pressure Changes to Intermittent Noise*

The test of blood pressure changes to intermittent noise has been evaluated for 4 types of sounds i.e. sound of airplanes, trains, ambulance sirens, and exploding bombs. The result of sound pressure level for each sound can be shown by the table 1 and the result of blood pressure changes after exposure can be shown by the figure 2.

**Table 1.** The sound pressure level of intermittent noises

| | Sound Pressure Level (dBA) | |
|---|---|---|
| Sounds | Minimum | Maximum |
| Airplane | 53.3 | 90.4 |
| Exploding Bomb | 81.6 | 93.9 |
| Ambulance Siren | 90.4 | 100.2 |
| Trains | 80.2 | 91.3 |

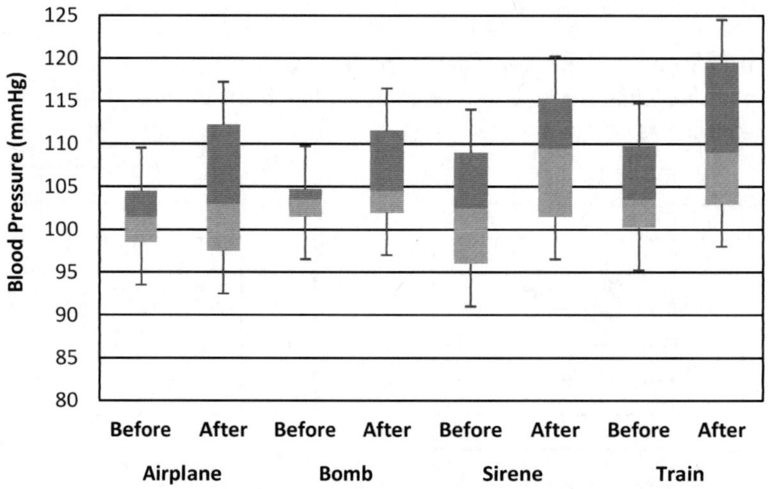


**Figure 2.** The average of blood pressure test (systole).

The Pilot test was conducted to evaluate and analyse the general response of participants after doing some testing as a feasibility test for this research. The 3 of participants (Number 6, 14, and 23) was tested in the experiment using 2 sources of noise i.e station and traffic noise.

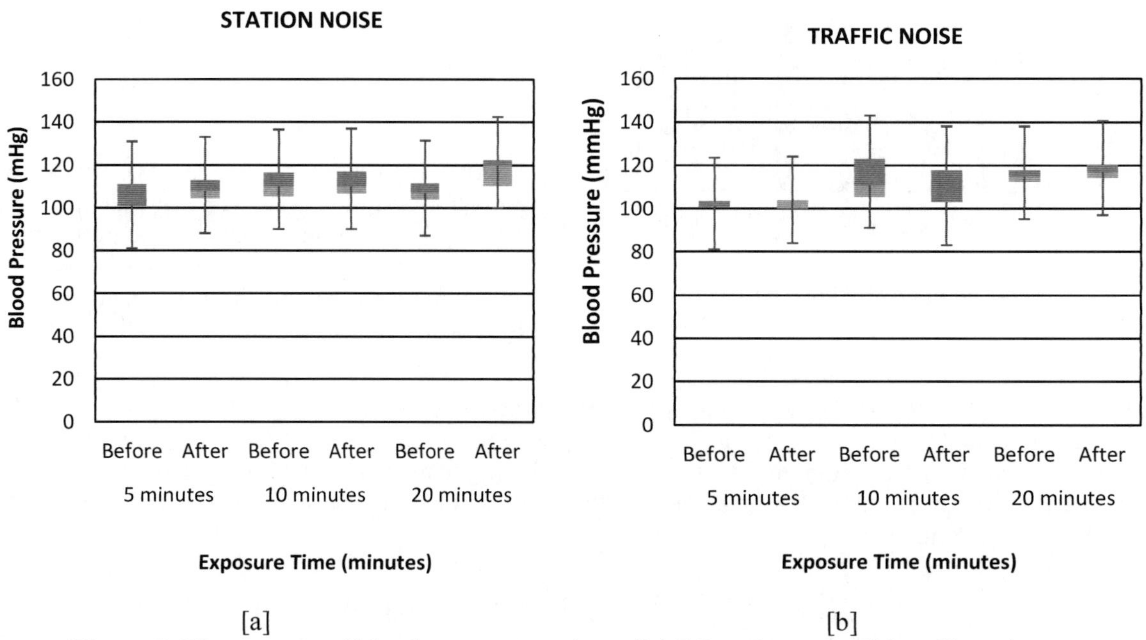


**Figure 3.** The average of blood pressure test (systolic) [a] station noise [b] traffic noise.

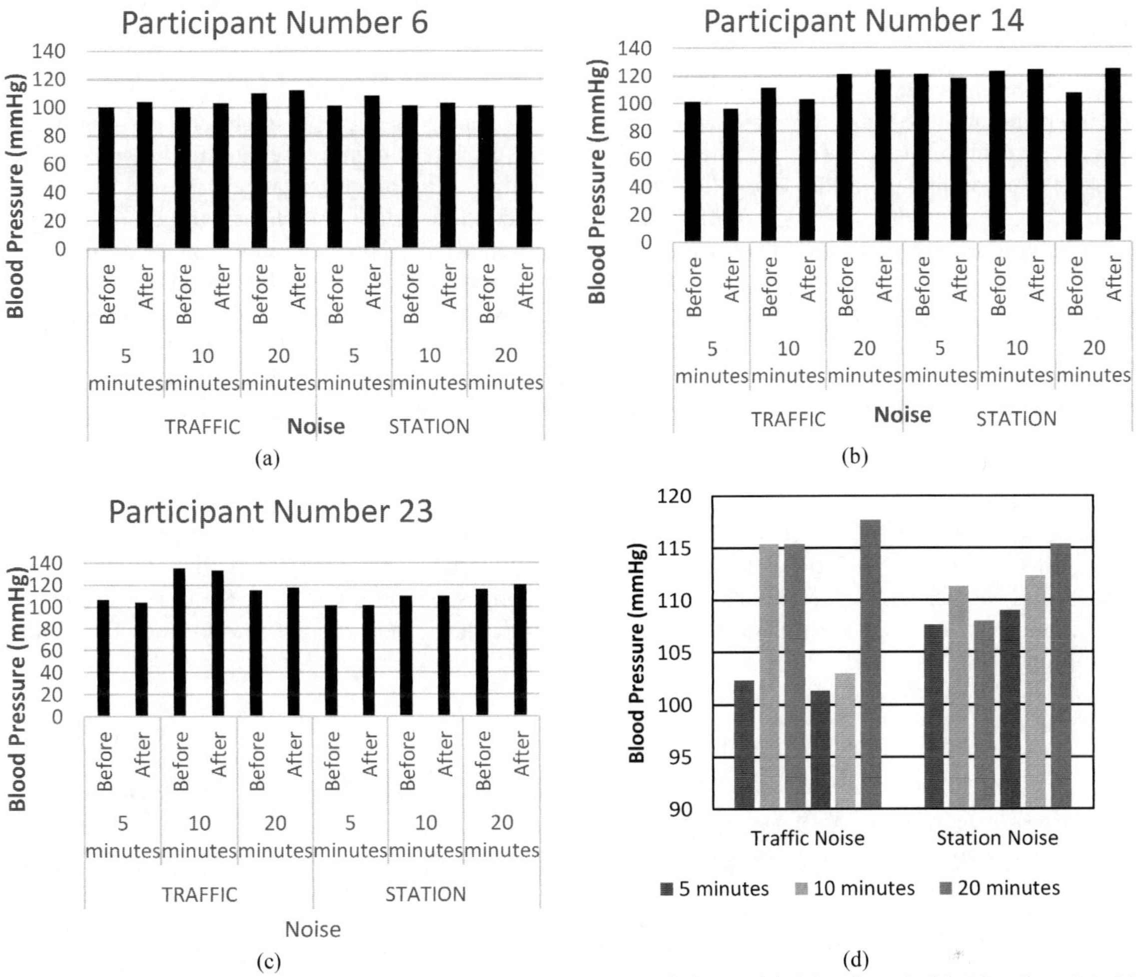


**Figure 4.** Pilot test blood pressure changes on some participant (a) Number 6 (b) Number 14 (c) Number 23 (d) The result of blood pressure changes after exposure.

## 4. Discussion

The continuous noise increase blood pressure changes and cause hypertense. The sensation is sensed by auditory sensor and hippocampus. The noise is defined as distraction and cause hyper-activity for HBF. It stimulates production of stress hormone (epinefine) and increase metabolism of glucose, protein, and neuron system. The metabolism changes cause heart beat changes then blood pressure changes [7].

### *4.1. Continuous Noise*

The test was conducted to analyze the impact of noise to blood pressure changes. From the result, there is no positive value for blood pressure changes. The theory said that exposure noise causes the changes of blood pressure with positive value. It means blood pressure will increase after exposure by noises. These results therefore seem to contradict the theory. However, White noise and pink noise have a constant sound spectrum to frequency. Therefore, participants can relax after exposed by the noises. Beside that, noise dose of noise exposure that give to the participants is low.

The conclusion of the results of continuous noise test can be shown by figure 1c. The blood pressure changes decrease at 74 dB(A) sound pressure level. For 80 dB(A) sound pressure level, the blood pressure increases after exposed by pink noise. And both of the noises increase after exposed at 86 dB(A) sound pressure level.

### *4.2. Intermittent Noise*

The intermittent noise was conducted to prove that continuous noise gives less effect than intermittent noise. The result of blood pressure changes after exposed by intermittent noise can be shown by figure 2. All of the condition increases the blood pressure changes. The biggest change occurred on aircraft noise exposure with the value is 8 mmHg. It means its condition does not change the blood pressure changes significantly. The blood pressure changes significantly if it changes more than 10 mmHg [8].

The result of test using station and traffic noise can be shown by figure 3. Increased blood pressure generally occurs when exposed noise is station noise. However, the traffic noise does not impact blood pressure significantly. Traffic noise conditions are very common and experienced by humans. The noise exposes continuously and affects the background noise of the environment. Therefore, the noise is not defined as a threat by the body. It means the traffic noise does not increase blood pressure significantly.

## 5. Conclusion

The experiment of the impact of several noise variety to blood pressure changes has been conducted. The results show that variation of continuous and intermittent noise increase the blood pressure changes not significantly. Increased blood pressure changes by the intermittent noise is more give impact than continuous noise. In addition, future research should add more training time. To prolong intermittent noise exposure.

## References

[1] A. T., "Noise Effect on Mental Health," Shahid Beheshti University of Medical Sciences, Tehran.

[2] M. Basner, W. Babisch, A. Davis, M. Brink, C. Clark, S. Janssen and S. Stansfeld, "Auditory and non-auditory effects of noise on health," *National Center for Biotechnology Information,* vol. 383, no. 9925, p. 1325–1332, 2013.

[3] P. Nasiri, M. M. Esmaeelpour, A. R. Foroushani, H. Ebrahimi and Y. Salimi, "Occupational noise exposure evaluation in drivers of bus transportation of Tehran City," *Iranian Journal of Health & Environment,* vol. 2, p. 124–131, 2009.

[4] S. Kalantary, A. Dehghani, M. S. Yekaninejad, L. Omidi and M. Rahimzadeh, "The effects of occupational noise on blood pressure and heart rate of workers in an automotive parts industry," *ARYA Atherosclerosis,* vol. 11, no. 4, pp. 215-219, 2015.

[5] S. O. Ismaila and A. Odusote, "Noise exposure as a factor in the increase of blood pressure of workers in a sack manufacturing industry," *Beni-Suef University Journal of Basic and Applied Sciences,* vol. 3, no. 2, pp. 116-121, 2014.

[6] H. Deghnan, M. T. Bastami and B. Mahaki, "Evaluating combined effect of noise and heat on blood pressure changes among males in climatic chamber," *Journal of Education and Health Promotion,* vol. 6, no. 39, 2017.

[7] M. Monsefi, A. Bahoddini, S. Nazemi and G. A. Dehghani, "Effects of Noise Exposure on the Volume of Adrenal Gland and Serum Levels of Cortisol in Rat," *IJMS,* vol. 31, no. 1, pp. 5-8, 2006.

[8] M. Ulyah, KDPK: Keterampilan Dasar Pabrik Klinik untuk Kebidanan, Health Book Publishing, 2008.

## Acknowledgments

This study was done in the semi-anechoic chamber of Department of Engineering Physics, Institut Teknologi Sepuluh Nopember, Indonesia, as a part of requirements of the study, and the authors are grateful to all those who participated in the study and supported its procedure.

# Improvement acoustic quality room live music P-Two cafe Surabaya

**I B A Pangestu[1], D N Fauzi[1] and W A Asmoro[1]**

[1]Dept. of Engineering Physics, Faculty of Industrial Technology, Institut Teknologi Sepuluh Nopember, Kampus ITS Sukolilo, Surabaya 60111, Indonesia

E-mail: indrabagusaji@gmail.com

**Abstract.** Music room needs good sound clarity to make the intonation tone can be delivered clearly. Room acoustic parameter which used are Reverberation Time, Clarity (C50), and Clarity (C80). The object of this research is P-Two cafe Surabaya. The purpose of this research is to redesign P-Two cafe room and to make an appropriate room for live music according to the international standard. The design is done by giving insulators and absorbers to the room. Method of interpretation data is Impulse Response. Besides, simulation from calculated results done by using EASE 4.3. The result of calculated sound pressure level has been able to reduce the sound pressure level for about 42 dB for wall which is made by glass. The simulation result of alternative redesign which used gypsum board material, wood floor carpet, concrete smooth finish and wood surface showed 1,43 second for room RT the result before are 0,6 second smaller. Simulation result of Clarity (C50) is -1 and Clarity (C80) is 1,67 that also showed the room is according to the standard which means the speech and live music of room are getting better.

## 1. Introduction

One of the most challenging problems that is faced by acoustic architect is to design a room whose purpose is used for meetings or musical performances with good acoustic quality. Thus, the problem of acoustic design for venues such as town hall, live music cafe and concert venues have been attracted observers of acoustics to do some studies and preliminary research. Design of a building show, cafetorium, should to consider the acoustic parameters that support a good room to perform its function at moment it's used, so that the sound can be heard clearly between the music player and the audience in the room. improving acoustic quality can be done by eliminating the usual hum known as reverberation time or RT [1,2,3]. To provide the good acoustic quality of the room such as reducing buzzing and noise is done by paying attention to an interior element. It is very supportive when designing room when an acoustic level of comfort and level clarity of voice are to be considered. The purpose of these studies and research are to analyse the sound acoustic quality in the cafe room P-Two Surabaya by using the simulation based on standard of acoustic parameters. At the time of data retrieval Using EASE simulation in reverberation Time at this cafe room of 0.5 seconds. For the room which is designed for music is recommended RT for more than 1.4 seconds.

IOP Conf. Series: Journal of Physics: Conf. Series **1075** (2018) 012082 doi:10.1088/1742-6596/1075/1/012082

## 2. Experiment Setup

The research focuses on redecoration of acoustic quality of live music room of P-Two café, Surabaya. Below are some descriptions of data retrieval methods, analysis and the discussion that will be applied in this research, among others:

### *2.1. Designing Simulation Objects*

The object used is the P-Two café which can be used for live music and conversation. Cafe located on the street Medokan Asri no 30 facing the street with an outside background noise of 65 dB SLM position is facing the entrance of P-Two café.

**Figure 1.** P-Two cafe room with room dimension 13.9 meters length, 8.64 meters width, and 3 meters high. The volume of the room is 347.33 meters$^3$.

### *2.2. Testing Using Simulation*

The simulation test using EASE 3.0 software is the first step to know the acoustic condition of room based on the acoustic parameter among others that is:

1) Reverberation Time. RT simulation test that is done by inserting value of room volume, humidity value of room and air pressure.
2) Clarity C80 For Music. The C80 simulation test is done to know the clarity of the room
3) Clarity C50 For Conversation. The C50 simulation test is done to know the clarity of the room.

### *2.3. Data Retrieval and Data Analysis*

Data retrieval is done in two stages measuring noise background data noise and impulse response. For noise background noise data retrieval done at the time of no activity at around 17.00-22.50 in room P-two café. The equipment used is the SLM (Sound Level Meter) Solo 01dB brand that records the background noise every ten minutes for one hour. SLM position is placed in the room directly opposite the music player with distance of 120 cm from the lip of the stage.

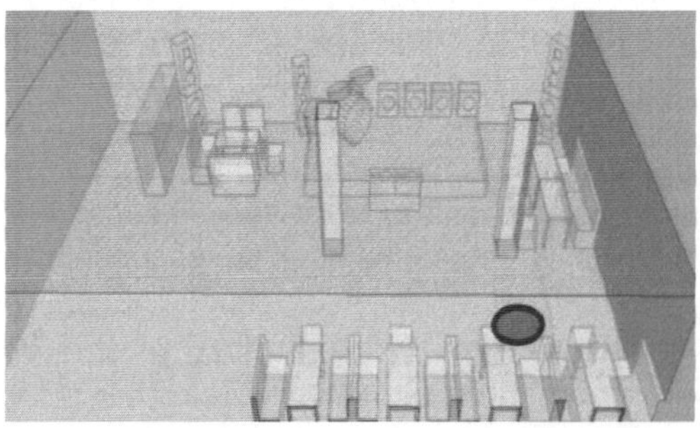

**Figure 2.** Point of data measurement

The second data retrieval process is impulse response. This measurement requires the source of impulse sound generated by balloon. The sound source of the impulse is recorded using SLM brand Solo 01dB.

## 3. Data Analysis

This chapter describes the data that has been taken will be processed with the simulation calculations to improve the acoustic quality of the room at the P-Two café to match the needs of live music and conversation:

### *3.1. Background Noise*

Measured data such as background noise and pulse response are required to support the improvement of acoustic quality. These data can be simulated with the software to compare the measured data with simulation. Supported by retrieving the data manually to know the background noise of the room and generating an impulse response to find out the reverberation time of the P-Two café room. Data is taken during the afternoon hours before the operation of P-Two café to ensure optimal noise value. The duration of continuous data retrieval is recorded every 10 (ten) minute intervals for 1 (one) hour inside and outside the room directly related to the surrounding environment. Here is background noise data from inside space.

**Table 1.** Background Noise

| Data to- | Indoor (dB A) |
|---|---|
| 1 | 45,75 |
| 2 | 44,63 |
| 3 | 43,6 |
| 4 | 40,93 |
| 5 | 40,38 |
| 6 | 40,38 |
| 7 | 80,35 |
| 8 | 39,71 |
| 9 | 38,41 |
| 10 | 82,93 |
| Leq | 49,71 |
| NC | 45,88 |

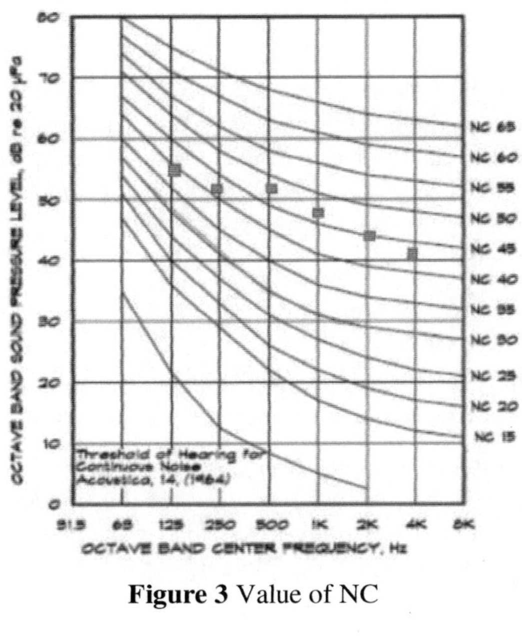


**Figure 3** Value of NC

IOP Conf. Series: Journal of Physics: Conf. Series **1075** (2018) 012082 doi:10.1088/1742-6596/1075/1/012082

*3.2. SRI Wall Data*

SRI data is processed by comparing the measurement and calculation results. If the measurement result is smaller than the calculation result, then the insulation ability of the wall of partition has been able to reduce the noise coming from the sound source. In this case the background noise, of the measurement results are greater than the calculation results. Here is the calculation of the SRI value of the back wall compared to the measured result.

**Table 2**. SRI Room Data Before Repair

| Element | SRI material (dB) | Transmission Coefficient τ |
|---|---|---|
| The front wall of the stage | 20,18 | 9,584 x$10^{-4}$ |
| Side walls of the stage | 24,96 | 3,189 x$10^{-4}$ |
| Backstage wall | 27,92 | 1,614 x10-4 |
| Wall of the gate | 20 | 0,01 |

*3.3. Indoor Insulation*

Indoor insulation is required for the sound generated, in this case the loudspeaker, the loudspeaker specification used, namely JBL VP7212-64, has a value of SPL output (Sound Pressure Level) ] can be achieved equal to 100,78 dB. When reviewed the SRI value that only reduces the front wall of the stage and the gate wall 20.18 dB and 20 dB on the wall is required wall insulation to fit the circumstances surrounding the room that has a background noise of 49.71 dB. Insulation background linked to music to reduce noise that is eligible under KEP-MENLH regulations 48, 1996.

Appropriate insulation treatment is required for the SRI value to reduce the sound source from indoors. For the walls of the room, the required SRI values are as follows:

**Table 3**. SRI Repair

| Element | SRI material (dB) | Transmission Coefficient τ |
|---|---|---|
| The front wall of the stage | 48 | 1,585x$10^{-5}$ |
| Side walls of the stage | 47 | 1,995x$10^{-5}$ |
| Backstage wall | 48 | 1,585x$10^{-5}$ |
| Wall of the gate | 42 | 6,309x$10^{-5}$ |

*3.4. Reverberation Time Data*

**Table 4** Reverberation time manual calculations before acoustic treatment

| Frekuensi (Hz) | 125 | 250 | 500 | 1000 | 2000 | 4000 |
|---|---|---|---|---|---|---|
| $\sum$ S.α | 39,15 | 44,7 | 40,5 | 35,4 | 34,1 | 32,2 |
| RT- 60 | 0,67 | 0,76 | 0,69 | 0,6 | 0,58 | 0,55 |

*3.5. Impulse Response Measurement Data*

The measurement data of the room impulse response is taken at 1 point of the room with volume size 362,68 $m^3$. Result of measurement at frequency 125 Hz - 4000 Hz using SLM Solo is as follows:

**Table 5.** Reverberation Time Data from SLM Solo

| Frekuensi (Hz) | RT (Second) |
|---|---|
| 125 | 0,6 |
| 250 | 0,62 |
| 500 | 0,6 |
| 1000 | 0,53 |
| 2000 | 0,7 |
| 4000 | 0,63 |

*3.6. Simulation Results Before Given Acoustic Treatment*

To find out the initial state of the room before being given the proper acoustic treatment, it is necessary to simulate a room that the condition of the room as described in Chapter III. The image of the simulation results of the room before being given acoustic treatment is as follows:

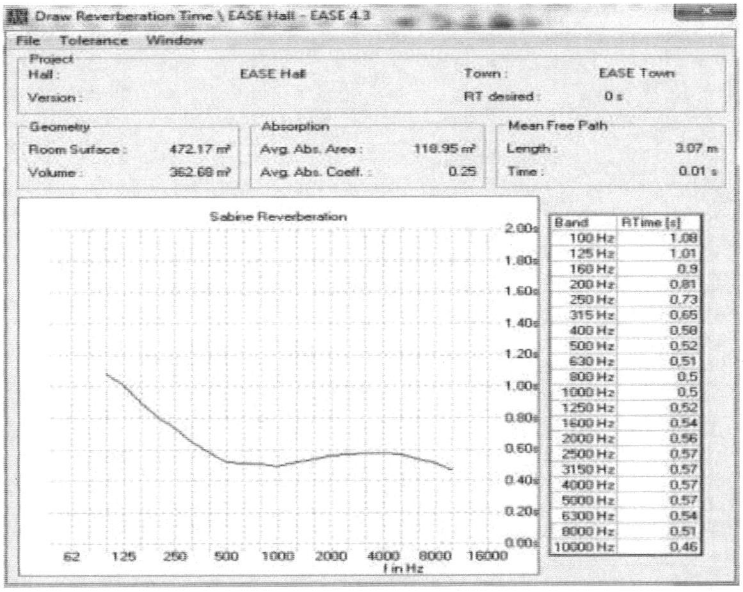


**Figure 4.** Simulation of Room Reverberation Time Before Acoustic Treatment

### *3.7. Clarity Simulation Results (C80)*

The simulation C80 is a simulation to know the room according to the criteria of music shows at a frequency of 125 Hz - 10 kHz according to the simulation sound clarity occurs at that frequency does not meet the criteria then the simulation needs to be given a treatment to achieve clarity value of the room around -2 < C80 <4.

### *3.8. Clarity Simulation Result (C50)*

Simulation C50 is a simulation to know the criteria of conversation in the room of P-Two café. It show that C50 not yet full fill criteria on - 3 <C50> 6 in frequency 125 Hz. The result is -2,9 dB.

### *3.9. Alternative Design*

Considered in the redesign, free variable that can be changed is the value of absorption coefficient Sabine (A) which is distributed in interior indoors. The optimized absorption coefficient value of Sabine is the value of A that follows the standard Reverberation time of the music room of 1.45 seconds for volume 362.68 m³.

**Table 6.** Manual Calculation of Reverberation Time After Acoustic Treatment

| Frequency (Hz) | 125 | 250 | 500 | 1000 | 2000 | 4000 |
|---|---|---|---|---|---|---|
| $\sum$ S. α | 68,8 | 55,9 | 55,7 | 57 | 59,2 | 53 |
| RT 60 | 1,2 | 1,0 | 1,0 | 1,0 | 1,0 | 0,9 |

### *3.10. Re-Design Results*

In the simulation using the acoustic software of EASE 4.3 space, Clarity (C80 and C50) meets the criteria at the middle frequency of 1dB and -1 dB. For the Reverberation Time also meets the criteria at the middle frequency of 1 second.

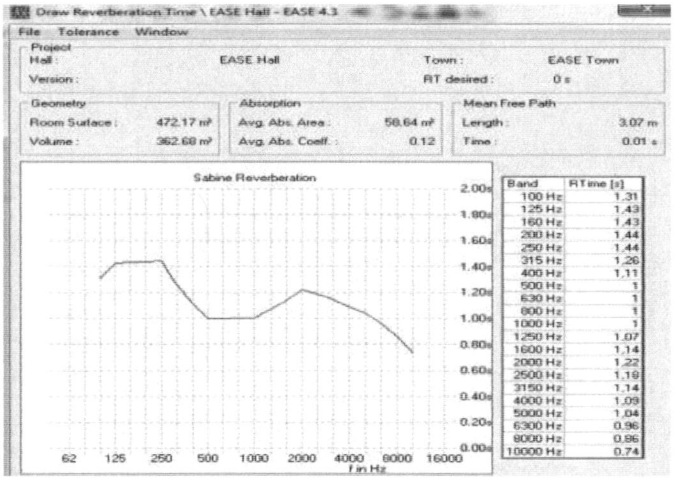


**Figure 5.** Simulation of Reverberation Time After Acoustic Treatment

## 4. Discussion

From background noise measurements in the room can be see that noise entering the room still does not meet the NC30 criteria recommended on the NC chart. With the value of NC is worth 45 still need to decrease the noise level. For SRI calculation results of the front wall of the stage and gate walls consisting of plaster of brick, wood, foam, leather and glass are not yet able to reduce sound pressure level by 20.18 dB. It needs additional room insulation. For indoor insulation treatment SRI calculation of the front wall of the stage is enough to reduce the maximum sound pressure level coming out of the loudspeaker by 48 dB. While the SRI value must be reduced by the insulation wall when the sound source is worth 100.78 dB is 42 dB. Then for the result of the impulse response the room Reverberation time has the RT value at the middle frequency of 1.0 seconds. While the RT for the music room with the volume of P-Two Cafe of 362.68 m3 is 0.53 seconds.

## 5. Conclusion

From the research that has been done, the conclusions that can be drawn from this research are as follows : the acoustic quality of the room at P-Two cafe Surabaya for the initial condition is not sufficiently crippled with the RT value at the middle frequency of 0.6 seconds, C50 in the middle frequency is 0.96 dB and C80 in the middle frequency is 1.71 dB; after redesigning the interior room of P-Two cafe obtained quality with better objective acoustic parameter values of RT for 1 second in middle frequency, C50 in middle frequency 0,96 dB and C80 in middle frequency equal to 1.71 dB; and redesigning the interior space P-Two cafe includes replacing the material on the wall of P-Two cafe room that is on the front wall and backstage using gypsum board, side walls of the stage using concrete and on the floor using wood.

## References

[1] Ribeiro, M.R.S. 2002. Room Acoustic Quality of A Multipurpose Hall:A Case Study, Centro de Estudos do Departamento de Engenharia Civil, Faculdade de Engenharia deUniversidade do Porto. Portugal

[2] Szokolay, Steven V. 2008. Introduction to Architectural Scoence: The Basis of Sustainable Design 2nd edition. USA: Elsevier Ltd.

[3] Everest, F. Alton. and Pohlmann, Ken C. 2009. Master Handbook of Acoustics 5thedition. USA: Mc GrawHill, Inc

[4] Long, Marshall. 2006. Architectural Acoustics. San Diego, California: Elsevier Academic Press.

[5] Mediastika, E. Christina. 2005. Akustika Bangunan. Indonesia: Erlangga

[6] Danarjati Wisnu Wardhana ,Desain Ulang Meeting Room P3AI ITS untuk Perbaikan Kualitas Akustik Video Conference Jurusan Teknik Fisika, Fakultas Teknologi Industri, Institut Teknologi Sepuluh Nopember, ITS.

[7] De Salis, M.H.F., Oldham, D.J., Sharples, S. 2002. Noise contol strategiesfor naturally ventilated buildings. The Journal of Building and Environment 37 (2002) 471-484.

[8] Doelle, L.L. 1972. Environtmental Acoustic. McGrawHill Publishing Company, New York

IOP Conf. Series: Journal of Physics: Conf. Series **1075** (2018) 012083 doi:10.1088/1742-6596/1075/1/012083

# Traffic noise control in SMAN 5 Surabaya

**F A Winarno[1], M P Liana[2] and T Dhanardono[2]**
[1]PT Pembangkitan Jawa Bali, Harun Tohir street Gresik 61112 East Java, Indonesia
[2]Departement of Physics Engineering, Institut Teknologi Sepuluh Nopember Surabaya, Keputih, Sukolilo 60111 East Java, Indonesia

E-mail: fradita10@gmail.com

**Abstract.** The development of transportation technology is now very rapidly with increasing of number vehicles that causes a traffic jam. The location nears a crowded highway that causes a noise pollution in SMAN 5 Surabaya. Based on noise standards of Minister of Environment of Indonesia (KEP-48/MENLH/11/1996), the permitted sound pressure level in educational environments is 55 dBA. The SPL in the environment of SMAN 5 Surabaya is 87 dBA. It is necessary to design noise barrier to reduce noise. To decrease the noise level 32 dBA, it can use noise barrier with 12 m height. The most effective method is using Maekawa based on direct calculations. This method is more accurate. The other methods used in this research are graph method, double screen and Nomograph. The best material of noise barrier planning is steel with thickness 1,25 mm that decrease noise about 32,32 dBA.

## 1. Introduction

Recently, the main problem on the street is noise which caused by traffic jam. Traffic noise is the second biggest environmental problem according to WHO after air pollution and it is affecting health the most [1]. Traffic noise arises from vehicle engine (engine operation, exhaust or exhaust system). This noise can be heard up to several meters from the source of noise [2]. Frequency of this traffic noise has the same range of frequencies audible to humans, but it can impair human comfort. Traffic noise is very disturbing to the human comfort who lives near the highway. It is very disturbing not only the human comfort who lives the highway but also human activities such as such as in the office environment, hospital environment, trade and service environment, and educational environment [1]. In this case, the authors will examine the control of noise levels due to traffic to the educational environment in SMAN 5 Surabaya. Based on the traffic noise data, it can be known the noise level in the environment and evaluate the standard of noise level for the educational environment according to the regional allocation based on the Decree of Minister of Environment No. 48 / MENLH / 1996. November 25, 1996. It designs to reduce traffic noise as well as providing recommendations of some appropriate Noise Barriers.

## 2. Research Methodology

### *2.1. Noise*

Noise Level of traffic can be described with graph of $L_{eq}$ value. $L_{eq}$ (*Equivalent Continuous Noise Level*) is a specific noise value from fickle noise (fluctuating in certain time equal with steady noise level) in the same time. $L_{eq}$ is measured directly by integrated sound pressure level. $L_{eq}$ is the level of energy average from some sound level Varian. It is not direct interference measurement. $L_{eq}$ can be calculated using equation 1 [3].

$$Leq = 10\log\left(\frac{1}{T}\sum t_i 10^{\frac{L_i}{10}}\right) \tag{1}$$

With $t_i$ as Length of time with noise $L_i$ and T as $\sum t_i = t_1 + t_2 + t_3 + \dots$. To determine $L_{eq}$ from ($L_{day}$ = Noise level in the noon) can be calculated from the equation 2 and equation 3 using ($L_{night}$ = Noise level in the night). To know the noise, exceed of noise level or not, we must determine the value of $L_{dn}$ from measurement of field data. $L_{dn}$ is calculated with equation 4.

$$L_{day} = 10\ \log\frac{1}{16}\{T1 \cdot 10^{0.1L1} + \dots + T4 \cdot 10^{0.1L4}\} \tag{2}$$

$$L_{night} = 10\ \log\frac{1}{8}\{T5 \cdot 10^{0.1L5} + \dots + T7 \cdot 10^{0.1L7}\} \tag{3}$$

$$L_{dn} = 10\ \log\frac{1}{24}\{16 \cdot 10^{0.1L_{day}} + 8 \cdot 10^{0.1(L_{night}+5)}\} \tag{4}$$

*2.2. Noise standard*

The Indonesian Government through of the Environment Minister Decree No: 48/ MENLH/ XI/ 1996 contains noise level limit criteria in residence and educational area which have maximum noise less than 55 dBA [4].

*2.3. Noise barrier.*

Noise Barrier is a barrier wall or partition used to control airborne noise transmissions, where the wall is located between the source and the receiver. The function of Noise Barrier is to provide a shadow zone or area where it has a quieter noise at the receiver [3].

*2.3.1. Maekawa method.* Maekawa method is used to reduce sound pressure level using noise barrier. Maekawa method usually uses graph method. In this method, we can determine the reduction of sound pressure level which depend on sound frequency and distance between source to barrier. In single screen barrier, path length difference can be determined from equation 5. Equation 6 shows double screen barrier [5].

$$\delta = A + B - d \tag{5}$$

$$\delta = \left[(d_{ss} + d_{sr} + e)^2 + a^2\right]^{\frac{1}{2}} - d \tag{6}$$

Where, $\delta$ as path length difference (m), $d_{ss}$ distance from source to the first tip barrier(m), $d_{sr}$ as distance from the second tip to receiver(m), $a$ as distance between two barriers (m), $e$ is distance between barrier(m) and $d$ distance between source of noise and receiver(m).

Noise Reduction (NR) is the different value of sound pressure level from source to receiver. The equation of noise reduction is shown in equation 7. NR is noise reduction(dB), $SPL_1$ as sound pressure level of source(dB), and $SP_2$ as sound pressure level of receiver(dB). It can be calculated using equation 8 which assumed after value of path length difference is obtained. With $B$ is barrier attenuation (dBA), $\delta$ is path length difference (m), and $\lambda$ is wave length.

$$NR = SPL_1 - SPL_2 \tag{7}$$

IOP Conf. Series: Journal of Physics: Conf. Series **1075** (2018) 012083 doi:10.1088/1742-6596/1075/1/012083

$$B = 10\log\left(3 + \frac{40\delta}{\lambda}\right) \tag{8}$$

*2.3.2. Nomograph method. Nomograph method is used to determine the attenuation value on noise barrier using graph of Nomograph. We must know about barrier break, barrier position value, source noise with receiver and angle subtended.*

*2.3.3. Graph method. Graph method is one of method used to determine of attenuation on noise barrier. This method used of Fresnel Number for finding noise barrier. Fresnel number is obtained by calculating of Path Length Difference as similar with Maekawa method. Equation 9 show of Fresnel Number. We can describe of N as Fresnel Number.*

$$N = \frac{2\delta}{\lambda} \tag{9}$$

*2.4. Object measurement*

The object measurement in this research is SMAN 5 Surabaya. This area is in the center traffic of Surabaya city near Kusumabangsa Street and Ambengan Street. The measurement was conducted on two measurement points. The first point (measurement point 1, ●) is placed on 10,5 meters from Kusumabangsa Street and the second point (measurement point 2, ●) on 6,5 meters from Ambengan Street. The position of measurement point in SMAN 5 Surabaya is shown in figure 2, 3, and 4.

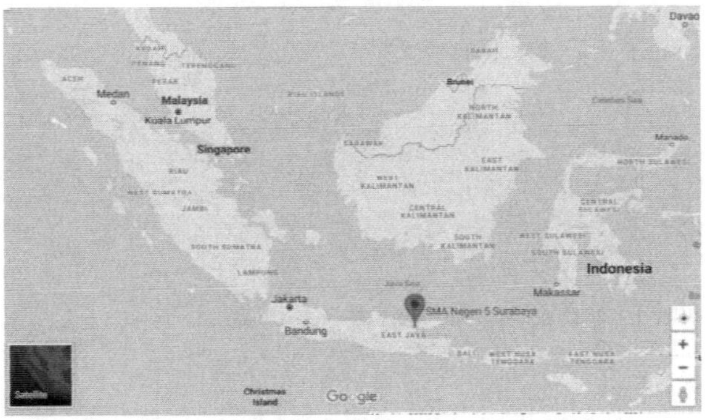


**Figure 1.** The location of SMAN 5 Surabaya on Google Map.

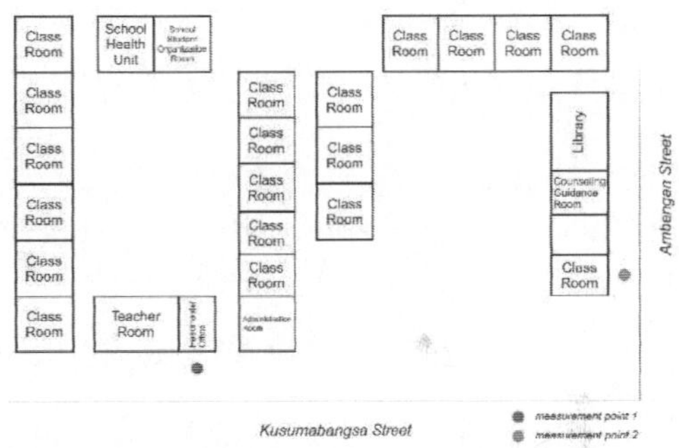


**Figure 2.** The layout and position of all of measurement points.

**Figure 3.** The position of first measurement point.

**Figure 4.** The position of second measurement point.

IOP Conf. Series: Journal of Physics: Conf. Series **1075** (2018) 012083 doi:10.1088/1742-6596/1075/1/012083

## 3. Data Analysis

*3.1. Data measurement of environmental noise*
Noise data measurement of the first measurement point which located near Kusumabangsa street is taken every hour on 12.00 am – 09.00 pm (Indonesia Time Zone). From the data calculation of $L_{eq}$, it can be determined the value of noise level in the noon ($L_{day}$), noise level in the night ($L_{night}$), and average noise day night ($L_{dn}$) using equation 2, 3, and 4. The value of $L_{day}$ is 84.31 dBA, $L_{night}$ is 86.26 dBA and $L_{dn}$ is 87.96 dBA. Meanwhile, noise data measurement of the second measurement point which located near Ambengan Street is 71.22 dBA for $L_{day}$, 65.76 dBA for $L_{night}$ and $L_{dn}$ is 71.07 dBA.

Figure 5 shows the comparison of the value of $L_{eq}$ between Kusumabangsa Street and Ambengan Street on time domain. From the figure 5 we know that both of $L_{eq}$ between first and second measurement points are exceeding the standard from Minister of Environment.

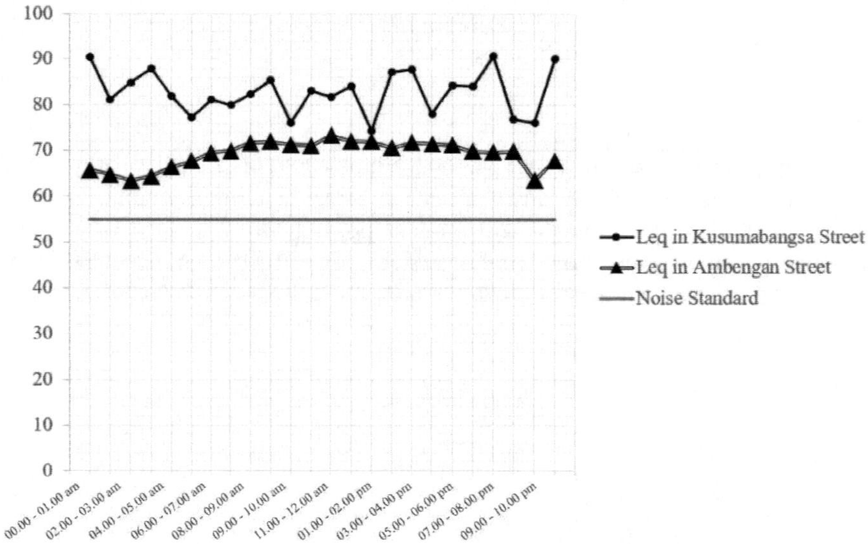


**Figure 5.** Measurement of Leq and Standard Leq

*3.2. Noise Barrier Plan in SMAN 5 Surabaya*
We will use the noise barrier planning several methods that will be applied in SMAN 5 Surabaya.

*3.2.1. Maekawa method.* From the Figure 6, we can design the noise barrier. The assumption of sound between source and receiver is same. The data of c is 10.5 m, $c_1$ is 7.5 m, $c_2$ is 3 m, f is 1000 Hz, v is 340 m/s, and $\lambda$ is 0.34. So, using the equation 8, its design can reduce the noise about 32,76 dBA with the height of noise barrier about 12 meters. Meanwhile, different noise barrier design shown in Figure 7 can reduce noise about 21.19 dBA.

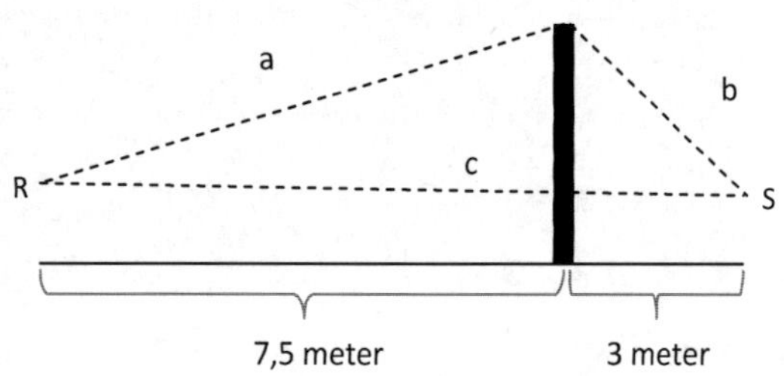


**Figure 6.** Noise barrier design with the height with similar assumption.

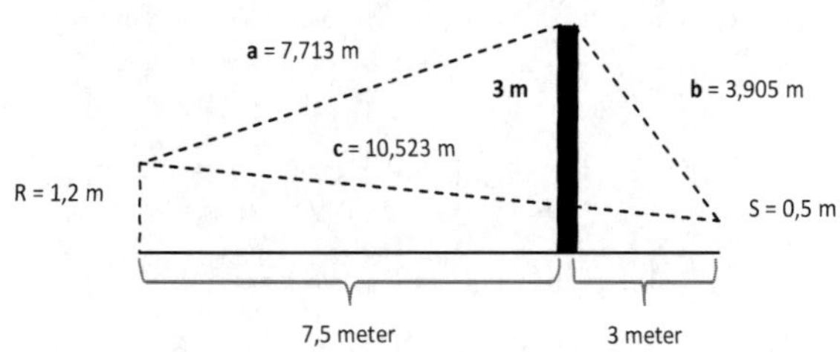


**Figure 7.** Noise barrier design with the height of source and different receiver.

IOP Conf. Series: Journal of Physics: Conf. Series **1075** (2018) 012083 doi:10.1088/1742-6596/1075/1/012083

*3.2.2. Double Screen Method.* It can reduce noise about 28,121 dBA shown in Figure 8 and table 1.

**Table 1.** Design variation of height of different source and receiver

| *Var* | $h_S$ | $h_R$ | $x_1$ | $x_2$ | $h$ | $a$ | $b$ | $c$ | $NR$ |
|---|---|---|---|---|---|---|---|---|---|
| 1 | 0,5 | 1,2 | 3 | 7,5 | 3 | 7,71 | 3,91 | 10,52 | 21,19 |
| 2 | 0,5 | 1,2 | 3 | 7,5 | 6 | 8,91 | 6,27 | 10,52 | 27,40 |
| 3 | 0,5 | 1,2 | 3 | 7,5 | 12 | 13,15 | 11,89 | 10,52 | 32,33 |
| 4 | 0,5 | 1,2 | 2 | 8,5 | 3 | 8,69 | 3,20 | 10,52 | 22,14 |
| 5 | 0,5 | 1,2 | 4,5 | 6,5 | 3 | 6,26 | 5,15 | 10,52 | 20,32 |

*3.2.3. Graph Method.* This method can determine the barrier attenuation used Fresnel Number from equation 9. The design of noise barrier with graph method is shown in Figure 9.

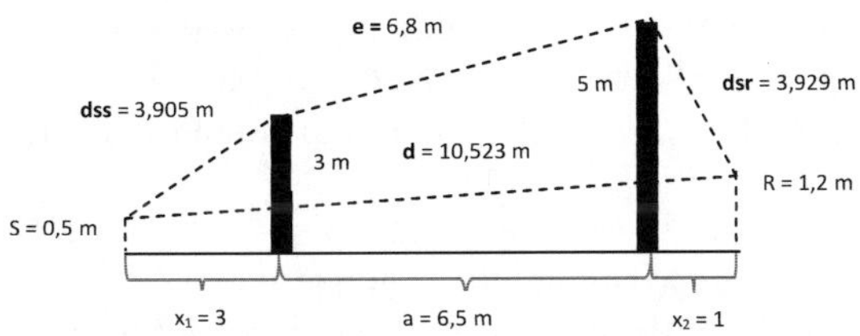


**Figure 8.** Design of *Noise Double Screen Barrier.*

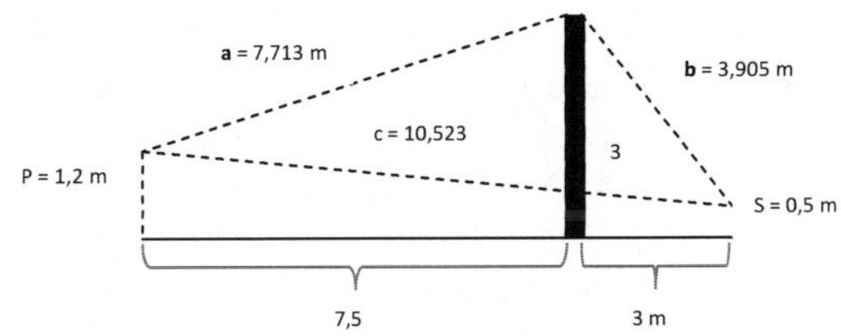


**Figure 9.** Design of Noise Barrier with graph method.

*3.2.4. Nomograph Method.* It can be determined by x is barrier position about 3.1594 m and t is barrier break about 2.295 m. Length of gate SMAN 5 Surabaya in Kusumabangsa street is 154 m. Measurement point is 7.5 m and measurement angle is $166.2^0$ look at picture 12. After the data is plotted in Nomograph form, we can get attenuation barrier value about 12.8 m (see the table 2).

**Table 2.** Barrier attenuation based on frequency

| *f (Hz)* | *λ (m)* | *N* | *A (dBA)* |
|---|---|---|---|
| 125 | 2,72 | 0,805 | 11,5 |
| 250 | 1,36 | 1,61 | 14 |
| 500 | 0,68 | 3,221 | 16 |
| 1000 | 0,34 | 6,441 | 20,3 |
| 2000 | 0,17 | 12,882 | 24,8 |
| 4000 | 0,085 | 25,77 | 26,4 |

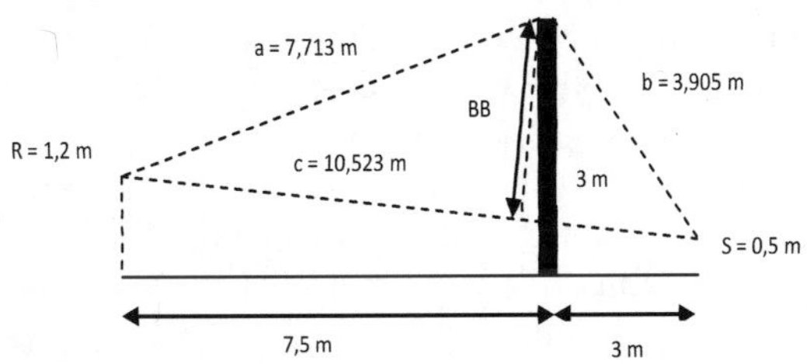


**Figure 10.** Noise graph in Nomograph method.

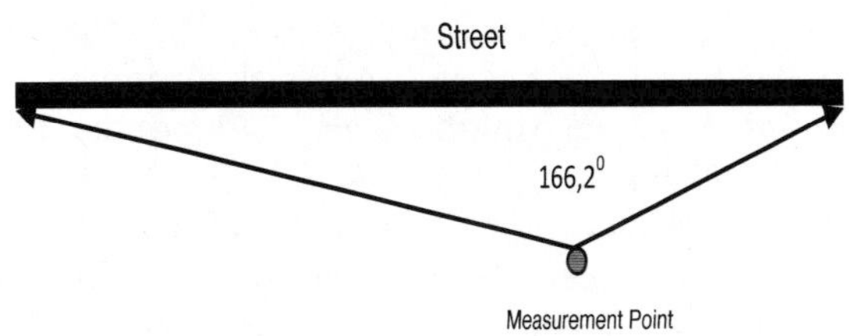


**Figure 11.** Angle Subtended.

**4. Discussion**

Based on measurement data of the noise near Kusumabangsa street, the highest $L_{eq}$ value in 00.00-01.00 is 90.42 dBA. It is caused by vehicle which through the street like truck, motorcycle, etc. In the night driver got the height speed than day because decrease volume of vehicle. So, it caused noise. Meanwhile in the noon at 14.00-15.00, it happened the lowest noise. In the other hand, on Ambengan street got the highest noise at 12.00-13.00 about 73.32 dBA. At this time, learning activities were occurring and many classrooms were located near Ambengan street. Absolutely, it can disturb the student concentration. This area doesn't obey the noise standard from MENLH No.48 /MENLH /11 /1996 which gave the maximum value of noise about 55 dB.

Design plan is using Maekawa method with some treatments such as change the height of noise barrier, position or distance of noise barrier with source of noise (Source) and with receiver. Based on the calculation of data, this method can reduce noise up to 32 dB at the height of 12-meter barrier noise. In the other variations, if the distance of the noise barrier shifted 1 meter near the noise source will increase the noise attenuation by 0.945 dB. When the noise barrier shifted 1 meter away from the noise barrier, it will reduce the noise damping by 0.881 dB. This is matching with the Maekawa method, which the higher the noise barrier will dampen the sound greater

Design plan used the double screen method which the height of the main barrier is fixed as high as 3 meters and added the height variation and the distance position on the screen barrier. The design results made at the height of the screen barrier 5 meters and positioned 1 meter from the classroom wall which noise will reduce of 28,121 dBA. If the high screen barrier made 3 meters with the same position, it will reduce noise by 26.219 dBA. If the high screen barrier made 5 meters with a position of 2 meters from the class wall, it will dampen the noise of 27.325 dBA. Thus, the screen barrier is influenced by the height and position placement.

Design used the Nomograph method have results that the biggest noise reduction is at the height of the noise barrier 6 meters is 13.8 dBA. The lowest is 12 dB at the height of noise barrier 3. Noise reduction (NR) obtained with this method is not too large because the scale of attenuation barrier on graph maximum up to 20 dB. This method is not effective to apply for this case study for the graph method obtained attenuation barrier at each frequency. However, high accuracy is needed in the determination of attenuation on the graph because the scale on the Fresnel Number is not the same for each range of values. So, it is very difficult to get the right attenuation of the barrier.

All the methods have tried in designing the noise barrier. The most effective method of noise control in SMAN 5 Surabaya is the Maekawa method because this method obtained the NR value based on the direct calculation so that the value is more accurate. Basically, the design of the most emphasized noise barrier is the high barrier, position or distance barrier with the source noise and receiver, and the noise barrier material itself. The higher and closer the noise barrier distance with the noise source will produce a large NR. The noise barrier material should be selected that has a high density so that it can reduce the noise is greater but also pay attention to the artistic aspect of the noise barrier. While the best material in the design of noise barrier is Steel material with noise attenuation of 32,32 dBA at layer thickness 1.25 mm.

## 5. Conclusion

Based on theoretical and experimental results, it can be concluded several things: Lsm value near kusuma bangsa street is 87,96 dBA whereas near ambengan street is 71.07 dBA. It is too higher noise and not fullfil of standart value MENLH. The area of SMAN 5 Surabaya need noise barrier plan to reduce the noisethat caused by traffic. The effective plan is used the Maekawa method with the height of noise barrier 12 m that can decrease the noise (NR) about 32,76 dBA. The best material of noise barrier plan is steel with thickness 1,25 mm that decrease noise about 32,32 dBA.

## 6. References

[1] ScienceDaily, "Science News: Traffic noise is dangerous for your health: Solutions exist for dense cities," Chalmers University of Technology, 1 July 2014. [Online]. Available: https://www.sciencedaily.com/releases/2014/07/140701085328.htm.

[2] W. H. Close and J. E. Wesler, "VEHICLE NOISE SOURCES AND NOISE-SUPPRESSION POTENTIAL," [Online]. Available: http://onlinepubs.trb.org/Onlinepubs/sr/sr152/152-001.pdf.

[3] Barron. Randall F, "Industrial Noise Control and Acoustics", New York: Marcel Dekker (2001) 5.12 dan 7.7

[4] "Keputusan Menteri Negara Lingkungan Hidup Nomor: Kep.Men-48/MENLH/XI/1996 Tentang Baku Tingkat Kebisingan", Menteri Negara Lingkungan Hidup (1996, November)

[5] Z. Maekawa, "Noise Reduction by Screens", Faculty Of Engineering Kobe University: Japan, (1968, January) 157 – 173

IOP Conf. Series: Journal of Physics: Conf. Series **1075** (2018) 012084 doi:10.1088/1742-6596/1075/1/012084

# Time-Frequency Independent Component Analysis for Multi-Damage Detection on a Rotating Machine

**D Arifianto**

Department of Engineering Physics, Institut Teknologi Sepuluh Nopember, Surabaya 60111 Indonesia

E-mail: dhany@ep.its.ac.id

**Abstract**. Maintenance of the plant machinery plays a crucial factor for maintaining continuity of industrial processes. This paper reported a development of an acoustic-emission-based (AE) technique of identifying multi-damage to the machine remotely using two sensors. In implementation, we emphasized in the separation of sound signals emitted by multiple machines using Time Frequency Independent Component Analysis (TFICA) recorded with a microphone array using the technique of mixing assuming a single source. Overall this study aimed to identify the unbalance, misalignment and bearing faults. Each machine had simultaneously two different damages, i.e. the bearing fault and unbalance, unbalance and misalignment, and bearing fault with misalignment. Separation process was performed using several types of techniques, namely, time domain ICA, frequency domain ICA. Observations FDICA superior in separation rather than TDICA with high MSE values are: 1.7 x$10^{-5}$. From the experimental results showed that the distance between the microphone so the shorter the distance the smaller the spatial aliasing occurs.

## 1. Introduction

Blind Source Separation (BSS) is a method that can be used to separate a signal from a mixture without knowing information sources and without knowing how the mixing process [1]. This method was originally used as a signal separation technique of the human voice (speech) or often referred to as cocktail party problem [2, 5]. From the research that the study of BSS was developed again but with a different object of research is the engine pump. Research on the engine pump has been done by [7], but the algorithms used are BSS ICA (Independent Component Analysis) where the assumption in this method is to have the characteristics of independence and non-Gaussian for the signal source. The purpose of this research is the BSS algorithm to be used is tfBSS (Time-Frequency Blind Source Separation) where the base used is the TFD (Time Frequency Distribution) with SWVS estimation approach (Spatial Wigner Ville Spectrum). The fundamental difference between [7] was the configuration of the sensor and source. Another difference lies in the assumption of the signal mixing techniques. Mixing assumptions on the signal [7] is a linear instantaneous mixture or superposition, while [3, 4] assuming the mixing is convolutive mixtures (mixing convolutive). This research was carried out two experiments: first take convolutive mixtures and the second assumption is instantaneous linear mixtures. Mixing is instantaneous linear mixtures for tfBSS is based on the paper [3]. While the assumption of convolutive mixtures has not been tested so that the necessary

IOP Conf. Series: Journal of Physics: Conf. Series **1075** (2018) 012084 doi:10.1088/1742-6596/1075/1/012084

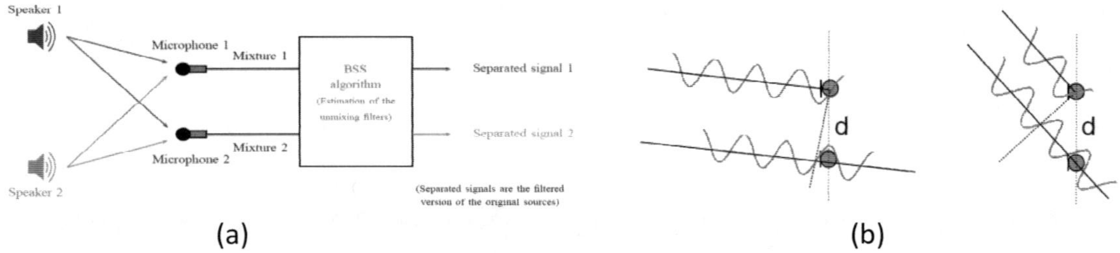


**Figure 1.** Data acquisition schematic, (a) mixture of sounds recorded by two microphone array and (b) two sensors separation to avoid spatial aliasing

research to find out how this influences tfBSS algorithm if the mixing is carried out in convolutive. The second aim is to test the hypothesis whether the algorithm is tfBSS mixed signal (convolutive mixtures) of several emission engines with different characteristics can be separated.

This paper is organized as follows. In Section II, we formulated the BSS problem and list all required assumptions. In Section III, we shown methodology on this research. In Section IV, we present the result of baseline and mixed signals, separation of signals with tfBSS, Voice Signal Convolutive Mixture and Instantaneous Linear, comparing with ICA and tfBSS methods. In Section V discusses the results of the separation of many sources of engine noise tfBSS obtained with this method compared with the results of sound source separation of many machines with the ICA method.

## 2. Measurement Method

### *2.1. Mixture separation*

The aim is to obtain the original sound of sources $s_1(t)$ and $s_2(t)$ whilst $Z^T$ is the output of the measurements $X^T$ [1, 2],

$$x_i(t) = \boldsymbol{A} * z_n(t) + n(t), \qquad (1)$$

where $\mathbf{x} = [x_1, x_2, ..., x_m]^T$ represents the measured data, and $\mathbf{x}_i$, $\mathbf{z} = [z_1, z_2, ..., z_n]$ was the vector of sources. $\boldsymbol{A}$ is a full rank matrix $\boldsymbol{A}$ $(A=[\boldsymbol{a}(\alpha_1)...\boldsymbol{a}(\alpha_N)])$ and the aim is to find $\boldsymbol{A}^{-1}$ , where $\boldsymbol{A}^{-1}\boldsymbol{x} = \mathbf{z}$. The * notation indicates that the convolutive mixtures was assumed.

Furthermore [5],

$$\begin{bmatrix} x_1(t) \\ x_2(t) \end{bmatrix} = \begin{bmatrix} a_{11} & a_{12} \\ a_{21} & a_{22} \end{bmatrix} \begin{bmatrix} z_1(t) \\ z_2(t) \end{bmatrix} + \begin{bmatrix} n_1(t) \\ n_2(t) \end{bmatrix} \qquad (2)$$

and,

$$\mathbf{x_1(t)=a_{11}z_1(t)+a_{12}z_2(t)+ n_1(t)} \qquad (2a)$$
$$\mathbf{x_2(t)=a_{21}z_1(t)+a_{22}z_2(t)+ n_2(t)} \qquad (2b)$$

The problem is to estimate $\boldsymbol{A}$ with information from the number of the sensors. With uncorrelated sources $\mathbf{z}$, we may recast the problem into

$$\boldsymbol{s(t)} = \boldsymbol{Wx(t)},$$

where $\boldsymbol{W} = \boldsymbol{A}^{-1}$ then mathematically the problem became

$$\begin{bmatrix} w_{11} & w_{12} \\ w_{21} & w_{22} \end{bmatrix} = \begin{bmatrix} a_{11} & a_{12} \\ a_{21} & a_{22} \end{bmatrix}^{-1} \quad and \quad \begin{bmatrix} s_1(t) \\ s_2(t) \end{bmatrix} = \begin{bmatrix} w_{11} & w_{12} \\ w_{21} & w_{22} \end{bmatrix} \begin{bmatrix} x_1(t) \\ x_2(t) \end{bmatrix}$$

We used the Time Domain Independent Component Analysis (TDICA) and Frequency Domain Independent Component Analysis (FDICA) in speech sound separation [3]. The schematic of both techniques was shown in the Fig. 2 (a) and Fig. 2 (b), respectively.

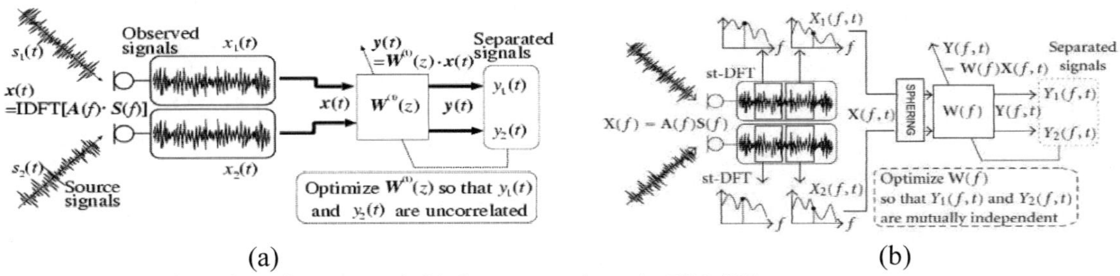


(a) (b)

**Figure 2.** (a) time domain ICA and (b) frequency domain ICA [3].

### *2.2. Measurement Setup*

*2.2.1. Measured Samples.* In total, six water pumps were used. On each pump was intentionally damaged with two damages of unbalance, bearing fault and misalignment as depicted in Table 1. First of all, we measured the so-called baseline. The baseline was basically one pump with one damage only. The pump was individually in normal, bearing fault, misaligned and unbalanced, respectively. The baseline was used to validate the sound pattern of the damages.

**Table 1.** One Pump with two damages

| Experiment | Number of Microphones | Damages type |
|---|---|---|
| 1 | 1 | unbalance (4 gr) |
| | 2 | bearing fault |
| 2 | 1 | unbalance (8 gr) |
| | 2 | misalignment (1 mm) |
| 3 | 1 | Bearing fault |
| | 2 | misalignment (3 mm) |
| 4 | 1 | unbalance (4 gr) misalignment (2 mm) |

At the Table 1, there were four experiments were set with different types of damages on one water pump. For example, experiment 1 the damages were one pump with 4 grams additional mass added to the impeller to introduce unbalance, and the bearings were intentionally damaged by hammer. The misaligned in mm were from the eccentricity of the rotating shaft. The number of sensors in the array was either one microphone and two microphones to compare the accuracy of the separated mixtures.

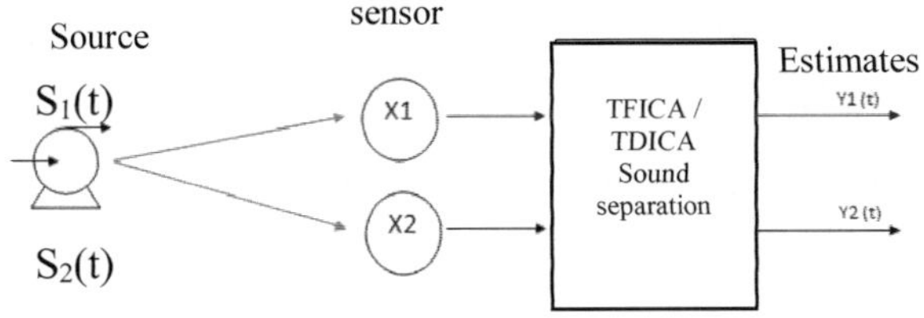


**Figure 3.** Schematic of the experiment

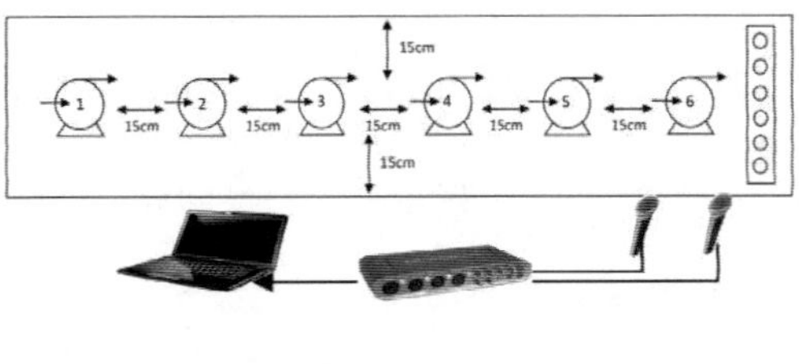


**Figure 4.** The schematic of the mini plant. **Only** pump no. 2, 3, 4, and 5 were used. Pump no. 1 and 6 were standing-by.

Fig. 3 depicts the experiment procedures of which the one pump emitted two distinct sounds of damages simultaneously $S_1(t)$ and $S_2(t)$ and recorded by one or two sensors. Using either TFICA and TDICA the estimates of $y_1(t)$ and $y_2(t)$ were obtained. When two microphones were used, the spatial aliasing limit were also tested by placing the microphone distance to each other from 10, 20 and 30 cm, respectively.

*2.2.2. Laboratory Arrangement.* The mini plant was set up inside a semi-anechoic chamber as shown the the Fig. 5.

**Figure 5.** General arrangement of the measurement setup.

*2.2.3. Measurement Procedure.* Sound data from Panasonic GP - 129 water pump with 2950 rpm rotational frequency with 4 different mounted on the table (shown in Fig.5). Then Behringer XM1800S microphone as a sound sensor with characteristics and cable jack female type, while the interface used as an analog data converter to digital data is USB DAC Multi Channel (M-Audio Fast Track Ultra) with 2 inputs. Adobe Audition 1.5 were used to record the digital samples and saved with .wav (PCM) format. We set the recording with mono, 24-bit (recording depth), 24-bit (audio mix down), and 44100 Hz sampling frequency. To reduce computational complexity the audio files were downsampled to 16000 Hz. For separation, we used natural gradient algorithm with 100 iteration number, 0.00001 step and 30 process-block.

*2.2.4 Mean Square Error measurement.* MSE is a method used in statistics as a comparison between the estimated spectrum of the estimated signal with the signal source (baseline signal). The equation used in calculating the value of MSE is as follows:

$$\mathrm{MSE} = \frac{1}{n}\sum_{i-1}^{n}(S - S_c)^2 \qquad (3)$$

where,

| | | |
|---|---|---|
| MSE | = | Mean Square Error |
| n | = | number of samples |
| S | = | original signals (baseline) |
| Sc | = | signal estimation |

## 3. Measurement Results

### *3.1. Baseline*

Measurement results showed that the spectral based patterns of baseline were in good agreement with ISO 13373-2:2016 [6].

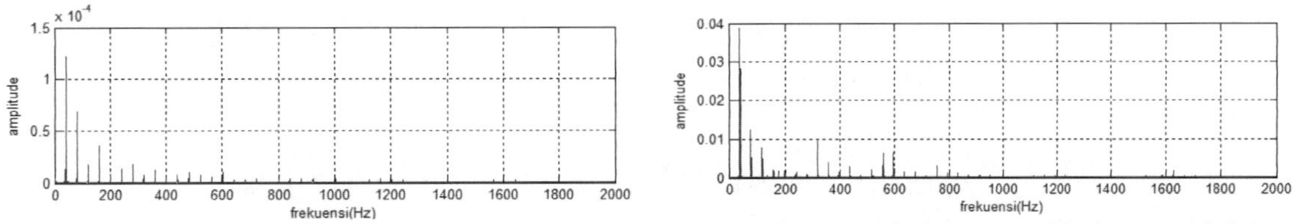


**Figure 6.** Spectral pattern of normal pump (left) and unbalance with 4 grams added mass (right).

As shown in Fig. 6 the normal pump spectrum is at 50 Hz which is in agreement to the rotation speed of the pump. The amplitude of the normal pump is also low which implies that the normal pump vibration is low. On the other hand, the amplitude of the unbalanced increased significantly compared to the normal. The location of the main spectrum remains the same at 50 Hz.

For the configuration of 2 sensors and one source with two simultaneous damages was conducted. The microphones position placed 15 cm to the pumps and distance between microphone is 10 cm, 20 cm, 30 cm, respectively. This is done in order to test the limit of spatial aliasing. The results were tabulated in the Table 2 and Table 3, respectively.

**Table 2.** MSE FDICA

| Pump # | Damages | 10 cm | 20 cm | 30 cm |
|---|---|---|---|---|
| **2** | ***Unbalance*** **(4 gr)** | $5.05 \times 10^{-5}$ | $5.7 \times 10^{-5}$ | $9.3 \times 10^{-5}$ |
| | ***Bearing fault*** | $2.2 \times 10^{-5}$ | $4.5 \times 10^{-5}$ | $6.5 \times 10^{-5}$ |
| **3** | ***Unbalance*** **(8 gr)** | $4.4 \times 10^{-5}$ | $4.9 \times 10^{-5}$ | $5.3 \times 10^{-5}$ |
| | **Misalignment 1 mm** | $5.3 \times 10^{-5}$ | $5.8 \times 10^{-5}$ | $6.6 \times 10^{-5}$ |
| **4** | ***Bearing fault*** | $4.4 \times 10^{-5}$ | $5.4 \times 10^{-5}$ | $6.7 \times 10^{-5}$ |
| | **Misalignment 3 mm** | $3.2 \times 10^{-5}$ | $6.4 \times 10^{-5}$ | $7.2 \times 10^{-5}$ |
| **5** | ***Unbalance*** **(4 gr)** | $1.8 \times 10^{-5}$ | $3.4 \times 10^{-5}$ | $5.7 \times 10^{-5}$ |
| | **Misalignment 2 mm** | $1.7 \times 10^{-5}$ | $3.3 \times 10^{-5}$ | $6.3 \times 10^{-5}$ |

Table 2 shows the MSE of the estimates of two sensors and compared to the baseline. Interestingly, the MSE of the estimates were significantly low. These results suggest that the technique may have accurately estimate the simultaneous damage on one pump.

Table 3 shows the MSE of the estimates of two sensors and compared to the baseline. In this experiment, TDICA showed lower accuracy in term of MSE compared to the TFICA, particularly the estimate of two simultaneous damages. Therefore, the frequency domain separation outperforms the time domain technique.

IOP Conf. Series: Journal of Physics: Conf. Series **1075** (2018) 012084 doi:10.1088/1742-6596/1075/1/012084

**Tabel 3.** MSE TDICA

| Pump # | Damages | 10 cm | 20 cm | 30 cm |
|---|---|---|---|---|
| 2 | ***Unbalance*** **(4 gr)** | $8{,}7 \times 10^{-3}$ | $8{,}7 \times 10^{-3}$ | $9{,}4 \times 10^{-3}$ |
| | ***Bearing fault*** | $5{,}2 \times 10^{-3}$ | $5{,}2 \times 10^{-3}$ | $6{,}8 \times 10^{-3}$ |
| 3 | ***Unbalance*** **(8 gr)** | $7{,}9 \times 10^{-3}$ | $7{,}9 \times 10^{-3}$ | $7{,}9 \times 10^{-3}$ |
| | **misalignment 1 mm** | $22 \times 10^{-3}$ | $24 \times 10^{-3}$ | $27 \times 10^{-3}$ |
| 4 | ***Bearing fault*** | $22 \times 10^{-3}$ | $22 \times 10^{-3}$ | $7{,}4 \times 10^{-3}$ |
| | **Misalignment 3 mm** | $1{,}1 \times 10^{-2}$ | $1{,}1 \times 10^{-2}$ | $1{,}2 \times 10^{-2}$ |
| 5 | ***Unbalance*** **(4 gr)** | $1{,}3 \times 10^{-2}$ | $1{,}3 \times 10^{-2}$ | $1{,}7 \times 10^{-2}$ |
| | **Misalignment 2 mm** | $1{,}2 \times 10^{-2}$ | $1{,}2 \times 10^{-2}$ | $1{,}6 \times 10^{-2}$ |

## 4. Conclusion

Acoustic emission (AE) based vibration measurement was performed in semi-anechoic chamber by simulating a rotating machine with two simultaneous damages. The time-domain and the frequency-domain independent component analysis techniques that previously used for speech mixture separation were proposed to separate mixtures with an array of two microphones. The results showed that the TFICA accuracy in MSE outperformed the TDICA significantly. The results are encouraging for field test in the on-going measurement in the presence of background noises. This may be used for alternative measurement where contact vibration measurement poses a hazardous threat to the human operator.

## References


[1] C. Fevotte, C. Doncarli, 2004 **Two Contribution to Blind Source Separation Using Time-Frequency Distribution**. IEEE Signal Processing Letters vol.11, no.3, pp 386–389

[2] A. Belouchrani and M. G. Amin 1998 **Blind Source Separation Based on Time-Frequency Signal Representations.** *IEEE Trans. Signal Processing*, vol. 46, pp. 2888–2897

[3] T. Nishikawa, H. Saruwatari and K. Shikano 2002 **Comparison of Time Domain ICA, Frequency Domain ICA and Multistage ICA for Blind Source Separation**. Proc. *EUSIPCO* pp 154-157

[4] L. Parra and C. Spence 2000 **Convolutive blind separation of non-stationary sources**. *IEEE trans. Speech and Audion Processing* Vol 8 No 3, pp.320-327

[5] E. Oja, J. Karhunen, A. Hyvarinen. 2001 **Independent Compoenent Analysis.** *Wiley Interscience*

[6] BSI. BS EN ISO 13373-2:2016 Condition monitoring and diagnostics of machines — Vibration condition monitoring Part 2: Processing, analysis and presentation of vibration data 2016.

[7] D Arifianto 2011 **Source separation using independent component analysis techniques for machine fault detection in the presence of background noise.** *Proc. ICA* pp. 274-278

# Noise exposure hazard at supporting facilities of Dr. Soetomo General Hospital, Surabaya, Indonesia

**D Arifianto[1], A Nadiroh[1], R D Kartika[1] and N Purnami[2]**

[1]Department of Engineering Physics, Institut Teknologi Sepuluh Nopember, Sukolilo Campus, Surabaya 60111 Indonesia
[2]Department of Otorhinolaryngology Head and Neck Surgery, Faculty of Medicine, Universitas Airlangga-Dr. Soetomo General Hospital, Surabaya 60231 Indonesia

E-mail: dhany@ep.its.ac.id

**Abstract.** Noise exposure in a hospital is known to have considerable impact not only to the healthy workers, but affects the ability for speedy recovery to the patients as well. However, recent investigations showed that the noise emission in a hospital tends to increase from year to year, for example John Hopkins Hospital. Dr. Soetomo General Hospital is the largest hospital in eastern part of Indonesia serving 2,500 outpatients and 1,500 inpatients per day as of January 2017, tripled from 2014. This also implies that the activities also increase significantly, from supporting services such as laundry to clinical services. In this paper, we present our recent study on the noise emission focusing on the supporting activities of the hospital such as water pump station, boiler, central kitchen, incinerator and workshop. The aim is to investigate the noise dose exposure to the workers of each supporting facility. We measured the noise emission level of the facility during its respective working hours. The results showed that all the facilities emitted higher level (in decibel) than allowed by the WHO. For the continuous noise, the average noise level for the water pump station was about 89.3 dB(A) and the workers were exposed almost 100 times more than that of the allowed noise exposure daily dose. The crushers at the incinerator facility were the highest of all intermittent noise sources at 98.4 dB(A). Furthermore, the noise level difference between outside and inside the water pump station was only reduced 1 dB, which suggests that either the sound enclosures was no longer effective. To summarize, the overall results showed that either the continuous or intermittent noise source the workers were exposed to high dose of noise daily.

## 1. Introduction

Generally, the main requirement of a hospital is to be located in a quiet area for an obvious reason which is better rest of the patients for speedy recovery [1]. However, the hospital itself constitutes people, equipments and their activities in the hospital that consequently creates noise in busy health care facilities. For most spaces in hospital, the maximum loudness of sound in the night time should not exceed 40 dB(A) indoors [2]. This was explained from the study by Busch-Vishniac that the night-day noise level in the hospital did not comply with the WHO guidelines, in fact increases from year to year [3]. In previous study, noises in the hospital may affect directly to the patients such as sleep disturbance, annoyance, and communication interference [2], [4].

Most of research work has conducted on areas that are disturb the patient directly such as intensive care units (ICUs), wards including medical and surgical, and emergency unit where patients might stay for treatment and/or recovery [3], [5], [6], [7], [8], [9]. However, it does not mean that research of noise which conducted outside the area where the patients stay is not important to be explored. Tziaferi *et. al.* in their published work obtained that noise level was measured high in the boiler room and laundry in the general hospital. This is reinforced by the result of measurement in oncology hospital which shows that the following departments had the highest levels of noise are the boiler unit, ironing room, and sterilization unit located next to the boiler [10]. It is indicate that noise levels in supporting facilities of hospital is necessary to be investigated especially in risk assessment of physical hazards of the workers in those areas.

Dr. Soetomo General Hospital, Surabaya is the largest hospital in eastern part of Indonesia. As one of the central tertiary referral hospitals in Indonesia, Dr. Soetomo General Hospital is an expansive healthcare institution with over 1500 beds and 26 departments [11] which serving 2,500 outpatients and 1,500 inpatients per day as of January 2017. With the large number of patients handled indicates that Dr. Soetomo General Hospital is a hospital with busy activities which means it also affects the rising of activities in supporting facilities of the hospital. The aim of this study is to measure the noise in supporting facilities of Dr. Soetomo General Hospital Surabaya which might to be highest noise source level in this hospital. The noise exposure on the workers of each facility was measured to obtain personal noise dose to assess the noise hazard to the workers.

## 2. Methods

The noise measurement used the criteria for a recommended standard of OSHA (Occupational Safety and Health Act), standard of the State Minister for The Environment of Republic of Indonesia KEP-48/MENLH/11/1996 for noise threshold level, and standard of the Ministry of Manpower of Republic of Indonesia KEP–51/MEN/I999 for occupational noise threshold level. During the day of the survey, measurement points were determined throughout the supporting facilities, in this study we measured water pump station, boiler, central kitchen, incinerator, and workshop at Dr. Soetomo General Hospital, Surabaya, Indonesia. The measurements were conducted for 2 consecutive days.

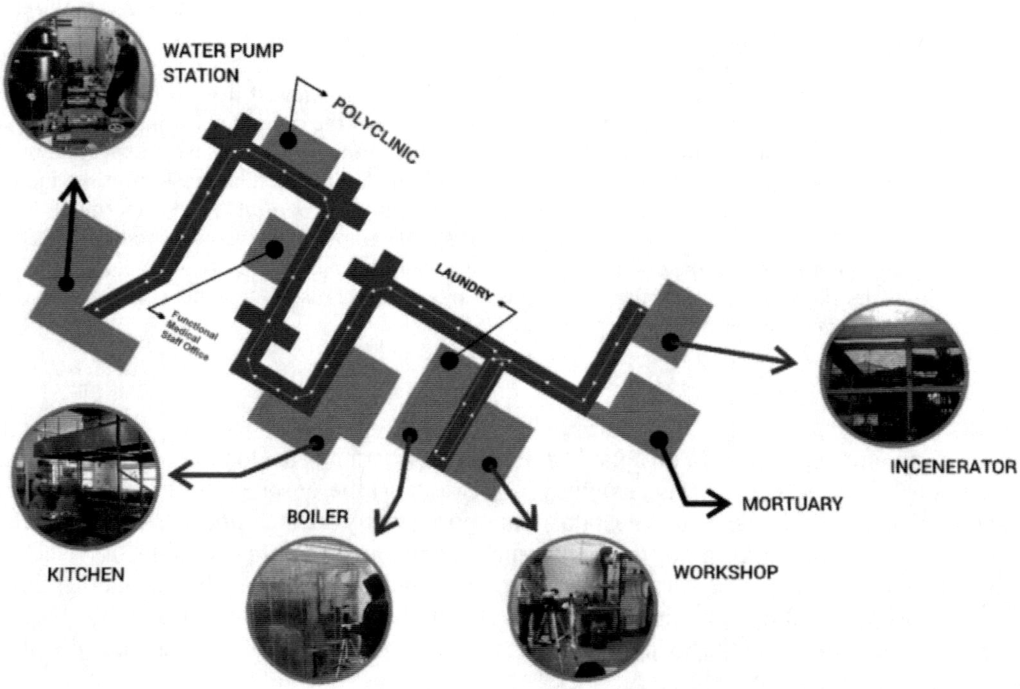


**Figure 1.** The measurement area.

IOP Conf. Series: Journal of Physics: Conf. Series **1075** (2018) 012085 doi:10.1088/1742-6596/1075/1/012085

### *2.1. Noise level measurements*

Noise measurements was performed in daytime and nighttime with integrated sound level mater placed on tripod at 1.2 m from the floor. The meter was calibrated on site and capable to measure the value of equivalent continuous sound level ($L_{eq}$) over the whole day, day-night sound level can be calculated. The following settings also recorded 1/3 octave band spectrum to determine source characteristics (frequency dominant).

### *2.2. Noise dose measurements*

The noise dose was calculated to observe the impact of source to the workers in measurement area. Reference and work hour were collected to determine the noise dose according to the Ministry of Manpower of Republic of Indonesia Number: KEP–51/MEN/I999. In addition, personal dosimeters were used to assess personal exposure on some of employees. The daily personal noise exposure levels are defined in terms of the noise dose exposure experienced by a worker during a working day of 8 hours.

## 3. Results

### *3.1. Noise levels*

*3.1.1. Water pump station.* Personal noise dose results are shown in Table 1. Five of pumps working simultaneously in this area, with day-night sound level of 89.28 dB(A) for indoor and 88.40 dB(A) for outdoor. All condition, indoor-outdoor sound level was measured 1 meters from source position. The difference level between indoor-outdoor sound level is only 1 dB(A). It means, noise enclosure around station is not effective to reduce noise from water pump station.

*3.1.2. Boiler.* Personal noise dose results are shown in Table 1. Sound level in this area ranged between 75 – 102 dB(A). Measurement points were taken to the front of boiler room, laundry, and outdoors facing the boiler room and laundry. The boiler room produced the highest noise level of between 91 – 102 dB(A) when the machine was running. It is indicated that boiler is noise source of this area. The measurement was conducted at 1 meters that face the laundry, it is purpose to know sound transmission from boiler into laundry. The result of measurement point near to laundry was much quieter at between 85 – 95 dB(A). The outdoor measurement was also conducted to observe how noise source affecting the environmental area. The meter was positioned at 1.5 meters from the wall facing the boiler room and laundry. The outdoor point measurement that facing the boiler room produced the lowest noise level of between 75 – 87 dB(A) and the other outdoor measurement produced the noise level between 76 – 93 dB(A). From all results of day-night sound level, the difference noise level between boiler room and laundry is 5.53 dB(A). Even though the boiler is noise source in this area, the enclosure of boiler room is good enough to reduce the noise machine that indicated by the difference of day-night sound level between indoor-outdoor is 13.8 dB(A) than the difference in laundry's point measurement of 6.32 dB(A).

*3.1.3. Central kitchen.* Personal noise dose results are shown in Table 1. Sound level in this area ranged between 93 – 106 dB(A). Many of the kitchen equipment in this area produced high noise levels, all noise level exceeds 90 dB(A), with $L_{eq}$ of 97.5 dB(A) when nearby stove, 92.2 dB(A) when nearby vegetable washing machine, 100.5 dB(A) when nearby cooker hood 1, and 99.2 dB(A) when nearby cooker hood 2. From the results, we obtain the cooker hood is noise source in this area. Outdoor measurement was not conducted for this area, this is because this area is not the main road and is rarely passed by patients and their families.

*3.1.4. Incinerator.* Personal noise dose results are shown in Table 1. We measured the sound level for four incinerator and one crusher at position of 2.5 meters from the machines. Sound level in this area ranged between 77.9 – 107.1 dB(A). The main source emanated from working machines in this area,

IOP Conf. Series: Journal of Physics: Conf. Series **1075** (2018) 012085 doi:10.1088/1742-6596/1075/1/012085

with noise level ($L_{eq}$) of 97.8 dB(A) for incinerator and noise level ($L_{eq}$) of 98.4 dB(A) for crusher. There is no significant level of noise between two main sources. Due to the incinerator position is far from health care activities, the impact of main sources is only recorded for inside the area.

*3.1.5. Workshop.* Personal noise dose results are shown in Table 1. There are two main rooms of workshop in the hospital with position adjacent to each other. The first one consists of grinder and compressor with noise level ranged between 73.3 – 102.9 dB(A) and at the other side of the room, the sound source emanated from lathe and grinder, with noise level between 72.2 – 90.4 dB(A) being recorded. We were not measured the outside sound level in this area. The measurement points of grinder and compressor produced noise level ($L_{eq}$) of 93.6 dB(A) and 77.4 dB(A) for lathe and grinder (room 2).

### *3.2. Noise source characteristics*

Even this case in this study is in hospital areas, the environmental noise in supporting facilities in the hospital is due to power of machine. Berglund in Guidelines for Community Noise categorize this noise as an industrial noise. It is describing as noise from fixed installations, such as factories or construction sites, heat pumps and ventilation systems on roofs, typically affect nearby communities [2]. In this study, noise sources were characterized by its frequency content (see Figure 2). This can be assessed by contributions of the frequency components to the total noise. Based on figure 2, the results show that the dominant sound pressure level is in frequency range of 1000 Hz – 8000 Hz except in stove and crusher which in ranged between 250 Hz – 8000 Hz. It means the highest sound level tends to have the dominant frequency in the wider band. The peak of the loudness level with respect to frequency may also suggest that the noise should affect the hearing threshold within 2000 Hz to 4000 Hz [3, 15].

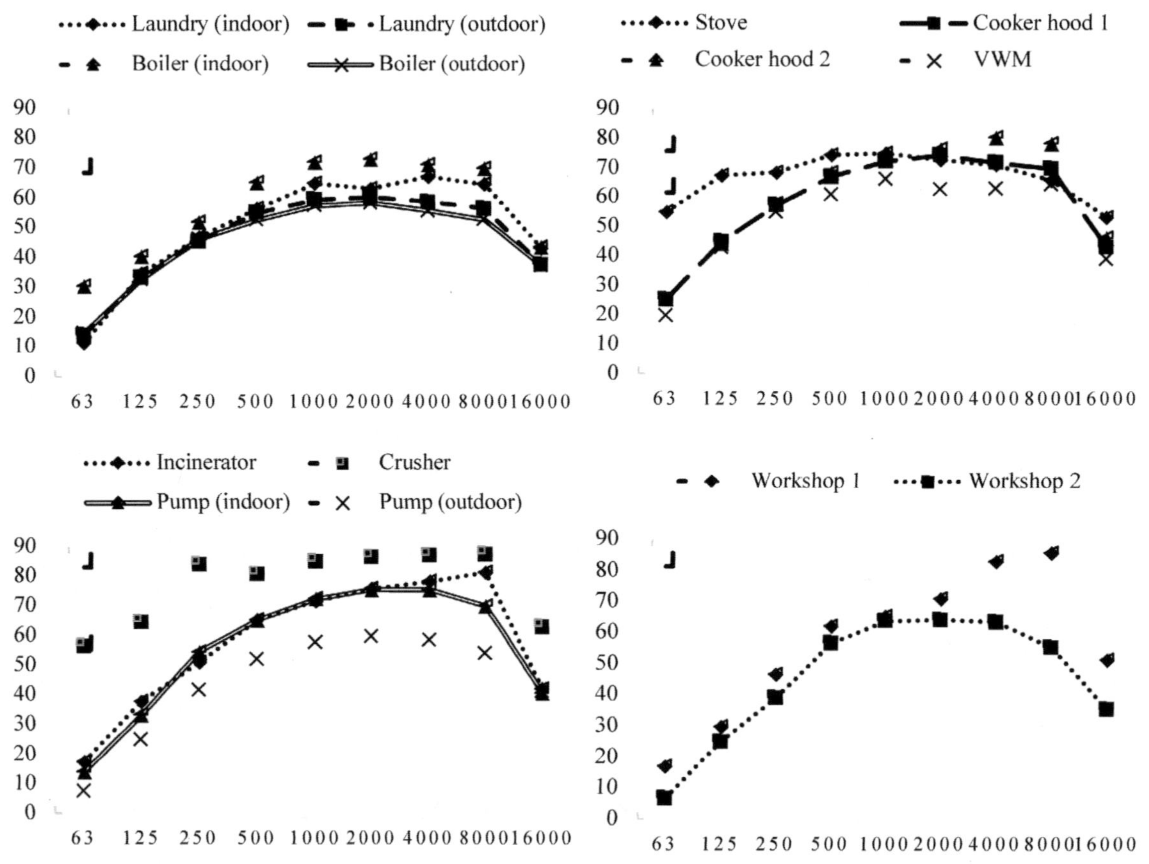


**Figure 2.** Octave bands noise level measured in various supporting facilities location.

IOP Conf. Series: Journal of Physics: Conf. Series **1075** (2018) 012085 doi:10.1088/1742-6596/1075/1/012085

### *3.3. Personal noise dose*

Personal noise dose is used to assess the noise exposure which emanated to the workers in this area. Measurement points at each room are adjusted with the position of the operator is located.

**Table 1.** Personal noise dose.

| Area/ Operator/ Machine | Time Exposure (hrs) | Dose (%) |
|---|---|---|
| **Water Pump Station** | | |
| Operator A | 10 | 335 |
| Operator B | 9 | 302 |
| **Boiler** | | |
| Operator A – Boiler | 7 | 1222 |
| Operator B – Laundry | 5 | 873 |
| **Central Kitchen** | | |
| Operator A – Stove | 7 | 2235 |
| Operator B – Vegetable washing machine | 7 | 2235 |
| Operator C – Cooker Hood | 5 | 1597 |
| **Incinerator** | | |
| Operator A – Incinerator | 2 | 476 |
| Operator B – Crusher | 2 | 547 |
| **Workshop** | | |
| Operator A – Room 1 | 2 | 181 |
| Operator B – Room 2 | 4 | 9 |

Based on previous results, noise levels around hospital support facilities shows a high average value that exceeds 85 dB(A). The maximum dose is 85 $dBL_{eq}$ over 8 hours. It means that the workers were exposed to 88 dB(A) for 4 hours, 91 dB(A) for 2 hours and so on [12]. Table 1 shows the measurement results on the personal noise dose to all workers in supporting facilities at Dr. Soetomo General Hospital, Surabaya. The workers were exposed to noise that far exceeded the recommended exposure limits except the workers in workshop area (room 2). According to OSHA, the noise dose exposed to a worker for eight-hour per one working day should be less than or equal to 50% [13]. According to [12] an intermittent noise may also induce hearing impairment damage if it exceeds the limit time exposure. Therefore, the workers who are exposed to noise levels in excess of the limit time exposure are required to wear ear protection equipment [14].

However, the noise characteristics were different. In term of loudness level, the noise shown in Fig.2 was higher than that of the John Hopkins Hospital [3, 15]. In term of frequency characteristics, the noise was dominantly within 2000 Hz to 4000 Hz region whereas the John Hopkins Hospital showed dominantly in the lower frequency region (below 500 Hz). The hazard of the noise exposure at the Dr. Soetomo General Hospital may have more severe impact on a normal hearing worker.

## 4. Conclusion

Identification of noise hazard in supporting facilities at Dr. Soetomo General Hospital was conducted. Noise levels and personal noise dose are used to assess noise hazard which expose the workers. The results showed that all workers in supporting facilities is indicated to enter in dangerous area due to high noise exposure. The safe area is only in the workshop-room 2 where the lathe and grinder operate. From those results, the use of ear protection equipment shall be applied to reduce the impact of hazards from noise exposure to the workers. Future research investigating physical hazards of workers is important to be conducted and observing the impact of wear ear protection equipment to compare of those condition.

In addition, the influence of supporting facilities on the ward or building where patients stay is also necessary to observe to know those facilities affects patients comfort in the hospital or not.

## 5. References

[1] A. Pierdait, "A Source of Design Reference Standards: Handbook to Build an Hospital," French Red Cross, Male, 2006.

[2] B. Berglund, T. Lindvall and D. H. Schwela, "Guidelines for Community Noise," World Health Organization, Geneva, 1999.

[3] I. J. Busch-Vishniac, J. E. West, C. Barnhill, T. Hunter, D. Orellana and R. Chivukula, "Noise levels in Johns Hopkins Hospital," *J. Acoust. Soc. Am.,* vol. 118, no. 6, pp. 3629-3645, 2005.

[4] defra, "Environmental Noise: Valuing impacts on sleep disturbance, annoyance, hypertension, productivity, and quiet," Open Goverment Licence: The National Archives, Kew, London, 2014.

[5] B. Shield, N. Shiers and R. Glanville, "The acoustic environment of inpatient hospital wards in the United Kingdom," *The Journal of the Acoustical Society of America ,* vol. 140, no. 3, pp. 2213-2224, 2016.

[6] A. McLaughlin, B. McLaughlin, J. Elliott and G. Campalani, "Noise levels in a cardiac surgical intensive care unit: a preliminary study conducted in secret," *Intensive and Critical Care Nursing,* vol. 12, no. 4, pp. 226-230, 1996.

[7] C. Tsiou, D. Eftymiatos, E. Theodossopoulou, P. Notis and K. Kiriakou, "Noise sources and levels in the Evgenidion hospital intensive care unit," *Intensive Care Medicine,* vol. 24, no. 8, pp. 845-847, 1998.

[8] D. J. MacKenzie and L. Galbrun, "Noise level and noise sources in acute care hospital wards," *Building Services Engineering Research and Technology,* vol. 28, no. 2, pp. 117-131, 2007.

[9] E. E. Ryherd, M. Moeller and T. Hsu, "Speech intelligibility in hospitals," *The Journal of the Acoustical Society of America,* vol. 134, no. 586, pp. 586-595, 2013.

[10] S. G. Tziaferi, P. Sourtzi, A. Kalokairinou, E. Sgourou, E. Koumoulas and E. Velonakis, "Risk Assessment of Physical Hazards in Greek Hospitals Combining Staff's Perception, Experts' Evaluation and Objective Measurements," *Safety and Health at Work,* vol. 2, pp. 260-271, 2011.

[11] A. Joyal, "Indonesia's Dr. Soetomo Academic Medical Center Hospital Uses UpToDate," Wolters Kluwer, 8 December 2015. [Online]. Available: https://wolterskluwer.com/company/newsroom/news/2015/12/indonesia%E2%80%99s-dr.-soetomo-academic-medical-center-hospital-uses-uptodate.html. [Accessed 2018].

[12] D. M. Howard and J. A. S. Angus, "Introduction to Hearing," in *Acoustics and Psychoacoustics 4th Edition,* Oxford, Focal Press, 2009, pp. 73-119.

[13] Occupational Safety and Health Act, "Occupational Noise Exposure: Revised Criteria 1998`," National Institute for Occupational Safety and Health, Cincinnati, 1998.

[14] Health and Safety Executive, "Controlling noise at work - Guidance on the Control of Noise at Work Regulations," HSE Books, 2005.

[15] Jonathan M. Kracht, Ilene J. Busch-Vishniac, and James E. West, "Noise in the operating rooms of Johns Hopkins Hospital," *The Journal of the Acoustical Society of America,* vol. 121, no. 586, pp. 2673- 2680, 2007.

**Acknowledgments**
Authors wishing to thank to member of VibrasticLab, Department of Engineering Physics, Institut Teknologi Sepuluh Nopember who had assisted in taking data at the hospital.

IOP Conf. Series: Journal of Physics: Conf. Series **1075** (2018) 012086 doi:10.1088/1742-6596/1075/1/012086

# Crack Detection of Propeller Shaft on board Marine Ship using Microphone Array

**D Arifianto**

Department of Engineering Physics, Institut Teknologi Sepuluh Nopember, Surabaya 60111 Indonesia

E-mail: dhany@ep.its.ac.id

**Abstract.** *Vibration due to rotation of a shaft of a ship propeller may cause crack due to lack of proper maintenance. It is considered to be hazardous area to conduct a direct measurement on rotating shaft using accelerometer. In this paper, acoustic emission using linear microphone array is proposed to collect acoustical data emitted from the rotation of the shaft. In a semi-anechoic chamber, the proposed technique using single microphone may have detected a crack in propeller shaft by observing the peak of the main spectrum. In engine room on a ship with highly noisy background, we propose microphone co-linear array to increase microphone sensitivity. First, the normal shaft sound was recorded using single sensor, thereafter used as baseline. The cracks were intentionally introduced to the shaft with depth 0.25, 0.5, and 0.75 diameter of the shaft respectively. The sound mixture was separated using frequency domain independent component analysis (TFICA) and then, compared to the accelerometer. The results showed that the measurement accuracy was increased from one sensor to the array and comparable to the direct measurement. These results suggest that the proposed technique may be used as an easy-to-deploy vibration monitoring in a noisy and hazardous engine room.*

## 1. Introduction

Recently, acoustic emission (AE) receives considerable attention for non-contact mechanical failure on a rotating machine because it may operationally simpler than that of the direct measurement [1]. However, the main problem is the difficulty to extract a crack sound of a failed rotating part of the machine due to highly adverse environment, particularly in an engine room. In this study, we propose a blind source separation (BSS) technique using microphone array to separate a crack signal of a rotating shaft from a mixture without knowing information sources and without knowing how the mixing process. The combination of loads bending, longitudinal and torsional vibrations in rotating shafts had been studied by [2], wherein the frequency of torsional excitation force was equal to the natural frequency of the bending load on the rotation axis of the crack.

The focus of this research is on the overdetermined case because with the increasing number of sensors, information signals are overcomplete. The purpose of this research is to use tfBSS (Time-Frequency Blind Source Separation) with Spatial Wigner-Ville Spectrum (SWVS) estimation approach. When compared with the underdetermined case, one of the advantages of this overdetermined case is in

IOP Conf. Series: Journal of Physics: Conf. Series **1075** (2018) 012086 doi:10.1088/1742-6596/1075/1/012086

its data processing or computing. For the overdetermined case, the settlement matrix easier where multiplication is not square $m \times n$ matrices where $m \geq n$ can be solved by joint diagonalization. Another characteristic in this case, namely the redundancy due to the number of sensors is more than the amount of the source. So that the necessary reduction of such redundancy. This is so that the output of the signal estimate equal to the number of sources and can be identified according to the original signal source.

The contribution of this paper is the following.

- The first goal to be achieved in this research is able to reduce redundancy signal (same signal) to the joint diagonalization in this tfBSS algorithm that estimates the output signal has the same amount to its source and has the same characteristics with the source identification .
- The second aim is to test the hypothesis whether the algorithm is tfBSS mixed signal (convolutive mixtures) of several emission engines with different characteristics can be separated.

This paper is organized as follows. In Section II, we formulate the BSS problem and list all required assumptions. In Section III, we outlined the methodology on this research. In Section IV, we present the result of baseline and mixed signals, separation of signals with tfBSS.

## 2. Measurement Method

### 2.1. BSS (*Blind Source Separation*)

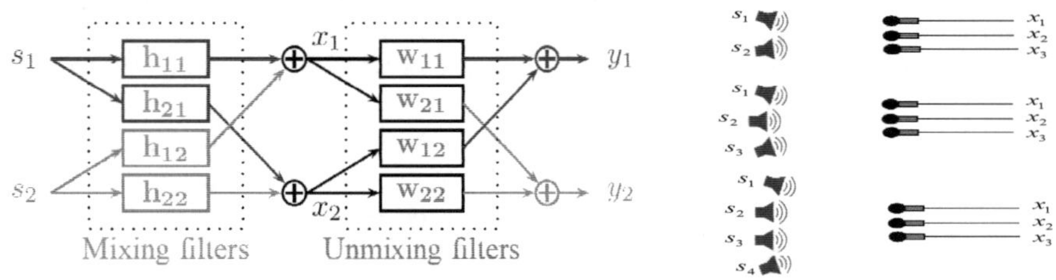


**Figure 1.** Schematic of Blind Source Separation (left), overdetermined, determined and underdetermined array (right)

Blind Source Separation Method is a method used to separate the mixed signals from multiple sources blindly (without knowing the mixing process) [1]. This method can be used for the overdetermined case / number of sensors is more than equal to the number of sources ($m \geq n$), determined cases / number of sensors is less than the number of sources ($m < n$) and the cases determined / number of sensors equals the number of source ($m = n$). In this study, the BSS is focused on solving the overdetermined case by taking tfBSS algorithm (time-frequency Blind Source Separation). Algorithm other in the often discussed in this BSS is ICA. BSS method that addresses the algorithm ICA (Independent Component Analysis), can be found in references [11,12,13]. In this section, the BSS method described by taking the case determined that the BSS method of explanation can be understood more easily. As seen in Fig. 1, there are two sources of sound from the speaker 1 and speaker 2 will be received by the microphone 1 and microphone 2 so happened that the sound mixing mixture 1 and mixture 2. In this there are two sound mixing techniques and mixing is convolutive mixtures of linear instantaneous mixtures. The results of this mixing will be processed by the BSS to be separated in order to be like the original single sound signal.

The results of the separation of sound signals or output signals of this BSS are separated signal 1 and separated signal 2. The BSS scheme can be seen in Fig. 2, showing that there are 4 filters mixture of $h_{11}$ (the source signal 1 received by microphone 1), $h_{21}$ (source signal 2 received microphone 1), $h_{12}$ (the source signal 1 received microphone 2), and $h_{22}$ (source signal 2 received microphone 2 ). Further mixing was separated and denoted by the symbol $W_{11}$, $W_{21}$, $W_{12}$, and $W_{22}$. The results of this separation are called the signal estimate and denoted is $y_1$ and $y_2$. From Fig. 1, tfBSS will be applied on the unmixing filters BSS algorithm.

IOP Conf. Series: Journal of Physics: Conf. Series **1075** (2018) 012086 doi:10.1088/1742-6596/1075/1/012086

### 2.2. Time-Frequency Blind Source Separation Steps

1. Whitening Data

The first stage in the separation of mixed signals that whitening the data from the sensor signal or the signal observation. In illustration, a data whitening process is illustrated in Fig. 5. The mixed signal that has not undergone a process called the original distribution whitening. From the figures can be seen the results of the original two-dimensional distribution of the oval-shaped or oval. After the whitening process is applied then the image that originally had an oval, a smaller image circle (shown in red). Image circle is a representation of the signal that has undergone a process of whitening. Blind source separation works well under Gaussianity and Independent. Subsequently, after data whitening process is completed then forwarded to the rotation or rotation.

2. Calculating STFD (Spatial Time Frequency Distribution) from the sensor signal which has undergone a process of whitening.

Here is the equation SWVS

$$SWVS_{XX}(t,f) = E\{D_{xx}^{WV}(t,f)\} \tag{2.1}$$

where is SWVD of the sensor signal which has undergone a process of whitening. SWVD equation itself is defined as follows

$$D_{xx}^{WV}(t,f) \stackrel{\text{def}}{=} \int_{-\infty}^{\infty} x\left(t+\frac{\tau}{2}\right) x^{H}\left(t-\frac{\tau}{2}\right) e^{-j2\pi\pi f\tau}\, d\tau \tag{2.2}$$

From the above calculation, the results will be obtained based on the estimated STFD is as shown in fig.7. In the figure, the frequency and time can be displayed directly to a specific frequency can be determined at any given time.

3. Calculating Criterion and Collecting *Single Auto Term.*

Before determining the location of Single Auto Term, required calculation of the criterion first. The calculation of this criterion can be obtained from equation (2.3).

$$C(t,f) \stackrel{\text{def}}{=} \frac{\max_{k} \lambda_k^{(t,f)}}{\sum_{k=1}^{n} | \lambda_k^{(t,f)}} = 1 \tag{2.3}$$

From equation (2.3), non-zero diagonal matrix obtained should only be one. So that non-zero eigenvalues to be obtained was only one. To obtain this value, then the above equation approximated by equation (2.4)

$$C(t,f) \approx \left(\frac{\max\left(\left|\mathrm{eig}\left(\underline{D}_{xx}^{\emptyset}(t,f)\right)\right|\right)}{\sum\left|\mathrm{eig}\left(\underline{D}_{xx}^{\emptyset}(t,f)\right)\right|}\right) \tag{2.4}$$

where the criterion is the ratio between the maximum eigen value of the diagonal matrix with the total number of eigenvalues diagonal matrix itself. Physical meaning of the calculation of this criterion is actually to get the energy of each signal. The highest mean energy value close to the original signal are compared.

$$\text{If } \left(\frac{\max\left(\left|\mathrm{eig}\left(\underline{D}_{xx}^{\emptyset}(t,f)\right)\right|\right)}{\sum\left|\mathrm{eig}\left(\underline{D}_{xx}^{\emptyset}(t,f)\right)\right|}\right) > \varepsilon \tag{2.5}$$

IOP Conf. Series: Journal of Physics: Conf. Series **1075** (2018) 012086 doi:10.1088/1742-6596/1075/1/012086

then (t, f) is the position of *Single Auto Term*.
From equation (2.5), the threshold value of ε in the interval [0,1]. For values closer to 1, it will be selected as a candidate in the joint diagonalization. From these intersection between the two signal sources is a diagonal matrix so that the necessary selection criterion for determining the position of *Single Auto Term*. At these selections there were two assumptions, namely instantaneous mixture and superposition. Election SAT (Single Auto Term) with additive term may be less complex than the superposition (sum signal) because the combination to find a mate with each other less than superposition. This needs to be done because the process is required permutation BSS method, sign, and scaling (scaled).

4. Joint Diagonalization (JD)

*Joint Diagonalization* technique is the most important step in the separation of the voice signal from the sensor is larger than the number of sources (m > n), because in this technique, the original matrix is not square can be a square matrix. Here is the equation to solve the joint diagonalization

$$\underline{D}^{\emptyset}_{xx}(t,f) \approx UD^{\emptyset}_{ss}(t,f)U^{H} \qquad (2.6)$$

where

$$\underline{D}^{\emptyset}_{xx}(t,f) \stackrel{\text{def}}{=} \widehat{W}(D^{\emptyset}_{xx}(t,f) - \hat{\sigma}^{2}\, I_{m})\widehat{W}^{H} \qquad (2.7)$$

From equation (2.7), $I_m$ an identity matrix while σ is the variance. If the value of the variance here is zero, then the multiplication between the variance and the identity matrix is zero as well. While is a hermitian matrix estimation in which this matrix when multiplied by another matrix will produce an identity matrix. Thus, obtained by multiplying the identity matrix with diagonal matrix , the result is a diagonal matrix itself. So, to find the inverse matrix can be easily done.

### 2.3. Reduction of Redundancy

Reduction of redundancy required in the selection of code where some information will be stored and some others deliberately discarded but does not reduce the required information (Barlow, 2001). The main focus of this idea, but actually for the computational efficiency without losing much information. Therefore, the necessary reduction of redundancy if the information is transmitted to other locations like. This is done for memory efficiency. This is an aspect of Shannon's formula, published in a clear and important in communication engineering in which the captured information obtained from the detection sensor redundancy in the form of a small and dense. The reduction of redundancy can be further subdivided and displayed in the form of levels (Barlow, 1959). This can be illustrated as in fig.9. In reference Scarpiniti Michele (2010) is said also that one of the assumptions and limitations of the BSS is an unknown mixture matrix is a square matrix. If the matrix is not square then the matrix is the result of redundant mixture. In this case, redundant this mixture can be removed so that the matrix can be a square matrix. The technique eliminates this redundancy is mathematically described in a paper Lars Omlor, Martin Giese (2007).

## 3. Measurement Setup

In total, six water pumps were used. On each pump was intentionally damaged with two damages of unbalance, bearing fault and misalignment. First of all, we measured the so-called baseline. The baseline was basically one pump with one damage only. The pump was in normal, bearing fault, misaligned and

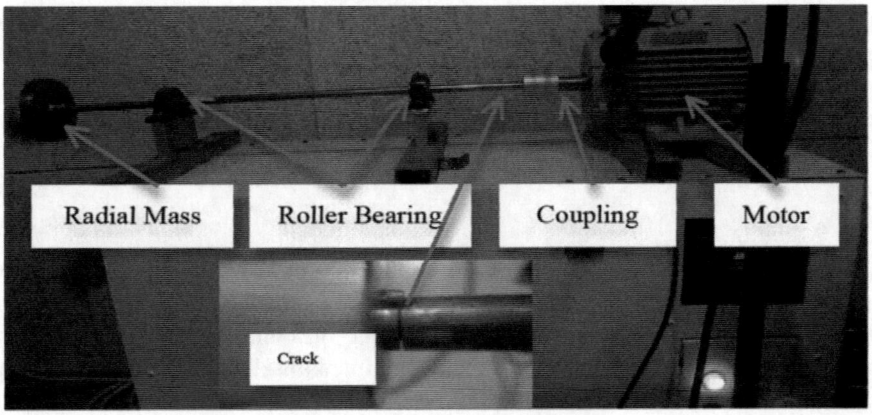


**Figure 2.** Experimental Setup

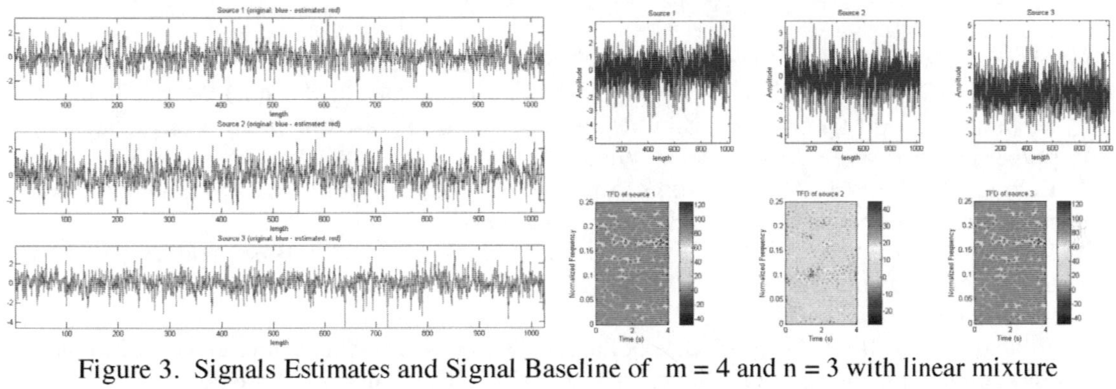

Figure 3. Signals Estimates and Signal Baseline of m = 4 and n = 3 with linear mixture

unbalanced, respectively. The baseline was used to validate the sound pattern of the damages. In recording of the baseline signal, the sampling frequency used is 44100 Hz. The baseline signal was downsampled to 256 Hz with data length of 1024 to reduce computational complexity. The results of this baseline recording signal will serve as the ground truth for MSE calculation with the signal estimation result.

## 4. Results and Discussion

### 4.1. Mixed Signal Data

Furthermore, retrieval of data recording performed for the number of sensors is greater than the amount of engine noise signal sources. Configuration used to take a maximum of 4 sensors and 3 sources. By positioning the microphone sensors placed parallel with the engine when recording sound source sound signal. This meant that the sound signal from a microphone right in front of it can receive signals from a source with a strong engine but still receive the signal of other sound nearby in order to avoid the phenomenon of spatial aliasing. In this research, the configuration used is the number of sensors (microphones) is greater than or equal to the number of sources (engine pump).

### 4.2. Instantaneous Linear Mixture Separation

The results of the estimated signal and the TFD of 4 sensors and 3 sound sources of this pump can be seen in Fig. 4. In the figure, the frequency and time is plotted in a 2D (two dimensional) where the frequencies are shown the frequency normalization. In the figure, the color shows the magnitude of the amplitude (dB). As the colour goes toward red means the power is getting higher, while for the blue, then the amplitude gets smaller. To determine the frequency of normalization of the estimated pump, characterized by the greatest amplitude (red) are depicted resemble a horizontal line.

The window analysis was also conducted where we compared between the Hamming and Blackman Harris window. The results of the normalized frequency of the pump bearing fault 0.7, 0.02 normal pump, and pump unbalance 0.1, respectively. For the Hamming window had loss of 1.78 while the Blackman-Harris 0.83.

## 5. Conclusion

Mixed sound signals were successfully separated by the method of time frequency blind source separation and redundant signals from the sensors successfully reduced to only one signal at the output signal due to the estimated number of sensors greater than the number of sources. With this tfBSS method, separation of mixed sound signals (instantaneous linear mixture) is better than the mixed sound signal (convolutive mixtures) with MSE score. The results obtained from the TFD estimates was with the MSE of the Bearing Fault 0.17, 0.1 Normal pumps and pump unbalance 0.16, respectively.

## References

[1] Fevotte, Cedric., Doncarli, Christian. "**Two Contribution to Blind Source Separation Using Time-Frequency Distribution".** IEEE Signal Processing Letters vol.11, no.3, March 2004.

[2] A. Belouchrani and M. G. Amin, "**Blind Source Separation Based on Time-Frequency Signal Representations**". *IEEE Trans. Signal Processing*, vol. 46, pp. 2888–2897, Nov 1998.

[3] A. Belouchrani, K. Abed-Meraim, M. G. Amin, A. M. Zoubir. **"Joint Anti-Diagonalization For Blind Source Separation"**. in *Proc. ICASSP*, vol.5, 2001, pp. 2789–2792.

[4] A. Belouchrani, K. Abed. Meraim, J-F. Cardoso dan E. Moulines, "**A Blind Source Separation Technique Based on Second Order Statistics**," *IEEE Trans. Signal Processing*, vol.45, pp. 434-444, Feb 1997.

[5] A.Holobar, C. Févotte, C. Doncarli, dan D. Zazula, "**Single Autoterms Selection for Blind Source Separation in Time-Frequency Plane**," in *Proc. EUSIPCO*, vol.1, 2002, pp 565-568.

[6] Barlow, H.B. "**Sensory Mechanism, The Reduction of Redundancy, and Intelligence**," NPL Symposium on the Mechanization of Thought Process. No.10, pp. 535-539, HM Stationery Office, London, (1959).

[7] Barlow, H.B. "**Redundancy Reduction Revisited**," Network : Comput. Neural Syst. 12 (2001) 241-253, Institute of Physics Publishing.

[8] L. Omlor, M. Giese. 2006. "**Blind Source Separation for Over-Determined Delayed Mixtures**." Laboratory for Action Representation and Learning. Department of Cognitive Neurology. University of Tübingen, Germany.

[9] J. P. Nadal, N. Parga. 1997. "**Redundancy Reduction and Independent Component Analysis : Conditions on Cumulants and Adaptive Approches**." Laboratoire associé au C.N.R.S. (U.R.A. 1306), à PENS,et au Universités Paris VI et Paris VII.

[10] L.Cohen. "**Time Frequency Analysis**." Prentice Hall PTR Englewood Cliffs, New Jersey, 1995.

.

**Institute of Physics**
Dirac House, Temple Back
Bristol BS1 6BE UK

ISSN: 1742-6588
ISBN 978-1-5108-7370-4
9 781510 873704